REVIEWS in MINERALOGY   Volume 32

STRUCTURE, DYNAMICS AND PROPERTIES OF SILICATE MELTS

EDITORS:

J.F. Stebbins, Stanford University

P.F. McMillan, Arizona State University

D.B. Dingwell, Universität Bayreuth

Front cover: Sketch of a possible mechanism for the exchange of silicon species. Oxygens: open circles; network-modifying cations: cross-hatched circles; silicons: solid circles. From Farnan and Stebbins (1994; see Fig.17, p. 222)

Series Editor: **Paul H. Ribbe**
Department of Geological Sciences
Virginia Polytechnic Institute & State University
Blacksburg, Virginia 24061 U.S.A.

Mineralogical Society of America
Washington, D.C.

COPYRIGHT 1995

MINERALOGICAL SOCIETY OF AMERICA

Printed by BookCrafters, Inc., Chelsea, Michigan.

REVIEWS IN MINERALOGY

(Formerly: SHORT COURSE NOTES)
ISSN 0275-0279

Volume 32

Structure, Dynamics and Properties of Silicate Melts

ISBN 0-939950-39-1

ADDITIONAL COPIES of this volume as well as those listed on page vi
may be obtained at moderate cost from:

THE MINERALOGICAL SOCIETY OF AMERICA
1015 EIGHTEENTH STREET, NW, SUITE 601
WASHINGTON, DC 20036 U.S.A.

STRUCTURE, DYNAMICS AND PROPERTIES OF SILICATE MELTS

FOREWORD

This the thirty-second volume in the *Reviews in Mineralogy* series is the result of contributions from fifteen authors overseen by three editors, Jonathan Stebbins (Stanford), Paul McMillan (Arizona State), and Don Dingwell (Bayreuth), who have worked diligently in selecting the topics and authors and in thoroughly reviewing and cross-referencing the twelve chapters. The result is a definitive treatise on the *Structure, Dynamics and Properties of Silicate Melts.* We apologize that there is rarely time, under the pressure of deadlines and late manuscripts, to assemble an index for most the volumes in this series. But the Table of Contents is detailed, and hopefully the reader will be able to "make do" with this as a means to find his or her way through this vast accumulation of information. There are two other volumes of *Reviews in Mineralogy,* one on *Spectroscopic Methods* (Vol. 18, 1988) and another entitled *Volatiles in Magmas* (Vol. 30, 1994), which are referenced often and are particularly relevant to the subject matter of this book. These are still available at modest cost from the MSA office (address on the opposite page)

The Mineralogical Society of America sponsored a short course for which this was the text at Stanford University December 9 and 10, 1995, preceding the Fall Meeting of the American Geophysical Union and MSA in San Fransisco, with about 100 professionals and graduate students in attendance.

I thank Margie Strickler for secretarial help and Jodi and Kevin Rosso for assistance with software problems.

Paul H. Ribbe
Series Editor
Blacksburg, VA

EDITORS' PREFACE

A silicate melt phase is the essential component of nearly all igneous processes, with dramatic consequences for the properties of the Earth's interior. Throughout Earth history and continuing to the present day, silicate melts have acted as transport agents in the chemical and physical differentiation of the Earth into core, mantle and crust. The occurrence of such magmatic processes leads to the definition of our planet as "active," and the resulting volcanism has a profound impact on the Earth's atmosphere, hydrosphere and biosphere. Although near-surface melts are observed directly during volcanic eruptions, the properties of magmas deep within the Earth must be characterized and constrained by laboratory experiments. Many of these experiments are designed to aid in developing an atomic level understanding of the structure and dynamics of silicate melts under the P-T conditions of the Earth's crust and mantle, which will make extrapolation from the laboratory results to the behavior of natural magmas as reliable as possible.

Silicate melts are also the archetypal glass-forming materials. Because of the ready availability of raw materials, and the ease with which molten silicates can be vitrified, commercial "glass" has necessarily implied a silicate composition, over most of the history of glass technology. The properties of the melt, or "slag" in metallurgical extractions, determine the nature of the glass formed, and the needs of the glass industry have provided much of the impetus for understanding the structure-property relations of molten silicates as well as for the glasses themselves. It is now recognized that any liquid might become glassy, if cooled rapidly enough, and understanding the thermodynamic and kinetic aspects of the glass transition, or passage between the liquid and glassy states of matter, has become a subject of intense interest in fundamental physics and chemistry.

Glasses have also been studied in many geochemical investigations, often as substitutes for the high temperature melts, with the results being extrapolated to the liquid state. In many cases, in situ techniques for direct investigation of these refractory systems have only recently become available. Much valuable information concerning the melt structure has been gleaned from such studies. Nevertheless, there are fundamental differences between the liquid and glassy states. In liquids, the structure becomes progressively more disordered with increasing temperature, which usually gives rise to major changes in all thermodynamic properties and processes. These changes must, in general, be investigated directly by in situ studies at high temperature. Studies of glass only represent a starting point, which reflect a frozen image of the melt "structure" at the glass transition temperature. This is generally hundreds of degrees below the near-liquidus temperatures of greatest interest to petrologists.

Since the early 1980s, a much deeper understanding of the structure, dynamics, and properties of molten silicates has been developed within the geochemical community, applying techniques and concepts developed within glass science, extractive metallurgy and liquid state physics. Some of these developments have far-reaching implications for igneous petrology. The purpose of this Short Course and volume is to introduce the basic concepts of melt physics and relaxation theory as applied to silicate melts, then to describe the current state of experimental and computer simulation techniques for exploring the detailed atomic structure and dynamic processes which occur at high temperature, and finally to consider the relationships between melt structure, thermodynamic properties and rheology within these liquids. These fundamental relations serve to bridge the extrapolation from often highly simplified melt compositions studied in the laboratory to the multicomponent systems found in nature. This volume focuses on the properties of simple model silicate systems, which are usually volatile-free. The behavior of natural magmas has been summarized in a previous Short Course volume (Nicholls and Russell, editors, 1990: *Reviews in Mineralogy,* Vol. 24), and the effect of volatiles on magmatic properties in yet another (Carroll and Holloway, editors, 1994: Vol. 30).

In the chapters by Moynihan, by Webb and Dingwell, and by Richet and Bottinga, the concepts of relaxation and the glass transition are introduced, along with techniques for studying the rheology of silicate liquids, and theories for understanding the transport and relaxation behavior in terms of the structure and thermodynamic properties of the liquid. The chapter by Dingwell presents applications of relaxation-based studies of melts in the characterization of their properties. Chapters by Stebbins, by Brown, Farges and Calas, and by McMillan and Wolf present the principal techniques for studying the melt structure and atomic scale dynamics by a variety of spectroscopic and diffraction methods. Wolf and McMillan summarize our current understanding of the effects of pressure on silicate glass and melt structure. Chapters by Navrotsky and by Hess consider the thermodynamic properties and mixing relations in simple and multicomponent aluminosilicate melts, both from a fundamental structural point of view and empirical chemical models which can be conveniently extrapolated to natural systems. The chapter by Chakraborty describes the diffusivity of chemical species in silicate melts and glasses, and the chapter by Poole, McMillan and Wolf discusses the application of computer simulation methods to understanding the structure and dynamics of molten silicates. The emphasis in this volume is on reviewing the current state of knowledge of the structure, dynamics and physical properties of silicate melts, along with present capabilities for studying the molten state under conditions relevant to melting within the Earth, with the intention that these techniques and results can then be applied to understanding and modeling both the nature of silicate melts and the role of silicate melts in nature.

This book is the result of the combined efforts of many people to whom we express our heartfelt thanks: the authors themselves; the reviewers who conscientously gave detailed and useful comments on the chapters, often on extremely short notice; and the Series Editor, Paul H. Ribbe, who made all of this possible with his usual efficiency and aplomb.

Jonathan F. Stebbins
Stanford University, USA

Paul F. McMillan
Arizona State University, USA

Donald B. Dingwell
Universität Bayreuth, Germany

ADDITIONAL COPIES of this volume as well as those listed below may be obtained at moderate cost from the MINERALOGICAL SOCIETY OF AMERICA, 1015 EIGHTEENTH STREET, NW, SUITE 601, WASHINGTON, DC 20036 U.S.A.

STRUCTURE, DYNAMICS AND PROPERTIES OF SILICATE MELTS

TABLE OF CONTENTS, VOLUME 32

Chapter 1 — C. T. Moynihan

STRUCTURAL RELAXATION AND THE GLASS TRANSITION

Chapter 2 — D. B. Dingwell

RELAXATION IN SILICATE MELTS: SOME APPLICATIONS

Chapter 6   **P. C. Hess**

THERMODYNAMIC MIXING PROPERTIES AND
THE STRUCTURE OF SILICATE MELTS

Chapter 7   **J. F. Stebbins**

DYNAMICS AND STRUCTURE OF SILICATE AND OXIDE
MELTS: NUCLEAR MAGNETIC RESONANCE STUDIES

Chapter 8   P. F. McMillan & G. H. Wolf

VIBRATIONAL SPECTROSCOPY OF SILICATE LIQUIDS

Chapter 9   G. E. Brown, Jr., F. Farges & G. Calas

X-RAY SCATTERING AND X-RAY SPECTROSCOPY
STUDIES OF SILICATE MELTS

Chapter 10   **S. Chakraborty**

DIFFUSION IN SILICATE MELTS

xiii

Chapter 11   **G. H. Wolf & P. F. McMillan**

PRESSURE EFFECTS ON SILICATE MELT
STRUCTURE AND PROPERTIES

Chapter 12   **P. H. Poole, P. F. McMillan & G. H. Wolf**

COMPUTER SIMULATIONS OF SILICATE MELTS

Chapter 1

STRUCTURAL RELAXATION AND THE GLASS TRANSITION

Cornelius T. Moynihan

Materials Science and Engineering Department
Rensselaer Polytechnic Institute
Troy, NY 12180 U.S.A.

INTRODUCTION: THE NATURE OF STRUCTURAL RELAXATION

In spite of the enormous activity and interest in the subject over the past thirty years or so, the nature of the glass transition and the structural relaxation process in liquids is frequently poorly understood by otherwise knowledgeable scientists or engineers. The purpose of this chapter is to explicate the glass transition, the structural relaxation process and the phenomenology associated therewith in a fairly basic fashion aimed more at fundamental perceptions than at quantitative details. Extensive treatments of much of this material can be found in the books of Brawer (1985), Scherer (1986a) and Nemilov (1995) and in various overview papers (Moynihan et al., 1976; Dingwell and Webb, 1989, 1990; Scherer, 1990; Simmons, 1993; Moynihan, 1994; Hodge, 1994; Angell, 1995).

At the macroscopic thermodynamic level, structural relaxation and the glass transition can be viewed as a phenomenon in which the state of the system depends not only on variables such temperature T, pressure P, shear stress, electric field, etc., but also on what are commonly called order parameters, internal parameters or (in chemistry) progress variables (Moynihan and Gupta, 1978). If one has such a system initially in equilibrium and makes a sudden change in one of the variables (change in T, P, chemical composition, etc.), the system will respond and seek a new equilibrium state consistent with the imposed change. However, the adjustment of the order parameters to new equilibrium values may be kinetically impeded and, in the extreme, be so slow as to be unobservable on any practicable experimental timescale.

Examples of this sort of kinetically impeded adjustment of order parameters to imposed changes abound in our everyday world. Consider, for example, the chemical reaction

$$CO(g) + 1/2\,O_2(g) \Leftrightarrow CO_2(g) \tag{1}$$

where the order parameter describes the relative amounts of reactants and products. At ordinary temperature (25°C) in ordinary Earth atmosphere ($PO_2 = 0.21$ atm, $PCO_2 = 3.3 \times 10^{-4}$ atm) the partial pressure of CO at chemical equilibrium is extremely small ($PCO,eq = 8 \times 10^{-49}$ atm). If we perturb this system by adding additional CO at ordinary temperature, we might expect most of the added CO to be consumed by reaction with oxygen. This of course will not occur in real time, as demonstrated by the ease with which the CO content in air can rise to and exceed the toxic level ($PCO = 5 \times 10^{-5}$ atm). The reaction is too slow at ordinary temperatures. However, this is not the case at higher temperatures, where flammable CO reacts quite readily with O_2.

Equilibrium constants for chemical reactions are in general temperature dependent. This means that if one starts with a mixture at chemical equilibrium and lowers the temperature, the relative amounts of reactants and products (the order parameter) will also change, provided that the reaction kinetics are sufficiently rapid. Rates of chemical reactions, however, decrease with decreasing temperature. If one continues to lower the temperature, eventually a point will be reached where the reaction kinetics are unable to keep pace with the imposed temperature changes. Below this temperature regime the chemical composition will remain constant, effectively "frozen" at a value characteristic of equilibrium at a higher temperature. On reheating, the reacting system will re-equilibrate in the same temperature regime in which it initially fell out of equilibrium during prior cooling and subsequently remain in equilibrium when heated to still higher temperatures. An excellent experimental study of this type of phenomenon is to be found in the paper of Barkatt and Angell (1979), in which the chemical reaction was the dimerization of an organic dye in dilute solution in the polyalcohol sorbitol, $C_6H_8(OH)_6$.

There exist other, more "physical" examples of these sorts of order parameter related phenomena. For example, the concentration of vacancies in a metal at equilibrium decreases with decreasing temperature. However, on cooling the vacancy concentration will eventually become "frozen" at a constant value when the rate at which the number of vacancies can decrease by diffusion from the interior of the metal to the surface becomes too slow to keep pace with the temperature change and concomitant shift in the equilibrium vacancy concentration.

The order parameter related phenomenon of interest in this chapter is the change in what is referred to loosely as the "structure" of a liquid or melt in response to temperature (or pressure) changes. In most cases, although we may have a moderately accurate picture of the average structure of a liquid or melt, the changes in that structure that accompany temperature changes are much more subtle and difficult to characterize experimentally. For example, it is universally accepted that the structure of SiO_2 melts and glasses is a disordered, 3-dimensional network of interconnected SiO_4 tetrahedra. Changes in temperature affect some of the detailed features of this disordered network. Best indications, based primarily on infrared and Raman vibrational spectroscopy studies, are that with increasing temperature changes in the average structure of SiO_2 melts are manifested by a decrease in the average Si-O-Si bond angle and an increase in the small number of $(SiO)_3$ siloxane rings in the network (McMillan et al., 1994). As another example, in their study of liquid sorbitol, $C_6H_8(OH)_6$, Barkatt and Angell (1979) found that with decreasing temperature there was a growth in the intensity of an infrared absorption band due to O-H vibrations of -OH groups hydrogen bonded to one another, accompanied by a decrease in the intensity of another band due to non-hydrogen bonded -OH groups. Thus the changes in the "structure" of liquid sorbitol with decreasing temperature are manifested as an increase in the average degree of hydrogen bonding between the molecules.

Rearrangement of the average structure of a liquid is known as *structural relaxation* and has a number of important features:

(1) Thermodynamic specification of the state of a liquid involves not only specification of T, P, etc., but also of the average structure. That is, like a chemically reacting system or a metal containing vacancies, liquids require an order parameter description.

(2) Structural relaxation appears to involve breaking and remaking of the bonds between the atoms or molecules, for example, primary Si-O bonds in liquid SiO_2 and intermolecular hydrogen bonds in liquid $C_6H_8(OH)_6$.

(3) At equilibrium at a given T and P the average structure of a liquid or melt is constant with time. However, this is a dynamic equilibrium, much like a chemically reacting system where at equilibrium the forward and reverse reactions

are occurring at equal rates in opposite directions. At equilibrium in a liquid the local structure is continually rearranging with time, but any changes in one small region are compensated by opposite changes in some other small region, so that the structure averaged over the entire specimen remains constant.

(4) This continual structural rearrangement is in large part responsible for the fluid character of liquids. Since it involves breaking and remaking of the interatomic or intermolecular bonds, it provides a mechanism whereby atoms or molecules can move past one another during viscous flow or diffusion. For example, diffusion of O^{-2} ions through liquid SiO_2 appears to involve breaking of the Si-O bonds, which is also part and parcel of the structural relaxation process (Dingwell and Webb, 1989, 1990; McMillan et al., 1994).

(5) Since the structural relaxation process involves breaking and making of bonds, it will contribute to changes in the enthalpy H, entropy S and volume V of a liquid as a function of T and P.

(6) Just like rates of chemical reactions or vacancy formation in metals, the rate at which structural relaxation occurs decreases with decreasing temperature. At a sufficiently low temperature this rate will become extremely slow, so that the structure of the liquid is "frozen" on an experimental time scale. In this situation the liquid will cease to exhibit fluid behavior and will take on the mechanical and thermodynamic properties of a solid. When this occurs we no longer refer to the amorphous system as a *liquid.* Rather, we refer to it as a *glass.*

PHENOMENOLOGY OF STRUCTURAL RELAXATION

Isothermal relaxation

Experimentally the most straightforward way to detect the structural relaxation process is to carry out observations in a temperature range where structural relaxation is very slow compared to the timescale over which properties of the liquid can be determined. One allows the liquid to equilibrate on this experimental time scale, subjects the specimen to a rapid change in temperature (or pressure), and then follows the evolution with time of some experimentally measureable macroscopic property (enthalpy H, volume V, refractive index n) at the new temperature (or pressure). The results of such an experiment are shown schematically in Figure 1 for the evolution of H or V of a melt initially in equilibrium after a step change in temperature from T_1 to T_2 imposed at time t = 0. The melt initially exhibits a "fast" or glass-like change in H or V associated primarily with the vibrational degrees of freedom. This is followed by a "slow" or kinetically impeded further change in H or V associated with the structural relaxation process. Structural relaxation progresses until equilibrium is reached at the new temperature. At the top of Figure 1 the molecular events attending these two processes are conceptualized for a network oxide liquid. In the fast or glass-like process there is a decrease in the amplitude of the atomic vibrations, very similar to what would occur in the corresponding crystal if the temperature were decreased. This leads to a net decrease in the distance of separation of the atoms, but no change in their relative positions. In the subsequent slow structural relaxation there is a change in the liquid structure involving breaking and remaking of bonds, suggested in Figure 1 as a decrease in the average ring size.

Results of experiments of this sort are frequently described in terms of changes in the fictive temperature, T_f. T_f is defined as the contribution of the structural relaxation process to the property of interest (H or V in Fig. 1) expressed in temperature units and may be considered as a measure of the order parameter associated with the structural relaxation process. For the equilibrium liquid, $T_f = T$, and during heating or cooling $dT_f/dT = 1$. For

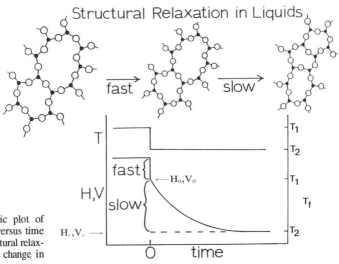

Figure 1. Schematic plot of enthalpy and volume versus time during isothermal structural relaxation following a step change in temperature.

a glass whose structure is frozen, T_f = constant, and during heating or cooling $dT_f/dT = 0$. As shown in Figure 1, during the course of structural relaxation following a step change in temperature from T_1 to T_2, T_f varies from T_1 to T_2 in parallel with the changes in H or V. It should be emphasized that T_f is not a quantity of fundamental significance, but is merely an auxiliary variable defined for computational and conceptual convenience. Moreover, the value of T_f is not necessarily a unique or complete specification of the structural state frozen into a glass. For example, T_f values calculated from different properties (e.g. H and V) for a glass formed by rate cooling will in general be slightly different.

The rate of structural relaxation may be described by a characteristic structural relaxation time, τ. As a crude first approximation, we might assume that the rate of approach of, say, enthalpy H to equilibrium in the experiment of Figure 1 will be proportional to the deviation of H from the equilibrium value H_e approached at long times at temperature T_2:

$$d(H - H_e)/dt = -k(H - H_e) \tag{2}$$

where k is a rate constant. Integrating this expression between time 0 and time t we get:

$$\phi(t) \equiv \frac{H - H_e}{H_0 - H_e} \equiv \frac{T_f - T_2}{T_1 - T_2} = \exp(-kt) \equiv \exp(-\frac{t}{\tau}) \tag{3}$$

where $\phi(t)$ is called the relaxation function and varies from 1 at t = 0 to 0 at very long times, H_0 is the enthalpy at time t = 0 immediately following the fast glass-like response of the liquid, and the structural relaxation time τ is defined as the inverse of the rate constant ($\tau = 1/k$). For small departures from equilibrium and over a short temperature range the temperature dependence of τ can be approximated by an Arrhenius expression:

$$\tau = \tau_0 \exp(\Delta H^*/RT) \tag{4}$$

where τ_0 is a pre-exponential constant, ΔH^* an activation enthalpy, and R is the ideal gas constant. Because ΔH^* is positive, τ increases and the rate of structural relaxation decreases rapidly with decreasing temperature.

Relaxation during cooling and heating

Cooling or heating a liquid or glass at a rate $q = dT/dt$ can be thought of as a series of small temperature steps ΔT followed by isothermal holds of duration $\Delta t = \Delta T/q$. In Figure 2a is shown the behavior of the enthalpy H of a glassforming liquid during stepwise cooling followed by stepwise reheating over the same temperature range. [Volume V and other properties would behave similarly.] The dashed line represents both the temperature T and the equilibrium enthalpy H_e. The solid line represents the experimentally measured enthalpy H. Following the first downward step in temperature, the relaxation time τ is sufficiently short compared to the time interval Δt ($\Delta t \gg \tau$) that the system is able to equilibrate and exhibits a liquid-like response for the enthalpy. Following the second downward step, however, the relaxation time is now longer at the lower temperature, and the system is unable to equilibrate completely ($\Delta t \sim \tau$). The extent of equilibration in time Δt becomes less and less after each subsequent downward temperature step, so that after the last two such steps virtually no structural relaxation occurs in time interval Δt ($\Delta t \ll \tau$), the system exhibits only the fast glass-like change in H (compare Fig. 1), and the relaxational part of the enthalpy is frozen at a value much greater than the equilibrium value. At this point the system is behaving as a glass.

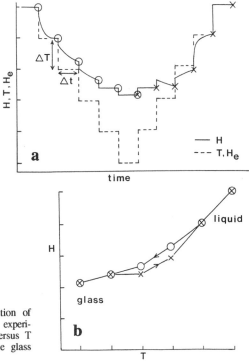

Figure 2. Schematic plots of (a) variation of temperature T, equilibrium enthalpy H_e and experimental enthalpy H with time, and (b) H versus T during stepwise cooling and reheating in the glass transition region.

Following the first upward step in temperature during reheating in Figure 2a, the relaxation time τ is still too long to allow structural relaxation in time interval Δt. Hence the enthalpy exhibits only the fast glass-like change, but now in an upward direction. Following the second upward step the system can exhibit partial relaxation. Since, however, H is above the equilibrium value at this point, it exhibits a downward relaxation, even though the system is now being heated. Following the third and subsequent upward temperature steps, H is below the equilibrium value H_e. Hence it now relaxes upwards and is eventually able to equilibrate in time Δt at the highest temperatures, so that the system has returned to liquid-like behavior.

The enthalpy H at the beginning and end of each time/temperature step in Figure 2a is plotted versus temperature in Figure 2b. The intermediate temperature region over which the system passes from liquid-like behavior to glass-like behavior and in which $\Delta t \sim \tau$ is called the *glass transition* region. Note that in the glass transition region the H vs. T cooling curve is different from the H vs. T heating curve, so that there is hysteresis between the two curves when a liquid is cooled and subsequently reheated through the transition region. This hysteresis is in no way unique to structural relaxation in liquids and the glass transition. Rather it is expected for any chemical or physical process requiring an order parameter description (e.g., a chemical reaction or disappearance and formation of vacancies in a metal crystal) when the equilibrium state of the system and the rate of approach to equilibrium are both temperature dependent. Note also that the glass transition, as explicated here, is entirely kinetic in origin and is not to be confused with thermo-dynamic transitions, such as melting of a crystal or vaporization of a liquid, which can take place at equilibrium.

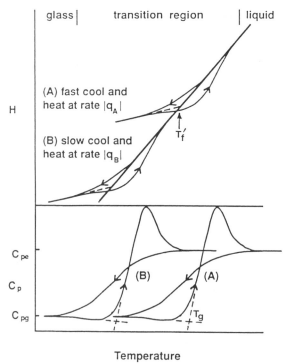

Figure 3. Schematic plots of enthalpy H and heat capacity C_P versus temperature during cooling and reheating through the glass transition region at two different rates.

Shown schematically in the upper part of Figure 3 are plots of H vs. T similar to that in Figure 2b for cooling and reheating through the glass transition region at two different rates, q_A and q_B ($q_A > q_B$) (Moynihan et al., 1974; Moynihan, 1994). During cooling the H vs. T plot changes slope monotonically as one passes from liquid to glass through the glass transition region. If instead of monitoring H directly one monitored the heat capacity C_P

(= dH/dT = slope of H vs. T curve), as is commonly done using differential scanning calorimetry (DSC) or differential thermal analysis (DTA), one would obtain the sigmoidal C_p vs. T cooling curves shown in the lower part of Figure 3. Because of the hysteresis between the H vs. T curves during cooling and reheating, the slope of the H vs. T curve on reheating passes through a maximum near the upper end of the glass transition region. This leads in turn to maxima in the C_p vs. T reheating curves, as shown in the lower part of Figure 3. Note that these maxima in the C_p reheating curves, which are often taken to be an essential or definitive feature of the glass transition, are in fact an artifact of fairly straight-forward kinetics for the structural relaxation process coupled with a continual shift in the parameters characterizing the rate of the process (e.g., k or τ in Eqns. 2 and 3) and in the departure of the system from equilibrium due to the continually changing temperature.

Although the glass transition covers a finite range in temperature, this range is fairly narrow and is commonly demarcated by a glass transition temperature Tg. Tg may be taken as any characteristic point on the cooling or reheating curve for a measured property or the temperature derivative of that property. If, for example, the property is enthalpy H, Tg might be taken as the extrapolated onset of rapid increase of the C_p vs. T reheating curve, as shown in the bottom part of Figure 3. Alternatively, Tg might be taken as the extrapolated point of intersection, denoted T_f' in the upper part of Figure 3, of the liquid and glass H vs. T curves measured during cooling. [T_f' as defined in Figure 3 is the limiting fictive temperature attained by cooling the liquid to well below the glass transition region.]

Shown in Figure 4 are plots of C_p (= dH/dT) and of the temperature derivative of the volume dV/dT of a $Na_2Si_2O_5$ glass vs. T measured during rate heating through the glass transition region at 5°C/min after prior cooling at 5°C/min (Webb, 1992). Both heating curves exhibit the shapes expected from the lower part of Figure 3 (rapid rise followed by a maximum in C_p and dV/dT) and serve to demarcate the glass transition region. Also worth

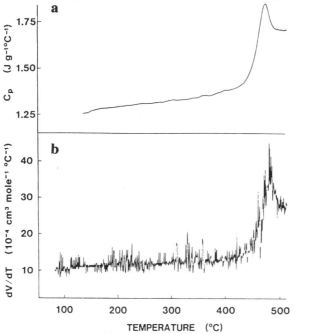

Figure 4. Heat capacity C_p and temperature derivative of volume dV/dT of $Na_2Si_2O_5$ glass during reheating at 5 K/min after prior cooling at 5 K/min through the glass transition region (Webb, 1992).

noting in Figure 4 is the close agreement between the C_p and dV/dT curves in terms of, for example, the temperatures and relative heights of the maxima. This suggests that the kinetics of the structural relaxation process monitored in terms of enthalpy H are very similar to those monitored in terms of volume V for silicate melts (Knoche et al., 1992). This correlation, however, is neither expected nor found experimentally to hold true for all types of glassforming liquids (Moynihan and Gupta, 1978).

Dependence of Tg on cooling and heating rate

As discussed above, during cooling and subsequent reheating the glass transition is observed in a temperature regime where the characteristic time scale $\Delta t = \Delta T/q$ becomes comparable to the structural relaxation time τ. Since a change in the cooling or heating rate q alters this time scale, it follows that the relaxation time τ in the glass transition region will be similarly altered. Since τ in turn is temperature dependent (cf. Eqn. 4), it then follows that a change in cooling or heating rate will change the location (temperaturewise) of the glass transition region and the value of Tg. In particular, as shown schematically in Figure 3, increasing the cooling or heating rate will shift the transition region to higher temperatures and increase the measured value of Tg. This is illustrated in Figure 5, which shows DSC traces for heating of As_2Se_3 glass through the glass transition at three different rates after cooling through the transition region at the same respective rates (Moynihan et al., 1974). The DSC output is proportional to the heat capacity C_p. Aside from some distortion of the DSC curves at low heating rates due to baseline curvature, the three DSC curves agree with the schematic curves in Figure 3. The glass transition temperatures, which might be taken from any characteristic points on the curves such as Tg_1, Tg_2 or Tg_3, shift uniformly upward in temperature with increasing heating rate.

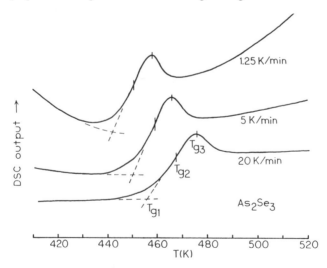

Figure 5. Differential scanning calorimeter traces for reheating of As_2Se_3 glass through the glass transition region at various rates after cooling through the transition region at the same respective rates.

It can be shown (Moynihan et al., 1974) that experiments of the type in Figures 3 and 5 (reheating rate proportional to cooling rate) can be used to obtain the activation enthalpy ΔH^* of Equation (4), which specifies the temperature dependence of the structural relaxation time in the transition region:

$$d \ln q_h/d(1/Tg) = -\Delta H^*/R \qquad (5)$$

where q_h is the heating rate. If one defines the glass transition temperature as the limiting fictive temperature $T_f{}'$ attained on cooling through the transition region, one can analogously obtain ΔH^* from the dependence of $T_f{}'$ on cooling rate q_c (Moynihan et al., 1976):

$$d \ln |q_c|/d(1/T_f{}') = -\Delta H^*/R \qquad (6)$$

(These last two expressions remain valid even when the nonlinear and nonexponential features of structural relaxation described in a subsequent section are taken into account.) Note that, since the glass and equilibrium liquid property vs. T curves are the same for cooling and subsequent reheating, $T_f{}'$ may be determined from the reheating curves. Any reheating rate may be used for this determination and will presumably be chosen to optimize experimental accuracy and precision. If, as in a DSC experiment, the property monitored is enthalpy H and the instrumental output is proportional to the temperature derivative of the property ($C_P = dH/dT$), determination of the extrapolated intersection of the glass and equilibrium H vs. T curves corresponds to a matching of areas under the measured C_P vs. T plots. Determination of $T_f{}'$ in this fashion is illustrated in Figure 6 for a B_2O_3 glass reheated at 10 K/min after being previously cooled through the glass transition region at 10 K/min. The equation governing the area match is

$$\int_{T \gg T_g}^{T_f{}'} (C_{Pe} - C_{Pg})dT = \int_{T \gg T_g}^{T \ll T_g} (C_P - C_{Pg})dT \qquad (7)$$

where C_{Pg} is the glass heat capacity observed below the transition region and C_{Pe} is the equilibrium liquid heat capacity observed above the transition region. Shaded area I in Figure 6 corresponds to the left side of this expression and shaded area II to the right side. Note from Figure 6 that, when the prior cooling and reheating rates are the same, $T_f{}'$ obtained from the C_P reheating curve is very close to Tg in Figure 3 or Tg_1 in Figure 5 defined as the extrapolated onset of rapid rise of the C_P vs. T reheating curve.

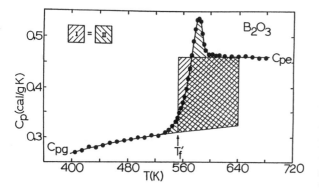

Figure 6. Heat capacity versus temperature for B_2O_3 glass during heating at 10 K/min following a rate cool at 10 K/min through the glass transition region. $T_f{}'$ is the limiting fictive temperature attained by the glass on cooling at 10 K/min.

In Figure 7 is shown an Arrhenius plot of $\log|q_c|$ vs. $10^3/T_f{}'$ (K) for an As_2Se_3 glass (Easteal et al., 1977). The $T_f{}'$ values were obtained from C_P vs. T curves during reheating at 10 K/min after prior cooling through the transition region at rates ranging from -0.31 to -20 K/min. The slope of the plot gives via Equation (6) an activation enthalpy ΔH^* of 342 kJ/mol for structural relaxation of As_2Se_3 glass in the glass transition region. This is a large activation enthalpy, so that although the cooling rate has been varied by a factor of 65 (= 20/0.31) in the experiments of Figure 7, the glass transition temperature as measured by $T_f{}'$ has only changed by about 20 K (from 435 K to 455 K). Note from Equation (4) that this also means that the characteristic structural relaxation time τ also decreases by a factor of 65 between 435 K and 455 K.

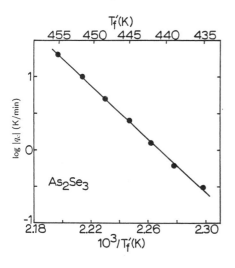

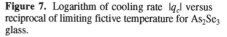

Figure 7. Logarithm of cooling rate $|q_c|$ versus reciprocal of limiting fictive temperature for As_2Se_3 glass.

KINETICS OF STRUCTURAL RELAXATION

Isothermal relaxation

Although the simple first order kinetic expressions of Equations (2) and (3) give an accurate qualitative description of the progress of the structural relaxation process with time following a step change in temperature from equilibrium, they fail to give an acceptable quantitative description. A quantitative description requires that two additional features be taken into account. The first of these is the *nonlinear* character of the process (in the sense that the rate of relaxation cannot be described by a linear differential equation such as Eqn. 2). This is illustrated in Figure 8 for the isothermal relaxation of the density ($\propto 1/V$) of a soda lime silicate glass (Scherer, 1986b). In the upper curve the glass was initially equilibrated at 500°C (so that its initial fictive temperature $T_f = 500°C$), then upquenched to 530°C and allowed to relax at that temperature. In the lower curve the glass was initially equilibrated at 565°C (initial $T_f = 565°C$), then downquenched to and allowed to relax at 530°C. At long times both samples have come to equilibrium (final T_f = annealing temperature = 530°C) and exhibit the same density. [The densities in Figure 8 were actually measured at room temperature after dropping the samples out of the furnace following annealing for various times at 530°C. Hence the fast, glass-like contribution to the density change (cf. Fig. 1) has effectively been subtracted from all the data points.]   Even though

Figure 8. Isothermal relaxation at 530°C of the density of a soda lime silicate glass following an upward step change in temperature from 500°C (upper curve) and a downward step in temperature from 565°C (lower curve). Solid lines are calculated from Equations (8), (14) and (15) using the parameters $\tau_0 = 3.9 \times 10^{-37}$ s, $\Delta H^* = 607$ kJ/mol, $x = 0.45$ and $\beta = 0.62$.

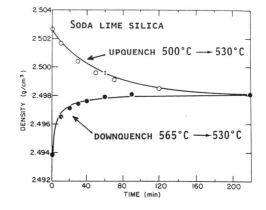

the initial magnitude of departure of the density from equilibrium is nearly the same for both specimens and even though they are relaxing at the same temperature (530°C), the downquenched sample with the higher initial T_f (= 565°C) clearly relaxes more quickly than the upquenched sample with the lower initial T_f (= 500°C). In terms of Equations (3) and (4) this means that the structural relaxation time τ depends not only on temperature T, but also on the instantaneous structure or fictive temperature. One common way of modifying Equation (4) to account for this is to use the so-called Tool-Narayanaswamy (TN) equation:

$$\tau = \tau_0 \exp[\frac{x\Delta H^*}{RT} + \frac{(1-x)\Delta H^*}{RT_f}] \tag{8}$$

where x $(0 \leq x \leq 1)$ is the nonlinearity parameter. Note that Equations (4) and (8) become identical in the limit of small departures from equilibrium $(T_f \rightarrow T)$.

Another way of accounting for the nonlinearity is via the Adam-Gibbs equation (Scherer, 1984; Hodge, 1994; Richet and Bottinga, this volume):

$$\tau = \tau_0 \exp[\frac{B}{TS_c(T_f)}] \tag{9}$$

where B is a constant and the configurational entropy $S_c(T_f)$ is presumed to be a function of the fictive temperature:

$$S_c(T_f) = \Delta C_p \ln(T/T_K) \tag{10}$$

ΔC_p $(= C_{Pe} - C_{pg})$ is the difference between the liquid and glass heat capacities, and T_K is the temperature at which the configurational entropy of the liquid would vanish at equilibrium. Over a short temperature range for moderate departures from equilibrium, Equations (8) and (9) cannot be distinguished functionally from one another (Scherer, 1986; Hodge, 1994).

The second of the aforementioned features of the structural relaxation process is its *nonexponential* character. This means that even for relaxation very close to equilibrium, where $T_f \approx T$ and nonlinear effects are unimportant, isothermal relaxation cannot be described by the simple exponential function of Equation (3). To account for this feature a distribution of relaxation times is incorporated into the relaxation function

$$\phi(t) = \sum_i g_i \exp[-\int_0^t dt'/\tau_i] \tag{11}$$

where the g_i are temperature independent weighting coefficients $(\Sigma_i g_i = 1)$ for the contributions from the various relaxation times τ_i. Each τ_i is given by an expression of the form of Equation (8):

$$\tau_i = \tau_{i0} \exp[\frac{x\Delta H^*}{RT} + \frac{(1-x)\Delta H^*}{RT_f}] \tag{12}$$

or of Equation (9)

$$\tau_i = \tau_{i0} \exp[\frac{B}{TS_c(T_f)}] \tag{13}$$

Hence the τ_i differ only in their pre-exponential factors τ_{i0}. This, along with the presumed temperature independence of the g_i, leads to a condition known as thermorheological simplicity. The need for an integral over time in Equation (11) is due to the variation of T_f and hence of the τ_i with time during relaxation.

The microscopic interpretation of a distribution of relaxation times in the relaxation function of Equation (11) remains something of an open question. To return for a moment to the chemical reaction analogy to the structural relaxation process, imagine a system in which two or more independent chemical reactions could take place. If one monitored the evolution of the enthalpy or volume of such a system as it equilibrated isothermally, one would expect nonexponential behavior, since each reaction would make an independent contribution to the changes in H and V with time, but it would be very unlikely that the rate constants k_i or relaxation times τ_i for the different reactions would be the same. This suggests in turn that during structural relaxation in liquids there may be a number of different molecular events involved which occur at different rates. Alternatively, some microscopic regions of the liquid may differ from other microscopic regions, so that the various regions relax at different rates. These differences might arise from thermally generated differences (fluctuations) in the configurational entropy S_c from microregion to microregion, which would lead via Equation (9) to differences in the local relaxation times (Moynihan and Schroeder, 1993).

The number of adjustable parameters in Equations (11) and (12) or (13) may be considerably reduced if one selects a continuous distribution or spectrum of relaxation times, the shape of which can be specified by a single parameter and the location of which on a logarithmic timescale can be specified by a reference relaxation time τ (e.g. the most probable relaxation time). One then need worry only about the T and T_f dependence of the reference relaxation time τ, which is presumed to be of the form of Equation (8) or (9). A common choice for the shape of the spectrum of relaxation times, which seems to give a good fit to most data, is that corresponding to the so-called Kohlrausch-Williams-Watts (KWW) or stretched exponential relaxation function:

$$\phi(t) = \exp[-(\int_o^t dt'/\tau)^\beta] \tag{14}$$

where β $(0 < \beta \leq 1)$ is the nonexponentiality parameter. Hence, to describe structural relaxation in response to temperature changes for substantial departures from equilibrium one requires a minimum of four adjustable parameters. In Equations (8) and (14) these are τ_0, ΔH^*, x and β. [For most glasses a "substantial departure from equilibrium" means a difference between T_f and T greater than about 2 K.] The solid lines in Figure 8 are fits to the data using Equations (8) and (14), where the relaxation function for density ρ and the fictive temperature T_f is defined by (compare Fig. 1 and Eqn. 3):

$$\phi(t) = \frac{\rho(t) - \rho_e}{\rho_0 - \rho_e} = \frac{T_f(t) - T_2}{T_1 - T_2} \tag{15}$$

Relaxation during cooling and heating

Structural relaxation in response to a complicated thermal history, e.g., a temperature change ΔT_1 at time t_1, ΔT_2 at time t_2, ..., ΔT_m at time t_m, may be dealt with by using the Boltzmann superposition principle in conjunction with Equations (8) or (9) and (14). Here we assume that the total response of the system is just the sum of its responses to each of one of the temperature steps starting at the time that temperature step is imposed. In this situation the fictive temperature $T_f(t)$ at time t $(>t_m)$ is given by

$$T_f(t) = T_0 + \sum_{j=1}^{m} \Delta T_j[1 - \phi(t, t_j)] \tag{16}$$

where T_0 is an initial temperature at which the sample is at equilibrium, and

$$\phi(t, t_j) = \exp[-(\int_0^{t-t_j} d(t' - t_j)/\tau)^\beta] \tag{17}$$

An example of such a complicated thermal history is cooling at rate q_c from a temperature T_0 above the glass transition region to well below the transition region and then

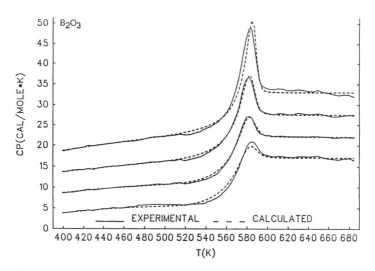

Figure 9. Heat capacities of B_2O_3 glass measured during heating at 10 K/min after cooling through the transition region at rates of 0.62, 2.5, 10 and 40 K/min (top curve to bottom curve). Heat capacity scale is correct for top curve; other curves have been displaced downward for clarity. Solid lines are experimental data. Dashed lines are calculated from Equations (8), (16), (17) and (18) using the parameters $\tau_0 = 1.3 \times 10^{-33}$ s, $\Delta H^* = 377$ kJ/mol, x = 0.39 and β = 0.62.

reheating at a (possibly different) rate q_h to above the transition region, as schematically shown in Figures 2 and 3. In Figure 9 are shown some actual data of this sort for B_2O_3 glass, where the solid lines are plots of the heat capacity C_p measured by DSC during rate heating at 10 K/min following cooling through the transition region at a variety of rates ranging from -0.62 to -40 K/min (DeBolt et al., 1976; Moynihan et al., 1991). The dashed lines were obtained from a fit to the C_p heating data using Equations (8), (16) and (17) and the τ_0, ΔH^*, x and β parameters given in the figure caption. T_0 in Equation (16) was taken as a temperature well above the transition region ($T_0 = 630$ K) where the relaxation time τ is short enough that the liquid remains in equilibrium during the cooling and heating rates employed. Equation (16) was used to calculate T_f and dT_f/dT during the initial cool through the transition region, followed by the subsequent reheat. C_p was then obtained from the expression (compare Eqn. 7):

$$dT_f /dT = (C_{P} - C_{P_g})/(C_{P_e} - C_{P_g}) \tag{18}$$

As might be expected, changes in prior thermal history (cooling rate in the case of Fig. 9) affect the subsequent relaxational behavior of the system (during reheating in the case of Fig. 9). In Figure 9 this is manifested mainly as an increase in the maximum in C_p during reheating at the same rate with decreasing prior cooling rate. This in turn reflects via Equation (6) the decrease in T_f' with decreasing cooling rate and the effect of this via Equation (8) on the relaxational kinetics during subsequent reheating. More drastic changes in the prior thermal history can have even more pronounced effects on the relaxational behavior during subsequent reheating (Moynihan, 1994). This is illustrated in Figure 10 for a ZBLA glass, where prolonged sub-Tg annealing causes a substantial upward shift in the temperature region for structural relaxation and in the value of Tg measured during reheating at a given rate.

The calculated density and C_p curves in Figures 8 and 9 agree with the measured curves within experimental error. This is generally found to be the case when the departure from equilibrium during actual relaxation is not extremely large. When this condition is not met, e.g. during annealing of a glass at temperatures very far below Tg, the quantitative agreement between the experimental data and best fits using Equations (8), (9), (14), (16) and (17) deteriorates to varying degrees, but all of the qualitative features of the experimental data are accurately reproduced in the calculated curves (Scherer, 1986; Gupta and Huang, 1992; Ducroux et al., 1994). It thus seems fair to conclude that the structural relaxation process and the glass transition are well understood from a macroscopic, phenomenological standpoint.

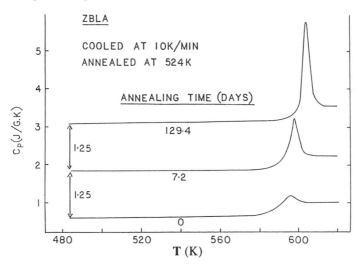

Figure 10. Heat capacities of $58 \cdot ZrF_4 \text{-} 33 \cdot BaF_2 \text{-} 5 \cdot LaF_3 \text{-} 4 \cdot AlF_3$ (ZBLA) glass measured during heating at 10 K/min after cooling through the transition region at 10 K/min and sub-Tg annealing at 524 K for times shown in the figure. Heat capacity scale is correct for the bottom curve. Other curves have been displaced upwards for clarity.

CORRELATIONS BETWEEN STRUCTURAL RELAXATION AND SHEAR VISCOSITY

Listed in Table 1 are structural relaxation and shear viscosity data in the glass transition region for a variety of inorganic glasses with Tg well above room temperature. These are mostly network oxide glasses, but examples of non-oxide glasses (As_2Se_3 and ZBLA) are also included. The data were compiled from a number of sources (Napolitano and Macedo, 1968; Tauber and Arndt, 1987; Knoche et al., 1992; Dingwell et al., 1993; Moynihan, 1993, 1994; Moynihan et al., 1995). Tg is the glass transition temperature defined as in Figure 3 and measured by DSC during heating at 10 K/min after cooling at a comparable rate. [Values of Tg for the glasses in Table 1 were actually measured at a variety of heating rates ranging from 5 to 20 K/min, but have been corrected where necessary to a standard heating rate of 10 K/min using Equation (5).] In the third column of Table 1 are listed values of $\log\langle\tau\rangle$, the logarithm of the mean or average equilibrium structural relaxation time at Tg determined from analysis of C_p data as in Figure 9. This mean relaxation time in terms of Equations (11) or (14) is defined respectively by

$$\langle\tau\rangle = \sum_i g_i \tau_i \quad \text{or} \quad (\tau/\beta)\Gamma(1/\beta) \tag{19}$$

where Γ signifies the gamma function. By "equilibrium" relaxation time we mean that this is the value of $\langle\tau\rangle$ corresponding to $T = T_f = Tg$ in Equations (8), (9), (12) or (13).

Table 1. Glass transition temperatures Tg measured by DSC at 10 K/min heating rate, logarithms of mean equilibrium enthalpy structural relaxation times $<\tau>$ and shear viscosities η at Tg, and activation enthalpies ΔH^* and ΔH_η^* for structural relaxation and viscous flow in the glass transition region.

Glass	Tg(K)	log$<\tau>$(s)	logη(Pa·s)	ΔH^* (kJ/mol)	ΔH_η^* (kJ/mol)
As$_2$Se$_3$	454	2.4	10.8	342	322
B$_2$O$_3$	557	2.6	11.4	385	385
ZBLA	587	2.8	11.6	1400	1140
lead silicate (NBS 711)	714	2.6	11.9	374	411
0.25Na$_2$O-0.75SiO$_2$	748	2.7	11.6	410	435
GeO$_2$	810	—	11.5	—	303
alkali lime silicate (NBS 710)	832	2.6	11.8	612	612
alkali borosilicate (BSC)	836	—	12.1	615	615
diopside (CaMgSi$_2$O$_6$)	973	—	12.7	—	965
anorthite (CaAl$_2$Si$_2$O$_8$)	1109	—	12.6	—	1084

As noted earlier, during cooling or reheating the glass transition region occurs in the temperature range where there is approximate correspondence between the time scale of the experiment and the characteristic relaxation time ($\Delta t \sim \tau$). For rate-cooling and -heating experiments this time scale is set by the cooling and heating rates q_c and q_h. Consequently, it is to be expected that the value of $\langle\tau\rangle$ at Tg observed at comparable prior cooling and reheating rates should be virtually the same for all glasses, as is borne out by the data in Table 1. For reheating at 10 K/min after cooling at a comparable rate, $\langle\tau\rangle$ at Tg is approximately $10^{2.6}$ s $= 400$ s. This is in line with the frequently made assertion that the characteristic relaxation time at Tg is of the order of 10^2 s.

Also listed in Table 1 are the logarithms of the shear viscosities η at Tg, along with the structural relaxation activation enthalpies ΔH^* (obtained as in Fig. 7) and the corresponding shear viscosity activation enthalpies ΔH_η^* in the glass transition region, where

$$d \ln \eta/d(1/T) = \Delta H_\eta^*/R \qquad (20)$$

As suggested in the initial section of this chapter, the microscopic mechanism by which structural relaxation occurs should be intimately connected with the mechanism of viscous flow under a mechanical stress. The data in Table 1 bear this out. The shear viscosities at Tg of these inorganic glasses are all very similar and lie in the range typically 10^{11} to 10^{12} Pa s. Indeed, the temperature at which $\eta = 10^{12}$ Pa s is often used as an indicator of Tg.

Likewise, the activation enthalpies ΔH^* and ΔH_η^* which characterize the temperature dependences of the structural relaxation time and the shear viscosity appear to be identical within experimental error.

Over an extended temperature range Arrhenius plots of log η vs. $1/T$ for glassforming melts commonly exhibit curvature or non-Arrhenius behavior to varying degrees. A good method, popularized by Angell (Angell, 1991 and 1995), of comparing this non-Arrhenius viscosity behavior for various melts is to plot log η vs. a normalized reciprocal temperature Tg/T, where Tg is taken as the temperature where $\eta = 10^{12}$ Pa s. An example of such a plot is shown in Figure 11 (Lee et al., 1993). Melts such as SiO_2 whose log η vs. Tg/T plots are nearly linear have been termed "strong" liquids by Angell, in the sense that they retain their high viscosities over a large range as the temperature increases relative to Tg. At the other extreme are melts such as ZBLA, whose viscosities exhibit a highly non-Arrhenius temperature dependence and decrease rapidly to very fluid values as the temperature rises above Tg; these are termed "fragile" liquids.

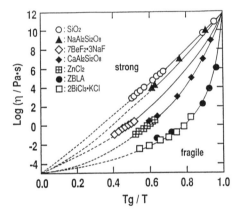

Figure 11. Plots of the logarithms of the shear viscosities of several melts versus normalized inverse temperature Tg/T.

In terms of the viscosity activation enthalpies defined in Equation (20), the slopes of the plots in Figure 11 are given by

$$d \log \eta / d(Tg/T) = \Delta H_\eta^*/2.3RTg \tag{21}$$

Moreover, as pointed out by Angell, the log η vs. Tg/T curve for a particular liquid in Figure 11 is determined by its slope $\Delta H_\eta^*/2.3RTg$ at Tg. Hence strong liquids exhibit an Arrhenius viscosity temperature dependence and have low values of $\Delta H_\eta^*/Tg$ at Tg. Fragile liquids exhibit a highly non-Arrhenius temperature dependence and a large value of $\Delta H_\eta^*/Tg$ at Tg.

The reason why some melts show an Arrhenius and others a non-Arrhenius viscosity temperature dependence over a long range in temperature and viscosity can be explicated using the Adam-Gibbs theory, as originally pointed out in a paper on the viscosity of the strong liquid BeF_2 (Moynihan and Cantor, 1968). The Adam-Gibbs expression (cf. Richet and Bottinga, this volume, and Eqns. 9 and 10) for the shear viscosity of an equilibrium melt ($T_f = T$) above Tg is given by

$$\eta = \eta_0 \exp[\frac{B_\eta}{T S_c(T)}] = \eta_0 \exp[\frac{B_\eta}{T \Delta C_P \ln(T/T_K)}] \tag{22}$$

where η_0 and B_η are constants. The melt viscosity will be roughly Arrhenius and the log η vs. $1/T$ plot roughly linear if the temperature dependence of the configurational entropy

$S_c(T)$ and hence the value of ΔC_P are small. A highly temperature dependent $S_c(T)$ and a large value of ΔC_P will give highly non-Arrhenius behavior. Hence in Angell's terms strong liquids have small ΔC_P values and fragile liquids have large ΔC_P values, a correlation that appears to be quite well borne out experimentally when comparing liquids of the same general type (e.g. comparing network oxide melts) (Angell, 1991 and 1995). Since the fragility of a liquid also corresponds to a large value of $\Delta H_\eta^*/Tg$ at Tg, one also expects a positive correlation between ΔC_P and $\Delta H_\eta^*/Tg$ at Tg. Such a correlation has been demonstrated very clearly in a recent series of papers by Minami and coworkers (Lee et al., 1993, 1994 and 1995) and is exemplified in Figure 12.

The heat capacity difference ΔC_P between liquid and glass is a measure of the contribution of the structural changes in the liquid to the enthalpy H as a function of temperature, the evolution of which with time is monitored in structural relaxation experiments. Hence a strong liquid with a small ΔC_P has a structure which maintains its "integrity" fairly well as the temperature increases above Tg, while a fragile liquid with a large ΔC_P has a structure which "falls apart" rapidly with increasing temperature.

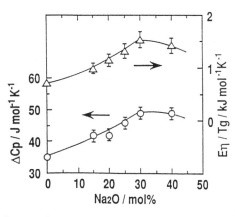

Figure 12. Composition dependences of ΔC_P and H_η^*/Tg (= E_η/Tg) at Tg for $x \cdot Na_2O \cdot (100-x) \cdot B_2O_3$ melts (Lee et al., 1995).

CONCLUSIONS

The glass transition, as explicated here, appears to be a strictly kinetic phenomenon and can be described in much the same way as the kinetics of chemical reactions are described. Presently unanswered or only partly answered questions with regard to the glass transition are of two (not unrelated) types. The first has to do with the nature of the structural relaxation process itself, that is, how the equilibrium structures of liquids change with changing temperature and pressure and what detailed molecular mechanisms are involved in the structural rearrangement. There are obviously no completely generic answers to these questions; each different liquid will have its own unique structure and mode of structural rearrangement. Obtaining detailed answers for specific liquids will depend, among other things, upon improvements in the precision and accuracy of spectroscopic and diffraction characterization and computer simulation of melt structures and dynamics. The second type of question has to do with the kinetic expressions used to describe the structural relaxation process, the physical reasons behind the nonlinear and nonexponential relaxational behavior, and the proper way to model structural relaxation occurring very far from equilibrium where the treatments described in this chapter are no longer quantitatively adequate.

On a more positive note, there has been sizeable progress in the past thirty years in our perception of and ability to model the detailed behavior of melts in the glass transition region. Some of these advances remain to be exploited. For example, as seen in Figures 9 and 10, the behavior of glasses on reheating through the transition region are strongly dependent on their previous thermal history, e.g. prior cooling rate and annealing time and temperature. This means that one should be able to infer this previous history from reheating experiments. Obvious candidates for such studies are naturally occurring geological glasses.

REFERENCES

Angell CA (1991) Relaxation in liquids, polymers and plastic crystals–strong/fragile patterns and problems. J Non-Cryst Solids 131-133:13-31
Angell CA (1995) Formation of glasses from liquids and biopolymers. Science 267:1924-1935
Barkatt A, Angell CA (1979) Optical probe studies of relaxation processes in viscous liquids. J Chem Phys 70:901-911
Brawer S (1985) Relaxation in Viscous Liquids and Glasses. American Ceramic Soc, Columbus, OH
DeBolt MA, Easteal AJ, Macedo PB, Moynihan CT (1976) Analysis of structural relaxation using rate heating data. J Am Ceram Soc 59:16-21
Dingwell DB, Webb SL (1989) Structural relaxation in silicate melts and non-newtonian melt rheology in geologic processes. Phys Chem Minerals 16:508-516
Dingwell DB (1990) Relaxation in silicate melts. Eur J Mineral 2:427-449
Dingwell DB, Knoche R, Webb SL (1993) A volume temperature relationship for liquid GeO_2 and some geophysically relevant derived parameters for network liquids. Phys Chem Minerals 19:445-453
Ducroux J-P, Rekhson SM, Merat FL (1994) Structural relaxation in thermorheologically complex materials. J Non-Cryst Solids 172-174:541-553
Easteal AJ, Wilder JA, Mohr RK, Moynihan CT (1977) Heat capacity and structural relaxation of enthalpy in As_2Se_3 glass. J Am Ceram Soc 60:134-138
Gupta PK, Huang J (1992) Enthalpy relaxation in glass fibers. In: The Physics of Non-Crystalline Solids. Pye LD, LaCourse WC, Stevens HJ (eds) Taylor and Francis, Washington, DC, p 321-326
Hodge IM (1994) Enthalpy relaxation and recovery in amorphous materials. J Non-Cryst Solids 169:211-266
Knoche R, Dingwell DB, Webb SL (1992) Temperature-dependent thermal expansivities of silicate melts: the system anorthite-diopside. Geochim Cosmochim Acta 56:689-699
Lee S-K, Tatsumisago M, Minami T (1993) Transformation range viscosity and thermal properties of sodium silicate glasses. J Ceram Soc Japan 101:1018-1020
Lee S-K, Tatsumisago M, Minami T (1994) Fragility of liquids in the system Li_2O-TeO_2. Phys Chem Glasses 35:226-228
Lee S-K, Tatsumisago M, Minami T (1995) Relationship between average coordination number and fragility of sodium borate glasses. J Ceram Soc Japan 103:398-400
McMillan PF, Poe BT, Gillet Ph, Reynard B (1994) A study of SiO_2 glass and supercooled liquid to 1950 K via high temperature Raman spectroscopy. Geochim Cosmochim Acta 58:3653-3664
Moynihan CT, Cantor S (1968) Viscosity and its temperature dependence in molten BeF_2. J Chem Phys 48:115-119
Moynihan CT, Easteal AJ, Wilder J, Tucker J (1974) Dependence of the glass transition temperature on heating and cooling rate. J Phys Chem 78:2673-2677
Moynihan CT, Easteal AJ, DeBolt MA, Tucker J (1976) Dependence of fictive temperature of glass on cooling rate. J Am Ceram Soc 59:12-16
Moynihan CT et al (1976) Structural relaxation in vitreous materials. Ann NY Acad Sci 279:15-35
Moynihan CT, Gupta PK (1978) The order parameter model for structural relaxation. J Non-Cryst Solids 29:143-158
Moynihan CT, Crichton SN, Opalka SM (1991) Linear and non-linear structural relaxation. J Non-Cryst Solids 131/133:420-434
Moynihan CT, Schroeder J (1993) Non-exponential structural relaxation, anomalous light scattering and nanoscale inhomogenieties in glass-forming liquids. J Non-Cryst Solids 160:52-59
Moynihan CT (1993) Correlation between the width of the glass transition region and the temperature dependence of the viscosity of high-Tg glasses. J Am Ceram Soc 76:1081-1087
Moynihan CT (1994) Phenomenology of the structural relaxation process and the glass transition. In: Assignment of the Glass Transition, Seyler RJ (ed) ASTM STP 1249, ASTM, Philadelphia, PA, p 32-49
Moynihan CT, Lee S-K, Tatsumisago M, Minami T (1995) Estimation of activation energies for structural relaxation and viscous flow from DTA and DSC experiments. Thermochim Acta, in press
Napolitano A, Macedo PB (1968) Spectrum of relaxation times in GeO_2 glass. J Res NBS 72A:425-433

Nemilov SV (1995) Thermodynamic and Kinetic Aspects of the Vitreous State, CRC Press, Ann Arbor, MI

Richet P, Bottinga Y, Rheology and configurational entropy of silicate melts. This volume

Scherer GW (1984) Use of the Adam-Gibbs equation in the analysis of structural relaxation. J Am Ceram Soc 67:504-511

Scherer GW (1986a) Relaxation in Glass and Composites, Wiley-Interscience, New York

Scherer GW (1986b) Volume relaxation far from equilibrium. J Am Ceram Soc 69:374-381

Scherer GW (1990) Theories of relaxation. J Non-Cryst Solids 123:75-89

Simmons JH (1993) Viscous flow and relaxation processes in glass. In: Experimental Techniques in Glass Science, Simmons CJ, El-Bayoumi OH (eds) Am Ceram Soc, Westerville, OH, p 383-432

Tauber P, Arndt J (1987) The relationship between viscosity and temperature in the system anorthite-diopside. Chem Geol 62:71-81

Webb SL (1992) Shear, volume, enthalpy and structural relaxation in silicate melts. Chem Geol 96:449-457

Chapter 2

RELAXATION IN SILICATE MELTS: SOME APPLICATIONS

D. B. Dingwell

Bayerisches Geoinstitut
Universität Bayreuth
95440 Bayreuth, Germany

INTRODUCTION

In the past decade a significant body of research has been generated in the geosciences that has referred to relaxation in silicate melts. Consideration of the relaxation timescale has provided criteria for the clear distinction between equilibrium and nonequilibrium behavior of silicate melts, led to the identification and correction of errors in previous studies of the structure and properties of silicate melts, and inspired new experiments to measure melt properties. Yet the treatments of the concepts of structural or property relaxation in molten silicates that have emerged in recent years are not, as a rule, new. Many are to be found in contributions to the fields of physical chemistry of liquids and properties of glass melts. The enhanced recognition of the potential applications of relaxation-based studies of silicate melts in the fields of geochemistry and petrology/volcanology in the past decade serve at least two purposes. First, relaxation-based studies have enhanced our ability to quantify the P-V-T equation of state, thermodynamic and transport properties of geo-relevant melts improving our ability to estimate magma properties that serve as vital parameters in igneous processes. Second, the range of melt compositions covered by igneous processes describes an otherwise poorly-investigated area of melt physics and chemistry. By uncovering the behavior, for example, of highly silicic, water-rich, tectosilicate melts, we are able to build up structure-property relationships for melts that are not only of great theoretical importance for models of geological melt behavior but also contribute in a significant way to the general decription of liquids in the physical sciences.

This chapter is concerned with some recently developed applications of the nature of relaxation in silicate melts in the description of bulk melt properties. Further discussion of the relaxational aspects of certain property determinations are covered in the chapters by Richet and Bottinga and by Webb and Dingwell in this volume. Discussions of relaxational aspects of spectroscopic and computer simulation studies are also included in the chapters by McMillan and Wolf, by Stebbins and by Poole et al. A discussion of the relationship of the glass transition to diffusion processes is in the chapter by Chakraborty. Recent advances in the area of interaction timescales for multiphase systems (not covered here) have recently been reviewed by Dingwell et al. (1993a). The emphasis here is on the contribution that relaxation studies have made to the development of new experimental pathways to the description of melt properties and melt behavior during magmatic processes.

FUNDAMENTALS

Phenomenology and significance of the glass transition

One of the most fundamental aspects of the physical properties of silicate melts is that they can exhibit both liquid-like and solid-like behavior. We denote solid-like behavior in amorphous materials as glassy and the transition from liquid-like behavior to glassy behavior as the glass transition. Virtually all melt properties are affected by the glass transition. Some properties, such as viscosity and shear modulus are drastically affected. Thus a description of the nature of the glass transition is an important first step towards the description of silicate melts themselves.

The glass transition has been traditionally defined in terms of its phenomenology as "that phenomenon... in which a solid amorphous phase exhibits with changing temperature a more or less sudden change in the derivative thermodynamic properties such as heat capacity and thermal expansivity from crystal-like to liquid-like values." (Wong and Angell, 1976). This behavior is illustrated schematically in Figure 1. The curve labeled p records the value of a melt property, P, measured during cooling of a silicate melt at a constant rate. Two segments of the curve are apparent. The inflection point is termed the glass transition of the sample material and the temperature at which the inflection occurs is the glass transition temperature recorded by the experiment (T_g). The temperature-dependent value of p at temperatures above T_g corresponds to liquid-like behavior. The temperature-dependent value of p at temperatures below T_g is solid-like or glassy behavior. If we record the temperature dependence of property p at a differing cooling rates and concentrate on the behavior in the vicinity of the inflection point of Figure 1a then we observe the behavior illustrated in Figure 1b; a path dependent hysteresis in the variation of the derivative properties in the glass transition interval.

The location of the inflection point shifts to higher values of temperature with increasing cooling rate and to lower values of temperature with decreasing cooling rate. This behavior can be used to describe a relationship between cooling rate and inflection or glass transition temperature. Such a relationship is illustrated in Figure 2. Intuitively, it can be appreciated that a slower cooling rate provides more time for the structure and properties of the silicate melt to relax to temperature-induced changes during cooling. We will see below that this can be quantified, but first we need to introduce the concept of a relaxation time.

If a silicate melt is subjected to an instantaneous change (i.e. step function) of an intensive thermodynamic variable such as temperature, the resulting change in the structure and extensive properties of the liquid is not an instantaneous one. Instead, the structure and properties of the melt "relax" towards the new equilibrium values of the thermodynamic properties as a function of time. The time taken to accomplish this re-equilibration is the relaxation time. The cooling ramp, implicit in Figure 1b, can be considered to be a series of

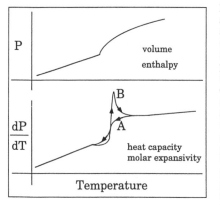

Figure 1. A schematic illustration of the variation of an arbitrary melt property, P (e.g. volume, enthalpy) and the temperature derivative of the property, dP/dT (e.g. expansivity, heat capacity) as a function of temperature. The temperature dependence of the melt property is divided by the glass transition temperature into a low temperature segment formed from the locus of values for the glassy state and a high temperature segment formed from the locus of values for the liquid state. The derivative properties illustrate (in an expanded temperature scale) that the glass transition is a region of finite width in which transient values of the derivative properties are recorded during heating and cooling transects. These transient values of the properties are path dependent and influenced by the kinetic material parameters, the temperature scanning rate and direction and (for the case of heating) the thermal history of the glass.

small steps in temperature separated by short isothermal time intervals during which the sample can approach or relax to the new equilibrium values of its properties. A log unit change in the cooling rate can be approximated by a log unit change in these isothermal time intervals and thus in the time available for the relaxation process during cooling. The relationship between relaxation time and cooling rate is thus seen to be a linear one. The trace of T_g versus time, in Figure 2, illustrates a fundamental point concerning the transition from liquid to glass. The glass transition is a curve in temperature-time space which divides the dynamic behavior of silicate melts into two fields of response; a liquid field at long timescales and high temperatures and a solid-like or glass field at short timescales and low temperatures. In Figure 2 these relations are generalized into a plot of time versus (reciprocal absolute) temperature. All processes concerning the dynamic response of silicate melts can be plotted in Figure 2 given knowledge of their characteristic timescales or rates and of their temperatures. Such processes include all experiments designed to measure the properties or structure of silicate melts and all petrogenetic processes involved in the formation of igneous rocks. Our description of the temperature dependence of silicate melt properties below will be repeatedly placed in the reference frame provided by Figure 2.

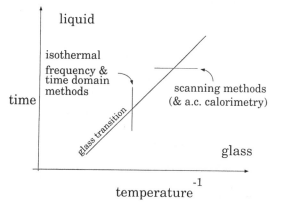

Figure 2. A schematic representation of the glass transition in reciprocal temperature-time space. The relationship between timescale of observation and resulting transition temperature yields an "activation energy" for relaxation processes. The glass transition may be investigated by (1) transects in which the timescale of the investigation is variable (e.g. time and frequency domain methods) and the relaxation of the melt to a state of equilibrium following a perturbation of the system is measured isothermally or by (2) transects in which the timescale of the observations is held constant and the temperature is scanned such that the temperature dependence of the melt properties and structure are the source of the perturbation from equilibrium. Both types of experiments yield comparable quantification of the relaxation time in silicate melts.

As noted above, the glass transition can be phenomenologically defined as the temperature where the properties of the melt change from those of the liquid-like state to those of the solid-like state (Wong and Angell, 1976; Dingwell and Webb, 1989; 1990). This is perhaps best pictured as a rheological glass transition between a dominantly viscous and a purely elastic response of the melt to an applied stress; that is, a transition between viscous dissipation and elastic storage of strain energy. Using the example of shear deformation, which, in the case of sustained, externally applied, stress also yields the brittle-ductile transition with its catastrophic consequences for volcanism, the timescale of the transition can be approximated via the Maxwell relationship for shear:

$$\tau = \frac{\eta_N}{G_\infty} \tag{1}$$

where η_N is the relaxed Newtonian shear viscosity, G_∞ is the unrelaxed elastic shear modulus and τ is the relaxation time. In this sense the shear viscosity and shear modulus can be thought of as energy *loss* and *storage moduli*, respectively. The shear viscosity is exponentially dependent on temperature and the shear modulus is essentially independent of temperature and composition and can be usefully approximated for this purpose to be a constant 10 GPa (Dingwell and Webb, 1990). The relaxation time is therefore represented by a time-temperature curve for each melt composition. The temperature dependence of the Maxwell relaxation time for any silicate melt can be readily approximated from the measured viscosity-temperature relationship together with either a measured or estimated value for the shear modulus to within a factor of two. Thus the resulting curves look very similar to the viscosity-temperature relationships of these melts. Normally, the viscosities of multicomponent melts can be calculated to within a similar uncertainty at superliquidus temperatures. Thus the shear relaxation time of melts of geological interest can be estimated for the relatively high temperatures and low viscosities relevant to eruptive temperatures.

Unfortunately, in this regard, the relationship between cooling rate and relaxation time presented below illustrates that the relaxation times corresponding to the glass transition in natural processes, and in most experiments, are much longer than those corresponding to the high temperatures and low viscosities mentioned above. Viscosity data in the range of 10^9 to 10^{15} Pa s, relatively high viscosities for experimental determination, are those that are needed in order to predict the shear relaxation time and, in combination with cooling rate data, the glass transition temperature for natural processes and for experimental quenching. The determination of viscosities in this range is possible with a number of methods (Dingwell, 1995) but most of them are more time consuming than high temperature, concentric cylinder viscometry. For this and other reasons, such as the metastability of most if not all silicate melts in this viscosity range, relatively few data are presently available, although this area is receiving considerable attention at present (see chapter by Richet and Bottinga). The shear relaxation time, derived from the Maxwell relation with the assumption of a temperature-independent G_∞, is plotted in Figure 3 for several examples of melts for which relatively complete viscosity-temperature relationships exist. The non-Arrhenian temperature dependence of these curves is the reason why low range viscosity data cannot be used, in general, to extrapolate the estimation of shear relaxation time to lower temperatures. Figure 3 illustrates well the range of relaxation time that is exhibited by silicate melts due to their temperature dependence and their composition dependence of viscosity. In this calculation, the temperature dependence of viscosity contributes to a variation in shear relaxation time of up to 12 orders of magnitude between the glass transition temperature of industrial processes and the practical onset of liquid instability due to melt volatilisation. The composition dependence of the viscosity leads to a variation in the "classical" glass transition temperature (that defined by industrial glassmaking processes), from approximately 1300°C for very pure SiO_2 to approximately 450°C for alkali silicates. In Figure 4 recent data for what is probably the most intensely investigated silicate glassforming system, Na_2O-SiO_2, are presented (Knoche et al., 1994). The greatest variation in glass transition temperature is accomplished with minor contamination of SiO_2 by water or alkalis leading to a drop of over 800°C with the addition of a few mole % of Na_2O (Fig. 5). The exact functional form of the decrease of Tg with composition near SiO_2 is not very well known and yet the sheer magnitude of the effect makes its explanation one of the challenges to all theories of relaxation.

The pressure dependence of the glass transition is very poorly known. The measurements are difficult to perform. Differential thermal analysis measurements offer an opportunity for high pressure measurements yet only one study has been conducted. Data for albite, diopside and sodium trisilicate glasses are reproduced from the study of Rosenhauer et al. (1979) in Figure 6 for pressures up to 7 kbar. The variation in glass transition temperature is very minor and positive for diopside and sodium trisilicate but

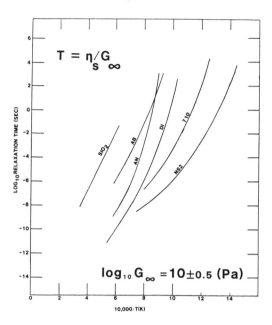

$$T = \eta_s / G_\infty$$

$$\log_{10} G_\infty = 10 \pm 0.5 \text{ (Pa)}$$

Figure 3. Quantification of the relaxation time-temperature relationship for several silicate melts through the use of shear viscosity data and the approximation of a constant value of the shear modulus together with the Maxwell relation. (AB = albite, AN = anorthite, DI = diopside, 710 = NBS SRM 710 soda lime glass, NS2 = Na-disilicate). For sources of viscosity data see Dingwell and Webb (1990 from whom this figure is reproduced).

negative for albite. These data, combined with evidence for a significant decrease in the high temperature viscosity of silicic melts from falling sphere investigations (Kushiro, 1978a,b; Dingwell, 1987) lead to the inference of greater melt fragility at higher pressure.

Figure 4. Variation in character and temperature of the glass transition in the Na_2O-SiO_2 system determined using scanning calorimetry. Note that the signal of the glass transition becomes almost immeasurably small with increasing SiO_2. Comparison of these data with estimates of the glass transition of pure SiO_2 reveal a catastrophic drop of the glass transition with the addition of a few percent of alkali oxide to the base composition. Variation of T_g with composition in the high SiO_2 range is complex and is a primary challenge to theories of relaxation in silicate melts. Reproduced with permission from Knoche et al. (1994).

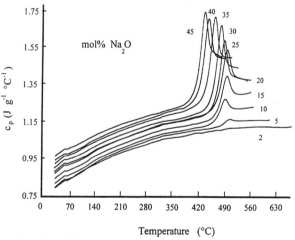

Experimental timescales and relaxation times

All experimental investigations of the properties and structure of silicate melts have timescales associated with them. The timescales are imposed by different types of experiments in different ways. Experiments that employ sinusoidal oscillations of an extrinsic field have a frequency which can be inverted to obtain a timescale. Such frequency-domain experiments are powerful probes of the structure and properties of silicate melts because the frequency, and thus the timescale, of the experimental probe can be varied over several log units to study the frequency-dependence of certain melt properties. Examples of such experiments designed to measure physical properties are the methods of ultrasonic wave propagation, specific heat spectrometric, a.c. conductivity and torsion pendulum studies. Similarly, stress-strain experiments sample melt properties on timescales that can be calculated as the inverse of the strain rate during sample deformation. Examples include constant strain-rate or constant stress determinations of viscosity, such as concentric-cylinder and fiber elongation, respectively.

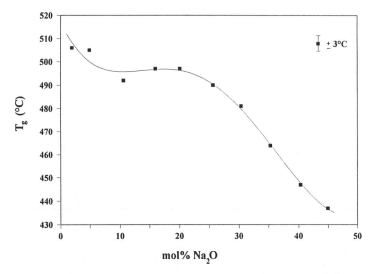

Figure 5. Variation in the glass transition temperature of melts in the system Na_2O-SiO_2. The T_g of SiO_2 lies well above 1000°C. In contrast, the total variation in T_g from 20 to 45% Na_2O is a mere 60°C. Reproduced with permission from Knoche et al. (1994).

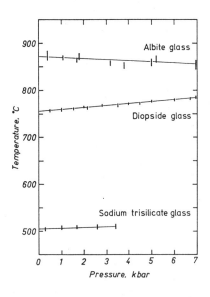

Figure 6. The variation of the glass transition temperature of silicate melts of albite, diopside and sodium trisilicate composition from 1 bar to 7 kbar determined using differential thermal analysis. The variations in this pressure range are slight. Comparison with available data on the higher temperature viscosities of albite melt at similar pressures indicates a decrease in activation energy at low viscosity which can only be reconciled with the relative invariance of T_g with pressure (pictured here) if albite melt becomes more fragile with increasing pressure. Reproduced with permission from Rosenhauer et al. (1979).

Experiments that employ a temperature scan, such as scanning calorimetric or ·dilatometric methods, operate on timescales that can be calculated from a comparison of stress relaxation times with the temperature dependence of the inflection in properties marking the glass transition. Each cooling/heating rate corresponds to a defined relaxation timescale for the experiment (quantified below).

With knowledge of the timescale of the experimental probe, the location of a particular experiment with respect to the relaxation timescale of the silicate liquid can be determined. Depending on the temperature and timescale either relaxed liquid or unrelaxed glassy melt properties will be recorded by the experiment. Most experiments are confined to certain distinct ranges of temperature (and pressure). In addition each experimental set-up is constrained in terms of the magnitude and dynamic range over which data can be acquired. Thus the absolute temperature and the temperature dependence of the melt properties themselves control the range of temperature over which property measurements can be made using a given technique. A compilation of experimental and relaxation timescales is provided in Figure 7.

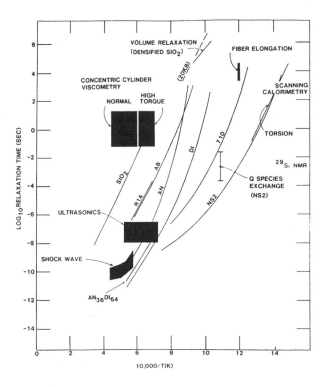

Figure 7. Comparison of the timescales commonly accessed by several experimental methods for the study of melt properties compared with the relaxation times calculated as in Figure 3 for several melt compositions. The sources of experimental timescales are discussed in Dingwell and Webb (1989). Reproduced from Dingwell and Webb (1989).

Volume versus enthalpy relaxation

Volume and enthalpy record significant inflections in their temperature dependence at the glass transition. The temperature dependence of volume and enthalpy and their derivative properties, expansivity and heat capacity, was illustrated schematically in Figure 1. The temperature dependence can be divided into a low temperature segment reflecting the glassy or solid-state properties of the melts, a high temperature segment corresponding to the liquid state properties of the melt and an intermediate temperature region of complex property variations across the glass transition. As we will see below actual calorimetric and dilatometric data for the heat capacity and expansivity of silicate melts respectively differ in the magnitudes of their temperature dependence and of their behavior in the glass transition interval. Nevertheless, to a very good approximation, the glass transition temperatures derived from both types of scanning experiments yield the same glass transition temperatures as, for example read from the peak temperatures of the derivative property curves. This is briefly illustrated in Figure 8a,b where heat capacity and expansivity data together with a normalization of the heat capacity and expansivity scans of a sample (in this case a 50:50 anorthite:diopside composition melt) with identical cooling history and heating rate has been performed such that the glassy temperature-dependent heat capacity equals 0 and the peak maximum equals 1. This normalization serves to correct out the differences in *magnitude* of variations of the two properties, heat capacity and expansivity, which are distinct for most melts, from the *relative* variations in the transient values of those properties in the glass transition interval, which are determined solely by the temperature-dependent *kinetic* properties of the sample and the experiment. The tight overlap of the normalized heat capacity and expansivity curves up to the peak temperature (where viscous deformation invalidates the expansivity trace) indicates that the peak position, the form and thus the underlying kinetics controlling the relaxation of both enthalpy and volume are very similar and possibly identical. This equivalence of volume and enthalpy relaxation provides a simplification which serves below as a powerful tool for the investigation of melt and magma properties.

The equivalence of relaxation behavior and timescale for such disparate properties as enthalpy and volume begs the question of whether the shear stress relaxation time is also equivalent to that of volume and enthalpy relaxation.

Volume versus shear relaxation

The determination of volume relaxation times for silicate melts may be directly carried out by annealing samples of glass and serially measuring their densities following perturbation of the system by a pressure or a temperature step (e.g. Ritland, 1954; Höfler and Seifert, 1984). Relatively few such data exist. An alternative approach, for which relatively many data exist, is the ultrasonic determination of sound velocities and attenuation from which viscosity may be determined. From the identity relating volume, shear and longitudinal moduli of isotropic substances,

$$M = K + (4/3)G \qquad (2)$$

we can write a viscous equivalent

$$\eta_L = \eta_v + (4/3)\eta_s \qquad (3)$$

where η_L, η_v, and η_s are the longitudinal, volume and shear viscosities, respectively.

Compendia of the elastic constants of silicate glasses (e.g. Bansal and Doremus, 1986) indicate that the shear (or rigidity) modulus and the bulk modulus of silicate glasses are similar in magnitude. Studies of the propagation of longitudinal ultrasonic waves in a small number of natural silicate melts (Sato and Manghnani, 1984) have also demonstrated that concentric-cylinder data and (ultrasonic) longitudinal wave attenuation data yield shear and longitudinal viscosity data that are also similar in magnitude. Dingwell and Webb (1989) later demonstrated that the extensive longitudinal wave attenuation data set of Rivers and Carmichael (1987) was consistent with the proposal that the volume and shear viscosities of the investigated melts were equal. Figure 9 illustrates this comparison using literature data for shear viscosities (see Dingwell and Webb, 1989, for data sources), the longitudinal viscosity data noted above, and Equation (3), to obtain volume viscosity by difference.

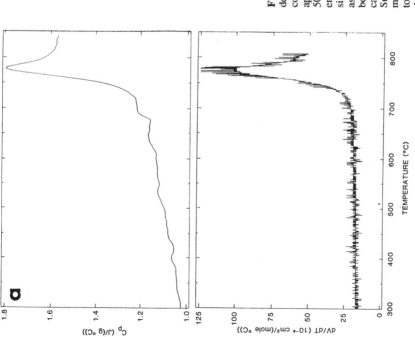

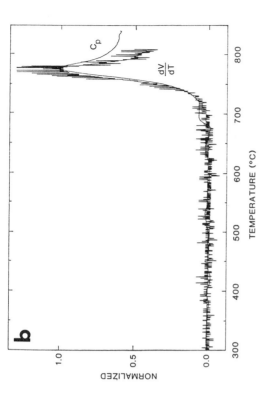

Figure 8. (a) A comparison of the dilatometric and calorimetric data for the temperature derivative properties, expansivity and heat capacity, for a melt of 50:50 anorthite-diopside composition. The general similarity of the glass transition temperature and the peak shape is apparent. (b) The normalised comparison of the expansivity and the heat capacity data for the 50:50 anorthite-diopside glass composition. The identical glass transition behavior of both enthalpy and volume is apparent in this normalisation, up to the point (the peak) where significant viscous deformation begins to distort the dilatometric signal. With the assumption of the complete equivalence of the kinetics factors underlying the relaxation behavior of volume and of enthalpy, the high temperature liquid value of the heat capacity can, through the normalised comparison, be used to infer the liquid value of the expansivity. Such expansivity data, combined with high temperature volume-temperature data, provide a much more complete volume temperature relation for each liquid from the glass transition up to temperatures above the liquidus than previously possible. Amongst other aspects, such determinations of the low temperature liquid expansivity have demonstrated for the first time the significant temperature dependence of the expansivity of some silicate melts. Reproduced with permission from Knoche et al. (1992a).

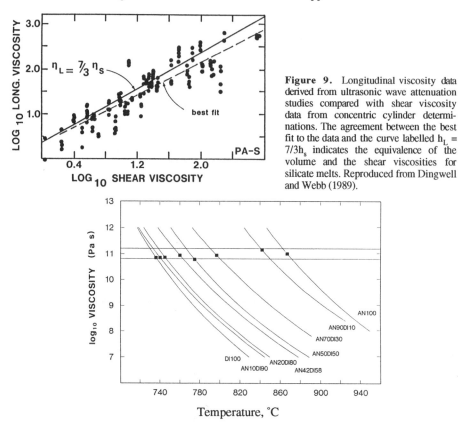

Figure 9. Longitudinal viscosity data derived from ultrasonic wave attenuation studies compared with shear viscosity data from concentric cylinder determinations. The agreement between the best fit to the data and the curve labelled $h_L = 7/3h_s$ indicates the equivalence of the volume and the shear viscosities for silicate melts. Reproduced from Dingwell and Webb (1989).

Temperature, °C

Figure 10. A comparison of calorimetrically determined glass transition temperatures and viscosity-temperature relationships for the system anorthite-diopside. The calorimetrically determined glass transition temperatures represent the peak temperatures from the transient values of heat capacity during heating in a scanning calorimeter. These T_g values correspond to viscosity values for each of the investigated melts that are equal within the uncertainties of the data sets. Thus the composition-dependence of the glass transition temperature can be described by an isokom or locus of points of equal viscosity for those compositions. Reproduced with permission from Knoche et al. (1992a).

If the volume and shear moduli and viscosities are similar or equal, then it seems inevitable that the relaxation times for shear and volume relaxation must be similar if not identical. Direct comparison of shear and volume relaxation times is not as simple as the comparison of volume and enthalpy discussed above. Nevertheless, the wealth of high quality shear viscosity data, together with our approximations regarding the shear modulus presented above, and the observations of the enthalpy-volume equivalence, allow a "short-cut" to the comparison of shear and volume relaxation times by comparison of the shear viscosity of various melts at a consistently defined glass transition temperature from scanning calorimetry. In short, if the shear relaxation time is identical to the volume and enthalpy relaxation times then, under the approximation of invariant shear modulus, the calorimetric glass transition temperature must represent an isokom or locus of constant viscosity for all melt compositions. This turns out to be a good approximation over wide ranges of melt composition. An example of this comparison is provided in Figure 10 for the system anorthite-diopside. The individual viscosity-temperature relations of the melts are each marked by a cross representing the calorimetric glass transition temperature. The crosses are distributed across Figure 10 at values of viscosity for the individual compositions which do not vary outside the error of the determinations, describing instead an isokom of viscosity.

The demonstration of constant viscosity at a consistently defined calorimetric (or dilatometric) glass transition temperature implies that the shear relaxation time (which is proportional to viscosity through the Maxwell relation), is closely related to the enthalpy and volume relaxation times. A more exact comparison of the shear with the volume and enthalpy relaxation times is difficult because, as mentioned above, our choice of the definition of the glass transition as the peak temperature on the scanning traces is an arbitrary selection of a single feature which, although certainly a useful approximation to the glass transition temperature, does not exactly represent the fictive temperature of the melt as frozen in during cooling to form a glass. However, we have one more possibility to fix, for our purposes, the relation between the enthalpy and the shear stress relaxation times.

Studies of enthalpy relaxation in silicate glass melts have indicated that the activation energy of enthalpy relaxation is identical to that of viscous flow (e.g. Scherer, 1984). This leads to the following relationship between T_g and viscosity, which is independent of melt composition;

$$-\log_{10}|q| = A_{DTA} + \frac{E_{DTA}}{T_g} \tag{5}$$

$$\log_{10} \eta_s \text{ (at } T_g) = A_\eta + \frac{E_\eta}{T_g} \tag{6}$$

(Moynihan et al., 1976) as the activation energies viscous flow E_η and enthalpy relaxation E_{DTA} are identical

$$\log_{10} \eta_s \text{ (at } T_g) = \text{constant} - \log_{10}|q| = \log_{10} t \text{ (at } T_g) + \log_{10} G \tag{7}$$

for shear relaxation time, t (s) and shear modulus G (Pa). Scherer (1984) found the constant to be 11.3 for η_s in $\log_{10}$ Pa s and quench rate $|q|$ in °C s^{-1}. This relationship between viscosity and quench rate can be used to estimate melt viscosity at the glass transition. Stevenson et al. (1995) have performed rheological and calorimetric experiments to determine the value of the constant of Equation (7) for a number of rhyolitic obsidians. Their results are presented in Figure 11.

The viscosity-temperature and cooling rate-glass transition temperature relationships are Arrhenian for the obsidian compositions investigated by Stevenson et al. (1995). The activation energies obtained for viscous flow and enthalpy relaxation are identical for each sample. The value of the constant required in Equation (7) is 10.49 ± 0.13 (1σ = 0.31). Equation (7) can now be used to relate the cooling rate of volcanic melts to the viscosity at the glass transition.

As a first approximation, the shift factor K derived by Stevenson et al. (1995) (10.49 ± 0.13; 1 σ = 0.31) can be used for all silicic volcanic compositions. This shift factor is a revision of the zeroth order approximation of Dingwell and Webb (1990) based on an average shear modulus of 10 GPa. On closer inspection the data in Figure 11 raise the possibility that a slight compositional dependence of the shift factor may exist. Figure 12 displays the shift factor obtained by Stevenson et al. (1995) versus the agpaitic index of those melts. The shift factor appears to increase with decreasing agpaitic index. This minor variation in shift factor is within the 1σ estimate of fit quality. Nevertheless the variation might reflect a melt structure influence on the relationship between the enthalpic relaxation time and the temperature of the peak in the Cp curve. Clearly a larger range of compositional variation needs to be investigated to address this possibility.

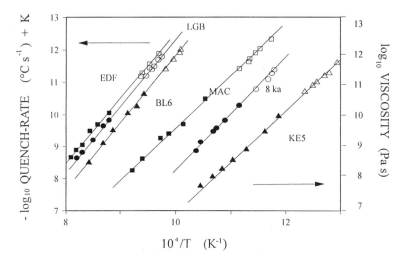

Figure 11. A comparison of the temperature-dependence of shear viscosity and the quench rate-dependence of the calorimetric (peak) glass transition temperature for six silicic volcanic obsidians. The identical activation energies exhibited by enthalpy relaxation (open symbols) and shear stress relaxation (solid symbols) together with several other lines of evidence lead to the inference that the same thermally activated process lies behind both phenomena. The two curves can be related through a logarithmic "shift factor" which then together with the Maxwell relation, explicitly relates cooling rate to relaxation time. Reproduced with permission from Stevenson et al. (1995). See that paper for identification of samples.

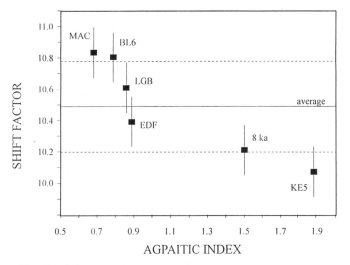

Figure 12. The shift factor derived in Figure 11 plotted versus the agpaitic index of the investigated obsidians. The shift factor very slightly to lower values with increasing peralkalinity. The slight variation may indicate a subtle variation with composition of the relation between viscosity and calorimetric glass transition temperature. It is however very small in comparison to the uncertainties in the other parameters which must be combined with the shift factor to apply it to the estimation of the location of the glass transition during volcanic cooling. See the text for further discussion. Reproduced with permission from Stevenson et al. (1995).

Also worth noting in Figure 11 is the activation energy range obtained for these volcanic obsidians. The values of the slope of viscosity versus reciprocal temperature are relatively low. The activation energies exhibited by the compositions in Figure 11 yield a decadic variation in the viscosity over a temperature range of some 30°C. We will see later that this leads to a large variation in the glass transition temperature experienced by these samples in nature. The relatively low activation energy of relaxation for these melts is typical of relatively "strong" melts (melts for which the viscosity-temperature relationship is not very non-Arrhenian) in the high viscosity region (Angell, 1984). One of the greatest limitations of the concept of a glass transition temperature applied to geological melts is that, in comparison to most other classes of liquids, they are quite strong. For very fragile liquids, the non-Arrhenian viscosity-temperature relationship leads to such a large activation energy at glass transition timescales corresponding to moderate cooling rates that the glass transition approximates a single temperature value over a range of timescales. That this is clearly not the case for silicate melts is demonstrated in Figure 11. We are faced with the prospect that the glass transition temperature in volcanological processes must be specified for each cooling rate with all the implications that will have for volcanic facies-specific cooling rate variations.

An important distinction regarding the comparison between calorimetric and shear stress relaxation times presented above is that calorimetric and viscometric data (peak temperature and viscosity, respectively) are not necessarily monotonically proportional to the calorimetric and shear stress relaxation times, respectively. Why not? Firstly, the viscosity is related to the shear relaxation time (τ) through the Maxwell relation $\tau = \eta/G$, where G is the shear modulus. Tabulated data for the (low temperature) shear moduli of silicate glasses (Bansal and Doremus, 1986) indicate that significant differences in the shear modulus do exist but that such differences are unlikely to give rise to a total variation in shift factor of more than 0.1. Secondly, the "peak" temperature in a DTA trace is a complex function of the melt structure and the enthalpic relaxation time distribution (see below). It is thus possible that minor shifts in the calorimetric peak temperature are due to differences in the relaxation time distribution and the melt structure between different melt compositions such that the 1:1 relationship between the peak temperature and calorimetric relaxation time is lost as a function of melt structure. The result is a variable shift factor. This is not the equivalent of saying that the viscous relaxation time (or a single relaxation time approximation to the relaxation behavior yields a value that) is varying independently of the calorimetric relaxation time. As shown by the present agreement between the activation energies of viscous relaxation and enthalpic relaxation, the use of a fixed point on the calorimetric curve to define T_g results in the calculation of a correct (consistent) value for the activation energy (i.e. identical with that of shear relaxation), but suggests that the relaxation time associated with this point on the curve might vary as a function of the composition and the resulting kinetic parameters of the melt.

The fact that the shift factor Equation (7) for the relationship between enthalpy and shear stress relaxation explicitly includes a cooling rate can be used to determine the temperature at which a silicate melt will intersect the glass transition for a given cooling rate and to determine, from the activation energy of the viscosity, the sensitivity of the glass transition temperature experienced by the cooling sample to the rate at which it is cooled. The former calculation gives us the effective relaxation time corresponding to a given cooling rate and the latter allows us to evaluate the range of glass transition temperature likely to be experienced by a given volcanic melt during cooling in nature at differing rates in differing volcanic facies. The relationship between cooling technique for a variety of experimental methods and the log of the relaxation time for each case is illustrated in Figure 13.

Figure 13. A schematic description of the "hierarchy" of cooling rates normally obtained using a variety of techniques for the cooling or quenching of melts in research and industry. The slowest effective cooling rates achievable are obtained by programmed rate cools with electronic controllers. The fastest rates are achieved with so-called spin quench methods. The entire range of quench rate available to experimental labs is some eight orders of magnitude. The lower cooling rate end can be extended by simulating slower cooling rates using annealing experiments. The higher cooling rate end can be simulated in the quench of molecular dynamic simulations of melts.

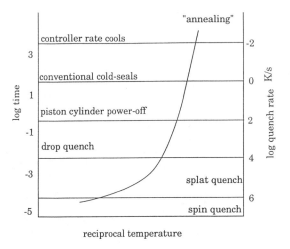

The equivalence of enthalpic and volume relaxation times also lies behind several applications of relaxation to the study of silicate melts. Extensive data for the calorimetric and for the dilatometric relaxation of silicate melts now points clearly to identical peak temperatures during scanning measurements (calorimetric and dilatometric) on a wide range of melt compositions, an example of which has been provided in Figure 8. The equality of these peak temperatures indicates that the kinetic parameters controlling the relaxation time for the two investigated properties, enthalpy and volume, are indistinguishable from each other. This equivalence of relaxation time, combined with the observation presented above for the equivalence of shear stress and enthalpic relaxation times, appears to yield a relatively robust simplification of the estimation of the glass transition for the three important physical properties of igneous melts, viscosity, enthalpy and volume. Prediction of the glass transition for a given experiment or process in nature is then fully transferable between calorimetric, volumetric and rheological studies of silicate melts. This is a very powerful simplification and one which apparently does not hold for glassforming liquids in general (Angell, 1991). Nevertheless the tests provided above and the examples of applications provided below, illustrate that in addition to greatly simplifying the evaluation of the role of the glass transition in petrology and volcanology, it opens up new paths in the experimental investigation of silicate melt properties under difficult conditions.

Secondary relaxations

It has long been known that certain aspects of the structure of silicate melts induce relaxations at conditions well away from the glass transition in temperature-time space. These relaxations have been variably described as beta, secondary, detached, or decoupled relaxations. For silicate melts there is no evidence of such relaxations occurring in the liquid field above the glass transition at longer times and higher temperatures. Instead such relaxations occur invariably at lower temperatures for a given timescale or at shorter timescales for a given temperature, than the glass transition and are thus in the glass field defined in Figure 14. Such relaxations have lower activation energies than the glass transition. They are Arrhenian at temperatures lower than the glass transition for the timescale of the investigations. Under temperature-time conditions, however, where the melt structure experiences sufficient time to relax to the selected temperature on the timescale of the investigation of the secondary relaxation, the temperature dependence is non-Arrhenian. This is because it is influenced as it is by the temperature dependence of the melt structure. One of the best investigated examples of such a relaxation is the case of

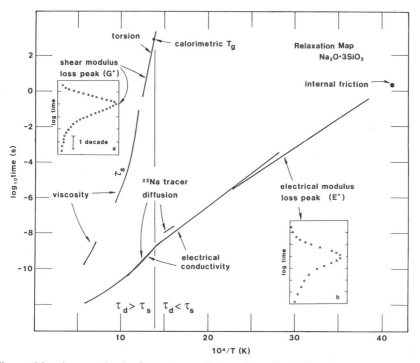

Figure 14. An example of a "relaxation map" for a silicate liquid. The diagram summarizes in an Arrhenian plot the location in time and in temperature of various relaxation modes in the silicate melt structure. Data on the temperature-timescale relationship for these relaxations come from calorimetric, dilatometric, viscometric, and electric property determinations. A detailed discussion of the data contributing to this diagram can be found in Dingwell (1990) from which this figure is reproduced.

alkali mobility in alkali silicate melts. Figure 14 illustrates a summary of data available for two relaxation modes in an alkali trisilicate melt. The secondary relaxation has a measurable expression in conduction of electricity as well as internal friction studies (see references in Dingwell, 1990). The electrical effect of alkali mobility is quite large as in other alkali-bearing melts whereas the internal friction effect is very small compared to the internal friction at the glass transition. Below, this presentation of relaxation modes as a relaxation "map" is used in the discussion of hydrous melts.

APPLICATIONS

Relaxation geospeedometry

Above it has been emphasized that great care must be taken to obtain identical thermal histories of melts before sensible comparisons of enthalpic, volume and shear stress relaxation are permitted. This sensitivity of the hysteresis in the glass transition interval to thermal history should be a potential source of information on the thermal history of glasses where it is not known. This hysteresis of melt properties in the glass transition interval stems from the path-dependence of the properties of the melt in passing from the glass to the liquid. Models for the empirical description of the transient variation or hysteresis of enthalpy in the glass transition interval have been proposed in order to quantify the relationship between hysteresis, material properties, thermal history and heating rate of measurement. These models form the basis of the application of scanning calorimetric data to the determination of the cooling history of the volcanic glass, i.e. relaxation geospeedometry.

As noted above, if the cooling or heating rate, $|q|$, is viewed as a series of jumps of temperature ΔT which are followed by isothermal holds of duration Δt (i.e. $|q|$ = $|\Delta T/\Delta t|$), it follows that a certain cooling rate corresponds to a certain time available for relaxation of the system. If the duration of Δt is relatively short (i.e. if the value of Δt approaches that of τ) then the time available for equilibration (relaxation) is no longer sufficient for relaxation (in this case backreaction of the structure with falling temperature) to permit further changes during cooling. Thus, when a glass is cooled below the glass transition a configuration (structure) is frozen into the glass, which can be correlated with a temperature at which the frozen structure corresponds to the equilibrium state. This is termed the fictive temperature, T_f, of the glass (Tool, 1946). The fictive temperature can be considered an order parameter for the description of the state of the system. Although subsequent experiments (e.g. Ritland, 1954) have demonstrated that complex thermal histories require more than a single order parameter to completely describe the relaxation behavior of glasses, the fictive temperature approximation yields remarkably good predictions of the bulk properties of quenched glasses. Therefore, this frozen structure, defining the fictive temperature, can be indirectly monitored by the study of its expression in any structurally-dependent property such as volume or (in the present case) enthalpy.

The onset of the glass transition upon cooling involves a change from a metastable equilibrium state ($T_f = T$) to a disequilibrium state ($T_f > T$) in a way which is illustrated with Figure 15. In Figure 15 the variation of the structural state of the liquid (indicated by variation in T_f) is illustrated for cooling and heating paths. At high temperature the melt is in equilibrium, described by $T_f = T$, whereas at low temperature the melt is not in equilibrium. During cooling, the path of T_f across the glass transition interval is marked by a departure from the equilibrium $T_f = T$ to a point where T_f becomes unaffected by temperature and the structure becomes frozen. When melts are rapidly quenched this departure from equilibrium occurs at a higher temperature and consequently a higher fictive temperature is frozen into the glass structure. For more slowly cooled glasses the melt remains in equilibrium to lower temperatures and thus lower values of fictive temperature are achieved. The glass structure (as represented by the fictive temperature, T_f) is therefore quench-rate-dependent with more rapid quench rates yielding higher values of T_f.

A hysteresis is observed between heating and cooling curves (Fig. 15). Cooling results in a gradual deviation from the equilibrium condition ($T_f = T$) until a constant value of fictive temperature (T_f = constant) below the transition, is reached. Reheating results in a

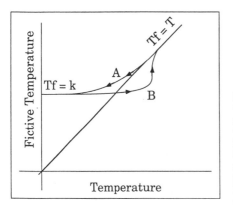

Figure 15. The variation of fictive temperature (T_f) with temperature (T) during the cooling and reheating of a melt. The derivative of the fictive temperature with respect to temperature goes from a value of 0 (T_f is constant) in the glassy state to a value of 1 ($T_f = T$) in the liquid state. This derivative is the normalisation basis for the comparison of the dilatometric and calorimetric data for the determination of liquid expansivity. Redrawn from Narayanaswamy (1971).

return to equilibrium via a path that is dependent on the heating rate, the previous cooling rate, the temperature dependence of structure and the temperature dependence of the relaxation time in the liquid. The detailed behavior of properties within the glass transition is a function of these parameters (Narayanaswamy, 1971; Moynihan et al., 1976a; Angell, 1988; Scherer, 1990; Angell, 1991).

If the value of some property, p (in our case enthalpy), is used to determine T_f, the value of T_f at temperature T' is defined as;

$$\frac{\Delta T_f}{\Delta T} = \frac{\left[(dp/dT) - (dp/dT)_g\right]\Big|_{T'}}{\left[(dp/dT)_e - (dp/dT)_g\right]\Big|_{T_f}} \tag{6}$$

where the subscripts "e" and "g" refer to the equilibrium liquid and the glass respectively (Scherer, 1984). T_f can be used to describe the property during changes in temperature (T) and time (t). In order to calculate T_f as a function of increasing temperature it is necessary to define the form of the relaxation equation and the form of the relaxation time as a function of temperature. An empirical expression for the relaxation process in silicate melts is the Kohlrausch-William-Watts "stretched exponential" function (KWW, Scherer, 1984)

$$p = p_0 \exp[-(t/\tau_{0k})^\beta] \tag{7}$$

where τ_{0k} is the characteristic relaxation time and ß is a constant ($0 < ß \le 1$). The KWW function has been found to give an excellent fit to a wide range of relaxation processes (DeBolt et al., 1976). Lower values of ß indicate a broader spectrum of relaxation times. The evolution of the fictive temperature and the related property can be written as

$$T_{fm} = T_0 + \sum_{j=1}^{m} \Delta T_j \left[1 - \exp-\left(\sum_{k=j}^{m} \Delta T_k / |q_k| \tau_{0k}\right)^\beta\right] \tag{8}$$

for temperature T_0 above the glass transition and temperature steps ΔT, heating (or cooling) rate $|q|$, and relaxation time τ_{0k} (DeBolt et al., 1976). This is the basic equation used in the present study to model the evolution of T_f and enthalpy across the glass transition interval.

Evidence has been reviewed above supporting the statement that the relaxation of enthalpy, volume, and shear stress all have the same timescales. This is supported by a number of investigations (DeBolt et al., 1976; Scherer 1984a,b, 1986, 1990; Stevenson et al., 1995; Hess et al., 1995). Following Narayanaswamy (1971, 1988) an Arrhenian form of the relaxation time equation with terms for the (temperature-dependent) structural contribution and the thermal contribution to the relaxation is used

$$\tau_p = \tau_0 \exp\left[\frac{x H}{RT} + \frac{(1-x) H}{R T_{fp}}\right] \tag{9}$$

where H is the activation enthalpy, R is the gas constant and x is a constant with $0 \le x \le 1$. H is equal to the activation energy for shear viscosity in the glass transition interval. When x=1 there is only a thermal contribution to the relaxation time. As x approaches zero there is an increasing contribution of the temperature dependence of the structure to the calculated relaxation time. This description of the temperature dependence of the relaxation time

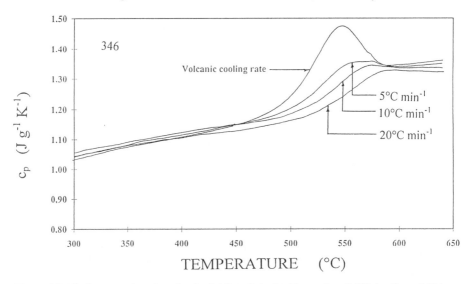

TEMPERATURE   (°C)

Figure 16. The heat capacity of a volcanic obsidian obtained with samples of differing thermal history. The known cooling rates of 20, 10 and 5°C/min were generated within the scanning calorimeter and the unknown cooling rate is continues the trend of the known cooling rates to a lower apparent cooling rate. The quantification of the hysteresis in the transient segment of the heat capacity at the glass transition is the method discussed in the text for the characterisation of the thermal history of the natural glass in terms of an apparent linear cooling rate across the glass transition using the concept of relaxation geospeedometry. Reproduced from Wilding et al. (1995).

results in a four parameter structural relaxation model; with x and ß unknown, τ_0 estimated from Equation (1), and H the activation energy calculated from Equation (6).

Of the four natural compositions investigated by Wilding et al. (1995), one sample, a phonolite from Montaña Rajada, Tenerife, is a massive obsidian from one of the more recent flows. Two compositions from lava flows within the Greater Olkaria Volcanic Complex (GOVC), Kenya, are peralkaline rhyolites with agpaitic indicies (A.I.) of 1.0 and 1.3 and the fourth composition is a pantellerite from the Eburru Complex, adjacent to the GOVC. The pantellerite occurs as a thin band of glass 0.5 m thick and 10 m in lateral extent within a poorly structured, proximal air fall deposit.

The experimental details of the present application of the experimental determination of cooling rate by scanning calorimetry have been presented in full by Wilding et al. (1995). They are summarized briefly here. The heat capacity of each raw sample was determined to ~50°C above the glass transition temperature at a heating rate of 5°C min^{-1}. The heat capacity for a natural glass is shown in Figure 16. This heat capacity data was subsequently used to determine the cooling rate of the natural glass. Central to the successful application of this method to volcanic glasses is a reliable estimate of their material specific parameters. These are obtained to be input into the fitting of calorimetric data for the determination of cooling rate by the adoption of an internal calibration procedure. Such an internal calibration is performed by recording the heat capacity curves of the sample with matched, known cooling and heating rates of 4, 5, 8, 10, 16, 20 and 32°C min^{-1}. Sample integrity during measurement was confirmed by bulk chemical analyses, ^{57}Fe Mössbauer spectroscopy (oxidation state), and infrared absorption spectroscopy (water content). Examples of the heat capacity data for the sample "346" obtained under such conditions are included in Figure 16. The temperature at which the peak in the heat capacity curve occurs increases by more than 40°C with this increase in matched cooling and heating rates. The activation energy of enthalpy relaxation for each sample was then calculated from the known heating rate and measured peak temperature T_g;

$$-\log_{10}|q| = -\log_{10}|q_0| + \frac{H}{RT_g}.\tag{10}$$

This treatment enables the derivation of the activation energy for enthalpic relaxation (H) through the relationship between quench rate, $|q|$, and reciprocal T_g.

The modeled c_p curves are expressed in terms of a normalized diagram (Fig. 17a,b,c) which makes reference to the fictive temperature in the glass and liquid fields. In this sample, "DP1", the modeled cooling rate of 0.8×10^{-3} °C s^{-1} is shown in comparison with the c_p trace for the raw glass on reheating (Fig. 17a). In addition modeled heat capacity curves which differ by 1 $\log_{10}$ unit are shown i.e. 0.8×10^{-2} °C s^{-1} (Fig. 17b) and 0.8 $\times 10^{-4}$ °C s^{-1} (Fig. 17c). The results of the modeled cooling rates are demonstrated in Figure 18. Each modeled cooling rate is shown with the calibration range, that is the range of quench rates in the subsequent calorimetry measurements. The slope of each line is related to the activation energy of enthalpic relaxation..

Four very different cooling rates have been obtained for the four different compositions. The most slowly cooled natural sample is the most rhyolitic and has the highest viscosity. The two samples of the more peralkaline rhyolite, "346", have different modeled cooling rates which we interpret as differences in the thermal geometry of the lava flow. The difference of a factor of 10 is greater than the reproducibility of the measurements and model (Fig. 18) and for this relatively viscous sample cooling may have varied over relatively small distances.

Both the samples of the phonolite, "DP1" have comparable cooling rates, as do the samples of the pantellerite, "KE5". The latter sample has a very slow modeled cooling rate of the order of 2×10^{-5} °C s^{-1}, equivalent to 2°C d^{-1}. The field occurrence of this sample suggests emplacement as a welded horizon within an air fall pumice, Wilding et al. (1995) have interpreted the relatively slow cooling rate as the result of slow dissipation of heat from a relatively thick volcanic pile possibly resulting from a reheating of the rapidly cooled sample due to ist primary burial within hot pyroclastics. This last aspect raises the important point that should volcanic samples be reheated at conditions sufficiently above the glass transition that total relaxation occurs, then all evidence of the primary cooling rate may be lost. The recorded state of the glass structure would in the case of such a double cooling cycle (or a multiple cooling cycle) which involves excursions above the glass transition, only record information on the last cooling cycle. If the annealing of a volcanic sample involves partial relaxation near the glass transition then the result may be a very complex signal from more than one cycle of the cooling history.

In further analysis of facies-specific samples of a very wide range of phonolite samples from Tenerife, Wilding et al. (in prep.) have identified a range in effective cooling rate of over four orders of magnitude from less than a degree per day to ten degrees per second, or 0.0001°C/s to 10°C/s. A very similar range (0.0003°C/s to 5.5°C/s) of effective cooling rates has been obtained by Zhang et al. (1995) for calcalkaline rhyolites from California based on entirely different data. This equivalence of cooling rates estimated from property relaxation (Wilding et al., 1995) and from hydrous species relaxation (Zhang et al., 1995) serves to underline the point (see also below) that the relaxation of properties and structures in silicate melts are intimately linked.

Equation of state: liquid expansivity and volume

The next application of melt relaxation involves the premise that not only the relaxation times of various properties are equivalent but also the details of the relaxation

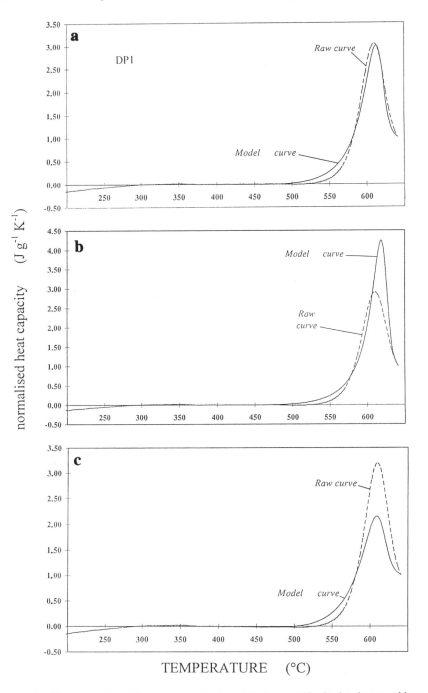

Figure 17. The results of modeling the transient values of the heat capacity in the glass transition range using the algorithm and parameters described in the text. An adjustment of the cooling rate value (representing predicted thermal history) by 1 order of magnitude lower (b) or higher (c) poduces a serious mismatch, the severity of which is an indication of the sensitivity of the calorimetric hysteresis to the cooling rates obtained. Reproduced with permission from Wilding et al. (1995).

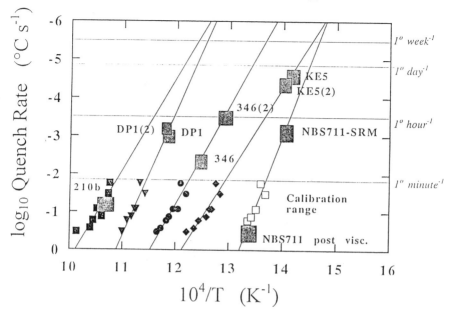

Figure 18. The logarithm of the quench rate versus the reciprocal of the absolute temperature for five investigated volcanic obsidians. The calibration range of the range of heating/cooling rates accessed by the scanning calorimeter is indicated in the small symbols for each composition. From the variation of the calorimetric glass transition temperature with heating/cooling rate, the enthalpy of the relaxation process can be estimated. This estimation provides the preexponential and activation energy parameters for the temperature dependence of the relaxation time. The modelled cooling rates for the naturally cooled glasses are the vertical positions of the large boxes. The temperatures which correspond to the intersections of these cooling rates and the slopes of each composition represent the glass transition temperature that each melt experienced during cooling in nature. The range of cooling rates estimated form these compositions originating from differing volcanic facies is four logarithmic units (from faster than 1 K/min to approximately 1 K/day). Reproduced with permission from Wilding et al. (1995).

process in the glass transition interval. The theory of obtaining relaxed-liquid molar expansivity data from a combination of scanning calorimetry and dilatometry, the application presented here, has its origins in that premise. It was introduced by Webb et al. (1992) and is described in full in that study. Here we review the method and its results to date.

The comparison of dilatometric and calorimetric measurements is based on the principle that the relaxation of melt properties in the glass transition region can be described by a universal set of parameters. These universal parameters can be derived from the relaxation of one property (in this case enthalpy) and then applied to predict the behavior of a second property, such as volume. Heat capacity is measured through the glass transition and molar expansivity up to the glass transition using differential scanning calorimetry and dilatometry, respectively. The normalized calorimeter trace is then applied to the molar expansivity curve to extend it into the liquid region.

It has been emphasized above that silicate glasses quenched from liquids preserve a configuration that can be approximated to the equilibrium structure of the liquid at some fictive temperature, T_f (Tool and Eichlin, 1931). To describe in general, the relaxed (liquid) or unrelaxed (glassy) properties of a silicate melt it is necessary to specify the temperature and the fictive temperature of the melt. The temperature derivative of the physical properties of a glass and a liquid (e.g. molar heat capacity (dH/dT) and molar thermal expansivity (dV/dT)) can be used to describe the temperature-derivative of the fictive temperature. To do this, the temperature-derivative of any property in the glass transition interval (*e.g.* enthalpy, volume) is normalized with respect to the temperature-derivative of the liquid and glassy properties. Assuming the equivalence in relaxation time and activation energy for volume and enthalpy relaxation for melts with identical thermal histories, the temperature derivatives of the fictive temperature T_f, the enthalpy (H) and volume (V) can be related by:

$$\frac{c_p(T') - c_{pg}(T')}{c_{pe}(T_f) - c_{pg}(T_f)} = \frac{dT_f}{dT}\bigg|_{T'} = \frac{\left(\dfrac{dV(T)}{dT} - \dfrac{dV_g(T)}{dT}\right)\bigg|_{T'}}{\left(\dfrac{dV_e(T)}{dT} - \dfrac{dV_g(T)}{dT}\right)\bigg|_{T_f}} \tag{11}$$

Thus, in the glass transition region, the behavior of any temperature-dependent property of a melt can be predicted from the known behavior of another temperature-dependent property if the relaxation of the two properties is equivalent. In the above equation, which relates c_p and thermal expansivity dV/dT, the only unknown parameter is the thermal expansivity of the relaxed liquid at temperature T' in the glass transition interval.

Due to the lack of relaxed thermal expansivity data (viscous flow of the melt above T_g results in the inability to determine volume as a function of temperature), the liquid molar thermal expansivity must be calculated from the dilatometric trace by normalizing both the scanning calorimetric and dilatometric data:

$$P'(T) = \frac{P(T) - P_g(T)}{P_p - P_g(T)} \tag{12}$$

where the subscripts "p" and "g" refer to peak and unrelaxed, glassy values. The relaxed value of thermal expansivity can now be generated from the peak and extrapolated glassy values of thermal expansivity; the volume and coefficient of volume thermal expansion α_V [1/V(dV/dT)] of the melt can be calculated.

Small changes in composition or fictive temperature of a silicate melt can strongly influence relaxation behavior. Thus the above method only can be applied to calorimetric and dilatometric data obtained on the same sample using identical experimental conditions and thermal histories. It is only this internal consistency that permits the use of the assumption of the equivalence of the enthalpy and volume relaxation behavior.

The investigation of a variety of synthetic melt compositions using the methods of determination of the expansivity and volume of the melt just above the glass transition have yielded results which may be generalized to the following conclusions. The volume and expansivity data obtained near the glass transition are consistent with the volume and expansivity data obtained with immersion methods at higher temperatures. This is not to say however that the volume and expansivity data are similar in magnitude. Figure 19 illustrates the point well for the example of GeO_2. The curve is a combined fit to the volume and expansivity data of Dingwell et al. (1993c) and those of Sekiya et al. (1980). Clearly the low and high temperature data are reconcilable with a smooth nonlinear volume-temperature relationship. Note that the segment superimposed on the curve near the glass transition is the expansivity value determined from the low temperature technique described above. The agreement between that expansivity value and the fitted curve is evident.

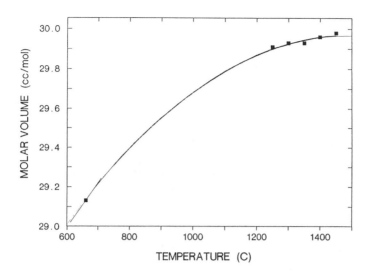

Figure 19. Temperature dependence of volume of GeO_2 liquid determined by the dilatometric/calorimetric method described in the text as well as the double-bob buoyancy method. The nonlinearity of the curve fitted to both expansivity data sets is corroborated by the expansivity segment near the glass transition illustrating the value derived for the liquid just above the glass transition. The agreement between the dilatometric/calorimetric estimate of the expansivity, the buoyancy-derived value, and the values derived from the volume-temperature fit to both data sets, is excellent. Reproduced from Dingwell et al. (1993c).

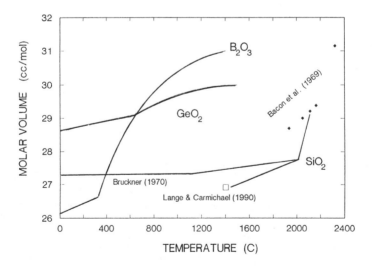

Figure 20. A summary of the estimates of volume-temperature relationships available for a number of simple oxide liquids stable at high temperatures. B_2O_3 and GeO_2 exhibit strongly decreasing expansivities with increasing temperature such that the high temperature liquid expansivities of the liquids are comparable to the glassy expansivities. SiO_2 is poorly constrained. The glassy expansivity and indirect estimates of the liquid expansivity at low temperature (Knoche et al., 1994, see text) indicate very low values. The high temperature single bob buoyancy data of Bacon et al. (1960) point to a higher expansivity for SiO_2 which would generate a volume temperature relation intersecting the low temperature glassy expansivity curve at reasonble estimates of the glass transition temperature (see text for discussion). Reproduced from Dingwell et al. (1993c).

The temperature dependence of the expansivity of melts is, in general, nonzero. Figure 20 illustrates the temperature dependence of the volume for B_2O_3, GeO_2 and SiO_2. The data for B_2O_3 are all obtained in the stable liquid state and indicate a strong decrease in the expansivity with temperature, the GeO_2 data behave similarly. The available data for the temperature dependence of the volume of SiO_2 are of extreme theoretical and practical importance to geochemistry but very poorly constrained. The glassy expansivity is very low (Brückner, 1970), whereas the only liquid data available indicate a significantly strong expansivity at higher temperatures. One might argue that an extraordinary effect at the glass transition temperature could yield high liquid expansivity data of SiO_2 even at temperatures just above the glass transition, but at least one strong argument exists against it. Knoche et al.'s (1994) study of melts in the system Na_2O-SiO_2, ranging in composition up to 98% SiO_2 indicate, by partial molar arguments, that the liquid expansivity of SiO_2 is indeed very low at low and at high temperatures (Fig. 21). Whether the expansivity of SiO_2 liquid increases with temperature remains an important question for future investigation.

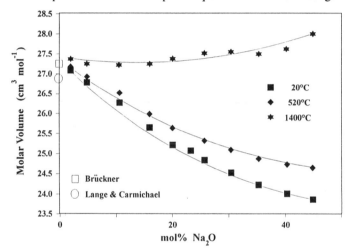

Figure 21. Melt volumes in the Na_2O-SiO_2 system at temperatures of 20°C (glass) as well as 520° and 1400°C (liquid). The data range from 45 to 2 wt % Na_2O and clearly indicate the very low expansivity of SiO_2 liquid in the temperature range from 520° to 1400°C but the expansivity of these liquids does not go to zero at a finite SiO_2 concentration as suggested by the work of Bockris et al. (1955). Such derivations of the relaxed partial molar volumes of inaccessible liquids (e.g. SiO_2 at 520°C) are a result of the dilatometric/calorimetric determination of melt volume and expansivity. Reproduced with permission from Knoche et al. (1994).

As can be seen from Figure 20, the expansivity of the liquid phase can be comparable to that of the glass phase. Data for the albite-anorthite-diopside system (Fig. 22) illustrate this well. The glassy and the high temperature liquid expansivity data are similar whereas the liquid expansivity data just above the glass transition temperature are significantly higher than either. This increasing contraction of the melt volume with falling temperature approaching the glass transition appears, where it is strong, almost as if it were recording an impending structural catastrophe for the liquid which is warded of by the intervention of the glass transition temperature - a volume equivalent of a Kauzman catastrophe?

The example of the Na_2O-SiO_2 system above brings us to the next application of the determination of melt expansivity using the dilatometric/calorimetric method. High precision, high temperature determinations of the densities of silicate liquids have thus been the subject of several experimental studies in the geosciences (Lange and Carmichael, 1987; Dingwell et al., 1988, Dingwell and Brearley, 1988; Dingwell, 1992) and empirical predictive schemes have been proposed (Lange and Carmichael, 1987; 1990; Kress and

Carmichael, 1991). Up to the present, such experimental studies have been focused on the determination of melt density at 1 atm pressure and superliquidus temperatures using double bob Archmedean methods (e.g. Dingwell et al., 1988) as well as high pressure studies using Stokesian falling sphere or sink-float densitometry (Kushiro, 1978a,b; Scarfe et al., 1987). Notable exceptions are the shock-wave Hugoniot density studies (Rigden et al., 1984; 1988; 1989), ultrasonic compressibility studies (Rivers and Carmichael, 1987; Kress et al., 1988; Webb and Dingwell, 1994) and inferences from the slopes of melting curves (e.g. see Lange, 1994), although in some of these studies the question of melt relaxation is not adequately addressed (see discussion by Dingwell and Webb, 1989).

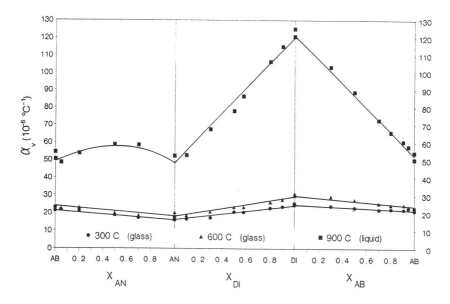

Figure 22. The expansivities of melts in the system albite-anorthite diopside derived by the methods outlined in Figure 16 as well as buoyancy methods at higher temperatures. The composition-dependence of the expansivity is largest at 900°C, just above the glass transition of melts in this system. The high temperature expansivities of the liquid match are similar to the glassy values of expansivity in this system. The "anomalous" high expansivities just above the glass transition may in fact be a reflection of structural constraints leading to the onset of the glass transition. Reproduced with permission from Knoche et al. (1992c).

The vast majority of the above work has concentrated on basic to intermediate melt compositions at the temperatures and pressures relative to the petrogenesis of basalts, andesites and derivative composition. Another branch of igneous petrogenesis, giving rise to intrusive granitic complexes and to more evolved pegmatitic is relatively under-investigated (cf. Knoche et al., 1992b; Dingwell et al., 1993b).

Knoche et al. (1995) have generated a low temperature multicomponent model for the densities of leucogranitic and pegmatitic melts. The densities and thermal expansivities of 39 haplogranitic silicate melts have been experimentally determined. The compositions represent the additions of selected oxide components Al_2O_3, Cs_2O, Rb_2O, K_2O, Na_2O, Li_2O, BaO, SrO, CaO, MgO, TiO_2, Ta_2O_5, Nb_2O_5, F_2O_{-1}, P_2O_5, B_2O_3 and WO_3 to a base composition of haplogranitic (HPG8) composition. The example of the effects of the addition of alkalies and alkaline earths to the volume of leucogranitic melt at 750°C are illustrated in Figure 23.

MOLAR VOLUME OF THE MELTS

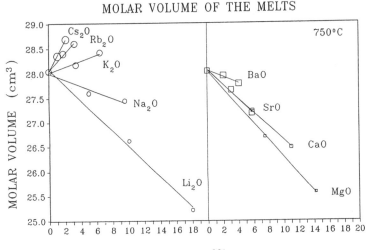

Figure 23. Plots illustrating the effects of the addition of alkalies and alkaline earths to the molar volume of a haplogranitic melt. The variation in molar volume has been derived for an isotherm of 750°C using a linear approximation to the melt expansivity in the temperature range from the glass transition to 750°C and employing the value of the expansivity obtained at the glass transition using the dilatometric/calorimetric method. Such volume data have been incorporated into the multilinear regression of oxide volumes for the determination of partial molar volumes of oxide components of leucogranitic and pegmatitic melt densities by Knoche et al. (1995).

The partial molar volumes were obtained by a least squares fits to a multicomponent linear dependence on the oxide concentrations. The results of the fit are the set of partial molar volumes for the oxide components at 750°C and expansivities of these partial molar volumes in Knoche et al. (1995). The residuals on the fit indicate that the root mean squared deviation is 0.24 cm^3/mole or approximately 0.3% of the specific volumes of these melts. This quality of fit indicates that further terms involving nonlinear oxide contributions to the volume are not required. The partial molar volumes range from 10.53±0.29 cm^3/mole for MgO to 69.09 ± 1.82 cm^3/mole for Ta$_2$O$_5$. Compared on the basis of one oxygen per mole they range from 10.53 ± 0.29 cm^3/mole (MgO) to 55.38 ± 1.69 cm^3/mole (Cs$_2$O).

The partial molar volumes provided by Lange and Carmichael (1990) from high temperature determinations have been compared at 750°C. Considering that this is a gross extrapolation of the Lange and Carmichael (1990) model, the agreement is not bad with one exception, TiO$_2$. The origin of the discrepancy for TiO$_2$ lies almost certainly in the composition dependence of the structural role of TiO$_2$ (Mysen, 1988; Dingwell et al, 1994). Several spectroscopic studies of Ti-rich silicate glasses indicate that the average coordination number of Ti decreases in the presence of Na and K. The consequences of the shift in coordination number include anomalous heat capacity (Richet and Bottinga, 1986; Lange and Navrotsky, 1993); anomalous compressibility-volume systematics (Webb and Dingwell, 1994) and a variable partial molar volume for TiO$_2$ in high temperature alkali and alkaline earth silicate melts (Dingwell, 1993). Thus granitic melt chemistry may stabilize an lower mean coordination number for Ti than is the case for alkali- and alkaline earth silicate melts (Dingwell et al., 1994; Paris et al., 1994a). In fact more recent X-ray absorption spectroscopic data on the effect of the addition of Al to alkali silicate melts indicates a growth in spectral features associated with a tetrahedral coordination of Ti, that is a lower average coordination than that observed in the simple Al-free melt compositions observed to date (Paris et al., 1994b).

Negative expansivities are obtained for TiO_2, Al_2O_3, WO_3, P_2O_5 and Nb_2O_5. These are all high field strength elements. Melt compositional dependence of the structural roles of Al_2O_3, TiO_2 and P_2O_5 have all been repeatedly proposed in the past based on spectroscopy, phase equilibria, thermochemistry or physical property behavior (see summary in Mysen, 1988). Recent high temperature density data for melts in a large compositional range of the $CaO-Al_2O_3-SiO_2$ system (Courtial and Dingwell, 1995) also indicate a negative expansivity of the Al_2O_3 component when treated without excess terms. The compositional variation of the volumetric behavior of TiO_2 is discussed above. Diffusivity and viscometry data (Chakraborty and Dingwell, 1993; Toplis and Dingwell, 1994) together with spectroscopic observations by Gan and Hess (1992) and solubility data for apatite and other phosphate minerals (Rapp and Watson, 1986; Montel, 1986; Pichavant et al., 1992) all indicate that the structural role of P_2O_5 is variable. High-field strength elements appear in general to be capable of adopting variable coordination based on the available complexing cations. Although little is known for WO_3 or Nb_2O_5, recent X-ray spectroscopic data for Nb_2O_5 indicate a variable coordination of Nb as well in silicate glasses, but, interestingly, not for Ta_2O_5 (Paris, pers. comm.), which here exhibits a positive expansivity.

Relaxation and rheology

The observation of Newtonian viscosity implies that the thermodynamic state of the melt is independent of the applied stress. The simplest microscopic interpretation of this behavior is that the self diffusive motion of atoms due to random thermal fluctuations in the melt is greater than that imposed by the experimental strain rate. With decreasing temperature the self diffusivities of the melt components decrease and, if a sufficiently high strain rate can be maintained experimentally, the relative motion of atoms required by this experimental strain rate approaches that due to self diffusion. The immediate result is a decrease in the measured viscosity with increasing strain rate, i.e. non-Newtonian flow. Experimentally, non-Newtonian flow of silicate melts has been observed using frequency domain (e.g. ultrasonic) and time domain (e.g. dilatometry) methods. The frequency domain methods operate at very low strains and are capable of mapping the viscoelastic response of silicate melts up to fully elastic behavior. They are discussed in the chapter by Webb and Dingwell. Here we turn our attention to the latter dilatometric studies.

Several types of dilatometric experiments have been performed which record the onset of non-Newtonian viscosity in silicate melts. The first example is fiber elongation dilatometry. Li and Uhlmann (1970) suspended long silicate melt fibers through a tube furnace with a load hanging from the lower end. They determined the onset of non-Newtonian flow for a Rb-silicate melt. At certain stresses and temperatures, the strain rate of elongation increased as the fiber drove into non-Newtonian rheology. This effect was observed to be runaway. Webb and Dingwell (1990a,b) extended these observations using a slightly modified fiber elongation technique in which the entire fiber was contained in the hot zone of the dilatometer furnace. Webb and Dingwell (1990a,b) surveyed a range of natural and synthetic silicate melt compositions using this method. The viscosity results for synthetic and natural melts are presented in Figures 24 and 25. The strain-rate deviates from the linear dependence of slope 1 on stress at approx. 10^8 Pa (Fig. 24). The calculated viscosities are presented in Figure 25 as functions of the strain rate. The decrease in viscosity (non-Newtonian flow) begins at $\sim 10^{-4}$ s^{-1}. Webb and Dingwell (1990a) demonstrated that the onset of non-Newtonian flow for the entire range of silicate melt compositions investigated could be reduced to s single curve of viscosity versus "reduced" strain rate $\varepsilon'/\varepsilon_r$ where ε_r is the inverse of the shear relaxation time

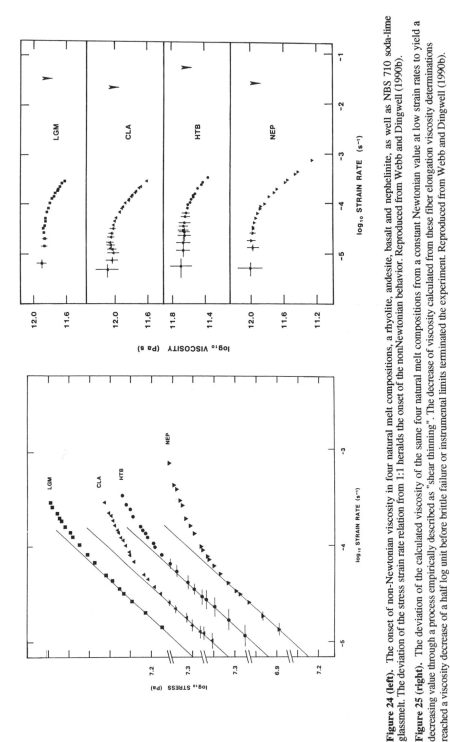

Figure 24 (left). The onset of non-Newtonian viscosity in four natural melt compositions, a rhyolite, andesite, basalt and nephelinite, as well as NBS 710 soda-lime glassmelt. The deviation of the stress strain rate relation from 1:1 heralds the onset of the nonNewtonian behavior. Reproduced from Webb and Dingwell (1990b).

Figure 25 (right). The deviation of the calculated viscosity of the same four natural melt compositions from a constant Newtonian value at low strain rates to yield a decreasing value through a process empirically described as "shear thinning". The decrease of viscosity calculated from these fiber elongation viscosity determinations reached a viscosity decrease of a half log unit before brittle failure or instrumental limits terminated the experiment. Reproduced from Webb and Dingwell (1990b).

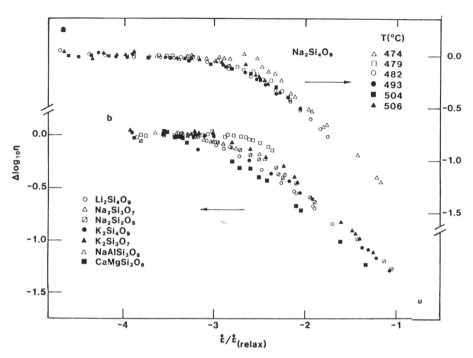

Figure 26. A normalisation of the onset of nonNewtonian viscosity based on the ratio of the measured to the Newtonian value of shear viscosity and the ratio of the experimental strain rate to the relaxation strain rate (reciprocal of the relaxation time). This normalisation, based on the idea that the structural relaxation time of the liquid is controlling the onset of the nonNewtonian viscosity is successful in unifying all the available data for the onset of nonNewtonian viscosity. Reproduced from Webb and Dingwell (1990a).

The same normalization appears to remove the composition dependence of the onset of non-Newtonian viscosity for a wide range of melts. In Figure 26 it is clear that the non-Newtonian onset occurs 2.5 to 3 log units of strain rate lower than the calculated shear relaxation strain rate. This is at a lower strain rate than would be estimated from the consideration of viscoelasticity in the glass transition for example from internal friction measurements (Mills, 1974). The large finite strains involved in fiber elongation mean that we can expect nonlinear effects due to structural relaxation as well as viscoelasticity resulting from shear stress relaxation. Simmons and Simmons (1989) suggest that the difference between the onset of viscoelasticity and the onset of non-Newtonian behavior is due to structural relaxation in response to the large strains employed and describe this effect of structural relaxation as "shear thinning".

Experiments in which the strain rate is defined and the resultant stress is recorded illustrate this behavior as well. Constant strain rates are produced by a dynamometer and stresses are measured using load gauges. Such methods have been used in tensile (fiber) and shear (cone-and-plate) geometries by Simmons and co-workers (Simmons et al., 1982; Simmons et al., 1988) and in uniaxial compression (parallel-plate) by Brückner and co-workers (Manns and Brückner, 1988; Hessenkemper and Brückner, 1990). Simmons and Simmons (1989) have attempted to model the two characteristic features of these experiments, stress overshoot and shear thinning. By comparing the results of different experimental geometries which combine to span measurements over 7 log units of viscosity, Simmons and Simmons (1989) have generalized the results. These authors identify a maximum shear stress developed in the experiments as the strain rate goes to infinity and propose its interpretation as the cohesive shear strength of the melt. Its temperature dependence is obtained by comparing the fitted values from the cone-and-plate and fiber experiments analysed by Simmons and Simmons (1989). Their parameterization essentially implies that the shear modulus alone is insufficient to describe the temperature dependence of the onset of non-Newtonian viscosity. The known temperature dependence of the shear modulus of several silicate glasses and some liquids (Bansal and Doremus, 1986) is certainly too low to account for the variations indicated in the range of cohesive shear strength values (Simmons and Simmons, 1989). In summary, the dynamometer experiments are complementary to and consistent with the dilatometer experiments in predicting the onset of non-Newtonian rheology in silicate melts. The non-Newtonian onset occurs at lower strain rates and stresses than predicted from purely viscoelastic considerations. This observation, along with the observation of a stress overshoot, has been have been interpreted as evidence for significant structural reorganization of the melt, which relaxes at a slower rate that the shear stress itself (viscoelastic effects). The much greater range of viscosity of the dynamometer studies reveals that the non-Newtonian onset cannot be simply normalized to the shear relaxation time but rather requires an additional adjustable parameter. The equivalent behavior of the parallel plate compression, cone-in-plate and fiber elongation experiments in the onset of non-Newtonian rheology is very significant. It implies that the pressure-dependence or density-dependence of the viscosity has no noticeable effect on the onset of non-Newtonian viscosity because these three types of experiments employ shear stress associated with a volume stress which is either positive (parallel plate), zero (cone-in-plate) or negative (fiber elongation).

Dissipation and failure. The measurement of viscosity at very high driving stresses or strain rates raises the question of heating of the sample by viscous dissipation. This effect requires a comparison of the energy input due to work done on the sample and the specific thermal conductivity of the sample. Clearly the effect will be more severe for more massive sample geometries. Hessenkemper and Brückner (1988) addressed this problem for the parallel plate geometry in a direct way. They embedded a NiCr thermocouple in a glass that was subjected to dynamometer deformation. The results of that study indicate that, for the lowest deformation rate of their study, the measured dissipative heating is far less than the adiabatic calculation and contributes a negligibly small viscosity decrease to the observed flow curve, whereas for the highest deformation rate the viscosity decrease the measured and calculated dissipative heating-viscosity decrease are the same, and thus the dissipative contribution dominates the flow behavior. The relatively thin geometry and lower strain rates of the fiber elongation investigations of non-Newtonian flow indicate that the onset of non-Newtonian flow in those experiments is unrelated to viscous heating (Webb and Dingwell, 1990a).

The general picture emerging from this studies is one of silicate melts undergoing an essential phase of non-Newtonian "shear thinning" prior to brittle failure. This scenario is quite likely to play a role in the brittle fragmentation of very viscous magmas which succumb to volatile overpressure at low melt water contents and low temperatures and experimental studies of the fragmentation process of magmas must of neccessity be concerned with the rheological details of such materials. The surest solution is to experimentally fragment magma under magmatic conditions (e.g. Alidibirov and Dingwell, 1995).

The tensile stresses in fiber elongation studies can be large enough to drive strain rates into the viscoelastic region of response near the glass transition. The absolute strains in fiber elongation experiments are usually less than 1 and the strain rates used in studies of non-Newtonian behavior range from 10^{-6} to 10^{-3}. The stresses range up to $10^{8.5}$ Pa and approach the tensile strength of the silicate fibers. This combination of stress and strain rate places the viscosity of the investigated silicate melts into the range of 10^{10} and higher.

Flow birefringence. If a silicate melt is exposed to a high differential stress during cooling across the glass transition then an anisotropic structure will be frozen in. Brückner and coworkers (Wäsche and Brückner, 1986a,b; Stockhorst and Brückner, 1987) have investigated such anisotropy produced during the production of borosilicate (E-glass) glass fibers. The specific birefringence has been measured on fiber bundles and the optical anisotropy is reported by Stockhorst and Brückner (1982) for a number of glasses and forming conditions. They plot the specific birefringence versus drawing speed and drawing stress. Two factors controlling the anisotropy are the cooling rate (up to 10^5 K/s) and the drawing stress. The anisotropy of the glass structure is small, (in the parts per mil range). Although the structural anisotropy is very low, the frozen-in "anisometry" or differential stress is approximately 10% of the theoretical strength of the fibers. Thus the actual flow anisotropy in silicate melts can be expected to be very low.

Isostructural viscosity. To this point we have emphasized that the viscosity of a silicate melt displays a non-Arrhenian temperature dependence. It has been noted elsewhere (see chapter by Richet and Bottinga) that the non-Arrhenian nature of the temperature dependence can be parameterized using an additional parameter, the configurational entropy (S_{conf}) of the silicate melt (Adam and Gibbs, 1965). The resulting inference is that the non-Arrhenian temperature dependence of silicate melt viscosities is due to the changing value of S_{conf} with temperature. The temperature dependence of the configurational entropy of the liquid records the temperature-dependent changes in the distribution of cations in the melt, i.e. order-disorder processes leading from low temperature, well ordered amorphous structures to higher temperature, disordered structures. The decreasing activation energy of silicate melt viscosities with increasing temperature can be interpreted as a result of this disordering with increasing temperature. The disordering leads to smaller units of cooperative relaxation and lower activation energies of viscous flow.

The viscosity of a silicate liquid in thermal equilibrium can be termed appropriately enough, the equilibrium viscosity. Equilibrium viscosities refer then to an individual composition but not to an individual structure. If the configurational entropy is the source of the non-Arrhenian behavior of viscosity then the temperature dependence of a silicate melt of constant structure might be Arrhenian. The temperature dependence of such an "isostructural" viscosity would then provide the activation energy of viscous flow corresponding to a certain melt structure or configuration and the proportion of the equilibrium viscosity activation energy due to the temperature dependence of configurational entropy could be obtained by difference.

Experimentally, the problem is how to measure "relaxed" stress strain relationships in a silicate melt without allowing significant relaxation of the structure. In this way, the temperature dependence of viscosity for a melt structure can be constructed. These conditions necessary for such measurements have been outlined by Mazurin et al. (1979) and experiments have been performed by Zijlstra (1963) and by Mazurin and co-workers (Mazurin et al., 1979). The essential feature of the experiments can be explained with the aid of Figure 27 (see also chapter by Richet and Bottinga). Figure 27 describes the time domain strain response of a silicate melt which has been loaded under a constant stress. We have seen in the chapter by Webb and Dingwell that the viscoelastic response of a silicate melt consists of three segments: instantaneous elastic, delayed elastic and viscous, which are included in Figure 27. The instananeous and delayed elastic components are defined to be recoverable. That is to say that upon subsequent removal of the load (stress) from the melt, the sample will recover the instantaneous and delayed elastic components of stress.

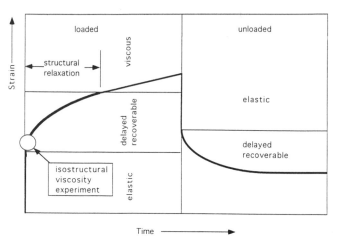

Figure 27. Schematic representation of the determination of the isostructural viscosity of silicate melts. The viscous nonrecoverable strain obtained in a loading-unloading cycle of a creep experiment is recorded for an experiment where the structural relaxation after the application of the stress is interrupted before a significant fraction of the structure is relaxed. The results yield temperature-viscosity data where the structure of the melt has not changed in response to temperature. These quasi-"isostructural" viscosity-temperature relationships yield Arrhenian values of the activation energy considerably lower than the equilibrium viscosity activation energy. The intersection of such an Arrhenian viscosity-temperature segment with the equilibrium viscosity curve marks the fictive temperature of the quenched liquid. Redrawn from DeBast and Gilard (1963).

The strain resulting from viscous flow is not recoverable, remaining in the sample after the removal of the load. If the experimenter waits until the delayed elastic strain relaxes out and the purely viscous flow region is reached then the sample has completely relaxed its structure to the new conditions of temperature and stress and the experiment is measuring the equilibrium viscosity (because the shear, volume and structural relaxation times are all related). This is common procedure for dilatometric determinations of melt viscosities using for example fiber elongation. In such experiments it is essential that the delayed elastic component is removed by waiting long enough before recording the strain rate to calculate viscosity. For the measurement of isostructural viscosity we want to prevent the structural relaxation of the sample so we cannot wait until the delayed elastic component relaxs out of the sample deformation. It is important to note that the viscous deformation of the sample begins immediately upon the application of the load to the sample. Thus viscous strain (non-recoverable) is in the sample at all times after loading. The key to measuring the isostructural viscosity is to measure the small viscous strains without allowing any significant structural change as recorded by the delayed elastic component. The method used by Mazurin and coworkers (Mazurin et al., 1979) was to load the sample for a period of time long enough to produce a an observable viscous strain, then to unload the sample and wait long enough for the delayed elastic component to recover its strain. The strain difference between the original and final, approximately time invariant, values of the samples dimensions can be divided by the time duration of the loaded condition to obtain the strain rate. The viscous stress strain rate relation is thus recorded. This method is an approximation. Clearly some delayed elastic component indicates some structural relaxation and the resultant viscosities are average values for the range of structure explored by the sample at that temperature during the experiment. The important point is that the structural changes can in practice be limited to very small changes so that a glass of a single chemical composition can yield viscosity data for a number of structures.

The results of such experiments on silicate melts (including soda lime 710) taken from Mazurin et al. (1979), indicate that activation energies of the isostructural viscosities are approximately one-half the activation energy of the equilibrium viscosity (Fig. 28).

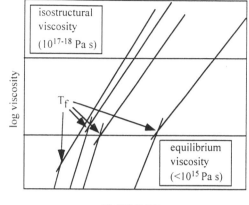

Figure 28. Isostructural viscosity data for a series of glass melts. The low temperature Arrhenian segments are the isostructural viscosity relationships which intersect the higher temperature equilibrium curves at the fictive temperatures of the glasses. Redrawn from Mazurin et al. (1979).

The isostructural experiments are performed in the range of 10^{16} to 10^{17} Pa s. The equilibrium and isostructural viscosity values are constrained by the requirements of obtaining and preventing, respectively, structural relaxation of the sample. Measurements of the equilibrium viscosity attempted above 10^{14} Pa s involve unreasonably long waits to avoid the measurement of an elastic strain component and that the practical upper limit for equilibrium viscosity measurements lies near this viscosity value. If this condition is not met, then the result will be erroneously low viscosities. In a similar way, the experimenter measuring isostructural viscosities must wait long enough after the removal of strain to fully recover the delayed elastic component. If not, again the result will be erroneously low viscosity values. The results of Zijlstra (1963), which illustrate an increasing deviation from equilibrium with decreasing temperature, may have been affected in this way. Scherer (1986a) and Rekhson (1985) suggest that the trend towards zero activation energy produced in Zijlstra's experiments (1963) is incorporating this effect. Trends of viscous activation energy towards lower values with lower temperature are, for the above reasons, always suspect.

Relaxation of fluid inclusions in melts

Determination of T_g of fluid-saturated melts. Some of the most common glass transition measurements are dilatometric determination of the expansivity (Tool and Eichlin, 1931; Knoche et al., 1992a,b,c; 1995; Webb et al., 1992) and calorimetric determination of the heat capacity (Martens et al., 1987). The application of such methods to geologically relevant melt compositions has been pursued in the past few years (Dingwell and Webb, 1989; Webb and Dingwell, 1990a,b; Dingwell, 1993). The dilatometric method at 1 atm has even been applied to wet rhyolite glasses (Taniguchi, 1981) although water loss is a potential problem. The calorimetric method has also been applied by Rosenhauer et al. (1979) to dry melts at high pressure. Hydrous melts at high pressure have not yet been investigated although in situ spectroscopy of water speciation in hydrous melts offers the promise of direct T_g determinations in the future. Romano et al. (1994) have developed a new method for the determination of the glass transition temperature in the high pressure, vesicular, hydrous glasses using the principles of structural relaxation applied to the relaxation of the melt surrounding vesicles that contain fluid inclusions.

The experimental method presently employed for the determination of T_g using fluid inclusions is illustrated schematically in Figure 29. The P-T trajectories of the H_2O liquid-vapor coexistence curve from some low temperature up to the critical point and of two representative isochores (lines of constant water density) are as indicated. Initially the experimental charge is allowed to dwell for a time sufficient to allow the chemical equilibration of a water-saturated silicate melt at pressure and temperature. The melt contains bubbles because the viscosity of the liquid (initially a powder) is high enough to prevent the escape of vesicles during the experiment. The horizontal line extending down temperature from the dwell point represents an isobaric quench path of the experiment. The two curves with negative slopes represent hypothetical glass transition curves corresponding to different cooling rates. The negative slope of the curves is based on the observation that the (isothermal) viscosity of water-saturated melts decreases with increasing pressure, due to increasing water content.

If water is contained in the vesicles of an equilibrated, water-saturated silicate melt then an isobaric quench (sufficiently rapid that no significant exchange of water occurs between melt and fluid) yields as a mechanical response to the quench a simple contraction of melt and bubbles, by volume relaxation. This viscous contraction continues with dropping temperature until a temperature point where the exponentially increasing viscosity, reflecting a structural relaxation time for the melt, becomes too high. This point is the glass transition temperature for the cooling rate used. It is the temperature at which the structure of the equilibrium liquid most closely resembles the structure of the quenched glass: and it is here that the densities of the ideally non-reacting fluid inclusions are frozen in.

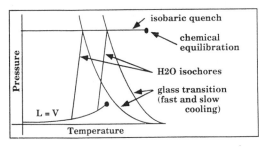

Figure 29. The principal behind the determination of the glass transition temperature of fluid saturated melts at elevated pressure using the densities of the fluid inclusions preserved in vesicular samples. See discussion in the text. Reproduced with permission, from Dingwell (1993).

Subsequent determination of the bulk densities of fluid inclusions trapped in the quenched glasses via measurement of their liquid-vapor homogenization temperatures identify isochores (Fig. 29) that must intersect the isobaric quench path at T_g. The results presented below, indicate that under certain conditions fluid inclusion analysis can be used to successfully predict the glass transition temperature.

Glass transition temperatures for the binary $NaAlSi_3O_8$-$KAlSi_3O_8$ system, calculated from fluid inclusion homogenization temperatures using volumetric data for pure water are shown in Figure 30a. The mean values of glass transition temperature range from 515° to 416°C. The $KAlSi_3O_8$ glass has a significantly higher transition temperature than the albite glass. The compositions reported in Figure 30a are not truly binary as the water content varies from 6.0 wt % in albite melt to 5.1 wt % in orthoclase melt (Romano et al., 1995a). The higher glass transition of orthoclase melt is consistent with the higher viscosity of the dry melt as well as the lower water content of the saturated melt. The composition dependence of the glass transition along this binary join shows a significant negative deviation from additivity. This deviation from additivity is a common feature of the transport properties of low temperature, highly viscous, silicate melt binaries. Well-known examples of this effect are described in particular for the case of mixing of the alkalies under the term "mixed alkali effect" (e.g. electrical conductivity, Isard, 1969; viscosity, Richet, 1984).

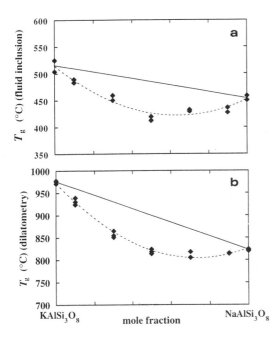

Figure 30. (a) The glass transition temperature of water-saturated feldspathic liquids (albite-orthoclase) quenched from 2 kbar and 1100°C. The variation in water contents across the join are approximately linear (Romano et al. (1994) and the variation in glass transition temperature across this join can be understood in terms of a contribution of the entropy of mixing of the alkalies to the total system entropy. The relationship between configurational entropy and viscosity leads to a viscosity minimum and thus a glass transition minimum at intermediate composition. (b) For comparison, the variation of the glass transition with composition in the water-free system at 1 atm is presented. The drop in glass transition temperature of over 400°C is due to the tremendous fluxing effect of water on the feldspar melt structure. Reproduced with permission from Romano et al. (1994).

The dry starting glass compositions yield peak temperatures of the expansivity curves (glass transition temperatures) that are presented in Figure 30b. Again, both a higher glass transition temperature for the orthoclase composition, as well as a negative deviation from additivity along the binary join are apparent. The effect of water on the glass transition temperature is enormous. The glass transition temperatures of the water saturated glasses are approximately 400°C below those of the dry glasses of equivalent composition. [The fact that the effective quench rates for the dilatometric experiments were considerably lower than the rapid quench of the hydrothermal glasses actually increases the difference between the T_g of dry and wet glasses by perhaps 30° to 40°C.] This is, in part, a reflection of the efficiency of water in reducing the volume relaxation time of the melts and is completely consistent with the drastic effect of water on the shear viscosity (and shear relaxation time) of the melts (Shaw, 1963). The intrinsic effect of pressure on the relaxation time of feldspathic melts can be estimated from the glass transition measurements of Rosenhauer et al. (1979) and viscosity measurements of Kushiro (1978b) to be minor.

Viscosity and fragility of hydrous melts. Romano et al. (1994) have used the equivalence of the temperature dependence of viscosity and the variation of T_g with the log of the quench rate to estimate the viscosity of these hydrous melts at the glass transition temperature recorded by the fluid inclusions. Their estimate is based on the relationship of Scherer (1984). This relationship differs from that of Stevenson et al. (1995) due to the differing conventions of picking the T_g as either the extrapolated onset temperature (Scherer, 1984) or the peak temperature (Stevenson et al., 1995). The peak temperature is recommended for viscosity calculations based on scanning dilatometry or calorimetry due to the high precision of peak temperature assignment. The extrapolated onset temperature is however a closer approximation to the fictive temperature of the system as it is there that T_f finally freezes out (Fig. 15).

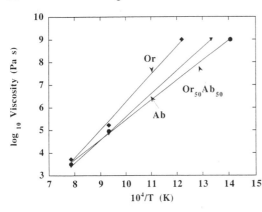

Figure 31. Viscosities estimated from the relationships between cooling rate, relaxation time and viscosity are presented for the water-saturated melts of albite, orthoclase and albite-orthoclase 50:50 composition. The estimation of the viscosities of these melts from the calculational scheme of Shaw (1972) is included for comparison. The Shaw calculations and the fluid inclusion-based estimates of melt viscosity are consistent with an Arrhenian temperature dependence of the viscosity of the water-saturated feldspathic melts from 10^4 to 10^9 Pa s. Reproduced with permission from Romano et al. (1994).

Figure 32. The variation of density with composition for dry and water-saturated feldspathic glasses quenched at constant cooling rates from 1 bar and 2 kbar respectively. The non-linear variations in density can be related through the data of Figure 30a to variations in fictive temperature of these glasses. Comparison of glass transition temperature and density deviations from the linear yield expansivity estimates for the dry and water-saturated melts. Reproduced with permission from Romano et al. (1994).

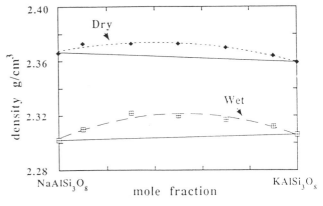

The quench rate pertaining to the synthesis of the hydrous glasses can be estimated at 200°C/s. Using the relationship between quench rate and effective relaxation time available during the quench (Scherer, 1984a) together with the Maxwell relation between viscosity and shear relaxation time for a quench rate of $10^{2.3}$ °C/s, we obtain a relaxation time of $10^{-1.7}$ (using a shear modulus of 25 GPa) and for such a relaxation time we obtain a viscosity of 10^9 Pa s. This viscosity-temperature data point compares favorably with the calculated viscosity-temperature relationship for each melt using the method of Shaw (1972) (Fig. 31). Thus the viscosity-temperature relationship of the water-rich feldspathic melt is Arrhenian within error over the investigated range. This result is somewhat unexpected as the usual consequence of adding a depolymerising agent to a polymerised composition is to make the viscosity temperature relationship more strongly non Arrhenian, that is more "fragile" in the sense of Angell (1984, 1988, 1991). It is however possible that the viscosity range over which the comparison of Figure 5 is being made is not large enough or does not extend to low enough temperature to observe the non-Arrhenian behavior. In this regard it is informative to note that the results of such synthetic fluid inclusion—based T_g determinations are in excellent agreement with Tg trends recently estimated from viscometry of hydrous melts (Dingwell et al., 1995).

Expansivity of hydrous melts. The densities of bubble-free glasses quenched from the same conditions at the same rates have been determined for the dry and wet joins. These data are presented in Figure 32. The deviation from additivity is evident in both cases. Assuming a linear volume composition relation for these joins and similar expansivities for the glasses, and ignoring, in the case of the wet melts, the slight

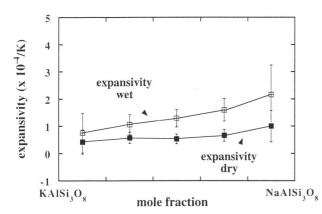

Figure 33. The expansivities of melts estimated from the non-linear variations on density and glass transition temperature described in the text. The water-saturated melts have considerably higher expansivities than the dry melts, consistent with the observations of Burnham and Davis (1974). Reproduced with permission from Romano et al. (1994).

Figure 34. Variation of concentrations of hydrous species (molecular water and hydroxyl groups) along the albite-orthoclase join. Characteristic deviation of the spectral intensities from linearity is also caused by glass transition variations in this system discussed above in the context of property variations. Fluid inclusion-based determinations of fictive temperatures of these glasses form the basis of the normalisation of spectral intensities through the equilibrium constant of Equation (14) presented in Figure 36. Data from Romano et al. (1994).

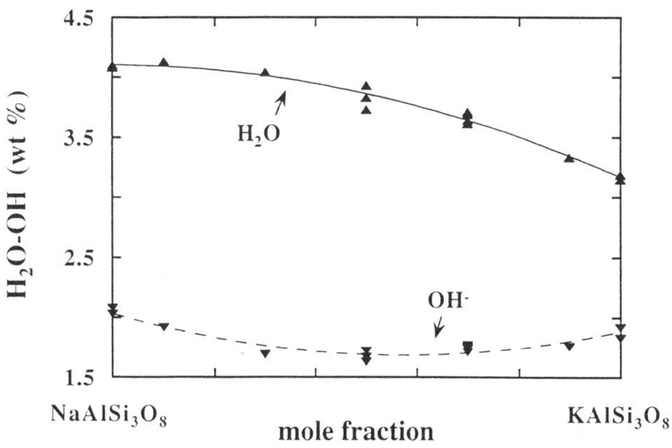

composition dependence of the water solubility, then we can estimate from the deviation of density from additivity, the expansivity of the melt. Romano et al. (1994) simply divided the deviation from additivity in the density by that of the glass transition temperature. The result yields values of expansivity which are presented in Figure 33. The expansivity of the wet melts can be seen to be higher than those estimated for the dry melts and are in reasonable agreement with the predictions from the study of Burnham and Davis (1971).

Water speciation.   The method of fluid inclusion investigation for the determination of glass transition temperatures discussed above allows us to return to the issue of the speciation of water in melts inferred from their quenched glasses (see review by McMillan, 1994). Romano et al. (1995a,b) have demonstrated that infrared spectroscopic investigation of melts quenched along the albite-orthoclase join reveal a nonlinear variation in water speciation across the Ab-Or join, yielding a minimum in OH and a maximum in H_2O content at an intermediate composition (Fig. 34) that correlates with the nonlinearity in other melt and glass properties noted above.

Demonstration of a quench-rate dependence of the speciation data is reproduced in Figure 35. Such a dependence of the speciation data on quench rate is the hallmark of a temperature-dependent equilibrium whose kinetics are being frozen in at differing temperatures of last equilibration. Above it was demonstrated that the deviations in density along the water-saturated Ab-Or join were entirely accounted for by correction for the

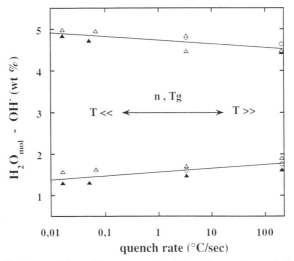

Figure 35. An illustration of the problem of quenching information on the speciation of water in silicate melts that is caused by the strong quench-rate- (or temperature-) and water concentration-dependence of the fictive temperatures of investigated melts. The inherent temperature-dependence of the speciation of water yields a quench-rate dependence of the structure through the relationship between quench rate and relaxation time. Additionally, the tremendous effect of water on the glass transition temperature of silicate melts generates a variation in fictive temperature of the glasses used for the estimation of concentration-dependence of the speciation of water of several hundred degrees. Reproduced with permission from Romano et al. (1995a).

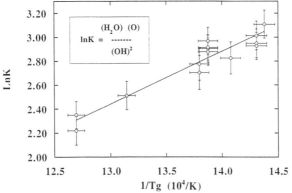

Figure 36. Variation of the equilibrium constant of Equation (14) with composition within the water-saturated albite-orthoclase system quenched from 2 kbar. The data are plotted versus the fictive temperatures of the quenched glasses. This correction of the spectra for the fictive temperature of the glasses yields a linear temperature dependence of the speciation. The conversion of a highly non-linear composition variation of the speciation data from infrared spectroscopy to a linear array, through the consideration of the fictive temperature dependence of the spectra, argues strongly for a strong temperature dependence of the speciation of water in this system combined with weak to no dependence of the water speciation on the identity of the alkali cation. See text for further discussion and comparisons. Reproduced with permissionfrom Romano et al. (1995a).

fictive temperature variations along the join. Romano et al. (1995b) have applied the same logic to the variations in the speciation data to test the proposal of Dingwell and Webb (1989) by seeing to what extent a fictive temperature dependence of the speciation might account for the variations observed in Figure 34 and the quench-rate dependence of Figure 35. The higher molecular water and lower hydroxyl contents observed at intermediate composition are consistent with a fictive temperature dependence where lower fictive temperatures at intermediate compositions yield higher molecular water contents. This trend to higher molecular water contents with lower fictive temperature has been demonstrated previously (Stolper, 1989; Dingwell and Webb, 1990).

The speciation data versus the reciprocal of the temperature are plotted in Figure 36. The temperature dependence of the water speciation data is cast as

$$K = (H_2O_m)(O_m^{2-})/(OH^-)^2 \qquad (13)$$

where K is the equilibrium constant of the equation

$$H_2O + O_m^{2-} \Leftrightarrow 2\,OH^- \qquad (14)$$

in mole fractions of molecular water, dry oxygens (in the melt but not associated with any H atoms) and hydroxyl groups dissolved in the melts.

The melt is modeled as an ideal mixture of water molecules, hydroxyl groups, and oxygens, in which all the different types of oxygens available for reaction with the water molecules (Al-O-Si, Al-O-Al, Si-O-Si) are energetically indistinguishable.

Assuming ideal mixing of species, activities in Equation (14) are replaced by molar fractions. The value of *ln* K varies linearly with reciprocal absolute temperature

$$ln\,K = -\Delta G/RT \qquad (15)$$

where ΔG is the free energy of the reaction and R is the gas constant. The standard state enthalpy of reaction is 36.5 ± 5 kJ/mol) from the data of Figure 36. This enthalpy compares with the estimate of 25 ± 5 kJ/mol by Dingwell and Webb (1990) for a rhyolitic melt. To the extent that the trend in Figure 6 is linear no further compositional aspects of a solution model are required to explain the present data. Thus the temperature dependence of reaction (13) does appear to dominate the speciation variations along the Ab-Or join. Perhaps surprisingly the exchange of Na for K produces little to no observable effect on the speciation (except indirectly through the total water concentration). The water solubility is however noticeably affected by the exchange. The solution model for water in these melts must therefore explain the variation in water solubility along the Ab-Or join in the absence of a variation in speciation.

The above analysis of the fictive temperature dependence of the speciation of water in melts along the Ab-Or join demonstrates that the observed deviations from linearity in the hydroxyl and molecular water content as a function of composition are completely explained by a fictive temperature dependence along the $NaAlSi_3O_8$-$KAlSi_3O_8$ join and that there is no observed residual of the composition dependence of water speciation along the system $NaAlSi_3O_8$-$KAlSi_3O_8$ which could be attributed to a true isothermal relaxed composition dependence of water speciation in this system. There is no evidence from Figure 36 that the water speciation is alkali specific in this system.

Very recently, the kinetics of the exchange of molecular water and hydroxyl groups in silicate melts has been revisited by Zhang et al. (1995). Comparison of those results with the original Dingwell and Webb (1989) analysis of Silver et al.'s (1990) data on the quench rate dependence of the speciation of water has been performed by McMillan (1994). The importance of such a comparison should not be underestimated. The study of Zhang et al. (1995) is a spectroscopic investigation of the kinetics of equilibration of hydrous species based on the spectra of quenched glasses in time series annealing experiments. Thus the information of that study is fundamentally microscopic and structural and independent of assumptions regarding the physical properties of such systems. In contrast, the earlier analysis of Dingwell and Webb (1989) was based on the premise that the structural relaxation of the melt, the relaxation of viscous stress in the liquid and the water speciation, were all linked through the same kinetics. Only the use of this premise allowed the analysis of Dingwell and Webb (1989) to be performed. This premise is essentially tested by the study of Zhang et al. (1995) (see discussion by McMillan, 1994) whose comparison of the results reveals a striking agreement between the initial estimates of Dingwell and Webb (1989) and the latter evaluation of Zhang et al. (1995; their Fig. 5). As noted by Dingwell and Webb (1989) utilisation of their premise regarding the equivalence of macroscopic property relaxation and microscopic structural relaxation requires estimation of the shear viscosity. Dingwell and Webb (1989) were forced to calculate it using the method of Shaw (1972). The result was an estimate that was well-constrained at 5 wt % water but increasingly poorly constrained at lower water contents. It is at these low water contents where a potential discrepancy between the estimates of temperature dependence of speciation of Dingwell and Webb (1989) and Zhang et al. (1995) appears. The possibility of such a discrepancy very likely finds its source in the Shaw (1972) viscosity calculation which fails badly at low water contents. A recent study of the viscosity of hydrous haplogranitic melts at 1 bar and water contents up to 2 wt % (Dingwell et al., 1995) demonstrates clearly that Shaw's (1972) viscosity calculation method seriously overestimates melt viscosity of silicic melts with low water contents. Similar conclusions can be drawn from extrapolation of the results of high pressure falling sphere viscometry of Schulze et al. (1995). Improved estimates of the viscosity of the water-poor viscosity of the rhyolite studied by Zhang et al. (1995) should result in even better agreement between the estimates of the temperature dependence of water speciation provided by the spectroscopic annealing studies and the quench-rate-dependence analysis using shear relaxation times.

The conclusion to be drawn from the above is a powerful one. To the best of our ability to determine them to date, the relaxation times for the the macroscopic response of properties such as volume, shear stress and enthalpy as well as that required for equilibration of hydrous species (and other aspects of the melt structure) are identical. The inference drawn here is that the macroscopic and microscopic investigations of the relaxation behavior point to a single structural relaxation behavior which controls both the relaxation of essential aspects of melt speciation (including water speciation) and properties which should be viewed as derivatives of the structure. This equivalence or interconvertability of the kinetics of structure and property relaxation holds great promise for the future investigation of melt kinetics under conditions of difficult experimental constraints that are unavoidable for the characterisation of transport properties of melts in the petrological range of P-T-X space.

Relaxation timescales of hydrous species

From the above discussion of the fictive temperature dependence of the distribution of hydroxyl versus molecular water dissolved in silicate melts it is clear that the kinetics of the water speciation reaction are significantly linked to those of the glass transition itself. This does not seem surprising in light of the fact that the reaction implied by the equilibrium constant presented in Figure 37 involves bridging oxygen bonds—the same bonds whose exchange with nonbridging oxygens is though to lie at the heart of the NMR observations of motional averaging presented in the chapter by Stebbins. One might anticipate that the self-diffusivity of oxygens that are linked into the network as bridging or nonbridging oxygens would describe a temperature dependence that follows the viscous activation energy. As noted above, the recent study of Zhang et al. (1995) on the kinetics of

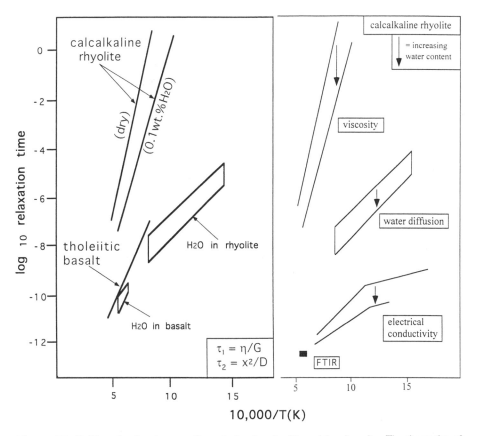

Figure 37 (left). A relaxation map for water-bearing rhyolite and basalt melts. The timescales of viscous relaxation, of molecular water mobility and of heat conduction are compared. The fundamental problem of relaxation of viscous stresses in highly viscous liquids during the diffusion-fed growth of waterfilled vesicles is illustrated here.

Figure 38 (right). Extended relaxation map of a water-rich silicic melt. At least three relaxation modes of hydrous species can be seen in silicic melts. (1) the mobility of network-bonded hydroxyl groups at the viscous relaxation, (2) the mobility of molecular water and (3) the mobility of protons. Sources of data discussed in the text.

Reaction (14) has confirmed the equivalence of the results of microscopic (structural) and macroscopic (property) relaxation timescales (McMillan, 1994).

Abundant data on the chemical diffusivity of water in silicate melts including several studies on water in rhyolite (e.g. Zhang et al., 1991) illustrate that the mobility of water in silicic melts is much higher than predicted by a mechanism associated with the glass transition. This long known observation can be quantified for comparison in a relaxation map as is illustrated in Figure 37. The available experimental data on the diffusivity of water in rhyolite melts falls within the box labeled "H_2O in rhyolite." The inference from the studies of others on this topic is that the mobility of molecular water dominates the diffusivity of the hydrous component determined in low temperature studies of water diffusion which exhibit a low activation energy characteristic of a secondary relaxation in the melt (Zhang et al., 1991). The comparison of the viscous and molecular water mobility timescales raises a fundamental point about the mobility of water in volcanic systems.

Water can, under appropriate boundary conditions, diffuse into vesicles at a higher rate than the vesicles can viscously relax to accommodate their physical growth. The ultimate consequence of such dehydration is the intersection of the melt with the glass transition. The case for basalt is quite different because the diffusivity of water lies virtually superimposed on the viscous relaxation curve (Fig. 37). Experimental and volcanological implications of the brittle behavior of vesicle walls have been discussed recently by Romano et al. (1995c) and Mungall et al. (1995).

A third aspect of the mobility of hydrous species has been discussed by Keppler and Bagdassarov (1994) on the basis of high temperature infrared spectra obtained on a rhyolite glass with a couple of tenths of a weight percent of dissolved water. In a study in which they took care to reverse the experiments and demonstrate stability of their samples during the heating cycles, they observed a feature in the infrared spectra which they interpreted as the loss of a distinction between hydroxyl and molecular water at temperatures above 1300°C. This feature can be interpreted as representing the chemical exchange of the mobile species at the temperature of measurement and the timescale corresponding to the resonant frequency of the structural probe. As shown in Figure 38 this timescale is much faster than that provided by the viscosity data for the exchange of oxygens in the structure. It is even faster than that provided by the available water diffusivity data. It thus appears to represent a third distinct aspect of species mobility in hydrous melts. One addition clue as to its nature comes from the inclusion in Figure 38 of relaxation timescale for electrical conductivity in hydrous granitic melts (Lebedev and Khitarov, 1964). The extrapolation of this relaxation mode to the temperature of the infrared spectroscopic result clearly indicates that the mechanism of electrical conductivity in these samples, most likely the mobility of dissociated protons within the structure, is the mechanism which has intersected the vibrational timescale at the conditions noted by Keppler and Bagdassarov (1994). This would imply that it is merely the mobility of the protons on the structural sites within the liquid structure (one may say a tertiary relaxation) which is responsible for the motional averaging in the infrared spectra and not the original notion of the loss of distinct oxygen sites within the melt. The oxygens, whose mobility is given by the shear relaxation time, are exchanging at a timescale corresponding roughly to a viscosity of 10^8 Pa s, or 10^{-2} s, 10 orders of magnitude lower than the proton mobility.

OUTLOOK

Many new vistas have been opened up in the study of silicate melts using the kinetic standpoint of the glass transition. Several misconceptions concerning the structure and properties of silicate melts have been cleared up using relaxation considerations. Relaxation of the structure of silicate melts has been shown to be directly linked to the kinetics of property determinations. Such links have generated new experiments that have already yielded a considerably improved description of melt properties and behavior. Although substantial, these contributions are likely a precursor of more sophisticated studies to come. The investigation of further fundamental aspects of the PVT-equation of state of melts, the pressure and temperature dependence of their transport properties, and the nature and kinetics of structural changes occuring in silicate melts will continue to be greatly assisted in the future by a proper application of the knowledge of melt relaxation gained in the recent past.

ACKNOWLEDGMENTS

Most of the work described would not have been possible without the enthusiastic and skillful cooperation of several colleagues. I am especially indebted to M. Alidibirov, N. Bagdassarov , H. Behrens, K.-U. Hess, F. Holtz, K. Klasinski, R. Knoche, J. Mungall,

C. Romano, R. Stevenson, S.L. Webb and M. Wilding. The research performed in Bayreuth has been generously supported by the European Commission, the Deutsche Forschunsgemeinschaft (DFG), the Alexander-von-Humboldt Stiftung and the Deutsche Akademischer Austauschdienst (DAAD).

REFERENCES

Adam G, Gibbs JH (1965) On the temperature dependence of cooperative relaxation properties in glass-forming liquids. J Chem Phys 43:139-146

Alidibirov MA, Dingwell DB (1995) Experimental facility for studies of magma fragmentation during rapid decompression. EUGVIII Proc, Strasbourg, Blackwell Scientific Publ

Angell CA (1984) Strong and fragile liquids. In: Ngai KL, Wright GB (eds) Relaxation in Complex Systems. Office of Naval Research and Technical Information Service Lab, Arlington, VA, 345 p

Angell CA (1988) Perspectives on the glass transition J Phys Chem Solids 49:863-871

Angell CA (1991) Relaxation in liquids, polymers and plastic crystals-strong / fragile patterns and problems. J Non-Cryst Solids 131:13-31

Bacon JF, Hasapis AA, Wholley JW Jr (1960) Viscosity and density of molten silica and high silica content glasses. Phys Chem Glass 1:90-98

Bansal N, Doremus R (1986) Handbook of Glass Properties. Academic Press, New York, 680 p

Bockris JO`M, Tomlinson JW, White JL (1956) The structure of liquid silicates: partial molar volumes and expansivities. Trans Faraday Soc 52:299-310

Brückner R (1971) Properties and structure of vitreous silica II. J Non-Cryst Solids 5:177-216

Burnham CW, Davis NF (1971) The role of water in silicate melts: I. P-V-T relations in the system $NaAlSi_3O_8$-H_2O to 1 Kilobar and 1100°C. Am J Sci 270:54-79

Chakraborty S, Dingwell DB (1993) Viscous flow and multicomponent diffusion in melts in the system Na_2O-K_2O-Al_2O_3-SiO_2-P_2O_5. EOS, Trans Am Geophys Union 74:617

Courtial P, Dingwell DB (1995) Non-linear composition dependence of melt volume in the CaO-Al_2O_3-SiO_2 system. (In press, Geochim Cosmochim Acta)

DeBast J, Gilard p (1963) Variation of the viscosity of glass and the relaxation of stresses during stabilisation. Phys Chem Glass 4:117-128

DeBolt MA, Easteal AJ, Macedo PB, Moynihan CT (1976) Analysis of structural relaxation in glass using rate heating data. J Amer Cer Soc 59:16-21

Dingwell DB (1987) Melt viscosities in the system $NaAlSi_3O_8$-H_2O-F_2O_{-1}. In: Magmatic Processes: Physicochemical Principles. BO Mysen (ed) Geochemical Society Spec Pub 1:423-433

Dingwell DB (1990) Effects of structural relaxation on cationic tracer diffusion in silicate melts. Chem Geol 182:209-216

Dingwell DB (1992) Density of some titanium-bearing silicate liquids, and the composition dependence of the partial molar volume of TiO_2. Geochim Cosmochim Acta 56:3403-3407

Dingwell DB (1993) Experimental strategies for the determination of granitic melt properties at low temperature. Chem Geol 108:19-30

Dingwell DB (1995) Viscosity and Anelasticity of melts and glasses. Mineral Physics and Crystallography A Handbook of Physical Constants AGU Reference Shelf 2. T Ahrens (ed) Am Geophys Union, p 209-217

Dingwell DB, Brearley M (1988) Melt densities in the CaO-FeO-Fe_2O_3-SiO_2 system and the compositional-dependence of the partial molar volume of ferric iron in silicate melts. Geochim Cosmochim Acta 52:2815-2825

Dingwell DB, Webb SL (1989) Structural relaxation in silicate melts and non-Newtonian melt rheology in igneous processes. Phys Chem Minerals 116:508-516

Dingwell DB, Webb SL (1990) Relaxation in silicate melts. Eur J Mineral 12:427-449

Dingwell DB, Brearley M, Dickinson Jr JE (1988) Melt densities in the Na_2O-FeO-Fe_2O_3-SiO_2 system and the partial molar volume of tetrahedrally-coordinated ferric iron in silicate melts. Geochim Cosmochim Acta 152:2467-2475

Dingwell DB, Bagdassarov N, Bussod G, Webb SL (1993a) Magma Rheology Mineralogical Assoc Canada Short Course on Experiments at High Pressure and Applications to the Earth´s Mantle, p 131-196

Dingwell DB, Knoche R, Webb SL (1993b) The effect of fluorine on the density of haplogranitic melts. Am Mineralseral 78:325-330

Dingwell DB, Knoche R, Webb SL (1993c) A volume temperature relationship for liquid GeO_2 and some geophysically relevant derived parameters for network liquids. Phys Chem Minerals 119:445-453

Dingwell DB, Paris E, Seifert F, Mottana A, Romano C (1994) X-ray absorption study of Ti-bearing silicate glasses. Phys Chem Minerals 21:501-509

Dingwell DB, Romano C, Hess K-U (1995) The effect of water on the viscosity of a haplogranitic melt under P-T-X- conditions relevant to silicic volcanism. (submitted to Contrib Mineral Petrol)

Gan H, Hess PC (1992) Phosphate speciation in potassium aluminosilicate glasses. Am Mineral 77:495-506

Hess K-U, Dingwell DB, Webb SL (1995) The influence of alkaline earth oxides on the viscosity of granitic melts: systematics of non-Arrhenian behavior. (In Press, Eur J Mineral)

Hessenkemper H, Brückner R (1988) Load-dependent flow behavior of silicate melts. Glastech Ber 61:312-320

Höfler S, Seifert F (1984) Volume relaxation of compacted SiO_2 glass: a model for the conservation of natural diaplectic glasses. Earth Planet Sci Lett 67:433-438

Isard JO (1969) Mixed alkali effect in glass. J Non-Cryst Solids 1:235-261

Keppler H, Bagdassarov N (1994) High temperature FTIR spectra of water in rhyolite melt to 1300°C. Am Mineral 78:1324-1327

Knoche R, Webb SL, Dingwell (1992a) A partial molar volume for B_2O_3 in haplogranitic melts. Can Mineral 130:561-569

Knoche R, Dingwell DB, Webb SL (1992b) Temperature-dependent thermal expansivities of silicate melts:The system anorthite-diopside. Geochim Cosmochim Acta 56:689-699

Knoche R, Dingwell DB, Webb SL (1992c) Temperature dependent expansivities for silicate melts: the system albite-anorthite-diopside. Contrib Mineral Petrol 111:61-73

Knoche R, Dingwell DB, Seifert FA, Webb S (1994) Nonlinear properties of supercooled liquids in the Na_2O-SiO_2 system. Chem Geol 116:1-16

Knoche R, Dingwell DB, Webb SL (1995) Leucogranitic and pegmatitic melt densities: partial molar volumes for SiO_2, Al_2O_3, Na_2O, K_2O, Rb_2O, Cs_2O, Li_2O, BaO, SrO, CaO, MgO, TiO_2, B_2O_3, P_2O_5, F_2O_{-1}, Ta_2O_5, Nb_2O_5, and WO_3. (In press, Geochim Cosmochim Acta)

Kress VC, Carmichael ISE (1991) The compressibility of silicate liquids containing Fe_2O_3 and the effect of composition, temperature and oxygen fugacity and pressure on their redox states. Contrib Mineral Petrol 108:82-92

Kress VC, Williams Q, Carmichael ISE (1988) Ultrasonic investigation of melts in the system Na_2O-Al_2O_3-SiO_2. Geochim Cosmochim Acta 52:283-293

Kushiro I (1978a) Density and viscosity of hydrous calkalkaline andesite magma at high pressures. Carnegie Inst Wash Yearb 77:675-677

Kushiro I (1978b) Viscosity and structural changes of albite ($NaAlSi_3O_8$) melt at high pressures. Earth Planet Sci Lett 41:87-90

Lange RA (1994) The effect of H_2O, CO_2 and F on the density and viscosity of silicate melts. Rev Mineral 30:331-369

Lange RA, Carmichael ISE (1987) Densities of Na_2O-K_2O-CaO-MgO-FeO-Fe_2O_3-Al_2O_3-TiO_2-SiO_2 liquids: new measurements and derived partial molar properties. Geochim Cosmochim Acta 53:2195-2204

Lange RA, Carmichael ISE (1990) Thermodynamic properties of silicate liquids with emphasis on density, thermal expansion and compressibility. Rev Mineral 24:25-64

Lange RA, Navrotsky A (1993) Heat capacities of TiO_2-bearing silicate liquids. Evidence for anomalous changes in configurational entropy with temperature. Geochim Cosmochim Acta 57:3001-3011

Lebedev EB, Khitarov NE (1964) Beginning of granitic melting and its melt electrical conductivity at high pressure. Geokhymia 3:195-201.

Li J, Uhlmann D (1970) The flow of glass at high stress levels. I Non-Newtonian behavior of homogeneous 0.08 Rb_2O-0.92SiO_2 glasses. J Non-Cryst Solids 3:127-147

Manns P, Brückner R (1988) Non-Newtonian flow behavior of a soda-lime silicate glass at high rates of deformation. Glastech Ber 61:46-56

Martens RM, Rosenhauer M, Büttner H, Von Gehlen K (1987) Heat capacity and kinetic parameters in the glass transformation interval of diopside, anorthite and albite glass. Chem Geol 62:49-70

Mazurin OV, Smartsev YK, Polutseva LN (1979) Temperature dependences of the viscosity of some glasses at a constant structural temperature. Sov J Glass Phys Chem 5:69-79 (English translation).

McMillan PF (1994) Water solubility and speciation models. Rev Mineral 30:131-152

Mills JJ (1974) Low frequency storage and loss moduli of soda-silica glasses in the transformation range. J Non-Crystal Solids 14:255-268

Montel, J-M (1986) Experimental determination of the solubility of Ce-Monazite in SiO_2-Al_2O_3-Na_2O-K_2O melts at 800°C, 2 kbar under water-saturated conditions. Geol 14:659-662

Moynihan, CT, Easteal, AJ, Wilder, J (1974) Dependence of the glass transition temperature on heating and cooling rate. J Phys Chem 78:2673-2677

Moynihan CT, Easteal AJ, DeBolt MA, Tucker J (1976a) Dependence of fictive temperature of glass on cooling rate. J Am Ceram Soc 59:12-16

Moynihan CT, Easteal AJ, Tran DC, Wilder JA, Donovan EP (1976b) Heat capacity and structural relaxation of mixed alkali glasses. J Am Ceram Soc 59:137-140

Mungall J, Romano C, Bagdassarov N, Dingwell DB (1995) A mechanism for microfracturing of vesicle walls in glassy lava: implications for explosive volcanism. (Submitted to J Volc Geotherm Res)

Mysen BO (1988) Structure and Properties of Silicate Melts. Elsevier, Amsterdam, 354 p

Narayanaswamy OS (1971) A model of structural relaxation in glass. J Am Ceram Soc 54:491-498

Narayanaswamy OS (1988) Thermorheological simplicity in the glass transition. J Am Ceram Soc 71:900-904

Paris E, Dingwell DB, Seifert F, Mottana A,, Romano C (1994a) Pressure-induced coordination change of Ti in silicate glass: a XANES study. Phys Chem Minerals 21:510-515

Paris E, Giuli G, Dingwell D, Seifert F, Mottana A (1994b) The influence of aluminum on titanium coordination in silicate melts: a XANES study. EOS, Trans Am Geophys Union 44:705

Pichavant M, Montel J-M, Richard LR (1992) Apatite solubility in peraluminous liquids: experimental data and an extension of the Harrison-Watson model. Geochim Cosmochim Acta 56:3855-3861

Rapp RP, Watson EB (1986) Monazite solubility and dissolution kinetics implications for the thorium and light rare earth chemistry of felsic magmas. Contrib Mineral Petrol 94:304-316

Rekhson S (1985) Viscoelasticity of glass. Glass Science Technology 3:1-117

Richet P (1984) Viscosity and configurational entropy of silicate melts. Geochim Cosmochim Acta 48:471-484

Richet P, Bottinga Y (1986) Thermochemical properties of silicate glasses and liquids. Rev Geophys 24:1-26

Rigden SM, Ahrens TJ, Stolper EM (1984) Density of liquid silicates at high pressure. Science 226:1071-1074

Rigden SM, Ahrens TJ, Stolper EM (1988) Shock compression of molten silicate: results for model basaltic composition. J Geophys Res 93:367-382

Rigden SM, Ahrens TJ, Stolper EM (1989) High pressure equation of state of molten anorthite and diopside. J Geophys Res 94:9508-9522

Ritland H N (1954) Density phenomena in the transformation range in a borosilicate crown glass. J Am Ceram Soc 37:370-378

Rivers ML, Carmichael ISE (1987) Ultrasonic studies of silicate melts. J Geophys Res 92:9247-9270

Romano C, Dingwell DB, Sterner SM (1994) Kinetics of quenching of hydrous feldspathic melts: quantification using synthetic fluid inclusions. Am Mineral 79:1125-1134

Romano C, Dingwell DB, Behrens H, Dolfi D (1995a) Solubility of water in melts along the joins $NaAlSi_3O_8$-$KAlSi_3O_8$, $NaAlSi_3O_8$-$LiAlSi_3O_8$ and $LiAlSi_3O_8$-$KAlSi_3O_8$. (In press, Am Mineral)

Romano C, Dingwell DB, Behrens H (1995b) The temperature dependence of the speciation of water in $NaAlSi_3O_8$-$KAlSi_3O_8$ melts: an application of fictive temperatures derived from synthetic fluid inclusions. (In press, Contrib Mineral Petrol)

Romano, C., Bagdassarov, N., Dingwell, D.B. and Mungall, J. (1995c) Strength and explosive behavior of vesicular glassy lavas: experimental constraints. (Submitted to Am Mineral)

Rosenhauer M, Scarfe CM, Virgo D (1979) Pressure dependence of the glass transition in glasses of diopside, albite and sodium trisilicate composition. Carnegie Inst Wash Yearb 78:556-559

Sato H, Manghnani M (1984) Ultrasonic measurements of Vp and Qp: relaxation spectrum of complex modulus on basalt melts. Phys Earth Planet Int 41:18-33

Scarfe CM, Mysen BO, Virgo DL (1987) Pressure dependence of the viscosity of silicate melts. Geochem Soc Spec Pub 1:59-67

Scherer GW (1984) Use of the Adam-Gibbs equation in the analysis of structural relaxation. J Am Ceram Soc 67:504-511

Scherer GW (1986a) Relaxation in Glass and Composites. Wiley, New York, 331 p

Scherer GW (1986b) Volume relaxation far from equilibrium J Am Ceram Soc 69:374-381

Scherer GW (1990) Theories of relaxation. J Non-Cryst Solids 123:75-89

Schulze F, Behrens H, Holtz F, Roux J, Johannes W (1995) The influence of water on the viscosity of a haplogranitic melt. (Submitted to Am Mineral)

Sekiya K, Morinaga K, Yanagase T (1980) Physical properties of Na_2O-GeO_2 melts. J Japan Ceram Soc 88:367-373

Shaw HR (1963) Obsidian-H_2O viscosities at 1000 and 2000 bars in the temperature range 700 to 900°C. J Geophys Res 68:6337-6343

Shaw HR (1972) Viscosities of magmatic silicate liquids: an empirical method of prediction. Am J Sci 272:870-889

Scherer GW (1984) Relaxation in Glass and Composites. Wiley, New York, 331 p

Silver LA, Ihinger PD, Stolper EM (1990) The influence of bulk composition on the speciation of water in silicate glasses. Contrib Mineral Petrol 104:142-162

Simmons J, Simmons C (1989) Nonlinear viscous flow in glass forming. Bull Cer Soc Am 11:1949-1955

Simmons J, Mohr R, Montrose C (1982) Non-Newtonian viscous flow in glass. J Appl Phys 53:4075-4080

Simmons J, Ochoa R, Simmons K, Mills J (1988) Non-Newtonian viscous flow in soda-lime-silica glass at forming and annealing conditions

Stevenson RJ, Dingwell DB Webb SL, Bagdassarov NS (1995) The equivalence of enthalpy and shear stress relaxation in rhyolitic obsidians and quantification of the liquid-glass transition in volcanic processes. (In press, J Volc Geotherm Res)

Stockhorst H, Brückner R (1982) Structure sensitive measurements on E-glass fibers. J Non-Cryst Solids 49:471-484

Taniguchi H (1981) Effects of water on the glass transition temperature of rhyolitic melt. J Jap Assoc Mineral Petrol Geochem 76:49-57

Tool AQ Eichlin CG (1931) Variations caused in the heating curves of glass by heat treatment. J Am Ceram Soc 14:276-308

Tool AQ (1946) Relation between inelastic deformability and thermal expansion of glass in its annealing range. J Am Ceram Soc 29:240-253

Toplis M, Dingwell DB (1994) High field strength cations in melts of variable alkali/aluminum ratio: the variable influence of P_2O_5 on melt viscosity. EOS, Trans Am Geophys Union 75:704.

Wäsche R, Brückner R (1986a) Flow birefringence and structure of alkali phosphate melts. Phys Chem Glass 27:80-86

Wäsche R, Brückner R (1986b) The structure of mixed alkali phosphate melts as indicated by their non-Newtonian flow behavior and optical birefringence. Phys Chem Glass 27:87-142

Webb SL, Dingwell DB (1990a) The onset of non-newtonian rheology of silicate melts. A fiber elongation study. Phys Chem Minerals 17:125-132

Webb SL, Dingwell DB (1990b) Non-newtonian rheology of igneous melts at high stresses and strain rates: experimental results for rhyolite, andesite, basalt, and nephelinite. J Geophys Res 95:695-701

Webb SL, Knoche R, Dingwell DB (1992) Determination of silicate liquid thermal expansivity using dilatometry and calorimetry. Eur J Mineral 4:95-104

Webb SL, Dingwell DB (1994) The compressibility of titanium-bearing alkali silicate melts. Contrib Mineral Petrol 118 157-168

Wilding M, Webb SL, Dingwell DB (1995) Evaluation of a relaxation geothermometer for volcanic glasses. (In press, Chem Geol)

Wong J, Angell CA (1976) Glass Structure by Spectroscopy. Dekker, New York, 864 p

Zhang Y, Stolper EM, Wasserburg G (1991) Diffusion of water in rhyolitic glasses. Geochim Cosmochim Acta 55:441-456

Zhang Y, Stolper EM, Ihinger PD (1995) Kinetics of the reaction $H_2O + O = 2$ OH in rhyolitic and albitic glasses: preliminary results. Am Mineral 80:593-612

Zijlstra AL (1963) The viscosity of some silicate glasses in connection with thermal history. Phys Chem Glass 4:143-151

Chapter 3

RHEOLOGY AND CONFIGURATIONAL ENTROPY
OF SILICATE MELTS

P. Richet and Y. Bottinga

Laboratoire de Physique des Géomatériaux, URA CNRS 734
Institut de Physique du Globe
4, place Jussieu, 75252 Paris cedex 05, France

INTRODUCTION

Throughout geological history the viscosity of silicate melts has been a major property controlling mass and heat transfer within the Earth's interior. The viscosity is thus a key parameter in modeling igneous petrological and volcanological processes. Unfortunately, prediction of the viscosity is made difficult by extremely strong dependences on temperature and chemical composition as illustrated by the experimental data of Figure 1 for a variety of homogeneous silicate melts at room pressure. The viscosity generally varies by more than 10 orders of magnitude between superliquidus conditions and the glass transition range. Likewise, at constant temperature one can observe similar differences as a function of composition since at 1500 K, for example, the viscosity of pure SiO_2 is about 10 orders of magnitude higher than that of molten diopside ($CaMgSi_2O_6$). As a matter of fact, the viscosity range spanned by the data is wider still at lower than at higher temperatures where they tend to converge (Fig. 1).

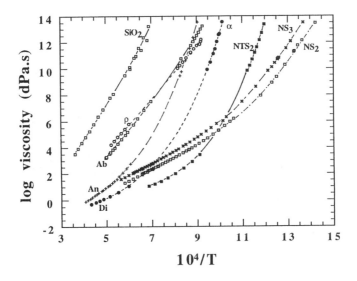

Figure 1. Viscosity-temperature relationship for some silicate melts. Data sources as follows (see Richet, 1984, for a review of the available data for most of these compositions). SiO_2: Urbain et al. (1982) and Hetherington et al. (1964); $NaAlSi_3O_8$ (Ab): Urbain et al. (1982) and Taylor and Rindone (1970); $CaAl_2Si_2O_8$ (An): Urbain et al. (1982) and Cukierman and Uhlmann (1973); $CaMgSi_2O_6$ (Di): Urbain et al. (1982) and Neuville and Richet (1991); rhyolite (ρ) and andesite (α) melts: Neuville et al. (1993); $Na_2TiSi_2O_5$ (NTS$_2$): Bouhifd and Richet (1995); $Na_2Si_2O_5$ (NS$_2$) and $Na_2Si_3O_7$ (NS$_3$): Bockris et al. (1955), Fontana and Plummer (1979) and Poole (1948).

Of course, magmas are not homogeneous liquids. They carry bubbles and crystals whose influence on the rheology becomes complex at high volume fractions, say about 40% or more. As long as the volume fraction of these inclusions is not large, however, their physical effects on the viscosity are small compared to chemical effects resulting from the changing composition of the melt (e.g. Ryerson et al., 1988; Spera et al., 1988; Bagdassarov and Dingwell, 1992; Stein and Spera, 1992; Lejeune and Richet, 1995). In any case, the rheology of real magmas has to be considered in the light of the viscosity of the liquid phase which is dealt with in this review. For obvious experimental reasons, most available data for homogeneous melts concern simple compositions, free of dissolved volatiles, at a pressure of 1 atm and temperatures higher than 1500 K. However, petrological and volcanological processes of interest to earth scientists generally take place at temperatures much lower than 1500 K (at low pressures), or at pressures much greater than 1 atm (at high temperatures). In addition, melts relevant to volcanological processes contain various amounts of volatiles, mainly H_2O, CO_2 and sulfur-bearing species. Hence, there is a considerable mismatch between the data actually needed for understanding igneous processes and available melt viscosities (e.g. Bottinga and Weill, 1972; Urbain et al., 1982; Scarfe et al., 1987; Ryan and Blevins, 1987).

Extrapolation of available data could be a way out of this predicament. For instance, one could make use for this purpose of the well-known Arrhenius equation:

$$\eta = A_a \exp (E_a/RT) , \tag{1}$$

which is often chosen to account for the temperature dependence of the viscosity. In this equation, obtained from the absolute rate theory of Eyring, R is the gas constant, A_a an adjustable pre-exponential factor and the activation energy E_a represents potential energy barriers overcome by atoms when they move from one site to another in the melt (see Glasstone et al., 1941). In fact, the curvature of the viscosity data plotted in Figure 1 shows that activation energies cannot be constant. As discussed by Richet et al. (1986), they appear much too large to represent actually the energy barriers involved in viscous flow. Furthermore, it is unlikely that these energy barriers decrease by a factor of 5 or more between 1800 and 1000 K, as do apparent activation energies obtained from the experimental viscosity data. Extrapolation of equations like (1), not containing physically well-defined parameters, is thus beset by considerable difficulties.

A physical understanding of the temperature, pressure, composition and even time dependences of the viscosity is required for geochemical purposes. As far as we are aware, only one theory has been shown capable of addressing all these effects. This is the configurational entropy theory of relaxation processes in viscous liquids proposed by Adam and Gibbs (1965), to which this review will thus be devoted. The relevance of configurational entropy to the viscosity of liquids, i.e. to their inability to oppose stress, stems from the simple fact that viscous flow takes place through the availability of a great many different configurational states. In this respect, the inherently cooperative nature of these rearrangements makes the Adam and Gibbs theory fundamentally different from Arrhenius equations, which rely invariably on the hopping of a single atom over a "constant" energy barrier as the fundamental step of fluid flow.

The versatility of Adam-Gibbs theory stems in part from the fact that its key parameter, the configurational entropy, is not a fitting parameter, but is amenable to calorimetric measurement and simple modeling. The qualitative merits of this theory have long been discussed (e.g. Chang et al., 1966; Angell and Sichina, 1976). Its application to relaxational processes has been described more recently by Scherer (1984), and the temperature and composition dependences of silicate melt viscosities were discussed by Richet (1984) and Richet and Neuville (1992). Since then, the Adam-Gibbs theory has been used to elucidate the development of nanometric inhomogeneities observed in liquids close to the glass transition (Moynihan and Schroeder, 1993). It has also been used to interpret the occurrence of viscous thinning in silicate liquids (Bottinga, 1994a,b) and to explain qualitatively the pressure dependence of silicate melt viscosities (Bottinga and Richet, 1995).

In this chapter we will not deal with all these aspects in a comprehensive way. We will first review briefly the notions of viscosity and configurational entropy. With specific examples, we will then illustrate salient aspects of the Adam and Gibbs (1965) theory as applied to the viscosity of silicate melts in the geochemically relevant range 1 to 10^{13} dPa s. In doing so, we will avoid as much as possible duplication of published material on topics closely related to the viscosity such as the glass transition and relaxation properties. For a broader perspective on some of the points considered here the readers could complement their reading by other chapters of this book and available reviews on relaxation (Moynihan et al., 1976; Dingwell and Webb, 1990; Scherer, 1990), viscosity regimes (Bottinga et al., 1995), glass transition (Angell, 1988), thermochemistry of the glass transition (Richet and Bottinga, 1983, 1986), and configurational properties (Richet and Neuville, 1992).

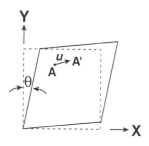

Figure 2. Displacement of a point A in the X,Y plane under a shear stress. Due to homogeneous shear, the initial square is deformed into a parallellogram and point A moved to A'.

VISCOSITY AND RELAXATION TIMES

Shear viscosity

In moving liquids, one observes that shear stresses resulting from interatomic forces yield velocity gradients perpendicular to the stream direction. If the liquid is Newtonian, the shear stresses are proportional to the velocity gradients. A two-dimensional picture of the deformation due to homogeneous shear is shown in Figure 2. In the plane perpendicular to the Y direction, the shear force per unit area associated with fluid flow at a rate v_X in the X direction is the XY component (σ_{XY}) of the shear stress tensor, viz:

$$\sigma_{XY} = \eta \, \partial v_X / \partial y = \eta \, \partial/\partial t \, (\partial u_X/\partial y) . \qquad (2)$$

In this equation, t is the time, u_X the X component of the displacement vector (u) of a point A in the liquid due to shear motion, and η a proportionality constant, known as the coefficient of viscosity or more shortly as the viscosity. An analogous expression can be written for σ_{XZ}, whereas σ_{XX} is a compression (or extension) stress related to the volume viscosity discussed below. Because the angle θ is very small (Fig. 2), $u_X = y \tan \theta = y\theta$, where y is the value of the Y coordinate of point A before shearing took place. Hence for homogeneous shear strain, u_X is proportional to the initial value of y, whereas the XY component of the strain tensor (ε) is defined as:

$$\varepsilon_{XY} = \theta = u_X/y . \qquad (3)$$

For a non-homogeneous shear Equation (3) is replaced by:

$$\varepsilon_{XY} = \partial u_X / \partial y , \tag{4}$$

and with Equation (2) one obtains for each tensor component a relation between the stress and the rate of deformation:

$$\sigma = \eta \, \partial \varepsilon / \partial t . \tag{5}$$

In other words, the viscosity is Newtonian as long as η is independent of the stress and deformation rate. The breakdown of this approximation is discussed later on in this chapter.

Maxwell model

In spite of its shortcomings, the Maxwell model is a convenient point of departure for introducing the concept of relaxation time in a discussion of the viscosity. The application of a shear stress to a body causes instantaneously a relative deformation or strain, which is elastic because it is recovered when the stress is released. The relation between the stress and strain may be written (Hooke's Law):

$$s = G_\infty \varepsilon , \tag{6}$$

where G_∞ is the high-frequency shear modulus, also known as the shear rigidity. Therefore the stress rate is:

$$d\sigma / dt = G_\infty \, d\varepsilon / dt . \tag{7}$$

In addition to this elastic deformation, liquids have a non-recoverable viscous deformation that tends to relax the stress built up macroscopically through strained bond lengths and angles. Such a relaxation occurs even when the deformation is kept constant because the instantaneous elastic deformation is progressively transformed into a viscous deformation. Experimentally (see de Bast and Gilard, 1965, for careful measurements on silicate melts), one observes this relaxation by exerting a stress on a viscous liquid and registering how it diminishes with time while keeping constant the initially elastic deformation. Assuming that the rate of stress relaxation is proportional to the stress itself, Maxwell replaced Equation (7) by:

$$d\sigma / dt = G_\infty \, d\varepsilon / dt - \sigma / \tau , \tag{8}$$

where τ is the stress relaxation time which characterizes the timecale of this process. In practice, the stress relaxation time is about the same as the structural relaxation time and both terms are used synonymously. If the deformation rate is constant, integration of Equation (8) gives the instantaneous stress:

$$\sigma = \eta \, d\varepsilon / dt + C \, e^{-t/\tau} \tag{9}$$

where C is a constant.

The great merit of the Maxwell model is its simplicity. To examine its validity for silicate melts, we have plotted in Figure 3 the data of de Bast and Gilard (1965) for the stress relaxation in a window glass at a constant temperature of 784 K for which the equilibrium viscosity is 10^{14} Pa s. The initial stress $\sigma_0 = 22.7$ MPa is progressively reduced in such a way that the deformation is constant. If Equation (9) were exact, $\ln (\sigma / \sigma_0)$ should vary as $-(t/\tau) \ln C$ and plot along a straight line in Figure 3. The actual curvature of the data of de Bast and Gilard (1965) thus illustrates the limitations of the Maxwell model.

For stresses applied at low frequencies, $d\sigma / dt$ in Equation (8) becomes very small and with Equation (5) one obtains:

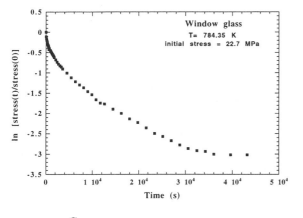

Figure 3. Stress relaxation at constant deformation of a window glass, as given by the ratio of the instaneous stress at time t over the initial stress. Data from de Bast and Gilard (1965).

$$\eta = G_\infty \tau. \tag{10}$$

The validity of this equation is fairly general for viscous liquids. For instance, Equation (10) holds for polymers (de Gennes, 1979) as well as for liquid silicates even though these melts have structurally very little in common (see Stebbins et al., 1992). Concerning silicate melts, the limited available data suggest that G_∞ is about 10^{10} Pa and varies by less than a factor of ten with either temperature or composition (Dingwell and Webb, 1989). Compared to the tremendous variations of the viscosity apparent in Figure 1, the parameter G_∞ can thus be considered approximately as a constant. If the viscosity is known, then Equation (10) yields straightforward estimates of structural relaxation times.

Volume viscosity

Besides the shear terms ε_{XY}, there are volumetric strains ε_{XX} resulting from the aforementioned stress component σ_{XX}. Again, one distinguishes an instantaneous, recoverable, elastic component and a viscous deformation which takes place through changes in the internal structure. The isothermal compressibility of a liquid, $\kappa = -1/V$ $(\partial V/\partial P)_T$ can thus be split into two terms:

$$\kappa = \kappa_\infty + \kappa_r \tag{11}$$

where κ_∞ is the compressibility at infinite frequency and κ_r its relaxed part, representing the elastic and viscous components, respectively. The bulk modulus is the inverse of the compressibility, $K = 1/\kappa$. Defining $K_\infty = 1/\kappa_\infty$ as the bulk modulus at infinite frequency, one rewrites Equation (11) as:

$$K = K_\infty - K_r \tag{12}$$

where $K_r = \kappa_r K_\infty K$. At low frequencies the volume viscosity is defined as:

$$\eta_V = K_r \tau_V \tag{13}$$

where τ_V is the relaxation time at constant volume. In addition, there is a volume relaxation time at constant pressure:

$$\tau_P = \tau_V K_\infty/K . \tag{14}$$

The shear and volume viscosities differ generally by less than a factor of ten. The ratio η_V/η is nearly temperature independent for many liquids (Litovitz and Davis, 1965) and the magnitudes of K_r and G_∞ are frequently similar. In Earth sciences, the shear viscosity is far more important than the volume viscosity. Therefore we will consider only the former in the following, referring the reader interested in the ramifications of volume viscosity to Litovitz and Davis (1965).

Elongational viscosity

Most of the viscosities higher than 10^9 poises represented in Figure 1 have been obtained from the rates of stretching (or shortening) of samples under a constant extensive (or compressional) stress. By analogy with the shear viscosity, the proportionality constant relating these deformation rates to the stress is called the elongational viscosity:

$$\sigma = \eta_{el} \, \partial \varepsilon_{el}/\partial t , \qquad (15)$$

where $\varepsilon_{el} = \Delta l/l_0$ is the relative change in the length of a sample with an initial length l_0. In addition to this viscous response, the elastic deformation must be considered. It is expressed as:

$$\sigma = E \, \varepsilon_{el} , \qquad (16)$$

where E is Young's modulus. Tending to counterbalance the change in length of the sample, the cross section also varies. Let $\Delta w/w_0$ be the relative change in diameter of a cylindrical sample with an initial diameter w_0. The relative changes in section and length of the sample are then related to Poisson's number μ with:

$$\mu = - (\Delta w/w_0) / (\Delta l/l_0) . \qquad (17)$$

For an isotropic body, the moduli E and G, for deformations due to elongational stress (Eqn. 16) and under shear stress (Eqn. 10), respectively, are geometrically related to the Poisson number by:

$$E = 2 (1 + \mu) \, G . \qquad (18)$$

If the stretching or shortening takes place at constant volume, then μ approaches the limiting value of 0.5. According to Equation (18), in this case $E = 3 \, G$, and one has therefore:

$$\eta_{el} = 3 \, \eta . \qquad (19)$$

This relationship thus provides a simple means of measuring elevated Newtonian viscosities under either extensional or compressional stress.

CONFIGURATIONAL ENTROPY

Glass transition

In glass-forming liquids, atoms can be viewed as lying in potential energy wells separated from neighboring wells by barriers with variable heights and shapes. At a given temperature, the configurational state is determined by the distribution of atoms over available potential energy minima. With increasing temperatures, configurational states of higher energy become available because part of the heat supplied to the liquid is converted in potential energy. Bond lengths and angles, nearest and next-nearest neighbor distances, coordination numbers and other parameters that can be used to describe the structure are not single-valued; they are distributed over ranges of values which generally become wider with increasing temperatures. In thermochemical terms, the energy needed for these temperature-induced structural changes is represented by the configurational heat capacity (C_p^{conf}) and the measure of the corresponding spreading of configurational states is the configurational entropy (S^{conf}). These are mutually related by:

$$dS^{conf} = C_p^{conf}/T \, dT . \qquad (20)$$

The structural relaxation times, which characterize the kinetics of these temperature-induced changes, scale as the viscosity according to Equation (10), $\eta = G_\infty \tau$. When they become long with respect to the timescale of an experiment, time-dependent properties are observed, corresponding to a slower and slower approach to internal thermodynamic equilibrium when the temperature is decreased. Eventually, at a point that depends on the cooling rate, the structure can no longer adjust to the new temperature (or new conditions of other intensive parameters) and the liquid then transforms into a glass, i.e. a solid with the frozen in disordered atomic arrangement of the liquid. Since the glass transition temperature depends on the cooling rate, properties of glasses depend on thermal history. Thus glasses are nonequilibrium phases.

With usual laboratory cooling rates of the order of 10 K/s, the glass transition temperature (T_g) is close to the temperature at which the viscosity is 10^{13} dPa s. From Equation (10), this viscosity corresponds to relaxation times of the order of 100 s, which indeed represent typical timescales of macroscopic calorimetry or dilatometry measurements. With the freezing in of the structure at the glass transition, the configurational heat capacity vanishes and the heat capacity decreases abruptly (Fig. 4). Similar decreases are oberved for the thermal expansion coefficient and the compressibility because the configurational changes no longer contribute to the dilation and compression mechanisms (see Richet and Neuville, 1992). As a matter of fact, these rapid changes in second-order thermodynamic properties are the most dramatic features of the glass transition. But we must stress that the glass transition is a nonequilibrium phenomenon, inherently dependent on the rate of temperature change, even though it is reproducible experimentally.

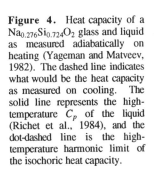

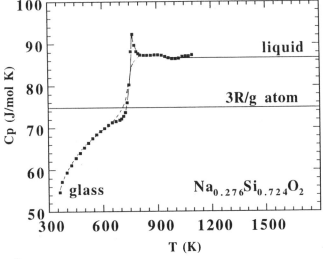

Figure 4. Heat capacity of a $Na_{0.276}Si_{0.724}O_2$ glass and liquid as measured adiabatically on heating (Yageman and Matveev, 1982). The dashed line indicates what would be the heat capacity as measured on cooling. The solid line represents the high-temperature C_p of the liquid (Richet et al., 1984), and the dot-dashed line is the high-temperature harmonic limit of the isochoric heat capacity.

Configurational heat capacity

On the glass side, the heat capacity is essentially determined by the distribution of kinetic energy among vibrational energy levels. The heat capacity is thus termed vibrational. Integrated from 0 K to a given temperature T, it gives simply the vibrational entropy of the glass with:

$$S_g(T) = \int_0^T C_{pg} / T \, dT \tag{21}$$

For silicates, a simplifying feature of considerable practical importance is that the glass transition takes place on heating when the heat capacity is close to the Dulong-and-Petit limit of $3R$ per g atom/K, where R is the gas constant (Fig. 4). This correlation first pointed out by Haggerty et al. (1968) has been found to hold without exception for the scores of silicates subsequently investigated (cf. Richet and Bottinga, 1986; Martens et al., 1987). This is illustrated in Figure 5 by the heat capacity data for some of the compositions whose viscosities are plotted in Figure 1. The main consequence of this correlation is that, on a gram atom basis, the configurational heat capacity of silicate liquids can be approximated by:

$$C_p{}^{conf} = C_{pl} - C_{pg}(T_g) = C_{pl} - 3R , \qquad (22)$$

where C_{pl} and C_{pg} are the heat capacities of the liquid and glass, respectively. A detailed justification of Equation (22) has been given by Richet et al. (1986). Here, we will just note that although $C_p{}^{conf}$ is not zero below T_g for some non-silicate substances (see Goldstein, 1976, 1977; Gujrati and Goldstein, 1980), calorimetric evidence rules out any significant configurational contributions to the relative entropy of silicate glasses well below the glass transition range (see Richet et al., 1982, 1986, 1993).

The data plotted in Figure 5 also show that, as given by Equation (22), the configurational heat capacity may or not vary with temperature, being constant for alkali-silicate melts, generally increasing for aluminosilicate melts, or decreasing markedly for alkali-titanosilicate melts. These specific variations contribute to a strong composition dependence of $C_p{}^{conf}$ which, as a result of Equation (22), is the same as that of C_{pl}. As reviewed by Richet and Neuville (1992) and Courtial and Richet (1993), in systems like $Na_2O\text{-}SiO_2$ or $MgO\text{-}Al_2O_3\text{-}SiO_2$ the configurational heat capacity varies linearly with composition. Nonlinear variations are observed in other systems like $Na_2O\text{-}Al_2O_3\text{-}SiO_2$, pointing to more complex interactions with aluminum of alkali than alkaline-earth cations. Even though nonideal models are actually needed to represent all the data (Richet and Bottinga, 1985; Courtial and Richet, 1993), the fact that additive models of calculation of the heat capacity as a function of composition have been proposed (Stebbins et al., 1984;

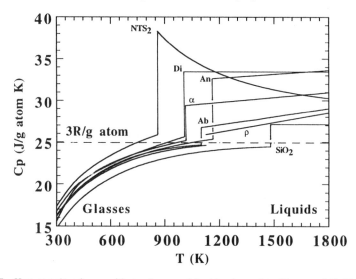

Figure 5. Heat capacity of some silicate glasses and liquids whose viscosities are plotted in Figure 1. Data for SiO_2: Richet et al. (1982); $NaAlSi_3O_8$ (Ab): Richet and Bottinga (1984a); $CaAl_2Si_2O_8$ (An) and $CaMgSi_2O_6$ (Di): Richet and Bottinga (1984b); rhyolite (ρ) and andesite (α) melts: Neuville et al. (1993); $Na_2TiSi_2O_5$ (NTS$_2$): Richet and Bottinga (1985).

Lange and Navrotsky, 1992) indicates that such complexities have not major thermo-chemical consequences, except for the extreme case of configurational rearrangements taking place in melts like alkali-titanosilicates (Richet and Bottinga, 1985; Lange and Navrotsky, 1993; Navrotsky, this volume).

Calorimetric determination of the configurational entropy

With Equations (20) and (22) the variation of the configurational entropy of silicate melts can be calculated from calorimetric data. To obtain absolute values of S^{conf}, it suffices to note that the residual entropy of a glass at 0 K represents the configurational entropy frozen in at the glass transition. The next step is to calculate this residual entropy from the thermodynamic cycle shown in Figure 6 in the case of $CaMgSi_2O_6$. Because it starts from the third-law entropy of a crystalline phase, this calculation is possible only for compositions for which the equilibrium congruent crystallization temperature and the enthalpy of crystallization are known. Analytically, one has:

$$S^{conf}(T_g) = S_g(0 \text{ K})$$
$$= \int_0^T C_{pc}/T dT + \Delta S_f + \int_{T_f}^{T_g} C_{pc}/T dT + \int_{T_g}^0 C_{pg}/T dT , \qquad (23)$$

where T_f and ΔS_f stand for the equilibrium temperature and entropy of fusion, and C_{pc} is the heat capacity of the crystal. We list in Table 1 all residual entropies obtained from Equation (23) for silicate glasses, including for comparison recent data for GeO_2 and B_2O_3. Of course, these residual entropies depend on thermal history since T_g in Equation (23) depends on the cooling rate at which the glass was formed. For this reason, all the T_g data of Table 1 refer to glasses rapidly quenched in a similar way. Unfortunately, the limited nature of these data prevent determinations of the composition dependence of $S^{conf}(T_g)$.

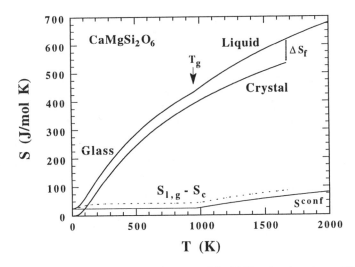

Figure 6. Entropy of diopside and $CaMgSi_2O_6$ glass and liquid. Note that the entropy difference between diopside and $CaMgSi_2O_6$ glass and liquid, shown as the dotted curve, differs by about 50 % from the actual configurational entropy. Data from Krupka et al. (1985), Richet and Bottinga (1984b), Richet et al. (1986) and Richet and Fiquet (1991).

Table 1. Configurational entropies determined
by calorimetry (J/mol K) [a]

Composition	T_g (K)	$S^{conf}(T_g)$
B_2O_3	543	11.2 ± 1.0
GeO_2	980	6.3 ± 1.0
SiO_2	1480	5.1 ± 1.0
$CaSiO_3$	1065	8.5 ± 3.0
$MgSiO_3$	874	8.7 ± 5.0
$CaMgSi_2O_6$	1005	24.3 ± 3.0
$NaAlSi_3O_8$	1100	36.7 ± 6.0
$NaAlSi_2O_6$	1130	16.0 ± 5.0
$NaAlSiO_4$	990	9.7 ± 2.0
$KAlSi_3O_8$	1221	28.3 ± 6.0
$Mg_3Al_2Si_3O_{12}$	1035	56.3 ± 13.0
$CaAl_2Si_2O_8$	1160	36.8 ± 4.0

[a] Data from de Ligny et al. (1995), Richet (1984),
Richet et al. (1982, 1986, 1990, 1993) and Téqui et
al. (1991).

Combining Equations (20), (22) and (23), one finally obtains the configurational
entropy at any temperature:

$$S^{conf}(T) = S^{conf}(T_g) + \int_{T_g}^{T} C_p^{conf} / T dT. \tag{24}$$

This equation will be extensively used in the next section.

VISCOSITY AND CONFIGURATIONAL ENTROPY

Adam-Gibbs theory

For a hypothetical liquid with a zero configurational entropy, there would be no way
to move any structural entity from one place to another and the structural relaxation time
would thus be infinite. If two configurations only were available for the whole liquid, then
structural rearrangement would require a simultaneous displacement of all entities. The
probability for such a cooperative event would be extremely small, but not zero, and the
relaxation time would be extremely long, but no longer infinite. When the configurational
entropy increases, cooperative rearrangements of the structure can take place independently
in smaller and smaller regions of the liquid. Correlatively, the relaxation time thus
decreases when the configurational entropy increases. Extending earlier work (Gibbs and
DiMarzio, 1958), Adam and Gibbs (1965) thus explained the relaxational properties of
glass-forming liquids by assuming that structural relaxation times are determined by the
probability of cooperative configurational rearrangements within microscopic regions of the
liquid.

Quantitatively, Adam and Gibbs (1965) calculated from a simple statistical mechanical
model that the probability for a cooperative rearrangement is:

$$w(T) = A \exp(-B_e/TS^{conf}), \tag{25}$$

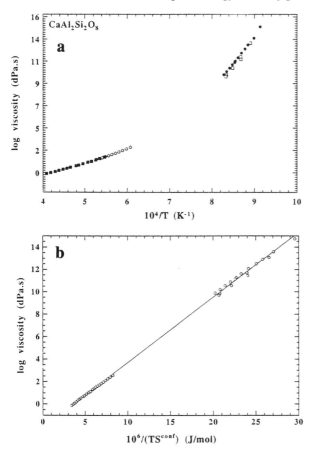

Figure 7. Viscosity of liquid $CaAl_2Si_2O_8$ as a function of reciprocal temperature (a) and as a function of $1/TS^{conf}$ (b), where S^{conf} is the calorimetrically determined configurational entropy. Data from Urbain et al. (1982; solid squares), Scarfe et al. (1983; open circles), Hummel and Arndt (1985; open squares) and Sipp (1993; solid circles).

where A is a pre-exponential term and B_e is approximately a constant proportional to the Gibbs free-energy barriers hindering the cooperative rearrangements. (Note that the temperature dependence of both of these material-dependent parameters is negligible with respect to that of the product TS^{conf}.) Since the structural relaxation time is inversely proportional to the average probability of a structural rearrangement, one has:

$$\tau = A_\tau \exp (B_e/TS^{conf}) . \tag{26}$$

Finally, the viscosity is proportional to the structural relaxation time, see Equation (10). It is thus given by:

$$\eta = A_e \exp (B_e/TS^{conf}) , \tag{27}$$

where A_e is an pre-exponential constant.

Temperature dependence of the viscosity

The quantitative validity of Equation (27) was assessed for silicate melts by Richet (1984) and Richet et al. (1986) who used calorimetrically determined configurational entropies (Eqn 24) for fitting it to laboratory observations. The quality of fit provided by Equation (27) will be illustrated by just one example, that of the strongly non-Arrhenian viscosity of $CaAl_2Si_2O_8$ (Fig. 7a). When plotted as a function $1/TS^{conf}$ (Fig. 7b), the data plot linearly as predicted by Equation (27). The viscosity expression obtained from the

Adam-Gibbs theory is thus remarkable in that it contains only two adjustable parameters instead of three as embodied in empirical expressions like the so-called Tamman-Vogel-Fulcher (TVF) equation:

$$\log \eta = A + B/(T - T_1) , \tag{28}$$

which has long been used to interpolate reliably experimental data, in the viscosity range 1 to 10^{13} dPa s, with adjustable parameters A, B and T_1.

As noted above, configurational entropies can be determined calorimetrically for a limited number of compositions. When C_p^{conf} is known, the quantitative validity of Equation (27) in fact allows one to use it to determine not only A_e and B_e, but also $S^{conf}(T_g)$ from a least-squares fit to the experimental viscosity data. Of course, with this procedure Equation (27) at first sight resembles other equations with three adjustable parameters which reproduce viscosities to within the error margins of the measurements. Its distinctive feature, however, is that $S^{conf}(T_g)$ has a clear physical meaning and that its values determined in this way agree very well with the calorimetric data, as apparent from the comparisons made in Figure 8. This agreement between these independently determined values justifies determinations of configurational entropies from viscosity data when these are available over wide enough temperature intervals.

In the range 1 to 10^{13} dPa s, the viscosity of silicate melts is thus quantitatively accounted for by Equation (27). Turning now to the significance of the pre-exponential constant A_e, we note that its average value is 0.03 dPa s for the compositions for which the residual entropies at 0 K are known (Table 1). From Equation (10), one has $A_e = G_\infty A_\tau$, whence the conclusion that A_τ is about $3 \cdot 10^{-13}$ s, i.e. a value compatible with vibrational frequencies of the silica network. We also recall that Angell and Sichina (1976) and other authors remarked that the TVF equation is derived from the Adam-Gibbs expression if C_p^{conf} is proportional to $1/T$. All available calorimetry data indicate that this proportionality does not hold for silicates. The validity of the TVF equation is thus strictly empirical for silicate melts.

As a result of its strong increase with temperature, at some point the configurational entropy becomes high enough that the availability of configurational states does not restrict any longer the fluidity of the melt. For the reasons expounded by Goldstein (1969), description of the structure of the liquid in terms of potential energy barriers becomes problematic for viscosities smaller than about 10 dPa s. At these low viscosities, the size of

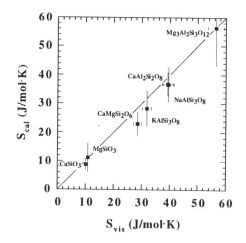

Figure 8. Comparison between configurational entropies at the glass transition determined from either calorimetric or viscosity measurements.

cooperatively rearranging regions has become so small that the Adam-Gibbs theory is no longer adequate and other mechanisms will determine the strain rate of the stressed liquid. In this low-viscosity regime, the Arrhenius equation has still some appeal in the rheological literature. As discussed by Bottinga et al. (1995), however, the power law derived from mode coupling theory (Götze, 1991) reproduces better the observed data for silicate melts than the Arrhenius equation.

Viscosity and structural relaxation

Below the usual glass transition temperature, the loss of ready access of the system to available configurational states is not really abrupt. Because of the exponential character of Equation (26), relaxation times become so large that the glass transition range is actually a somewhat narrow temperature interval where slow kinetics toward equilibrium results in time-dependent macroscopic properties. To characterize the state of a sample in or below the glass transition range, the fictive temperature (T') has been defined as the temperature at which its actual configuration would represent the equilibrium configuration of the liquid (Tool and Eichlin, 1931). As shown in Figure 9, the approach to thermodynamic equilibrium can be readily determined in viscosity measurements through changes in fictive temperature within the glass transition range. In Figure 9, the viscosity of two samples of window glass is plotted as a function of time. These samples were previously equilibrated at 774 K ($\eta = 10^{15.52}$ dPa s) and 892 K ($\eta = 10^{10.28}$ dPa s), respectively, for periods of time long enough that the fictive temperature was equal to the anneal temperature. When the fictive temperature is initially higher than the actual temperature of the measurement, the viscosity increases with time. In contrast, it decreases when the fictive temperature is lower than the actual temperature. As shown in Figures 9, equilibrium is reached when the fictive temperature is equal to the run temperature and the viscosity becomes time-independent.

If measurements are actually performed too rapidly to allow for structural changes, then the configurational entropy does not vary and Equation (27) reduces to an Arrhenius law. Only under these circumstances does the viscosity of silicate melts follow an Arrhenius law. This effect has been observed by Mazurin (1979) for the standard glass NBS 710 and was discussed qualitatively by Scherer (1984) in the light of the Adam-Gibbs theory. We stress that this analysis can be quantitative, as shown in Figure 10 where we have plotted the data of de Bast and Gilard (1965), Mazurin et al. (1979) and Sipp et al. (1995) for window glasses with nearly the same composition. From new heat capacity measurements, $S^{conf}(T_g)$ and the parameters A_e and B_e of Equation (27) could be determined by Sipp et al. (1995) from the equilibrium viscosity data. Using these parameters,

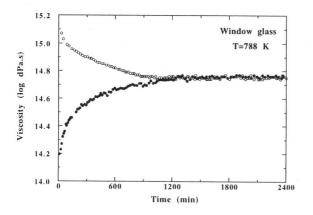

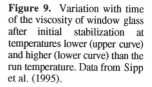

Figure 9. Variation with time of the viscosity of window glass after initial stabilization at temperatures lower (upper curve) and higher (lower curve) than the run temperature. Data from Sipp et al. (1995).

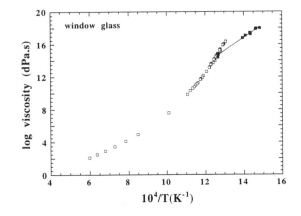

Figure 10. The viscosities of several window glasses. The open squares (de Bast and Gilard, 1965; Mazurin et al., 1979; Sipp, 1993) refer to equilibrium viscosities, even at the lowest temperatures; the solid squares (Mazurin et al., 1979), indicate the viscosity of samples whose configuration has been frozen in at T = 777 K. The solid line represents the values predicted from Equation (27), without any adjustable parameter.

the viscosities of a sample with a constant fictive temperature of 777 K is then readily calculated as the solid line shown in Figure 10. Without any adjustable parameters, the agreement with the experimental data of Mazurin et al. (1979) is excellent.

The glass transition separates an upper temperature range where configurational changes take place, during a given timescale, from a lower temperature interval where the configuration is fixed. In passing, we thus note that the data of plotted in Figure 10 do show that the viscosity changes at the glass transition. The reason for the opposite assertion commonly made is simply that viscosity measurements near 10^{13} dPa s are usually made with long timescales, implying that the reported data refer to equilibrium values for a relaxed configuration. We also emphasize that Adam-Gibbs equations obtained from measurements made above the glass transition range extrapolate accurately for viscosities as high as 10^{15} dPa s, provided these are actually equilibrium data determined in the way shown in Figure 9.

In the rest of this section we will thus concentrate on the non-equilibrium, time-dependent viscosities and outline how these features could also be treated within the Gibbs theory. Under conditions that the fictive temperature T' differs from the actual temperature T, the structural relaxation time depends on both of these temperatures and can be written:

$$\tau(T,T') = A_\tau \exp [B_e/TS^{conf}(T')] . \tag{29}$$

Using the definition of a relaxation time :

$$T'(t) = [T'(0) - T'(\infty)] \exp (- t/\tau) + T'(\infty) , \tag{30}$$

one obtains an expression for the fictive temperature at time t. Resorting to numerical methods to solve Equations (29) and (30) simultaneously for τ and T', then one could evaluate very simply η with Equation (10). We are however unaware of solutions of these equations for glasses or liquids for which A_τ, B_e, $S^{conf}(T_g)$, and C_P^{conf} are known.

Relaxation has been discussed by Scherer (1990) who reviewed various empirical functions which are known to be acceptable only when the fictive temperature is not too different from the actual temperature. At $T > T_g$, the relaxation time will be very small hence $T' = T$, and Equation (30) is irrelevant. Time dependences cannot be ignored, however, when one studies the reaction of a liquid to a very high frequency signal, such that the signal period is comparable to the structural relaxation time. Elastic measurements by Brillouin scattering is a case in point. The timescale of Brillouin scattering being of the order of 10^{-10} s, it is thus for viscosities lower than 1 Pa s that relaxed compressibilities could be obtained (see Askarpour et al., 1993). In such a case, usage of Equation (30) is required.

Composition dependence of the viscosity

This subject has certainly an utmost importance in geochemistry, but it is also probably the most difficult given the very strong composition dependence of the viscosity apparent in Figure 1. Quantitative predictions of the viscosity as a function of temperature and composition were initiated by Bottinga and Weill (1972). Relying on the distinction between network-former and network-modifier elements, Bottinga and Weill (1972) broke down available data for simple silicate systems into different SiO_2 ranges where the viscosity could be considered as an additive function of composition. This model and that by Shaw (1972) immediately proved useful in a wide variety of magmatic contexts. The main limitations of these models are that they deal with superliquidus conditions only and assume Arrhenian temperature dependences of the viscosity, which is followed approximately over wide intervals only by pure SiO_2 and some SiO_2-rich melts like molten alkali feldspars (Fig. 1). For less SiO_2-rich melts, Arrhenian extrapolations of data can result in considerable errors. In the case of $CaMgSi_2O_6$ liquid, for example, these errors would amount to 8 orders of magnitude when extrapolating data down from the liquidus or up from the glass transition range (Fig. 1).

Prediction of the viscosity as a function of composition thus requires experimental data over wide viscosity ranges, which are currently available for limited geochemically relevant compositions. As discussed by Neuville et al. (1993), empirical estimates of the viscosity of molten lavas from measurements on component mineral compositions are possible. Of course, this approach is too qualitative and its main merit is just to show that even at high viscosities, where the composition dependence is the strongest, predictions from simpler systems are indeed possible for complex compositions.

To show how viscosity variations with composition can be treated rigorously within the framework of the Adam-Gibbs theory, we will consider the data plotted in Figure 11 for (Ca,Mg) garnet compositions. Measurements show that mixing of Ca and Mg results in a very strong nonlinear variation of the viscosity just above the glass transition, with a minimum of about two orders of magnitude (Fig. 11). When the temperature increases, the depth of this minimum decreases and η is eventually a linear function of composition at

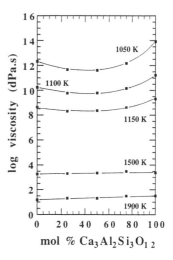

Figure 11. Viscosity as a function of composition for $(Ca,Mg)Al_2Si_3O_{12}$ melts at different temperatures. Data from Neuville and Richet (1991).

mol % $Ca_3Al_2Si_3O_{12}$

high fluidities. To interpret such temperature-dependent variations, consider the configurational entropy of an intermediate composition. As a function of the entropies of the endmembers (S_i^{conf}), it can be expressed as :

$$S^{conf}(\text{T}) = \sum x_i S_i^{conf}(\text{T}) + S_{mix} \tag{31}$$

where the temperature dependence of the S_i^{conf} is given by Equation (24). If mixing is ideal, then the entropy of mixing is given by:

$$S_{mix} = - n\, R \sum x_i \ln x_i \tag{32}$$

where x_i is the molar fraction of entity i, R the gas constant and n the number of entities exchanged per formula unit (e.g. $n = 3$ and $x = 0.5$ for the (Ca,Mg) exchange in $Ca_3Al_2Si_3O_{12}$-$Mg_3Al_2Si_3O_{12}$ considered in Fig. 11). This S_{mix} term is constant and contrasts in this respect with the $S_i^{conf}(T)$ terms of the endmembers which increase considerably with temperature. At lower temperatures, when the $S_i^{conf}(T)$ are small, intermediate compositions have a strong excess entropy with respect to the endmembers and their viscosity is considerably lowered. With increasing temperatures, the entropy-of-mixing term is then progressively overwhelmed and its contribution to the total configurational entropy is eventually slight enough that the viscosity varies linearly with composition. The lowering of the viscosity, shown as a solid curve in Figure 11, resulting from the excess entropy of the intermediate compositions is thus most effective at lower temperatures, i.e. at high viscosities.

The good agreement found between the observed viscosities and the values calculated with Equations (27) and (31) points to essentially ideal mixing of calcium and magnesium in molten garnets. Similar conclusions were obtained for (Ca,Mg) exchange in molten pyroxenes (Neuville and Richet, 1991) and for (Na,K) exchange in alkali-silicates (Richet, 1984) giving rise to the well known mixed alkali effect. In the latter case, however, mixing of pairs of alkali cations (and not of individual cations) had to be assumed to obtain entropies consistent with the viscosity data. A two-lattice entropy model was found adequate by Hummel and Arndt (1985) and Tauber and Arndt (1987) for calculating the viscosity of molten plagioclases and of melts along the join anorthite-diopside. This illustrates the fact that entropy modeling becomes more difficult because of the general lack of reliable mixing models for compositionaly complex silicate melts. In turn, however, one can expect that entropy modeling will be strongly constrained by extensive viscosity data when they are available down to the glass transition range.

Pressure dependence of the viscosity

The pressure dependence of the viscosity of silicate melts depends sensitively on composition. It can be either positive, as for diopside liquid (Scarfe et al., 1976), or negative, as for some andesite or basalt melts (Kushiro et al., 1976). Assuming that both A_e and B_e do not depend on pressure, Richet (1984) calculated from Equation (23) and a Maxwell relation that:

$$(\partial \ln \eta / \partial P)_T = - (B_e/TS^{conf\,2})\,(\partial S^{conf}/\partial P)_T = (B_e/TS^{conf\,2})\,(\partial V^{conf}/\partial T)_P \tag{33}$$

where V^{conf} is the configurational volume. With the approximation $(\partial V^{conf}/\partial T)_P = [(\partial V_l/\partial T)_P - (\partial V_g/\partial T)_P]$, this equation predicts that the viscosity should increase with pressure because thermal expansion coefficients are greater for liquids than for glasses (e.g. Knoche et al., 1992). This conclusion is thus at variance with the viscosity decreases observed for a variety of compositions. Assuming that B_e varies with pressure would resolve the difficulty (see Richet, 1984), but this requires to consider $(\partial B_e/\partial P)_T$ as just a fitting parameter.

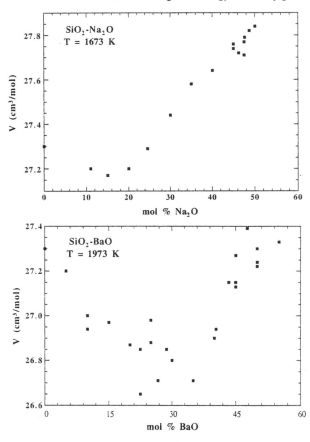

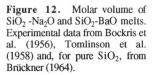

Figure 12. Molar volume of SiO_2-Na_2O and SiO_2-BaO melts. Experimental data from Bockris et al. (1956), Tomlinson et al. (1958) and, for pure SiO_2, from Brückner (1964).

It is more satisfactory to evaluate the variations of the configurational entropy due to pressure-induced structural changes. The basis for this approach is the observation that bridging oxygens (BO) have a higher molar volume than nonbridging oxygens (NBO). As discussed by Bottinga and Richet (1995), this feature is borne out by crystallographical data for a variety of silicates where BO and NBO coexist, as well as by available measurements of the molar volumes of SiO_2-rich binary melts which show an initial volume decrease when a BO is transformed into a NBO (see Fig. 12). At magmatic temperatures, these data indicate that the difference in molar volume between BO and NBO is about 0.7 cm^3/mol, which also corresponds to the difference between the molar volume of pure SiO_2 (Brückner, 1964) and the partial molar volume of SiO_2 in silicate melts with a SiO_2 mol fraction in the range 0.2 to 0.5 (Bottinga et al., 1982, 1983). Defining most simply the degree of polymerization of the simple binary melts considered in this section as $\xi = BO/(BO+NBO)$, it follows from Le Châtelier's principle that ξ must decrease with increasing pressure.

Unfortunately, we do not know a priori how the configurational entropy depends on ξ. Formally, however, one can split the configurational entropy into two parts:

$$S^{conf}(P,T,\xi) = S^{conf*}(P,T) + S^{conf**}(\xi) , \qquad (34)$$

where $S^{conf*}(P,T)$ does not depend on ξ, i.e. is that part of the configurational entropy usually considered. The further assumption made by Bottinga and Richet (1995) is that, close to the glass transition, the configurational entropy associated with polymerization is given by the entropy of mixing of bridging and non-bridging oxygens:

$$S^{conf**}(\xi) = -R\,N_O\,[\xi \ln \xi + (1-\xi) \ln (1 - \xi)], \tag{35}$$

where N_O is the relevant number of oxygen atoms, i.e. the number of oxygen atoms per mole of oxide components. For instance, $N_O = 2$ and $5/3$ for SiO_2 and $Na_2Si_2O_5$ melts, respectively. Note that Equation (35) is only a first approximation for alkali-silicates because of the structural reasons discussed by Xue et al. (1991, 1994), for instance, and it is invalid for alkaline-earth silicates because divalent cations prevent random mixing of two NBOs (see also Gaskell et al., 1991) Configurational entropies for liquid wollastonite, enstatite, and diopside (see Richet and Neuville, 1992) per oxide mole are thus significantly lower than the values for the system SiO_2-Na_2O. The degree of polymerization is independent of temperature for the simple binary silicates discussed here because of the very large Si-O bonding energy.

Differentiating S^{conf}, one can write:

$$dS^{conf} = (\partial S^{conf}/\partial P)_{T,\xi}\,dP + (\partial S^{conf}/\partial T)_{P,\xi}\,dT +$$

$$(\partial S^{conf}/\partial \xi)_{P,T}\,[(\partial \xi/\partial P)_T\,dP + (\partial \xi/\partial T)_P\,dT], \tag{36}$$

and one obtains finally (see Bottinga and Richet, 1995):

$$(\partial \ln \eta/\partial P)_T = (B_e/TS^{conf\,2})\,\{(\partial V^{conf}/\partial T)_{P,\xi}$$

$$+ R\,N_O \ln [\xi/(1 - \xi)]\,(\partial \xi/\partial P)_T\}, \tag{37}$$

where S^{conf} is the configurational entropy at T,P and ξ.

At $T = T_g$ and P = 1 bar, S^{conf} is known (Eqn. 23) and S^{conf**} can be calculated with Equation (35). As noted above, the volume difference between bridging and nonbridging oxygens implies that $(\partial \xi/\partial P)_T < 0$. When $\xi > 0.5$, the configurational entropy increases with pressure (Eqn. 35), resulting in a lowering of the viscosity (Eqn. 37). In contrast, if $\xi < 0.5$ a pressure-induced decrease of the degree of polymerization results in a decrease of the configurational entropy and hence in a viscosity increase conform to the behavior of ordinary liquids. These conclusions are in qualitative agreement with the data compiled by Scarfe et al. (1987) on the pressure dependence of the viscosity of 14 silicate melts with ξ larger as well as smaller than 0.5, see Table 2. At high temperature and for $\xi > 0.5$, the anomalous pressure effect tends to disappear, while for $\xi < 0.5$ the normal viscosity increase with pressure should become less important (Eqn. 37). The reason is that high temperatures mean large S^{conf}, and with large values of T and S^{conf} the absolute value of $(\partial \ln\eta/\partial P)_T$ decreases. As ξ decreases with pressure or with addition of network modifiers to the melt, the viscosity will eventually change to a "normal" pressure dependence.

Density fluctuations

Nanometric density inhomogeneities in liquids close to the glass transition have been observed spectroscopically and discussed by several groups (Malinovsky and Sokolov, 1986; Mazurin and Porai-Koshits, 1989; Duval et al., 1990; Sokolov et al., 1993). These density fluctuations are not due to incipient unmixing or to compositional fluctuations. Confirming one of the conjectural aspects of the Adam-Gibbs theory, Moynihan and Schroeder (1993) have shown instead that they can be interpreted as configurational entropy fluctuations. Denoting as ΔS and ΔV the amplitudes of entropy and volume fluctuations, the ensemble average $<\Delta S\,\Delta V>$ is given by:

Table 2. Observed changes of the viscosity of silicate melts due to increases of pressure, compiled by Scarfe et al. (1987).

Composition	$\xi^{(c)}$	Viscosity change (dPa s)	Pressure interval (GPa)	Temp. (°C)
GeO_2	1.0	6270;1160[a]	0 - 1	1425
$Na_2O.SiO_2$	0.33	2; 15	0 - 1	1300
$Na_2O.2SiO_2$	0.60	151; 90	0 - 2	1200
$Na_2O.3SiO_2$	0.71	485; 170	0 - 2	1175
$CaMgSi_2O_6$	0.33	3; 11	0 - 1.5	1640
$K_2MgSi_5O_{12}$	0.67	2890;1000	5 - 2	1300
$NaCaAlSi_2O_7$	0.71	36; 26	0 - 1.5	1450
$NaAlSi_2O_6$	1.0	64700;5090	0 - 2.4	1350
$NaAlSi_3O_8$	1.0	113000;18400	0 - 2	1400
Ol.nephel.	0.71	10; 6	0 - 2	1450
Alk.ol.basalt	0.79	10; 8	0 - 1.5	1400
Ol.tholeiite	0.81	35; 17	0 - 2	1400
Abyssal thol.	0.84	161; 80	0 - 1.2	1300
Andesite	0.93	1820; 810	0 - 2	1350
Obsidian	0.97	-80%[b]	0 - 2	1400

[a] First value measured at low pressure, second value at high pressure.
[b] The observed decrease of the viscosity is given in % of initial value at $P = 0.1$ MPa; no other details available.
[c] $\xi = BO/$Total oxygen, see text; estimated from the stoichiometry.

$$< \Delta S\, \Delta V > \ = \ k_B\, T\, (\partial V/\partial T)_P \tag{38}$$

(see Landau and Lifshitz, 1958). From Equation (38) one derives (Robertson, 1978):

$$< \Delta S^{conf} \Delta V > \ = \ k_B\, T\, [(\partial V_l/\partial T)_P - (\partial V_g/\partial T)_P] . \tag{39}$$

As already noted, liquids and glasses with the same composition show invariably that $\alpha_l > \alpha_g > 0$, where $\alpha = 1/V\, (\partial V/\partial T)_P$. Hence the fluctuations ΔS^{conf} and ΔV are positively correlated and the presence of density inhomogeneities implies the presence of configurational entropy inhomogeneities as well. Moynihan and Schroeder (1993) have pointed out that introduction of these configurational entropy fluctuations into Equation (26) explains the observed non-exponential structural relaxation kinetics in liquids in the vicinity of the glass transition. Detailed descriptions of the relaxation kinetics close to the glass transition can be found in Macedo and Litovitz (1965) for B_2O_3, in Napolitano and Macedo (1968) for GeO_2, and in Webb (1991) and Bagdassarov et al. (1993) for some silicates. The occurrence of entropy fluctuations just above T_g is consistent with the heat capacity change associated with the glass transition. The heat capacity is related to entropy fluctuations (Landau and Lifshitz, 1958) by:

$$< (\Delta S)^2 > \ = \ k_B\, C_P \tag{40}$$

and this gives (Robertson, 1978):

$$< (\Delta S^{conf})^2 > \ = \ k_B\, (C_{pl} - C_{pg}) = k_B\, \Delta C_P. \tag{41}$$

Close to T_g, the aforementioned inhomogeneities become observable because of their large size which ranges from 2 to 5 nm. Moynihan and Schroeder (1993) have calculated a volume of about 12 nm^3 for inhomogeneities observed in liquid B_2O_3, which would contain about 400 formula units.

It has long been known that glasses, and in particular silica glass, have anomalous thermal and mechanical properties at very low temperature (see Krause and Kurkjian, 1968). In the present context, the observed excess heat capacity (Pohl, 1981) with respect to the Debye value and the anomalous vibrational properties observed between 2 and 20 K are of great interest. Low-temperature vibrational properties of silica glass have been observed by Buchenau et al. (1984) by neutron scattering experiments and were interpreted to be due to acoustic phonons and low-frequency modes associated with coupled rotations of SiO_4 tetrahedra. Buchenau et al. (1984) have shown that the heat capacity of silica calculated from their neutron scattering data is in qualitative agreement with the heat capacity anomaly around 15 K. These results support the suggestion that inhomogeneities observed in melts close to T_g may also contribute to the anomalous C_P of glasses at low temperature (Moynihan and Schroeder, 1993); or, more explicitly, the low-temperature C_P anomalies could be caused by density fluctuations frozen in at $T = T_g$. The observation that the heat capacity anomaly of silica glass around 2 K is affected by the fictive temperature (Fagaly and Bohn, 1981) supports this interpretation. Richet et al. (1986) have also observed a fictive temperature dependence of the low temperature C_P anomaly of vitreous diopside (Fig. 13). No satisfactory explanations for these heat capacity anomalies and their dependence on fictive temperature have yet been published. (We stress, however, that these structural inhomogeneities have calorimetric effects in temperature ranges too low to affect in a significant way the vibrational entropy of glasses at and above room temperature.)

Newtonian vs. non-Newtonian viscosity

The last topic reviewed in this chapter is the non-Newtonian viscosity observed at high stress or strain rates by Li and Uhlmann (1970), Simmons et al. (1982), Guillemet and Gy (1990) and Webb and Dingwell (1990a,b). These observations can be understood quantitatively within the context of the Adam and Gibbs theory (Bottinga 1994a,b). In the fiber-elongation measurements which have evidenced non-Newtonian rheology, the process leading to the observed viscous thinning can be split up schematically into recurring time steps Δt, during which three consecutive events take place: (1) under the influence of a force F, a viscous fiber is elastically elongated by an amount dl; (2) the

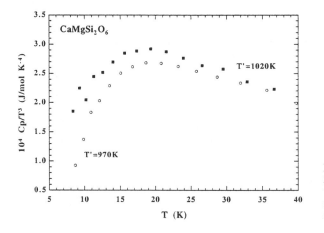

Figure 13. Low-temperature heat capacity of diopside glasses with the fictive temperatures indicated. Data from Richet et al. (1986).

elastic work done by F on the fiber generates internal energy (dU) and concomitantly configurational entropy; and (3) subsequently, the elastic strain relaxes viscously while the deformation remains constant. During the next time step this sequence is repeated, and so on, until the charge is removed (F becomes zero) or, alternatively, until the steady state regime is not maintained and rupture occurs.

Entropy is produced and consumed during each time step; the rate of entropy production is equal to:

$$T \, dS/dt = dU/dt - F \, dl/dt .$$ (42)

Integrated over the time interval Δt, the internal energy change in the stretched fiber should be zero. The steady state character of the deformation process prohibits the net accumulation of internal energy during a time step Δt. Hence:

$$\Delta U = \int (dU/dt) \, dt = 0 ,$$ (43)

and the integration of Equation (43) over the time interval Δt should be equal to the elastic work done by F. Therefore:

$$T \, \Delta S(T,\sigma) = - F \, \Delta l = \sigma^2 \, V_m / (2 \, E) ,$$ (44)

where E and V_m denote the Young's modulus and the molar volume of the silicate liquid, respectively. Because the applied stress is relatively small ($\sigma < 10^{8.5}$ Pa) and the temperature of the stretched fiber did not change (Simmons et al., 1982; Webb and Dingwell, 1990a,b), possible changes in vibrational entropy will be very small compared to the configurational entropy variations. Under steady state conditions, the configurational entropy produced by stretching of the fiber at constant temperature is thus given by:

$$\Delta S^{conf}(T,\sigma) = \sigma^2 \, V_m / (2 \, E \, T) .$$ (45)

Using Equations (1) and (45) one obtains for the reduced viscosity η/η_0:

$$\ln (\eta/\eta_0) = B_e \{ 1/[T \, (S^{conf}(T)+\Delta S^{conf}(T,\sigma))] - 1 / [TS^{conf}(T)]\} .$$ (46)

Because $|\Delta S^{conf}(T,\sigma)| << |S^{conf}(T)|$ the reduced viscosity should depend on the square of the applied stress as follows:

$$\ln (\eta/\eta_0) = - B_e \, \Delta S^{conf}/TS^{conf \, 2} = - B_e V_m \, \sigma^2/2E \, (Ts^{conf})^2 .$$ (47)

During a recurring time step Δt, the elastic work done by the stress on the fiber thus causes accumulation of an elastic strain and consequently a change in the configurational entropy. This entropy increase given by Equation (47) speeds up the viscous deformation of the fiber and is thus the reason for viscous thinning. The viscous deformation relaxes the elastic strain and annihilates the configurational entropy produced by the applied stress. Consequently, no net changes in internal energy or configurational entropy occur during this cyclic steady state deformation process. When the stretching experiment is terminated by removing the stress, the elastic deformation is recovered, but the viscous deformation remains, as observed by de Bast and Gilard (1965). The model predicts that the reduced viscosity depends on the squared value of the applied stress during steady-state non-Newtonian deformation of silicate liquids. This turns out to be the case (Bottinga, 1994a,b) and is well illustrated by a plot of the data of Webb and Dingwell (1990b) for liquid CaMgSi$_2$O$_6$ (Fig. 14).

Silicate liquids are Newtonian as long as the increase in configurational entropy, given by Equation (45), is significantly smaller than the configurational entropy of the nonstressed liquid. The stress dependence of the viscosity becomes apparent when ΔS^{conf} becomes no longer negligible with respect to S^{conf}. When the temperature increases and the Newtonian viscosity decreases, the minimum stress needed for the onset of non-

Newtonian rheology becomes larger. This deduction is confirmed by the observations of Simmons et al. (1988) at different temperatures on the standard glasses NBS710 and E-4.

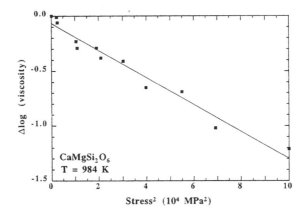

Figure 14. Stress-dependent viscosity of liquid $CaMgSi_2O_6$, where $\Delta\log \eta = \log (\eta/\eta_0)$, η being the observed viscosity and η_0 the Newtonian viscosity. Experimental data (squares) from Webb and Dingwell (1990b) and least-squares linear fit as given by Equation (47).

EPILOGUE

It is well known that magmas containing significant amounts of crystals or bubbles may be non-Newtonian. Such inhomogeneous liquids have not been discussed in this chapter. These liquid suspensions are quite common in nature, but one first needs to know the rheology of the bubble- and crystal-free liquids before studying how they react to stress. Under natural conditions magmas or lavas experience shear rates varying from 10^{-8} s-1 to 30 s-1 (Spera et al., 1988). Doubt has been expressed about the validity of Equation (5) for such a large range of strain rates, in particular when the stress is small (Spera et al., 1982, 1988; McBirney and Murase, 1984; Mazurin, 1986). However, careful laboratory work has shown that bubble- and crystal-free silicate liquids, even when they are metastable at subliquidus temperatures, are Newtonian at strain rates as low as 10^{-8} s-1 and stresses as low as 10^4 dPa. According to Mazurin (1986), the breakdown of Equation (5) at low stress or strain rate has never been well documented.

For silicate liquids the Adam and Gibbs (1965) theory reproduces better than any other two-parameter model available observations up to magmatic temperatures. It is at this moment the most comprehensive theory available to understand the relaxation and viscosity of volatile-free silicate liquids at magmatic temperatures. It is unlikely that this theory will be applicable at very low viscosities ($\eta < 1$ Pa s), or for silicate liquids very rich in volatiles, such as pegmatitic liquids. The Adam-Gibbs theory has shown its power by furnishing quantitative explanations for the deviations from the Arrhenius viscosity equation and for the phenomenon of viscous thinning shown by silicate melts, and it allows a qualitative understanding of the pressure dependence viscosity of these melts and of the structural inhomogeneity of silicate glasses. For the time being, the compositional dependence of the configurational entropy is insufficiently understood. It will require a major effort to obtain the needed observational data to close this enormous gap in our knowledge of the properties of petrologically and volcanologically important silicate melts.

In all these respects, application of the Adam-Gibbs theory to silicate melts is free from practical difficulties met with other classes of glass-forming liquids for which one observes frequently that rotational or librational relaxation becomes active at much lower temperatures than shear relaxation. For instance, in an ionic liquid like $Ca_{0.4}K_{0.6}(NO_3)_{1.4}$ structural, volume, shear and NO_3 rotational relaxation times differ considerably at low temperatures (Grimsditch and Torell, 1989). Likewise, infrared spectra of organic chain polymers indicate the presence of rotations or librations of groups of atoms below the glass transition. As a consequence, only that part of the total configurational entropy which is related to shear relaxation should be taken into account. This invalidates the approximations described in the preceding sections for determining configurational heat capacities and entropies by calorimetric methods.

In contrast, polymerized silicate liquids are characterized by strongly bonded three-dimensional entities made up of SiO_4 (or AlO_4) tetrahedra. Hence the structural fluctuations which render the relaxation of shear strain possible are the same as those responsable for volume and enthalpy relaxation. Relaxation times for shear, volume, and enthalpy excitations in $Na_2Si_2O_5$ (glass and liquid, from 800 to 1450 K) are equivalent (Webb, 1992). In short, the entropy associated with these processes is the cause of the difference between C_{pl} and C_{pg}, the approximation $C_P^{conf}(T < T_g) = 0$ is justified, and the residual entropy at 0 K of silicate glasses represents adequately the configurational entropy frozen in at the glass transition. Without these features, it would be impossible to evaluate $S^{conf}(T)$ and thus to verify the quantitative applicability of Equations (22) and (23).

ACKNOWLEDGMENTS

Our work mentioned in this contribution has been benefitted over the years from financial support of CNRS-INSU-DBT programs and the enjoyable participation of M.A. Bouhifd, A.M. Lejeune, D. de Ligny, D.R. Neuville and A. Sipp. We also thank A. Navrotsky and J. Stebbins for their reviews of this chapter.

REFERENCES

Adam G, Gibbs JH (1965) On the temperature dependence of cooperative relaxation properties in glass-forming liquids. J Chem Phys 43:139-146

Angell CA (1988) Perspectives on the glass transition. J Phys Chem Solids 49:863-871

Angell CA (1991) Relaxation in liquids, polymers, and plastic crystals—strong/fragile patterns and problems. J Non-Cryst Solids 131-133:13-31

Angell CA, Sichina W (1976) Thermodynamics of the glass transition: empirical aspects. Annals New York Acad Sci 279:53-67

Askarpour V, Manghnani MH, Richet P (1993) Elastic properties of diopside, anorthite, and grossular glasses and liquids: a Brillouin scattering study up to 1400 K. J Geophys Res 98:17,683-17,689

Bagdassarov NS, Dingwell DB (1992) A rheological investigation of vesicular rhyolite. J Volcan Geotherm Res 50:307-322

Bagdassarov NS, Dingwell DB, Webb SL (1993) Effect of boron, phosphorus and fluorine on the shear stress relaxation in haplogranitic melts. Eur J Mineral 5:409-425

de Bast J, Gilard P (1965) Rhéologie du verre sous contrainte dans l'intervalle de transformation. Centre R Rech IRSIA 1:9-188

Bockris JO'M, Tomlinson JW, White JL (1956) The structure of liquid silicates: partial molar volumes and expansivities. Trans Faraday Soc 53:299-310

Bottinga Y (1994a) Configurational entropy and the non-Newtonian rheology of homogeneous silicate liquids. Phys Rev B49:95-100

Bottinga Y (1994b) Non-Newtonian rheology of homogeneous silicate melts. Phys Chem Minerals 20:454-459

Bottinga Y, Weill DF (1972) The viscosity of magmatic silicate liquids: a model for calculation. Am J Sci 272:438-475

Bottinga Y, Richet P (1995) Silicate melts: the "anomalous" pressure dependence of the viscosity. Geochim Cosmochim Acta 59:2725-2731

Bottinga Y, Weill DF, Richet P (1982) Density calculations for silicate liquids, I. Revised method for aluminosilicate compositions. Geochim Cosmochim Acta 46:909-920

Bottinga Y, Richet P, Weill DF (1983) Calculation of the density and thermal expansion coefficient of silicate liquids. Bull Minéral 106:129-138

Bottinga Y, Richet P, Sipp A (1995) Viscosity regimes of homogeneous silicate melts. Am Mineral 80:305-318

Bouhifd MA, Richet P (1995) Viscosity and configurational entropy of alkali titanosilicate melts. To be submitted to Geochim Cosmochim Acta.

Brückner R (1964) Charakteristische physikalische Eigenschaften der oxydischen Hauptglasbildner und ihre Beziehungen zur Struktur der Gläsern II: Mechanische und optische Eigenschaften als Funktion der thermische Vorgeschichte. Glastechn Ber 37:459-475

Buchenau U, Nücker N, Dianoux AJ (1984) Neutron scattering study of the low-frequency vibrations in vitreous silica. Phys Rev Lett 53:2316-2319

Chang SS, Bestul AB, Horman JA (1966) Critical configurational entropy at the glass transformation. Proc 7th Int'l Congress on Glass, Brussels, 1965, 1:26.1-26.15, Gordon and Breach, New York

Courtial P, Richet P (1993) Heat capacity of magnesium aluminosilicate melts. Geochim Cosmochim Acta 57:1267-1275

Cukierman M, Uhlmann DR (1973) Viscosity of liquid anorthite. J Geophys Res 86:7951-7956

Dingwell DB, Webb SL (1989) Structural relaxation in silicate melts and non-Newtonian melt rheology in geological processes. Phys Chem Minerals 16:508-516

Dingwell DB, Webb SL (1990) Relaxation in silicate melts. Eur J Mineral 2:427-449

Duval E, Boukenter A, Achibat T (1990) Vibrational dynamics and the structure of glasses. J Phys: Condensed Matter 2:10,227-10,234

Fagaly RL, Bohn RG (1981) The effect of heat treatment on the heat capacity of vitreous silica between 0.3 and 4.2 K. J Non-Cryst Solids 28:67-76

Farnan I, Stebbins JF (1990) A high temperature ^{29}Si NMR investigation of solid and molten silicates. J Am Chem Soc 112:32-39

Fontana EH, Plummer WA (1979) A viscosity-temperature relation for glass. J Am Ceram Soc 62:367-369

Gaskell PH, Eckersley MC, Barnes AC, Chieux P (1991) medium-range order in the cation distribution of a calcium silicate glass. Nature 350:675-677

de Gennes PG (1979) Scaling concepts in polymer physics. Cornell Univ Press, Ithaca, New York

Gibbs JH, Di Marzio EA (1958) Nature of the glass transition and the glassy state. J Chem Phys 28:373-383

Glasstone S, Laidler KJ, Eyring H (1941) The theory of rate processes. 486 p McGraw-Hill, New York

Goldstein M (1969) Viscous liquids and the glass transition: a potential energy barrier picture. J Chem Phys 51:3728-3739

Goldstein M (1976) Viscous liquids and the glass transition. V. Sources of the excess specific heat of the liquid. J Chem Phys 64:4767-4774

Goldstein M (1977) Viscous liquids and the glass transition. VII. Molecular mechanisms for a thermodynamic second-order transition. J Chem Phys 67:2246-2253

Götze W (1991) Aspects of structural glass transisitions. In: Liquids, Freezing and Glass Transition. Hansen JP, Levsque D, Zin-Justin J (eds) 1:292-503. Elsevier Science Publishers B V, Amsterdam

Grimsditch M, Torell LM (1989) Opposite structural behaviour in glassformers; a Brillouin scattering study of B_2O_3 and $Ca_{0.4}K_{0.6}(NO_3)_{1.4}$. In: Dynamics of Disordered Materials. Richter D, Dianoux AJ, Petry W, Teixeira J (eds) Springer Proc in Physics 37:196-210, Springer-Verlag, Berlin

Guillemet C, Gy C (1990) Non-Newtonian viscous flow and tensile strength of silicate glass at high deformation rates. Riv Staz Sper Vetro 6:221-224

Gujrati PD, Goldstein M (1980) Viscous liquids and the glass transition. 9. Nonconfigurational contributions to the excess entropy of disordered phases. J Phys Chem 84:859-863

Haggerty JS, Cooper AR, Heasley JH (1968) Heat capacity of three inorganic glasses and supercooled liquids. Phys Chem Glasses 5:130-136

Hetherington G, Jack KH, Kennedy JC (1964) The viscosity of vitreous silica. Phys Chem Glasses 5:130-136

Hummel WJ, Arndt J (1985) Variation of viscosity with temperature and composition in the plagioclase system. Contrib Mineral Petrol 90:83-92

Johari GP, Goldstein M (1970) Viscous liquids and the glass transition. II. Secondary relaxations in glasses of rigid molecules. J Chem Phys 53:2372-2388

Knoche R, Dingwell DB, Webb SL (1992) Temperature dependent thermal expansivities of silicate melts: the system anorthite-diopside. Geochim Cosmochim Acta 56:689-699

Krause JT, Kurkjian CR (1968) Vibrational anomalies in inorganic glassformers. J Am Ceram Soc 51:226-227

Krupka KM, Robie RA, Hemingway BS, Kerrick DM, Ito J (1985) Low-temperature heat capacities and derived thermodynamic properties of antophyllite, diopside, enstatite, bronzite, and wollastonite. Am Mineral 70:249-260

Kushiro I, Yoder HS Jr, Mysen BO (1976) Viscosities of basalt and andesite melts at high pressures. J Geophys Res 81:6351-6356

Landau LD, Lifshitz EM (1958) Statistical Physics, Pergamon Press, London

Lange RA, Navrotsky A (1992) Heat capacities of Fe_2O_3-bearing silicate liquids. Contrib Mineral Petrol 110:311-320

Lange RA, Navrotsky A (1993) Heat capacities of TiO_2-bearing silicate liquids: evidence for anomalous changes in configurational entropy with temperature. Geochim Cosmochim Acta 57:3001-3011

Lejeune AM, Richet P (1995) Rheology of crystal-bearing silicate melts: an experimental study at high viscosities. J Geophys Res 100:4215-4229

Li JH, Uhlmann DR (1970) The flow of glass at high stress levels, 1. Non-Newtonian behavior of homogeneous 0.08 Rb_2O-0.92 SiO_2 glasses. J Non-Cryst Solids 33:127-147

de Ligny D, Richet P, Westrum EF, Jr (1995) Low-temperature heat capacity of vitreous GeO_2 and B_2O_3: thermochemical and structural implications. Abstract, 5th Int'l Silicate Melt Workshop, la Petite Pierre.

Litovitz TA, Davis CM (1965) Structural and shear relaxation in liquids. In: Physical Acoustics. Mason WP (ed) p 282-349, Academic Press, San Diego

Macedo PB, Litovitz TA (1965) Ultrasonic viscous relaxation in molten boron trioxide. Phys Chem Glasses 6:69-80

Malinovsky VK, Sokolov AP (1986) The nature of the boson peak in Raman scattering in glasses. Solid State Comm 57:757-761

Martens RM, Rosenhauer M, Büttner H, von Gehlen K (1987) Heat capacity and kinetic parameters in the glass transformation interval of diopside, anorthite and albite glass. Chem Geol 62:49-70

Mazurin OV (1986) Glass relaxation. J Non-Cryst Solids 87:392-407

Mazurin OV, Porai-Koshits EA (1989) Inhomogeneity in monophase glass-forming melts and glasses. In: The Physics of Non-Crystalline Solids. Pye LD, LaCourse WC, Stevens HJ (eds) p 22-25. Taylor and Francis, Washington, DC

Mazurin OV, Startsev YK, Potselueva LN (1979) Temperature dependence of the viscosity of some glasses at a constant structural temperature. Sov J Glass Phys Chem 5:68-79

McBirney AR, Murase T (1984) Rheological properties of magmas, Ann Rev Earth Planet Sci 12:337-357

Moynihan CT, Schroeder J (1993) Non-exponential structural relaxation, anomalous light scattering inhomogeneities in glass-forming liquids. J Non-Cryst Solids 160 52-59

Moynihan CT, Macedo PB, Montrose CJ, Gupta PK, De Bolt MA, Dill JF, Dom BE, Drake PW, Easteal AJ, Elterman PB, Moeller RP, Sasabe M, Wilder JA (1976) Structural relaxation in vitreous materials. Annals New York Acad Sci 279:15-35

Napolitano A, Macedo PB (1968) Spectrum of relaxation times in GeO_2 glass. J Res Nat'l Bureau Standards 72A:425-433

Navrotsky A (1995) Energetics of silicate melts. This volume.

Neuville DR, Richet P (1991) Viscosity and mixing in molten (Ca, Mg) pyroxenes and garnets. Geochim Cosmochim Acta 55:1011-1019

Neuville DR, Courtial P, Dingwell DB and Richet P (1993) Thermodynamic and rheological properties of rhyolite and andesite melts. Contrib Mineral Petrol 113:572-581

Pohl RO (1981) Low-temperature specific heat of glasses. In: Topics in Current Physics. Phillips WA (ed) 24:27-52, Springer-Verlag, Heidelberg

Poole JP (1948) Viscosité à basse température des verres alcalino-silicatés. Verres Réfract 2:222-228

Richet P (1984) Viscosity and configurational entropy of silicate melts. Geochim Cosmochim Acta 48:471-483

Richet P, Bottinga Y (1983) Verres, liquides, et transition vitreuse. Bull Minéral 106:147-168

Richet P, Bottinga Y (1984a) Glass transition and thermodynamic properties of amorphous SiO_2, $NaAlSi_nO_{2n+2}$ and $KAlSi_3O_8$. Geochim Cosmochim Acta 48:453-470

Richet P, Bottinga Y (1984b) Anorthite, andesine, wollastonite, diopside, cordierite, and pyrope: thermodynamics of melting, glass transitions, and properties of the amorphous phases. Earth Planet Sci Lett 67:415-432

Richet P, Bottinga Y (1985) Heat capacity of aluminum-free liquid silicates. Geochim Cosmochim Acta 49:471-486

Richet P, Bottinga Y (1986) Thermochemical properties of silicate glasses and liquids: a review. Rev Geophys 24:1-25

Richet P, Neuville DR (1992) Thermodynamics of silicate melts: Configurational properties. In: Adv Phys Geochem. Saxena S (ed) 11:132-160. Springer-Verlag, Heidelberg

Richet P, Bottinga Y, Deniélou L, Petitet JP, Téqui C (1982) Thermodynamic properties of quartz, cristobalite and amorphous SiO_2: drop calorimetry measurements between 1000 and 1800 K and a review from 0 to 2000 K. Geochim Cosmochim Acta 46:2639-2658

Richet P, Bottinga Y, Téqui, C (1984) Heat capacity of sodium silicate liquids. J Am Ceram Soc 67:C6-C8

Richet P, Robie RA, and Hemmingway BS (1986) Low-temperature heat capacity of diopside glass ($CaMgSi_2O_6$): A calorimetric test of the configurational-entropy theory applied to the viscosity of liquid silicates. Geochim Cosmochim Acta 50:1521-1533

Richet P, Robie RA, Rogez J, Hemingway BS, Courtial P. (1990) Thermodynamics of open networks: ordering and entropy in $NaAlSiO_4$ glass, liquid, and polymorphs. Phys Chem Minerals 17:385-394

Richet P, Robie RA, Hemmingway BS (1991) Thermodynamic properties of wollastonite and $CaSiO_3$ glass and liquid. Eur J Mineral 3:475-484

Richet P, Robie RA, Hemmingway BS (1993) Entropy and structure of silicate glasses and melts. Geochim Cosmochim Acta 57:2751-2766

Robertson RE (1978) Effect of free-volume fluctuations on polymer relaxation in the glassy state. J Polymer Sci: Polymer Symp 63:173-183

Ryan MP, Blevins JYK (1987) The viscosity of synthetic and natural silicate melts and glasses at high temperatures and 1 bar pressure and at higher pressures. U.S. Geol Survey Bull 1764

Ryerson FJ, Weed HC, Piwinskii AJ (1988) Rheology of subliquidus magmas. 1. Picritic compositions. J Geophys Res 93:3421-3436

Scarfe CM, Mysen BO, Virgo D (1979) Changes in viscosity and density of melts of sodium disilicate, sodium metasilicate, and diopside composition with pressure. Yearbook, Carnegie Inst Washington 78:547-551

Scarfe CM, Cronin DJ, Wenzel JT, Kaufman DA (1983) Viscosity-temperature relationships at 1 atm in the system diopside-anorthite. Am Mineral 68:1083-1088

Scarfe CM, Mysen BO, Virgo D (1987) Pressure dependence of the viscosity of silicate melts. In: Magmatic Processes: Physicochemical Principles. Mysen BO (ed) Spec Publ 1:59-67, The Geochemical Society

Scherer GW (1984) Use of the Adam-Gibbs equation in the analysis of structural relaxation. J Am Ceram Soc 67:504-511

Scherer GW (1990) Theories of relaxation. J Non-Cryst Solids 123:75-89

Shaw H (1972) Viscosities of magmatic silicate liquids: an empirical method of prediction. Am J Sci 272:870-893

Simmons JH, Mohr RK, Montrose JC (1982) Non-Newtonian viscous flow in glass. J Appl Phys 53:4075-4080

Simmons JH, Ochoa R, Simmons KD, Mills JJ (1988) Non-Newtonian viscous flow in soda-lime-silica glass at forming and annealing temperatures. J Non-Cryst Solids 105:313-322

Sipp A (1993) Relaxation structurale dans les silicates. Rapport DEA, Université Paris 6.

Sipp A, Neuville DR, Richet P (1995) Viscosity and configurational entropy of window and borosilicate melts. To be submitted to J Non-Cryst Solids

Sokolov AP, Kisliuk A, Soltwisch M, Quintmann D (1993) Low-energy anomalies of vibrational spectra and medium range order in glass. Physica, A201, 295-299

Spera FJ, Yuen DA, Kirschvink SJ (1982) Thermal boundary layer convection in silicic magma chambers: Effects of temperature-dependent rheology and implications for thermogravitational chemical fractionation. J Geophys Res 87:8755-8767

Spera FJ, Borgia, A, Strimple J, Feigenson M (1988) Rheology of melts and magmatic suspensions, 1, Design and calibration of concentric viscometer with application to rhyolitic magma, J Geophys Res 93:10,273-10,294

Stebbins JF, Carmichael ISE, and Moret LK (1984) Heat capacity and entropies of silicate liquids and glasses. Contrib Mineral Petrol 86:131-148

Stebbins JF, Farnan I, Xue X (1992) The structure and dynamics of silicate liquids: A view from NMR spectroscopy. Chem Geol 96:371-386

Stein DJ, Spera FJ (1992) Rheology and microstructure of magamtic emulsions: theory and experiments. J Volcan Geotherm Res 49:157-174

Tauber P, Arndt J (1987) The relationship between viscosity and temperature in the system anorthite-diopside. Chem Geol 62:71-81

Taylor TD, Rindone GE (1970) Properties of soda aluminosilicate glasses. V, low-temperature viscosities. J Am Ceram Soc 53:692-695

Téqui C, Robie RA, Hemingway BS, Neuville DR, Richet P (1991) Melting and thermodynamic properties of pyrope ($Mg_3Al_2Si_3O_{12}$). Geochim Cosmochim Acta 55:1005-1010

Tomlinson JW, Heines MSR, Bockris JO'M (1958) The structure of liquid silicates, pt. 2. Trans Faraday Soc 54:1822-1833

Tool AQ, Eichlin CG (1931) Variations caused in the heating curves of glass by heat treatment. J Am Ceram Soc 14:276-308

Urbain G, Bottinga Y, Richet P (1982) Viscosity of liquid silica, silicates and aluminosilicates. Geochim Cosmochim Acta 46:1061-1071

Webb SL (1991) Shear and volume relaxation in $Na_2Si_2O_5$. Am Mineral 76:451-456

Webb SL (1992) Shear, volume, enthalpy and structural relaxation in silicate melts. Chem Geol 96:449-458

Webb SL, Dingwell DB (1990a) Non-Newtonian viscosities of geologic melts at high stresses: experimental results for rhyolite, andesite, basalt and nephelinite. J Geophys Res 95:15,695-15,701

Webb SL, Dingwell DB (1990b) Onset of non-Newtonian rheology of silicate melts: A fiber elongation study. Phys Chem Minerals 17:125-132

Xue X, Stebbins JF, Kanzaki M, McMillan PF, Poe B (1991) Pressure-induced silicon coordination and tetrahedral structural changes in alkali oxide—silica melts up to 12 GPa: NMR, Raman, and infrared spectroscopy. Am Mineral 76:8-26

Xue X, Stebbins JF, Kanzaki M (1994) Correlations between ^{17}O NMR parameters and local structure around oxygen in high-pressure silicates: Implications for the structure of silicate melts at high pressure. Am Mineral 79:31-42

Yageman VD, Matveev GM (1982) Heat capacity of glasses in the system SiO_2-Na_2O.SiO_2. Fiz Khim Stekla 8:238-245

Chapter 4

VISCOELASTICITY

Sharon L. Webb and Donald B. Dingwell

Bayerisches Geoinstitut
95440 Bayreuth
Germany

INTRODUCTION

The structure of a silicate melt requires a certain amount of time after a perturbation in its environment (e.g. temperature, pressure, electric field) in order to reach its equilibrium state. Using a simple gas model, Maxwell (1867) calculated that, in response to a change in the shear stress, the structure of a fluid requires an equilibration time τ which is a function of the Newtonian shear viscosity and the elastic shear modulus of the fluid. Recent contributions have highlighted the role of viscoelasticity of silicate melts in the context of melt structure and rheology. The utility of viscoelastic investigations of silicate melts or relevant to the glass industry has been recently treated by Zarzycki (1991), and the contribution of melt viscoelasticity to the investigation of the rheology of natural melts has been discussed by Dingwell et al. (1993). Viscoelasticity plays a central role in a host of experimental investigations of silicate melts. As often as not it is essential to avoid viscoelastic effects, remaining totally within either the fully relaxed viscous response field or within the fully unrelaxed elastic field. Performing this prerequisite for the determination of Newtonian viscosities or for elastic moduli is not always easy and cannot always be demonstrated in silicate melt studies. Nevertheless the attempt must be made to classify experimental timescales appropriately, and the path to this goal is a complete as possible description of viscoelasticity in silicate melts (Dingwell and Webb, 1989). The investigation of viscoelasticity is important for another reason. The nature of the relaxation process in silicate melts lies behind a structural explanation of virtually all aspects of melt rheology and elasticity. If we are to understand the mechanism of viscous flow then we must be in a position to predict the frequency dependence of the loss modulus. Conversely, a full explanation of the viscoelasticity of melts will form an essential component of any structural explanation of flow processes.

The present discussion is confined to the equilibration of silicate melt structure in response to a mechanical perturbation; i.e. shear and compressional stresses and strains. The equilibration times required for the melt structure with respect to other types of perturbations are described in the Appendix.

Three types of deformation occur in a silicate melt (or any material) upon the application of a step-function in stress. As shown in Figure 1, there is the instantaneous recoverable (elastic) deformation, the time-dependent recoverable deformation and the time-dependent non-recoverable (viscous) deformation. The instantaneous recoverable deformation is the elastic deformation; this in combination with the time-dependent recoverable deformation is anelastic deformation; and this in combination with the time-dependent non-recoverable deformation is viscoelastic deformation. (Nowick and Berry, 1972). Therefore, as a function of time, different parts of the melt structure are deforming differently due to the applied stress. The structure at first deforms elastically and then viscously. This change in deformation mechanism is due to relaxation of the melt structure.

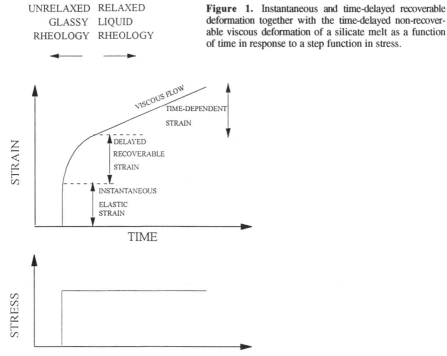

UNRELAXED RELAXED
GLASSY LIQUID
RHEOLOGY RHEOLOGY

Figure 1. Instantaneous and time-delayed recoverable deformation together with the time-delayed non-recoverable viscous deformation of a silicate melt as a function of time in response to a step function in stress.

For an ideal material, the structure relaxes (equilibrates) as an exponential function of time. The relaxation behavior can be described with a single valued time constant called the relaxation time τ. As pointed out by Maxwell (1867) this relaxation time for shear deformation is a function of the Newtonian (relaxed - long timescale) shear viscosity η_0, and the Hookean (unrelaxed - short timescale) elastic modulus G_∞ of the material;

$$\tau_s = \frac{\eta_0}{G_\infty} . \tag{1}$$

When deformation is measured a short time (time $< \tau$) after the application of a stress, the unrelaxed glassy strain response of the melt is observed. If the deformation is measured a long time (time $> \tau$) after the application of the stress, the relaxed liquid strain response of the melt is observed. The relaxed response includes of course the elastic component of deformation. For the cases of small stresses and strains, there is a unique, linear relationship between the applied stress and the resulting strain. For large stresses and strains (which can be employed for example in rheological creep experiments) the deformation behavior becomes non-linear as bonds are forced beyond the limits of the simple Hookean approximation of the stress-strain relationship.

In the linear regime simple Hookean physics can be used to describe the relationship between stress, strain and strain-rate. The modulus M of a material is the ratio of stress σ to strain ε:

$$M = \frac{\sigma}{\varepsilon} = J^{-1} \tag{2}$$

where J, the compliance is the ratio of strain to stress. The viscosity η of a material is the ratio of stress to strain-rate

$$\eta = \frac{\sigma}{\dot\varepsilon} = \phi^{-1} \tag{3}$$

where ϕ is the fluidity of the material.

In principle, the viscoelastic behavior of a melt can be determined as a function of time simply by applying a step function in stress and measuring the resultant deformation of the melt as a function of time. As shown in Figure 2, the modulus and viscosity determined depends upon the elapsed time after the application of the step function in stress that the deformation was measured.

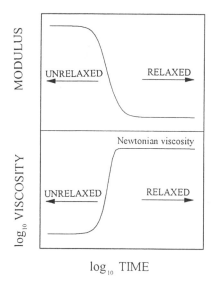

Figure 2. Modelled time-dependent, relaxed and unrelaxed modulus and viscosity measured after the application of a step function in stress.

For a melt of viscosity 10^0 Pa s and a modulus of 20 GPa, the application of 10^7 Pa stress results in an elastic deformation of 5×10^{-3}, total anelastic deformation of $\sim 10^{-2}$ occurring within 5 s and a viscoelastic deformation of 1 occurring within 1000 s (17 min). For a melt of viscosity 10^2 Pa s the time required for similar amounts of anelastic and viscoelastic deformation is 5 nsec and 10 μsec, respectively. The need to measure 3 orders of magnitude in deformation over short periods of time makes it difficult to observe the viscoelastic behavior of melts in the time domain. It is, however, relatively simple to measure the large viscous deformation as a function of time and therefore to determine the shear viscosity of the melt from the time-dependent deformation in response to an applied stress (Dingwell, 1995; see also chapters by Dingwell; and by Richet and Bottinga).

The viscoelastic behavior of melts can, however, be determined in the frequency domain by application of a sinusoidally varying stress wave. The timescales of deformation (period of the wave) can easily be varied over 3 (or more) orders of magnitude, and the elastic and viscous deformation due to the sinusoidally oscillating stress remain within an order of magnitude of each other.

The viscoelastic nature of silicate melts results in a frequency-dependent phase lag

$\delta(\omega)$ between the applied sinusoidal stress $\sigma^*(\omega) = \sigma_0 \exp\{i\omega t\}$ and the resulting sinusoidal strain $\varepsilon^*(\omega) = \varepsilon_0 \exp\{i[\omega t+\delta(\omega)]\}$ and the modulus and viscosity are described by the complex functions:

$$M^*(\omega) = \frac{\sigma^*(\omega)}{\varepsilon^*(\omega)} \tag{4}$$

and

$$\eta^*(\omega) = \frac{\sigma^*(\omega)}{\dot{\varepsilon}^*(\omega)}. \tag{5}$$

As discuss by Herzfeld and Litovitz (1959), viscoelastic materials display frequency-dependent behavior (see Fig. 3). If the frequency of the applied signal is high and the stress oscillation occurs on a timescale so short that viscous flow does not have time to occur within one oscillation, only the elastic behavior of the material is observed. If the frequency of the applied signal is lower, however, viscous flow of the melt is possible within the timescale of the stress oscillations and the viscoelastic response of the melt will be observed. The elastic response is the strain in-phase with the applied stress and is described by the real part of the complex modulus in Figure 3; whereas the viscous response is the strain out-of-phase with the applied stress which is described by the imaginary part of the modulus in Figure 3. The real and imaginary parts of the modulus describe the storage and loss, of energy, respectively. When $\delta(\omega) = 0$, there is only elastic deformation occurring in the melt, at the frequency of the stress wave. For silicate melts, which are linear viscoelastic materials under conditions of small stresses and strains, the volume [bulk, $K^*(\omega)$], shear [$G^*(\omega)$] and longitudinal [$M^*(\omega)$] moduli are frequency-dependent, as are the volume [$\eta_v^*(\omega)$], shear [$\eta_s^*(\omega)$] and longitudinal [$\eta_l^*(\omega)$] viscosities.

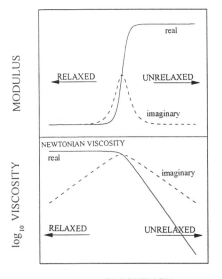

Figure 3. Modelled frequency-dependent real and imaginary components of the relaxed and unrelaxed modulus and viscosity as a function of frequency; the solid lines are the real part of the modulus and viscosity, the dotted lines the imaginary part.

The shear modulus as a function of frequency of applied stress can be described as:

$$G^*(\omega) = G_\infty \frac{\omega^2\tau_s^2}{1+\omega^2\tau_s^2} + iG_\infty \frac{\omega\tau_s}{1+\omega^2\tau_s^2} \tag{6}$$

and the bulk modulus as:

$$K^*(\omega) = (K_\infty - K_0) + K_0 \frac{\omega^2 \tau_v^2}{1 + \omega^2 \tau_v^2} + iK_0 \frac{\omega \tau_v}{1 + \omega^2 \tau_v^2} \tag{7}$$

where G_∞ and K_∞ are the elastic shear and volume moduli (i.e. the values for infinite frequencies, or the instantaneous [zero time after application of the stress] values of the moduli) and K_0 is the relaxational volume modulus (Herzfeld and Litovitz, 1959). The shear and volume relaxation times τ_s and τ_v are functions of the low frequency viscosity and the high frequency elastic modulus;

$$\tau_s = \frac{\eta_s}{G_\infty}$$

and

$$\tau_v = \frac{\eta_v}{K_0}. \tag{8}$$

The longitudinal modulus is

$$M = K + (4/3) \cdot G . \tag{9}$$

As viscosity is a function of stress and strain-rate, the viscosity is also described by complex equations

$$\eta_s^*(\omega) = \frac{G^*(\omega)}{i\omega} = G_\infty \tau_s \frac{1}{1 + \omega^2 \tau_s^2} - i G_\infty \tau_s \frac{\omega \tau_s}{1 + \omega^2 \tau_s^2}$$

$$= \eta_{0s} \frac{\tau_s}{1 + \omega^2 \tau_s^2} - i \eta_{0s} \frac{\omega \tau_s}{1 + \omega^2 \tau_s^2} \tag{10}$$

and

$$\eta_v^*(\omega) = \frac{K^*(\omega)}{i\omega} = K_0 \tau_v \frac{1}{1 + \omega^2 \tau_v^2} - i K_0 \tau_v \left(\frac{\omega \tau_v}{1 + \omega^2 \tau_v^2} + \frac{K_\infty - K_0}{K_0 \omega} \right)$$

$$= \eta_{0v} \frac{1}{1 + \omega^2 \tau_v^2} - i \eta_{0v} \left(\frac{\omega \tau_v}{1 + \omega^2 \tau_v^2} + \frac{K_\infty - K_0}{K_0 \omega} \right) \tag{11}$$

where the real and imaginary parts of the viscosity describe the loss (viscous flow) and storage (elastic deformation) of energy in the melt respectively.

Thus the measured viscosity of silicate melts depends upon the frequency of the applied stress wave; or the timescale on which the stress changes. Newtonian viscosity is independent of strain-rate (and stress) and therefore independent of the frequency of the applied stress. At low frequencies the frequency-independent Newtonian viscosity of the melt is measured and at high strain-rates the non-Newtonian rate-dependent viscosity is determined.

The Equations (1) through (11) are for an ideal liquid. The boundary conditions for these equations are such that at low frequencies the liquid exhibits only shear flow and the shear modulus is zero, while at high frequencies the liquid exhibits a shear modulus. In the case of the volume modulus which describes homogeneous compression of a liquid, both the low and high frequency modulus have non-zero values. The low frequency volume modulus must be non-zero as it is not physically reasonable to observe continuous compression for a constant applied pressure. On the other hand the shear modulus at low

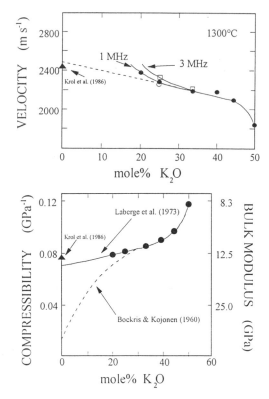

Figure 4. (a) Frequency-dependent (1 and 3 MHz) compressional wave velocity determined by Bockris and Kojonen (1960) and Laberge et al. (1973) in the K_2O-SiO_2 system. The velocity for SiO_2 is calculated from the data of Krol et al. (1986). (b) Extrapolation of relaxed [Laberge et al. (1973)] and partially relaxed [Bockris and Kojonen (1960)] data to the compressibility of SiO_2 melt at 1300°C. The velocity for SiO_2 is calculated from the data of Krol et al. (1986). (a) and (b) redrawn after Laberge et al. (1973).

frequencies may be zero as it is possible to have an infinite amount of shear deformation for a fixed applied shear stress, with no volume change. Comparison of compressibility (or shear) data obtained on different composition melts at the same temperature and frequency of signal; or on the same composition at different deformation rates can result in a mixture of relaxed and partially relaxed data which can lead to misinterpretations of the rheological properties of melts.

The study of Laberge et al. (1973) illustrates the problems associated with comparison of compressibility data for different composition melts at the same temperature and frequency. As shown in Figure 4a, the curvature in the measured compressional wave velocities as a function of composition in the system K_2O-SiO_2 is largely due to the frequency dependence of the measured velocity as a function of the change in structural relaxation time with changing composition. Laberge et al. (1973) assumed that the relaxed compressibility data could be extrapolated linearly and calculated a compressibility for SiO_2 melt at 1300°C of 14 GPa in comparison with the original extrapolation of partially unrelaxed data of Bockris and Kojonen (1960) to 77 GPa (Fig. 4b). The extrapolation of Laberge et al. (1973) agrees well with the Brillouin scattering calculation of the 13 GPa for the relaxed compressibility of SiO_2 melt at 1300°C (Krol et al., 1986).

All methods of determining the stress - strain - strain-rate behavior of melts [shock wave (e.g. Rigden et al. 1984,1988, Miller et al., 1991); ultrasonics (e.g. Macedo et al., 1968; Simmons and Macedo, 1970; Rivers and Carmichael, 1987; Secco et al., 1991; Webb, 1991); Brillouin scattering (e.g. Krol et al., 1986; Askarpour et al., 1993); light scattering (e.g. Siewert and Rosenhauer, 1994); torsion deformation (e.g. Kurkjian, 1963; Mills, 1974; Webb, 1992b); parallel plate deformation (e.g. Stevenson et al., 1995); beam-bending deformation (e.g. Cooper, 1990); and micropenetration (e.g. Dingwell et al., 1992), fibre elongation (e.g. Li and Uhlmann, 1970; Webb and Dingwell, 1990a,b) and concentric cylinder (e.g. Spera et al., 1988; Dingwell, 1992) viscosity measurements] involve a rate of deformation, or of stress application. Therefore, all such data must be critically inspected to separate the effects of viscoelasticity of the melt from the composition-, pressure- and temperature-dependence of the measured stress and strain.

In addition to the main high temperature structural relaxation responsible for viscous flow in silicate melts, a second relaxation, associated with the motion of alkalies within the glass structure has been widely documented (Zdaniewski et al., 1979; Dingwell, 1990). It is discussed in chapter x and not treated further here. The lifetime of the Si-O bond has been observed at high temperatures from ultrasonic (MHz frequencies: Rivers and Carmichael, 1987; Webb, 1991) and NMR (~10 Hz: Liu et al., 1988; Hz - MHz: Farnan and Stebbins, 1994) measurements; and at lower temperature via mechanical torsion (mHz to Hz: Mills 1974; Webb, 1992b; Bagdassarov et al., 1993) and viscosity measurements (deformation rates of 10^{-3} s^{-1}: Li and Uhlmann, 1970; Simmons et al., 1982; Webb and Dingwell, 1990a,b).

The Maxwell relationship reproduces the lifetime of Si-O bonds in silicate melts fairly well (see discussion by Stebbins, this volume). In the case of $Na_2Si_2O_5$ melt it can be seen in Figure 5 that the relaxation time calculated from Equation (1) using the measured shear viscosity for this melt, and an assumed G_∞ of 10 GPa agrees with the structural relaxation observed in ultrasonic, NMR, heat capacity, thermal expansion and torsion deformation experiments across a 10^9 to 10^{-5} Hz range in frequency (deformation rate) and 400° to 1150°C in temperature and 10^1 to 10^{15} Pa s in viscosity. To the extent that this equivalence

$$Na_2Si_2O_5$$

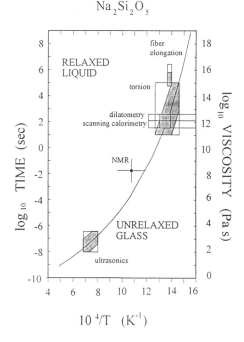

Figure 5. Relaxation time curve for Na₂Si₂O₅ calculated from Equation (1) divides timescale-temperature space into regions of relaxed and unrelaxed rheology. The rectangles illustrate the timescale-temperature regimes in which a variety of techniques is used to determine melt rheology. The hatched areas indicate the regimes in which shear (fibre elongation, torsion, ultrasonics), volume (dilatometry, ultrasonics) and enthalpy (scanning calorimetry) relaxation has been observed. The NMR datum is from Liu et al. (1988) and indicates the lifetime of Si-O bonds at temperature (redrawn after Webb, 1992a).

can be demonstrated we are confronted with the powerful conclusion that the species involved in the co-operative rearrangements of viscous flow involve the rearrangement of Si-O bonds within the melt structure (Liu et al., 1988; Dingwell and Webb; 1989).

VISCOELASTICITY

There have been a number of experimental studies to investigate the low frequency viscoelastic rheology of silicate melts (Mills 1974; Perez et al., 1981; Webb, 1992b) and also the frequency dependence of the longitudinal moduli (Sato and Manghnani, 1985; Manghnani et al., 1986) and shear moduli (Macedo et al., 1968; Webb, 1991; Bornhöft and Brückner, 1994) at high frequencies. Rivers and Carmichael (1987) demonstrated that a wide range of silicate melt compositions displayed frequency dependence of the longitudinal modulus. They showed that the frequency at which this phenomenon occurred was proportional to the relaxation time calculated from the Maxwell relationship. The volume viscosities calculated by Rivers and Carmichael (1987) were, however, negative, indicating that there was a further structural relaxation occurring in the frequency range between 0 and 1 MHz. This dilemma caused considerable confusion in the late 1980s because an additional relaxation mechanism at timescales longer than the viscous relaxation of these melts would have implied extended species in the melt structure (complexes or polymers) whose lifetime exceeded that of viscous flow deformation rates and might have been held responsible for the role of "flow units" in silicate melts. The calculation of negative volume viscosities was, as pointed out by Dingwell and Webb (1989), a flaw in the calculations of Rivers and Carmichael (1987) due to a factor of 100 error in the shear viscosity data of Rivers and Carmichael (1987) (possibly due to an error in the conversion of viscosity from 10 poise = 1 Pa s). Using the correct shear viscosity data, the longitudinal relaxation observed by Rivers and Carmichael (1987) occurs at the timescale calculated from the Maxwell relationship for shear relaxation, and the calculated volume viscosities are positive and equal to the shear viscosities [as had been previously purported (e.g. Simmons and Macedo, 1970; Sato and Manghnani, 1984)]. This observation removed the need for a further structural relaxation at lower frequencies. With this vital simplification of silicate melt relaxation at long timescales an array of microscopic and macroscopic data on silicate melt structure and properties fell into place in a consistent description of the time-temperature relationship of structural relaxation in silicate melts of all chemistries providing a powerful example of the link between microscopic and macroscopic expressions of the melt dynamics (see chapter by Dingwell for further discussion). The implication of this observed agreement of the longitudinal relaxation time with the calculated shear relaxation time is that the shear and volume relaxation times are identical, i.e. $\tau_v = \tau_s$.

Given the general applicability of the Maxwell relationship to calculate the timescale of structural relaxation, the deformation conditions under which shear and volume relaxation are to be expected in silicate melts can be estimated. In the case of shear relaxation, especially, the question of non-Newtonian rheology arises because the large strains involved in igneous processes are, except for the case of vesicular materials, dominated by the shear component. It has been generally assumed, for example in calculational schemes of silicate melt rheology (see chapter by Dingwell), that all silicate melts are Newtonian and that is there is a unique linear relationship between stress and strain-rate, independent of strain-rate. For each temperature, however, the structure of a melt has a relaxation (equilibration) time. If the melt is sheared at a rate faster than the relaxation rate, the effects of structural relaxation are observed and the stress necessary to achieve the deformation rate decreases and non-Newtonian viscosity is determined (Li and Uhlmann, 1970; Simmons et al., 1982; Webb and Dingwell, 1990a,b). The systematics of the results of time domain investigations of the onset of non-Newtonian flow in silicate melts are discussed in the chapter by Dingwell.

Time- versus frequency-domain measurements

A melt of shear viscosity η_s has a shear relaxation time $\tau_s = \eta_s/G_\infty$, or a relaxation strain-rate of $\dot{\varepsilon}_r = (\eta_s/G_\infty)^{-1}$. The shear viscosity of a melt can be determined in time-space by the application of a constant stress and the measurement of the resulting strain-rate (or by application of a constant strain-rate and the measurement of the necessary stress). In order to reach the relaxation strain-rate, a stress of $\eta_s\dot{\varepsilon}_r$ must be achieved (based on Newtonian viscosity calculations). For a shear viscosity of 1 Pa s, non-Newtonian viscosity should occur at a strain-rate of $\sim 10^{10}$ s^{-1} and a stress of $\sim 10^{10}$ N m^{-2}. For a melt of Newtonian shear viscosity 10^{11} Pa s, a strain-rate of $\sim 10^{-1}$ s^{-1} and a stress of $\sim 10^{10}$ N m^{-2} is required to match this non-Newtonian condition [as $G_\infty \cong 10$ GPa (Dingwell and Webb, 1990), the stress required to reach the relaxation strain-rate is also 10 GPa (1 Pa = 1 N m^{-2}); as $\tau = \dot{\varepsilon}_r = G_\infty/\eta = G_\infty/(\sigma/\dot{\varepsilon})$]. For a strain-rate of 10^{-1} s^{-1}, a strain of one occurs within a measurement time of 10 s. Due to the approximately (for present purposes) exponential nature of structural relaxation, the calculated onset on viscoelastic shear rheology should occur at strain-rates 2 orders of magnitude less than the relaxation strain-rate at stresses greater than 100 MPa. Most techniques commonly used to determine shear viscosity as a function of observation time cannot reach these extreme conditions of stress, and the range of strain rates versus temperature within which non-Newtonian flow occurs form a relatively small "window" within the total available experimental measurement range. Therefore very few observations of non-Newtonian viscosity have been documented. A stress of 100 MPa is a mass of 0.3 kg on a melt fibre of diameter 0.2 mm; or a mass of 200 kg on a melt "fibre" of 5 mm diameter.

Shear and volume relaxation can also be measured in the frequency domain. The relaxation of strain can be observed directly by applying a sinusoidal stress and observing the phase shift between the applied stress and the resulting sinusoidal strain. This method involves the application of stresses of ~ 100 kPa and strains of 10^{-5}. Shear and volume relaxation can be observed in an indirect manner by determining the velocity of a shear or longitudinal wave travelling through the melt as a function of the frequency of the sinusoidal deformation. This involves stresses of several $\sim$kPa and strains of $\sim 10^{-8}$.

SHEAR RHEOLOGY

Forced torsion

Forced torsion oscillation is a method of observing frequency-dependent viscoelastic rheology in melts as a function of strain-rate. There have been a large number of such studies; all of which are based on the design of a simple torsion pendulum. This consists of a cylindrical sample of melt which is held fixed at one end, and which has a sinusoidally oscillating stress applied to the other (free) end (see Fig. 6). Most of these investigations cover the frequency range of mHz to Hz with some measurements going up to kHz. The angle of twist of the melt is $\sim 10^{-5}$ rad, with the applied torque being $\sim 10^{-3}$ N m. As this is an oscillating deformation, there are two components of deformation in the melt. There is the deformation in-phase with the applied stress and the deformation 90° out-of-phase with the applied stress. These are the instantaneous recoverable elastic deformation, and the time-dependent recoverable anelastic deformation, and non-recoverable viscous deformation of the melt, respectively. The amplitude of the in- and out-of-phase components of deformation vary as a function of the frequency of the applied stress, with the out-of-phase (imaginary) component increasing in the vicinity of the relaxation frequency (see Fig. 3). The viscoelastic shear rheology of a range of silicate melts have been determined by forced oscillation (Mills, 1974; Perez et al., 1981; Webb, 1992b; Bagdassarov et al., 1993). The shear viscosity is seen to become non-Newtonian with increasing strain-rate (increasing frequency of the applied deformation).

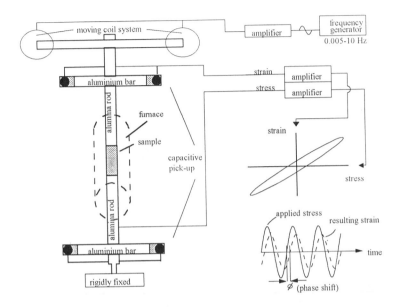

Figure 6. Torsion machine used to measure the frequency dependence of the shear modulus and viscosity of silicate melts. For an elastic response there is no phase difference between the applied stress and the resulting strain. In the frequency range in which there is a phase lag between the stress and strain, a plot of strain as a function of stress is an ellipse, instead of a straight line. The melt sample is attached to two alumina rods. One rod is rigidly fixed, while a sinusoidal stress is applied to the other. The resulting stress-strain curve is determined (redrawn after Webb, 1992b).

Thermorheological simplicity

As seen in the data of Mills (1974) in Figure 7a,b the forced oscillation technique covers a frequency range of some 3 orders of magnitude. The viscoelastic behavior of silicate melts, however, occurs over 6 or more orders of magnitude in rate of deformation. Due to the large range of frequencies required for the description of the complete frequency dependence of the viscoelastic relaxation peak, a practical alternative has been adopted which involves the assumption that a log-linear shift factor can be applied to the segments of relaxation modulus data which are obtained over the same restricted frequency range at differing temperatures. This assumption is equivalent to a temperature invariance of the shape or distribution of energy loss with frequency. This assumption constitutes the approximation of thermorheological simplicity (Narayanaswamy, 1971, 1988; Mazurin, 1986) and results in all of the data being plotted upon one master curve, with the frequency being normalised to the calculated relaxation frequency of the melt at each temperature (Fig. 7a,b).

The viscoelastic rheology of bubble- and crystal-free synthetic granitic melt compositions have been determined over a temperature range 500° to 1000°C by Bagdassarov et al. (1993). The shear moduli as a function of composition are plotted against the normalised frequency $\omega\tau_s$ in Figure 8. The relaxation time τ_s was calculated at each temperature from Equation (1) using the measured shear viscosity and shear modulus for these melt compositions. The infinite frequency shear modulus was found to range from 21.5 to 26.5 GPa for these melt compositions at these temperatures. These data cannot be fit by a single shear relaxation time theory where the modulus is described by

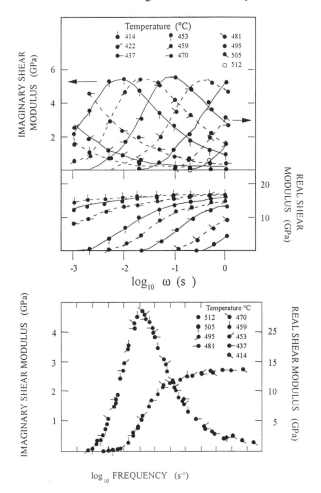

Figure 7. (a) The frequency-dependent shear modulus of $Na_2Si_2O_5$ melt over a range of temperatures (redrawn after Mills, 1974). (b) The frequency dependent shear modulus of $Na_2Si_2O_5$ melt plotted on a master curve assuming thermorheological simplicity (redrawn after Mills, 1974).

Equation (6). The distribution in relaxation times that is required to describe the equilibration of the melt structure is due to the range of topologically different Si-O bonds within the melt, together with the range of relaxation times for each of these different Si-O bonds. The frequency-dependent real and imaginary components of the shear modulus of haplogranite composition melts with added B_2O_3, P_2O_5 and F_2O_{-1} from 500° to 1000°C were fit to a curve with a distribution in relaxation time of the form;

$$G^*(\omega, \tau_s) = \sum_j a_j \, G^*\left(\omega, \, 10^{\left[\log_0(\tau_s) + j\right]}\right)$$ (12)

At these high viscosities, a distribution in relaxation times is necessary to describe the observed shear relaxation. The distribution of relaxation times is described by an asymmetric peak with a long tail extending to short times and a sharp cut-off at long times (Fig. 9). This is similar to the relaxation time spectra observed by other authors for silicate melts at high viscosities (i.e. DeBast and Gilard 1963; Kurkjian, 1963; Mills, 1974). Each

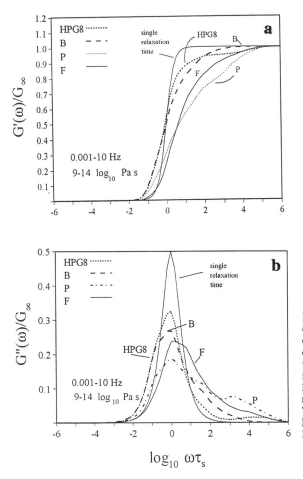

Figure 8. The frequency-dependent real (a) and imaginary (b) components of the shear modulus of haplogranite composition melts with added B_2O_3, P_2O_5 and F_2O_{-1} from 500° to 1000°C. The dotted line is calculated for a single relaxation time from Equation (6). The frequency is normalised to the Maxwell relaxation time (data from Bagdassarov et al., 1993).

study of the relaxation of shear deformation in high viscosity silicate melts employs a different equation to describe the relaxation time spectrum. It is therefore very difficult to compare the fine details of the relaxation spectra between studies. In all cases, however, the resulting spectrum can be described as "an asymmetric peak with a long tail extending to short times and a sharp cut-off at long times." This corresponds to the fundamental observation of non-exponential relaxation behavior in time domain studies. Further comparison of the resulting spectra shows that the long tail extending to short times extends ~6 to 7 orders of magnitude away from the calculated Maxwell relaxation time and the sharp cut-off at long times occurs 2 orders of magnitude away from the Maxwell relaxation time. This distribution in relaxation times contrasts with the single relaxation time behavior observed in silicate melts at lower viscosities using high frequency ultrasonic techniques. Despite the non-exponential nature of relaxation observed in these studies, such determination of the strain-rate dependent deformation of silicate melt at high viscosities and small strains ($<10^{-5}$) shows that the silicate melts display viscoelastic shear rheology with the shear relaxation time being well approximated by the Maxwell relationship.

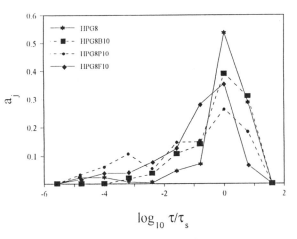

Figure 9. The relaxation spectrum for haplogranite composition melts with added B_2O_3, P_2O_5 and F_2O_{-1} calculated from Equation (12). Redrawn after Bagdassarov et al. (1993).

The viscoelastic behavior of the shear viscosity of a boron-bearing haplogranitic composition melt is presented in Figure 10a,b. At low frequencies of deformation the real part of the viscosity is frequency-independent and equal to the Newtonian shear viscosity determined by micropenetration viscometry on this sample. At high frequencies, the

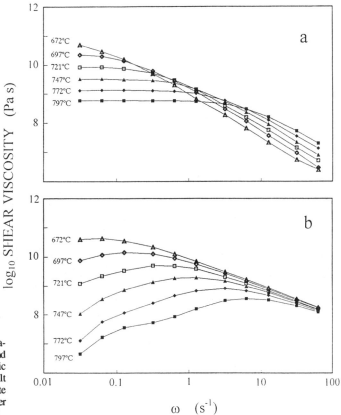

Figure 10. The temperature-dependent real (a) and imaginary (b) viscoelastic shear viscosities for a melt of B-bearing haplogranite composition (redrawn after Bagdassarov et al., 1993).

measured frequency-dependent shear viscosity is up to 5 $\log_{10}$ Pa s less than the Newtonian shear viscosity. As shown in Figure 8, changes in the melt composition result in small but significant changes in the frequency dependence of the stress/strain/strain-rate behavior. Such measurements, performed on rhyolitic melts with varying amounts of added water would answer the important question of whether water can significantly alter the description of the frequency dependence of the shear loss modulus.

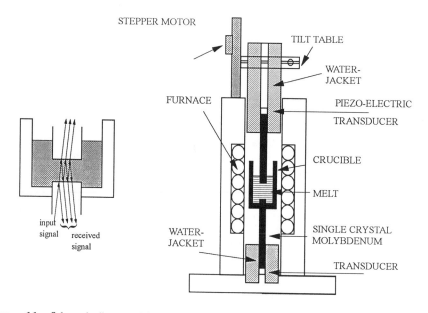

Figure 11. Schematic diagram of the ultrasonic apparatus for measuring frequency-dependent velocity and attenuation in melts (redrawn after Webb, 1991).

Ultrasonics

Ultrasonic measurements involve strains of 10^{-8} and stresses of several kPa and result in the observation of the frequency dependence of the velocity of a wave travelling through the melt. The ultrasonic set-up is shown in Figure 11. It consists of a crucible through the bottom of which passes a buffer rod. This rod is cooled at the free end; to which a piezo-electric transducer is glued. The upper buffer rod can be moved vertically, and it also has a transducer attached to the water-cooled end. A sinusoidal longitudinal or shear wave is created by the transducer. This signal travels through the buffer rod and is transmitted into the melt. The wave echoes within the melt, between the ends of the two buffer rods and is transmitted back to the sending transducer, and also to the receiving transducer on the other buffer rod. The amplitude of the signal reaching the transducer is measured as a function of the melt thickness as the upper buffer rod is moved vertically. The interference pattern that results is shown in Figure 12. Knowing the frequency of the propagating wave, the velocity of propagation can be determined from the interference pattern as the distance between each minima is 0.5 wavelength. The shear and longitudinal moduli are a function of the melt density ρ and the shear and compressional wave velocity v_s and v_p respectively;

$$G(\omega) = \rho\, v_s^2(\omega)$$

(13)

and

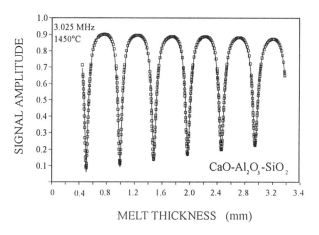

Figure 12. Ultrasonic interferometric data at a frequency of 3.025 MHz and temperature of 1450°C as a function of travel distance through the melt - between the two buffer rods for a CaO-Al$_2$O$_3$-SiO$_2$ melt (redrawn from Webb and Courtial, 1995).

$$M(\omega) = \rho\, v_p^2\, (\omega). \tag{14}$$

The viscoelastic rheology of silicate melts as a function of shear deformation has been determined by a number of authors. Macedo et al. (1968) and Simmons and Macedo (1970), and Webb (1991) have determined the frequency dependence of the shear modulus and the shear viscosity of a number of silicate melts in the frequency ranges 2 to 25 MHz and 5 to 150 MHz respectively; and Newtonian viscosity ranges 10^1 to 10^5 Pa s and 10^1 to 10^2 Pa s respectively. The shear viscosity determined by this ultrasonic technique is identical to that determined by conventional concentric cylinder techniques for a melt of the same composition (method described by Ryan and Blevins, 1987). The measured frequency-dependent shear modulus and viscosity relaxation can be described by Equations (6) and (10) for a single relaxation time process (see Figs. 13 and 14). The relaxation spectrum observed here is identical to that of a single relaxation time.

The onset on non-Newtonian rheology observed in Na$_2$Si$_2$O$_5$ melt (Webb, 1991) at ultrasonic frequencies at timescales two orders of magnitude slower than the Maxwell relaxation time is in agreement with observation made on the frequency-dependent rheology of silicate melts observed at high viscosities by forced oscillation techniques (Kurkjian, 1963; Mills, 1974; Webb, 1992b). The forced oscillation data of Mills (1974) on a melt of the same composition (Na$_2$Si$_2$O$_5$), however, observed at high viscosities, that the relaxation time distribution had a long tail extending to times faster than the Maxwell relaxation time. The single relaxation time data of Webb (1991) for Na$_2$Si$_2$O$_5$ melt at shear viscosities of $10^{1.3}$ to $10^{2.3}$ Pa s in combination with the relaxation time distribution data of Mills (1974) in the viscosity range 10^8 to 10^{15} Pa s, would indicate that the extent to which the viscoelastic regime extends to timescales faster than the calculated Maxwell relaxation time depends upon the viscosity of the melt. This suggests that for high viscosity melts, there exists a range of relaxation times which collapse to one relaxation time at lower viscosities. This is an important observation of general relevance. The explanation for such a collapse of the relaxation time distribution is not clear. This points to either a range of Si-O bonds or bond strengths or activation energies of Si-O bond exchange at low temperature/high viscosity conditions which are no longer distinguishable at low viscosities (high temperatures). It might also be related to the increasing detachment or decoupling of secondary relaxation processes from the main viscous relaxation with

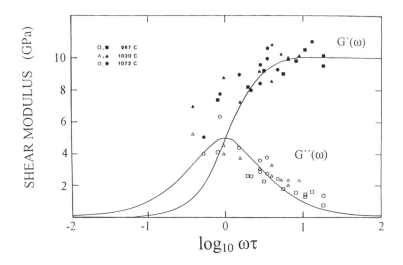

Figure 13. Frequency-dependent real and imaginary components of the shear modulus of $Na_2Si_2O_5$ melt in the viscosity range 10^1 to $10^{2.5}$ Pa s determined by ultrasonic techniques at frequencies of 5 to 150 MHz (redrawn after Webb, 1991).

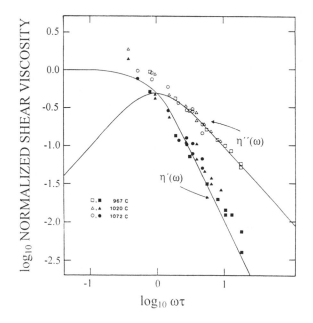

Figure 14. The frequency-dependent real and imaginary components of the shear viscosity $\eta_s(\omega) = \eta'_s(\omega) + \eta''_s(\omega)$ for $Na_2Si_2O_5$ melt as a function of normalised angular frequency in the viscosity range 10^1 to $10^{2.5}$ Pa s determined by ultrasonic techniques at frequencies of 5 to 150 MHz. The shear viscosity is also normalised to the Newtonian viscosity of the melt. The curve is calculated from Equation (10) (redrawn after Webb, 1991).

decreasing temperature (chapter by Dingwell). The onset on non-Newtonian rheology in geological processes as a function of strain-rate will be uninfluenced by the absolute value of shear viscosity. The effects of structural relaxation in cooling melts will, however, extend over large ranges of temperature below the glass transition temperature into the glassy region of melt behavior.

Shear viscosity

In general, shear viscosities of silicate melts are determined by concentric cylinder techniques at low viscosities in the range 10^0 to 10^5 Pa s. There are a range of techniques which allow the determination of very high viscosities in the range 10^8 to 10^{14} Pa s. The techniques which can be employed in this viscosity range have always been plagued by the problem of structural relaxation of the melt, which results in partially unrelaxed (i.e. lower) viscosities being determined. As shown above (and in the following) viscosity is a measure of the timescale on which the Si-O structure of a melt relaxes. Therefore the temperature and composition dependence of shear viscosity is an important indicator for changes in melt structure, and therefore in other physical properties of silicate melts. At low viscosity conditions it has been observed that the viscosity is not very sensitive to changes in composition, but at high viscosity conditions, the observed viscosity is very sensitive to small changes in composition (Hess et al., 1995). A wide range of geological processes occur at very high viscosities (e.g. crystal-melt fractionation in granitic composition melts at $\eta \geq 10^9$ Pa s at ~580°C) and magma degassing leading to explosive eruptive activity in calc-alkaline, dacitic and rhyolitic systems (Sparks et al., 1994). Therefore it is necessary to be able to determine relaxed viscosities greater than 10^8 Pa s; and to recognize the physical effects of the structural relaxation of the melt on the measured viscosity.

Micropenetration (Dingwell et al., 1992), parallel plate (Stevenson et al., 1995) and fibre elongation (Li and Uhlmann, 1970; Tauber and Arndt, 1987; Webb and Dingwell 1990a,b) are some of the time domain methods of measuring viscosities in the range 10^9 to 10^{12} Pa s. Parallel plate and torsion (viscosities in the range 10^9 to 10^{12} Pa s) techniques are also applicable for crystal and bubble bearing melts (Bagdassarov et al., 1994; Lejeune and Richet, 1995). Frequency-dependent torsion measurements of shear viscosity and modulus provide additional information about the timescale of various deformation mechanisms in crystal- and bubble-bearing melts, e.g. crystal and bubble growth, grain-boundary sliding.

VOLUME RHEOLOGY

Viscoelastic volume rheology can be determined at constant temperatures by ultrasonic measurements of the frequency dependence of the longitudinal and bulk moduli. Studies in which the longitudinal modulus was determined (Rivers and Carmichael, 1987; Tauke et al., 1968) indicate that the relaxation time for volume relaxation is identical to that of shear relaxation;

$$\tau_s = \frac{\eta_s}{G_\infty} = \frac{\eta_v}{K_0} = \tau_v \tag{15}$$

and therefore

$$\frac{\eta_s}{\eta_v} = \frac{G_\infty}{K_0} . \tag{16}$$

This means that if η_s and η_v are identical, the relaxational components of the shear and volume moduli are also identical. Rivers and Carmichael (1987) observed a factor of 5 increase between relaxed and unrelaxed longitudinal modulus for all of their silicate melt compositions. Assuming $25 < (K_\infty - K_0)$ GPa < 5, and $25 < G_\infty$ GPa < 5; this results in $K_0 = \sim 3\ G_\infty$. Webb (1991) observed $K_0 = 0.5\ G_\infty$ for a Na_2O-SiO_2 melt [$G_\infty = 10$ GPa and $K_0 = 5$ GPa]. It is expected that there is at most a factor of 5 difference possible between K_0 and G_∞. This would lead to a $\pm 0.7\ \log_{10}$ unit different in the shear and volume viscosity, if the relaxation times were identical. Ultrasonic studies in which both shear and longitudinal moduli of silicate melts were determined, allowed the calculation of the bulk modulus as a function of frequency, with the result that the relaxation times for shear and volume are found to be the same, within error of the measurements (Webb, 1991; Bornhöft and Brückner, 1994; Siewert and Rosenhauer, 1994; Webb and Knoche, 1995), at these temperatures. This type of measurement also gives information about volume viscosity. In contrast to shear viscosity which can occur for an infinite strain (without change in volume) volume viscosity must occur over a limited strain. Logically it is expected that anelastic (as opposed to viscoelastic) volume deformation occurs in silicate melts. The imaginary part of the volume modulus is then treated mathematically as a viscosity term.

Anelasticity

In the case of volumetric deformation, silicate melts (and their magmas; in the absence of crystals and vesicles) behave as anelastic materials. Only instantaneous and delayed recoverable deformation occur, with no time-dependent non-recoverable deformation being observed. Both the instantaneous and long timescale observation of melt deformation result in the determination of Hookean rheology (strain-rate independent rheology). However, if the deformation of the melt is observed on a timescale similar to that on which the delayed recoverable deformation takes place linear anelastic time-dependent (strain-rate dependent) rheology of the melt will be observed.

In the cases where volume relaxation time has been determined, the frequency-dependent volume (bulk) modulus behavior can be described by Equation (7) and the Maxwell relationship defines the relaxation time for volume deformation for the "volume viscosity" η_V and the relaxation component of the bulk modulus K_0 of the melt. The volume viscosity can be determined from the frequency-dependent volume modulus via Equation (11). It has been shown that for silicate melts in general the approximation $\eta_V \approx \eta_S$ holds (Macedo et al., 1968; Dingwell and Webb, 1990; see chapter by Dingwell).

The compressibility of silicate melts is a subject of great importance to the understanding of magma ascent and eruption within the Earth and terrestrial planets. Accordingly several experimental studies of the elastic properties of silicate melts have been carried out over the past decade (e.g. Rigden et al. 1984,1988; Manghnani et al. 1986; Agee and Walker, 1988; Miller et al. 1991; Webb and Dingwell, 1994; Webb and Courtial, 1995). The relaxed bulk modulus of silicate melts has been studied for a wide variety of silicate melt compositions and calculational models have been derived for the estimation of multi-component melt compressibilities (Rivers and Carmichael, 1987; Kress et al., 1988; Kress and Carmichael, 1991; Webb and Courtial, 1995).

Most ultrasonic studies of silicate melts are conducted at the high temperature ($900°$ to $1500°C$) conditions and low frequencies (3 to 22) MHz required to observe the relaxed (frequency-independent) longitudinal (= volume) modulus of the melt. In the cases where the experimental conditions approached the relaxation frequency of the melt (e.g. Manghnani et al., 1981; Sato and Manghnani 1985; Rivers and Carmichael, 1987) frequency-dependence of the longitudinal modulus has been observed. In studies where the frequency of the applied signal has been greater than that of the relaxation frequency, shear wave propagation has been observed (Macedo et al., 1968 (Na_2O-B_2O_3-SiO_2 melt); Webb, 1991 ($Na_2Si_2O_5$ melt)).

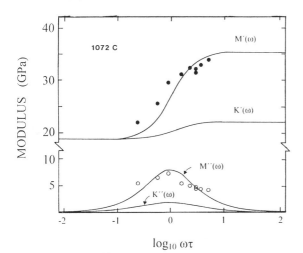

Figure 15. Frequency-dependent real and imaginary components of the longitudinal and volume moduli of $Na_2Si_2O_5$ melt at 1072°C determined by ultrasonic techniques at frequencies of 5 to 150 MHz. The curves are calculated from Equations (7) and (9) for a single relaxation time for both volume and longitudinal deformation (redrawn after Webb, 1991).

Figure 16. The frequency-dependent volume: $\eta_v(\omega) = \eta_v'(\omega) + i\eta_v''(\omega)$ and longitudinal viscosity:

$$\eta_l^*(\omega) = \eta_l'(\omega) + i\eta_l''(\omega)$$

for an $Na_2Si_2O_5$ melt at 1072°C as a function of angular frequency. Curves are calculated from Equations (11) and (9) for a single relaxation time for both volume and longitudinal deformation (redrawn after Webb, 1991).

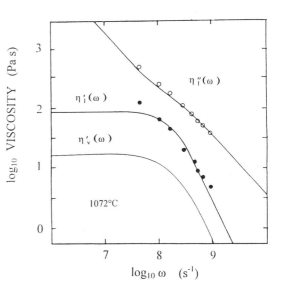

In the study of Webb (1991) the viscosity of the melt (10^1 to 10^3 Pa s) and the frequency of the ultrasonic signal (5 to 150 MHz) were chosen such that both the relaxed and unrelaxed longitudinal (M) and shear (G) moduli and viscosities of $Na_2Si_2O_5$ could be measured with the volume modulus and viscosity being determined from Equation (11). The frequency-dependent volume modulus and viscosity (see Figs. 15 and 16) of $Na_2Si_2O_5$ melt could then be determined from the measured shear and longitudinal moduli and viscosities. The measured longitudinal viscosity and calculated volume viscosity of $Na_2Si_2O_5$ melt are illustrated in Figure 16. As no data was obtained at viscosities greater than $10^{2.3}$ Pa s for this composition melt the volume relaxation time could not be resolved unambiguously. The relaxed volume modulus is 19 GPa with the unrelaxed modulus being 22 GPa. No distribution in relaxation time could be resolved from the frequency-dependent modulus and viscosity data, with all indications being that the data could be fit with a single relaxation time curve for the shear and longitudinal data and the calculated volume data.

RELAXED COMPRESSIBILITY OF SILICATE MELTS

Elasticity systematics

The general approach to calculating the elastic properties (relaxed bulk modulus) of a silicate melt is borrowed from the density literature in which a linear summation of the partial molar volumes of the component oxides with/without excess terms is used to calculate the molar volume of a melt. There are a number of possible equations to calculate the partial molar compressibility of the oxide components of a melt (Rivers and Carmichael, 1987; Kress et al., 1988; Kress and Carmichael, 1991; Webb and Dingwell, 1994). All of these are based upon the assumption that, to a first approximation, the individual oxides add together in the melt as isolated structural units. As Equation (17) is successful in calculating the partial molar volume of the oxide components of silicate melts, it is expected that Equation (19) should be used to calculate the partial molar compressibility of the oxide component of silicate melts;

$$V = \sum_{i=1}^{n} X_i \overline{V}_i + \sum_{i,j=1}^{n} X_i X_j \overline{V}_{ij} \tag{17}$$

for the partial molar fractions X_i and the partial molar volumes $\overline{V}_{ij}$. Given that:

$$K^{-1} = \beta = -\frac{1}{V} \cdot \frac{\partial V}{\partial P} . \tag{18}$$

it follows that

$$K^{-1} = -\frac{1}{V} \cdot \left[\sum_{i=1}^{n} X_i \frac{\partial \overline{V}_i}{\partial P} + \sum_{i,j=1}^{n} X_i X_j \frac{\partial \overline{V}_{ij}}{\partial P} \right] . \tag{19}$$

This equation results in the calculation of negative compressibility of some oxide components in some melts (Kress et al., 1988; Kress and Carmichael, 1991; Webb and Courtial, 1995). At present Equation (19) appears to be successful in the interpolation of compressibility data within an already investigated composition range, but it does not produce physically meaningful data, and therefore should not be used to predict the compressibility of melts outside the composition range of investigation (Kress et al., 1988).

Iso-structural melts

The bulk modulus of a melt can be determined by ultrasonic techniques at frequencies lower than the frequency of volume relaxation of the melt. In this way the composition dependent structure of melts can be investigated as a function of anomalous increase or decrease in relaxed bulk modulus. An example is the bulk modulus of titanosilicate melts. In order to investigate the structural changes occurring in these melts as a function of composition the bulk modulus has been plotted in Figure 17 as a function of the molar volume of the melt. A semi-empirical relationship between the bulk modulus, K, and the molar volume, V for iso-structural crystalline materials has been developed from the classical ionic model;

$$K \propto V^{-4/3} \tag{20}$$

(Bridgman, 1923; Anderson and Nafe, 1965). This relationship has been found to apply in general to iso-structural oxides and silicates. Rivers and Carmichael (1987) and

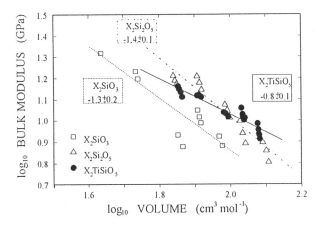

Figure 17. $KV^{4/3}$ relationship for X_2SiO_3, $X_2Si_2O_5$ and X_2TiSiO_5 melts (X = Li, Na, K, Rb, Cs) (redrawn after Webb and Dingwell, 1994).

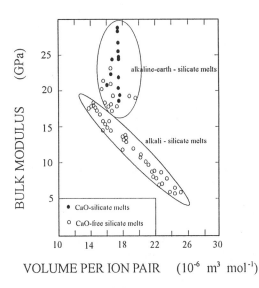

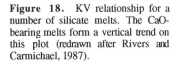

Figure 18. KV relationship for a number of silicate melts. The CaO-bearing melts form a vertical trend on this plot (redrawn after Rivers and Carmichael, 1987).

Herzberg (1987) have applied this relationship in discussions of the compressibility of silicate melts. Both of these studies pointed out that this relationship is applicable to simple binary alkali-silicate melts but not to alkaline-earth compositions. The compressibility of CaO-bearing melts (Webb and Courtial, 1995) and alkaline-earth bearing melts in general (Rivers and Carmichael, 1987) is anomalously low, with respect to other silicate melt compositions as shown in Figure 18. This indicates that the structure of alkaline-earth bearing melts should be different to alkali-bearing silicate melts.

Figure 17 is a log/log plot of the relaxed bulk modulus versus molar volume for alkali-metasilicate (950° to 1400°C) alkali-disilicate (1100° to 1500°C) and alkali-titano-silicate melts (950° to 1600°C). The slope of the straight lines fitted to the alkali-metasilicate (-1.3 ± 0.2) and alkali-disilicate (-1.4 ± 0.1) melt compositions is $-4/3$, implying that the melts in each of these series are iso-structural. The slope of the straight line fitted to the alkali-titanosilicate data is -0.8±0.1, implying that the structure of the melts changes as a function of composition. The structural parameter causing the shift in melt modulus/volume systematics is likely to be the shift in average co-ordination number of Ti inferred from X-ray absorption studies of glasses quenched from these melts (Dingwell et al., 1994). For melts with the same structure (co-ordination) a plot of $\log_{10}$ K against $\log_{10}$ molar volume will fall on a line of slope -4/3. In the case of the alkali titanosilicate melts the data fall on a line of lower slope, indicating that these melts are not iso-structural. The low slope of the line points to there being an increasing amount of the compressible octahedrally co-ordinated titanium in the melts with the smaller alkali cations, with respect to the less compressible tetrahedrally co-ordinated titanium.

OUTLOOK

The investigation of viscoelasticity of silicate melts represents a focal point for several applications of fundamental importance to the understanding of deformation and flow in silicate melts and for the determination of relaxed values of both the viscous and elastic properties of silicate melts as functions of temperature, pressure and composition. Several outstanding problems remain in a full description of the viscoelasticity of silicate melts.

The transition from exponential shear relaxation behavior inferred from ultrasonic data at low viscosities to demonstrably non-exponential shear relaxation at high viscosities is a feature of silicate melts whose explanation may hold the key to a more fundamental understanding of relaxation processes of materials in general, the current goal of much effort in physics and physical chemistry.

The connection between the microscopic relaxation of melt species and the macroscopic property relaxation has yielded one of the most powerful structure-property relations in silicate melts. It has led to a number of conceptual revisions of melt structure. Yet we must refine this relationship to the point where the potential existence of larger flow units, in a small timescale window between Si-O exchange frequency information from NMR studies and viscous relaxation times from viscoelasticity studies is either fully ruled out or reproducibly quantifiable. Only then will we have the definitive statement on the polymer nature of silicate melts.

The relaxed moduli of silicate melts determined by ultrasonic methods at 1 atm pressure represent an incomplete description of the composition dependence of the elastic moduli as emphasized in recent evaluations of the role of Al_2O_3 and TiO_2 in simple and multi-component melts. Clearly further compositionally defined experiments are needed.

The pressure dependence of the bulk modulus of silicate melts, the relative magnitudes and contributions of K and K' to the pressure-dependence are only known from a few preliminary experiments in a reconnaissance fashion. K-K' systematics are not available.

We must also evaluate whether a connection can be made between the fragility of the temperature-dependence of melt viscosity and the nature of the loss modulus as recorded by viscoelastic measurements. In particular, the proposal that increasing pressure may make strong liquid structures weak requires experimental evaluation by viscometric and viscoelastic methods.

For all of these reasons, and because the key to understanding the dynamics of silicate melts lies in the complete description of their relaxation behavior, i.e. their viscoelasticity, the collection of high quality frequency domain data on the viscous relaxation of silicate melts is an essential component of progress in understanding viscous deformation in natural systems.

REFERENCES

Agee CB, Walker D (1988) Static compression and olivine flotation in ultrabasic silicate liquids. J Geophys Res 93:3437-3449

Anderson OL, Nafe JE (1965) The bulk modulus-volume relationship for oxide compounds and related geophysical problems. J Geophys Res 70:3951-3963

Askarpour V, Manghnani MH, Richet P (1993) Elastic properties of diopside, anorthite, and grossular glasses and liquids: a Brillouin scattering study up to 1400K. J Geophys Res 98:17683-17689

Bagdassarov NS, Dingwell DB, Webb SL (1993) Effect of boron phosphorus and fluorine on shear stress relaxation in haplogranitic melts. Eur J Mineral 5:409-425

Bagdassarov NS, Dingwell DB, Webb SL (1994) Viscoelasticity of crystal- and bubble-bearing rhyolite melts. Phys Earth Planet Int 83:83-99

Bockris JO´M, Kojonen E (1960) Compressibilities of certain molten alkali silicate and borates. J Am Chem Soc 82:4493-4497

Bornhöft H, Brückner R (1994) Ultrasonic measurements and complex elastic moduli of silicate glass melts in the viscoelastic and viscous range. Glastech Ber 67:241-254

Bridgman PW (1923) The compressibility of thirty metals as a function of pressure and temperature. Proc Am Acad Arts Sci 58:165-242

Cooper RF (1990) Differential stress-induced melt migration: an experimental approach. J Geophys Res 95:6979-6992

DeBast J, Gilard P (1963) Phys Chem Glasses 4:117

Dingwell DB (1990) Effects of structural relaxation on cationic tracer diffusion in silicate melts. Chem Geol 182:209-216

Dingwell DB (1992) Shear viscosity of alkali and alkaline earth titanium silicate liquids. Am Mineral 77:270-274

Dingwell DB (1995) Viscosity and Anelasticity of melts and glasses. Mineral Physics and Crystallography A Handbook of Physical Constants AGU Reference Shelf 2. T. Ahrens (ed) American Geophysical Union, 209-217.

Dingwell DB, Webb SL (1989) Structural relaxation in silicate melts and non-Newtonian melt rheology in igneous processes. Phys Chem Minerals 116:508-516.

Dingwell DB, Webb SL (1989) Structural relaxation in silicate melts and non-Newtonian melt rheology in geologic processes. Phys Chem Minerals 16:508-516

Dingwell DB, Webb SL (1990) Structural relaxation in silicate melts. Eur J Mineral 2:427-449

Dingwell DB, Knoche R, Webb SL, Pichavant M (1992) The effect of B_2O_3 on the viscosity of haplogranitic liquids. Am Mineral 77:457-461

Dingwell D, Bagdassarov N, Bussod G, Webb SL (1993) "Magma Rheology" Mineralogical Association of Canada Short Course on Experiments at High Pressure and Applications to the Earth´s Mantle. 131-196.

Dingwell DB, Paris E, Seifert F, Mottana A, Romano C (1994) X-ray absorption study of Ti-bearing silicate glasses. Phys Chem Minerals 21:501-509

Farnan I, Stebbins, JF (1994) The nature of the glass transition in a silica-rich oxide melt. Science 265:1206-1209

Herzberg CT (1987) Magma density at high pressure Part 1: The effect of composition on the elastic properties of silicate liquids In Magmatic Processes: Physicochemical Principles. BO Mysen (ed) The Geochemical Society Spec Publ 1:25-46

Herzfeld KF, Litovitz TA (1959) Absorption and Dispersion of Ultrasonic Waves. Academic Press, New York, 535 p

Hess K-U, Dingwell DB, Webb SL (1995) The influence of excess alkalies on the viscosity of a haplogranitic melt. Am Mineral 80:297-304

Kress VC, Carmichael ISE (1991) The compressibility of silicate liquids containing Fe_2O_3 and the effect of composition temperature oxygen fugacity and pressure on their redox states. Contrib Mineral Petrol 108:82-92

Kress VC, Williams Q, Carmichael ISE (1988) Ultrasonic investigations of melts in the system Na_2O-Al_2O_3-SiO_2. Geochim Cosmochim Acta 52:283-293

Krol DM, Lyons KB, Brawer SA, Kurkjian CR (1986) High temperature light scattering and the glass transition in vitreous silica. Phys Rev B 33:4196-4202

Kurkjian CR (1963) Relaxation of torsional stress in the transformation range of a soda-lime-silica glass. Phys Chem Glasses 4:128-136

Laberge NL, Vasilescu VV, Montrose CJ, Macedo PB (1973) Equilibrium compressibilities and density fluctuations in K_2O-SiO_2 glasses. J Am Ceram Soc 56:506-509

Lejeune A-M, Richet P (1995) Rheology of crystal-bearing silicate melts: an experimental study at high viscosities. J Geophys Res (in Press)

Li JH, Uhlmann DR (1970) The flow of glass at high stress levels I Non-Newtonian behavior of homogeneous $0.08Rb_2O$ $0.92SiO_2$ glasses. J Non-Cryst Solids 3:127-147

Liu S-B, Stebbins JF, Schneider E, Pines A (1988) Diffusive motion in alkali-silicate melts: an NMR study at high temperature. Geochim Cosmochim Acta 52:527-538

Macedo PB, Simmons JH, Haller W (1968) Spectrum of relaxation times and fluctuation theory: ultrasonic studies on an alkali-borosilicate melt. Phys Chem Glasses 9:156-164

Manghnani MH, Rai CS, Katahara KW, Olhoeft GR (1981) Ultrasonic velocity and attenuation in basalt melt. In Anelasticity in the Earth. Geodynamics Series AGU 4:118-122

Manghnani M, Sato H, Rai CS (1986) Ultrasonic velocity and attenuation measurements on basalt melts to 1500°C: role of composition and structure in the viscoelastic properties. J Geophys Res 91:9333-9342

Maxwell JC (1867) On the dynamical theory of gases. Phil Trans Roy Soc London Ser A 157:49-88

Mazurin OV (1986) Glass relaxation. J Non-Cryst Solids 87:392-407

Miller GH, Stolper EM, Ahrens TJ (1991) The equation of state of a molten komatiite 1. shock wave compression to 36 GPa. J Geophys Res 96:11831-11848

Mills JJ (1974) Low frequency storage and loss moduli of soda silica glasses in the transformation range. J Non-Cryst Solids 14:255-268

Narayanaswamy OS (1971) A model of structural relaxation in glass. J Am Ceram Soc 54:491-498

Narayanaswamy OS (1988) Thermorheological simplicity in the glass transition. J Am Ceram Soc 71:900-904

Nowick AS, Berry BS (1972) Anelastic relaxation in crystalline solids. Academic Press, New York

Perez J, Duperray B, Lefevre D (1981) Viscoelastic behavior of an oxide glass near the glass transition temperature. J Non-Cryst Solids 44:113-136

Rigden SM, Ahrens TJ, Stolper EM (1984) Densities of liquid silicates at high pressures. Science 226:1071-1074

Rigden SM, Ahrens TJ, Stolper EM (1988) Shock compression of molten silicate: results for a model basaltic composition. J Geophys Res 93:367-382

Rivers ML, Carmichael ISE (1987) Ultrasonic studies of silicate melts. J Geophys Res 92:9247-9270

Ryan MP, Blevins JYK (1987) The viscosity of synthetic and natural silicate melts at high temperatures and 1 bar (10^5 Pa) pressure and at higher pressures. US Geol Surv Bull 17864, 563 p

Sato H, Manghnani MH (1985) Ultrasonic measurements of v_p and Q_p: relaxation spectrum of complex modulus on basalt melts. Phys Earth Planet Int 41:18-33

Secco RA, Manghnani MH, Liu T-C (1991) The bulk modulus-attenuation-viscosity systematics of diopside-anorthite melts. Geophys Res Lett 18:93-96

Siewert R, Rosenhauer M (1994) Light scattering in jadeite melt: strain relaxation measurements by photon correlation spectroscopy. Phys Chem Minerals 21:18-23

Simmons JH, Macedo (1970) Viscous relaxation above the liquid-liquid phase transition in some oxide mixtures. J Chem Phys 53:2914-2922

Simmons JH, Mohr RK, Montrose CJ (1982) Non-Newtonian viscous flow in glass. J Appl Phys 53:4075-4080

Sparks RSJ, Barclay J, Jaupart C, Mader HM, Phillips JC (1994) Physical aspects of magmatic degassing I. Experimental and theoretical constraints on vesiculation. Rev Mineral 30:413-445

Spera FJ, Borgia F, Strimple J, Feigenson M (1988) Rheology of melts and magmatic suspensions 1. design and calibration of concentric cylinder viscometer with application to rhyolitic lava. J Geophys Res 93:10273-10294

Stevenson RJ, Dingwell DB, Webb SL, Bagdassarov NS (1995) The equivalence of enthalpy and shear stress relaxation in rhyolitic obsidians and quantification of the liquid-glass transition in volcanic processes. J Volcanol Geotherm Res (in press)

Tauber P, Arndt J (1987) The relationship between viscosity and temperature in the system anorthite-diopside. Chem Geol 62:71-82

Tauke J, Litovitz TA, Macedo PB (1968) Viscous relaxation and non-Arrhenius behavior in B_2O_3. J Am Ceram Soc 51:158-163

Webb SL (1991) Shear and volume relaxation in $Na_2Si_2O_5$. Am Mineral 76:1449-1454

Webb SL (1992a) Shear volume enthalpy and structural relaxation in silicate melts. Chem Geol 96:449-458

Webb SL (1992b) Low-frequency shear and structural relaxation in rhyolite melt. Phys Chem Minerals 19:240-245

Webb SL, Dingwell DB (1990a) Non-Newtonian rheology of igneous melts at high stresses and strain rates: Experimental results for rhyolite, andesite, basalt and nephelinite. Scarfe Volume, J Geophys Res 95:15695-15701

Webb SL, Dingwell DB (1990b) The onset of non-Newtonian rheology of silicate melts: A fiber elongation study. Phys Chem Minerals 17:125-132

Webb SL, Dingwell DB (1994) Compressibility of titanosilicate melts. Contrib Mineral Petrol 118:157-168

Webb SL, Courtial P (1995) Compressibility of melts in the CaO-Al$_2$O$_3$-SiO$_2$ system. Geochim Cosmochim Acta (in press)

Webb SL, Knoche R (1995) The glass transition, structural relaxation and shear viscosity of silicate melts. Chem Geol (submitted)

Zarzycki J (1991) Glasses and the Vitreous State. Cambridge University Press, Cambridge, 505 p

Zdaniewski WA, Rindone GE, Day DE (1979) Review: The internal friction of glasses. J Mat Sci 14:763-775

APPENDIX

Each silicate melt has a equilibration time for the structure in response to a mechanical, electrical, thermal, etc. perturbation. By analogy to the mechanical relaxation time τ_m, the electrical τ_e and thermal τ_t equilibration times can be calculated.

Mechanical relaxation time

$$\tau_m = \frac{\eta_0}{M_\infty} = \frac{\sigma\,(\text{stress}) \,/\, \dot{\varepsilon}\,(\text{strain}-\text{rate})}{\sigma\,(\text{stress}) \,/\, \varepsilon\,(\text{strain})} \quad . \tag{A1}$$

Electrical relaxation time

$$\tau_e = \frac{e_0\,(\text{permittivity of free}-\text{space})\,\varepsilon_\infty\,(\text{dielectric cons}\tan t)}{\sigma_0\,(\text{conductivity})} \tag{A2}$$

where the electric modulus $M_\infty = \varepsilon_\infty^{-1}$; and

$$\sigma_0\,(\text{conductivi}\,t\,y) = \frac{J\,(\text{current density})}{E\,(\text{electric field})} = \frac{\dfrac{\Delta Q(\text{ch}\arg\text{e})}{\Delta t(\text{time})}\Big/\text{area}}{\text{voltage}\,\big/\text{length}}$$

and

$$e_0\varepsilon_\infty\,(\text{permittivit}\,y) = \frac{D\,(\text{ch}\arg\text{e density})}{E\,(\text{electric field})} = \frac{\text{ch}\arg\text{e}\big/\text{area}}{\text{voltage}\,\big/\text{length}}$$

Thermal relaxation time

$$\tau_t = \frac{Z}{c_p^{-1}} = \frac{T\,(\text{temperature})\big/\dot{q}\,(\text{rate of heat}-\text{flow})}{T\,(\text{temperature})\big/q\,(\text{heat}-\text{flow})} \tag{A3}$$

for impedence Z, and heat modulus = $(c_p)^{-1}$.

Chapter 5

ENERGETICS OF SILICATE MELTS

Alexandra Navrotsky

Department of Geological and Geophysical Sciences
Princeton Materials Institute
Princeton University
Princeton, New Jersey 08544 U.S.A.

WHY STUDY ENERGETICS?

The physical and chemical properties of silicate melts reflect their structure and speciation dictated by molecular scale interactions. The P-T-X regions of stability of the melt (phase equilibria) result from competition between energetic (ΔE or ΔH), entropic ($T\Delta S$), and volumetric ($P\Delta V$) terms for crystalline and molten phases, again related to bonding and interactions on the atomic scale. Thus thermochemical data are useful in two different contexts. First, thermodynamic data enable one to systematize, predict, store and retrieve information about melting and crystallization sequences, compositions of coexisting phases, major, minor, and trace element distributions, and other petrologic information. A knowledge of the equilibrium states of an evolving magmatic system is essential to determine when deviations from equilibrium occur, and this knowledge is the starting point for kinetic studies. Second, though thermodynamics is by its nature a macroscopic description, the magnitudes of the various parameters (heats, entropies and volumes of fusion, of mixing, etc.) give considerable insight into structure and speciation on the microscopic scale, especially when one compares these parameters in a systematic way for different compositions, pressures, or temperatures. Furthermore, changes in thermodynamic parameters near the glass transition provide insights into the differences between glass and melt and into the restructuring that occurs within and above the glass transition interval as temperature is varied.

This review summarizes current knowledge and active research areas in the thermodynamic properties of silicate melts, with an emphasis on calorimetric measurements made directly on molten samples. Although quenched glasses are good models for melts in many respects, and much of the earlier thermochemical studies were done on the glassy state, there are significant differences between glass and melt. Properties, such as heat capacities and heats of mixing, can depend on temperature in a more pronounced and complicated manner for melts than for glasses. These differences relate to changes in speciation in both a static and dynamic sense, and the simultaneous application of calorimetry and of high temperature spectroscopic techniques (see other chapters in this volume) offer complementary evidence to be integrated in ongoing efforts to understand the molten state.

METHODS OF STUDYING ENERGETICS

Calorimetry, the measurement of heat effects associated with heating or cooling a sample or with a chemical reaction, forms the most direct means for studying energetics. For silicate melts, calorimetric measurements can be grouped into two categories: on the one hand, measurement of heat capacity, heat content, heat of fusion, and, on the other, measurement of enthalpy associated with a chemical reaction, such as the mixing of components in a melt or the dissolution of a crystal or glass in an appropriate solvent.

For measurement of heat capacities above room temperature, several methods are available. In the range 300 to 1000 K, a number of commercial differential scanning calorimeters (DSC) can be used, with careful calibration, to obtain C_p with an accuracy of about ±1%. This upper temperature limit is set, not so much by technical limitations, as by the marketplace; the major use of DSC is in polymer science, and higher temperatures are not required. Because many silicate glasses have glass transition temperatures, T_g, of 950 to 1100 K the upper limit of most commercial DSC instruments brings one tantalizingly close to the molten state. Conventional DSC has been used to obtain heat capacities of glasses in a number of systems of petrologic importance, see, for example, Stebbins et al. (1984). Two commercial DSC instruments extend the temperature range of accurate C_p measurements: the Setaram DSC 111 series instruments to about 1100 K and the Netzsch 404 to about 1673 K. Our experience with both instruments has been quite satisfactory, and accuracy in C_p of ±2% at 800 to 1100 K and ±3 to 4% at 1300 to 1673 K appear attainable, provided that extreme care is taken in reproducing sample and crucible geometry and placement and in appropriate calibration. Our experience in testing a number of other high temperature units, which are basically differential thermal analysis (DTA) instruments (sometimes claimed by their manufacturers to have quantitative capability), has been disappointing, with heat capacities obtained to no better than ±20%. Nevertheless, such DTA instruments, some of which operate to well over 2000 K, are very useful for the qualitative detection of melting and/or phase transitions. Both DTA and DSC instruments typically use samples in the 10 to 100 mg range.

Heat capacities at high temperature are also obtained by differentiating the heat content ($H_T - H_{298}$) curve obtained in either of two ways. The first is by conventional drop calorimetry; a molten sample at high temperature is dropped into a calorimeter at room temperature, and the heat released is measured. Because the final state is a quenched sample, one must be concerned about the reproducibility of this state, both with respect to fictive temperature, if the sample is a glass, and with respect to phase assemblage and degree of order, if the sample quenches to partially or totally crystalline products. If these complications are well controlled, and fairly massive samples (1 to 10 g) are used, accuracy on the order of ±0.5 to ±1% is attainable, but the measurements are slow and painstaking. The calorimeters used for this purpose appear to be mainly home-built, and several groups have been obtaining high quality data on heat capacity of silicate melts and heats of fusion, see for example; Adamkovicova et al. 1980; Stebbins et al. 1984; Richet and Bottinga, 1985; Richet et al. 1990.

The second method is transposed temperature drop calorimetry, in which a sample is dropped from room temperature into a hot calorimeter and the heat absorbed measured. Varying the temperature of the calorimeter allows one to measure the heat content and heat of fusion. We have had considerable success using Setaram HT 1500 calorimeters for this purpose (Ziegler and Navrotsky, 1986; Navrotsky et al. 1989; Tarina et al. 1994). With some modification of sample crucibles, software, computer control and operating procedure, we have been able to use the Setaram HT 1500 calorimeter as a quantitative scanning calorimeter, especially in step-scanning mode, where the change in enthalpy for every 5 K or 10 K step delineates the heat capacity (Lange et al. 1991; Lange and Navrotsky, 1992; Lange and Navrotsky, 1993; Lange et al. 1995). The accuracy attainable on C_p is about ±3% at 1400 K and ±8% at 1750 K. It is fair to caution, however, that any such heat capacity measurements at high temperature must be done very carefully, with proper attention to calibration and reproducibility, and to the chemical problems intrinsic with dealing with corrosive silicate melts. It is overoptimistic to think that one can buy a commercial calorimeter, have it installed, undergo a brief training period, and begin to immediately produce reliable data. High temperature calorimetry remains as much an art as a science.

Up to the late 1980's, heats of mixing in silicate melts (except for some data for low melting alkali silicates) were generally obtained by obtaining solution calorimetric data for silicate glasses either in aqueous hydrofluoric acid near room temperature (Hovis, 1984) or in molten lead borate near 973 K (Navrotsky et al. 1980; Capobianco and Navrotsky, 1982; Navrotsky et al. 1982; Navrotsky et al. 1983; Hervig and Navrotsky, 1984; Roy and Navrotsky, 1984; Hervig and Navrotsky, 1985; Hervig et al. 1985; Wright and Navrotsky, 1985; Geisinger et al. 1987; Geisinger et al. 1988; Navrotsky et al. 1990; DeYoreo et al. 1990; Maniar et al. 1990). For oxide melt solution calorimetry, home built calorimeters of the Tian Calvet type have been used (Navrotsky, 1977; Navrotsky, 1990). Commercial Calvet-type calorimeters have also been used for drop solution calorimetric studies (Kiseleva and Ogorodova, 1984; Akaogi and Ito, 1993), although their sample chambers are too small for introducing samples and stirring devices for solution calorimetry.

The development of calorimeters and techniques for drop-solution calorimetry to 1773 K, using the Setaram HT 1500 calorimeter and a home built "hybrid" calorimeter in our laboratory (Topor and Navrotsky, 1992) has opened new possibilities for direct studies of enthalpies of mixing in melts. So far, two types of experiments have been tried. The first consists of dropping into the calorimeter at high temperature (1773 K) a mechanical mixture of crystalline phases which melt to form a homogeneous liquid. By varying the composition of the melt produced, both enthalpies of fusion and enthalpies of mixing in the molten state are obtained. This method has been applied to the systems diopside - anorthite - albite and anorthite - forsterite (Navrotsky et al. 1989) and diopside - anorthite - wollastonite (Tarina et al. 1994). The second method involves dropping small pellets (10 to 30 mg) of a crystalline oxide into a crucible containing several grams of a silicate melt of relatively low viscosity in the calorimeter at high temperature. In this way, the enthalpy of solution of the oxide is measured as a function of melt composition and the concentration of the dissolving oxide. So far, this technique has been applied only to rutile (TiO_2) in a potassium and a calcium aluminosilicate melt (Gan and Navrotsky, 1994). The method shows great promise at mapping out a good approximation to the partial molar enthalpy of mixing of oxides in silicate melts, and of correlating the variation in Δh with composition, and with changes in structure and speciation inferred from structural and volumetric studies. The accuracy of both these methods in delineating mixing parameters appears to be in the $\pm 10\%$ range. Because the measurements are made directly at high temperature, with no need for other corrections or extrapolations, the data are unique and useful.

Thus the emphasis in calorimetry has shifted from the study of glasses to direct *in situ* measurements in the molten state by both calorimetric and spectroscopic methods. Such calorimetric measurements can probe temperature-dependent changes in melt structure and speciation which are not retained, or only partially retained, when the melt is cooled rapidly and a snapshot of the structure at T_g is "frozen in". As the following sections will show, the glass is an imperfect energetic model for the melt, especially in cases where considerable restructuring occurs above T_g, i.e. the liquid tends toward "fragility" (Angell 1985, 1988).

Other less direct methods can be used to infer melt energetics. The temperature dependence of free energy measurement (vapor pressure, solid cell e.m.f., phase equilibria) can be used to infer enthalpies of reaction. Similarly, the temperature dependence of speciation equilibrium constants, determined, for example, by ^{29}Si NMR, provide enthalpies for the given equilibrium. The accuracy of such estimates is hard to judge, both because they often are somewhat model-dependent, and because of the limited temperature and/or composition range over which such data usually exist.

FACTORS AFFECTING MELT ENERGETICS

Major acid-base interactions and polymerization equilibria

Figure 1 shows the enthalpy and free energy of mixing (formation from molten oxides) of several metal oxide-silica systems in the molten state. Figure 2 shows inferred speciation in alkali silicate glasses from NMR data (Maekawa et al., 1991). It is clear that the large exothermic heats of mixing (about 60 to 80% as large in magnitude as the heats of formation of the corresponding crystalline compounds) reflect the acid-base reaction between silica (or other acidic oxide) and alkali or alkaline earth (basic) oxide. For silicates, the enthalpy and free energy of formation are most exothermic near the orthosilicate composition, where the dominant change in speciation upon mixing can be written as

$$SiO_2 + 2\,M_xO = SiO_4^{4-} + 2x\,M^{2+/x} \tag{1}$$

or

$$Q_4 + 2\,O^{2-} = Q_0 \tag{2}$$

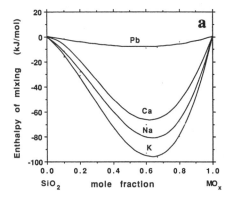

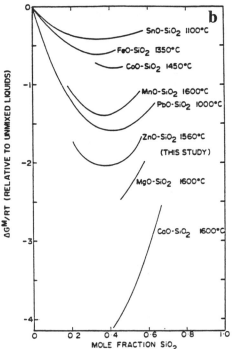

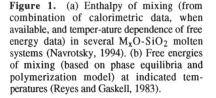

Figure 1. (a) Enthalpy of mixing (from combination of calorimetric data, when available, and temper-ature dependence of free energy data) in several M_xO-SiO_2 molten systems (Navrotsky, 1994). (b) Free energies of mixing (based on phase equilibria and polymerization model) at indicated tem-peratures (Reyes and Gaskell, 1983).

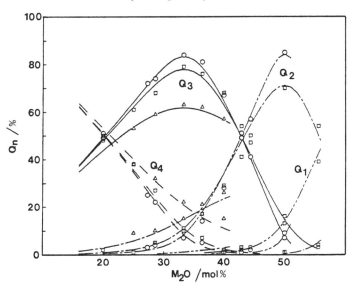

Figure 2. Speciation as a function of composition in alkali silicate glasses. Triangle = Li, square = sodium, circle = potassium, dashed curve = Q_4, solid curve = Q_3, dot-long dash = Q_2, double-dot dash = Q_1, dot-short dash = Q_0. Curves are calculated from equilibrium constants, points are NMR data. From Maekawa et al. (1991), used with permission.

This reaction becomes more exothermic as the difference in acid-base character between the oxides increases, that is, as the ionic potential (z/r) of the cation of the network modifying oxide decreases for a given network former oxide, or as the ionic potential of the network forming cation increases for a given network modifier. These interactions are of the order of -30 to -100 kJ/mol; based on heats of mixing in the molten state (see Fig. 1). Energetics of other speciation reactions also are associated with substantial free energy (and presumably also enthalpy) effects (see Table 1). Thus a small change in speciation can have a significant energetic effect.

The application of polymer theory to silicate melts (Masson, 1965; Masson et al., 1970; Gaskell, 1973, 1977; Hess, 1977, 1980; see also Hess, this volume) has led to quantification of equilibrium constants for reactions of the type

$$2 \, SiO_4^{4-} = Si_2O_7^{6-} + O^{2-} \tag{3}$$

or

$$2 \, Q_0 = 2 \, Q_1 + O^{2-} \tag{4}$$

In simple polymerization models, such equilibrium constants are representative of the growth of a chain to any length, and thus of the equilibria:

$$SiO_4^{4-} + (Si-O-Si)_n = (Si-O-Si)_{n+1} + O^{2-} \tag{5}$$

One may also consider disproportionation reactions of the sort

$$Q_{n-1} + Q_{n+1} = 2 \, Q_n \tag{6}$$

Energetics of such reactions, derived from NMR and other spectroscopy of glasses and application of polymer theory to free energy of mixing data in alkali, alkaline earth, and transition metal silicate melts, are shown in Table 1. The melts containing very basic cations (e.g. K^+) have equilibria (5) and (6) shifted far to the right, and have the sharpest changes in speciation with composition (see Fig. 2) and are the least viscous. In general, viscosities for liquids with higher field strength cations (e.g. Li vs. K) are expected to be smaller, especially when normalized to T/T_g, because such liquids are more "fragile", and thus their viscosity drops more rapidly with increasing T. This may be related to the greater configurational entropy of mixing Q species (in turn associated with the larger equilibrium constants for reactions such as those above) when field strength is greater for the modifier (Stebbins et al., 1992, see also Dingwell (this volume) for a discussion of rheology and relaxation processes, and Richet and Bottinga (this volume) for a discussion of configurational entropy and viscosity).

Table 1. Equilibrium constants and free energy changes for speciation reactions in melts and glasses

System	Reaction	Equilibrium constant	$\Delta G°$ (kJ)	State, Temperature
$Li_2O - SiO_2$	$2 Q_3 = Q_4 + Q_2$	0.08^b	21.0	glass[a]
	$2 Q_2 = Q_3 + Q_1$	0.30^b	10.0	glass[a]
	$2 Q_1 = Q_2 + Q_0$			
$Na_2O - SiO_2$	$2 Q_3 = Q_4 + Q_2$	0.02^b	32.5	glass[a]
	$2 Q_2 = Q_3 + Q_1$	0.16^b	15.2	glass[a]
	$2 Q_1 = Q_2 + Q_0$	0.14^b	22.9	glass[a]
$K_2O - SiO_2$	$2 Q_3 = Q_4 + Q$	0.01^b	38.3	glass[a]
	$2 Q_2 = Q_3 + Q_1$	0.01^b	38.3	glass[a]
	$2 Q_1 = Q_2 + Q_0$	0.02^b	32.5	glass[a]
$CaO - SiO_2$	$2 Q_0 = Q_1 + O^{2-}$	0.0016^c	100.2	melt, ~1873 K
$FeO - SiO_2$	$2 Q_0 = Q_1 + O^{2-}$	0.25^c	18.4	melt, ~1600 K
$MnO - SiO_2$	$2 Q_0 = Q_1 + O^{2-}$	0.70^c	4.7	melt, ~1600 K
$CoO - SiO_2$	$2 Q_0 = Q_1 + O^{2-}$	2.0^c	-9.2	melt, ~1600 K
$PbO - SiO_2$	$2 Q_0 = Q_1 + O^{2-}$	0.20^c	17.0	melt, ~1273 K
$SnO - SiO_2$	$2 Q_0 = Q_1 + O^{2-}$	2.55^c	-9.9	melt, ~1273 K
$CaO-Na_2O-SiO_2$	$2 Q_3 = Q_2 + Q_4$	0.03^d	30.0	glass
$Na_2O-Al_2O_3-SiO_2$	$Al^{IV} = Al^{VI}$	0.1^e	30.1	melt, ~1573 K

(a)   all equilibrium constants for glasses presumably refer to temperatures near glass transitions where speciation equilibria become "frozen", assumed to be near 1000 K.
(b)   Maekawa et al. (1991)
(c)   Masson et al. (1970)
(d)   Brandriss and Stebbins (1988)
(e)   Stebbins and Farnun (1992)

Interactions in geologically relevant silicate melts

Whereas the large exothermic effects associated with the reactions above, which can be viewed as acid-base neutralizations, dominate in binary silicate systems, most geologically relevant silicate melts differ from the above simple systems in that they contain aluminum and their silica content is typically restricted to a fairly narrow range, roughly 50 to 70 mol %. This means that, rather than being dominated by exothermic acid-base interactions, magmas exhibit more complex thermodynamic behavior, with inter-actions which are smaller in magnitude, closely balanced, and both positive and negative.

Charge coupled substitutions. The substitution of aluminum, except possibly in a few very peraluminous melts, occurs by the charge coupled mechanism, with aluminum entering tetrahedral coordination as a network former:

$$Si^{4+} = Al^{3+} + (1/n) M^{n+} \qquad (7)$$

The energetics of this reaction have been studied extensively in glasses (Roy and Navrotsky, 1984: Navrotsky et al., 1985) by measuring heats of solution in molten test borate at 973 K of glasses with compositions along $SiO_2 - M^{n+}_{1/n}AlO_2$ binary joins. The slope of enthalpy of solution curve, related to the enthalpy of the reaction

$$SiO_2 - M^{n+}_{1/n}AlO_2 \text{ (dilute solution in lead borate)} + SiO_2 \text{ (in silicate glass)} =$$

$$SiO_2 - M^{n+}_{1/n}AlO_2 \text{ (in silicate glass)} + SiO_2 \text{ (dilute solution in lead borate)} \qquad (8)$$

is a proxy for the enthalpy of substitution (Reaction 7). Maximum stabilization occurs near Al/(Al + Si) = 0.5; see Figure 3. The reaction becomes more exothermic with increasing basicity (decreasing ionic potential, z/r) of the cation, M; see Figure 4. This implies that the stabilization is largely electrostatic in nature. This pattern has also been interpreted as expressing the competition of the cation, M, and silicon or aluminum, for bonding to oxygen. Molecular orbital calculations on appropriate aluminosilicate clusters, in which a cation approaches the bridging oxygen, support this view (Navrotsky et al., 1985). There is indication that the tetrahedral framework in the glass is more perturbed, in terms of bond angle and ring size distribution, as the field strength of the network modifier increases. Furthermore, the change in heat capacity at the glass transition, the deviation from Arrhenius behavior in viscosity, and the ease of crystallization, all increase in the order silica, albite, anorthite. It is generally assumed that charge compensation in the framework must occur with the M cation spatially fairly close to the aluminum. NMR data suggest a significantly non-random Al-Si distribution in framework aluminosilicate glasses, indicative of at least partial Al avoidance, (Murdoch et al., 1985). Presumably these general trends hold for the melts as well (but Stebbins, this volume, for a discussion of dynamics). These observations are important to structural models and the calculation of entropies of mixing in silicate melts.

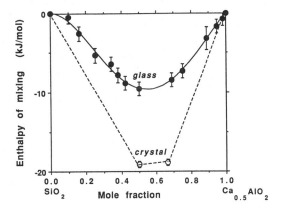

Figure 3. Comparison of energetics of crystals and glasses along the join $SiO_2-Ca_{0.5}AlO_2$ (Navrotsky, 1994).

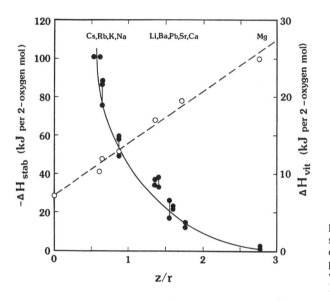

Figure 4. Energy of coupled substitution (Reaction 7) and of vitrification versus ionic potential (z/r) of nonframework cation (Navrotsky et al., 1985).

Reaction (7) may also be viewed as a complexation reaction. The coupled substitution represents an equilibrium which may be viewed, in the nomenclature of Hess and coworkers, (Hess, 1977, 1986, 1987; Ellison and Hess, 1986; Gan, 1994; Hess, this volume) as the formation of a complex between aluminum and the charge balancing cation, written, for example, as K-O-Al, implying that the alkali and aluminum share a common oxygen, but not implying any structural details such as bond distance or coordination number. The greater exothermicity of the reaction in the order Mg, Ca, Na, K implies a more stable complex (larger equilibrium constant) in that order. Thus one is comfortable, and probably correct, in thinking that, for example, K-O-Al complexes will form in preference to Ca-O-Al complexes in a multicomponent melt. A similar formalism (see below) is very useful in considering equilibria involving other highly charged cations, e.g. Ti^{4+}

There are several advantages of the formalism of homogeneous equilibria. Although stoichiometry is assumed, knowledge of detailed structure of species is neither required nor implied. The model is easily parameterized in terms of temperature, pressure and composition, and can be used to calculate complexation equilibria and thermodynamic activities from phase equilibrium data. Such variation can imply equilibria which shift continuously and strongly with temperature, consistent with viscosity changes, density changes, and fragile behavior. However, it is easy to use very many parameters, and questions of uniqueness of fit and physical significance need to be asked. As discussed below, simultaneous application of equilibria and calorimetry offers very promising new insights.

Speciation, clustering, and phase separation. In binary alkali and alkaline earth silicates, positive heats of mixing at silica-rich compositions reflect a tendency toward phase separation, which can occur in the stable liquid, subliquidus in the metastable supercooled liquid, or in the glass (see for example, Hess and Tewhey, 1979). Generally the exsolving compositions represent a silica-rich phase, predominantly Q_4 in linkage, and a silica-poor phase, often Q_2 to Q_3 in average degree of polymerization (Sen et al., 1994; Sen and Stebbins, 1994). In geologic systems, especially magmas rich in potassium, iron, and titanium, phase separation has also been documented (see for example, Philpotts, 1979). A consequence of such phase-separation is that minor and trace elements can be strongly partitioned into one of a pair of coexisting liquids (Watson, 1976; Ryerson and Hess, 1978).

In certain multicomponent systems, such as the join albite-diopside, the glasses show positive heats of mixing (Navrotsky et al. 1980). This implies that forming a glass with average degree of polymerizatrion intermediate between Q_2 (diopside) and Q_4 (albite) is energetically unfavorable. Such a glass, annealed for several weeks below T_g, shows signs of phase separation (Henry et al., 1983).

From the point of view of thermodynamics, these observations suggest that the equilibrium

$$2\,Q_3 = Q_2 + Q_4 \tag{9}$$

will depend on temperature, and that even nominally homogeneous melts and glasses may contain clusters or regions of larger and smaller than average degree of polymerization. If Reaction (9) is exothermic, equilibrium will shift to the right with decreasing temperature, and heterogeneity in speciation will be greater at lower temperatures. If ΔH is endothermic, as suggested by some NMR data (Brandriss and Stebbins, 1988; Stebbins et al., 1992) a more heterogeneous distribution of species (more Q_2 and Q_4) may be favored at higher temperature. Subtle changes in melt properties with temperature may reflect changes in the extent and scale of such clusters. Such effects may be very important to the structure and dynamics of melts and glasses.

Mixed alkali and mixed cation effects. It is well known from glass science that glasses in which two alkalis substitute for each other in an otherwise constant composition often show exothermic heats of mixing (see Table 2), nonlinear variation of T_g with composition, lowered electrical conductivity, and other anomalous properties. Such behavior is known collectively as the mixed alkali effect, (Day, 1976; Kawamoto and Tomozawa, 1981), and it appears most pronounced at lower temperatures. Mixed molten salts, such as NaCl-KCl also show small exothermic heats of mixing, (Kleppa, 1965), though these are smaller in magnitude than those seen in oxide glasses.

Similar effects probably occur when other nonframework cations (e.g. Mg and Ca) are mixed. Thus a number of glassy joins, where the predominant change is in the nonframework cation composition, show exothermic heats of mixing, see Table 2.

The mixed alkali effect is probably smaller in melts than in glasses. Thus glasses along the $NaAlSi_3O_8$-$KAlSi_3O_8$ join at 973 K show definite exothermic heats of mixing (Hervig and Navrotsky, 1984) while melts at 1673 K show ideal mixing (Rammensee and Fraser 1982; Fraser et al., 1983, 1985; Fraser and Bottinga, 1985). This would imply excess heat capacities of mixing between 973 and 1673 K, but the relevant C_p data do not exist.

APPLICATIONS TO MELTS OF GEOLOGIC COMPOSITION

Heat capacities and heats of fusion

In most aluminosilicate glasses, the heat capacity between liquid nitrogen temperature and the glass transition can be treated as an additive function of oxide components (see for example, Richet et al., 1993). As vibrational modes are

Table 2. Energetics of mixed cation glasses and liquids

System	$T(°C)$ maximum	State	Regular Sol'n Parameter[a] (kJ per mole cations mixed)
Orthosilicate			
$1/2(Mn_2SiO_4\text{-}Fe_2SiO_4)$	1550	Liquid	-50.6[b]
$1/2(Mg_2SiO_4\text{-}Fe_2SiO_4)$	1550	Liquid	-8.4[b]
$1/2(Fe_2SiO_4\text{-}Co_2SiO_4)$	1450	Liquid	-48[b]
$1/2(Ca_2SiO_4\text{-}Mn_2SiO_4)$	1660	Liquid	-86[b]
$1/2(Ca_2SiO_4\text{-}Fe_2SiO_4)$	1150	Liquid	-60[b]
$1/2(Ca_2SiO_4\text{-}Mg_2SiO_4)$	1600	Liquid	-65[b]
Alkali Silicates and Aluminosilicates			
$1/2(Na_2SiO_3\text{-}K_2SiO_3)$	1100	Liquid	$+7.2$[c]
$1/2(Na_2Si_2O_5\text{-}K_2Si_2O_5)$	1100	Liquid	$+3.2$[c]
$1/2(Na_2Si_4O_9\text{-}K_2Si_4O_9)$	1100	Liquid	$+2.0$[c]
$NaAlSi_3O_8\text{-}KAlSi_3O_8$	50	Glass	-7.3[d]
	700	Glass	-10.1[e]
	700	Glass	-14.5[f]
	1400	Melt	~ 0[g]

a.　in approximation that ΔG_{mix}^{xs} (or ΔH_{mix}^{xs}) $= WX_1X_2$, where X_1 and X_2 are mole
　　fractions. Uncertainty is generally about $\pm 20\%$.
b.　Belton et al. (1973)　　　　c.　　Choudary et al. (1977)
d.　Hovis (1984)　　　　　　　e.　　Brousse et al. (1982)
f.　Hervig and Navrotsky (1984)　g.　　Rammensee and Fraser (1982)

Table 3. Heat capacities of components in magmatic liquids

Oxide	Cp*(liq)* $= \sum X_iCp_i$ (*J/g f w-K*)		
	Lange & Navrotsky (1992)	Stebbins et al. (1984)	Courtial & Richet (1993)
SiO_2	82.6 ± 1.2	80.0 ± 0.9	81.4
TiO_2	109.2 ± 8.9	111.8 ± 5.1	
Al_2O_3	170.3 ± 5.1	157.6 ± 3.4	$130.2 \pm 0.036\,T$
Fe_2O_3	240.9 ± 7.9	229.0 ± 18.4	
FeO	78.8 ± 4.6	78.9 ± 4.9	
MgO	94.2 ± 4.3	99.7 ± 7.3	85.8
CaO	89.8 ± 3.1	99.9 ± 1.9	
Na_2O	97.6 ± 3.1	102.3 ± 1.9	
K_2O	98.5 ± 5.5	97.0 ± 5.1	

progressively excited, C_p increases. At the glass transition, there is a jump in the heat capacity (see Fig. 5), the magnitude of which reflects the fragility of the liquid (Angell, 1985, 1988). In the liquid, in many cases, the heat capacity is almost independent of temperature. Thus, a set of partial molar heat capacities of oxide components in magmatic liquids, with C_p independent of temperature and composition, is a useful approximation, (see Table 3, above). If a broader range of temperatures and/or compositions is chosen, or accuracy in calculated C_p values of better than about ±3% is sought, then the effects of temperature and composition (so called nonideality in heat capacities) must be considered. Courtial and Richet (1993) have argued that, in the system MgO-Al_2O_3-SiO_2 at 900 to 1800 K, a temperature dependence of the partial molar heat capacity of Al_2O_3, coupled with C_p of SiO_2 and MgO having temperature independent values identical to those in Al_2O_3-free melts, is a better description. They also point out that alkali aluminosilicates show even greater heat capacity nonideality than magnesium aluminosilicate melts, and that this may reflect the strength of alkali-aluminum complexation equilibria. This tendency appears even more strongly in alkali titanosilicate melts, see below.

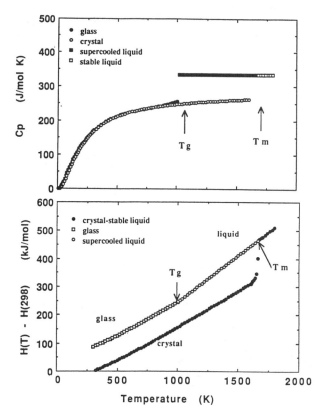

Figure 5. Heat capacity and enthalpy of crystal, glass, and liquid for $CaMgSi_2O_6$ (Navrotsky, 1994).

For a given composition, the heat capacities of glass and crystal are generally similar, but that of the liquid is generally higher (see Fig. 5). This means that the heat of fusion increases with temperature, sometimes by as much as a factor of two over a

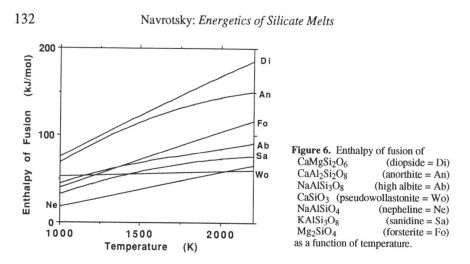

Figure 6. Enthalpy of fusion of

$CaMgSi_2O_6$	(diopside = Di)
$CaAl_2Si_2O_8$	(anorthite = An)
$NaAlSi_3O_8$	(high albite = Ab)
$CaSiO_3$ (pseudowollastonite = Wo)	
$NaAlSiO_4$	(nepheline = Ne)
$KAlSi_3O_8$	(sanidine = Sa)
Mg_2SiO_4	(forsterite = Fo)

as a function of temperature.

1000 K interval (see Fig. 6). This is geologically important because minerals in multi-component systems crystallize over a temperature interval of a hundred degrees below the melting point of the pure compound. Thus both for appropriate computation of phase equilibria and for thermal modelling, the temperature dependence of the heat of fusion must be included (see Table 4). This term can be more important than the heat of mixing in the multicomponent melt because the latter tends to be quite small in many systems (see below). Expressions for heats of fusion for some common minerals are shown in Table 4.

Table 4.　Equations for enthalpy of fusion of some minerals (kJ/mol), valid at approximately 1000 to 2000 K

Pseudowollastonite, $CaSiO_3$ (modified from Adamkovicova et al., 1980)

$\Delta H(fusion) = 47.1 + 0.0056T$

Diopside, $CaMgSi_2O_6$ (Stebbins et al., 1983)

$\Delta H(fusion) = -29.0 + 0.109T - 5.34 \times 10^{-6}T^2$

High albite, $NaAlSi_3O_8$ (Stebbins et al., 1983).

$\Delta H(fusion) = -28.3 + 0.089T - 1.64 \times 10^{-5}T^2$

Nepheline, $NaAlSiO_4$ (modified from Stebbins et al., 1983)

$\Delta H(fusion) = -21.3 + 0.039T$

Sanidine, $KAlSi_3O_8$ (Stebbins et al., 1983)

$\Delta H(fusion) = -51.6 + 0.106T - 2.19 \times 10^{-5}T^2$

Forsterite, Mg_2SiO_4 (calculated from Navrotsky et al., 1989; valid at approximately 1500 to 2400 K)

$\Delta H(fusion) = -24.6 + 0.0641T$

Anorthite, $CaAl_2Si_2O_8$ (modified from data in Henry et al., 1982)

$\Delta H(fusion) = -72.7 + 0.172T - 2.97 \times 10^{-5}T^2$

These equations and the corresponding curves in Figure 6 represent simplifications of the equations given in the original papers in order to avoid maxima in the heat of fusion extrapolated to 2000 K and also to avoid excessive curvature in the heat of fusion curves at 1000 to 1200 K in the supercooled liquids. If one wishes to use heats of fusion to calculate the energetics of melting in multicomponents eutectic and hydrous systems, then the 1000 to 1200 K range becomes important. If one wishes to calculate melting relations at high pressure, then both the 1500 to 2000 K range and the $P\Delta V$ term, not discussed here (but see Chapter by Wolf and McMillan, this volume) become important. The equations given here are meant to be simple, useful, and reasonably accurate for this wide range of applications.

Table 5. Enthalpy of mixing parameters for several liquids and glasses

System	W_{xy}[a]	W_{xz}	W_{yz}	W_{xyz}
diopside-albite-anorthite				
$CaMgSi_2O_6$-$NaAlSi_3O_8$-$CaAl_2Si_2O_8$				
x y z				
glass[b]				
melt[c]	124.8	0	-85.0	0
diopside-forsterite-anorthite				
$CaMgSi_2O_6$-Mg_2SiO_4-$CaAl_2Si_2O_8$				
x y z				
glass[d]	0	-23.9	-16.7	0
melt[c]	0	0 .	0	0
diopside-wollastonite-anorthite				
$CaMgSi_2O_6$-$CaSiO_3$-$CaAl_2Si_2O_8$				
x y z				
melt[e]	-60.6	0	-44.0	0
melt[e]	0	0	0	0
gehlenite-wollastonite-anorthite				
$Ca_2Al_2SiO_7$-$CaSiO_3$-$CaAl_2Si_2O_8$				
x y z				
melt[f]				

(a) $\Delta H_{mix} = xy\,W_{xy} + xz\,W_{xz} + yz\,W_{yz} + xyz\,W_{xyz}$	(b) 973 K, Hon et al. (1981)
(c) 1773 K, Navrotsky et al. (1989)	(d) 973 K, Navrotsky et al. (1990)
(e) 1773 K, Tarina et al. (1994)	(f) 1830 K, Kosa et al. (1992)

Heats of mixing of major components

Table 5 lists mixing parameters for glasses and melts in several systems. In general, the following trends are seen. When heats of mixing in the glasses are exothermic (indicative of some stabilizing complexation), those in the melt are less exothermic or zero, suggesting the decreasing importance of such specific ionic associations. This is seen along the anorthite-diopside join and also in the much less pronounced mixed alkali effects in melts than in glasses. When heats of mixing are endothermic in the glass (indicative of clustering and unfavorable interactions), they can become more endothermic in the melt, consistent with the energetically unfavorable (but entropically favorable) cost of breaking large clustered regions into smaller ones. Thus, for both the breakdown of ordered and clustered regions, increasing temperature shifts equilibrium in the energetically endothermic direction (as predicted by LeChatelier's Principle) and to a state of higher entropy. Note that this means that the heat of mixing becomes more endothermic (either less negative or actually more positive) and that athermal mixing ($\Delta H_{mix} = 0$) need not be the limit approached at high temperature. Nevertheless at 1200 to 1500°C, heats of mixing in many melts of geologic composition, (with respect to molten mineral end-members) are often small. In systems in which complex species (with negative ΔH_{mix}) decompose to end-members with unfavorable interactions ΔH_{mix} may change from negative at lower temperature to positive at higher. The possibility of such behavior led Navrotsky (1992) to speculate that a lower consolute temperature (LCST) could exist in silicate or borate melts analogous to that seen in organic systems and that it is thermodynamically possible for a magma to unmix with increasing temperature.

Application of the two-lattice model to entropies of mixing in silicate melts

In the early 1980s, Navrotsky and coworkers began to calculate phase equilibria in pseudoternary aluminosilicate systems such as albite-anorthite-diopside by combining measured enthalpies of mixing and of fusion with a "two lattice" model of the entropy of mixing (Weill et al., 1980; Hon et al., 1981; Henry et al., 1982; Navrotsky, 1986). This model calculates the configurational entropy of mixing silicon and aluminum on tetrahedral sites and of mixing mono- and divalent cations (defining a nonframework sublattice). Framework and nonframework distributions are each assumed random and independent of each other (no aluminum avoidance: Al-O-Al, Si-O-Si, and Si-O-Al linkages are assumed to exist in statistical proportions). The model neglects inequivalency of oxygen sites, as might arise from different degrees of polymerization (Q speciation), differences in tetrahedral linkages (Si-O-Si, Si-O-Al, or Al-O-Al), or differences in bonding to nonframework cations (e.g. Si-O-K vs. Si-O-Mg). They found that this model could be used to adequately calculate the pyroxene saturation surface in Ab-An-Di (Hon et al., 1981), but that consideration of aluminum avoidance in the melts must be included to obtain satisfactory results for the plagioclase loop in Ab-An, (Henry et al., 1982). These calculations were done using heats of mixing in glasses with correction terms arising from estimated ΔC_p terms.

The availability of better heat of fusion data and heat of mixing data directly for melts permits further assessment of the range of validity of the two lattice model. Tarina et al. (1994) have shown that heats of mixing of melts in the system diopside-anorthite-wollastonite at 1500°C are very small (see Table 5), and that the two lattice model combined with a zero heat of mixing (athermal mixing) permits a calculation of the diopside saturation surface with an accuracy of about ±10° (see Fig. 7). The combination of athermal mixing and two-lattice model also appears to work moderately well for the diopside saturation surface in diopside-anorthite-forsterite, diopside-gehlenite, diopside-leucite, diopside-anorthite-enstatite, and diopside-nepheline-akermanite (Tarina and Navrotsky, unpublished). The assumption of athermal mixing is not warranted in diopside-albite-anorthite or probably in other systems containing an alkali aluminosilicate (e.g. diopside-nepheline-SiO_2, diopside-albite-wollastonite). A tentative conclusion appears to be that athermal mixing plus the two lattice model appears to be a useful first approximation for melts with Al/(Al + Si) < 0.25, Ca/(Mg + Ca) < 0.5, and Na/(Na + Mg + Ca) < 0.3. For melts richer in alkali aluminosilicate, nonzero heats of mixing must be considered, and for melts rich in aluminum, the two-lattice model breaks down because of aluminum avoidance in the melt. For melts of orthosilicate composition, where the degree of polymerization depends significantly on the nature of the divalent cation (see Table 2), neither athermal mixing nor the two lattice model apply.

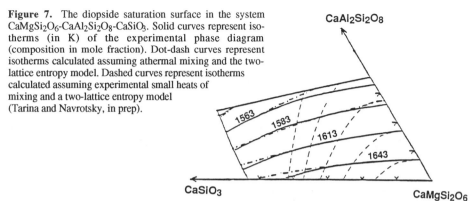

Figure 7. The diopside saturation surface in the system $CaMgSi_2O_6$-$CaAl_2Si_2O_8$-$CaSiO_3$. Solid curves represent isotherms (in K) of the experimental phase diagram (composition in mole fraction). Dot-dash curves represent isotherms calculated assuming athermal mixing and the two-lattice entropy model. Dashed curves represent isotherms calculated assuming experimental small heats of mixing and a two-lattice entropy model (Tarina and Navrotsky, in prep).

Energetics of minor components — particularly TiO_2

Petrologically important silicate melts, in addition to the major components SiO_2, MgO, CaO and Al_2O_3, contain iron, manganese and titanium (and water) in minor amounts (commonly up to several weight percent) and the rest of the periodic table in trace amounts (parts per million to several tenths of a percent). TiO_2 is particularly interesting because the Ti^{4+} ion, although larger than Si^{4+} or Al^{3+} in the same coordination, is marginally capable of entering the tetrahedral silicate network, and lies on the borderline between a network former and network modifier. Thus its behavior in melts might be expected to be very sensitive to changes in temperature, pressure, and composition, and its effects on density (Dingwell, 1992a), viscosity (Dingwell, 1992b) and melt energetics (Gan, 1994; Gan and Navrotsky, 1994) have indeed been found to be complicated. TiO_2-containing melts appear to change structure with increasing temperature above T_g and represent much more fragile liquids than their TiO_2 free analogues.

The heat capacities of several TiO_2-bearing silicate glasses and liquids containing Cs_2O, Rb_2O, Na_2O, K_2O, CaO, MgO, or BaO have been measured to 1100 K using a differential scanning calorimeter and to 1800 K using a Setaram HT-1500 calorimeter in step-scanning mode (Lange and Navrotsky, 1993). The results for liquids of M_2O-TiO_2-$2SiO_2$ composition (M = Na, K, Cs), compared to those for liquids M_2O-$3SiO_2$ composition (Lange and Navrotsky, 1993; Richet and Bottinga, 1985; Richet et al., 1984), show that the presence of TiO_2 has a profound influence on the heat capacity of simple three-component silicate liquids over the temperature range 900 to 1300 K. Specifically, replacement of Si^{4+} by Ti^{4+} leads to doubling of the magnitude of the jump in C_p at the glass transition (T_g); this is followed by a progressive decrease in liquid C_p for over 400 K, until C_p eventually becomes constant and similar to that in Ti-free systems (see Fig. 8). The large heat capacity step at T_g in the TiO_2-bearing melts suggests significant configurational rearrangements in the liquid that are not available to TiO_2-free silicates. In addition, these "extra" configurational changes apparently saturate as temperature increases, implying the completion of whatever process is responsible for them, or the attainment of a random distribution of structural states. Above 1400 K, where the heat capacities of TiO_2-bearing and TiO_2-free alkali silicate liquids are similar, their configurational entropies differ by ~3.5 J/g f w-K. The observed excess heat capacity in alkali titanosilicate melts, which persists for several hundred degrees above T_g and produces a much larger and more spread-out anomaly than a simple glass transition, suggests a gradual change in melt structure and/or titanium speciation over that temperature range.

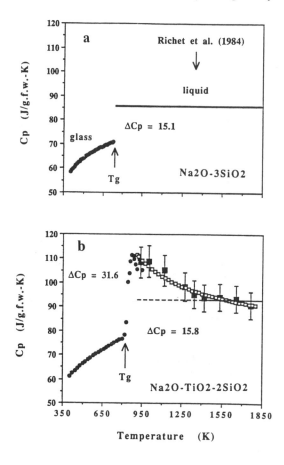

Figure 8. (a) Heat capacity vs. temperature of Na_2O-$3SiO_2$ liquid and glass from Richet et al. (1984). ΔC_p refers to the jump in heat capacity at the glass transition (T_g). (b) Heat capacity vs. temperature of Na_2O-TiO_2-$2SiO_2$ glass and liquid from Lange and Navrotsky (1993). Data obtained using the DSC and the HT-1500 are shown by solid circles and squares respectively. The open squares represent the fitted curve of Richet and Bottinga (1985) based on their drop calorimetric measurements for this composition.

We have recently used drop solution calorimetry in a Setaram HT-1500 calorimeter at 1756 K to probe the question of the composition dependence of the energetics of rutile dissolution in silicate melts (Gan and Navrotsky, 1994; Gan et al., in prep.). The experiments consist of dropping rutile pellets (~30 mg) sequentially into a silicate melt, (~5 g) and thus measuring the sum of heat content plus heat of solution as a function of melt composition and TiO_2 content. These experiments represent the first direct measurement of the partial molar enthalpy of solution of an oxide in a silicate melt above 1273 K. Results for melts of disilicate composition, $K_2O\cdot2SiO_2$, $CaO\cdot2SiO_2$, and $0.5K_2O\cdot0.5CaO\cdot2SiO_2$ are shown in Figure 9. The enthalpy of solution of rutile in potassium disilicate melts is less endothermic than in calcium disilicate melts, and the solubility higher (Gan, 1994). The heat of solution in K-bearing melts becomes more endothermic with increasing TiO_2 content, while that in Ca-bearing and mixed Ca,K melts shows little dependence on TiO_2 concentration. These results suggest stabilization of Ti in melts with high K/Ti ratios, consistent with ideas of homogeneous equilibria and K,Ti complex formation developed from rutile solubility data (Gan, 1994).

The enthalpies of solution of rutile in silicate melts differ significantly from 66 kJ/mol, the heat of fusion of rutile at 2130 K (JANAF 1985). The heat of fusion of rutile

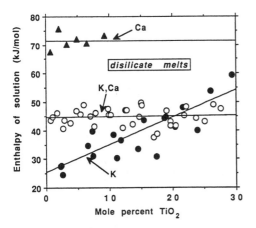

Figure 9. Enthalpy of solution at 1487°C of TiO_2 (rutile) in melt of composition $xTiO_2 - (1-x)K_2O \cdot 2SiO_2$. Each data point represents one measurement of heat of solution, plotted against average composition of melt before and after the addition of TiO_2 sample. From Gan, Wilding and Navrotsky (in prep.)

adjusted to 1756 K is estimated to be 60 kJ/mol (JANAF 1985). Thus the heat of solution of rutile in the calcium silicate melts is slightly more endothermic than its heat of fusion, implying a slightly positive (destabilizing) heat of mixing between $CaO \cdot 2SiO_2$ and supercooled liquid TiO_2, while the heat of solution of rutile in $K_2O \cdot 2SiO_2$ melts is substantially less endothermic than its heat of fusion, implying a negative (stabilizing) mixing term.

One can write speciation reactions involving the formulation of species (Ti-O-Ti, Ti-O-Si, Ti-O-K). Such a thermodynamic formalism does not uniquely imply the geometry or coordination environment of these complexes.

From the view point of thermochemistry, the rutile dissolution equilibria, as expressed by Gan (1994),

$$Ti\text{-}O\text{-}Ti \text{ (rutile)} = Ti\text{-}O\text{-}Ti \text{ (melt)} \tag{10}$$

$$Ti\text{-}O\text{-}Ti \text{ (melt)} + Si\text{-}O\text{-}Si = 2\,Ti\text{-}O\text{-}Si \tag{11}$$

$$Ti\text{-}O\text{-}Ti \text{ (melt)} + 2\,Si\text{-}O\text{-}K = 2\,Ti\text{-}O\text{-}K + Si\text{-}OS\text{-}i \tag{12}$$

can be characterized by the enthalpy of each reaction. Reaction (10) is endothermic since its enthalpy is essentially the heat of fusion at a given temperature. Reaction (11) is presumably endothermic too because the Ti-O-Si species concentration increases with increasing temperature (Gan, 1994). This could be due to the need to break strong Si-O-Si bonds. Reaction (12) is analogue to an acid-base reaction which should be exothermic and result in a stable Ti-O-K complex (e.g. a potassium titanate species). It is thus energetically favorable for Reaction (12) to strongly favor the right hand side as indicated by the very strong perakaline effect in the rutile solubility experiments. In addition, Raman spectra of Ti-bearing alkali silicate glasses also show that the band assigned to titanium species gains its intensity at the cost of that assigned to Si-O-K species (Furukawa and White, 1979). In rutile undersaturated melts, increasing temperature would shift the reaction

$$Ti\text{-}O\text{-}K + Si\text{-}O\text{-}Si = Ti\text{-}O\text{-}Si + Si\text{-}O\text{-}K \tag{13}$$

toward the right (in the endothermic direction). This reconstruction of Ti-O-Si and Ti-O-K species with changing temperature was suggested by Gan (1994) to explain the observed abnormally large heat capacity change in alkali titanate silicate melts (Lange and Navrotsky, 1993).

In addition to temperature, the TiO_2 concentration in silicate solvents also affects the homogeneous equilibria between the major Ti species. The initial addition of TiO_2 in potassium trisilicate melts will tend to complex K^+ to form Ti-O-K simply because Ti-O-K is energetically more stable (more exothermic). The combined total heat effect of solution of rutile will thus be much less endothermic than the heat of fusion of rutile at the same temperature. As more TiO_2 dissolves, relatively less additional Ti-O-K species and more Ti-O-Si species will form. The endothermic character of Reaction (11) will increase the partial molar enthalpy of solution of rutile. This trend continues and appears to result in an almost linear increase of the enthalpy of solution with the TiO_2 concentration in $K_2O\cdot2SiO_2$ and $K_2O\cdot3SiO_2$ melts. In contrast to the potassium silicate melts, the formation of Ti-O-Ca species in calcic silicate melts is not very favorable energetically (Dickinson and Hess, 1985; Gan, 1994). Rutile solubility increases only slightly with excess Ca^{2+} in percalcium melts, indicating that the reaction

$$\text{Ti-O-Ti} + 2 \text{ Si-O-Ca} = 2 \text{ Ti-O-Ca} + \text{Si-O-Si} \tag{14}$$

is not significant in affecting the total energy budget of rutile dissolution. Instead, Reactions (10) and (11) dominate and the enthalpy of mixing is endothermic.

We have used the drop solution calorimetric data to quantify the above equilibria. The enthalpy of Reaction (10) is assumed to be the enthalpy of fusion of rutile. The enthalpy of Reaction (11) is taken as the partial molar enthalpy of mixing of supercooled liquid TiO_2 with $CaO\cdot2SiO_2$ melt (the heat of solution minus the heat of fusion) equal to 12 kJ/mol, and that of Reaction (12) to be the partial molar enthalpy of mixing of supercooled liquid TiO_2 with $K_2O\cdot2SiO_2$ melt at low TiO_2 concentrations (where Ti-O-K complexes dominate), namely –34 kJ/mol. Using these energetics, we have calculated the temperature dependence of the speciation equilibria and modeled the heat capacity resulting from the shift in equilibria as a function of temperature (details are being prepared for publication). The results reproduce the heat capacity anomalies rather well, see Figure 10. Thus an equilibrium between titanium bearing species, which depends on temperature and composition, can be used in a thermodynamically consistent fashion to describe simultaneously enthalpies of solution of rutile, rutile solubility, and excess heat capacities.

The density measurements of Lange and Carmichael (1987) and Johnson and Carmichael (1987) indicate a strong compositional dependence to the partial molar volume of TiO_2. Specifically, at 1673 K, the fitted value for V_{TiO_2} in silicate liquids containing Cao and MgO is 23.61±0.35 cc/mol. In contrast, at the same temperature, the fitted values for V_{TiO_2} in sodic and potassic liquids are 29.59±0.57 and 34.14 ± 0.90 cc/mol, respectively. These results indicate that the partial molar volume of TiO_2 in silicate liquids increases quite substantially (more than 40%) as the field strength of the network modifying cation decreases and could reflect a composition-induced structural change. Density measurements on TiO_2-bearing liquids by Dingwell (1992) confirm this trend.

What, then, microscopically are the species involved? Because the heat capacity anomaly is measured in scanning mode, it appears that the species equilibrate rapidly, and a quenched glass will probably give information only about the structural state frozen

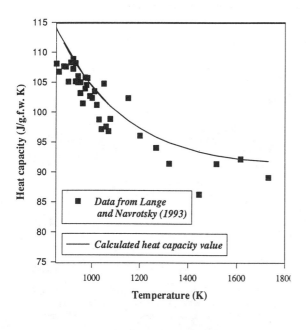

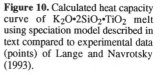

Figure 10. Calculated heat capacity curve of $K_2O \cdot 2SiO_2 \cdot TiO_2$ melt using speciation model described in text compared to experimental data (points) of Lange and Navrotsky (1993).

in at T_g. Spectroscopic studies on glasses have provided evidence for Ti in 4-fold, 5-fold and 6-fold coordination in titanosilicate glases (Farges et al., 1995; Dickinson and Hess, 1985; Dingwell et al., 1994; Hanada and Soga 1980; Mysen and Neuville, 1995; Sakka et al., 1986). The recent EXAFS data (Farges et al., 1995; Dingwell et al., 1994) favor 5-coordinated Ti as the dominant species in both glass and melt. The alkali titanate complex Ti-O-K presumably involves a strong interaction between a titanium-oxygen polyhedron and a potassium-oxygen polyhedron, probably, but not necessarily, involving a common oxygen. The Ti-O-Si species could bond the titanium either in the network or as a network modifier, strongly perturbing the bridging oxygens; indeed the distinction between these two cases may become dynamically blurred. The change in structure with increasing temperature may be more long-range and diffuse than just that affecting nearest neighbors, and the range of interactions and clustering on the meso-scale may depend on temperature. Plausibility arguments from physical property systematics, such as those made on the basis of density of melts (Lange and Navrotsky, 1993; Dingwell, 1992a) are useful, but *in situ* spectroscopic studies at high temperature are essential for structural insight into what is almost certainly a system with dynamics which change rapidly with temperature. Similar considerations probably apply to other minor component equilibria, especially those involving high field strength cations such as Zr, Mo, W, and the rare earths and actinides.

CONCLUSIONS

Silicate melts are structurally, energetically, and dynamically distinct from glasses. *In situ* high temperature calorimetric techniques increasingly allow us to study melts directly. Such work, combined with *in situ* spectroscopic studies and determination of other physical properties at high temperature, represents the present frontier. Work involving high pressure and the presence of volatile constituents such as H_2O and CO_2 represents future directions.

ACKNOWLEDGMENTS

This review summarizes work, largely from my own Laboratory, which has been supported by the National Science Foundation, the Department of Energy, Corning, Inc., and Sandia Laboratories over a fifteen-year period. I thank Hao Gan and Martin Wilding for the inclusion of data recently obtained in the calorimetry laboratory at Princeton.

REFERENCES

Adamkovicova K, Kosa L, Proks I (1980) The heat of fusion of CaSiO₃. Silikáty 24:193-201
Akaogi M, Ito E (1993) Refinement of enthalpy measurement of MgSiO₃ perovskite and negative pressure-temperature slopes for perovskite-forming reactions. Geophys Res Lett 20:1839-1842
Angell CA (1985) Strong and fragile liquids. In: Relaxation in Complex Systems. K Ngai, GW Wright (eds) p 3-11, Nat'l Tech Info Service, U S Dept Commerce
Angell CA (1988) Perspective on the glass transition. J Phys Chem Solids 49:863-871
Belton GR, Suito H, Gaskell DR (1973) Free energies of mixing in the liquid iron-cobalt orthosilicates at 1450°C. Metallurgical Trans 4:2541-2547
Brandriss ME, Stebbins, JF (1988) Effects of temperature on the structures of silicate liquids: ^{29}Si results. Geochim Cosmochim Acta 52:2659-2670
Brousse C, Rogez J, Castanet R, (1982) Enthalpie de melange du systeme vitreux albite-orthose (Na,K) AlSi₃O₈. Mat Res Bull 17:125-132
Capobianco C, Navrotsky A (1982) Calorimetric evidence for ideal mixing of silicon and germanium in glasses and crystals of sodium feldspar composition Am Mineral 67:718-724
Choudary UV, Gaskell DR, Belton GR (1977) Thermodynamics of mixing in molten sodium-potassium silicates at 1100 °C: The effect of a calcium oxide addition. Metallurgical Trans B 8B:67-71
Courtial P, Richet P (1993) Heat capacity of magnesium aluminosilicate melts. Geochim Cosmochim Acta 57:1267-1275
Day DE (1976) Mixed alkali glasses — their properties and uses. J Non-Cryst Solids 21:343-372
DeYoreo JJ, Navrotsky A, Dingwell DB (1990) Energetics of the charge-coupled substitution Si^{4+} --> Na^{1+} + T^{3+} in the glasses $NaTO_2$-SiO_2 (T = Al, Fe, Ga, B). J Am Ceram Soc 73:2068-2072
Dickinson JE, Hess PC (1985) Rutile solubility and titanium coordination in silicate liquids. Geochim Cosmochim Acta 49:2289-2296
Dingwell DB (1992a) Density of some titanium-bearing silicate liquids and the compositional dependence of the partial molar volume of TiO_2. Geochim Cosmochim Acta 56:3403-3407
Dingwell DB (1992b) Shear velocity of alkali and alkaline earth titanium silicate liquids. Am Mineral 77:270-274
Dingwell DB, Parise E, Seifert F, Mottana A, Romano C (1994) X-ray absorption study of Ti-bearing silicate glasses. Phys Chem Minerals 21:501-509
Ellison AJ, Hess PC (1986) Solution behavior of +4 cations in high silica melts: Petrologic and geochemical implications. Contrib Mineral Petrol 94:343-351
Farges F, Brown GE Jr, Navrotsky A, Gan H, Rehr JJ (1995) Coordination chemistry of Ti(IV) in silicate glasses and melts. Part I. Glasses at ambient temperature and pressure. Geochim Cosmochim Acta (submitted)
Fraser DG, Rammensee W, Jones RH (1983) The mixing properties of melts in the system $NaAlSi_2O_6$-$KAlSi_2O_6$ determined by Knudsen cell mass spectrometry. Bull Minéral 106:11-117
Fraser DG, Rammensee W, Hardwick A (1985) Determination of the mixing properties of molten silicates by Knudsen cell mass spectrometry–II. The systems (Na-K)$AlSi_4O_{10}$ and (Na-K)$AlSi_5O_{12}$. Geochim Cosmochim Acta 49:349-359
Fraser DG, Bottinga Y (1985) The mixing properties of melts and glasses in the system $NaAlSi_3O_8$-$KAlSi_3O_8$: Comparison of experimental data obtained by Knudsen cell mass spectrometry and solution calorimetry. Geochim Cosmochim Acta 49:1377-1381
Furukawa T, White WB (1979) Structure and crystallization of glasses in the Li_2SiO_3-TiO_2 system. Phys Chem Glasses 20:69-80
Gan H, Navrotsky A (1994) Enthalpy of solution of rutile in silicate melts: Direct calorimetric data and their implications to structure of titanosilicate melts. EOS, Trans Am Geophys Union 1994 Fall Mtg Supplement, p 726
Gan H (1994) The speciation and solution properties of high field strength cations P^{5+}, Ti^{4+}, and B^{3+} in aluminosilicate melts. PhD Thesis, Brown University, Providence, Rhode Island
Gaskell DR (1973) The thermodynamic properties of the masson polymerization models of liquid silicates. Metallurgical Trans 4:185-191

Gaskell DR (1977) Activities and free energies of mixing in binary silicate melts. Metallurgical Trans 8B:131-145

Geisinger KL, Ross NL, McMillan P, Navrotsky A (1987) $K_2Si_4O_9$: Energetics and vibrational spectra of glass, sheet silicate, and wadeite phases. Am Mineral 72:984-994

Geisinger KL, Oestrike R, Navrotsky A, Turner GL, Kirkpatrick RJ (1988) Thermochemistry and structure of glasses along the join $NaAlSi_3O_8$-$NaBSi_3O_8$. Geochim Cosmochim Acta 52:2405-2414

Hanada T, Soga N (1980) Coordination of titanium in sodium titanium silicate glasses. J Non-Crystal Solids 38-39:105-110

Henry DJ, Navrotsky A, Zimmermann HD (1982) Thermodynamics of plagioclase-melt equilibria in the system albite-anorthite-diopside. Geochim Cosmochim Acta 46:381-391

Henry DJ, Mackinnon IDR, Chan I, Navrotsky A (1983) Subliquidus glass-glass immiscibility along the albite-diopside join. Geochim Cosmochim Acta 47:277-282

Hervig RL, Navrotsky A (1984) Thermochemical study of glass in the system $NaAlSi_3O_8$-$KAlSi_3O_8$-Si_4O_8, and the join $Na_{1.6}Si_{2.4}O_8$-$K_{1.6}Al_{1.6}Si_{2.4}O_8$. Geochim Cosmochim Acta 48:513-522

Hervig RL, Navrotsky A (1985) Thermochemistry of glasses in the system Na_2O-B_2O_3-SiO_2. J Am Ceram Soc 11:284-298

Hervig RL, Scott D, Navrotsky A (1985) Thermochemistry of glasses along joins of pyroxene stoichiometry in the system $Ca_2Si_2O_6$-$Mg_2Si_2O_6$-Al_4O_6. Geochim Cosmochim Acta 49:1497-1501

Hess PC (1977) Structure of silicate melts. Can Mineral 15:162-178

Hess PC, Tewhey JD (1979) The two phase region in the CaO-SiO_2 system: Experimental data and thermodynamic analysis. Phys Chem Glasses 30:41-53

Hess PC (1980) Polymerization model for silicate melts. Ch 1 in Physics of Magmatic Processes, p 3-48

Hess PC (1986) The role of high field strength cations in silicate melts. In: Advances in Physical Geochemistry, Springer-Verlag, Berlin, Heidelberg, New York (in press)

Hon R, Henry DJ, Navrotsky A, Weill DF (1981) A thermochemical calculation of the pyroxene saturation surface in the system diopside-albite-anorthite. Geochim Cosmochim Acta 45:157-161

Hovis GL (1984) A hydrofluoric acid solution calorimetric investigation of glasses in the systems $NaAlSi_3O_8$-$KAlSi_3O_8$ and $NaAlSi_3O_8$-Si_4O_8. Geochim Cosmochim Acta 48:523-525

JANAF (1985) JANAF Thermochemical Tables, 3rd Ed'n, Part II. J Phys Chem Ref Data, Vol 14, Supp 1

Johnson T, Carmichael ISE (1987) The partial molar volume of TiO_2 in multicomponent silicate melts. Geol Soc Am Abstr Prog 19:719

Kawamoto Y, Tomozawa M (1981) The mixed alkali effect in the phase separation of alkali silicate glasses. Phys Chem Glasses 23:72-75

Kiseleva IA, Ogorodova LP (1984) High-temperature solution calorimetry for determining the enthalpies of formation for hydroxyl-containing minerals such as talc and tremolite. Geochem Int'l 21:36-46

Kleppa OJ (1965) The solution chemistry of simple fused salts. Rev Phys Chem 16:187-212

Kosa L, Tarina I, Adamkovicova K, Proks I (1992) Enthalpic analysis of melts in the $CaO \bullet SiO_2(CS)$-$CaO \bullet Al_2O_3 \bullet 2SiO_2(CAS_2)$-$2CaO \bullet Al_2O_3 \bullet SiO_2(C_2AS)$ system. Geochim Cosmochim Acta 56:2643-2655

Lange RA, Carmichael ISE (1987) Densities of Na_2O-K_2O-CaO-MgO-FeO-Fe_2O_3-TiO_2-SiO_2 liquids: New measurements and derived partial molar properties. Geochim Cosmochim Acta 51:2931-2946

Lange RA, DeYoreo JJ, Navrotsky A (1991) Scanning calorimetric measurement of heat capacity during incongruent melting of diopside. Am Mineral 76:904-912

Lange R, Navrotsky A (1992) Heat capacities of Fe_2O_3-bearing silicate liquids. Contrib Mineral Petrol 110:311-320

Lange RA, Navrotsky A (1993) Heat capacities of TiO_2-bearing silicate liquids: evidence for anomalous changes in configurational entropy with temperature. Geochim Cosmochim Acta 57:3001-3011

Lange RA, Cashman KV, Navrotsky A (in press) Direct measurement of latent heat during crystallization and melting of a ugandite and an olivine basalt. Contrib Mineral Petrol

Maekawa H, Maekawa T, Kawamura K, Yokokawa T (1991) The structural groups of alkali silicate glasses determined from ^{29}Si MAS-NMR. J Non-Cryst Solids 127:53-64

Maniar PD, Navrotsky A, Draper CW (1990) Thermochemistry of the amorphous system SiO_2-GeO_2: comparison of flame hydrolysis materials to hith temperature fused glasses. Mat Res Soc Symp Proc 172:15-20

Masson CR (1965) An approach to the problem of ionic distribution in liquid silicates. Proc Roy Soc London A287:201-221

Masson CR, Smith IB, Whiteway SG (1970) Activities and ionic distribution in liquid silicates: Application of polymer theory. Can J Chem 48:1456-1464

Murdoch JB, Stebbins JF, Carmichael ISE (1985) High resolution ^{29}Si NMR study of silicate and aluminosilicate glasses: The effect of network modifying cations. Am Mineral 70:332-343

Mysen B, Neuville D (1995) Effects of temperature and TiO_2 content on the structure of $Na_2Si_2O_5$-$Na_2Ti_2O_5$ melts and glasses. Geochim Cosmochim Acta 59:325-342

Navrotsky A (1977) Recent progress and new directions in high temperature calorimetry. Phys Chem Minerals 2:89-104

Navrotsky A, Hon R, Weill DF, Henry DJ (1980) Thermochemistry of glasses and liquids in the systems $CaMgSi_2O_6$-$CaAl_2Si_2O_8$-$NaAlSi_3O_8$, SiO_2-$CaAl_2Si_2O_8$-$NaAlSi_3O_8$, and SiO_2-Al_2O_2-CaO-Na_2O. Geochim Cosmochim Acta 44:1409-1423

Navrotsky A, Peraudeau G, McMillan P, Coutures JP (1982) A thermochemical study of glasses along the joins silica-calcium aluminate and silica-sodium aluminate. Geochim Cosmochim Acta 46:2039-2047

Navrotsky A, Zimmermann HD, Hervig RL (1983) Thermochemical study of glasses in the system $CaMgSi_2O_6$-$CaAl_2SiO_6$. Geochim Cosmochim Acta 47:1535-1538

Navrotsky A, Geisinger KL, McMillan P, Gibbs GV (1985) The tetrahedral framework in glasses and melts - inferences from molecular oribital calculations and implications for structure, thermodynamics, and physical properties. Phys Chem Minerals 11:284-298

Navrotsky A (1986) Thermodynamics of silicate glasses and melts. Mineralogical Assoc Canada Short Course in Silicate Melts, CM Scarfe (ed) 12:130-153

Navrotsky A, Ziegler D, Oestrike R, Maniar P (1989) Calorimetry of silicate melts at 1773 K: measurement of enthalpies of fusion and mixing in the system diopside-anorthite-albite and anorthite-forsterite. Contrib Mineral Petrol 101:122-130

Navrotsky A, Maniar P, Oestrike R (1990) Energetics of glasses in the system diopside-anorthite-forsterite. Contrib Mineral Petrol 105:81-86

Navrotsky A (1990) Calorimetry of phase transitions and melting in silicates. Thermochim Acta 163:13-24

Navrotsky A (1992) Unmixing of hot inorganic melts. Nature 360:306.

Navrotsky A (1994) Physics and Chemistry of Earth Materials. Cambridge University Press, Cambridge, UK

Philpotts AR (1979) Silicate immiscibility in tholeitic basalts. J Petrol 20:99-118

Rammensee W, Fraser DG (1982) Determination of activities in silicate melts by Knudsen cell mass spectrometry —I. The system $NaAlSi_3O_8$-$KAlSi_3O_8$. Geochim Cosmochim Acta 46:2269-2278

Reyes RA, Gaskell DR (1983) The thermodynamic activity of ZnO in silicate melts. Metallurgical Trans B 14B:725-731

Richet P, Bottinga Y, Tequi C (1984) Heat capacity of sodium silicate liquids. J Am Ceram Soc 67:C6-C8

Richet P, Bottinga Y (1985) Heat capacity of aluminum-free liquid silicates. Geochim Cosmochim Acta 49:471-486

Richet P, Robie RA, Rogez J, Hemingway BS, Courtail P, Tequi C (1990) Thermodynamics of open networks: ordering and entropy in $NaAlSiO_4$ glass, liquid, and polymorphs. Phys Chem Minerals 17:385-394

Richet P, Robie RA, Hemingway BS (1993) Entropy and structure of silicate glasses and melts. Geochim Cosmochim Acta 57:2751-2766

Roy BN, Navrotsky A (1984) Thermochemistry of charge-coupled substitutions in silicate glasses: the systems $M_1^{n+}/_n AlO_2$-SiO_2 (M = Li, Na, K, Rb, Cs, Mg, Ca, Sr, Ba, Pb). J Am Ceram Soc 67:606-610

Ryerson FJ, Hess PC (1978) Implications of liquid-liquid distribution coefficients to mineral-liquid partitioning. Geochim Cosmochim Acta 42:921-932

Sakka S, Miyaji F, Fumuki K (1986) Structure of binary K_2O-TiO_2 and Cs_2O-TiO_2 glasses. J Non-Crystal Solids 112:64-68

Sen S, Stebbins JF (1994) Phase separation, clustering, and fractal characteristics in glass: A magic-angle-spinning NMR spin-lattice relaxation study. Phys Rev B 50:822-830

Sen S, Gerardin C, Navrotsky A, Dickinson JE (1994) Energetics and structural changes associated with phase separation and crystallization in lithium silicate glasses. J Non-Cryst Solids 168:64-75

Stebbins JF, Carmichael ISE, Weill DE (1983) The high temperature liquid and glass heat contents and the heats of fusion of diopside, albite, sanidine and nepheline. Am Mineral 68:717-730

Stebbins JF, Carmichael ISE, Moret LK (1984) Heat capacities and entropies of silicate liquids and glasses. Contrib Mineral Petrol 86:131-148

Stebbins JF, Farnan I, Xue X (1992) The structure and dynamics of alkali silicate liquids: A view from NMR spectroscopy. Chem Geol 96:371-385

Stebbins JF, Farnan I (1992) Effects of high temperature on silicate liquid structure: A multinuclear NMR study. Science 255:586-589

Tarina I, Navrotsky A, Gan H (1994) Direct calorimetric measurement of enthalpies in diopside-anorthite-wollastonite melts at 1773 K. Geochim Cosmochim Acta 58:3665-3673

Topor L, Navrotsky A (1992) Advances in calorimetric techniques for high pressure phases. In: High Pressure Research: Application to Earth and Planetary Sciences. Y Syono, M Manghnani (eds) Tena Publishing Co, Tokyo, Japan, and Am Geophys Union, Washington, DC, p 71-76

Watson EB (1976) Two-liquid partition coefficients: Experimental data and geochemical implications. Contrib Mineral Petrol 56:119-134

Weill DF, Hon R, Navrotsky A (1980) The ignesous system $CaMgSi_2O_6$-$CaAl_2Si_2O_8$-$NaAlSi_3O_8$: Variations on a classic theme by Bowen. In: Physics of Magmatic Processes. RB Hargraves (ed) Princeton Univ Press, Princeton, New Jersey, p 49-92

Wright DP, Navrotsky (1985) A thermochemical study of the distribution of cobalt and nickel between diopsidic pryoxene and melt. Geochim Cosmochim Acta 49:2385-2393

Ziegler D, Navrotsky A (1986) Direct measurement of the enthalpy of fusion of diopside. Geochim Cosmochim Acta 50:2461-2466

Chapter 6

THERMODYNAMIC MIXING PROPERTIES AND THE STRUCTURE OF SILICATE MELTS

Paul C. Hess

Department of Geological Sciences
Brown University
Providence, RI 02912 U.S.A.

INTRODUCTION

A growing body of evidence shows that the solution properties of cations in multicomponent silicate melts depend not only upon the polymerization state of the liquid (i.e. the Si:O ratio) but also the identities and concentrations of the other cations, particularly the highly charged cations. An instructive example are the solution mechanisms for P_2O_5 which occurs in the form of $AlPO_4$ species in peraluminous melt but as K-phosphate chains in peralkaline melts (Gan et al., 1992; Mysen et al., 1981). Clearly, there is no single solution mechanism for P_2O_5 (or other oxides) in silicate melts. In general, a large part of the solution mechanism of a cation is the result of its interactions with its next-nearest neighbors, the other cations in the melt. The purpose of this paper is to bring order to the complex interactions that exist in silicate melts.

The key to understanding silicate melts is to establish a hierarchy of interactions between network-modifying cations, mainly monovalent and divalent cations but also certain trivalent cations, and the anionic complexes formed between highly charged cations and oxygen. The "peralkaline" effect, for example, is an expression of the hierarchies that exist between alkali silicate species and anionic complexes formed from highly charged cations such as P^{+5} or Ti^{+4}. When P_2O_5 is added to peralkaline silicate melts the homogeneous equilibrium

$$KOSi + POP = KOP + SiOSi$$

is displaced to the right with the formation of alkali phosphate species at the expense of alkali silicate species (Ryerson and Hess, 1980; Mysen et al., 1981). The phosphate species evolve from alkali orthophosphates at low P_2O_5 contents to chainphosphate species at high P_2O_5 contents whereas the alkali silicate sheet units, the Q^3 species [The Q^n designation refers to SiO_4 tetrahedra with n bridging oxygens.], are polymerized to form extended networks, or Q^4 species (Gan and Hess, 1992).

These equilibria establish that phosphates have a stronger affinity for network-modifying cations than do silicates. Can we predict or at least rationalize these hierarchies? What about the interactions in complex silicate melts with multiple network-forming cations and a range of anionic complexes? How are the network cations apportioned among the anionic complexes? The simple answer, of course, is that all species are apportioned in such a way to minimize the free energy of the melt. But how is this division accomplished? There are competing enthalpic and entropic effects to consider. What is needed is an understanding of how the microscopic state of the silicate melt, as expressed by the coordinations of cations and anions and their association as complex melt species, is related to the thermodynamic properties. Specifically, an equation of state that links the activities of melt oxide components to the activity coefficients of the melt species is desired. Unfortunately, the thermodynamic data base as obtained by calorimetry, and other techniques while growing at a smart pace, is still inadequate to meet this formidable objective.

The thermodynamic data base is much more extensive then widely recognized, however. The thermodynamic properties of silicate melts, particularly in the form of melt oxide activities, are contained in all crystal-melt phase diagrams.

The liquidus phase diagram described by the coordinates temperature, pressure and component mass (or mole) fractions (P,T,X) is a remarkable invention. The liquidus phase diagram, the workhorse in igneous petrology, has a number of practical applications. To some workers, it is a graphical record of a set of experimental results. But read correctly, it describes quantitatively the phase transitions that occur in response to changes in pressure and temperature, the paths of crystallization or melting and the resulting liquid lines of descent or ascent! What is often overlooked, however, is that the phase diagram is a storehouse of thermodynamic information.

Gibbs (1961) secured a solid theoretical foundation for the phase diagram. The P,T,X phase diagram records the phase or phase assemblage which minimizes the specific Gibbs free energy of the system at a given P,T,X. In addition, the conditions of heterogeneous equilibrium require that coexisting phases have the same temperature, pressures (assuming only hydrostatic conditions) and the chemical potentials of the phase components (a phase component is a component that can be independently varied in the phase). The thermodynamic properties of the silicate melts, specifically some of the chemical potentials and activities of the melt oxide components, are uniquely determined along the liquidi of unary crystalline phases (Appendix 1). All liquids coexisting with quartz, for example, are characterized by chemical potentials of SiO_2 which are functions only of P and T. Cotectic liquids coexisting with two or more unary crystalline phases uniquely fix a whole array of chemical potentials and activities of melt components and species. Cotectic liquids in equilibrium with forsterite and enstatite at a specific P and T for example, determine the chemical potentials not only of the oxide components MgO and SiO_2 in the melt but also the chemical potentials of the melt species Mg_2SiO_4, $Mg_3Si_2O_7$ and $MgSi_2O_5$ (or whatever MgO-SiO_2 species of interest) (Appendix 1)!

The main point of this discussion is that liquidus phase diagrams are a vast, relatively unexploited thermodynamic resource. The objectives of this chapter are to link the structure of silicate melts to their thermodynamic mixing properties and to their phase diagrams; these data are supplemented by data obtained from the spectroscopic investigations of the melts or glasses. The multifaceted approach is used to link the macroscopic thermodynamic properties of silicate melts to their microscopic states.

The plan of this chapter is somewhat unconventional. The first sections are devoted to the crystal chemistry, phase relations and thermodynamic properties, primarily the enthalpies of formation, of simple crystalline oxides. The rationale of this approach is based on the conviction that an understanding of the structural and the enthalpic characteristics of silicate melts is advanced by a consideration of the structure and enthalpic properties of their crystalline counterparts. This quasicrystalline approach clearly has its theoretical limits. After all, it is the entropic part of the Gibbs free energy that makes the liquid more stable than the solid! It is therefore dangerous to assume without justification that some of the structural and thermodynamic properties of the crystalline state are transferred to the molten state. Notwithstanding these real concerns, it is surprising that trends observed in the crystalline states, for example, trends in the enthalpies of formation, exist also in the molten state. The effectiveness of quasi-crystalline comparisons is founded on the fact that the interactions between cations and anions in condensed phases are dominated by short range forces (Gibbs, 1982). This hypothesis is one of the central themes of this paper.

CRYSTAL CHEMISTRY OF SIMPLE SILICATES

Observed crystalline silicate structures represent only a small fraction of those that are topologically possible (Dent Glasser, 1979). Explanations for the existence or non-existence of silicate structures are typically couched in crystal chemical language. There certainly is a large element of validity in this approach. But this approach considers only part of the problem. Whether a phase is stable is not simply dependent of the thermodynamic property of the phase in question but depends also on the thermodynamic properties of neighboring phases. $FeSiO_3$ does not appear in low pressure phase diagrams, for example, because fayalite and quartz are more stable and not because $FeSiO_3$ is unstable. It is useful, nevertheless, to investigate what crystal chemical limitations are placed on the chemical compositions of crystalline structures. To this end, unary crystalline compounds which exist either stably or metastably at ambient conditions in binary silicate systems are summarized in Table 1. The systems are arranged in terms of increasing ionic potential, z/r, of the network-modifying cation, where z is the formal charge and r is the radius of the cation in a given coordination (CN).

Several general observations are noteworthy. First, the number of crystalline compounds and the extent of polymerization of the silicate anion decrease as the ionic potential of the network-modifying cation increases. Compounds formed between small, highly charged cations such as Ti^{+4} or Nb^{+5} and silicate anionic groups generally do not exist. This observation is true quite generally. Compounds (X_xY_yO) formed between highly electronegative cations (X) and oxyanionic groups of highly electronegative cations (Y) are relatively uncommon (Wells, 1985); $ZrSiO_4$ and $Si_2P_2O_9$ are some of the notable exceptions (Levin et al., 1964). The greater the polymerization of the anionic group the more limited are the structures containing network-modifying cations of high ionic potential.

The most polymerized anhydrous silicate structures, the sheet silicates (Q^3) or "interrupted" networks (Q^3,Q^4) do not exist for network-modifying cations with ionic potentials greater than about 1.4. The $BaSi_2O_5$ sheet silicate is the only example of such a highly polymerized structure with other than a monovalent network-modifying cation. Liebau (1985) has argued that the stability of these structures are limited by the size of the sheet formed by the coordination polyhedra of the network-modifying cation. Whenever there is a misfit between the network-modifying cation sheet and the silicate sheet, it is the silicate sheet that adjusts to the steric constraints imposed by the network-modifying sheet. The silicate sheet, formed by corner sharing tetrahedra, has a range of alternative configurations of comparable energies and can be perturbed without a large energy penalty.

The cation-sheet, in contrast, is formed of edge or face sharing polyhedra that do not have these structural options available. The ideal fit between the silicate sheet and the network-modifying cation sheet is for the large monovalent Cs cation (Liebau, 1985). As the network-modifying cation becomes smaller and increases its charge, the silicate sheet must make progressively more extreme adjustments until the strain reaches the critical level for which the compounds are unstable. This critical limit is reached for divalent cations smaller than Ba^{+2}. Substitution of the larger AlO_4 tetrahedron for the SiO_4 tetrahedron should expand the sheet and thereby accommodate larger network-modifying cations. This prediction is confirmed as the alumino-silicate sheet silicates $M (AlSiO_4)_2$ with $M = Ca, Sr$ and Pb exist (Liebau, 1985). Cations of large ionic potential are also capable of forming anhydrous sheet silicate provided that large monovalent cations are included within the cationic sheet. Anydrous sheet silicates of $K_3Nd(Si_6O_{15})$ and $K_2(Ti,Zr)(Si_6O_{15})$ are stable whereas the alkali free counterparts are not. These features provide useful insights for silicate melts that will be discussed later.

The vacant regions that exist in the low SiO_2 regions of Table 1 are more difficult to rationalize. At the time Dent-Glasser (1979) wrote a paper on "Non-existent silicates" orthosilicates with K, Rb or Cs were unknown. Dent-Glasser argued that the large size of K (and also the larger Rb and Cs cations) and the Pauling radius-ratio rule required that K could not coordinate with less than six oxygens. This assumption means that six K atoms must be in contact with the non-bridging oxygen, a condition difficult to satisfy for an oxygen that is part of a silicate tetrahedron. Thus, packing considerations alone argue against the existence of the K_4SiO_4 compound. But, in fact, K_4SiO_4 has since been synthesized (Bernet and Hoppe, 1990)!

It has long been recognized but perhaps not widely known that the ionic radius ratio rules for coordination numbers are unreliable. Consider, as a relevant example, the structures of the alkali oxides, Li_2O, Na_2O, K_2O and Rb_2O. The ionic radii for octahedrally coordinated alkalis range from 0.59 $A°(Li)$ to 1.50$A°(Rb)$ (Shannon, 1976) yet the alkalis in these oxides are in four-fold coordination with oxygen (Wells, 1985), in clear violation of the radius ratio rules! Why did these rules fail so miserably in these simple oxides? The answer is simple. If the alkalis were in six-coordination with oxygen, then the oxygen must

Table 1. Unary crystalline silicates which exist stably
or metastably in binary SiO_2 systems.

Metal Oxide	CN	Z/r	SiO₄O	SiO₄	Si₂O₇	SiO₃	Si₂O₅	SiₓOᵧ
	CN	Z/r	SiO_4O	SiO_4	Si_2O_7	SiO_3	Si_2O_5	Si_xO_y
Cs	6	0.6			x	x	x	x
Rb	6	0.7			x	x	x	x
K	6	0.7		x	x	x	x	x
Na	6	1.0		x	x	x	x	x
Li	5	1.4		x	x	x	x	x
Ba	8	1.4		x		x	x	
Sr	6	1.7	x	x		x		
Ca	6	2.0	x	x	x	x		
Fe	6	2.6		x				
Mg	6	2.8		x		x		
La	7	2.8	x		x			
Lu	8	3.1	x		x			
Zr	8	4.8		x				
Al	6	5.7	x					
Ti⁺⁴	6	6.7						
Nb⁺⁵	6	7.8						
P	4	29.0					x	

z/r = ionic potential; z=formal charge of M_i r=ionic radius in coordination (CN) specified.
(from Levin et al., 1964)

be in 12-coordination with alkalis (i.e. the number of cations times their coordination number equals the number of anions times their coordination number). Such close-packings of cations would lead to cation-cation repulsions (O'Keeffe and Hyde, 1984) and contribute unfavorably to the stability of compound. Even a five coordinated alkali requires 10-coordinated oxygen. The four coordinated alkali, however, allows for an 8- to coordinated oxygen, a coordination that is an upper limit for oxygen in the vast majority of crystalline oxides. In fact, the coordination of oxygen typically is six or less in most crystalline silicates and oxides. It is the balance between oxygen-oxygen and cation-cation repulsions, not radius ratio rules alone, that control coordinations and the stability of ionic compounds (O'Keeffe and Hyde, 1984). This analysis places no constraints on the existence or nonexistence of Rb_4SiO_4 and Cs_4SiO_4.

Note that there exist crystalline compounds less polymerized than the orthosilicate, i.e. compounds with $X_{SiO_2} < 0.33$. These include the oxyorthosilicates $(M^{+2})_3SiO_4O$ with $M^{+2} = Ca^{+2}, Sr^{+2}, Eu^{+2}$ and Pb^{+2} and $(M^{+3})_2SiO_4O$ where M^{+3} are the trivalent rare earth elements (REE), Y (Felshe, 1973) and Al. The structures are "multidisperse" and contain two anionic groups: the anionic groups are the isolated SiO_4 (Q^o) tetrahedron and an additional M_xO_y complex where some of the oxygen ions are not bonded to silicon ("free oxygen" or O^{-2}). Alkali oxyorthosilicates are conspicuous by their absence. A simple rationalization is that alkali cations do not bond strongly with the free oxygens in the M_xO_y anionic groups. In contrast bonds between rare earth elements and the free oxygens are much shorter than the bonds between rare earths and non-bridging oxygen. These short bonds are less than the sum of the ionic radii, an indication of high bond strengths brought about by the strong polarizing forces of the rare earth cation (Felshe, 1973). The strong polarizing power of the rare earth cations, a consequence of the poor screening of the nuclear charge by f electrons, also results in extreme distortions of the individual SiO_4 tetrahedra. These considerations are consistent with our conclusion that cations of low ionic potential do not coordinate efficiently with free oxygen in these structures.

ENTHALPIES OF FORMATION OF SIMPLE SILICATES

The purpose of this section is to relate the energies of bond formation to the existence and nonexistence of silicate compounds. The enthalpy of formation is taken relative to the oxides. The enthalpy of formation of forsterite, for example, is the enthalpy change of the reaction.

$$2\ MgO + SiO_2 = Mg_2SiO_4 \tag{1}$$
$$\text{Periclase} \quad \text{Quartz} \quad \text{Forsterite}$$

The magnitude of enthalpy change must depend on a number of factors including the coordination changes of cations and anions and the reorganization of next nearest neighbors. Whereas the coordinations of Mg and Si are the same in forsterite as in periclase and quartz respectively, the coordination of oxygen is not; oxygen is coordinated with 6 Mg in periclase and 2Si in quartz but with three Mg and one Si in forsterite. The enthalpy change associated with the changes in next nearest neighbors can be described by the reaction

$$MgOMg + SiOSi = 2\ MgOSi \tag{2}$$

which is the longhand form of the "oxygen" reaction

$$O^{-2} + O^0 = 2\ O^- \tag{3}$$

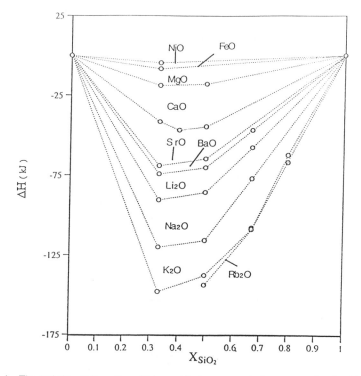

Figure 1. The enthalpy of formation of binary silicate compounds from one mole of oxides at 298 K. Thermodynamic data are from Barin et al. (1993) and Viellard and Tardy (1988).

commonly used to explain homogeneous equilibrium in silicate melts (Toop and Samis, 1962). We show later that the enthalpies of formation of the silicates relative to their oxides are comparable to the enthalpy of mixing of molten silicates relative to their molten oxides.

The enthalpies of formation (relative to the crystalline oxides) of a select number of stable binary crystalline silicates at 298°K are displayed in Figure 1. The thermodynamic data are chosen at this temperature to avoid the complication of melting or decomposition of the crystalline oxides. Fortunately, the enthalpies of formation relative to the oxides are not strongly temperature dependent. The enthalpy of formation of forsterite, for example, increases by only 7% from 298 to 1800 K (Robie et al., 1979). Whatever systematics are identified between the enthalpies of formation and bulk composition at one temperature will generally (with one important exception!) remain valid to temperatures extending up to the solidus. Note also that $\Delta \overline{H} \simeq \Delta \overline{G}$ at these temperatures, making the enthalpy and free energy of formation nearly equivalent.

Some care must be exercised when thermodynamic data obtained form the literature are plotted on this figure. Note that the enthalpies of the oxides are given on a per mole basis. Any representation of a molar property of a phase of intermediate composition must also be normalized to one mole of the oxides. As a consequence, the enthalpies recorded in Figure 1 are given in terms of one mole of the oxide components and not to one mole of the phase. The enthalpy of formation of forsterite, for example, is the enthalpy of forming 1/3 mole of Mg_2SiO_4 form one mole of the oxides or

$$2/3 \, MgO + 1/3 \, SiO_2 = 1/3 \, Mg_2SiO_4 \; ; \quad \Delta\overline{H} = 1/3 \, \Delta H^F \qquad (4)$$

where $\Delta \overline{H} F$ is the enthalpy of formation of one mole of forsterite.

The enthalpies of formation of the unary silicates are joined to form a concave upward series of segments which become progressively more negative in the order ("the sequence")

$$Ni < F\,e < Mg < Ca < Sr < Li < Ba < Na < K < Rb \, ,$$

a sequence that occurs partly or wholly in other systems and in silicate melts as well (Hess, 1991 and see below). The enthalpy of formation for a given network-modifying cation oxide (MO) is most negative at the orthosilicate composition or, in some instances, the sorosilicate composition. The enthalpies of formation are generally most negative for compounds with one to zero bridging oxygen per silicon. The minimum occurs near $X_{SiO_2} \approx 0.33$ for equivalent molten systems giving the enthalpy of mixing curves for binary silicate melts a decided asymmetry (Blander and Pelton, 1987). Some authors have associated this asymmetry with the maximum "degree of ordering" within the silicate liquid. Since the same asymmetry appears in the enthalpies of the crystalline compounds, it is obvious that the interpretation of the enthalpy minimum with a fully ordered structure is without meaning. The asymmetry, however, is simply related to the homogeneous reaction

$$SiOSi + MOM = 2 \, SiOM \qquad (5)$$

which leads to the formation of non-bridging oxygen. The orthosilicate composition in the crystalline compound and in the molten state represents the composition where all (crystalline) or nearly all (melt) bridging and free oxygen are converted to non-bridging oxygen. It is no mystery, therefor, why the minimum exists where it does.

The asymmetry can be changed, of course, by recasting the compounds into new species. Indeed a more logical choice of oxide components is to express them on an equal oxygen basis, for example, as Na_2O, $Si_{0.5}O$. The orthosilicate composition occurs at $X_{Si_{0.5}O} = 0.5$ and the asymmetry in the DH curves disappears!

The enthalpy curves display several interesting features. The number of stable crystalline compounds and the crystalline compound with the most polymerized silicate anion occur in the systems with the most negative (favorable) enthalpies. The most polymerized silicate compounds, however, are only marginally more stable (in terms of their enthalpies) than their neighboring phases. Indeed, the enthalpy of the sheet silicate with Na,Ba or Li is nearly the same as the enthalpy of an equivalent mixture of the respective chain silicate and quartz. It should come as no surprise, therefore, that highly polymerized melts in these systems become unstable (immiscible) with respect to their neighboring melts at comparatively high temperatures. Very large two liquid fields occur also in the SiO_2-rich regions of the SrO-,CaO-,MgO-, FeO- and NiO-SiO_2 systems where no crystals are stable in the subsolidus (see later).

This idea is completed by the observation that the enthalpies of orthosilicates or metasilicates are significantly more negative than the enthalpies of an equivalent mixture of neighboring phases. Melts less polymerized than metasilicate for these systems are stable with respect to unmixing at all temperatures investigated to date.

The enthalpies of formation are most negative for those silicates formed with the network-modifying cations of small ionic potentials (compare Table 1). This concept is developed more fully below; but generally the enthalpies of formation are most negative for compounds formed from "network-modifying" cations of low ionic potential with anionic groups formed by cations of high ionic potential. This observation has been noted before (Dowty, 1987, Navrotsky, 1982, Hess, 1980). What is even more interesting is that the interactions depend not only on the nature of the cations X, Y in the compound X_xY_yO but also on the state of polymerization of the anionic group.

The enthalpy of formation of a given silicate compound from its oxides can be related to the enthalpy change of forming one non-bridging oxygen, that is, the enthalpy change of the "oxygen" reaction

$$0.5 \, O^{-2} + 0.5 \, O^0 = O^-$$ (6)

The enthalpy of formation of forsterite on a per non-bridging oxygen basis is the molar enthalpy divided by four. Similarly, the enthalpy of formation of enstatite per non bridging oxygen (the "normalized enthalpy") is obtained by dividing the molar enthalpy by two and so on. The normalized enthalpies vary significantly with the extent of polymerization of the silicate anion.

The enthalpy of formation per non-bridging oxygen is most negative for compounds with the smallest number of non-bridging oxygen per silicon. These trends are best displayed for compounds in the K_2O-SiO_2 system. The enthalpy of formation of a mole of $K_2Si_2O_5$, a sheet silicate with only Q^3 tetrahedra, is -326 kj. Since each mole of $K_2Si_2O_5$ contains two non-bridging oxygen, the normalized enthalpy of forming a single Q^3 species is -163 kj. The enthalpy of formation per non bridging oxygen is -138 kj for the chain silicate K_2SiO_3 (Q^2) and -111 kj for the orthosilicate K_4SiO_4 (Q^0). There is a 52kj/O-stabilization of the non-bridging oxygen in the Q^3 tetrahedron over the non-bridging oxygen in the Q^0 tetrahedron. The enthalpy per non-bridging oxygen becomes less negative in the sequence $Q^3 \rightarrow Q^2 \rightarrow Q^1 \rightarrow Q^0$. These trends are displayed in Figure 2 where enthalpies

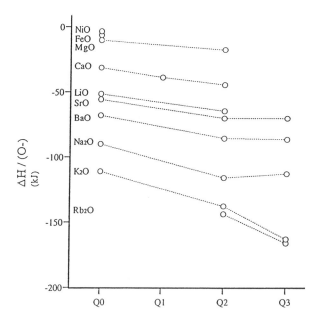

Figure 2. The enthalpy of formation of one mole of the compound from its oxides at 298 K divided by the number of non-bridging oxygen per mole. The Q^n notation refers to the number of bridging oxygens per SiO_4 tetrahedron.

of formation per non-bridging oxygen for binary silicates are plotted against their Q^n number.

The trends in the normalized enthalpies are duly reflected in the crystalline structures. It is well established that the length of the non-bridging Si-O bond decreases with increasing polymerization of the silicate anion (Liebau, 1985). The effect is to strengthen the SiO bond (Uchino et al., 1991). Thus, the non-bridging Si-O bond strengthens in the order of increasing polymerization of the silicate anion from Q^0 to Q^3 in the same system. In the same manner, the SiO non-bridging bond becomes shorter on a given silicate anion in the order Li,Na,K (Uchino et al., 1992). Thus, the SiO bond strengthens in the same order. Both these results are summarized in the normalized enthalpies of the alkali silicates.

These results have some useful implications for silicate melts if we assume that the trends in the normalized crystalline enthalpies carry over to their molten equivalents. Note that the difference between the normalized enthalpies between Q^0 and Q^3 species increase in the order Li,Na,K. The enthalpy change for the reaction, for example,

$$Q^2 + Q^4 = 2 Q^3 \tag{7}$$

is many times larger for the K_2O-SiO_2 system than for the Na_2O-SiO_2 and Li_2O-SiO_2 systems. It follows from this that the Q distribution in K_2O-SiO_2 melts should be more peaked around stoichiometric compositions than in either Na_2O or Li_2O-SiO_2 melts. This prediction is confirmed by NMR MAS spectra of alkali silicate glasses (Maekawa, et al., 1991). The Q^3 concentration in M_2O-SiO_2 glasses with $X_{M_2O} = 0.33$ is 86%, 79% and 63% respectively for M = K, Na or Li. (Table 2).

The very flat normalized enthalpy patterns for the alkaline earth silicates, and indeed, for silicates with either divalent or trivalent network-modifying cations, implies that the Q-distributions should be generally quite broad in these silicate melts. A high concentration of Q^3 species in these melts certainly is not favored since melts more SiO_2-rich than the metasilicate composition are unstable (immiscible) except at high temperatures. Raman spectra of $(Ca,Mg)O$- orthosilicate glasses reveal not only vibrations from the isolated silicate tetrahedra but also from non-bridging oxygen vibrations on dimers and even more polymerized silicate species (Cooney and Sharma, 1990). Both these results are compatible with the normalized enthalpies obtained from crystalline silicate.

Table 2. Distribution of Q^n species in binary M_2O-SiO_2 glasses.

	M_2O (mol %)	Q^4	Q^n % Q^3	Q^2
K_2O	20.0	49	51	--
	33.3	7	86	7
Na_2O	20.0	50	48	3
	33.3	11	79	10
Li_2O	25.0	38	53	9
	33.3	22	63	15

ENTHALPIC ELECTRONEGATIVITIES

The normalized enthalpies appear to be scaled to the difference in the electronegativities of the network-modifying and the network-forming cations. The normalized enthalpy of K_2SiO_3 is more negative than that of $MgSiO_3$ since the electronegativity difference $(X_{Si} - X_K)$ is greater than that of $(X_{Si} - X_{Mg})$ (Allen, 1993). But which electronegativity scales most effectively order the normalized enthalpies? The Pauling sale, which is based on the enthalpy of formation of binary compounds from their gaseous elements, fails to reproduce the normalized enthalpy sequence. Neither do the Mulligan or Allred Rochow scales which are based on the electronic properties of isolated atoms. None of these scales reproduce the K > Na > Ba > Li > Sr > Ca > Mg normalized enthalpy sequence. The scale that orders the normalized enthalpies is the spectroscopic electronegativity scale by Allen (1989).

The spectroscopic scale calculates the eletronegativity of a free atom from the ionization potentials of the atomic orbitals. The spectroscopic electronegativities of the monovalent and divalent cations are in the correct sequence. The spectroscopic electronegativities also order the highly charged cations correctly; for example, they correctly predict that the normalized enthalpies of K-bearing compounds are in the order S>P>B>Si. These results are unanticipated; why should atomic properties reproduce molecular properties? Part of the answer is that atomic properties are not totally erased during bond formation. The energetics of bond formation, nevertheless, cannot solely be the province of atomic properties. Molecular properties cannot quantitatively or even qualitatively be derived from the properties of the atom. Indeed, Duffy (1990) has demonstrated that the oxygen electronegativity, calculated after Pauling, cannot be assigned a fixed value if the cation electronegativities are fixed. Instead, the oxygen electronegativity increases with the electronegativity of the cation.

The electronegativity scales of atoms also miss the finer scale effects of bond formation. Recall that the normalized enthalpy of forming the non-bridging oxygen

$$1/2\ O^0 + 1/2\ O^{-2} = O^- \tag{8}$$

for the KOSi bond is not a constant but becomes more negative in the sequence $Q^0>Q^2>Q^3$. Constant atomic electronegativities are useless in discriminating between next-nearest neighbor effects. To be useful concept, an electronegativity scale must be designed to include these effects.

The approach followed here is to calculate group electronegativities rather than cation electronegativities for the most electronegative cation (see also Duffy and Ingram, 1976). A compound, in this approach, is divided into a relatively electropositive cation and an electronegative anionic group. The silicate compounds K_4SiO_4, K_2SiO_3 and $K_2Si_2O_5$, for example, are characterized by the electropositive cation, K^+, and the electronegative anionic groups (SiO_4^{-4}), (SiO_3^{-2}) and $(SiO_{2.5}^{-1})$; the latter are the expressions for Q^0, Q^2 and Q^3 tetrahedra respectively. Similarly, the group electronegativity for the compounds K_3PO_4, K_2SO_4 and $KAlSiO_4$ are the electronegativities calculated for the PO_4^{-3}, SO_4^{-2} and $AlSiO_4^-$ complexes, respectively.

The group electronegativities are calculated in the following way. The enthalpy of formation of the compound relative to the elements but normalized to the unit charge of the cation is expanded in a Taylor series as a function of the electronegativity difference (ΔX) of the anionic complex (X^-) and the cation (X^+)

$$H = H^0 + (\partial H/\partial \Delta X)(X^- - X^+) + 1/2\ (\partial^2 H/\partial \Delta X^{2)})(X^- - X^+)^2 \qquad (9)$$

where terms higher than quadratic are omitted. H° is the fictive enthalpy of the compound for which $\Delta X = 0$. The partial derivatives are evaluated where $\Delta X = 0$. Following Pauling (1960) H^0 is identified with the sum of enthalpy of the oxides per non-bridging oxygen. With this "arithmetic assumption" $(H - H^0)$ is the normalized enthalpy of formation of the compound. The normalized enthalpy of formation is an extremum at $\Delta X = 0$ so that the linear term in the expansion is zero. The normalized enthalpy of formation is therefore a quadratic function of the electronegativity difference between the cation and anionic complex.

$$\Delta H = 1/2\ a_2 (X^- - X^+)^2 \qquad (10)$$

I assume, for simplicity, that $a_2\ (= \partial^2 H/\partial \Delta X^2)$ is a constant and equal to -2, so that $\Delta H = -(X^- - X^+)^2$. The assigned value of a_2 has no consequences because it simply changes the X values by a constant.

The electronegativity scale, the "H-Scale", is obtained in a straightforward manner. The electronegativity of potassium is set to $X_K = 4.3$ as given by Allen's spectroscopic scale. The electronegativity of the anionic complex is calculated from the normalized enthalpy of formation of the K-compound. Given the electronegativity of the anionic complex and the normalized enthalpy of formation, the electronegativities of the remaining cations which form compounds with the given anionic complex are obtained. This process is repeated until the electronegativity scale is completed.

The results are given in Tables 3 and 4. By definition, the electronegativity of potassium is fixed and, by the procedure described above, the electronegativity of a given anionic complex is calculated only once. All inaccuracies resulting from the Taylor approximation are imbedded in the variations in the electronegativities of the cations. Note, for example, that the electronegativity of Ca ranges from 6.2 in $CaUO_4$ to 7.5 in $CaZrO_3$. The variation in the H value for Li is even wider, ranging from 5.1 in $LiPO_3$ to 7.1 in $LiHfO_3$. The H value for Li is therefore not a reliable index. These electronegativities are averaged so that one value is assigned to each cation (Table 3).

The electronegativities of the cations and anionic complexes are compared to the spectroscopic electronegativities in Table 3 and Table 4. The two electronegativity scales for the cations in Table 3 are in reasonable agreement with the notable exception of the Li,Sr sequence. In this case, the average electronegativities are misleading since the Li may have a smaller or larger electronegativity than Sr in a given compound. The spectroscopic scales, are of little use, however, in ordering the anionic electronegativities (Table 4). There are several reasons for this. The most obvious reason is that the group electronegativity depends not only on the central cation but also on the nature of the anionic group. The electronegativity of the silicate anion, for example, spans more than one electronegativity unit which is 22% of entire scale. Secondly, the coordination and charge of the central cation is a variable that atomic electronegativity scales do not consider.

The electronegativity scale summarized in Tables 3 and 4 is clearly an oversimplified model for the energetics of bond formation. First of all, it is solely an enthalpy based scale (as are most electronegativity scales) that is independent of temperature. Second, the scale

Table 3. H-scale and electonegativity scale for network-modifying cations.

Cation	H-Scale	Electro-negativity
Cs	4.1	3.9
Rb	4.2	4.2
K	4.3	4.3
Na	5.1	5.1
Li	6.3	5.3
Ba	5.6	5.2
Sr	6.1	5.7
Ca	6.8	6.1
Mg	8.0	7.7
Pb	8.1	--
Mn^{+2}	8.2	9.2
Fe^{+2}	8.3	9.9
Ni	8.6	11.0

Table 4. H-scale for anionic groups and electronegativity of the cation in the anionic group.

Anionic Group	H-Scale	Electro-negativity of cation
SeO_4	13.2	14.3
PO_3	13.0	13.3
SO_4	12.9	15.3
NO_3	12.6	18.1
CrO_4	11.6	8.6
PO_4	11.5	13.3
NbO_3	11.5	7.4
SeO_3	11.5	14.3
MoO_4	11.3	8.2
AsO_4	10.9	13.1
BO_2	10.9	12.1
CO_3	10.9	15.1
$SiO_{2.5}$	10.5	11.3
TiO_3	10.4	7.4
SiO_3	10.0	11.3
$SiO_{3.5}$	9.9	11.3
AlO_2	9.8	9.5
FeO_2	9.6	9.9
SiO_4	9.4	11.3
TiO_4	8.8	7.4

accounts for only nearest-neighbor effects. Third, the scale is not a function of the variations in the coordination of the cation or the anion coordinated by these cations Saying this, it is nevertheless true that the concept of cation and group anionic properties will prove of use in rationalizing or predicting thermodynamic and structural properties of melts and crystals.

The utility of the enthalpy scale is highlighted by the effective use of exchange potentials. Consider the following exchange equilibrium

$$K_2SiO_3 + 2\ Ca_{0.5}PO_3 = CaSiO_3 + 2\ KPO_3 \qquad (11)$$

which can be rewritten as

$$2\ Ca_{0.5}PO_3 - CaSiO_3 = 2\ KPO_3 - K_2SiO_3 \qquad (12)$$

The terms on either side of Equation (12) are the "exchange potentials". The exchange potentials are written so that their exchange enthalpies are negative. But more importantly the enthalpy of exchange potential for the RHS is significantly more negative that of the LHS. This result means that the RHS of the exchange equilibrium (11) is favored over that on the LHS. In general, the exchange potentials have the most negative enthalpies if they are formed from cations to the left of the "sequence"; cations of low enthalpic electronegativity will combine with anionic groups with the greatest enthalpic electronegativity.

ENTHALPY OF SIMPLE "MINERAL" MELTS

This section examines the hypothesis that the thermodynamic properties of crystalline oxides (including silicates) have a reasonably close relationship to the physical and thermodynamic properties of their melts. The correspondence between the crystalline and liquid state can't be pushed too far, of course. The free energy of an ordered crystalline phase relative to its oxides [the free energy of formation relative to an equivalent mixture of oxides] is largely due to its enthalpy of formation; that is, the enthalpy of formation of the compound from its oxides is much larger than the entropic contribution to the free energy. This fact remains true even at the melting point where the entropic contribution is maximum. The entropy of formation is relatively small because the vibrational states of the constituent atoms are not greatly changed as they are transferred from their oxide environment to that of the compound. In contrast, new species involving next nearest neighbors are formed in the compound and the additional stability of these new bonds are reflected in the free energy of formation.

The formation of a melt is entropy driven, however. Indeed, the entropic contribution to the free energy of melting is negative and exactly compensates the positive enthalpic contribution to the free energy of melting at the melting point. The gain in entropy is derived from the increase in volume on melting (except for those rare cases where ΔV melting < 0) and the destruction of the crystalline lattice. In statistical thermodynamic language, these two effects increase the number of microstates that are consistent with the macrostate of the phase. In thermodynamic language, the configurational and the vibrational entropy of the melt exceeds that of the corresponding crystalline phase. It is true, therefore, that only the enthalpies of formation of crystalline and melt phases are likely to be similar. To what extent are the systematics of the normalized enthalpies of the crystalline solids carried over to the molten state? Are the normalized enthalpies of formation of the molten compounds from their liquid oxides ordered in the same way as the normalized enthalpies of their crystalline counterparts?

The normalized enthalpy of formation (the enthalpy per non-bridging oxygen) of molten $Mg_2SiO_{4(L)}$ from its liquid oxides $MgO_{(L)}$ and $SiO_{2(L)}$ is the enthalpy change of the reaction

$$2/4\, MgO_{(L)} + 1/4\, SiO_{2(L)} = 1/4\, Mg_2SiO_{4(L)} \tag{13}$$

The following thermodynamic data are required: (1) the heat of fusion of periclase, (2) the heat of fusion of cristobalite, and (3) the heat of fusion of forsterite all at their respective melting points. The heats of fusion (ΔHFu) can be corrected to a common temperature from

$$\Delta HFu(T) = \Delta HFu(Tm) + \int_{Tm}^{T} \Delta C\rho\, dT \tag{14}$$

where $\Delta HFu(Tm)$ is the heat of fusion at the melting point, $\Delta HFu(T)$ the heat of fusion at the desired temperature and $\Delta C\rho$ is the heat capacity change on fusion. The enthalpies of fusion, $\Delta HFu(T)$, when added to the heats of formation of the crystalline oxides yield the desired enthalpies of the liquid oxides and compounds.

The enthalpy of mixing of simple binary silicate melts corresponds to the enthalpy change of the equilibrium (using the enthalpy of liquid Mg_2SiO_4 as an example)

$$2/3\, MgO + 1/3\, SiO_2 = 1/3\, Mg_2SiO_4 \tag{15}$$

where all terms refer to the liquid enthalpies of the corresponding components. Note that the ΔH^{mix} is for one mole of the oxides, not one mole of liquid Mg_2SiO_4! The enthalpies of fusion are given in Barin et al. (1993) but only at the respective melting temperatures. The heat capacities of fusion are used to bring the liquid enthalpies to the same temperature.

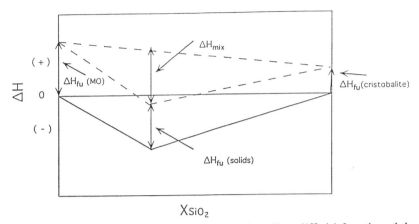

X_{SiO_2}

Figure 3. Steps to calculate the enthalpy of mixing of molten silicate (ΔH_{mix}) from the enthalpy of formation of the crystalline solid from its crystalline oxides.

These enthalpies are then substituted into the equation above to obtain the enthalpy of mixing of the appropriate liquid compound. Figure 3 describes the steps necessary to obtain the liquid enthalpy of mixing.

The diagram in Figure 3 highlights an important result. Note that the enthalpy of fusion of the oxides and silicate compound partially cancel. We can anticipate that the ordering of enthalpies of formation obtained for crystalline compounds will be similar to those obtained for the liquid state. Indeed, the magnitudes of the enthalpies of formation of the respective crystalline and liquid compounds should be similar, provided that the enthalpies are obtained at identical temperatures.

The liquid enthalpies at 1600 K of certain mineral compounds are given Figure 4. A few words of caution are in order. First, the enthalpies of fusion for the Ba and Sr silicates are not known. The enthalpies are estimated by equating the entropies of fusion of Ba and Sr compounds to the entropy of fusion of the typical ortho- and meta silicate and then multiplying the value by the melting temperature. Secondly, the listed enthalpies of fusion of some of the alkali silicates are suspect. The entropies of fusion of Li_4SiO_4 and Li_2SiO_3 are ~40% of the typical values for orthosilicate and metasilicates respectively. These low entropies suggest that the enthalpies of fusion are low and that the enthalpy of mixing of liquid Li_4SiO_4 and Li_2SiO_3 are too negative. Indeed, the liquid entropies of mixing at 1500 K of these liquids are sharply negative whereas the true entropies of mixing should be positive! Similar results are obtained for liquid Na_4SiO_4 whereas the enthalpy data for liquid K-, Rb- silicates and the $Li_2Si_2O_5$ and $Na_2Si_2O_5$ liquids do not appear to be grossly in error.

The liquid enthalpies are related, as was the case for the crystalline enthalpies of formation, to the reaction

$$O^{-2} + O^0 = 2\,O^- \tag{16}$$

where oxygen ions in one state of polarization are transferred to those in another state. The differences between this reaction in the crystalline and liquid state are important, however. We again use the liquid and crystalline states of Mg_2SiO_4 to illustrate the main differences.

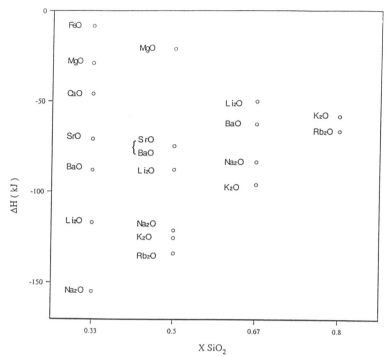

Figure 4. Enthalpies of mixing for certain mineral melts at 1600 K.

In the crystalline state, Reaction (16) is heterogeneous and the free- and bridging- oxygen are totally converted to non-bridging oxygen. In contrast, Raman spectra of Mg_2SiO_4 glass show that bridging oxygen and by mass balance the free oxygen, exist in homogeneous equilibrium with the non-bridging oxygen; the homogeneous reaction does not go to completion (Cooney and Sharma, 1990). In the crystalline state, the non-bridging oxygen is located solely on Q^0 silicate tetrahedra; the crystalline state is monodisperse. The Mg_2SiO_4 glass, and probably the corresponding melts, are polydisperse. The non-bridging oxygen are located not only on silicate monomers (Q^0) but a small percentage are parts of silicate dimers (Q^1) and probably small silicate chains (Q^2). The free oxygen must be associated with various $(MgO)x$ complexes, but there is no independent spectroscopic evidence to identify the nature of these species.

Notwithstanding these differences between crystalline and liquid states, it is noteworthy that the sequence of liquid enthalpies follows, with one exception, the sequence obtained for the crystalline enthalpies of formation. The obvious exception is that the enthalpies of liquid Li-silicates are more negative than those of liquid Ba-silicates whereas the opposite is true of the crystalline enthalpies obtained at 25°C. It is not clear whether this switch is real since it has already been noted that the liquid enthalpies of Li_4SiO_4 and Li_2SiO_3 are probably too negative. Moreover, the liquid enthalpies of the Ba-silicate are calculated by assuming "reasonable" enthalpies of fusion. The results are probably correct, however, the enthalpies of formation of the crystalline Li-orthosilicates and Li-metasilicates become more negative with temperature and become roughly equal to those of the corresponding Ba-silicates at temperatures in excess of 1400 K. Thus, the switch in the Li to Ba enthalpy sequence is anticipated even in the crystalline state, making the liquid enthalpy sequence less extraordinary. We conclude, therefore, that the liquid enthalpy series is, from least to most negative,

$$Mg < Ca < Sr < Ba < Li < Na < K < Rb \,, \tag{17}$$

a series that explains many of the phase properties of silicate systems.

ANALYSIS OF PHASE DIAGRAMS

Cristobalite-tridymite liquidi

The thermodynamic and microscopic properties of silicate melts can be inferred by examining how the liquidus fields of crystalline phases, particularly unary phases, are shifted owing to the addition of certain components to the melt (Hess, 1991; Ryerson, 1985; Hess, 1980; Kushiro, 1975). The liquidi of unary phases are particularly useful because such phases uniquely fix the chemical potential and activity of the phase component at the stated T,P. The liquidus of cristobalite, for example, is a surface of constant a_{SiO_2} at fixed T,P. The coexistence of two unary phases and silicate melt in a binary system provides even more thermodynamic constraints because such assemblages uniquely determine not only the chemical potentials of the phase components but also the chemical potentials of the system components and the chemical potentials of the melt species (see Appendix). Ryerson (1985) and others used these powerful ideas to determine the relationship between activity coefficients of SiO_2 (Fig. 5) and melt composition. The liquidus surface of cristobalite as given by the SiO_2 mole fraction of the liquid at fixed T,P decreases in the order (Ryerson, 1985)

$$Cs > Rb > K > Na > Li > Ba > Sr > Ca > Mg > Fe \,, \tag{18}$$

provided that the moles of the network-modifying cations are expressed on an equal oxygen basis (the oxides are N_2O, MO where N and M are monovalent and divalent cations

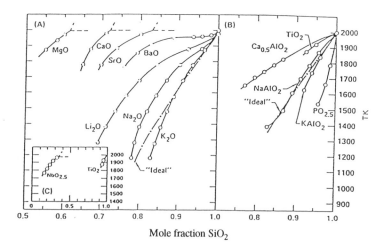

Figure 5. Experimentally determined silica polymorph liquidus surfaces (Ryerson, 1985).

respectively). But if the metal oxides are written on an equal cation basis as some suggest (Varshal, 1993), then Li and Ba switch places in the sequence. Which is the more "correct" sequence? Since the liquidus at fixed P,T is a surface of fixed a_{SiO_2}, the different ways of representing the liquid mole fractions simply require alternative ways of representing the liquid activity coefficient. The activity coefficient is a useful measure of the ideality, or lack therefore, of the liquid. Which sequence Li, Ba or Ba, Li more correctly describes the trend of non-ideality in the liquid? What is needed is an index of non-ideality that is independent of how the component is defined. The index exists and is the temperature of the critical point of stable or metastable two liquids fields in binary SiO_2 melts.

The critical temperatures of the stable and metastable two-liquid fields increase in the order (see later)

$$K < Na < Li < Ba < Sr < Ca < Mg \qquad (19)$$

(Hagemann and Oonk, 1986; Tewhey and Hess, 1980; Haller et al., 1974;). This index of non-ideality shows that the cation oxide should be expressed on an equal oxygen rather than on an equal cation basis. What we mean by this statement is that SiO_2-rich melts become more non-ideal when cations to the left of the sequence are replaced on a charge equivalent basis by a cation to the right of the sequence. This sequence is one of increasing activity coefficients of SiO_2; that is, the activity coefficient of SiO_2 at fixed P,T and X_{SiO_2} in high silica melts increases from K_2O-bearing to MgO-bearing melts (Ryerson, 1985). Are these relationships maintained in more complex systems? What other insights are gained by this analysis?

Cotectics and the activities of melt species

The power of this analysis is illustrated by examining how the forsterite-enstatite boundary curve is affected upon the addition (or substitution) of oxides of various network-forming and network-modifying cations (Fig. 6) (Ryerson, 1985). The thermodynamic properties of the cotectic melt can be expressed in several equivalent ways:

$$\underset{\text{forsterite}}{\mu_{Mg_2SiO_4}} + \underset{\text{melt}}{\mu_{SiO_2}} = \underset{\text{enstatite}}{2\mu_{MgSiO_3}} \qquad (20)$$

$$\underset{\text{forsterite}}{\mu_{Mg_2SiO_4}} = \underset{\text{enstatite}}{\mu_{MgSiO_3}} + \underset{\text{melt}}{\mu_{MgO}} \qquad (21)$$

When forsterite and enstatite are truly unary phases (there is negligible solid solution between these phases and the added components), the chemical potentials and the activities of Mg_2SiO_4 and $MgSiO_3$ are functions only of temperature and pressure. It follows from the equilibria described by (20) and (21) that the chemical potentials and the activities of SiO_2 and MgO in the melt are functions only of pressure and temperature. Since the activities are fixed, the shift of the liquidus boundary between forsterite and enstatite with the addition of other metal oxides records directly the changes in the activity coefficients of SiO_2 and MgO in these liquids after suitable corrections for the changes in temperature are made. Indeed, the substitution of MgO by equimolar amounts of K_2O, Na_2O, Li_2O or CaO shifts the forsterite-enstatite boundary curves towards higher X_{SiO_2} contents (Ryerson, 1985). This displacement of the boundary curve means that the activity coefficients of SiO_2 are shifted to smaller values. Lower activity coefficients mean that the substitution of MgO by metal oxide of this group results in lower activities of SiO_2 for the same mole fraction of SiO_2. The magnitude of the shift is in the order $K_2O > Na_2O > Li_2O > CaO$ (i.e. the activity coefficient of SiO_2 is lowest in the K_2O-MgO-SiO_2 system). Note that these effects

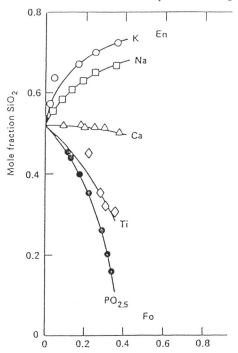

Figure 6. The displacement of the forsterite-enstatite cotectic due to the substitution of MgO by equimolar amounts of oxides.

are in the same order as obtained from the analysis of the cristobalite liquidus. These results are quite general. If we apply this analysis to the pseudowollastonite-silica boundary curve, for example, the substitution of CaO by K_2O or Na_2O decreases the activity SiO_2 whereas the substitution of MgO or FeO for CaO increases the activity of SiO_2 at constant X_{SiO_2}. The shifts in the activity coefficients of SiO_2 are in the order $K_2O > Na_2O > CaO > MgO > FeO$ again in agreement with the sequence.

The regularities in the shift of liquidus boundaries can be summarized in the following way; the liquidus boundaries are shifted towards SiO_2 when cations to the right of the sequence are substituted on a charge equivalent basis for cations to the left of the sequence. The net effect is that the liquidus field of the least polymerized silicate crystalline phase expands at the expense of the liquidus of the more polymerized silicate phase. These conclusions are important; can these rules be extended to include cations of higher charge?

The available phase equilibria are more limited for these ternary systems but the general observation is that cotectics in the MO-SiO_2 system are shifted to lower silica contents when oxides of cations of $\geq +3$ charge are substituted on an equal oxygen basis for oxides of the sequence. For example, the substitution of $P_{0.4}O$ for MgO causes a dramatic shift of the forsterite-enstatite cotectic to very low SiO_2 values. Consequently, the activity coefficient and activity of SiO_2 increases sharply in the direction of increasing $P_{0.4}O/MgO$. Indeed, the increase in non-ideality leads to the onset of silicate liquid immiscibility between SiO_2-rich and magnesian phosphate-rich melts (Levin et al., 1964).

Recall that the changes in the activity of SiO_2 obtained by exchanging the cation oxides of the sequence were predicted from the relative positions of the cristobalite liquidus in the appropriate metal oxide-silica system. These relationships are not true for liquids containing oxides of highly charged cations (Ryerson, 1985). The liquidus of cristobalite in the $P_{0.4}O$-SiO_2 system, for example, lies very close to the cristobalite liquidus of the K_2O-SiO_2 system. Yet the effect of adding K_2O or $P_{0.4}O$ to the MgO/SiO_2 system couldn't be more unlike; the addition of K_2O displaces the cotectics towards SiO_2, lowers activity coefficients of SiO_2 and eliminates the stable two liquid field that exists in the MgO-SiO_2 system. The addition of $P_{0.4}O$ does exactly the opposite! Indeed, the addition of $P_{0.4}O$ to the Na_2O-SiO_2 system generates a stable two liquid field in the ternary system when none exists in any of three binary joins (Levin et al., 1964).

The phase equilibria data, unfortunately, are not complete enough to develop a sequence for the oxides of highly charged cations. Such a sequence, if indeed one exists, is complicated by the various bonding opportunities available for highly charged cations. Phosphorus, again, serves as an illustrative example. Raman spectra of glasses along the P_2O_5-SiO_2 join indicate that PO_4 and SiO_4 tetrahedra copolymerize into an extended network (Mysen et al., 1981). In contrast, the speciation of P_2O_5 in fully polymerized subaluminous ($K = Al$ in moles) granitic melts is more complex; P_2O_5 exists as $AlPO_4$ and K-phosphate species ranging from monomers to small chains (Gan and Hess, 1992). In peraluminous melts ($Al > K$ in moles) and peralkaline melts ($K > Al$ in moles) of the same system phosphorus exists as $AlPO_4$ units and K-phosphate monomers and chains, respectively. There is no evidence for Si-O-P bonds indicative of a copolymerized phosphoro-silica network. It is not surprising, therefore, that the phase equilibria in the P_2O_5-SiO_2 system are inconsistent with the phase equilibria in more complex systems.

Analysis of cotectic shifts-concept of neutral species

How then can the phase equilibria systematics be explained? Why does the addition of an oxide of a high field strength cation to a MO-SiO_2 melt increase the activity coefficient of SiO_2? Hess (1977) proved that the activity of SiO_2 is proportional to the activity of the bridging oxygen within a melt of a given binary system (or the activity of the SiOSi species). We make the reasonable assumption that the activity of SiO_2 is proportional to the concentration of the SiOSi species. (This assumption cannot be true for coexisting immiscible melts since they have the same activities of SiO_2 but different concentrations of SiOSi species). Consider, then, the homogeneous equilibrium

$$SiOK + POSi = POK + SiOSi \qquad (22)$$

which describes the speciation within a K_2O-SiO_2-P_2O_5 melt. Recall that in peralkaline melts, POK species of various types were identified by spectroscopy but there was no evidence for the existence of POSi species (Gan and Hess, 1992). These data show that the right hand side of the homogeneous equilibrium is favored and that the corresponding equilibrium constant,

$$K = \frac{(POK)(SiOSi)}{(SiOK)(POSi)} \qquad (23)$$

is significantly greater than unity. It follows that the addition of P_2O_5 to MO-SiO_2 melts creates MOP species at the expense of MOSi species and in the process increases the activity (concentration) of SiOSi species. The response of the system is as observed; the cotectics in the MO-SiO_2 systems are displaced to lower SiO_2 concentrations, and therefore greater activities of SiO_2 at a given $XSiO_2$. The displacement of the cotectics away from SiO_2 is also consistent with a decrease in the activity coefficient and activity of K_2O. The decrease in the activity of K_2O is consistent with the formation of KOP species through Reaction (22).

We have used the available spectroscopic data to correlate the nature of melt speciation and the phase properties. Can the converse be done? What do the phase equilibria infer about the nature of melt speciation? Indeed, some useful information about melt speciation can be gleaned from the analysis of cotectics.

Consider the following thought experiment. Suppose that metal oxide MxOy is added to the forsterite-enstatite cotectic liquid with the result that the MgO/SiO_2 ratio of the cotectic liquid remains fixed at the original value. The cotectic is controlled by the homogeneous equilibrium

$$Mg_2SiO_{4(L)} + SiO_{2(L)} = 2\ MgSiO_{3(L)} \tag{24}$$

Because the cotectic is not shifted, we would argue that the activity coefficients of $Mg_2SiO_4(L)$, $MgSiO_3(L)$ and $SiO_2(L)$ are all affected similarly, all species are diluted equally with the addition of the metal oxide MxOy. For convenience, we call the metal oxide in this case a "neutral species" ie. the metal oxide does not interact preferentially with any of the melt species.

In general, the cotectics are displaced on the addition of the metal oxide. Our goal is to predict and also to explain these cotectic shifts by identifying the homogeneous equilibria that are responsible for the displaced heterogeneous equilibrium. To this end, we erect a tangent to the cotectic at the cotectic composition (where $X_{MxOy} \to 0$) and extend the tangent to one of the remaining two sides of the ternary diagram (Fig. 7). The composition obtained by this construction is the "neutral species" which when added to the cotectic liquid does not change the (normative) MgO/SiO_2 ratio of the cotectic melt. The composition of the "neutral species" identifies the complex that is formed when MxOy is added to the cotectic liquid.

To illustrate this hypothesis, consider the TiO_2-MgO-SiO_2 liquidus phase diagram at 1 bar (Fig. 8) (Levin et al., 1964). The addition of TiO_2 displaces the cotectics between periclase-forsterite, and forsterite-enstatite to lower SiO_2/MgO ratios. The tangents to these cotectics project close to the compositions Mg_2TiO_4 and $MgTi_2O_5$. These compositions are the "neutral species" defined earlier, and by hypothesis are the titanate species formed in

Figure 7. Concept of neutral species (see text).

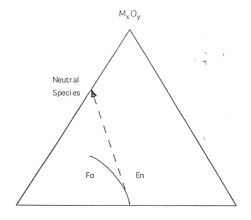

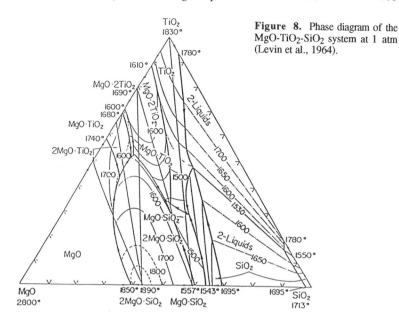

Figure 8. Phase diagram of the MgO-TiO₂-SiO₂ system at 1 atm (Levin et al., 1964).

the MgO-SiO₂ cotectic liquids. The heterogeneous equilibria at the two olivine-bearing cotectics are consistent with the following homogeneous equilibria (all are liquid species):

$$2\,MgO + TiO_2 = Mg_2TiO_4 \tag{25}$$

$$Mg_2SiO_4 + 2\,TiO_2 = MgTi_2O_5 + MgSiO_3 \tag{26}$$

The homogeneous equilibria describe the interactions of TiO_2 with the more MgO-rich species to produce a Mg-titanate species in (25) and in (26) a more polymerized Mg-silicate species. Mg_2TiO_4 and $MgTi_2O_5$ are the neutral species because these species do not combine with the most MgO-rich species, that is, the neutral species already have the requisite Mg-Ti ratio to preserve the homogeneous equilibrium.

This hypothesis provides a quick, if approximate, guide to the description of homogeneous equilibrium in silicate melts. Some results may not be as clear cut as above. It is possible, for example, that the tangent projects to points between known crystalline compositions. The composition identified by this projection might imply the existence of a liquid species that is not stable as a crystalline compound. Alternatively, more than one neutral species may be indicated. A survey of phase diagrams, however, shows that a surprisingly large number of "neutral species" compositions coincide with crystalline stoichiometric compositions. Conversely, the neutral species rarely correspond to compositions in binary systems which do not have stable compounds. In the CaO-SiO₂-Ta₂O₅ system (Levin et al., 1964), e.g., the cotectics to the CaO-SiO₂ binary system project to the CaO-Ta₂O₅ binary where there are a number of CaO-Ta₂O₅ compounds and not to the SiO₂-Ta₂O₅ binary where there are no stable unary Ta₂O₅-SiO₂ crystalline compounds near the solidus (Fig. 9). Moreover, stable single liquids in the Ta₂O₅-SiO₂ system are relegated to SiO₂- and Ta₂O₅-rich compositions. Instead, a large two liquid field occupies most of the phase diagram. (Ta,Si)xOy species apparently are not favored either in the crystalline or liquid states in this system as indeed implied by the compositions of the "neutral species"

Figure 9. Phase diagram of the CaO-Ta₂O₅-SiO₂ system at 1 atm (Levin et al., 1964).

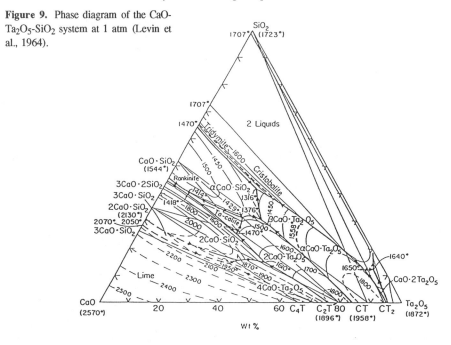

This analysis can be extended to all points on the cotectic curve as it passes into the ternary system. A curved cotectic implies that the melt speciation changes continuously with composition. The tangent to the cotectic between CaO and Ca_2SiO_4 in the CaO-SiO₂-P₂O₅ system (Levin et al., 1964), for example, projects first to the oxy-phosphate compositions and then with the addition of progressively larger amounts of P_2O_5, projects to orthophosphate and then to the metaphosphate phases (Fig. 10). The phosphate compositions represent the composition of liquidus crystalline phosphates. The increasing polymerization of the phosphate is consistent (Gan and Hess, 1992), with the addition of progressively larger amounts of P_2O_5 to peralkaline melts to generate phosphate species varying from ortho-phosphate to metaphosphate states of polymerization.

This analysis can also be used to describe the variation of the activity coefficient of the metal oxide when SiO_2 is replaced on an equal oxygen basis by oxides of highly charged cations. For example, the substitution TiO_2 for SiO_2 shifts the forsterite-enstatite cotectic away from MgO, requiring that the activity coefficient of MgO increases and the activity of MgO increases in liquids of constant mole fraction of MgO. In contrast, the substitution $P_{0.8}O_2$ for SiO_2 in the same equilibrium decreases the activity coefficient of MgO. The activity of MgO for fixed X_{MgO} increases in the order $P_{0.8}O_2$-SiO_2-TiO_2. But the effects are modest compared to the changes in the activity of SiO_2 when $Ti_{0.5}O$ and $P_{0.5}O$ are substituted for MgO.

Aluminosilicate systems

In order to bridge the gap towards more petrological melts, the liquidus and boundary curves are evaluated for aluminosilicate systems. Consider the position of the cristobalite liquidus in $MAlO_2$-SiO_2 systems (where M is a monovalent cation or one-half of a divalent cation). The positions of cristobalite liquidi in the $KAlO_2$-SiO_2 and $NaAlO_2$-SiO_2 systems

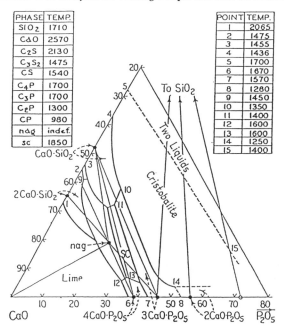

Figure 10. Phase diagram of a part of the CaO-P_2O_5-SiO_2 system at 1 atm (Levin et al., 1964).

more or less follow the ideal liquidus, that is, the activity of SiO_2 is nearly equal to the mole fraction of SiO_2 provided that the latter is calculated on an equal oxygen basis (Fig. 5) (Ryerson, 1985). In contrast, the liquidi in the $Ca_{0.5}AlO_2$ and $Mg_{0.5}AlO_2$ lie to significantly higher SiO_2 concentrations, and therefore, have activity coefficients of SiO_2 greater than unity. At a given temperature, the activity coefficients of SiO_2 increase in the order $KAlO_2 < NaAlO_2 < LiAlO_2 < Ba_{0.5}AlO_2 < Ca_{0.5}AlO_2 < Mg_{0.5}AlO_2$ which is parallel to the sequence (Levin et al., 1964). This sequence means that alkali subaluminous (M=Al in moles) melts are more ideal that alkaline earth subaluminous melts. Enthalpy of mixing curves for $MAlO_2$-SiO_2 glasses show the same trends; the enthalpy of mixing of $KAlO_2$-SiO_2 glasses are more exothermic than those of $Mg_{0.5}AlO_2$-SiO_2 glasses (Roy and Navrotsky, 1984). The enthalpy of mixing are more exothermic (favorable) in the order $Mg_{0.5}AlO_2 < Ca_{0.5}AlO_2 < Sr_{0.5}AlO_2 < Ba_{0.5}AlO_2 \leq LiAlO_2 < NaAlO_2 < KAlO_2$.

The complex thermodynamic and structural properties of aluminum are highlighted in granitic melts were the compositional variables are limited yet have extraordinary effects on these properties. Petrologists have long been aware that granitic melts of peraluminous, peralkaline or metaluminous compositions have widely different phase equilibria. It has been observed, for example, that the concentrations of high valence cations is significantly greater in peralkaline volcanic rocks than in either peraluminous or metaluminous volcanic rocks of comparable composition (see, for example, Gwinn and Hess, 1989). Watson (1979) showed by experiment that the solubility of zircon ($ZrSiO_4$) increased almost linearly with the K/Al ratio in peralkaline H_2O-saturated granitic melts. The solubility of zircon in peraluminous H_2O-saturated granitic melts was negligible by comparison. The K/Al mole ratio of approximately unity was a major divide separating regions of low solubility in peraluminous melts from those of high solubility in peralkaline melts. Since zircon is a unary phase in this system, the zircon liquidus is also a surface of fixed a_{ZrSiO_4}

(or μ_{ZrSiO_4}). The remarkable increase in solubility observed in peralkaline melts (the "peralkaline effect") records a remarkable decrease in the activity coefficient of $ZrSiO_4$ in these melts. It occurred to us that this system could sensibly be used to examine the thermodynamic properties of high valency cations by measuring the liquidus of unary phases containing these components.

The solubility of rutile (Dickinson and Hess, 1985; Gan and Hess (1991), cassiterite (Naski and Hess, 1985), zircon (Watson, 1979; Ellision and Hess, unpublished), Ce-monazite (Montel, 1986), La-phosphate (Ellison and Hess, 1988) hafnon (Ellison and Hess, 1986) and pseudobrookite (Gwinn and Hess, 1989) demonstrated clearly that not only is the peralkaline effect a general phenomenon, but also that there is also a peraluminous effect. Because peraluminous and peralkaline melts are very different it will be convenient to discuss their thermodynamic properties separately. But before we proceed, it is useful to give a brief discussion of the thermodynamic properties of the base (K_2O-Al_2O_3-SiO_2) melts themselves.

To focus the discussion, consider an isopleth of roughly 80 mole % SiO_2 in the K_2O-Al_2O_3-SiO_2 system. Whatever statements are true for this system will be qualitatively true of the corresponding Na_2O-bearing system. From the nature of the cristobalite liquidus (Levin et al., 1964), it is evident that the activity of SiO_2 is nearly constant to within a few percent variation in peralkaline melts. In contrast, the strong contraction of the cristobalite liquidus in peraluminous melts requires that the activity of SiO_2 decreases significantly with increasing Al_2O_3 - how much can't be quantified.

The activity of Al_2O_3 must increase with increasing Al_2O_3 contents from peralkaline to peraluminous melts. Indeed, reflecting this fact, the mullite liquidus underlies most of the peraluminous join. But a more useful index of the variation in the activity of Al_2O_3 is obtained from a crystal-melt partitioning study between pseudobrookite $(Fe,Al)_2TiO_5$ and melts along peraluminous-peralkaline isopleth (Gwinn and Hess, 1989). The crystal/liquid partition coefficient for Al_2O_3 varies from values greater than unity in the most peraluminous melts studies to approximately 0.05 in the most peralkaline melts studies. Whereas the partition coefficient varies by a factor of 20, the concentration of Al_2O_3 in these melts varies only by a factor of two! Assuming that the total variation in non-ideality is that of the liquid, then the activity coefficient of Al_2O_3 will have decreased by a factor of 20 from peraluminous to peralkaline liquids.

The activity of K_2O must certainly decrease from peralkaline to peraluminous melts but the magnitude of such a decrease can only be guessed. By symmetry, I expect that the changes be comparable in magnitude to those of Al_2O_3, but in the opposite sense, of course.

Rutile saturation surface

The application of phase equilibria to characterize the thermodynamic properties of various components in peraluminous to peralkaline granitic melts is illustrated below by examining the liquidus of rutile over a wide range of temperature, pressure and composition. Because rutile is a unary phase in this system, all melts saturated with rutile have a fixed activity of TiO_2 at constant T and P. We will set the $a_{TiO_2}=1$ meaning that the chemical potential of rutile is chosen as the standard state. Since at fixed T,P

$$a_{TiO_2} = \gamma_{TiO_2} X_{TiO_2} = 1$$

$$(27)$$

$$\gamma_{TiO_2} = \frac{1}{X_{TiO_2}}$$

it follows that the activity coefficient of TiO_2 in rutile saturated melts is equal to the reciprocal of the mole fraction of TiO_2 (X_{TiO_2}) in these melts. Indeed, the value of the activity coefficient of TiO_2 summarizes the non-ideal mixing properties of TiO_2 in these melts.

The generic phase diagram for rutile saturated melts is given in cartoon form in Figure 11. The phase diagram is obtained by adding sufficient TiO_2 to melts of initially fixed X_{SiO_2} but variable $K_2O/(K_2O+Al_2O_3)$ (moles = K^*) until rutile saturation is achieved. Note that the mole fraction of SiO_2 must vary inversely with the mole fraction of TiO_2; the rutile saturation surface is no longer an isopleth of constant X_{SiO_2}. It is a surface of fixed a_{TiO2}, however.

The rutile liquidus in the $K_2O-Al_2O_3-SiO_2$ melts shows the same qualitative dependence on the K^* ratio for temperatures from 1475°C to 927°C and in dry or H_2O-saturated melts up to 2 kbar (Gan and Hess, 1991; Dickinson and Hess, 1985). The solubility of rutile is roughly constant for peraluminous to subaluminous melts but then increases steadily with K^* in peralkaline melts. As deduced from our previous discussion, the activity coefficient of TiO_2 is roughly constant in peraluminous to subaluminous melts but then decreases significantly in more peralkaline melts. In rutile undersaturated melts of fixed TiO_2 (and X_{SiO_2}), the activity of TiO_2 must decrease from peraluminous to peralkaline melts. The rutile liquidus in this system is therefore a useful reference to examine the solution properties of TiO_2 in melts of varying composition. The results of several investigations are briefly summarized herein

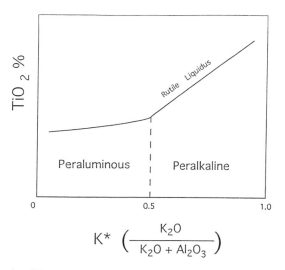

Figure 11. Cartoon describing the solubility of rutile as a function of $K^* = (K_2O/K_2O+Al_2O_3)$.

(1) Rutile solubility is not sensitive to the NaO/K_2O ratio (Ellison and Hess, 1986).

(2) Rutile solubility increases upon substitution of CaO or MgO for K_2O in peraluminous melts but it decreases with substitution of CaO or MgO for K_2O in peralkaline melts (Gan and Hess, in prep; Dickinson and Hess, 1985).

(3) The TiO_2 content of rutile saturated melts of slightly peraluminous composition increases with increases in the $(K,Na)AlO_2/SiO_2)$ ratio (Ellison and Hess, 1986). The changes in solution behavior are entirely due to changes in the nature of the fully polymerized aluminosilicate network.

(4) The solubility of rutile decreases with the addition of P_2O_5, Nb_2O_5 or Ta_2O_5 to highly peralkaline melts but the solubility increases with the addition of these oxides to peraluminous to slightly peralkaline melts (Gan et al., 1994b).

These observations are rationalized as follows:

(1) The solubility of rutile increases, albeit only slightly, in peraluminous melts ranging in composition from $K^* = 0.4$ to $K^* = 0.5$. In order to explain this trend, we consider three possible homogeneous equilibria

$$SiOSi + TiOTi = 2\ SiOTi \tag{28}$$

$$2\ KAlOSi + TiOTi = 2\ KAlOTi + SiOSi \tag{29}$$

$$AlOAl + TiOTi = 2\ AlOTi \tag{30}$$

Equation (28) accounts for the formation of Ti-Si next nearest neighbors, the second equation assumes the formation of Al-Ti nearest neighbors, where the Al is tetrahedrally coordinated with oxygen and is associated with K for charge balance and the third equation assumes the formation of Ti-Al nearest neighbors but the Al is not necessarily 4-coordinated with oxygen nor is it associated with a charge balancing cation. Each equilibrium creates Ti-species at the expense of TiOTi species, with the result that the activity coefficient of TiO_2 decreases (i.e. the a_{SiO_2} is assumed to be proportional to the concentration of TiO_2 species in the silicate melt (Hess, 1977)) and the solubility of rutile increases.

The third equilibrium, (30), cannot dominate the solution properties of TiO_2 because it predicts that the RHS is favored with increasing concentration (activity) of AlOAl. If true, the solubility of rutile would increase as peraluminous melts become more peraluminous. The opposite trend is observed. Either of the other two equilibria are consistent with experiment, however. We remarked earlier that the a_{SiO_2} in peraluminous melts increases with K^*. An increase in the a_{SiO_2} favors the RHS of (28) causing TiOSi species to form from TiOTi species. The solubility of rutile would increase as observed. But the rutile solubility data of Ellison and Hess (1986) show conclusively that the dominant equilibrium is given by (29); these authors found that the solubility of rutile increased as $(K,Na)AlO_2$ replaces SiO_2 in peraluminous melts. The effect in these peraluminous melts is modest, however.

(2) The peralkaline effect is consistent with the homogeneous equilibrium

$$SiOK + TiOTi = TiOK + SiOSi \tag{31}$$

where the pair of melt species on the RHS are favored above those on the LHS. Rearranging and inserting the appropriate chemical potentials

$$\mu_{SiOK} - \mu_{SiOSi} = \mu_{TiOK} - \mu_{TiOTi} \tag{32}$$

quantities analogous to the "exchange potentials" defined earlier in the paper are recovered. It is now evident that the peralkaline effect exists because the (standard state) exchange potentials of the Ti species are more negative then those of the Si species. In other words, it is the difference the free energies between Ti species relative to the free energy difference of Si species that determine the solution properties, not solely the free energies of formation of the SiOK and TiOK species.

(3) In peralkaline melts, the total to partial exchange of K (or any monovalent cation) by Ca (or any divalent cation) decreases the solubility of rutile implying that the RHS of the exchange equilibrium

$$TiOCa - SiOCa = TiOK - SiOK \tag{33}$$

is favored. These results follow from our previous analysis which demonstrated that the exchange enthalpy ($H_{TiOK} - H_{SiOK}$) of anionic species formed from cations of low enthalpic electronegativity must be more negative (favorable) than of those anionic species formed from cations of high enthalpic electronegativity. In the exchange equilibrium

$$SiOK + TiOCa = SiOCa + TiOK \tag{34}$$

the RHS is favored because the greatest stabilization is obtained by joining cations of lowest enthalpic electronegativity (K) with the anion of the highest enthalpic electronegativity (Ti) (see also Duffy, 1993).

In peraluminous melts, the total or partial exchange of K by Ca increases the solubility of rutile implying that the LHS of the exchange equilibrium

$$Ca_{0.5}AlOTi - Ca_{0.5}AlOSi = KAlOTi - KAlOSi \tag{35}$$

is favored over the RHS. Following the previous analysis, the increase in rutile solubility means that the $Ca_{0.5}AlOTi + KAlOSi$ species are favored over the pair $Ca_{0.5}AlOSi + KAlOTi$.

(4) In highly peralkaline melts, the addition of M_2O_5 ($M = P^{+5}, Nb^{+5}$ or Ta^{+5}) decreases the solubility of rutile (Gan et al., 1984). The results are consistent with the homogeneous equilibrium

$$2 TiOK + POP = 2 POK + TiOTi \tag{36}$$

where the RHS is favored over the LHS. The creation of TiOTi species at the expense of TiOK species increases the activity coefficient of TiO_2 and thereby reduces the solubility of rutile. This equilibrium is consistent with the rule that K should combine with the anionic complex with the greatest H-value (P).

The effect on the solubility of rutile by the addition of P_2O_5 to peraluminous melts is more complex. At low P_2O_5 concentrations the solubility change is negligable. We suggest that the P_2O_5 interacts with that part of the Al in excess of the charge balancing cation (K) according to the homogeneous equilibrium

$$AlOAl + POP = 2 AlOP \tag{37}$$

(see Gan and Hess, 1992). These species do not directly involve Ti and therefore have little effect on the activity coefficient of TiO_2. But when the P_2O_5 content exceeds the content of excess Al, the governing equilibrium becomes (Gan et al, 1994; Gan and Hess, 1992).

$$2 \text{ KAlOTi} + \text{POP} = 2 \text{ AlOP} + 2 \text{ TiOK} \qquad (38)$$

wherein more stable titinate species are created, thereby lowering the activity coefficient of TiO_2 and increasing the solubility of rutile.

Other network-forming species

Up to this point, the discussion has focussed on the relative roles of network-modifying cations and anionic complexes formed from highly charged cations in aluminosilicate melts. What are the roles of other network-forming (or potential network-forming species)?

DeYoreo and Navrotsky (1990) determined that the solution of $NaTO_2$ in SiO_2 glasses where $T = Al^{+3}$, Ga^{+3}, Fe^{+3} or B^{+3} became more exothermic (favorable) in the order NaAl > NaGa > NaFe > NaB. At first glance, this order appears to be at variance with the order implied by the H-scale which has $MBO_2 > MAlO_2 > MFe_2$. Unfortunately the H-scale for these species is irrelevant in this instance. The H-scale refers to compounds in which the next-nearest neighbors are TOT species; for example, the H-scale for $MAlO_2$, is obtained from compounds in which the species are AlOAl with Al in tetrahedral coordination with oxygen (Wells, 1985). In contrast, the calorimetric data of DeYoreo and Navrotsky (1990) are for aluminosilicate subaluminous melts in which the dominant species are surely AlOSi species (Taylor and Brown, 1979).

It is concluded that the true H values of the trivalent network-forming cations are not those of Table 4 but instead should increase in the order B<Fe<Ga<Al. It is difficult, however, to relate the H values of theses species to those of other highly charged cations since the necessary enthalpies of formation of the appropriate silicate compounds are not available. Much more work is needed to examine their interactions (see however, Gwinn and Hess, 1993 and 1989; Gan and Hess, 1992; Ellison and Hess, 1988; Dickenson and Hess, 1986, 1981).

CRITICAL MELTS

The most direct manifestation of non-ideal mixing of silicate melts is the widespread occurrence of stable (super liquidus) and/or metastable (subliquidus to subsolidus) immiscible liquids in simple and even natural occurring silicate melts. The location of the two liquid solvi is completely consistent with stability arguments based on quasi-crystalline concepts.

It was argued many years ago (Hess, 1971, 1977) that two liquid fields exist in compositional regions of phase diagrams in which no intermediate compounds were stable. The low-SiO_2 arms of the binodals in the two component M_xO_y-SiO_2 systems invariably are bounded, and probably are asymptotic at low temperatures, to the most SiO_2-rich stable compound of the given system. The high-SiO_2 arms are generally asymptotic to pure SiO_2 (Tewhey and Hess, 1979) (Fig.12). The widest two liquid fields occur in systems with no silicate compounds (TiO_2-, Cr_2O_3-, Fe_2O_3-SiO_2 for example) whereas metastable two liquid fields have not been detected in the Rb_2O-, Cs_2O or even K_2O-SiO_2 systems for which the most SiO_2-rich compounds have $X_{SiO_2} = 0.80$.

Because of these relations, the shapes of the miscibility gaps are decidedly asymmetric around the critical composition when plotted as a function of the molar

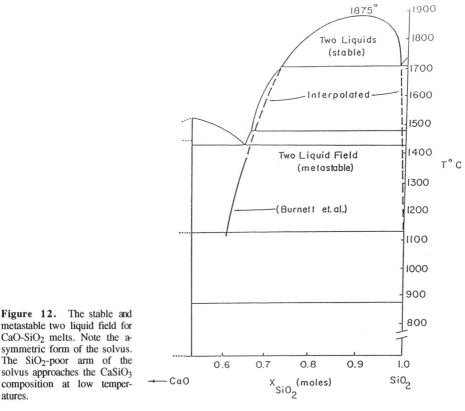

Figure 12. The stable and metastable two liquid field for CaO-SiO₂ melts. Note the asymmetric form of the solvus. The SiO₂-poor arm of the solvus approaches the CaSiO₃ composition at low temperatures.

concentrations of the oxides. The miscibility gaps can be symmetrized, however, by the proper choice of components. This formulation leads to a modified regular solution description not only of the phase boundaries but also of the thermodynamic mixing properties (Haller et al., 1974). In the Na₂O-SiO₂ system, for example, symmetrization is achieved by setting one component equal to the composition of the silicate compound bounding the miscibility gap on the low-SiO₂ side (Na₂O·3SiO₂) and the other equal to (SiO₂)₈ or Si₈O₁₆ (Haller et al., 1974). Taken at face value, this model implies that these species control the phase equilibria. Or perhaps, the stoichiometries do not represent single species but the average of two or more coexisting species. NMR spectra of glasses of appropriate compositions indicate the coexistence of Q^2, Q^3 and Q^4 species (Buckermann et al., 1992, Maekawa et al., 1991), an observation that supports the multiple species hypothesis. The main point here is that the framing of the miscibility gap by the compositions of bounding crystalline silicates is not a coincidence but provides quasi-crystalline evidence as to the nature of melt speciation.

It is also noteworthy that the critical temperatures of N₂O- or MO-SiO₂ melts follow the high temperature canonical sequence K < Na < Li < Ba < Sr < Ca < Mg. The critical temperatures, in fact, are approximate linear functions of the charge of the cation divided by the cation-oxygen bond distance of the network-modifying cation. The correlation of the critical temperature with charge and size of the network-modifying cation (Fig.13) indicates

Table 5. Comparison of observed
and theoretical critical temperatures.

System	$T_C(K)_{obs}$	$T_C(K)_{Theo}$
K_2O	(833)	990
Na_2O	1100	1193
Li_2O	1273	1306
BaO	1751	1919
SrO	2020	2040
CaO	2150	2212
MgO	2250	2613

that the forces driving melts towards instability have a strong coulombic character. Indeed, McGahay and Tomozawa (1989) have applied the Debye-Hückel theory of dilute ionic solutions to predict critical temperatures to within about 10% of their experimental values without the benefit of adjustable constants (Table 5). The Debye-Hückel theory treats the silicate melt as an infinitely dilute solution of "free" cations (the network-modifying cations) immersed in a continuum (the Si-O polymerized liquid) characterized by a low dielectric constant. The critical temperature obtained by this theory is given by

$$Tc = \frac{[Z^+Z^{-1}]e_o^2}{201 \ K_B \varepsilon_r \varepsilon_o r} \tag{39}$$

where Z^+ is the formal charge of the network-modifying cation, Z^- is the effective charge of the non-bridging oxygen (the part of the negative charge of the non-bridging oxygen that must be satisfied by the network-modifying cation, i.e. -1) eo is the charge on an electron, ε_0 the permittivity of free space, K_B Boltzmann's constant, ε_r the static dielectric constant of the solvent (ε_r is for SiO_2 glass) and r is the cation-oxygen bond distance.

The important variables in this expression are the charge of the network-modifying cation, the cation-oxygen bond distance and the static dielectric constant. Note that the critical temperature in this expression scales to the ratio of charge/bond distance, giving credibility to the observed correlation of critical temperature with the same variables (Fig. 13). Perhaps, as important, is the value of the static dielectric constant of the medium in which the cations are distributed. The static dielectric constant is a measure of the capacity of the solvent, here Si-O, to isolate the cations form each other. Since silica has one of the lowest dielectric constants known for the common oxides (Shannon, 1993), the onset of liquid phase separation arises because the network-modifying cations are not effectively shielded from each other's influence by the SiO medium. Strong coulombic repulsion between these cations render the melts unstable.

Note, however, that the concept of free ions in a silicate melt needs some elucidation. This analysis is particularly relevant since the Debye-Hückel theory is strictly true only in the dilute solution limit where the solute approaches zero concentration. Why then does the theory appear to embody much of the physics for silicate solutions that obviously are not dilute? Indeed most critical compositions contain about 7 to 12 mole % of the metal oxide (Hageman and Oonk, 1986; Tewhey and Hess, 1979; Haller et al., 1974).

The answer, I believe, rests in the difficulty of forming a coordination polyhedron of non-bridging oxygen around the network forming cation. In dilute solutions of Na_2O in

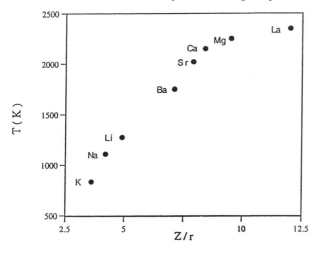

Figure 13. Critical temperatures as a function of the charge of the network-modifying cation (z) divided by the cation-oxygen distance (r).

SiO_2 melt, the Na^+, i.e. the non-bridging oxygen-cations, probably form in pairs (Charles, 1969):

$$Si\text{-}O\text{-} \underset{Na+}{\overset{Na+}{}} \text{-}O\text{-}Si$$

These cations are coordinated not only to non-bridging but to bridging oxygen as well. The bridging oxygen are part of the dielectric medium as described by the Debye-Hückel theory and serve as screens separating ions from each other. The cations that are partly or wholly coordinated by bridging oxygen are poorly screened from each other and interact strongly even at small concentrations. The concept of a poorly screened "free ion" is also supported by the thermal expansivities of SiO_2-rich melts. Molar expansitivies are negligible for SiO_2 melts with less than about 10 to 12% K_2O, Na_2O, Li_2O or BaO (Bockris et al., 1956; Tomlinson et al., 1958). The expansivities then undergo a rapid rise at greater contents of the network-modifying cation. Bockris et al. (1956) suggested that when the number of M_2O (or MO) added to the SiO_2 melt approaches 12%, cages formed from non-bridging oxygen and bridging oxygen are just able to completely enclose all network-modifying cations. Since the cage is made of strong Si-O-Si bonds having a negligible thermal expansion, the expansion associated with the weaker M-O bond is muted. If this model is correct, the ions enclosed in these cages are largely coordinated by bridging bonds as well as non-bridging bonds, and therefore are poorly screened from each other. This hypothesis is supported by the onset of phase separation, even at very low mole fractions of the network-modifying cation. Phase separation in $CaO\text{-}SiO_2$ melts, for example, occurs with only about 2 mole % CaO at T ~ 1700°C (Tewhey and Hess, 1979)! These poorly coordinated species are treated as free ions in the spirit of the Debye-Hückel theory.

As the metal oxide content is increased, the melt approaches a composition equivalent to crystalline stoichiometries wherein a significant fraction of the network-modifying cations develops crystal-like coordination polyhedra of non-bridging oxygens. These cations are efficiently screened from each other by non-bridging oxygen and are here called associated cations. The associated cations do not strongly contribute to the coulombic forces that drive phase separation and perhaps contribute to the dielectric properties of the medium. In this view, the critical point is the composition at which the concentration of the poorly screened "free" cations is at a maximum.

In this respect, it is interesting that the Debye-Hückel model fails to predict the critical temperatures in rare earth element oxide—SiO_2 melts. The critical temperatures for the trivalent rare earth oxide melts, for example, should be roughly 50% higher than the critical temperature of the solvus in the $MgO-SiO_2$ system or about 3300°C. Yet, critical temperatures range from 2273°C in Nd_2O_3 melts to 2598°C for Dy_2O_3 melts—more than 700°C below those predicted from the Debye-Hückel model. To be sure, there is considerable uncertainty in the experimental derived critical temperatures (see Tewhey and Hess, 1979; Hageman and Oonk, 1986). But these large discrepancies can not be explained away by the experimental uncertainties which appear to be in the ±100°C range.

McGahay and Tomozowa (1993) argue that the estimated critical temperatures can be brought to the expected range by requiring that the rare earth ions have an effective charge of +2 rather than the formal +3 charge of the free cation. This situation arises because the rare earth is closely associated with the non-bridging oxygen much like the non-bridging oxygen is closely associated with Si thus lowering the effective charge on oxygen from -2 to -1. There appears to be a grain of truth in this idea - highly charged cations should more effectively compete with Si^{+4} for the non-bridging oxygen. In fact, Hess (1977) used this idea to rationalize the extraordinarily wide miscibility gap that exists in the TiO_2-SiO_2 system. Phase separation exists because relatively few O are shared by Ti and Si. Instead, separate liquid domains of largely Ti-O-T and SiOSi species are created. A Debye-Hückel treatment of the TiO_2-SiO_2 system is clearly inappropriate because neither Ti nor Si can be treated as free cations in a dielectric continuum. This conclusion seems to be generally true for SiO_2 melts containing cations of +3 charge or greater.

REDOX EQUILIBRIA

It is now well documented that the oxidation-reduction equilibrium established by a single multivalent element in a silicate melt is a strong function of melt composition as well as the temperature, pressure and the oxygen fugacity (Schreiber et al., 1994 among others, Gwinn and Hess, 1989, 1993; Kilinc et al., 1983; Dickinson and Hess, 1981; Paul and Douglas, 1965). Redox equilibria are usually represented as an equilibrium between ions

$$M^{(x+n)+} + n/2\ O^{2-} = M^{x+} + n/4\ O_2 \qquad (40)$$

where n is the number of electrons transferred between the two cations, O_2 is the oxygen dissolved in the melt and O^{-2} refers to the oxygen ion in the melt that is directly involved in the equilibrium. The corresponding equation for the Fe redox equilibrium is

$$Fe^{+3} + 1/2\ O^{-2} = Fe^{+2} + 1/4\ O_2 \qquad (41)$$

where n = 1. The equilibrium constant is

$$K = \frac{[Fe^{+2}]\ fo_2^{1/4}}{[Fe^{+3}]\ [O^{-2}]^{1/2}} \qquad (42)$$

where the brackets refer to the activities of the indicated ions and fo_2 is the fugacity of oxygen. What the oxygen ion activity refers to is not clear; oxygen ion activity could refer to the activity of non-bridging, free or bridging oxygen or some combination of all three.

The use of oxygen ion activity as a measure of the electron donor power of the oxygen atom, is a concept borrowed from the Lewis acid-base theory (Duffy, 1993). In this theory, the state of the oxygen atom and its ability to donate some of its negative charge to the metal ion is a measure the "basicity". The basicity is at a maximum when the oxygen exists as a free oxygen ion removed from the influence of surrounding cations. Such a state is unstable (Cotton and Wilkinson, 1988) but is approximated when the coordinated cations are of low electronegativity (Duffy, 1993). When oxygen is coordinated to cations of low electronegativity such as Rb^{+1} or K^{+1} the basicity is high and the charge on oxygen approaches -2. But even these oxides are unstable at ambient conditions and tend to oxidize forming peroxides (K_2O_2) or superperoxides (KO_2)(Duffy, 1991). When an oxygen is coordinated with one (non-bridging) or two Si (bridging) its basicity is much less since more of the charge is drawn towards the Si atom(s). If the acid-base approach is correct, one predicts that the upper oxidation state ion benefits more from the electron donor capacity of oxygen so more"basic" melts favor oxidation of the redox couple This trend is observed; for example, the higher oxidation state for each of the redox couples Fe^{+2}-Fe^{+3}, Ce^{+3}-Ce^{+4}, Cr^{+6}-Cr^{+3} and U^{+6}-U^{+5} increases with the M_2O/SiO_2 in binary silicate melts (M = K, Na or Li) at constant T, P and fo_2 (Schreiber et al., 1994; Paul and Douglas, 1965; Nath and Douglas, 1965).

An apparent paradox is created, however, when the acid-base theory and the experimental results are compared to the redox equilibrium represented by Equation (40). Note that an increase in the "oxygen ion activity" favors the RHS, that is, higher oxygen ion activities stabilize the lower oxidation state rather than the upper oxidation states!

The paradox is created by the difficulty of defining the standard state free energy terms in the redox equilibrium in terms of these ionic species. Suppose we consider the iron redox equilibrium in K_2O-SiO_2 melts of varying K_2O/SiO_2 contents. The homogeneous equilibrium is given by (41) and (42). What are the standard states? If we choose liquid K_2O for the standard state of O^{-2}, that is, the oxygen ion activity is unity in these melts, then increasing the K_2O/SiO_2 ratio increases the activity of the "oxygen ion" activity, and drives the Fe^{+2}/Fe^{+3} ratio to lower values, a trend that is opposite to that observed by experiment. If we choose liquid SiO_2 for the standard state of the oxygen ion, then increasing the K_2O/SiO_2 ratio reduces the activity of the "oxygen ion" and thereby stabilizes the upper oxidation state as observed. But then how does this choice square with the definition of the electron donor power of oxygen as a measure of the basicity of the melt? Do lower oxygen ion activities mean more basic melts? There is nothing wrong with this choice - it is only a tautology. Yet it does violence with our chemical intuition. We would prefer that a basic melt has a large fraction of oxygen ions weakly coordinated with their cations - activities and mole fractions of "oxygen ions" should trend together.

An alternative and more instructive view of the role of melt composition in controlling redox equilibrium is to express the homogeneous equilibrium in terms of neutral melt species rather than on an ionic basis. The Fe redox equilibrium in very SiO-rich K_2O-SiO_2 melts, for example, is expressed as

$$2 Fe^{+2}Si_2 O_5 + K_2Si_2O_5 + 1/2 \ O_2 = 2 \ KFe^{+3}O_2 + 6 \ SiO_2 \qquad (43)$$

where the network-modifying cation (Fe^{+2}) in the lower oxidation state is oxidized to a network-forming cation (Fe^{+3} in tetrahedral coordination) in the upper oxidation state. An increase in the K_2O/SiO_2 ratio decreases the activity of SiO_2 of the melt and thereby favors the RHS. The result is to increase the Fe^{+3}/Fe ratio as is observed experimentally. The redox ratio Fe^{+3}/Fe increases in the order Li, Na, K (Paul and Douglas, 1965) a series consistent with the trends established from other equilibria. Note that the RHS is favored not only by the relatively strong bonds in $KFe^{+3}O_2$ but also by the comparatively weak bonds in $Fe^{+2}Si_2O_5$; recall that SiO_2-rich liquids in the FeO-SiO_2 system are immiscible and therefore unstable except at very high temperatures.

Equilibrium (43) applies only to very SiO_2-melts. At lower SiO_2 contents, the network-modifying cations are associated with less polymerized silicate species. Near the metasilicate composition ($X_{SiO_2} = 0.50$) the equlibrium is expressed as

$$2 \, Fe^{+2}SiO_3 + K_2SiO_3 + 1/2 \, O_2 = 2 \, KFe^{+3}O_2 + 3 \, SiO_2 \qquad (44)$$

At these and even less polymerzied silicate melts, we expect that the RHS would be progressively less favored. The reason is that since FeO-rich silicate melts are stable, it is likely that the FeOSi bond is relatively more stable in these melts than in high SiO_2 melts.

How then does this treatment of redox equilibria connect with the H-scale? To see this connection more clearly, consider the Cr^{+3}-Cr^{+6} redox equilibrium in high-SiO_2, K_2O-SiO_2 melts. The homogeneous equilibrium is written as

$$3/2 \, (Cr^{+3})_{2/3}SiO_3 + K_2Si_2O_5 + 3/4 \, O_2 = K_2Cr^{+6}O_4 + 7/2 \, SiO_2 \qquad (45)$$

where Cr^{+3} is a network-modifying cation coordinated with Q_2 silicate species and Cr^{+6} is a discrete potassium chromate species. The network-modifying role of Cr^{+3} is consistent with that of Y^{+3} or Yb^{+3} in K_2O-SiO_2 melts (Ellison and Hess, 1994; 1990) whereas the existence of the chromate anion was established by Brawer and White (1977). After rewriting the equation in terms of the appropriate exchange potentials

$$3/2 \, (Cr^{+3})_{2/3}SiO_3 - 7/2 \, SiO_2 + (3/4 \, O_2) = K_2CrO_4 - K_2Si_2O_5 \qquad (46)$$

it is not hard to accept that the exchange potential on the RHS is favored over that of the LHS. Indeed, no compounds formed between Cr^{+3} (or other large trivalent cations) and Q^2 species (or Q^3) are known, implying that the Cr-silicate species has only a marginal stability. Similar homogeneous equilibria can be invented to rationalize the redox equilibria of other cation pairs.

DISCUSSION

This chapter attempts to link the macroscopic phase and thermodynamic properties of silicate melts to their microscopic state. More specifically, we asked how network-modifying cations were apportioned between anionic groups to minimize the free energy of the melt. Because of a paucity of thermodynamic data, a quasicrystalline approach is used to assign H-values to cations and anionic groups. Quite generally, this enthalpy-based scale predicts that the most stable melt species are formed between cations and anionic groups for which the difference in H-values were a maximum. These predictions are not infallible, of course. The H-scale considers only enthalpic factors and totally neglects entropic contributions to the free energy. Secondly, no single number such as an H-value can accurately summarize the multibody interactions that surely occur in all condensed phases. It is of interest, nevertheless, to examine what additional insights to melt speciation, particularly in "petrological" melts, can be gleaned from such a simple approach. A useful data set to test these ideas is that of Toplis et al. (1994) who examined the effects of the progressive addition of P_2O_5 to a ferrobasalt melt saturated with a number of silicate and oxide phases. The effects were numerous and dramatic but we focus only on three results; (1) the disappearance of olivine from the liquidus (olivine exists only under reducing conditions. (2) the disappearance of magnetite (in more oxidized melts) and (3) the increased solubility of ilmenite.

The decrease in the abundance of olivine in P_2O_5-rich melts is consistent with the displacement of the olivine-pyroxene cotectic to lower SiO_2 contents and the stabilization of a more polymerized silicate phase at the expense of the less polymerized silicate phase. The homogeneous equilibrium associated with this shift in mineral stabilities is (all are melt species).

$$1.5 \, Mg_2SiO_4 + 2 \, PO_{2.5} = 2 \, Mg_{1.5}PO_4 + 1.5 \, SiO_2 \tag{47}$$

and in terms of exchange potentials

$$1.5 \, \mu_{Mg_2SiO_4} - 1.5 \, \mu_{SiO_2} = 2 \, \mu_{Mg_{1.5}PO_4} - 2 \, \mu_{PO_{2.5}} \tag{48}$$

It is clear that the RHS of (47) and (48) are favored. The addition of P_2O_5, therefore, decreases the activity of Mg_2SiO_4 and increases the activity of SiO_2 as required by the shift in mineral stabilities.

The disappearance of magnetite from the P_2O_5-bearing ferrobasalt liquidus is consistent with the homogeneous equilibria (all melt species):

$$1/2 \, Fe_2O_3 + PO_{2.5} = Fe^{+3}PO_4 \tag{49}$$

or

$$3 \, Fe_2O_3 + 4 \, PO_{2.5} = 4 \, Fe^{+2}_{1.5}PO_4 + 3/2 \, O_2 \tag{50}$$

wherein the activity of Fe_2O_3, and hence the activity of $FeFe_2O_4$, is reduced either by the formation of a ferric phosphate species or by the reduction of the ferric oxide to the ferrous phosphate. The first equilibrium is consistent with the stability of $AlPO_4$ complexes (Gan and Hess, 1992) given that Fe^{+3} and Al^{+3} have the same crystal chemical affinities. The $AlPO_4$ species, however, is most stable only in peraluminous melts; the $AlPO_4$ species should not be an abundant constituent in metaluminous melts such as the ferrobasalt. It is then more likely that the second equilibrium dominates the solution properties and controls the liquidus of magnetite. This interpretation is supported by the observed reduction of Fe^{+3} to Fe^{+2} upon the addition of P_2O_5 to ferrobasaltic melts (Toplis et al., 1994b), which is described by (50) and not (49).

The increased solubility of ilmenite in P_2O_5-bearing melts is very interesting because it agrees nicely with the predictions derived from the H-scale. The homogeneous equilibrium can be written as (all melt species)

$$1.5 \, FeTiO_3 + PO_{2.5} = Fe_{1.5}PO_4 + 1.5 \, TiO_2 \tag{51}$$

and in terms of exchange potentials

$$1.5 \, FeTiO_3 - 1.5 \, TiO_2 = Fe_{1.5}PO_4 - PO_{2.5} \tag{52}$$

Again, it is the RHS of (51) and (52) that are favored. The equilibrium implies not only that the activity coefficient of $FeTiO_3$ is reduced, hence the ilmenite solubility increases, but also that the activity coefficient of TiO_2 increases. If this homogeneous equilibrium is correct then rutile may replace ilmenite on the liquidus at high P_2O_5 contents.

This result is quite general. If the TiO_3 anionic group is replaced by an anionic group below that of the phosphate anion on the H-scale the addition of P_2O_5 will always tend to bring the phase corresponding in composition to the anionic group closer to the liquidus. The addition of P_2O_5 to alkali silicate melts, for example, brings cristobalite closer to the liquidus.

ACKNOWLEDGMENTS

Thanks to Jonathan Stebbins and an anonymous reviewer for some useful suggestions to improve the manuscript, to Alessandra Cazzaniga for help with the figures, to Gloria Correra for typing and formatting and to present and former colleagues who have proved to be fertile ground for good ideas. This work was supported by NSF grant EAR-9304046.

APPENDIX

Free energy of mixing

The thermodynamic mixing relations of a binary silicate melt will be developed at three different levels of sophistication. The most conventional approach is to express the solution properties in terms of the moles of the simple oxides. But how should the formula units of these oxides be given? The composition of the Na_2O-SiO_2 melt, for example, can be expressed on an equal oxygen basis (the components are Na_2O-$Si_{0.5}O$), an equal cation basis (the components are $NaO_{0.5}$-SiO_2) or in terms of the "conventional" oxides (the components are Na_2O-SiO_2). Which components order best the phase and thermodynamic properties of silicate systems? On what component basis are the solutions most ideal? It turns out, that the most efficient choice of components are those written on an equal oxygen basis (see "Critical Phases") but even these components leave much to be desired. The entropy of mixing, for instance, cannot be described by these components since the formula units have no relation to the physical entities that exist in the silicate melts.

The next most realistic approach is to express the thermodynamic mixing properties in terms of their melt species. Melt species are chemical formulas for physical entities that exist (or are thought to exist) in the melt. In analogy to the approach used to rationalize the enthalpy of formation of a crystalline silicate phase, the thermodynamic mixing properties of the silicate melt are first described as a homogeneous equilibrium between next-nearest cation neighbors. Consider, again, the properties of the Na_2O-SiO_2 binary melt. The composition and microstate of the silicate melt is described by the abundance and distribution of the three chemical entities NaONa, NaOSi and SiOSi (next nearest cation neighbors). These chemical species are identified by a shorthand which focuses solely on the oxygen ion; O^{-2} (free oxygen), O^- (non-bridging oxygen) and O^0 (bridging) oxygen respectively. Our task is to join the microstate view of the melt as given by the "oxygen" equation of state to the macrostate of the system as expressed by the oxide components. This task will be completed below.

The thermodynamic mixing properties can be further refined by organizing the oxygen species into polymeric functional groups. In this approach, the microstate of the melt is viewed as a homogeneous equilibrium between polyhedral complexes formed between silicate anionic groups and their network-modifying cations. The structure of a Na_2O-SiO_2 silicate melt is expressed in terms of the neutral species like Na_2O, Na_4SiO_4, $Na_3SiO_{3.5}$, Na_2SiO_3, $NaSiO_{2.5}$ and SiO_2. The silicate species are given in terms of one mole of Si but this mode of representation is not mandated. Indeed, perhaps a more physical representation is on an equal oxygen basis or roughly on an equal volume per mole of species basis. We write the species as above, however, so that the silicate species are in direct correspondence with the Q nomenclature; the Na_4SiO_4, $Na_3SiO_{3.5}$, Na_2SiO_3, $NaSiO_{2.5}$ and SiO_2 species are identified with the Q^0, Q^1, Q^2, Q^3 and Q^4 species respectively where the Q^n notation refers to a silicon tetrahedron with n bridging oxygen. The Q^n species refer to functional groups of the silicate monomer (n = 0), the silicate dimer (n = 1), the silicate chain (n = 2), the silicate sheet (n = 3) and the silicate network (n = 4). The thermodynamic mixing properties in the Q representation as well as in the oxide and oxygen representations are developed below.

The variation of the Gibbs free energy of Na_2O-SiO_2 melts at constant temperature and pressure is written in three equivalent ways:

$$(dG)_{P,T} = \mu_{SiO_2}dn_{SiO_2} + \mu_{Na_2O}dn_{Na_2O} \tag{A1}$$

$$(dG)_{P,T} = \mu_{O^\circ}dn_{O^\circ} + \mu_{O^-}dn_{O^-} + \mu_{O^{-2}}dn_{O^{-2}} \tag{A2}$$

$$(dG)_{P,T} = \mu'_{Na_2O}dn'_{Na_2O} + \sum_{n=1}^{4} \mu_{Q^n}dn_{Q^n} \tag{A3}$$

The free energy in (A1) is given as a function of the moles of the oxide components and their chemical potentials, in (A2) as a function of the moles and chemical potentials of the oxygen species and in (A3) as a function of the moles and chemical potentials of the Na_2O species and the various Q^n silicate species. Note that n'_{Na_2O} in (A3) refers to the number of moles of the chemical entity of Na_2O in the melt not to the moles of the oxide component that describes the composition of the melt.

The three representations of the Gibbs free energy of the binary melt are joined by applying a powerful theorem of Gibbs. Gibbs (1961) proved that any equation that links the composition of components/species is also an equation between their chemical potentials. The chemical potentials of the various oxide and Q^n silicate species are related in a straightforward way to the chemical potentials of the oxide components:

$$2\,\mu_{Na_2O} + \mu_{SiO_2} = \mu_{Na_4SiO_4} \qquad (Q^0) \tag{A4}$$

$$1.5\,\mu_{Na_2O} + \mu_{SiO_2} = \mu_{Na_3SiO_{3.5}} \qquad (Q^1) \tag{A5}$$

$$\mu_{Na_2O} + \mu_{SiO_2} = \mu_{Na_2SiO_3} \qquad (Q^2) \tag{A6}$$

$$0.5\,\mu_{Na_2O} + \mu_{SiO_2} = \mu_{NaSiO_{2.5}} \qquad (Q^3) \tag{A7}$$

$$\mu_{SiO_2} = \mu_{SiO'_2} \text{ (species)} \qquad (Q^4) \tag{A8}$$

$$\mu_{Na_2O} = \mu_{Na_2O'} \text{ (species)} \qquad (A9)$$

The chemical potentials of the "oxygen" components are similarly related. Note that the oxygen species are written on a one oxygen basis so that O^0 (SiOSi) contains 0.5 moles of Si, O^{-2} (NaONa) contains 2 moles of Na and O^- (NaOSi) contains one mole of Na and 0.25 mole of Si. The chemical potentials of the Q species are then

$$4\,\mu_{O^-} = \mu_{Na_4SiO_4} \qquad (Q^0) \tag{A10}$$

$$3\,\mu_{O^-} + 0.5\,\mu_{O^\circ} = \mu_{Na_3SiO_{3.5}} \qquad (Q^1) \tag{A11}$$

$$2\,\mu_{O^-} + \mu_{O^\circ} = \mu_{Na_2SiO_3} \qquad (Q^2) \tag{A12}$$

$$1\,\mu_{O^-} + 1.5\,\mu_{O^\circ} = \mu_{NaSiO_{2.5}} \qquad (Q^3) \tag{A13}$$

$$2\,\mu_{O^\circ} = \mu_{SiO'_2} \qquad (Q^4) \tag{A14}$$

It is evident that the chemical potentials of the chemical species can be expressed in terms of the chemical potentials of the chemical components or any other of the chemical species. Consider as an illustrative example the heterogeneous equilibrium between forsterite, enstatite and melt in the binary MgO-SiO_2 system at a given T and P (see text). The following relations are mandated by the conditions of equilibrium:

$$\mu_{Mg_2SiO_4} (Fo) = \mu_{Mg_2SiO_4} (L) \tag{A15}$$

$$\mu_{MgSiO_3} (En) = \mu_{MgSiO_3} (L) \tag{A16}$$

$$\mu_{Mg_2SiO_4} (Fo) + \mu_{SiO_2}(L) = 2 \, \mu_{MgSiO_3} (En) \tag{A17}$$

$$\mu_{Mg_2SiO_4} (Fo) = \mu_{MgSiO_3}(En) + \mu_{MgO} (L) \tag{A18}$$

where Fo and En refer to the appropriate crystalline phase and L refers to the chemical potential of the component or species in the liquid phase. Since at fixed T,P, the chemical potentials of μ_{Fo} and μ_{EN} are uniquely fixed it follows that all the chemical potentials of components and species in Equations (A15)-(A18) are uniquely fixed by the conditions of heterogeneous equilibrium. The conditions of homogeneous equilibrium (Eqns. A4-A9 and A10-A14) combined with the constraints provided by (A15)-(A18) fix uniquely all the chemical potentials of melt species. The chemical potential of $MgSi_2O_5$ (L) for example, is given by

$$0.5 \, \mu_{Mg_2SiO_4} + 1.5 \, \mu_{SiO_2} = \mu_{MgSi_2O_5} \tag{A19}$$

or by

$$\mu_{MgSiO_3} + \mu_{SiO_2} = \mu_{MgSi_2O_5} \tag{A20}$$

These relationships are clearly revealed in a $\overline{G}$-X diagram (Fig. A-1). The coexistence of forsterite, enstatite and liquid is obtained by constructing a common tangent to the three phases. The chemical potentials of Mg_2SiO_4 and $MgSiO_3$ are simply the specific Gibbs free energy of the appropriate compositions. The chemical potentials of μ_{SiO_2} and μ_{MgO} are obtained by extending the tangent to $X_{SiO_2} = 1$ and $X_{SiO_2} = 0$ respectively. Note, that by (A8) and (A9), the chemical potentials for the component and chemical species are identical. The chemical potentials of $Mg_3Si_2O_7$ and $Mg_2Si_2O_5$ are recorded by the value of the specific Gibbs free energy given by the tangent where it intersects the respective compositions.

The activities of the melt species and the melt components can be related in a straight forward manner. The Na_2O-SiO_2 melt is again utilized as an example. The activity of the melt components, Na_2O and SiO_2, are related to their chemical potentials

$$\mu_{Na_2O} = \mu^\circ_{Na_2O} (T,P) + RT \, ln \, \gamma_{Na_2O} X_{Na_2O} \tag{A21}$$

$$\mu_{Si_2O} = \mu^\circ_{SiO_2} (T,P) + RTv \, ln \, \gamma_{SiO_2} X_{SiO_2} \tag{A22}$$

where the standard state chemical potentials refer to the pure liquids at the pressure and temperature of the Na_2O-SiO_2 melt, X_{Na_2O} and X_{SiO_2} are the mole fractions of the melt components

$$X_{Na2O} = \frac{n_{Na_2O}}{n_{Na_2O} + n_{SiO_2}} \quad ; \quad X_{SiO2} = \frac{n_{SiO_2}}{n_{Na_2O} + n_{SiO_2}} \tag{A23}$$

(n_{Na_2O} moles of Na_2O; n_{SiO_2} moles of SiO_2) and γ are the respective activity coefficients. Note that $\gamma_{Na_2O} \to 1$ as $X_{Na_2O} \to 1$ and similarly for γ_{SiO_2}.

Consider again the same melt but express the composition of the melt in terms of the melt species rather than the melt components. For simplicity, the melt species are expressed

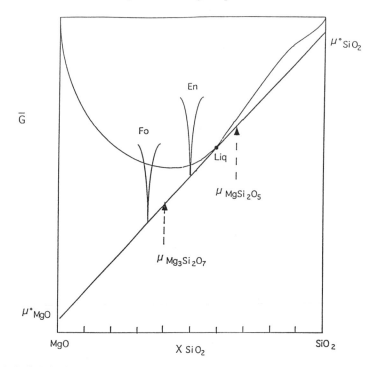

Figure A-1. Relation between the chemical potentials of components and species.

as Na_2O', SiO_2' and Q_n, where Q_n corresponds to Q_1, Q_2 and Q_3 melt silicate species. The chemical potentials of Na_2O' and SiO_2' are

$$\mu_{Na_2O'} = \mu^°_{Na_2O'}\,(P,T) + RT\,\mathit{ln}\,\gamma_{Na_2O'}\,X_{Na_2O'}$$

$$\mu_{SiO_2'} = \mu^°_{SiO_2'}\,(P,T) + RTv\,\mathit{ln}\,\gamma_{SiO_2'}\,X_{SiO_2'}$$

(A24)

where (') emphasizes that we are dealing with species. The species mole fractions are

$$X_{Na_2O'} = \frac{n_{Na_2O'}}{n_{SiO_2'} + \sum n_{Q_n}} \quad ; \quad X_{SiO_2'} = \frac{n_{SiO_2'}}{n_{SiO_2'} + \sum n_{Q_n}}$$

(A25)

where n refers to the number of moles of the given species. The standard state chemical potentials refer to the pure liquids, Na_2O and SiO_2, as before.

Since $\mu_{Na_2O'} = \mu_{Na_2O}$ and $\mu_{Si_2O'} = \mu_{Si_2O}$, it follows that the activity coefficient of the melt components are:

$$\gamma_{Na_2O} = \gamma_{Na_2O'}\,\frac{X_{Na_2O'}}{X_{Na_2O}} \quad ; \quad \gamma_{SiO_2} = \gamma_{SiO_2'}\,\frac{X_{SiO_2'}}{X_{SiO_2}}$$

(A26)

Suppose that melts can be treated as an "ideal associated" solution (Prigogine and Defay, 1954) which means that the species (but not the components) mix ideally or $\gamma_{Na_2O'} = \gamma_{SiO_2'} = 1$. This assumption is certainly incorrect but nevertheless pedagogically useful.

On substitution for the species activity, it is obvious that the component activity coefficients are equal to the ratio of species mole fraction to the component mole fractions

$$\gamma_{SiO_2} = \frac{X_{SiO_2}'}{X_{SiO_2}} \quad ; \quad \gamma_{Na_2O} = \frac{X_{Na_2O}'}{X_{Na_2O}} \tag{A27}$$

Consider now two melts, not necessarily binary, with the same mole fraction X_{SiO_2} but different species mole fractions $X_{SiO_2'}$. It follows that the melt with the lower species mole fraction has the lower component activity coefficient of SiO_2, and therefore, the lower component activity of SiO_2. The melt with the lower activity coefficient of SiO_2 is more undersaturated with respect to cristobalite than the melt with the higher activity coefficient. In this simple model, the activity of SiO_2 is simply related to the species mole fraction of SiO_2. In terms of the "oxygen model" the homogeneous equilibrium

$$SiOSi + KOK = 2 \, KOSi \tag{A28}$$

if driven to the RHS would lower the concentration of SiOSi species and thereby lower the activity of SiO_2 (Hess, 1977).

In general whenever the species mole fraction of a chemical entity is lowered we can expect that the component activity of that chemical entity is also lowered. This result explains the increased solubility of certain compounds in peralkaline melts. The solubility of rutile in peralkaline melts is much greater than in subaluminous melts because the RHS of the homogeneous equilibrium

$$SiOK + TiOTi = 2 \, TiOK \tag{A29}$$

is greatly favored. The generation of TiOK species reduces the concentration of TiOTi species and thereby lowers the component activity coefficient of TiO_2. At rutile saturation where the $a_{TiO_2} = 1$ (rutile standard state)

$$X_{TiO_2} = \frac{1}{\gamma_{TiO_2}} \tag{A30}$$

where X_{TiO_2} is the solubility of TiO_2 in rutile saturated melts. A lower γ_{TiO_2} increases the solubility of rutile.

Entropy of mixing

This paper has largely focused on the enthalpic mixing properties of silicate melts. A complete model must also address the entropic mixing properties. In general the entropy of mixing of a binary silicate melt (because of space constraints this discussion deals only with binary melts) is typically equated to the configurational entropy of mixing, a reasonable approximation given that the entropies of minerals can be obtained by summing the entropies of the equivalent oxides once corrections for volume and coordination changes are applied (Holland, 1989). But calculating the entropy of mixing is no small task. The assumption that the entropy of mixing is given by

$$\Delta S^{Mix} = -R \, (X_1 ln X_1 + X_2 ln X_2) \tag{A31}$$

where X_1 = mole fraction of the oxide of the network-modifying cation and $X_2 = X_{SiO_2}$ will be grossly in error since the structural entities cannot be solely represented by these species.

Charles (1969) argued that the configurational entropy of a binary silicate melt is largely due to permutations of the different "sites" occupied by oxygen ions in solution. The different oxygen sites (or species) are the free oxygen (O^{-2}), the bridging oxygen (O^0) and the non-bridging oxygen (O^-). Charles (1969) assumed with good justification that the non-bridging oxygen generally existed as pairs in such solutions (see also Araujo, 1983). With this assumption, the configurational entropy of mixing for one mole solution is

$$\Delta S = -R \left[\frac{O^-}{2} \ln \frac{O^-}{2W} + \left(O^0\right)\ln \frac{O^0}{W} + \left(O^{-2}\right)\ln \frac{O^{-2}}{W} \right] \qquad (A32)$$

where the brackets refer to the moles of the given oxygen species per one mole of solution and $W = (O^-)/2 + (O^0) + (O^{-2})$. The moles of the oxygen species are calculated from the homogeneous equilibrium (Toop and Samis, 1962)

$$O^0 + O^{-2} = 2\,O^- \qquad (A33)$$

the equilibrium constant

$$K = \frac{(O^0)(O^{-2})}{(O^-)^2} \qquad (A34)$$

and a material balance for the system (see also Hess, 1971, 1980).

Figure A-2 compares the entropy of mixing of a binary silicate melt calculated for several values of K. Small values of K mean that melts at stoichiometric compositions have

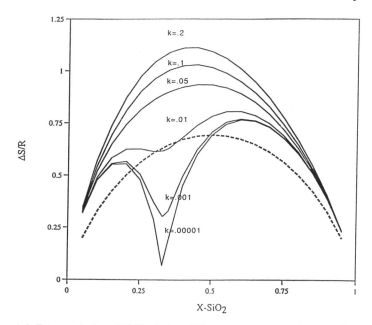

Figure A-2. Entropy of mixing ($\Delta S/R$) of binary SiO_2 melts for varying values of K (see text). Compare these curves to that of the ideal entropy of mixing of the oxides (dashed).

an oxygen content almost equivalent to that of the corresponding crystal (Hess, 1980). A melt with $X_{SiO_2} = 0.33$ and $K \rightarrow O$, for example, consists largely of non-bridging oxygen. In contrast, the same melt but with $K = 0.10$ contains about 60% (O^-) and 20% each of (O^{-2}) and (O^0). Note that for low K values, the entropy of mixing curves show two maxima and one minimum near $X_{SiO_2} \approx 0.33$. This result is not unexpected since melts more basic than this composition are solutions of largely (O^{-2}) and (O^-) and those more silicic are solutions mainly of (O^-) and (O^0). Indeed, in this model, the entropy of mixing must approach zero at $X_{SiO_2} \approx 0.33$ as K approaches zero.

This model is clearly oversimplified. First it tends to overcount the possible configurations. For example, bridging and free oxygen cannot be exchanged freely because, in certain instances, non-bridging oxygen are either created or destroyed in the process. Secondly, it takes no account of the association of oxygen ions into more complex anions such as rings, chains and/or sheets. Both these effects tend to reduce the configurational entropy of mixing given by the model. On the other hand, the excess entropy arising from the number of possible conformations of polymers should be positive. Nevertheless, it is likely that Charles' model should constitute an upper bound for the true entropy values.

A more realistic approach calculates the configurational entropy in terms of the permutations of Q^n species (Gurman, 1990; Maekawa et al., 1991, among others). These models have had only limited success because they include the assumption that there are no interactions between tetrahedra. The existence of fields of stable and/or metastable liquid immiscibility cannot be reconciled with the assumption of athermal mixing of Q^n species. Indeed, our emphasis on the varying energetics of next-nearest neighbors emphasizes the need to include enthalpic effects even between polymeric species. These enthalpic effects should express themselves most dramatically in the values of the entropy of mixing for systems with network-modifying cations of large H-values.

REFERENCES

Allen LC (1989) Electronegativity is the average one-electron energy of the valence-shell electrons in ground state free atoms. J Am Chem Soc 111:9003-9014

Allen LC (1993) Chemistry and electronegativity. Inter J Quant Chem 49:253-277

Araujo RJ (1983) A statistical mechanical interpretation for chemical disorder in interacting systems: Application to phase separation in alkali silicates, alkaline earth silicates and alkaline earth aluminosilicates. J Non-Cryst Solids 55:257-267

Barin I (1993) Thermochemical Data of Pure Substances. VCH, Germany

Bernet K, Hoppe R (1990) Zur Kristallstruktur von K_4SiO_4. Z Anorg Allg Chem 589:129-138

Blander M, Pelton AD (1987) Thermodynamic analysis of binary liquid silicates and prediction of ternary solution properties by modified quasichemical equations. Geochim Cosmochim Acta 51:85-96

Bockris JO'M, Tomlinson JW, White JL (1956) The structure of the liquid silicates: Partial molar volumes and expansivities. Trans Faraday Soc 52:299-311

Brawer SA, White WB (1977) Raman spectroscopic study of hexavalent chromium in some silicate and borate glasses. Mat Res Bull 12:281-288

Buckermann W-A, Müller-Warmuth W (1992) A further ^{29}Si MAS NMR study of binary alkali silicate glasses. Glast Ber 65:18-21

Burnett DG, Douglas RW (1970) Liquid-liquid phase separation in the soda-lime-silica system. Phys Chem Glasses 11:125-135

Charles RJ (1969) The origin of immiscibility in silicate solutions. Phys Chem Glasses 10:169-178

Cooney TF, Sharma SK (1990) Structure of glasses in the systems Mg_2SiO_4-Fe_2SiO_4,Mn_2SiO_4-Fe_2SiO_4,Mg_2SiO_4-$CaMgSiO_4$ and Mn_2SiO_4-$CaMnSiO_4$. J Non-Cryst Solids 122:10-32

Cotton FA, Wilkinson G (1988) Advanced Inorganic Chemistry. John Wiley, New York

Dent Glasser LS (1979) Nonexistent silicates. Z Kristallogr 149:291-305

DeYoreo JJ, Navrotsky A (1990) Energetics of the charge coupled substitution $Si^{+4} \rightarrow Na^+\ T^{+3}$ in the glasses $NaTO_2$-SiO_2 (T=Al,Fe,Ga,B). J Am Ceram Soc 73:2068-2072

Dickenson MP, Hess PC (1981) Redox equilibira and the structural role of iron in aluminosilicate melts. Contrib Mineral Petrol 78:352-357

Dickenson MP, Hess PC (1986) The structural role of Fe^{+3}, Ga^{+3} and Al^{+3} and homogeneous iron redox equilibria in $K_2O-Al_2O_3-Ga_2O_3-SiO_2-Fe_2O_3-FeO$ melts. J Non-cryst Solids 86:303-310

Dickinson JE, Hess PC (1985) Rutile solubility and Ti coordination in silicate melts. Geochim Cosmochim Acta 49:2289-2296

Douglas RW, Nath P, Paul A (1965) Oxygen ion activity and its influence on the redox equilibrium in glasses. Phys Chem Glasses 6:216-223

Dowty E (1987) Vibrational interactions of tetrahedra in silicate glasses and crystals: II Calculations on melilites, pyroxenes, silica polymorphs and feldspars. Phys Chem Minerals 14:122-138

Duffy JA (1993) A review of optical basicity and its application to oxidic systems. Geochim Cosmochim Acta 57:3961-3970

Duffy JA (1990) Bonding Energy Levels and Bands in Inorganic Solids. Longmans UK

Duffy JA, Ingram MD (1976) An interpretation of glass chemistry in terms of the optical basicity concept. J Non-cryst Solids 21:373-410

Ellison AJ, Hess PC (1986) Solution behavior of +4 cations in high silica melts: Petrologic and geochemical implications. Contrib Mineral Petrol 94:343-351

Ellison AJ, Hess PC (1988) Peraluminous and peralkaline effects upon monazite solubility in high silica liquids. EOS, Trans Am Geophys Soc 69:498

Ellison AJ, Hess PC (1990) Lanthanides in silicate glasses: A vibrational spectroscopic study. J Geophys Res 95:15717-15,726

Ellison AJ, Hess PC (1991) Vibrational spectra of high-silica glasses of the system $K_2O-SiO_2-La_2O_3$. J Non-crystal Solids 127:247-259

Ellison AJ and Hess PC (1994) Raman study of potassium silicate glasses containing Rb^+, Sr^{+2}, Y^{+3} and Zr^{+4}: Implications for cation solution mechanisms in multicomponent liquids. Geochim Cosmochim Acta 58:1877-1887.

Felsche J (1973) The crystal chemistry of the rare earth silicates. In: Structure and Bonding. JD Donitz et al. (eds) 13:100-197. Springer-Verlag, New York

Gan H, Hess PC (1991) Modelling activity coefficients of high field strength cations: Titanium. EOS, Trans Am Geophys Soc 72:547

Gan H, Hess PC (1992) Phosphate speciation in potassium aluminosilicate glasses. Am Mineral 77:495-506.

Gan H, Hess PC, Kirkpatrick RJ (1994) Phosphorus and boron speciation in $K_2O-B_2O_3-SiO_2-P_2O_5$ glass. Geochim Cosmochim Acta 58:4633-4648

Gan H, Hess PC, Horng W-S, (1994b) P, Nb and Ta interactions with anyhdrous haplogranite melts. EOS, Trans Am Geophys Soc 75:369

Gibbs JW (1961) The scientific papers of J Willard Gibbs: Vol. I. Thermodynamics. Dover, New York

Gibbs GV (1982) Molecules as models for bonding in silicates. Am Mineral 67:421-450.

Gurman SJ (1990) Bond ordering in silicate glasses: a critique and a re-solution. J Non- Cryst Solids 125:151-160.

Gwinn R and Hess PC (1993) The role of phosphorus in rhyolitic liquids as determined by homogeneous iron redox equilibria. Contrib Mineral Petrol 113:424-435.

Gwinn R, Hess PC (1989) Iron and titanium solution properties in peraluminous and peralkaline rhyolitic liquids. Contrib Mineral Petrol 101:326-338

Hageman VBM, Oonk HAJ (1986) Liquid immiscibility in the $SiO_2 + MgO$, $SiO_2 + SrO$, $SiO_2 + La_2O_3$ and $SiO_2 + Y_2O_3$ systems. Phys Chem Glasses 17:194-198

Haller W, Blackburn DH, Simmons JH (1974) Miscibility gaps in alkali-silicate binaries - Data and thermodynamic interpretation. J Am Ceram Soc 57:120-126

Hess PC (1971) Polymer model of silicate melts. Geochim Cosmochim Acta 35:289-306

Hess PC (1977) Structure of silicate melts. Can Mineral 15:162-178

Hess PC (1980) Polymerization model for silicate melts In: Physics of Magmatic Processes RB Hargraves (ed), p 1-48

Hess PC (1991) The role of high field strength cations in silicate melts In: Perchuk LL, Kushiro I (eds) Physical Chemistry of Magmas, Adv Geochem 9:152-191 Springer-Verlag

Holland, TJB (1989) Dependence of entropy on volume for silicate and oxide minerals: A review and predictive model, Am Mineral 74:5-13

Kilinc A, Carmichael ISE, Rivers ML, Sack RO (1983) The ferric-ferrous of natural silicate liquids equilibrated in air. Contrib Mineral Petrol 83:136-140

Kushiro I (1975) On the nature of silicate melt and its significance in magma genesis: Regularities in the shift of the liquidus boundaries involving olivine, pyroxene and silica minerals. Am J Sci 275:411-431

Levin EM, Robbins CR, McMurdie HF (1964) Phase diagrams for ceramists. Am Ceram Soc (supplement)

Liebau F (1985) Structural Chemistry of Silicates. Springer-Verlag, New York

Maekawa H, Maekawa T, Kawamura K, Yokokawa T (1991) The structural groups of alkali silicate glasses determined from ^{29}Si MAS-NMR. J Non-cryst Solids 127:53-64

McGahay V, Tomozawa M (1989) The origin of phase separation in silicate melts and glasses J Non-Cryst Solids 109:27-34

McGahay V, Tomozawa M (1993) Phase separation in rare-earth doped SiO_2 glasses J Non-cryst Solids 159:246-252

Montel,JC (1986) Experimental determination of the solubility of Ce-monazite in SiO_2-Al_2O_3-K_2O-Na_2O melts at 800°C, 2kbar under H_2O-saturated conditions. Geology 14:659-662

Mysen BO, Ryerson FJ, Virgo D (1981) The structural role of phosphorus in silicate melts. Am Mineral 66:106-117

Naski GC, Hess PC (1985) SnO_2 solubility: Experimental results in peraluminous and peralkaline high silica glasses. EOS, Trans Am Geophys Soc 66:412

Nath P, Douglas RW (1965) Cr^{+3}-Cr^{+6} equilibrium in binary alkali silicate glasses. Phys Chem Glasses 6:197-202

O'Keefe M, Hyde BG (1984) Stoichiometry and the structure and stability of inorganic solids. Nature 309:411-414

Paul A, Douglas RW (1965) Ferrous-ferric equilibrium in binary alkali silicate glasses. Phys Chem Glasses 6:207-211

Pauling L (1960) Nature of the Chemical Bond (3rd Edn) Cornell University Press, Ithaca, NY

Prigogine I, Defay R (1954) Chemical Thermodynamics. John Wiley, New York

Robie RA, Hemingway BS, Fisher JR (1979) Thermodynamic properties of minerals and related substances at 298.15 K and 1 bar pressure and at higher temperatures. U S Geol Surv Bull 1452

Roy BN, Navrotsky A (1984) Thermochemistry of charge coupled substitutions in silicate glasses: The systems $M^{n+}{}_{1/n}AlO_2$-SiO_2(M = Li,Na,K,Rb,Cs,Mg,Ca,Sr,Ba,Pb). J Am Ceram Soc 67:606-610

Ryerson FJ (1985) Oxide solution mechanisms in silicate melts: Systematic variations in the activity coefficient of SiO_2. Geochim Cosmochim Acta 49:921-932

Ryerson FJ, Hess PC (1980) The role of P_2O_5 in silicate melts. Geochim Cosmochim Acta 44:611-624

Schreiber HD, Kochanowski BK, Schreiber CW, Morgan AB, Coolbaugh MT, Dunlap TG (1994) Compositional dependence of redox equilibria in sodium siliate glasses. J Non-Cryst Solids 177:340-346

Shannon RD (1976) Revised effective ionic radii and systematic studies of interatomic distances in halides and chalcogenides. Acta Crystallogr A32:751-767

Shannon RD (1993) Dielectric polarizabilities of ions in oxides and fluorides. J Appl Phys 73:348-366

Taylor M, Brown GE Jr (1979) Structure of mineral glasses: I-The feldspar glasses $NaAlSi_3O_8$, $KAlSi_3O_8$ and $CaAl_2Si_2O_8$. Geochim Cosmochim Acta 43:61-75

Tewhey JD, Hess PC (1979) Silicate immiscibility and thermodynamic mixing properties of liquids in the CaO-SiO_2 system. Phys Chem Glasses 20:41-53

Tomlinson JW, Heynes MSR, Bockris JO'M (1958) The structure of liquid silicates: Molar volumes and expansivities. Trans Faraday Soc 54:1822-1833

Toop GWE, Samis CS (1962) Some new ionic concepts of silicate slags. Can Met Quart 1:129-156

Toplis MJ, Dingwell DB, Libourel G (1994) The effect of phosphorous on the iron redox ratio, viscosity and density of an evolved ferrobasalt. Contrib Mineral Petrol 117:293-304

Toplis MJ, Libourel G, Carroll MR (1994) The role of phosphorous in crystallization processes of basalt: An experimental study. Geochim Cosmochim Acta 58:797-810

Varshal BG (1993) Structural model of the immiscibility of glass-forming silicate melts. Fiz Khim Stekla 19:149-153

Viellard PH, Tardy Y (1988) Estimation of enthalpies of formation of minerals based on their refined crystal structures. Am J Sci 288:997-1040

Uchino T, Sakka T, Ogata Y, Iwasaki M (1991) Ab initio molecular orbital calculations on the electronic structure of sodium silicate glasses. J Phys Chem 95:5455-5462

Uchino T, Sakka T, Ogata Y, Iwasaki M (1992) Changes in the structure of alkali-metal silicate glasses with the type of network modifier cation: An ab initio molecular orbital study. J Phys Chem 96:2455-2463

Watson EB (1979) Zircon saturation in felsic liquids: experimental data and applications to trace element geochemistry. Contrib Mineral Petrol 70:407-419

Wells AF (1985) Structural Inorganic Chemistry (5th Edn) Clarendon, Oxford

Chapter 7

DYNAMICS AND STRUCTURE OF SILICATE AND OXIDE MELTS: NUCLEAR MAGNETIC RESONANCE STUDIES

Jonathan F. Stebbins

Department of Geological and Environmental Sciences
Stanford University
Stanford, CA 94305 U.S.A.

INTRODUCTION

Nuclear Magnetic Resonance (NMR) spectroscopy has been a standard tool for the characterization of organic molecules in liquid solution since at least the late 1960s. However, its application to crystalline and then to glassy silicates only began in earnest with the development of high-field superconducting magnets, pulsed Fourier transform methods, and the application of the magic-angle spinning (MAS) technique around 1980 (Lippmaa et al., 1980). Although NMR data have been collected at quite high temperatures by physicists working on molten salts and metals since at least the mid-1960s, molten silicates were not studied by the technique until much more recently (Stebbins et al., 1985). Since then, NMR has made major contributions in our understanding of glass and melt structure, of the effects of pressure and temperature, and, most uniquely, in the dynamical processes controlling diffusion and viscous flow.

The static, frozen-in structure of a glass is at least a good approximation of that of the liquid at the glass transition temperature. [See chapters in this volume by Richet and Bottinga, Dingwell, and Webb and Dingwell, for a complete discussion of the phenomenon of the glass transition, and the interrelationships between structure and viscosity.] One fundamental question for understanding of both the thermodynamic and transport properties of glass-forming liquids is thus the quantitative contribution of various types of disorder to the entropy differences between the glass and the crystal. The inherently quantitative nature of NMR spectra have lead to major progress over the last 10 to 15 years on this problem.

However, as shown in several other chapters in this volume (e.g. that by McMillan and Wolf on vibrational spectroscopy), glass structure is only the starting point for understanding liquids. Most oxide liquids of interest in geochemistry and in technology are at least somewhat "fragile" (Angell, 1985, 1990), in that their heat capacities, thermal expansivities, compressibilities, etc., all increase markedly on heating through the glass transition because of the onset of significant, progressive, structural disordering. Integrating up in temperature from the glass transition (typically $500°$ to $750°C$ for silicate components other than pure SiO_2) to liquidus temperatures (typically $900°$ to $1600°C$), leads to what can be large configurational contributions to enthalpy, entropy, and molar volume. These in turn are at the heart of the currently most successful approach to understanding melt viscosity (see chapter by Richet and Bottinga). If the effects of temperature-induced structural change in the liquid are ignored for diopside ($CaMgSi_2O_6$), for example, its calculated heat of fusion would be only half of that observed. Identifying and quantifying the nature of this structural change with temperature is thus a second critical issue. Although this problem is far from solved, NMR studies of glasses quenched with different transition temperatures, and at high temperature of liquids themselves, have begun to at least better define the details of the question.

A third fundamental question is the challenge posed by the enormous range of viscosities and diffusivities in melts (more than 10 orders of magnitude in geological systems) caused by variation in temperature, composition, and pressure. What atomic-scale mechanisms control these properties? Can useful predictions be made? In situ, high temperature NMR studies have again provided the first direct clues as to the relationship between the time scales of local bond-breaking and of macroscopic transport.

For all of these questions, which may seem rather esoteric to geochemists and petrologists closer to the "real world" of outcrops and volcanoes, it should be kept in mind that earth scientists face a peculiar set of challenges to solving our chosen problems. Because of the inherent limits in our ability to fully simulate nature in the laboratory (temperature, pressure, and, most importantly, long time and distance scales), we will always be forced to make long extrapolations from our data sets, rather than interpolating safely within them. Accurate extrapolation is only possible with models that are firmly based in fundamental understanding of process. It is with that spirit that this chapter, and this volume, have been composed.

Other sources of background information on NMR

In general, in an NMR measurement a signal from a single isotope is observed, and the information content of the signal is usually dominated by the local structure around the element being studied, for example by coordination number, first and second-neighbor bond distances, bond angles, and connectivity to the first few neighboring atoms. More sophisticated NMR experiments can sometimes select out information about connectivities and distances to specific types of atom neighbors (e.g. between ^{29}Si and ^{1}H). In crystalline solids, NMR is thus an important complement to diffraction methods, which, although they often reveal structure more directly, "see" only a relatively long-range average. In amorphous materials and in liquids, the same kind of short- to intermediate-range information is still available from NMR, even if no long-range structure exists to give a coherent diffraction pattern. In addition, in an ideal, properly-conducted experiment, the amount of signal observed in a given NMR peak is directly proportional to the concentration of the isotope being observed in the site giving rise to the peak, regardless of the structural details of the site. Such quantitation is not, however, always possible in practice.

In crystalline and glassy solids, and in liquids, NMR can provide what is often unique information on the dynamics of atomic or molecular motion that occur on time scales ranging from seconds to nanoseconds or shorter. These time scales are much longer than those of interatomic vibrations, and are often of great interest in defining mechanisms of displacive phase transitions, ionic conductivity, diffusion, and viscous flow. Because this chapter deals primarily with liquids, dynamical effects will receive considerable attention.

The theory of NMR, and reviews of recent applications to crystalline and glassy silicates, have been presented recently and in detail, and thus will not be covered at great length here. The basic principles needed to understand recent work on silicate melts will be summarized, and then applications will be described. The reader is encouraged to examine the latter sections first, then turn to the former to answer technical and theoretical questions.

Perhaps the most accessible, somewhat more detailed treatments for geochemists appear in a previous volume in this series (Kirkpatrick, 1988; Stebbins, 1988b). Discussion of both theory and applications to silicates and other geological materials can be found in several monographs (Engelhardt and Michel, 1987; Wilson, 1987), and NMR data for silicates have been recently tabulated (Engelhardt and Koller, 1994; Stebbins, 1995). The basics of NMR, usually applied primarily to organic liquids, are described in many chemistry texts (Derome, 1987; Sanders and Hunter, 1987), while the details of theory are covered in more advanced works (Abragam, 1961; Harris, 1983; Mehring, 1983; Slichter, 1990). The mysteries of real hands-on spectroscopy have been revealed (Fukushima and Roedder, 1981). State-of-the-art modern theory and applications to solids appear in several edited volumes (Dye and Ellaboudy, 1991; Grimmer and Blümich, 1994; Wind, 1991). NMR studies of "less common" nuclides have been reviewed (Mason, 1987; Sebald, 1994). Several extensive reviews of applications of NMR to glass structure have been recently published (Bray et al., 1991; Eckert, 1990; Kirkpatrick, 1988), as has a review of high temperature NMR studies (Stebbins, 1991b). Applications to silicate and oxide liquids have been summarized recently as well (Stebbins et al., 1992b; Stebbins and Sen , 1994; Stebbins et al., 1995b). Tables of NMR "accessible" nuclides appear in the geochemical literature (Kirkpatrick, 1988; Stebbins, 1995); a more complete tabulation may be found in (Harris, 1983). Relatively complete figures showing a variety of calculated solid-state NMR peak shapes resulting from chemical shift anisotropy, quadrupolar effects, magic angle spinning, etc., have recently been compiled (Dye and Ellaboudy, 1991; Engelhardt and Michel, 1987).

BASIC NMR CONCEPTS

Neutrons, protons, and electrons each possess a spin quantum number (I) of 1/2. By classical analogy to a spinning charged sphere, these spins give rise to a magnetic moment. The way that the nuclear spins pair up determines the net spin of a given isotope: a nuclide with an even number of protons and neutrons may or may not have a spin equal to zero, while a nuclide with an odd number of neutrons and protons must have $I > 0$. Just as closed shells of electrons give rise to chemically very stable neutral atoms (the noble gases), certain configurations of nucleons give rise to very stable, and thus relatively abundant, isotopes with $I = 0$ during nucleosynthesis in stellar interiors. It is for this reason that the NMR spectroscopist must face the inconvenient fact that typically, the most abundant isotopes of elements with even atomic numbers (e.g. ^{12}C, ^{16}O, ^{24}Mg, ^{28}Si, ^{32}S) are NMR-inactive. In these cases, lower abundance isotopes must be examined, sometimes requiring isotopic enrichment (as is generally done for ^{17}O). Fortunately, many odd-atomic-numbered elements have easily observable, high-abundance isotopes (e.g. ^1H, ^{11}B, ^{19}F, ^{23}Na, ^{27}Al, ^{31}P).

For a nuclide with $I = 1/2$ (such as ^1H, ^{13}C, ^{19}F, ^{29}Si, ^{31}P), the magnetic moment of the nucleus interacts with an externally imposed magnetic field to produce a pair of energy levels with a single transition between them. The higher energy level can be thought of as having the nuclear magnetic moment oriented in a direction opposite to that of the field; the lower level as being oriented with the field. The energy of the transition between the levels is directly proportional to the magnetic field (H) at the nucleus, with the corresponding resonant or Larmor frequency ν_L fixed by the gyromagnetic ratio, γ, which is characteristic of each nuclide:

$$\nu_L = \frac{|\gamma|}{2\pi} H \tag{1}$$

The phenomenon of nuclear spin transitions is useful to chemists (and geochemists!), because resonant frequencies are slightly perturbed by the location and density of electrons near the nucleus. In particular, the *local* magnetic field "seen" by the nucleus is slightly different than the external magnetic field H_0. It is these small perturbations that allows structural information to be deduced from NMR data.

For magnetic fields typically used in NMR (about 4 to 18 Tesla), all commonly observed Larmor frequencies are at radio frequencies (rf), typically a few to a few hundred MHz. In most modern NMR experiments, the nuclear transitions of a single isotope are excited by a pulse of rf energy. The re-emission of this energy is then observed in real time and Fourier-transformed to give the frequency-domain spectrum. Almost always, frequencies are reported relative to that of some standard material. Commonly, the "δ" scale is used, with results in parts per million (ppm):

$$\delta = 10^6 \times (\nu_{sample} - \nu_{standard})/\nu_{standard} \tag{2}$$

NMR spectra are generally plotted with the high frequency end at the left. This is a holdover from the first 30 years of NMR data collection, where spectra were generated by sweeping the magnetic field at constant observation frequency, rather than detecting a range of frequencies at a fixed external field.

In oxides without conduction electrons and without abundant unpaired *electronic* spins (such as those associated with transition metal cations), total ranges of δ in all oxide compounds may be as small as a few ppm (e.g. ^1H, ^6Li) to thousands of ppm (e.g. ^{207}Pb), with heavy, more electron-rich elements having larger ranges.

The "chemical shift" is that part of the relative NMR resonant frequency that is caused by the shielding of the nucleus from the applied external magnetic field by the surrounding electrons. It is constant relative to ν_L and is independent of external magnetic field (and hence of what laboratory is measuring it) when expressed as δ. However, the description of *any* NMR peak position that happens to be measured in terms of δ as the "chemical shift" is erroneous, because other physical effects (especially quadrupolar, as described below) can affect both peak shape and position.

The energy of a nuclear spin transition ($\Delta E = |\gamma| hH/2\pi$, where h is Planck's constant) is relatively small (compared to infrared, optical and x-ray photons), and the ratio of the population of spins in the high energy state (b) to that in the low (a) is therefore close to one, with $N_a/N_b = \exp(-\Delta E/kT)$, where k is Boltzmann's constant. A more relevant ratio can be approximated as:

$$\frac{N_b - N_a}{N_b + N_a} \approx \frac{|\gamma| hH}{4\pi kT} \tag{3}$$

Because the size of the observed NMR signal depends on the ratio in Equation (3), which for typical experiments is about 10^{-5} to 10^{-6}, the equation indicates that (1) NMR is of low sensitivity when compared to visible, IR, or x-ray spectroscopy; (2) signal strength is reduced at high temperature; (3) signal strength is higher for nuclides with higher Larmor frequency (making observation of "high-γ" nuclides much easier than "low-γ" nuclides); and (4) signal strength increases at greater magnetic fields. The latter is one reason for investing in bigger, more expensive NMR magnets. Barring certain technical limitations (e.g. very broad NMR peaks), this equation also says that the only factors controlling the relative amount of signal from a given nuclide is the number of nuclei in the sample, γ, H, and temperature. For this reason NMR can be highly quantitative, with intensities and calibrations independent of details of local structure and chemical environment. Sensitivity is improving with better technology. For example, useful solid-state ^{11}B NMR spectra have been obtained for calcite containing less than 100 ppm B (Sen et al., 1994); good [27]Al and [29]Si spectra (the latter with isotopic enrichment) can be obtained on high pressure samples of 1 to 10 mg in weight (Kanzaki et al., 1992; Kirkpatrick et al., 1991; McMillan et al., 1989; Phillips et al., 1992; Xue et al., 1989).

For atoms whose local environment is not of very high symmetry (i.e. less than that of a perfect sphere, tetrahedron, octahedron, cube, or dodecahedron), the chemical shift for a given nuclide in a given site depends on the orientation of the surrounding atoms and bonds with respect to the external magnetic field. This chemical shift anisotropy (CSA) can be described by the principle values ($\delta_{11}, \delta_{22}, \delta_{33}$) of the CSA tensor, which is a 3×3 matrix that can be thought of as describing an ellipsoid analogous to an optical indicatrix. A complete description requires in addition the two angles that relate the CSA tensor to the symmetry axes of the crystal or of the local site. This generally requires a single crystal study, which has been done only rarely for silicates (Spearing and Stebbins, 1989; Weiden and Rager, 1985).

In a powdered or an amorphous sample that contains a random collection of all possible orientations, a characteristic "powder pattern" defined by the CSA is mapped out. Because this pattern is relatively broad, and a fixed amount of signal is thus spread out over a wide frequency range, powder spectra for non-rotating solid samples are not commonly observed, although they may in fact contain unique and important information. Typical CSA patterns for [29]Si in a glass are shown in Figure 18 (below). Most commonly now, high resolution "magic angle spinning" (MAS) spectra are obtained (see below). CSA tensor values (but not orientations) for spin 1/2 nuclides can often be obtained by analysis of MAS "spinning side bands," as well (see below) (Smith et al., 1983). A complete view of the theory for NMR of spin 1/2 nuclides has been recently published (Wind, 1991).

Quadrupolar nuclides

Most nuclides with I > 0 also have I > 1/2. For such nuclides, the nucleus has a quadrupolar moment and thus can interact with any gradient that may be present in the electric field at the nucleus. The nuclear spin energy levels can be split into (2 I + 1) levels, with 2 I transitions allowed between adjacent levels only. This splitting occurs when the nucleus is placed in an unsymmetrical electric field that has a non-zero gradient (the change in field with distance, or EFG). The EFG is zero if the atom is at the center of a perfectly tetrahedral, octahedral, or cubic (etc.) site, and in this case (e.g. [23]Na in NaCl), the energies of the transitions are the same and all their intensity is observed at a single frequency. Most quadrupolar nuclides have I equal to an odd multiple of 1/2. (Geochemically interesting exceptions include ^{2}H and [6]Li, with I = 1.) In this case, a "central", 1/2 to −1/2 transition is present that is analogous to that of nuclides with I = 1/2. The frequency of this transition is perturbed somewhat by "second order" quadrupolar effects, which, like the CSA, are orientation-dependent and give rise to well-understood powder patterns (Fig. 1). These are characterized by the quadrupolar coupling constant, abbreviated "QCC" or C_Q, with

$$C_Q = e^2qQ/h \qquad (4)$$

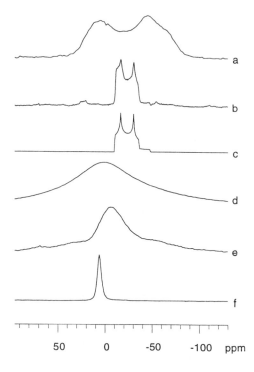

Figure 1. ^{23}Na spectra for crystalline, glassy, and molten silicates. (a) Static spectrum for crystalline low albite (NaAlSi$_3$O$_8$) (George and Stebbins, 1995). Note quadrupolar peak shape, broadened by Na-Na and Na-Al dipole-dipole couplings. (b) MAS spectrum for albite. (c) Best-fit calculated quadrupolar peak shape for albite. (d) Static spectrum for Na$_2$Si$_3$O$_7$ glass, which is severely broadened by disorder, quadrupolar, and dipole-dipole interactions. (e) MAS spectrum of same composition glass. (f) Liquid of same composition, showing fully-averaged peak. All spectra were acquired at a magnetic field of 9.4 T. Scales are relative to 1 M aqueous NaCl.

Here, e is the charge of the electron, eq is the principle value of the tensor describing the EFG, and Q is the nuclear quadrupole coupling constant, again a fixed parameter for a given nuclide. The deviation from cylindrical symmetry of the EFG tensor is described by the asymmetry parameter η_Q, the value of which can vary from 0 (highest symmetry) to 1 (lowest). The "quadrupolar frequency" v_Q is a convenient alternate variable for describing the size of quadrupolar effects, with

$$v_Q = \frac{3}{2 \cdot I(I-1)} C_Q \tag{5}$$

The factor in front of C_Q in Equation (5) is 1/2 for I = 3/2, and 3/20 for I = 5/2. Again when I is an odd multiple of 1/2 and $C_Q > 0$, there are (2 I – 1) additional transitions, half of which are shifted to higher frequency, half to lower frequency. These "satellite transitions" are shifted by v_Q, and their positions thus contain potentially useful information about the EFG and therefore about the structure. Because in many cases these shifts are relatively large, they may be inconvenient to observe even in a single crystal. Again, orientational effects often spread this part of the spectrum over a wide and nearly unobservable range for a powdered or amorphous sample. A detailed treatment of the theory of quadrupolar nuclides has recently appeared (Dye and Ellaboudy, 1991).

Dipole-dipole interactions

The other major effects on NMR peak shapes in inorganic solids are those of magnetic dipolar interactions. These may arise from neighboring nuclear spins of the same isotope (homonuclear dipolar coupling), from nuclear spins of other isotopes (heteronuclear coupling), or from unpaired electronic spins. The energy state of each adjacent spin is "felt" at a nucleus under observation as a perturbation in the local magnetic field. The effect on energy (and thus on the observed NMR frequency) scales as $1/r^6$, and thus is important only when the γ of the other spin is large and/or when distances are short. The effect of dipole-dipole interactions (for unlike spins A and B) on peak widths and on relaxation (see below) is proportional to $\gamma_A^2\gamma_B^2$, and thus is especially strong for a nuclide such as ^1H or ^{19}F with a high γ. In the case of an isolated pair of nuclear spins (e.g. the H_2O molecule in gypsum), the NMR signal from one is simply split by the other. In the more common case of many neighboring spins, each of which can have at least two different states, it is more common to observe a broadening of the NMR peak. This perturbation, which again is orientation dependent, can be as large as several tens of kHz, and can severely limit the information obtainable from strongly coupled spins in solid materials, for example from ^1H in hydroxides or ^{19}F NMR in fluorides.

Electrons have γ about 1000 times that of the largest value for a commonly observed nuclear spin (^1H) and thus unpaired electrons can be particularly effective in broadening NMR peaks. It is for this reason that NMR spectra of silicates with more than about 1% of Fe (or of many other transition metals or rare earth elements) are usually severely broadened, even to the point of being unobservable.

In contrast, spectra of "dilute" nuclides, such as ^{13}C (about 1% of natural C) and ^{29}Si (about 4.7% of natural Si), are not subject to homonuclear dipolar interacts and often very high resolution spectra can be obtained. Heteronuclear dipolar effects (e.g. from ^1H neighbors), if present, can generally be removed by MAS (see below) or by "decoupling", during which the energy levels of the "other" nuclide are saturated by long, high-power pulses at the appropriate frequency.

Motional averaging

Many of the issues raised above (chemical shift anisotropy, quadrupolar shifts and broadening, dipolar broadening) receive little or no attention in standard introductory texts on NMR. There is a very good reason for this: most NMR spectra that chemists collect are of liquids. In a liquid of relatively low viscosity, molecules (or other less well-defined "structural units") tumble rapidly, randomly sampling all spatial orientations. If this reorientation is rapid when compared to the total frequency range of the NMR spectrum of the equivalent "static" (immobilized) solid, then orientational effects on the NMR frequencies are averaged to zero. For example, the CSA is averaged so that only its "isotropic" value, δ_{iso}, is observed. The EFG is averaged to zero, making all 2I transitions for a quadrupolar nuclide equivalent in energy and concentrating their signal into a single peak. Dipole-dipole interactions are also averaged to zero, eliminating this source of broadening. At the expense of the loss of information that is contained in these orientation-dependent parameters, a sometimes enormous narrowing of the NMR spectrum is obtained (often a factor of 1000 or more). The ability to distinguish among different sites or molecules with slightly different values of δ_{iso} is thus greatly enhanced in a liquid. Again because the total signal is fixed by the number of NMR-active nuclei in a given sample, concentrating this signal into a narrow peak will give a correspondingly greater signal-to-noise ratio. Reorientation can occur not only through the true rotation of a structural unit, but can effectively be caused by the diffusion of an ion from one site to another with a series of different orientations, or by the rapid rearrangement of the atoms *around* a nuclide being observed.

The meaning of "liquid-like" behavior with respect to NMR depends on the width of the spectrum of interest. In ^{29}Si (I = 1/2) spectra of silicates, for example, the static peak width is usually dominated by the CSA, which is typically on the order of 100 ppm. At a typical magnetic field of 9.4 T, the ^{29}Si Larmor frequency is 79.5 MHz, and the CSA (and peak width without MAS) is thus 7950 Hz. Rotation of silicate "molecules" (if they existed in melts, see below) at a rate about 10 to 20% of this frequency would begin to affect its shape, while rotation at about 10 to 100 times this frequency would lead to a completely averaged, narrow spectral line.

"Liquid-like" behavior can even be observed for rapidly diffusing cations in crystalline solids, and partial motional averaging is common in "soft" molecular solids, where organic (e.g. benzene) or inorganic molecules (e.g. P_4) may rotate rapidly even well below the melting point (Stebbins, 1988b). Whether in a liquid or a solid, motional averaging is most commonly induced by heating, but can also be caused by the introduction of defects that speed diffusion (Stebbins et al., 1995c).

For a quadrupolar nuclide such as ^{23}Na (I = 3/2), more complex behavior will be observed. Recall that typically only the "central", 1/2 to –1/2 peak is observed in silicates, which is typically a few to a few tens of kHz wide. As shown in Figure 2, as Na mobility in a glass or liquid increases with increasing T, an initial narrowing may be observed, as reorientation leads to averaging of broadening effects on this central transition, which now include not only CSA and dipole-dipole couplings but second-order quadrupolar effects. Under these conditions, even if considerable narrowing is present, the position of the peak should still be shifted to lower frequency by the isotropic average of the second order quadrupolar coupling, δ_Q (Youngman and Zwanziger, 1994):

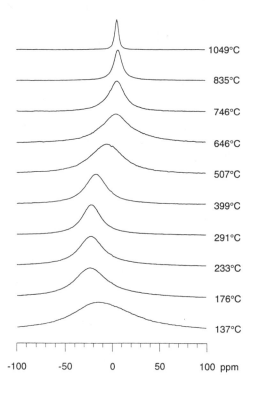

Figure 2. ^{23}Na spectra (static) for $Na_2Si_3O_7$ glass (T < 400°C) and liquid. Initial narrowing with increased temperature is caused by averaging of dipole-dipole and quadrupolar effects on the central, 1/2 to –1/2 transition. Broadening occurs on further T increase (between 300° and 600°C) because of the beginning of averaging of quadrupolar satellite transitions, which are not observed directly. Finally, complete averaging results when reorientation and exchange become much faster than the total quadrupolar peak width (George and Stebbins, in prep.). Spectra were acquired at a magnetic field of 9.4 T. Scales are relative to 1 M aqueous NaCl.

1049°C
835°C
746°C
646°C
507°C
399°C
291°C
233°C
176°C
137°C

-100   -50   0   50   100 ppm

$$\delta_Q = 25000 \left(\frac{C_Q}{\nu_L}\right)^2 \left(1 + \frac{\eta_Q^2}{3}\right) \tag{6}$$

All quadrupolar effects will not be fully averaged away until random, isotropic reorientation becomes rapid with respect to the full quadrupolar peak width, given by $2\nu_Q$ and in this example being several MHz. During partial averaging of the central and satellite transitions, the observed peak may actually become broader (Fig. 2). Final, truly "liquid-like", complete motional averaging will not be obtained until the reorientation frequency is roughly 10 to 100 times ν_Q. Only at this point does the peak position become the same as the isotropic chemical shift. During the approach to this state, all other factors remaining the same, the peak will move from lower to higher frequency as δ_Q is averaged to zero. Note that as discussed below, the shift of the peak in Figure 2 to *lower* frequency at higher temperature thus probably requires a structural change.

MAS, DOR, DAS: high resolution spectroscopy of solids

The key to high-resolution NMR of solid materials was the realization that many of the effects that broaden solid spectra (in particular, CSA and dipole-dipole couplings) can be averaged out not only by liquid-like random reorientation, but by mechanical rotation of the entire solid sample about a single axis. This is true because the equations for orientation dependence of these effects contain the multiplier $(3\cos^2\theta - 1)$, where θ is the angle between the external magnetic field and the largest component of the tensor describing the interaction, for example the long axis of the CSA ellipsoid. If a powdered sample is rotated about the correct θ such that this term is equal to zero ($54.7°$), and if that rotation is similar to or greater than the static peak width, then narrowing to nearly liquid-like resolution is often observed. Rotation of samples at the required rates (generally about 5 to 15 kHz or 300,000 to 900,000 rpm) is technically non-trivial, but has now become nearly routine. This "magic-angle-spinning" (MAS) technique is what has enabled most mineralogically interesting NMR to be done, beginning about 1980. It is applicable not only to solids, but, as will be shown below, to very viscous oxide melts.

An "artifact" of MAS is the formation of spinning sidebands in spectra, spaced at precisely the spinning frequency. For quadrupolar nuclides, a separate set of sidebands may be observed for each pair of satellite transitions. For $I = 5/2$, it turns out that second order quadrupolar effects are much smaller on the 1/2 to 3/2 transition than on the central, 1/2 to $-1/2$ transition. As a result, resolution among different sites may be better in the sidebands than in the central peak, and their analysis may give more accurate determinations of C_Q and δ_{iso} (Jakobsen et al., 1989; Massiot et al., 1990a; Skibsted et al., 1991).

For some nuclides that are very easy to observe in liquids, such as [207]Pb and [51]V, very large ranges of isotropic chemical shifts, and of CSA, often cover frequencies ranges far beyond possible sample spinning rates (e.g. thousands of ppm or hundreds of kHz for [207]Pb in a glass; Bray et al., 1988). In this case, MAS spectra of amorphous materials may resemble non-spinning spectra. Confusion on this issue has lead to significant mis-interpretation of NMR spectra in at least one very recent study of lead vanadate glasses (Hayakawa et al., 1995).

MAS does not completely average out second-order quadrupolar broadening, leaving sometimes inconveniently broad or complex peak shapes for nuclides such as [17]O, [23]Na, and [27]Al (Fig. 1). This broadening is particularly troublesome in glasses (and disordered crystals), where it can mask the "interesting" broadening caused by distributions of chemical shifts that could potentially characterize the amount of disorder.

It turns out that all significant orientational broadening effects can be essentially eliminated if the sample is spun about two axes simultaneously. This mechanically rather difficult task has been accomplished in the technique of "double rotation" (DOR) NMR, which has been shown to be capable of producing high resolution, narrow-peaked spectra for ^{17}O in crystalline silicates (Mueller et al., 1990). Its applicability remains somewhat limited by the relatively low spinning rates that can presently be obtained.

A second approach that can also eliminate quadrupolar broadening is "dynamic angle spinning" (DAS) NMR. In this experiment, the rotation angle is flipped back and forth between two carefully selected angles in as little as a few tens of ms. A complex rf pulse sequence and data acquisition scheme results in a two-dimensional NMR spectrum (see below), one dimension of which contains narrow, isotropically averaged peaks. The other dimension contains anisotropic peak shapes that may be analyzed to give quadrupolar parameters, CSA, etc. As in DOR, the positions of the isotropic peaks are controlled by the sum of the isotropic chemical shift and the averaged second-order quadrupolar shift. DAS NMR on ^{17}O in silicate glasses recently has allowed quantification of the distribution of Si-O-Si bond angles and of cation neighbors around non-bridging oxygens (Farnan et al., 1992; Vermillion et al., 1995). DAS NMR on ^{11}B in B_2O_3 glass has been used to detect multiple B sites (Youngman and Zwanziger, 1994).

Chemical exchange

A special kind of motional averaging occurs when an atom in an NMR-distinct site physically exchanges with a different site. This process can occur during diffusion in a crystalline or glassy solid, or may be tied to viscous flow in a liquid. In any case, the resulting effect on the spectrum is a merging of the peaks for the two sites and eventually an averaging into a single narrow peak as the exchange frequency becomes large relative to the frequency separation of the two resonances. This can be as small as a few Hz or can approach the maximum observable peak width, typically 0.1 to 5 MHz. This process is illustrated in Figure 3, which was created by a computer program that calculates effects

$v_{ex} / \Delta v =$

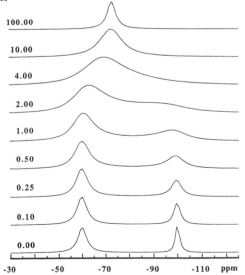

Figure 3. Simulation of exchange between two sites, each of which has a simple, Lorentzian-shaped peak. The exchange frequency relative to the frequency separation of the peaks is shown. Note that the position of the fully averaged peak is the weighted average of the two unaveraged peaks, and is closer to the left peak because of its greater area. All spectra are normalized to the same height. If spectra were plotted on the same absolute scale, heights of broadened peaks would be much reduced, as the area is conserved.

of exchange in a simple two-site system (Abragam, 1961). Similar approaches can be taken to simulate exchange among multiple sites, if the NMR peaks for those sites are each symmetrical and at least roughly Lorentzian or Gaussian in shape. This is often the case in MAS spectra, although complications may arise when the exchange rate approaches the sample spinning rate (Fenske et al., 1992). Correct simulation of experimental peak shapes in the region of partial chemical exchange can lead to accurate determination of exchange frequencies, as described below for ^{29}Si in melts. The more complex problem of simulating exchange among different sites and different orientations simultaneously in a static spectrum can be approached by starting with many "sites" (possibly hundreds) that map out a static CSA or quadrupolar powder pattern. Exchange among these can then be simulated, again yielding an overall average exchange rate for the system (Farnan and Stebbins, 1990a). A large body of literature exists on the interpretation of chemically exchanging systems, still based largely on observations of organic molecular systems (Schmidt-Rohr and Spiess, 1991).

Two-dimensional exchange experiments

A wide variety of two–dimensional (2-D) NMR experiments are routinely performed on liquid samples, and the concepts involved are described in a number of chemistry texts (Derome, 1987; Sanders and Hunter, 1987). In any 2-D NMR experiment, at least two rf pulses are given, separated by an "evolution time" t_1. The acquired data set consists of a series of spectra collected with t_1 values (often several hundred) incrementally increasing from zero. These spectra can be imagined as being stacked up to form a two-dimensional data array. The set made up of the first frequency point in each spectrum defines a new time-domain signal with increasing t_1, as does the set of second points, third points, etc. Each of these "interferograms" can then be Fourier transformed a second time, yielding a 2-D spectrum with two frequency axes.

The basis of a 2-D exchange experiment is a sequence of three rf pulses (Jeener et al., 1979). The spacing between the first two pulses is t_1 as described above. However, a second delay called the "mixing time," followed by a third pulse, is added. The pulse sequence is designed to make the two frequency axes in the 2-D spectrum essentially equivalent, yielding a particularly simple interpretation of results. Suppose that the material of interest contains two types of sites A and B that give rise to two distinct NMR peaks at v_A and v_B in a normal, 1-D spectrum. In the absence of any exchange, this spectrum appears on the diagonal in the 2-D plot, with peaks at the coordinates (v_A, v_A) and (v_B, v_B). However, if exchange between sites A and B takes place during the mixing time τ_{mix}, pairs of "cross-peaks" appear off the diagonal at (v_A, v_B) and (v_B, v_A). The intensities of the cross peaks increase with the probability of exchange during τ_{mix}, which can be varied over a wide range to investigate the time scale of the exchange, and to determine exchange rates quantitatively. This approach is especially useful for observing exchange at frequencies much slower than the width of the 1-D spectrum. Time scales almost arbitrarily long can be studied, as long as τ_{mix} remains short relative to the spin-lattice relaxation time (see below). An elegant, if non-oxidic, example of 2-D exchange spectroscopy is that of a molten salt mixture of $SnBr_4$ and $SnCl_4$, where NMR peaks for all five mixed species can by observed by ^{119}Sn NMR, and exchange among them can be quantified (Ramachandran et al., 1985) (Fig. 4).

An important technical consideration in this kind of 2-D experiment is that misleading results can result from a process called spin diffusion, where spin energy only is transferred from one nucleus to another of the same isotope, without physical exchange. If this transfer is from a nucleus in one type of site to one in a different type,

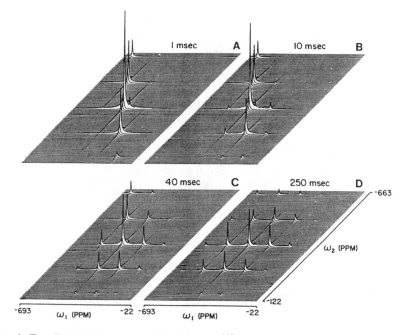

Figure 4. Two-dimensional exchange NMR spectra for ^{119}Sn in a molten mixture of $SnCl_4$ and $SnBr_4$ at 340 K at the mixing times as labeled. In (A), the five peaks along the diagonal from top to bottom correspond to $SnBr_4$, $SnBr_3Cl$, $SnBr_2Cl_2$, $SnBrCl_3$, and $SnCl_4$ groups. In (B) through (D), increasing mixing times allow increasing exchange among these species. The corresponding off-diagonal or cross peaks grow in intensity, while the on-diagonal peaks decrease. [Used by permission of the editor of *Journal of Magnetic Resonance*, from Ramachandran et al. (1985), Fig. 1, p. 137.]

off-diagonal peaks will appear. Spin diffusion requires significant homonuclear dipolar coupling, which can generally be eliminated by MAS or by isotopic dilution. Spin diffusion is also expected to not be strongly temperature dependent, because it is not an activated process.

Structural effects on chemical shifts

The development of high-resolution NMR techniques for solids, in particular MAS NMR, has focused attention on isotropic chemical shifts (δ_{iso}) and the structural information that they contain. The theoretical understanding of this problem is not complete, and, in general, values of δ_{iso} cannot be calculated accurately from first principles for complex structures. However, ab initio calculations on small molecular clusters that approximate the major structural features of silicates have recently shown great potential in quantifying at least relative effects of parameters such as coordination number, bond angles, and distances, etc. (Tossell, 1990; Tossell, 1992; Tossell and Vaughn, 1992). Most of what we know about chemical shifts is thus based on empirical correlations derived from known, crystalline structures. Some of the most important features of these correlations are summarized here. More complete discussion of these relationships can be found in earlier publications (Dupree et al., 1993; Engelhardt and Koller, 1994; Engelhardt and Michel, 1987; Kirkpatrick, 1988).

The largest structural influence on chemical shifts for Si and Al in inorganic oxides is the cation coordination number. SiO_4 groups ([4]Si) have δ_{iso} values between about −65 and −120 ppm relative to tetramethylsilane; SiO_6 groups ([6]Si) lie between about −180 and −220 ppm (Stebbins, 1995; Stebbins and Kanzaki, 1990) (Fig. 5). SiO_5 groups ([5]Si) appear to give signals at about −150 ppm in glasses (Stebbins and McMillan, 1993; Xue, et al., 1989), although this has not been rigorously confirmed due to the absence of this unit in known silicate crystal structures. [This group is, however, known in organic systems (Stebbins and McMillan, 1993; Swamy et al., 1990), and is suspected in a high pressure calcium silicate of unknown crystal structure (Kanzaki et al., 1991).] For AlO_4 groups ([4]Al), δ_{iso} general ranges from 85 to 60 ppm relative to aqueous Al^{3+}, AlO_6 groups ([6]Al) range from about 15 to 0 ppm, and again, five coordinate sites give signals about halfway in between (Fig. 6).

Coordination number also appears to be a major, sometimes predominant, influence on δ_{iso} for [6]Li (Xu and Stebbins, 1995), [23]Na (Koller et al., 1994; Maekawa, 1993; Xue and Stebbins, 1993), and [25]Mg in silicates (Fiske and Stebbins, 1994a) and for [11]B in borates (Bunker et al., 1990; Turner et al., 1986). In these cases also, higher coordination by oxygen leads to increased shielding (lower resonant frequency). The immediate cause of this coordination number effect, at least in the well-studied case of [29]Si, appears to be the increase in the ionicity and/or in the length of the bonds from the cation to its neighboring oxygens as its coordination number increases. A higher positive charge on the Si results. Somewhat counterintuitively, this results in greater shielding of the nucleus (and thus a lower δ_{iso}), because the occupancy in non-bonding, asymmetrical orbitals is reduced as the electron distribution around the Si becomes more spherical, reducing deshielding effects. This problem has been discussed at length in the theoretical literature (Engelhardt and Koller, 1994; Tossell and Vaughn, 1992). A corresponding effect on anions, with increased negative charge on the ion corresponding to more ionic, longer bonds, can contribute to *decreased* shielding for [17]O (Xue et al., 1991a) and [35]Cl (George and Stebbins, in preparation) in silicates. Chemical shift systematics are inherently complex, however, and a correlation established for one group of structures or ligands may be quite different for another group. For example, δ_{iso} for [29]Si, when plotted as a function of a wide range in net charge on the Si, displays a maximum, such that coordination number effects for non-oxygen ligands (e.g. Se) may have the opposite sign (Marsmann, 1981). Fortunately, all silicates are on the same side of this roughly parabolic curve (Engelhardt and Michel, 1987).

A second major structural effect, again best known for [29]Si and [27]Al, is that of the identity of first neighbor cations. Consider a Si cation Si_A surrounded by four oxygens O, each of which is also bonded to one or more cations X. As the degree of ionicity of the O–X bonds decreases, for example from Mg to Al to Si, δ_{iso} of the central Si_A decreases (becomes more shielded). This effect is thus similar in sign and probably in cause to that of coordination number, because in this same sequence the degree of ionicity of the Si_A–O bonds *increases*. Similar effects are known for Al and for Si in six-coordination (Dirken et al., 1992; Stebbins and Kanzaki, 1990). The effect of first neighbor cations is often described for SiO_4 groups by the label Q^n, where Q signifies "quaternary" (four bonds) and n the number of bridging oxygen bonds to other Si or Al (or sometimes B or P) cations. Chemical shift values for each of the Q^n species fall into distinct, if somewhat overlapping, regions, with each increment of n causing about a 10 ppm shift to more negative values (e.g. peaks for [4]Si in Fig. 5). Particularly in framework silicates where all Si sites are Q^4 groups, it is also well-established that substitution of each successive Si neighbor with Al increases the chemical shift (to less negative values) by about 5 ppm (Engelhardt and Michel, 1987).

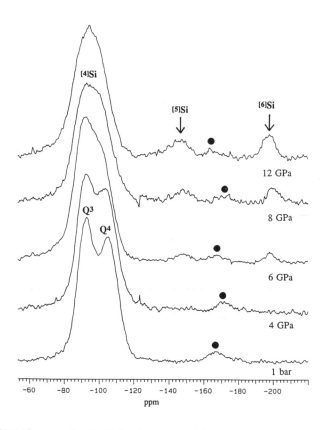

Figure 5. ^{29}Si MAS spectra for $Na_2Si_4O_9$ glass quenched from melts at pressures shown (Xue et al., 1991b). Four-, five-, and six-coordinated Si sites are shown. Spinning sidebands are marked by dots.

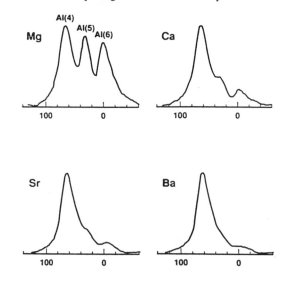

Figure 6. ^{27}Al MAS NMR spectra for glasses of composition 40% MO (M = Mg, Ca, Sr, Ba), 40% B_2O_3, 20% Al_2O_3. The Al coordination numbers are labeled. Scales are in ppm. [Used by permission of the American Ceramic Society, from Bunker et al. (1991), Fig. 6, p. 1434.]

Other structural effects are known to influence chemical shifts for ^{27}Al and ^{29}Si, in particular variations in bond distances and angles (Engelhardt and Michel, 1987; Phillips and Kirkpatrick, 1994). These are important in distinguishing multiple, but topologically similar, sites in minerals, and contribute to broadening of NMR peaks in glasses. For ^{23}Na, compositional effects, such as the ratio of non-bridging oxygens to tetrahedral cations, are also know to be important (Maekawa, 1993; Xue and Stebbins, 1993), probably again in causing variations in next neighbor cation populations.

Spin-lattice relaxation

Energy transfer out of a nuclear spin system to another form (such as thermal motion of a crystal lattice) is called "spin-lattice relaxation", and is required for the system to return to equilibrium after being excited by an rf pulse during an NMR experiment. This process is often characterized by a single exponential rate constant T_1, although in solids this may often be only a rough approximation. In the mineralogical literature, spin-lattice relaxation has been discussed most commonly because of its effect on the ease of acquiring usable spectra. NMR spectra are usually acquired by averaging data from many separate pulses in order to increase the signal to noise ratio. If pulses are given too rapidly relative to the relaxation rate, the spin system will saturate and the signal will disappear. Different sites in a material may have different relaxation rates, leading to non-quantitative spectra. In contrast, very rapid relaxation can lead to peak broadening, which in extreme cases can lead to unobservable NMR spectra.

On the other hand, measurements of spin-lattice relaxation can yield important data on dynamical processes, as well as on short-range to intermediate-range structure. Because the energies of nuclear spin transitions are relatively low, relaxation must be stimulated by fluctuations in the local magnetic field at the Larmor frequency, or by fluctuations in the EFG for quadrupolar nuclides. These fluctuations may in turn be caused by motion of the atoms of the element of interest, or of those around it, or, in some cases, by fluctuations in the state of unpaired electrons on paramagnetic ions such as transition metals and rare earth elements. If the relaxation process can be understood, then information on diffusion, polyhedral motion, site exchange, etc., may be obtainable.

Theories of relaxation may be complex, and relaxation processes, especially in solids, are not always well understood. Multiple relaxation mechanisms may be present, and distinguishing among them may be difficult. It is beyond the scope of this chapter to discuss these issues thoroughly, and useful descriptions appear in the literature at a variety of levels (Abragam, 1961; Fukushima and Roedder, 1981; Harris, 1983). However, several features are common to most descriptions of relaxation. Equations for relaxation contain the product of two kinds of terms. The first of these describes the interaction that is causing relaxation, and its magnitude in frequency units. For example, chemical exchange can cause relaxation because the chemical shift, and thus the local magnetic field, fluctuates during a jump from one site to another. Similarly, rotation of a molecule causes fluctuations in any orientation-dependent interaction that may be present, such as CSA, dipolar and quadrupolar couplings. Larger fluctuations in the predominant interaction term will lead to faster relaxation, everything else being equal.

The second kind of term describes the dynamics, and includes the correlation time τ_c (or times) that characterize the frequency of the fluctuation causing relaxation, and the Larmor frequency. The latter is generally taken as the angular frequency (the inverse of the time to rotate through one radian), with $\omega_L = 2\pi\nu_L$. In the simplest version of relaxation theory, as originally proposed by Bloembergen, Pound, and Purcell

(Bloembergen et al., 1948), the relaxation rate for a spin 1/2 nuclide is controlled by a single correlation time, and the correlation function itself is a simple exponential, corresponding to a random loss of positional correlation. In this case, the relaxation rate is proportional to a dynamical term as follows:

$$\frac{1}{T_1} \propto \frac{\tau_c}{1 + \omega_L^2 \tau_c^2} \tag{7}$$

This kind of relationship results in variations with temperature in T_1 that are common in most systems, regardless of the details of the modification of this basic model. At relatively low temperature, the dynamics of diffusion, site hopping and molecular rotation are slow, resulting in relatively long correlation times, with $\tau_c \gg 1/\omega_L$ (or $\tau_c \omega_L \gg 1$). In this case, $1/T_1$ becomes proportional to $1/\tau_c$. If, in turn, τ_c shows Arrhenian behavior (i.e. *ln* τ_c is proportional to E_a/RT, where E_a is an activation energy), then in this low temperature region a plot of *ln* $(1/T_1)$ versus inverse temperature will be linear with a slope of $-E_a$. At much higher temperature, where motional correlation times are short (generally true in low-viscosity liquids), the relationship $\tau_c \ll 1/\omega_L$ (or $\tau_c \omega_L \ll 1$) may hold. In this regime, $1/T_1$ becomes proportional to τ_c. Again, if τ_c is Arrhenian, a plot of *ln* $(1/T_1)$ versus inverse temperature will be a straight line with a slope of $+E_a$. If this entire curve can be observed, then at some intermediate temperature, τ_c will be equal to ω_L. At this point, relaxation will be most efficient and the $1/T_1$ curve will show a maximum (T_1 will show a minimum). It should be noted that in the systems considered here, normal interatomic vibrations in either liquids or solids are at such high frequencies (10^{13} to 10^{14} Hz) that they contribute little to spin-lattice relaxation.

Simple BPP behavior has been commonly observed for ^{1}H relaxation in organic liquids. In many other systems, a variety of complications are found. Two issues are most relevant here. The first of these applies most commonly to highly disordered systems, such as oxide glasses. Here, it is commonly observed that the absolute value of the slope on the low temperature side of the T_1 minimum is considerably less than that on the high temperature side (Fig. 7), with a systematic increase in slope with temperature. A large body of theory exists to explain this phenomenon and the 'correct' model is still a subject

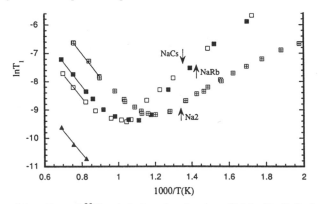

Figure 7. Natural logarithms of ^{23}Na spin-lattice relaxation times (T_1) for $Na_2Si_3O_7$ (crossed squares), $NaRbSi_3O_7$ (solid squares), and $NaCsSi_3O_7$ (open squares) glasses and liquids, plotted against inverse temperature in K. Inverses of peak widths for ^{87}Rb in the NaRb liquid are shown by triangles. Lines fitted to the highest temperature data all give apparent activation energies of 70 ± 3 kJ/mol. T_g's for typical laboratory heating rates are shown by the arrows. Redrawn from (Sen et al., 1995; Stebbins et al., 1995b).

of considerable debate. Some treatments assume that a distribution of correlation times is needed, perhaps with a corresponding distribution of activation energies. This makes good physical sense because of the disorder: a range in the height of energy barriers to motion *must* be present. Other treatments assume that the correlation function itself (i.e. the probability of a particle moving away from its original position as a function of time) is not a simple exponential decay, which again may be sensible in any complex, strongly interacting system. Regardless of details, both of these approaches may be describing the same basic physical phenomenon. Consider a mobile cation, such as Na^+, in a glass or melt. Recall that relaxation is caused by fluctuations at ω_L. For ^{23}Na, the most important fluctuations are likely to be those in the EFG caused by the Na^+ cation moving relative to its surrounding oxygen atoms. The motion of a cation is made up of a possibly complex distribution of low frequency oscillations within its site, and much less frequent diffusive jumps to other sites. At low temperature, the only part of this motion that is sufficiently probable at ω_L may be local motions inside its "cage" of oxygens. The energy involved in such motion is relatively low, and the apparent E_a for relaxation will also be low. In contrast, at high temperature, most motional frequencies are well above ω_L. In this case, the most infrequent motions, which are the high-energy diffusive jumps, will be closest to ω_L and will dominate relaxation. Thus, a relatively large apparent E_a will be observed. In between these extremes, of course, a range of types of motion, perhaps even a continuum from local to through-going, will participate in relaxation and intermediate behavior will be seen.

A second kind of complication arises in the study of quadrupolar nuclides. Because of the presence of multiple transitions, it may not be possible to characterize relaxation by a single T_1. For $I = 3/2$, for example, relaxation often appears to be well approximated by a double exponential fit of magnetization vs. time. In relatively mobile systems where motional averaging has "collapsed" the entire quadrupolar spectrum and all transitions contribute to a single observed peak, then relaxation often does behave as a single exponential decay, and fitting data in terms of single T_1 values is a sensible approach. This has generally been the case for ^{23}Na and ^7Li relaxation studies of oxide glasses and melts. In fully "liquid-like" spectra, on the high temperature side of the T_1 minimum where $\tau_c \ll 1/\omega_L$, quadrupolar relaxation may take on a relatively simple form (Abragam, 1961; Liu et al., 1987):

$$\frac{1}{T_1} = \frac{3}{40} \frac{2I+3}{I^2(2I-1)} \left[1 + \frac{\eta_Q^2}{3}\right] \left[\frac{C_Q}{2\pi}\right]^2 \tau_c \qquad (8)$$

This equation is appropriate for the value of C_Q actually "seen" by the nuclear spins. If C_Q is estimated from the positions and charges of the surrounding ions only (e.g. in a "point charge" calculation), then it must be multiplied by the term $(1 - F)$, where F is the "Sternheimer antishielding factor" (more often abbreviated γ_∞) (Dye and Ellaboudy, 1991). The Sternheimer factor accounts for the effect of the deformation of the electron cloud around the central atom, and thus varies from element to element, and, especially for high atomic numbers, may vary with the type of bonding. Values must generally be derived by comparing calculated field gradients to measured C_Q data.

Because the gyromagnetic ratio γ for an electron is so large (about 3300 times larger than γ for ^{29}Si), and magnetic dipolar interactions scale with γ^2, magnetic coupling between nuclear and the unpaired electronic spins of paramagnetic ions can be very effective in causing spin-lattice relaxation. This can occur in several ways. The field "seen" by the nucleus may fluctuate as above due to physical motion of the ions (relative to the paramagnetic center), or it may fluctuate because the state of the electron is changing. In the latter case, the correlation time of concern is that for the electron, not that for ionic motion. In either case, the relaxation may contain spatial information, because the dipolar coupling scales as $1/r^6$, where r if the distance from the nucleus of interest to the paramagnetic center. Paramagnetic relaxation is often predominant for dilute spin 1/2 nuclides (e.g. ^{29}Si) in materials without other abundant, high γ nuclides (e.g. ^1H). In this case, magnetization vs. time plots often show power-law, rather than exponential behavior. The slopes of such plots are related to the dimensionality of the distribution of spins, and the relaxation rate follows a predictable relationship to the concentration of the paramagnetic centers. Deviation from this behavior may be evidence of non-uniformity in the distribution of paramagnetic ions, often because of clustering (Sen and Stebbins, 1994; Sen and Stebbins, 1995).

A variety of approaches can be taken to distinguish among relaxation mechanisms, and to increase the range of dynamics sampled. One of the best of these is to collect data at a range of magnetic fields, which can determine if the relaxation depends on an interaction that is field independent (e.g. dipolar coupling) or field dependent (e.g. chemical shift fluctuations or CSA). Unfortunately, this approach has been made more difficult by the modern technology of superconducting NMR magnets, which generally are kept at a constant field, unlike older "iron" (non-superconducting) electromagnets. Isotopic substitutions may also be very useful to sample different frequencies and to explore the effects of quadrupolar or dipolar couplings: common (and relatively inexpensive) pairs of isotopes are ^1H and ^2H, ^6Li and ^7Li, ^{10}B and ^{11}B, and ^{35}Cl and ^{37}Cl.

Spin-lattice relaxation times are typically measured by a two-pulse "inversion recovery" sequence (Fukushima and Roedder, 1981). The first pulse inverts the magnetization. The second pulse, given after a delay time t, measures how much the magnetization M has returned to equilibrium. Fitting of M vs. t gives T_1, if the decay is actually a single exponential. This approach can be very time consuming if T_1 is long, because time must be allowed between pulse sequences for full relaxation to occur. Another approach, called "saturation-recovery," may be more suitable and faster. Here, a rapid train of pulses is given that saturates the magnetization, then, after a time t, a second pulse is given to assess how much magnetization has returned.

In low viscosity liquids, the peak width may be controlled only by the T_1, if other broadening effects (e.g. field heterogeneity, incomplete exchange, etc.) are negligible. A Lorentzian-shaped line should then be observed, with a full width at half height equal to $1/(\pi T_1)$. If possible, this assumption should be confirmed by direct T_1 measurements.

EXPERIMENTAL APPROACHES TO NMR OF OXIDE MELTS

The technical difficulties of NMR at high temperatures have been discussed in some detail recently (Stebbins, 1991b). These can be briefly categorized as signal strength problems and materials problems. In the former category are the inevitable decrease in the signal intensity caused by the Boltzmann factor in Equation (3), and, if the NMR receiver coil is heated along with the sample, the additional thermal noise that is generated and is picked up by the receiver electronics. The most unusual materials problem for high temperature NMR is that if a sample container is used, then it must be an electrical insulator in order to not to shield the rf signals into and out of the sample. This poses a significant problem for many molten oxides. Hexagonal boron nitride (BN) has been used with some success, but requires a reducing or a clean neutral atmosphere to prevent oxidation.

The most commonly used high temperature NMR probe design type has a static (non-spinning) sample, surrounded by the coil that both transmits and receives the rf signal, both contained within a small, resistively heated furnace (Fig. 8). Direct current power supplies and heater wires that are parallel to the external magnetic field are usually used to eliminate strong magnetic forces, and "windings" are generally made as non-inductive as possible to limit distortion of the field. This approach has been used to make low resolution NMR measurements on a wide variety of materials, generally below about 1000°C, but has been applied to studies as high as 1700°C (Kolem et al., 1990).

Recently, resistively-heated NMR probes have been refined to allow more accurate, high resolution measurements of isotropic chemical shifts in melts to precisions as high as a few tenths of a ppm (Maekawa, 1993; Stebbins and Farnan, 1992). Here, careful calibration or compensation of inevitable magnetic field effects of the heater are required.

High temperature MAS NMR has recently become possible to temperatures above

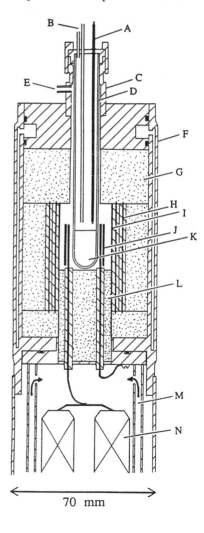

Figure 8. Typical resistively-heated high temperature NMR probe for vertical axis superconducting magnet, combining features of several probes developed in the author's laboratory. (A) Thermocouple, withdrawn during rf pulsing. (B) Tube for gas inlet to sample. (C) Seal. (D) Silica glass or alumina tube. (E) Gas inlet to heater container. (F) Water-cooled jacket. (G) Ceramic fiber insulation . (H) Heater wire in ceramic tubing. (I) Sheet metal (e.g. Mo) rf shield. (J) RF coil, shown as a Helmholz coil here. A horizontal solenoidal coil and boron nitride sample tube are often used instead. (K) Sample. (L) Support pedestal. (M) Tube for cooling air. (N) High-power tuning capacitors. [Used by permission of the editor of *Chemical Reviews*, from Stebbins (1991b), Fig. 1, p. 1356.]

70 mm

600°C, but remains very much non-routine (Farnan and Stebbins, 1994; Stebbins et al., 1989a). The major problems here are in the high temperature strength required of MAS rotor materials, the problem of heating the very large volumes of gas required to spin the sample, and difficulties in keeping the nearby probe electronics cool. Temperature calibration can also be difficult, but has been done by actually inserting a thermocouple into a MAS rotor spinning at high temperature.

A radically different approach has been pioneered by a group at the Centre de Recherches de Physique des Hautes Temperatures in Orléans, France (Coutures et al., 1990; Massiot et al., 1990b; Taulelle et al., 1989). In this system, a carefully designed gas jet levitates a small sphere (~3 to 5 mm in diameter) of sample, without a container, inside the coil of an NMR probe. A powerful laser or lasers, directed along the bore of the magnet, can heat the sample to temperatures well over 2500°C. As described below, a variety of interesting experiments have been done with this method, primarily on ^{27}Al, which provides intense, rapidly observable signals. Container problems are of course eliminated, and very refractory materials can be melted. Difficulties include uniformity, stability, and measurement (by optical pyrometry) of temperature, as well as vaporization of the sample into the rapidly flowing gas stream. In any high temperature NMR experiment in an open system, vaporization may be incongruent, possibly changing the sample composition.

Temperature effects can sometimes actually make the observation of NMR spectra in melts easier than in glasses or even in crystals. At least below T_1 minima, spin-lattice relaxation times usually decrease with increasing temperature, allowing faster data acquisition. More importantly, the peak narrowing caused by motion and exchange may be substantial. For example, static ^{29}Si spectra of glasses or crystals with natural isotopic abundance can take many hours or even days to collect. At high temperature in low-viscosity melts, however, narrow, peaks can be observed in a few minutes.

APPLICATION OF NMR TO GLASS STRUCTURE

As mentioned in the introduction, applications of NMR to glass structure have been recently and extensively reviewed. These results have been of major significance in defining our views of melt structure, because the energetically most significant features are frozen into the glass on cooling through the glass transition temperature. However, no attempt at a systematic review of the literature will be made here. Instead, only the major features of glass spectra that are needed to understand in situ NMR studies of melts, or that are most relevant to issues of melt structure and dynamics raised elsewhere in this volume, will be discussed. The emphasis will be on silicates and aluminates, neglecting the rich realms of borate and phosphate systems.

Silicon sites in glasses: Q species and thermodynamic models

^{29}Si MAS spectra for glasses are often broad and unresolved, because of the wide range of local structural environments for Si in these disordered materials. This is especially true in alkaline earth silicates and in aluminosilicates (Brandriss and Stebbins, 1988; Libourel et al., 1991; Merzbacher et al., 1990; Murdoch et al., 1985; Oestrike et al., 1987). In these systems, quantitation of species must often rely on model-dependent peak fitting. In simple binary compositions, however, especially alkali silicates, distinct, partially resolved peaks may be observed that can be unambiguously assigned to different Q^n species (Figs. 9, 10, 11) (Dupree et al., 1990; Dupree et al., 1986; Emerson et al., 1989; Hater et al., 1989; Maekawa et al., 1991b; Stebbins et al., 1992b). The areas of

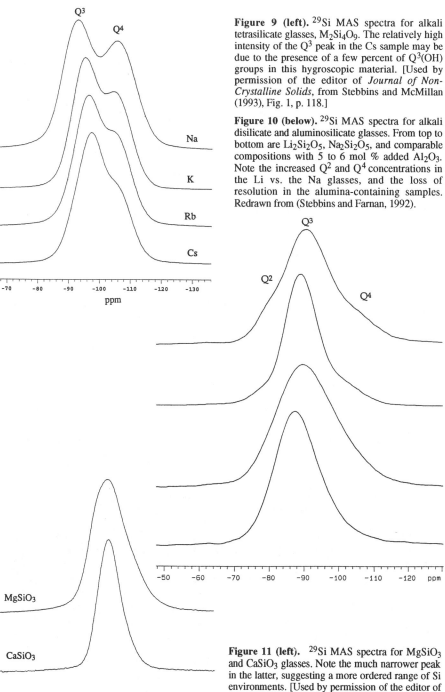

Figure 9 (left). ^{29}Si MAS spectra for alkali tetrasilicate glasses, $M_2Si_4O_9$. The relatively high intensity of the Q^3 peak in the Cs sample may be due to the presence of a few percent of $Q^3(OH)$ groups in this hygroscopic material. [Used by permission of the editor of *Journal of Non-Crystalline Solids*, from Stebbins and McMillan (1993), Fig. 1, p. 118.]

Figure 10 (below). ^{29}Si MAS spectra for alkali disilicate and aluminosilicate glasses. From top to bottom are $Li_2Si_2O_5$, $Na_2Si_2O_5$, and comparable compositions with 5 to 6 mol % added Al_2O_3. Note the increased Q^2 and Q^4 concentrations in the Li vs. the Na glasses, and the loss of resolution in the alumina-containing samples. Redrawn from (Stebbins and Farnan, 1992).

Figure 11 (left). ^{29}Si MAS spectra for $MgSiO_3$ and $CaSiO_3$ glasses. Note the much narrower peak in the latter, suggesting a more ordered range of Si environments. [Used by permission of the editor of *Journal of Non-Crystalline Solids*, from Stebbins and McMillan (1993), Fig. 2, p. 120.]

these peaks accurately reflect the relative proportions of the species. In some ranges of composition, Q species can be better distinguished in static (non-MAS) spectra, because they may have large differences in CSA. For example, Q^4 groups, because they have high local symmetry, have relatively small CSA's and give narrow peaks in static spectra (Fig. 18, below) (Stebbins, 1987). Q^3 groups, in contrast, locally resemble sites in sheet silicates, and the single non-bridging oxygen provides a natural symmetry axis. These groups thus tend to have broad CSA patterns displaying uniaxial symmetry.

A fundamental part of the disorder in an oxide glass is in the distribution of bridging oxygens (BO) and non-bridging oxygens (NBO). This is reflected to some extent in the distribution of Q species. For example, an equilibrium among species can be symbolized by the reaction:

$$2Q^3 = Q^4 + Q^2 \qquad (9)$$

This is analogous to reactions that have long been written to describe solution models for melts (chapter by Hess), as well as silicate species as observed by other spectroscopies (Mysen, 1988), for example. The equilibrium constant K for Reaction (9) can be approximated by (X is mole fraction):

$$K = X_{Q4}X_{Q2}/(X_{Q3})^2 \qquad (10)$$

If K is small, the system is highly ordered, because NBO's and BO's must be arranged in a way such that only one NBO is found in most tetrahedra (Q^3's). The mixing of these species therefore contributes little to the total configurational entropy. If K is larger, then more entropy can be generated. The implications of such reactions for solution models and energetics are discussed in the chapters by Hess and by Navrotsky.

The most complete study of Q species distributions in alkali silicate glasses (Maekawa et al., 1991b) has been used to constrain thermodynamic models of the liquids, where equilibria among species are controlled by reactions such as Reaction (9) (Greaves et al., 1991; Gurman, 1990). By fitting values of equilibrium constants for such reactions as a function of composition, enthalpies are derived that predict temperature-dependent variations in Q species proportions that are remarkably similar to those directly measured through fictive temperature studies (see below), with K as defined by Equation (10) increasing with temperature. Silicate speciation in glasses has long been studied by Raman spectroscopy, and extensive work has now been done on liquids at high temperature (see chapter by McMillan for review).

It has often hypothesized that the concentration of structural units resembling particular crystal structures has some bearing on the thermodynamic activity of the corresponding components in melts (Burnham, 1981; Burnham and Nekvasil, 1986). NMR studies of Q species distributions clearly show that K for Reaction (9) increases with the field strength of the modifier cation (defined in various ways, but related to the charge to radius ratio), leading to higher concentrations of Q^4 species (Maekawa et al., 1991b; Murdoch et al., 1985; Stebbins, 1988a). This correlates with the well-known increase in the activity coefficient of SiO_2 required by the systematic displacement of the silica liquidus to lower silica contents as the modifier field strength increases (Brandriss and Stebbins, 1988; Stebbins, 1988a) (see also the chapter by Hess). The relative increase with temperature in the concentration of Q^4 species (see below) is also similar to that predicted from solution models of both complex and simple silicate liquids at near-liquidus temperatures (Stebbins and Farnan, 1989).

Five- and six-coordinated silicon. In glasses quenched from melts at high pressure, NMR peaks for six– and five–coordinated Si have been identified in alkali disilicate and tetrasilicate glasses (Stebbins and McMillan, 1989; Xue et al., 1991b; Xue et al., 1989) (Fig. 5). By analogy with the effects of pressure on crystalline silicates, it has usually been assumed that at least [6]Si becomes important in the densification of molten silicates at high pressure; however [5]Si had only previously been "seen" in computer simulations, as described in the chapter by Poole and McMillan. At least [5]Si and probably both species also contribute to increasing melt "fragility" and decreasing viscosity at high pressure (Angell et al., 1983), although relaxation during depressurization can complicate interpretations of results for high pressure glasses (Wolf et al., 1990). Thermo-dynamically these concepts are linked through a negative "activation volume" for viscosity. The chapter by Wolf and McMillan gives a detailed account of pressure effects on glass and melt structure.

[5]Si has also been detected in small concentrations in samples quenched at ambient pressure, showing that this species is not only the result of high pressures (Stebbins, 1991a; Stebbins and McMillan, 1993) (Fig. 12). High coordinate Si has not been detected in high pressure samples of SiO_2 glass, and seems to be at a maximum concentration in alkali tetrasilicate compositions (Stebbins and McMillan, 1993; Xue et al., 1991b). A model in which non-bridging oxygens promote the formation of at least [5]Si and possibly [6]Si was suggested. Different mechanisms for densification of glasses and liquids that don't contain NBO's may be important, and may be rapidly reversible.

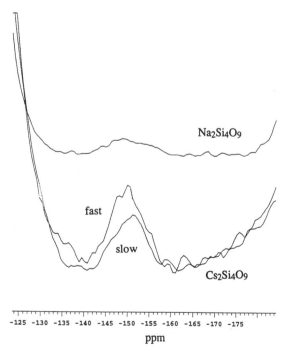

Figure 12. ^{29}Si MAS spectra for $Cs_2Si_4O_9$ and $Na_2Si_4O_9$ glasses, with vertical scale expanded greatly to show [5]Si peak. Note that this feature is more intense in the fast-quenched sample (higher fictive temperature). The steep slope to the left is the side of the SiO_4 peak; that to the right is the side of the spinning sideband. [Used by permission of the editor of *Journal of Non-Crystalline Solids*, from Stebbins and McMillan (1993), Fig. 3, p. 120.]

Peaks for [6]Si (but not for [5]Si) in sodium and lead silicophosphate glasses prepared at ambient pressure have been reported (Dupree et al., 1987; Dupree et al.,

1989; Prabakar et al., 1991). In these materials the high silicon coordination is due to the ability of the highly electronegative P ion to out compete Si for close tetrahedral oxygen neighbors. These have positions and widths similar to known crystalline [6]Si phosphates (Grimmer et al., 1986; Weeding et al., 1985), suggesting similar local structural environments.

Oxygen sites

Oxygen volumetrically dominates in silicate crystals and melts, and knowledge of oxygen site geometry and distribution is thus of considerable importance. Discussion of these aspects of structure has, however, been hindered by experimental limitations and, in part, by the classical mineralogical view that usually considers only *cation* polyhedra. NMR has provided a new way of approaching this problem.

Static, MAS, and DAS spectra for [17]O (I = 5/2) in silicate glasses have shown that bridging oxygen sites can often be readily distinguished from non-bridging oxygens (Bray et al., 1988; Bunker et al., 1990; Kirkpatrick, 1988; Schramm and Oldfield, 1984; Timken et al., 1987; Turner et al., 1987; Xue et al., 1994). Static and MAS peak shapes are usually dominated by second order quadrupolar effects. Because the EFGs at bridging oxygen sites are considerably larger than those at NBOs, NMR peak widths for the former are generally several times those for the latter (Fig. 23, below). In contrast, the positions of NBO peaks are quite sensitive to the types of network-modifying cations present, largely because of chemical shift variations. The widths of DAS [17]O peaks for NBO's in silicate glasses are thus a good measure of the extent of disorder of the modifier cations (Farnan et al., 1992) (see section on disorder, below). Correlations established between C_Q for [17]O and Si–O–Si angle may also be used to quantify angle distributions, and give somewhat different results from the well-studied case of pure SiO_2 glass (Farnan et al., 1992). This finding is not surprising given that many or most bridging oxygens have modifier cations in their first coordination shell, not just Si and Al. This ability to study the structure around a particular type of oxygen site seems to unique, as x-ray and neutron scattering techniques provide only averages over the entire structure.

The large effect of network modifier size and charge on [17]O δ_{iso} for NBO's is even larger for oxide ions that have *no* tetrahedral neighbors, as exemplified by studies of simple crystalline oxides, such as MgO, CaO, BaO, PbO, and TiO_2, (Turner et al., 1985). The existence of such oxygen species in glasses and melts of geological composition (as opposed to low-silica metallurgical slags) is controversial, and could imply clustering. This phenomenon is most likely for high field strength cations such as Ti^{4+}, and would have major implications for thermodynamic activities of minor components (see chapter by Hess). Initial ^{17}O NMR studies of TiO_2-rich glasses, indicating the presence of oxide ions, are thus particularly interesting (Fiske, 1993; Fiske and Stebbins, 1994b).

Aluminum coordination

Al is found in four-, five-, and six-coordination in silicate minerals, and high pressure favors minerals with higher coordination (e.g. Al-rich pyroxenes over feldspars) because of their reduced molar volumes. There has thus been considerable speculation about the presence of [5]Al and [6]Al in glasses and melts, and these species have been included in a number of solution models that treat pressure effects on phase equilibria (Boettcher et al., 1982; Burnham, 1981). The distribution of Al coordinations has been surprisingly difficult to quantify by spectroscopic and diffraction techniques (see chapter by Wolf and McMillan), but NMR has made a important new contributions to this old question.

Static and MAS spectra for ^{27}Al (I = 5/2) in glasses are commonly poorly resolved, both because of disorder and because of quadrupolar broadening. Spectra for most aluminosilicate glasses (with sufficient alkali or alkaline earth cations for charge balance), when collected at high magnetic fields and with high sample spinning rates, suggest that most or all of the Al is four-coordinated ([4]Al) (Kirkpatrick et al., 1986; Merzbacher et al., 1990; Oestrike et al., 1987). Reports of small amounts of [6]Al in a Na-Al silicate glass (Stebbins and Farnan, 1992), and of large concentrations of [6]Al in metaluminous Ca-Al silicate glasses (Engelhardt et al., 1985) have not been confirmed by other studies. However, in a variety of other aluminosilicate, aluminate, and aluminoborate compositions, MAS peaks for large (~10 to 50%) concentrations of five- and six-coordinated Al ([5]Al, [6]Al) are clearly observed (Bunker et al., 1991; Poe et al., 1992a; Risbud et al., 1987; Sato et al., 1991a; Sato et al., 1991b) (Fig. 13). Most of these glasses were very rich in Al_2O_3 (highly "peraluminous"), where Al may begin to play the role of a charge-balancing network modifier as well as a network-former. On the other hand, samples on the SiO_2–$MgAl_2O_4$ binary also contained high-coordinate Al (McMillan and Kirkpatrick, 1992). The role of high field strength modifier (or charge balancing) cations in promoting the formation of [5]Al and [6]Al (e.g. Mg^{2+} as opposed to Ca^{2+} or Na^+) is also seen in glasses in the MO–Al_2O_3-B_2O_3 ternary (Fig. 6), where high coordinate Al is systematically more abundant going from M = Ba to Sr to Ca to Mg (Bunker et al., 1991). A similar trend has been indicated by data on average isotropic chemical shifts for ^{27}Al in melts at high temperature (see below). One might therefore speculate that Mg- or Ca-rich magmas may contain significant amounts of [5]Al or [6]Al even at upper-mantle pressures.

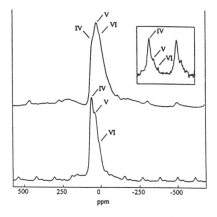

Figure 13. ^{27}Al MAS NMR spectra for a rapidly quenched glass with 30 mol % CaO and 70 mol % Al_2O_3, acquired at magnetic field strengths of 7 T (upper) and 11.7 T (lower). Al coordination numbers are labeled. Inset shows spinning sidebands. [Used by permission of the editor of *Journal of Non-Crystalline Solids*, from Poe et al. (1994), Fig. 4, p. 1835.]

High-coordinate Al was not unambiguously detected in $NaAlSi_3O_8$ glass quenched from liquids at pressures as high as 10 GPa (Stebbins and Sykes, 1990). No clear, separate [5]Al or [6]Al peaks were observed, although a shoulder was seen in the right frequency range. Spectra collected at two different magnetic fields were nearly identical, suggesting that this shoulder was largely caused by a real distribution in chemical shifts, not solely to quadrupolar broadening (see also chapter by Wolf and McMillan). More recently, clearly separated features for [5]Al and [6]Al were reported in samples halfway between $NaAlSi_3O_8$ and $Na_2Si_4O_9$ in composition quenched from pressures to 12 GPa

(Yarger et al., 1994). As for Si, it seems that non-bridging oxygens may be required to readily form high coordinate Al (or perhaps to preserve it during quench).

Analysis of satellite transition spinning sidebands for ^{27}Al in glasses has been particularly useful in resolving different coordination numbers in alkaline earth silicate and borate glasses, as well as in calculating δ_{iso} (Coté et al., 1992a,b; Jäger et al., 1993) (Fig. 13). Relative proportions of Al species may in fact be easier to estimate in sideband, rather than central, peaks.

Alkali and alkaline earth cations

NMR spectra for ^{7}Li (I = 3/2) and ^{23}Na (I = 3/2) are quite easy to collect for glasses, and data for ^{87}Rb (I = 3/2) and ^{133}Cs (I = 7/2) can be obtained with more difficulty. However, little structural information has been obtained, again because of peak broadening due to disorder and to quadrupolar effects, which prevent resolution of peaks for different sites. Recently, ^{6}Li (I = 1) has been shown to behave almost as if it were a spin 1/2 nuclide, because its quadrupole coupling constant is close to zero. In binary lithium silicate glasses, ^{6}Li peaks are dominated by ranges of chemical shifts and indicate that at least four-, five-, and six-coordinated Li are present (Xu and Stebbins, 1995). For ^{25}Mg, the resonant frequency is so low and the quadrupolar coupling constant so large that it is not practical at present to obtain glass spectra, although some useful data have been collected for crystalline silicates (Fiske and Stebbins, 1994a; MacKenzie and Meinhold, 1994). However, peak narrowing has allowed spectra to be observed in melts (see below).

Boron coordination

The longest-studied nuclide in inorganic glasses has been ^{11}B (I = 3/2), which has a high natural abundance and high resonant frequency, resulting in very intense NMR signals. One of the key structural issues in borate and borosilicate glasses is the relative concentrations of trigonally ($^{[3]}$B) and tetrahedrally ($^{[4]}$B) coordinated boron. The presence of these species in glasses was inferred from the structures of the corresponding crystals and documented by NMR (Bray et al., 1991; Bray et al., 1988; Bray and O'Keefe, 1963; Turner et al., 1986). The ratio of these species varies strongly as a function of composition and has a major influence on thermodynamic and transport properties. The two types of sites are readily distinguished by both static and MAS NMR, and can be accurately quantified. In the former, the large difference in EFG's at the two kinds of sites (analogous to the CSA's for Si in Q^3 and Q^4 sites, see above) allow clear and quantitative distinction. In MAS NMR, generally done at higher fields, both chemical shift differences and quadrupolar effects distinguish the two kinds of sites (Fig. 14). DAS NMR has given promising new, very high resolution results on boron sites in B_2O_3 glass (Youngman and Zwanziger, 1994).

Phosphorous in phosphate glasses

Because ^{31}P (I = 1/2) is abundant and has a high resonant frequency, NMR spectra are readily obtained. Like silicon in silicates, phosphorous sites in crystalline phosphates have distinct numbers of bridging and non-bridging oxygens, and characteristic CSA's and ranges of isotropic chemical shifts. In glasses, multiple sites can often be recognized and quantified. This observation has been exploited in a number of recent studies of speciation (Brow et al., 1991; Gan and Hess, 1992; Gan et al., 1994) which are in turn useful in constraining models of structure and thermodynamics (see chapter by Hess).

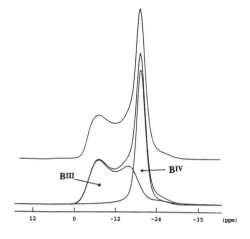

Figure 14. Typical ^{11}B MAS NMR spectrum for an alkali borate glass, with 30 mol % Na_2O and 70% B_2O_3 (Ellsworth, 1994; Ellsworth and Stebbins, 1994), acquired at a magnetic field of 9.4 T, an rf tip angle of 15°, and a spinning speed of about 10 kHz. Scale is in ppm relative to aqueous boric acid. A fit to a Gaussian peak for $^{[4]}$B (=B^{IV}) , and to a quadrupolar line shape for $^{[3]}$B (=B^{III}) are shown.

Fictive temperature studies: the effect of temperature on melt structure

One of the major issues addressed in this volume is the nature of both the structural and the dynamical differences between glasses and liquids. At the heart of this question is the way in which the melt structure changes with temperature, because this change is what gives rise to the often large changes in thermodynamic properties at the glass transition (see chapter by Richet and Bottinga). One approach to understanding this process is through in situ measurements in the liquids themselves; this is discussed in detail below and in other chapters. A second, complementary approach is to study (at ambient temperature) glasses made by quenching liquids at different rates, which record the structure of the melt at different temperatures. This effective "freezing in" temperature, which becomes a characteristic of a given glass until it is re-heated into the glass transformation range, is often called the fictive temperature T_f (see chapters by Dingwell and Webb and Dingwell). The higher the quench rate, the higher the value of T_f. It is possible that not all aspects of the structure of a given material, for example alkali cation positions or the relative proportions of H_2O and OH groups, have the same T_f (McMillan, 1994; Zhang and Stolper, 1993). In any case, if variations in speciation with temperature are not accounted for in the modeling of the high temperature (near to the solidus or liquidus) thermodynamics of primary interest in petrology, major systematic errors are likely.

Speciation in silicate glasses. In one of the few studies of fictive temperature effects on silicate melt structure, several silicate glasses were quenched with varying rates, and changes in K for Reaction (9) were noted (Brandriss and Stebbins, 1988). The reaction was displaced to the right at higher T_f, consistent with an increase in entropy. A reaction enthalpy of 30 ± 15 kJ/mol was derived by analyzing the change in K with T_f and applying the Van't Hoff equation, making the admittedly crude assumption that activity coefficients could be neglected. It was found that the change in speciation, although possibly very significant to thermodynamic activities of components (e.g. SiO_2, see above), probably does not make a major, direct contribution to the total change with T in configurational enthalpy and entropy, although mixing of bridging and non-bridging oxygens has an important role in some recent modeling (chapter by Richet and Bottinga). The effect of temperature on the above reaction has been confirmed by in situ, high temperature Raman spectra (chapter by McMillan and Wolf), where data can be obtained over the entire range from T_g to above the liquidus (McMillan et al., 1992; Mysen and Frantz, 1992; Mysen and Frantz, 1993).

In similar work on alkali tetrasilicate glasses, the effect of T_f on the concentration of five coordinated silicon ([5]Si) was measured. A higher T_f produces more [5]Si in $K_2Si_4O_9$, and the estimated enthalpy of the reaction to form the species was similar to that noted above for Reaction (9) (Stebbins, 1991a). The same conclusion was reached for $Cs_2Si_4O_9$ (Stebbins and McMillan, 1993) (Fig. 12). These melts therefore contain more [5]Si at higher temperature, an important factor in extrapolating measurements of high-coordinate species in high pressure glasses 1000°C or more up in temperature to the near-liquidus conditions of interest to geophysics. For example, extrapolation of data on the alkali tetrasilicates quenched from 10 GPa suggests that most of the Si in these melts might have a high coordination number at the melting point. The positive temperature effect on [5]Si concentration also supports the possible role of this species as a "transition complex" in viscous flow and diffusion (see chapters by Chakraborty and by Poole and McMillan, and Fig. 17, below).

In sodium silicophosphate glasses, more rapidly quenched samples showed much smaller [6]Si peaks relative to [4]Si (Dupree et al., 1989; Dupree et al., 1992).

Aluminum coordination: changes with T_f. Fast-quenched (10^2 to 10^3 K/s) and super-quenched (10^5 to 10^6 K/s) glasses in the SiO_2–Al_2O_3 binary have been studied with ^{27}Al MAS NMR (Poe et al., 1992a; Sato et al., 1991a). The former seemed to comprised mainly of [4]Al and [6]Al, the latter also contained a high proportion of [5]Al. These results may suggest that the latter species increases in concentration at higher temperature, although results were complicated by phase separation. As for Si, the possibility that [5]Al is significant as a "transition complex" in viscous flow is suggested. In an early study at a relatively low magnetic field of sodium aluminosilicate glasses, a decrease in the ^{27}Al MAS NMR signal intensity in faster quenched samples was noted, indicating a higher population in distorted sites with C_Q high enough so that signal was lost (Hallas and Hähnert, 1985). It is conceivable that these sites include Al in a high coordination state, but this remains to be confirmed by high-field NMR, which should be better at "seeing" such lost signals because second-order quadrupolar broadening is relatively less important.

Boron coordination: changes with T_f. As for silicates, relatively few studies of temperature effects on the structure of borate or borosilicate melts have been made. In CaO–Al_2O_3–B_2O_3 glasses, low field, "wideline" ^{11}B NMR showed that the fraction of four-coordinated boron (traditionally labeled N_4) did not change with respect to the fraction of three coordinated boron (N_3) on annealing of fibers, which have very high fictive temperatures (Bray, 1985). In contrast, if SiO_2 was added to the system, a strong T_f effect was found, with N_4 increasing at lower T_f. For the commercial glass fiber material "e-glass" (a complex alkali and alkaline earth borosilicate), a similar large effect was noted (Gupta et al., 1985). These findings have recently been confirmed and extended, using high-field MAS NMR (Ellsworth, 1994; Ellsworth and Stebbins, 1994). As shown in Figure 15, a ternary alkali borosilicate glass shows a similar effect of T_f on N_4. However, for several two component sodium borate glasses, no effect at all was observed (Fig. 16). This intriguing result may be rationalized by a mechanism in which non-bridging oxygens, which are present in the borosilicate but of low abundance in the borate glass, are required for this structural change to occur.

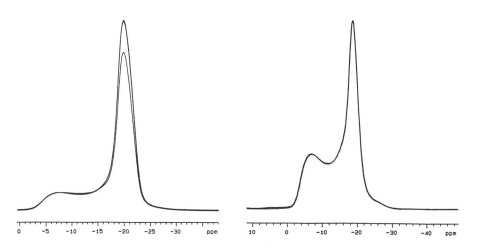

Figure 15 (left). ^{11}B MAS NMR spectra for a glass with 20 mol % Na_2O, 20% B_2O_3, and 60 mol % SiO_2 (Ellsworth, 1994; Ellsworth and Stebbins, 1994). The upper curve shows two superimposed spectra, the first collected on an annealed sample (low T_f), the other on reannealing after rapid quenching. The lower spectrum is the same sample after rapid quenching (high T_f). The narrow peak is primarily [4]B, the broad peak [3]B. Acquired as for Figure 14.

Figure 16 (right) ^{11}B MAS NMR spectra for a glass with 30 mol % Na_2O and 70% B_2O_3 (Ellsworth, 1994; Ellsworth and Stebbins, 1994). Two spectra are superimposed, one for a fast quenched sample, the other for an annealed sample. The two are essentially identical. Acquired as for Figure 14.

The extent of ordering in silicate glasses

The amount of order or disorder in glasses is a fundamental question in the nature of amorphous materials. For our purposes, it is significant because the quenched-in structure of a glass is the starting point for the structural changes that begin on heating through the glass transition. These changes in turn control the thermodynamic distinction between liquid and solid. Oxide glasses are far from random structures (Gaskell, 1992; Gaskell et al., 1991; Stebbins, 1987), despite oft-quoted "random network" models. Identifying how ordered a glass is thus can help decide what aspect of the structure may become disordered with increasing temperature.

As described above, one important view of disorder in silicate glasses has been the distribution of Q species, which in turn reflects that of bridging and non-bridging oxygens. A crude estimate of the magnitude of this entropy can be made by ideally mixing the observed mole fractions of species. In $Na_2Si_2O_5$ glass, this suggests that almost half of the entropy difference between the glass and the crystal could result from just this aspect of the structure (Brandriss and Stebbins, 1988). Despite this, the oxygen distribution is far from the fully randomized one discussed in the literature long before spectroscopic measurements were available (Lacey, 1965), and probably remains that way to liquidus temperatures. Models of mixing of oxide species or components need to account for this non-randomness, which should limit the applicability of both ideal and regular solution models. Similar conclusions about non-randomness have been made for alkaline earth silicate glasses (Libourel et al., 1991).

For alkali and alkaline earth glasses, both in binary systems and in aluminosilicates,

^{29}Si MAS peak widths increase systematically with the field strength of the modifier or charge balancing cation (Libourel et al., 1991; Merzbacher et al., 1990; Murdoch et al., 1985; Oestrike et al., 1987). This indicates an increasing disorder in the structure, probably including a displacement of reactions such as Reaction (9) to the right. In aluminosilicates, the following analogous reaction may also be shifted to the right, again leading to greater disorder (Murdoch et al., 1985). Here, k refers to the number of Al neighbors:

$$2Q^4[kAl] = Q^4[(k-1)Al] + Q^4[(k+1)Al] \qquad (11)$$

A direct look at disorder in the coordination environments of oxygen itself has been obtained with ^{17}O DAS NMR, where the number, position, and width of peaks for non-bridging oxygens has been studied in several glasses. In $KMg_{0.5}Si_4O_9$, for example, only a single, relatively narrow peak was seen, suggesting that most NBO's have the same coordination environment, perhaps one Si, two K, and one Mg cation (Farnan et al., 1992). In contrast, the NBO peak in $NaKSi_2O_5$ was broad enough to be comprised of the entire range of likely neighbor populations (e.g. 4Na, 3Na1K, 2Na2K, etc.) (Vermillion et al., 1995). These observations are analogous to known differences in crystal structures. Several ordered phases, including $K_2MgSi_6O_{14}$, exist in the $K_2O–MgO–SiO_2$ ternary, while K and Na silicates generally show continuous solid solution (Levin et al., 1964).

A long-standing question in aluminosilicate glasses is the extent of Al/Si ordering. It is likely that, as in high temperature feldspars, melts at high temperature are disordered. However, the extent of aluminum avoidance caused by the energetic unfavorability of Al–O–Al linkages remains to be evaluated. Answering this question precisely is difficult because the effects on ^{29}Si chemical shifts of varying the number of Al neighbors is comparable to that of bond angle variations, leading to unresolved spectra for most aluminosilicate compositions. However, the observation that overall peak widths reach a maximum at a Si/Al ratio of 2 or 3, and decrease markedly when the ratio is 1, suggests that Al-avoidance is important (Murdoch et al., 1985; Oestrike et al., 1987). Some further increase in Si/Al disorder with increasing temperature above T_g is thus possible, possibly contributing significantly to configurational entropy.

Intermediate range order. Intermediate range order (IRO) in glasses may be considered as non-randomness beyond the first coordination shell, and thus Q species distributions described above are part of IRO. For many questions, however, in particular for understanding nucleation, growth and phase separation, information about longer range structure, to nm or longer distances, may be required. This kind of information is notoriously difficult to obtain in glasses, and is a large and complex subject.

The connectivity among Q species in a silicate glass is perhaps the next step towards longer-range structure. 2-D NMR techniques have been applied to this problem (Knight et al., 1990), with the conclusion that there is relatively little clustering of Q^3 and Q^4 sites in sodium silicate glasses: most Q^3's have Q^4 neighbors and vice versa. This is consistent with models of viscous flow derived from high temperature species exchange data, as discussed below.

One aspect of IRO that is clearly important in understanding the differences between liquids at high temperature, and glasses that are most commonly studied in the laboratory, is incipient phase separation. Recent studies of ^{29}Si spin-lattice relaxation have shown that if the relaxation process is dominated by through-space coupling to the unpaired electronic spins of paramagnetic impurities (often the case in silicates), the observed non-exponential behavior can be interpreted in terms of the dimensionality of

the structure (Sen and Stebbins, 1994). In $Li_2Si_4O_9$ glass, which lies within the spinodal of the Li_2O-SiO_2 binary, this intermediate range structure was found to have a dimensionality of less than 3, indicating significant nm-scale, fractal-like compositional heterogeneity. This would presumably be absent in the liquid above the spinodal temperature. In similar work on SiO_2 glass doped with rare earth element cations, clustering was detected at concentrations as low as a few hundred ppm, of considerable significance to models of the activities of high field strength cations (see chapter by Hess) as well as to technologically important properties such as fluorescence (Sen and Stebbins, 1995)

Orientational disorder. In a truly polymeric liquid (such as molten polyethylene), large, long-lived molecules become partially aligned during flow, often leading to non-Newtonian rheology. If quenched to a glass during rapid shearing (e.g. during drawing of a fiber or film), some of this alignment is retained, as exemplified by the strong birefringence of transparent plastic tape. One test of the flow mechanism of oxide liquids is thus the extent of orientational ordering in glass fibers, films, and other highly sheared forms. Static NMR spectra are extremely sensitive to non-randomness in spatial orientation of certain kinds of small, very asymmetrical structural units. Examples are BO_3 (large C_0) and Q^3 (large CSA) groups, which both have strong uniaxial symmetry. In a single study of several types of silicate glass fibers, of B_2O_3 films (primarily BO_3 groups), and of splat quenched $Na_2Si_2O_5$ glass sheets (primarily Q^3 species), no trace of non-random orientation was observed, suggesting that the structural units rearrange at a rate comparable to that of local viscous relaxation (Stebbins et al., 1989b). This result thus agrees with interpretations of the mechanism of flow deduced from in situ NMR studies of species exchange rates (see below).

APPLICATIONS OF NMR TO OXIDE MELTS

In situ, high temperature NMR studies of silicate, aluminate, borate, and phosphate melts have provided both structural and dynamical information. The latter has been particularly important in a growing understanding of the atomic-scale mechanisms that control viscosity, diffusion, and structural change with temperature: it will be discussed first.

Chemical exchange in melts: silicate species and viscous flow

Since the first ^{29}Si NMR spectra of melts were published, it has been clear that exchange of Si among the various possible sites (especially Q^n species) must be rapid at temperatures higher than a few hundred degrees above the glass transition: only single, averaged peaks are observed (Stebbins et al., 1985; Stebbins et al., 1986). Here, "rapid" means faster than the total static spectral width, generally about 10 kHz. Because this exchange requires the breaking of Si-O bonds, it means that the average lifetimes of these bonds (the strongest in the system) are short. This observation has been confirmed in all compositions studied, ranging from a lithium silicate with only 40 mol % SiO_2 to $NaAlSi_3O_8$ liquid (Farnan and Stebbins, 1990a; Farnan and Stebbins, 1990b; Farnan and Stebbins, 1994; Fiske and Stebbins, 1994a; Liu et al., 1987; Liu et al., 1988; Shimokawa et al., 1990; Stebbins, 1988a; Stebbins and Farnan, 1992). High temperature Raman spectra (McMillan et al., 1992; Mysen and Frantz, 1993) confirm that at the same temperatures, Q species remain distinct on the much shorter time scale of interatomic vibrations (see also chapter by McMillan and Wolf). Thus a real exchange process, not the loss of energetic or structural distinction among sites, must be occurring. It is also clear that silicates are not "molecular" liquids in the sense of organic polymers. A

possible pathway for the exchange process, based on what is typically observed in molecular dynamics simulations (see chapter by Poole and McMillan), is shown in Figure 17. One hypothesized feature of this scenario is the involvement of a five-coordinated intermediate. At least the existence of this species has been shown by MAS NMR on glasses (see discussion of glass structure above). Another feature of this mechanism is the involvement of network modifiers. If one of the modifiers (e.g. K^+) diffuses away from a particular NBO, then the excess and temporarily uncompensated negative charge on the NBO may make it more likely to bond as a fifth oxygen to an adjacent Si. Similarly, the return of a modifier cation is likely to help break up this "transition complex."

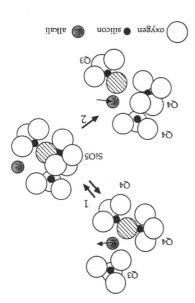

Figure 17. Sketch of a possible mechanism for exchange of silicon species. Oxygens are shown by open circles, network modifying cations as crosshatched circles, and silicons by solid circles. In step 1, a non-bridging oxygen in a Q^3 group approaches a Q^4 group as the modifier cation diffuses away, and the Q^4 silicon becomes 5-coordinated. Perhaps depending on additional modifier cation motion, this SiO_5 transition complex then relaxes back to a Q^4 group, either reforming the original configuration (self-exchange) or by going through step 2 to form a new Q^3 group in a completed exchange event. Only the latter contributes to macroscopic flow, and, at least at low temperatures, it is much less frequent than the self-exchange. Redrawn from (Farman and Stebbins, 1994; Stebbins and Farman, 1989).

oxygen ◯   silicon ●   alkali ▨

Chemical exchange and complete motional averaging in oxide melts are the rule at high temperature for every other isotope studied, including ^{17}O in $K_2Si_4O_9$ (Farman, 1991; Maekawa et al., 1991; Stebbins et al., 1991a; Stebbins et al., 1992b); ^{11}B in borates (Inagaki et al., 1993; Maekawa, 1993; Stebbins et al., 1995b; Xu et al., 1988; Xu et al., 1989); ^{27}Al in a variety of aluminosilicates and aluminates (Coté et al., 1992a; Coté et al., 1992b; Coutures et al., 1990; Massiot et al., 1994; Massiot et al., 1990b; Poe et al., 1992b; Poe et al., 1993; Poe et al., 1994; Stebbins and Farman, 1992; Taulelle et al., 1989); ^{23}Na (Fiske and Stebbins, 1994a; Liu et al., 1987; Liu et al., 1988; Maekawa, 1993; Stebbins and Farman, 1992), ^{25}Mg (Fiske and Stebbins, 1994a) and ^{87}Rb in silicates (Stebbins et al., 1995b); and ^{31}P in phosphates (Kirkpatrick et al., 1995). For each of these, multiple, energetically distinct sites (e.g. different coordination numbers) are expected, as for Si. Again, site "identity" is short-lived. These observations mean, however, that detailed structural information cannot be obtained from NMR under these conditions, although inferences about averages of parameters such as coordination number can be made (see below). However, with this loss of structural detail comes the unique possibility of actually quantifying the dynamics, mechanisms and rates of the exchange processes themselves.

In order to reach this goal, spectra must be collected over a temperature range where exchange, rotation, etc., are slow enough so that averaging is not complete. With typical NMR peak widths, this is the case up to about 300°C above the glass transition

temperature. For most compositions, this requires that data be collected on supercooled liquids well below their melting points, so that compositions must be chosen that do not crystallize readily, or experiments must be done quickly.

Figure 18 shows an example of this kind of study, for ^{29}Si in $K_2Si_4O_9$ liquid (Farnan and Stebbins, 1990a). Here, the static spectrum at low temperature is dominated by a Q^4 peak, which is narrow because of its small CSA, and a broad, uniaxial powder pattern for the Q^3 sites. By stoichiometry, the two types of site should make up half of the total (neglecting the small amounts of Q^2, etc., that are known to be present), and in fact, the peak areas have close to a 1:1 ratio. The spectrum remains about the same until somewhat above the glass transition T_g, which is at about 550°C for normal, slow heating rates. At higher temperatures, however, the peak begins to become narrower, and eventually appears as a single, narrow line.

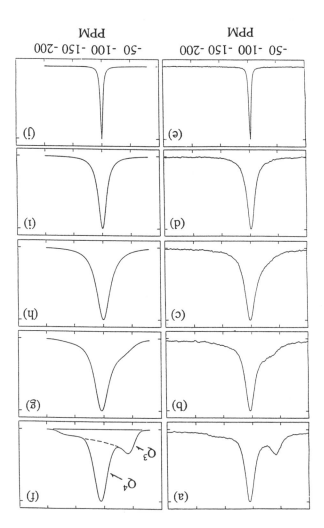

Figure 18. Static (i.e. non-MAS) ^{29}Si spectra of $K_2Si_4O_9$ glass and liquid (left) and multi-site exchange simulations (right). Temperatures: (a) 697°C; (b) 774°C; (c) 800°C; (d) 847°C; (e) 997°C. Exchange frequencies, in kHz: (f) 2; (g) 10; (h) 25; (i) 50; (j) 500. Q^4 and Q^3 static CSA patterns are labeled in the lowest temperature spectrum. [Used by permission of the editor of *Journal of the American Chemical Society*, from Farnan and Stebbins (1990a), Fig. 5, p. 36.]

These observations can be quantified by simulating the peak shapes with a model that has a single, adjustable exchange time τ_{ex}, and allows random exchange at a frequency of $1/\tau_{ex}$ among a large number of sites that map out the entire static spectrum. This approach clearly reproduces the observed peak shapes accurately, and allows determination of τ_{ex} by fitting (Fig. 18).

Several approaches can be taken that allow NMR exchange times to be compared with macroscopic dynamics, or even to predict bulk transport properties (Farman and Stebbins, 1990a; Farman and Stebbins, 1990b; Farman and Stebbins, 1994; Stebbins et al., 1995a). The first of these relies on the Maxwell relationship between viscosity η, the infinite frequency shear modulus G_∞ (typically about 1010 Pa), and the shear relaxation time τ_{shear} (Dingwell and Webb, 1989; Dingwell and Webb, 1990)(see also chapter by Webb and Dingwell):

$$\eta = \tau_{shear} \times G_\infty \qquad (12)$$

We can test a simple assumption that $\tau_{shear} \approx \tau_{ex}$ by calculating the viscosity using Equation (12) and values of τ_{ex} derived from NMR spectra. A second approach is to use the Einstein-Smoluchowski relation to calculate a diffusivity D from τ_{ex} and a hypothesized cation (or anion) jump distance d, here taken as 0.31 nm:

$$D = d^2/6\tau \qquad (13)$$

The factor of six in this equation reflects the number of possible sites that the ion can hop to, and is a reasonable approximation for a three-dimensional system. The viscosity can then be estimated from the Eyring equation, where k_B is Boltzmann's constant:

$$\eta = k_B T/(dD) \qquad (14)$$

As shown in Figure 19 (above), at relatively high temperatures the two approaches roughly bracket the experimental curve, with the Maxwell relation giving higher viscosities. At lower temperature, the viscosities predicted from 1-D spectra become systematically lower than measured values. The likely explanation for this is that the motional narrowing that is observed and simulated involves in part "self exchange", where the first step in Figure 17 reverses. No net diffusion occurs in this case, but the structural units may partially rotate. The difference between the rate of this unsuccessful exchange attempt and complete exchange probably increases at lower temperature and higher viscosity, resulting in the observed divergence. It is also important to point out that at present, only a single, average correlation time can be derived from fitting silicate

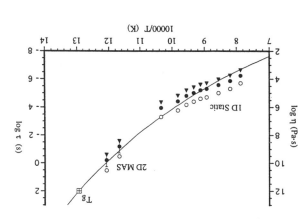

Figure 19. Plot of log₁₀ of viscosity and of shear relaxation time for $K_2Si_4O_9$ liquid vs. inverse temperature. The solid curve is a fit to observed viscosities and can be converted to mean shear relaxation time (right side scale) by using Equation (12). Open circles are viscosities calculated from NMR exchange times derived from 29Si NMR data; solid triangles are exchange times using an Eyring model (Eqs. 13 and 14, except with a factor of 6 instead of 2); solid circles are calculated using the Stokes-Einstein equation. Redrawn from (Farman and Stebbins, 1994).

NMR peak shapes. However, all data on mechanical relaxation show wide ranges of correlation times, only the peak of which is at the mean shear relaxation time (see chapter by Webb and Dingwell). The potential for high T NMR studies of chemical exchange to resolve more complex dynamics is illustrated by elegant recent work on phosphorus-selenium glasses and liquids (Maxwell and Eckert, 1994; Maxwell and Eckert, 1995).

In order to explore dynamics closer to the glass transition, 2-D NMR techniques are very useful because they can "see" exchange at frequencies much less than the peak width. The 2-D exchange experiment discussed in the introductory section has been done for ^{29}Si in $K_2Si_4O_9$ liquid, under both static and MAS conditions (Farnan and Stebbins, 1990b; Farnan and Stebbins, 1994) (Fig. 20). The latter allows even slower processes to be observed because of its higher resolution. Only complete exchange generates off-diagonal peaks, so there is no complication by "self exchange", although this may cause some peak broadening. Measured exchange rates can again be used to estimate viscosity. As in the high temperature results, the two equations again bracket the experimental curve. This indicates that at least in this material, the same flow mechanism dominates over a wide range in temperature and a span of at least 6 orders of magnitude in viscosity.

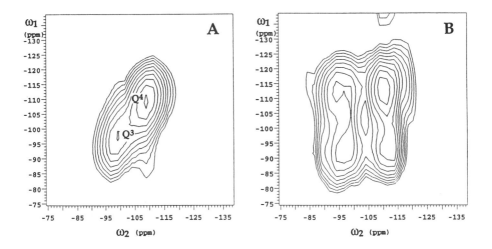

Figure 20. Contour plot of two-dimensional ^{29}Si MAS NMR exchange spectra of $K_2Si_4O_9$ liquid at 555°C (about 55°C above Tg). (A) Mixing time = 0.5 s. The normal 1D spectrum, with distinct peaks for Q^3 and Q^4 peaks, appears along the diagonal. (B) Mixing time = 4.0 s. Note the appearance of off-diagonal peaks in the latter, indicating substantial exchange of Q^3 and Q^4 species. Redrawn from (Farnan and Stebbins, 1994).

Recently, similar results have been obtained for a much lower silica content liquid, with 40 mol % Na_2O and 60 mol % SiO_2, using 1-D MAS NMR (Stebbins et al., 1995a). In this glass, the MAS spectrum has two clear peaks, for the Q^2 and Q^3 species that make up most of this composition (Fig. 21). As temperature was increased above T_g, exchange was noted as a merging and narrowing of the two peaks. The experiments were done rapidly (all data collected within a few minutes) to avoid crystallization, which began to be visible in the last spectrum shown. Data were analyzed using both Equations (12) and (14), and again, experimental viscosities are bracketed by the two methods (Fig. 22).

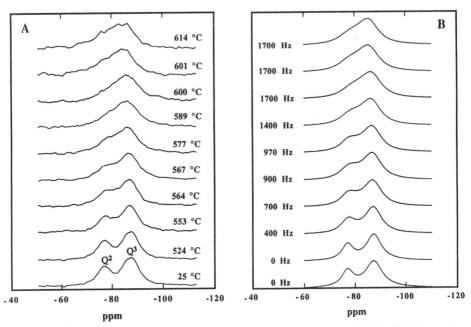

Figure 21. ^{29}Si MAS spectra of glass and liquid with 40 mol % Na_2O, 60 mol % SiO_2 (Stebbins et al., 1995a). (A) Experimental data; (B) simulation with two-site exchange model, with exchange frequencies shown. Small bumps in 614°C experimental spectrum are caused by the beginning of crystallization.

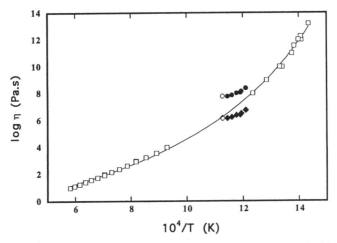

Figure 22. Plot of log_{10} of viscosity vs. inverse temperature for a sodium silicate liquid with 40% Na_2O (Stebbins et al., 1995a). Open squares are viscosity measurements; solid circles are calculated from NMR data using Equation (12); solid diamonds using Equations (13) and (14). Open circle and diamond are data for the sample after crystallization had begun, and thus may represent a slightly more silica-rich melt.

In $Na_2Si_3O_7$ liquid, ^{29}Si spectra demonstrating silicate species exchange have also been reported (Shimokawa et al., 1990), and were observed even at natural isotopic abundance.

The fundamental control on viscosity in all of these liquids thus seems to be a process that involves local structural rearrangements that require Si–O bond breaking. Explanation of the relatively minor differences among the various approaches for estimating viscosity from microscopic exchange rates of course awaits a detailed model of the exchange and diffusion process. It is clear that these kind of NMR observations detect only an average of what is likely be a wide distribution of exchange and other microscopic structural relaxation rates in the liquid. Somewhat different parts of this distribution may be sampled by other measurements. Nonetheless, certain kinds of processes are ruled out as having major influences on viscosity. For example, one could imagine relatively large domains or "molecules" that are strongly bonded (e.g. silica-like "icebergs" of Q^4 units or mica-like "sheets" of Q^3 units), which could move relative to one another as coherent, long-lived units, breaking primarily weak bonds between NBO's and modifier cations. This kind of scenario predicts that the average Si-O bond lifetime should be significantly longer than the time scale of viscous flow, and is thus clearly ruled out be the NMR data. It is likely that at even lower SiO_2 content, a transition in mechanism may occur, such that small silicate molecules (perhaps single or pairs of tetrahedra) have relatively long lifetimes. To speculate further, this might be especially likely in compositions that are highly clustered because of high contents of high field strength cations, e.g. Mg^{2+}. However, the point of transition to this behavior is still unknown.

Chemical exchange in melts: oxygen, boron and phosphate species

The exchange of silicon species also requires that oxygen species exchange, as illustrated by Figure 17. If this picture is sensible, then exchange between bridging and non-bridging oxygens should occur on the same time scale as Q^n species exchange. As shown by Figure 23, motional averaging of the ^{17}O NMR peaks for BO and NBO in $K_2Si_4O_9$ melts certainly occurs. Quantitative simulation of all features of an exchanging spin 5/2 line shape is still an incompletely solved problem; however, at the higher temperatures, simulation of partially averaged ^{17}O peaks (prior to re-broadening by the influence of quadrupolar satellites) does yield exchange rates essentially equal to those observed by ^{29}Si studies (Farnan, 1991; Stebbins et al., 1992b). Another study of ^{17}O in the same composition also shows averaging of BO and NBO (Maekawa et al., 1991a). However, averaging appeared to take place at a much lower temperature. The reasons for this discrepancy are unclear, but could involve lowering of the viscosity by contamination with H_2O or B_2O_3.

A number of measurements of ^{11}B NMR peak shapes in borate melts have been reported, and it has been clear that rapid reorientation eventually causes the spectrum to narrow to a single peak (Inagaki et al., 1993; Xu et al., 1988; Xu et al., 1989). However, the difference in isotropic chemical shifts between $^{[3]}B$ and $^{[4]}B$ is small (about 15 ppm), which was probably not resolvable at the low fields at which these experiments were done. It has thus not been obvious whether exchange among these two species takes place or whether these systems have "molecular" behavior (long persistence of local structural units relative to the time scales of macroscopic transport properties). Thus, higher resolution experiments at higher fields are useful. As shown in Figure 14, $^{[3]}B$ and $^{[4]}B$ peaks are clearly identifiable in MAS spectra of sodium borate glasses (Ellsworth, 1994; Ellsworth and Stebbins, 1994). As temperature is raised above the glass transition, motional averaging begins and single, narrow peaks begin to appear in samples with 10 and 30 mol % Na_2O. Note that as discussed above, the peak positions of interest here are probably not the isotropic chemical shifts, but the sums of the isotropic chemical shifts and the isotropic averages of the second order quadrupolar coupling (Eqn. 6). The latter is

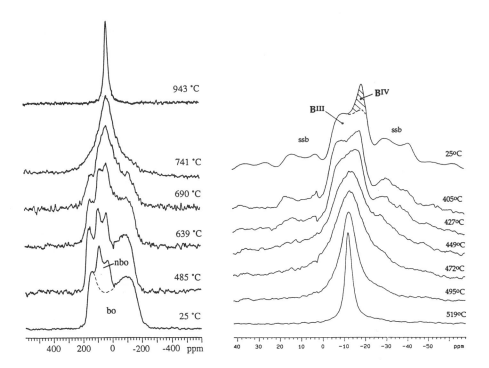

Figure 23 (left). Static (i.e. non-MAS) ^{17}O spectra for $K_2Si_4O_9$ glass and liquid. Quadrupolar doublets for non-bridging oxygen (nbo) and bridging oxygen (bo) are shown. Note the chemical exchange averaging into a single peak at high temperature. [Used by permission of the editor of *Chemical Geology*, from Stebbins et al. (1992b), Fig. 4, p. 377.]

Figure 24 (right). ^{11}B MAS NMR spectra for a glass and melt with 10 mol % Na_2O and 90 mol % B_2O_3 (Ellsworth, 1994; Ellsworth and Stebbins, 1994). Trigonal and tetrahedral boron sites are labeled. Data were collected at a relatively slow spinning speed of about 3 kHz at the temperatures shown. Spinning sidebands are labeled "ssb." Note collapse of two peaks due to chemical exchange and motional narrowing.

negligible for $^{[4]}B$, but will move the $^{[3]}B$ peak about 6 ppm to lower frequency, closer to the $^{[4]}B$ peak. More complete averaging was observed for the sample with 10% Na_2O and 90% B_2O_3, because of its lower T_g and viscosity (Fig. 24). In this case, significant averaging was observed by about 450°C, which suggests exchange at a rate roughly equal to the separation between the two peaks, or about 1000 Hz. Viscosity data, coupled with a value of G_∞ of 2.5×10^{10} Pa and Equation (12), give a mean τ_{shear} of about 1ms, equal to the inverse of the exchange frequency estimated from NMR. Again, it is clear that species exchange, which requires breaking of bonds between the network former (B) and oxygen, must be a critical part of viscous flow. A cartoon of a possible model that couples viscosity, diffusion, and borate species exchange is shown in Figure 25.

Studies of ^{31}P peak shapes in Na and Li metaphosphate glasses and melts suggest that P–O bond breaking, and phosphate group reorientation, are closely linked to the glass transition and to the time scale defined by viscous flow (Kirkpatrick et al., 1995).

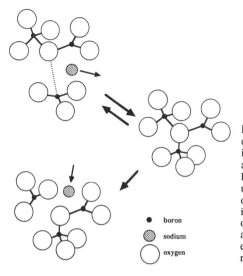

Figure 25. Cartoon of a possible mechanism for exchange among three- and four-coordinated boron in an alkali borate melt. The Na^+ cation diffuses away from the bridging oxygen in the top sketch, leaving the latter with a partial, temporarily uncompensated negative charge. The bridging oxygen is attracted to a three-coordinated boron, and, in the center sketch, forms a three-coordinated oxygen as a "transition complex." If this breaks up along the lower pathway, a net oxygen diffusion event has occurred. This may be triggered by the return of a modifier cation.

- boron
- sodium
- oxygen

Spin-lattice relaxation and dynamics in melts

Silicon. Studies of motional averaging and chemical exchange can be made only over limited ranges in temperature, because they require the observation of only partially averaged peak shapes. And, of course, there must be resolvable features in the spectra, which for many nuclides in many oxide glasses and liquids is not the case because of disorder-induced broadening. Measurements of spin-lattice relaxation can, in principle, provide dynamical information at almost any temperature and for any nuclide for which NMR signal can be obtained in a relatively short time, and have been (for decades) the mainstay of NMR studies of dynamics in polymers, ionic conductors, molten salts, and liquid metals. However, distinguishing among relaxation mechanisms can be difficult, and transforming relaxation time data into useful pictures of ionic or molecular motion can depend on complex and possibly under-constrained theory.

Relaxation time data for ^{29}Si in $K_2Si_4O_9$ and $NaAlSi_3O_8$ in glasses, for example, shows only minor changes with temperature up to T_g, then an abrupt steepening that begins just above T_g (Farnan and Stebbins, 1990a; Liu et al., 1987; Liu et al., 1988) (Fig. 26). A connection with the dynamics of the network is clearly indicated, but the exact nature of this tie is still uncertain. The apparent activation energy for relaxation in the $K_2Si_4O_9$ melt (about 180 kJ/mol) is similar to that observed for bulk viscosity (200 kJ/mol) and for Q species exchange (see above); for $NaAlSi_3O_8$ the E_a for T_1 (about 130 kJ/mol) is much less than that for viscosity (400 kJ/mol). Analysis of the magnetic field and compositional dependence of the relaxation, and estimates showing that ^{29}Si–^{29}Si and ^{29}Si–^{23}Na dipolar interactions are relatively small, suggested that coupling to paramagnetic centers, rather than chemical shift fluctuations, are important for relaxation even in the melts (Farnan and Stebbins, 1990a).

Because relaxation requires field fluctuation at the Larmor frequency, the abrupt turnovers in relaxation times near T_g initially appears surprising, because exchange frequencies in this range are many orders of magnitude smaller than ω_L. A significant spectral density of motion at ω_L is implied even at low exchange frequencies. One possible explanation is that the diffusion of paramagnetic cations (e.g. Fe^{2+}) suddenly

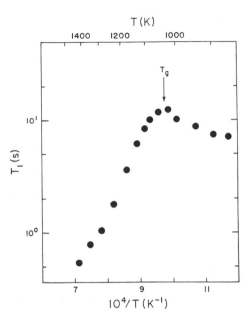

Figure 26. Spin-lattice relaxation times (T_1) vs. inverse temperature for ^{29}Si in $NaAlSi_3O_8$ glass and liquid. T_g is the glass transition for a normal slow laboratory heating rate. [Used by permission of the editor of *Physics and Chemistry of Minerals*, from Liu et al. (1987), Fig. 2, p. 158.]

becomes more rapid because of the change in network dynamics at T_g. Another, more likely, possibility is that the exchange process itself has a high frequency component. Suppose, for example, that exchange is occurring at an average frequency of 1 kHz. Two extremes in the way that this process occurs can be imagined. The first is that the exchange resembles a smooth oscillation or rotation with a period of 1 ms. In this case displacement and field fluctuations might be well-described by a sine wave with a single frequency of 1 kHz. However, given that the exchange and diffusive processes are known to be thermally activated and thus to require crossing an energy barrier, a much more likely scenario is that a 1 kHz exchange frequency implies only that the average time between jumps to a different configuration is 1 ms. In this case, the actual exchange event may take place much more rapidly, and the frequency spectrum needed to describe this motion is likely to contain a range of components much greater than 1 kHz, including a component at ω_L. A more thorough understanding of spin-lattice relaxation could potentially reveal much more about how exchange, and thus viscous flow and diffusion, actually operates.

Aluminum peak widths, relaxation and dynamics. Interpretation of spin-lattice relaxation may be simpler for quadrupolar nuclides in the fully-averaged, high temperature regime where $\tau_c << 1/\omega_L$, as indicated by Equation (8). In some cases in low viscosity melts, values of T_1 can be derived directly from measured peak widths (see above). This assumption has been made in a number of studies of ^{27}Al in aluminate melts, which are described in more detail in a later section. In the SiO_2–Al_2O_3 binary, peak widths increased systematically with higher silica content, indicating slower relaxation and longer motional correlation times (Poe et al., 1992a; Poe et al., 1992b). Using Equation (8), τ_c values were estimated and found to be of the same order of magnitude as the shear correlation times estimated from viscosity data (Eqn. 12). This suggested that, as for species exchange in silicate melts, there is a close link between local diffusive motion (which presumably controls T_1) and macroscopic flow. Similar

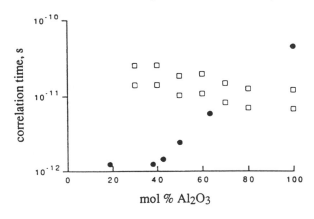

Figure 27. Correlation times for liquids in the $CaO–Al_2O_3$ binary. Open squares are derived from ^{27}Al peak widths in melts using Equation (8), with upper and lower bounds on C_Q taken from measurements on solids. Solid circles are shear correlation times derived from viscosity data using Equation (12). [Used by permission of the editor of *Journal of Non-Crystalline Solids*, from Poe et al. (1994), Fig. 5, p. 1835.]

results were obtained in the $CaO–Al_2O_3$ binary (Poe et al., 1994) (Fig. 27). In a recent study of Al_2O_3 melt during rapid cooling (Massiot et al., 1994), T_1 values derived from peak widths gave an activation energy of about 130 kJ/mol, close to that derived from viscosity (110 kJ/mol).

Alkali and alkaline earth cation diffusion. A number of studies of alkali motion in silicate and borates glasses have been made using spin-lattice relaxation, particularly for ^7Li and ^{23}Na (Göbel et al., 1979; Hendrickson and Bray, 1974; Jain et al., 1991; Kanert et al., 1991; Martin, 1989). The last three of these review recent results and theory. In some cases, measurements have been extended well above T_g (Liu et al., 1987; Liu et al., 1988; Maekawa, 1993; Sen et al., 1995; Stebbins et al., 1995b). As for the ^{27}Al studies described above, Equation (8) has been applied to ^{23}Na dynamics in silicate melts to calculate values of correlation times, activation energies, and C_Q (Liu et al., 1987; Liu et al., 1988). The latter were similar to those found in crystals, and were interpreted in terms of nearest ion neighbor distances to calculate Na-O bond lengths that agree reasonably well with those in known crystal structures. Thus, structural as well as dynamical data may be obtainable from relaxation studies.

As noted above and as illustrated by Figure 7, apparent activation energies (E_a) are greater on the high temperature sides of T_1 minima. In this region, E_a is usually similar to that for electrical conductivity and tracer diffusion, indicating that relaxation is being caused by the relatively high energy, infrequent diffusive jumps of the alkali cations from site to site. In a recent study of the "mixed alkali effect" (which included alkaline earth cations also) in disilicate and trisilicate liquids and glasses, these high temperature slopes were found to be nearly independent of the presence of "other" cations, and became similar to those for viscous flow (e.g. slopes for ^{23}Na T_1 in $Na_2Si_3O_7$ and in $NaCsSi_3O_7$ were the same within error). The apparent E_a for diffusion of Rb^+, derived from ^{87}Rb peak width measurements, was also the same within error (Stebbins et al., 1995b). These results indicate that in the low-viscosity melt, the energy barriers for diffusive motion are not strongly dependent on composition. On the other hand, mixing of other cations displaced T_1 minima to higher temperatures. Because these minima indicate the temperatures at which the mean $\tau_c = 1/\omega_L$, the addition of "other" cations obviously decreases the average diffusive hopping frequency for a given cation by somehow hindering its motion. At lower temperatures, through T_g and into the glass, apparent activation energies are systematically higher in the mixed cation compositions. These data were accurately fitted with a model, similar to one recently applied to sulfide and to chloride-borate glasses (Svare et al., 1993), in which distributions of energy barriers, and thus of correlation times, were included (Sen et al., 1995). Given that the systems are disordered, the existence of such distributions is realistic, although of course their shapes are not known a priori. The most obvious change caused by mixing cations is apparently a large drop in the number of available sites for a cation to hop to, indicating a strong energetic preference for sites previously occupied by the same cation. A given sites thus "remembers" what cation was in it even after the cation has jumped out of that site. This effect is also sensible, because at relatively low temperatures, the network structure physically relaxes at a rate much slower than that of the alkali cation hopping. The features of the structure that define the energy a cation will have when it jumps into the site (distribution of NBO's and BO's, coordination number, size of the site) thus remain fixed for relatively long periods of time. In the liquid at relatively high temperature, as exchange and Si–O bond breaking begin to occur at a rate similar to that of the alkali cation motion, this "memory effect" may be lost, resulting in the observed similarity of activation energies. Percolation theory was applied to the results of the modeling of spin-lattice relaxation, and predicted direct current conductivities with reasonably good accuracy.

In this, as well as in recent work on $NaAlSi_3O_8$ liquid (George and Stebbins, in preparation) (Stebbins et al., 1994), considerable care was taken to look for inflections in the T_1 curves for ^7Li and ^{23}Na at T_g, but none were detected. This makes sense in that the time scales for framework motion (which gives rise to the glass transition) and of alkali ion motion are strongly decoupled at these low temperatures. Earlier, less precise studies that reported such inflections (Liu et al., 1987; Liu et al., 1988) thus appear to be in error. In compositions where modifier cation content is too low to reach the percolation threshold, a stronger effect of T_g may be expected (Sen and Stebbins, in preparation).

In a study of peak widths and relaxation for ^{25}Mg and ^{23}Na in a Mg-Na silicate melt, activation energies were also reported that are similar to those expected for tracer diffusion of these cations (Fiske and Stebbins, 1994a).

Boron in borate melts. As described above, recent evidence suggests that exchange among borate species occurs at a time scale similar to that of viscous flow. Spin-lattice relaxation time measurements again have the potential to explore this relationship over wide ranges of temperature, because ^{11}B (I = 3/2) is very easy to observe by NMR, given its high natural abundance and Larmor frequency. However, both of these parameters also indicate that not just quadrupolar, but both hetero- and homonuclear dipolar couplings may contribute to relaxation, complicating its interpretation. A recent study reported ^{11}B T_1 data for several alkali borate melts to temperatures well above T_g (Inagaki et al., 1993). Results were fitted with a two-component relaxation model, possibly related to local and through-going diffusive motions. At the highest temperatures studied (about 1200°C), the correlation times for the slower motion closely approach those of the shear relaxation times calculated from viscosity, suggesting again a connection between local bond breaking processes and macroscopic dynamics (Stebbins et al., 1995b).

Average local structure in melts

In low viscosity molten oxides, generally well above their melting points, NMR peaks are generally fully narrowed by both reorientational and chemical exchange averaging, and their positions are a true measure of the average isotropic chemical shift. If correlations between δ_{iso} and structure are well established, then inferences about average structure in the liquid may be made. Non-structural effects of temperature on chemical shift, including changes in composition, bulk magnetic susceptibility, and electronic populations are also possible, and need to be investigated for structural conclusions to be robust.

Effects of temperature and melting on structure. Good examples of this approach are the extremely high temperature ^{27}Al NMR studies of a series of aluminate liquids (Coutures et al., 1990; Massiot et al., 1990b). For Al_2O_3 melt, a δ_{iso} value of 53 ppm indicated that most of the Al was four and/or five-coordinated Al ($^{[4]}Al$ or $^{[5]}Al$), probably related to the lower density of the liquid than the six-coordinated crystalline phase ($\delta_{iso} = 9$ ppm). A shift of about 70 ppm in $CaAl_2O_4$ liquid, on the other hand, indicated that it contained almost all $^{[4]}Al$, like the low-density crystalline solid ($\delta_{iso} = 75$ ppm). Similar studies (at much lower temperatures) on Na-Al fluoride melts indicated analogous results, with significantly lower average coordination numbers in the liquids than in the crystals (Stebbins et al., 1992a). In this system, data could be gathered over a wide enough range in temperature to indicate a significant further decrease in Al coordination with higher temperature. A more recent study of Al_2O_3 liquid reported data rapidly collected during a few seconds of cooling after shutting off of the heater laser, prior to crystallization of this non-glass forming composition (Florian et al., 1995; Massiot et al., 1994). Here, a significant effect on δ_{iso} of the oxidation state of the liquid was described, with δ_{iso} values varying between about 64 and 59 ppm. In this case, a slight decrease in shift with decreasing T ($d\delta_{iso}/dT > 0$) was attributed to changes in electronic structure accompanying thermal expansion, not to changes in coordination number. Similar effects have been noted in high T NMR of both ^{25}Mg and ^{17}O in crystalline MgO (Fiske et al., 1994).

In situ, high temperature ^{27}Al NMR has also allowed the comparison of δ_{iso} values in melts with those in glasses, which presumably represent the change in structure from that at T_g to that above the liquidus. For quadrupolar nuclides, such as ^{27}Al, either careful analysis of spinning sidebands (see above), or data collection at several magnetic fields, is required to obtain δ_{iso} for the glass. Using this approach, a systematic offset of about 5 ppm between melts (more shielded) and glasses in the $CaAl_2O_4$–SiO_2 binary was observed (Coté et al., 1992a; Coté et al., 1992b) (Fig. 28), although the two published data sets are somewhat different. A higher mean coordination number for Al at higher

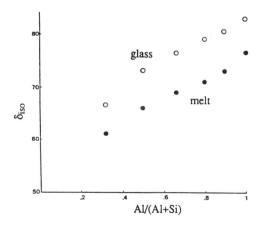

Figure 28. Isotropic chemical shifts for ^{27}Al in glasses and melts along the SiO_2–$CaAl_2O_4$ binary. Open circles (glasses) were derived from analysis of satellite spinning sidebands in MAS spectra; solid circles (melts) were direct measurements of fully averaged peaks at high temperature. [Used by permission of the publisher, from Coté et al. (1992b), Fig. 2, p. 755.]

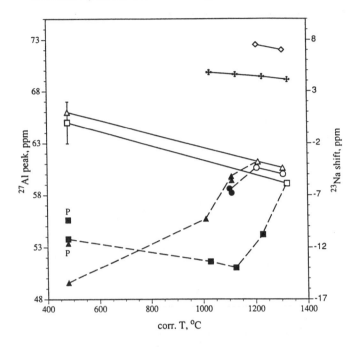

Figure 29. Peak positions and isotropic ^{27}Al and ^{23}Na chemical shifts for a variety of Na and Li silicate and aluminosilicate melts and glasses. See original reference for compositions. Crosses are for ^{23}Na and refer to scale at right; all other symbols are for ^{27}Al and refer to scale at left. For ^{27}Al, solid symbols and dashed lines are for data affected by quadrupolar shifts, while open symbols and solid lines are for isotropic chemical shifts. Three points were derived from each ^{27}Al ambient temperature MAS spectrum and are plotted at T_g: P indicates the MAS central peak maxima; other solid symbols are the centroids of the 1/2 to –1/2 transitions; open symbols are isotropic chemical shifts estimated on the basis of spinning side bands of satellite transitions. Note that peaks initially shift to higher frequency with increasing T as the quadrupolar effects are averaged, then shift slightly to lower frequency, possibly because of T effects on structure. The latter is consistent with the trend suggested by the isotropic shifts for the glasses. Redrawn from (Stebbins and Farnan, 1992).

temperatures is suggested. A similar observation was made on lower temperature alkali aluminosilicate melts (Stebbins and Farnan, 1992) (Fig. 29). This effect could be of dynamical significance if it is caused by an increase in the concentration of $^{[5]}$Al, and if this species plays a role as a "transition complex", as has been suggested for $^{[5]}$Si.

Studies of different nuclides in the same liquids have been useful to understand possible systematic errors introduced by non-structural effects on chemical shifts. Electronic effects, as noted above, can involve both changes in orbital overlap (covalency) and in electronic populations in excited states (Fiske et al., 1994), and seem generally to give slightly positive values of $d\delta_{iso}/dT$ (deshielding at higher temperature). Another possibility is that temperature can effect the bulk magnetic susceptibility of a liquid or glass, changing the magnetic field "seen" by all nuclei in a sample by the same amount. If this effect predominated, $d\delta_{iso}/dT$ should be the same for different isotopes. In a study of ^{23}Na, ^{27}Al, and ^{29}Si in several alkali silicate and aluminosilicate liquids, this was found not to be the case (Stebbins and Farnan, 1992) (Figs. 29 and 30). Relatively large positive values of $d\delta_{iso}/dT$ for ^{29}Si (confirmed by further studies; Maekawa, 1993) were attributed to lengthening of bonds between NBO's and modifier cations during

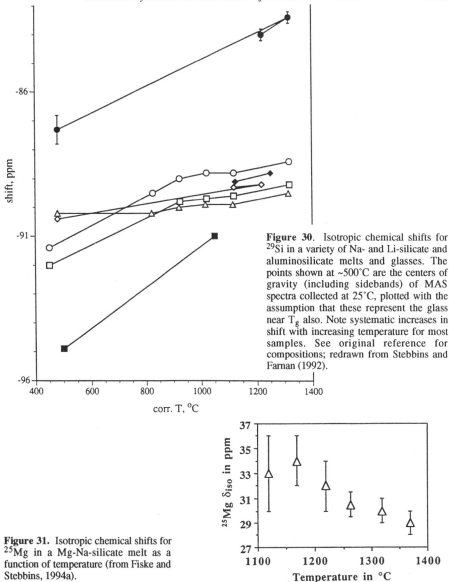

Figure 30. Isotropic chemical shifts for ^{29}Si in a variety of Na- and Li-silicate and aluminosilicate melts and glasses. The points shown at ~500°C are the centers of gravity (including sidebands) of MAS spectra collected at 25°C, plotted with the assumption that these represent the glass near T_g also. Note systematic increases in shift with increasing temperature for most samples. See original reference for compositions; redrawn from Stebbins and Farnan (1992).

Figure 31. Isotropic chemical shifts for ^{25}Mg in a Mg-Na-silicate melt as a function of temperature (from Fiske and Stebbins, 1994a).

thermal expansion, which may lead to more covalent Si–O bonds and deshielding. However, electronic effects are not yet well explored for Si shifts.

Structural effects on δ_{iso} for ^{25}Mg have not been well studied because of the difficulty of obtaining good spectra on known crystalline materials, but coordination number seems to be most important (Fiske and Stebbins, 1994a). In a case where spectra in a silicate melts are easier to observe than in a glass, a study of ^{25}Mg in a Na-Mg silicate showed a relatively large negative $d\delta_{iso}/dT$, which is probably the result of the average expansion of Mg sites at higher temperature, possibly leading to an increase in at least the time-averaged coordination number (Fig. 31). This may be the result of expansion of all Mg sites, or to the sampling of energetically unfavorable, large sites during cation diffusion. The latter could make a significant contribution to the configurational entropy of the liquid. In this study, ^{29}Si and ^{23}Na shifts were also measured, and varied considerably less with temperature relative to their total ranges in silicates. An alternate explanation of the data for ^{25}Mg is that it was caused by changes in the distribution of first neighbor cations, which has been shown to be significant for ^{23}Na (Xue and Stebbins, 1993). However, such a change would have probably produced a shift in the ^{29}Si peak greater than that observed.

In this study, the speculation was made that if in some general sense $d\delta_{iso}/dT$ reflects the overall change in structure around a given cation with temperature, then perhaps this should be small for tightly bonded cations such as Al and Si. For a very weakly bonded cation such as Na, much of the change in the averaged local environment might happen at much lower temperature, even below T_g, where cation diffusion may already be rapid. An alkaline earth cation at the temperatures studied (1100° to 1400°C) may be in an intermediate situation, experiencing a rapidly changing bonding environment.

This speculation may be in part supported by new data on ^{23}Na in glasses and liquids (Stebbins et al., 1994). As noted above, ^{23}Na peak positions, either static or MAS, are not the same as δ_{iso} except in the fully averaged liquid regime. However, as shown in Fig. 2, the peak progressively shifts to a more shielded (lower frequency) position with increasing temperature. This is opposite to any change that could be caused by averaging of quadrupolar effects, and suggests a significant increase in the coordination number of Na^+, probably again through diffusive sampling of large, high energy sites. Note that much of this change does indeed occur below T_g. Similar results have been observed in $NaAlSi_3O_8$ melt.

The large values of the partial molar thermal expansivities for network modifiers (largest for the lowest field strength cations) relative to that for SiO_2 (Lange and Carmichael, 1990) suggest that a major fraction of the total expansion may involve increases in the average volumes occupied by the modifier cations. Some in situ EXAFS data (Farges et al., 1994) suggest that the coordination number for Ni^{2+} may decrease with increasing temperature. Whether this is a composition-dependent phenomenon, or whether the two spectroscopies are "seeing" different aspects of structural change with temperature, will require further experimentation.

Compositional effects on melt structure. A number of studies of compositional effects on the average structure of Al sites in oxide melts have been made, primarily by laser-heated techniques on levitated samples. These have been particularly interesting because temperatures were high enough, and viscosities low enough, to do ion dynamics computer simulations at the same temperature. This has eliminated a common difficulty, where limits on computation time restrict such work to simulated temperatures much higher than those obtainable in the laboratory (see chapter by Poole and McMillan). In work on the $SiO_2–Al_2O_3$ binary, compositional effects on δ_{iso} for ^{27}Al were small (Poe et al., 1992b). This was explained by two compensating effects: a decrease in the mean coordination number of Al with decreasing Al_2O_3 content (which decreases shielding), and a first neighbor cation effect (substituting Si for Al increases shielding). Ion dynamics simulations showed these effects clearly. In work on the $MgAl_2O_4$ and $CaAl_2O_4$ liquids, the Al coordination number appeared to be slightly higher in the former, but was near 4 in both (Poe et al., 1993). This finding is intriguing because of the complete difference in the crystal structures, with mostly $^{[6]}Al$ in $MgAl_2O_4$ spinel and

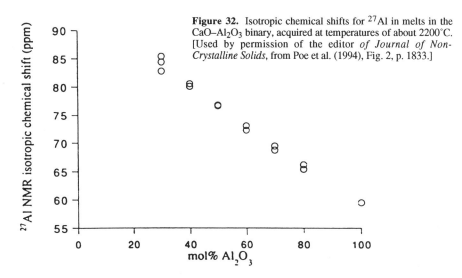

Figure 32. Isotropic chemical shifts for [27]Al in melts in the CaO–Al₂O₃ binary, acquired at temperatures of about 2200°C. [Used by permission of the editor *of Journal of Non-Crystalline Solids*, from Poe et al. (1994), Fig. 2, p. 1833.]

[4]Al in calcium aluminate. Again, ion dynamics simulations supported these observations, and indicated that the proportion of [5]Al should increase at higher temperature, as also suggested by the NMR data on liquids and glasses (see above). The increase in high-coordinate Al in Mg- vs. Ca-containing melts agrees with trend observed in glasses (see above).

In liquids in the CaO–Al₂O₃ binary, δ_{iso} for [27]Al decreases with increasing Al₂O₃ content (Poe et al., 1994) (Fig. 32). Again by comparison with ion dynamics simulations, this was correlated with an increase in the coordination number of Al, although first cation neighbor effects (substitution of Al for Ca) could again have been significant. MAS NMR showed the presence of multiple coordinations of Al in the glasses, as well (Fig. 13). High temperature Raman spectroscopy on the same liquids also indicated that AlOₙ polyhedra are persistent on the vibrational time scale.

The effect of composition on Mg sites in melts has begun with [25]Mg NMR on a Na-Mg silicate and a Ca-Mg silicate (Fiske and Stebbins, 1994a). The coordination number of Mg in the former appeared to be about 5, in the latter about 6. This result can be rationalized in terms of competition among the network modifiers for coordination by non-bridging oxygens, probably analogous to the effect of high field strength cations on Al coordination noted above.

Structural controls on δ_{iso} for [23]Na in silicates have now been well studied (George and Stebbins, 1994; George and Stebbins, 1995; Koller et al., 1994; Maekawa, 1993; Xue and Stebbins, 1993). Again, coordination number and Na–O bond length seem to dominate, although first neighbor cation effects, and the ratio of NBO's to tetrahedral cations, are also quite important. If these effects are taken into account, a variety of silicate and aluminosilicate melts seem to have similar coordination numbers of about 6 to 7, while liquids that nominally do not have abundant NBO's (e.g. NaAlSi₃O₈) have coordination numbers of 7 to 8 (Fig. 33). This is sensible in the less effective charge balance of Na+ by bridging instead of non-bridging oxygens.

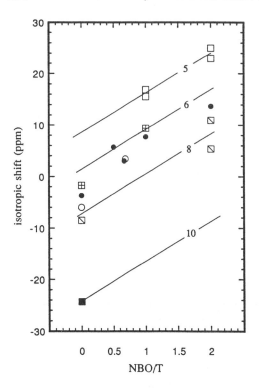

Figure 33. Isotropic chemical shifts for ^{23}Na vs. the ratio of non-bridging oxygens to tetrahedral cations for crystalline silicates (squares) and melts (circles), relative to 1 M aqueous NaCl at 25°C. The various symbols for the crystals are for different coordination numbers for Na, as marked by numbers labeling the solid lines. Redrawn from Stebbins et al. (1995b); see also Xue and Stebbins (1993).

CONCLUSIONS

In the introduction, three fundamental and general questions concerning the relationship of liquid and glass structure to macroscopic properties were posed. The contributions of NMR to answering these questions have been described above. Some of these will be summarized here. NMR has provided the first accurate quantitative measures of the real statistical distributions of some structural parameters in some oxide glasses, particularly those of simple composition. These include silicate anionic (Q^n) species, bridging and non-bridging oxygens, Si-O-Si angles in glasses other than SiO_2, and cation neighbors to NBOs. Systematic effects of network modifier cation field strength on this order have be noted that agree well with earlier inferences from Raman spectroscopy and the solution properties implied by phase diagrams. These findings have already been incorporated into newly constrained thermodynamic models. Useful, if still qualitative, data have been obtained on the old question of Si/Al disorder in glasses. Following decades of work quantifying the distribution of boron coordination numbers in glasses, the first unequivocal signatures (and concentration data) of 5- and 6-coordinated Al and Si in glasses have been reported, and pressure and compositional effects have been explored. The non-observation of high-coordinated Al in many glass compositions has been definitive enough to greatly restrain the range of speculation in structure-based thermodynamic modeling. Studies of 3-, 4-, and 5-component glasses, although generally less unambiguous, have begun to provide important new constraints to models that have more direct relevance to natural magmas. New approaches to and findings about the difficult problem of intermediate-range order have been made.

Recent NMR studies have focused attention on the differences between glasses and liquids, and the nature of structural change with temperature. Through fictive temperature studies, significant temperature effects on silicate speciation, on boron, aluminum, and silicon coordination, and on overall disorder have been observed, some of which have been confirmed (or even preceded) by high-T Raman studies. However, initial analyses of all of these data suggest that while these effects are important in understanding thermodynamic activities of components, they probably are not the dominant explanation of the overall configurational properties of the liquids. This remains therefore a critical problem for future research. In situ NMR studies are also beginning to provide results that can be interpreted in terms of mean coordination numbers for cations. Changes in these with temperature may reflect diffusive exploration by modifier cations of sites that are unoccupied at the glass transition. and thus may have important implications to diffusivity as well as overall energetics and to configurational complexity.

Finally, high temperature NMR has provided a unique window into the microscopic mechanisms that control viscous flow and the diffusion of network-forming cations. Results suggest that local effects of the breaking of strong network bonds predominate over larger-scale "molecular" mechanisms, but only a few compositions have been studied. As yet, only average motional correlation times have been derived, that just begin to characterize the true spectra of dynamical complexity known to be present from bulk measurements.

The complementary, and occasionally conflicting, contributions from other spectroscopies are described at length in other chapters in this volume, as is the wider context of the relationship of spectroscopic findings to bulk properties. It is hoped that the reader will consider all of these views, whether struggling with the types of "microscopic to macroscopic" connections posed by an earlier MSA short course (Kieffer and Navrotsky, 1985), or in attempting to formulate entirely new research directions.

ACKNOWLEDGMENTS

I am particularly grateful to the students, research associates, and scientific collaborators at U.C. Berkeley and Stanford who have made many of the studies discussed here possible, including Ian Carmichael, Alex Pines, Jim Murdoch, Erika Schneider, Shang-Bin Liu, Ian Farnan, Mark Brandriss, Xianyu Xue, Dane Spearing, Peter Fiske, Philip Grandinetti, Sabyasachi Sen, Zhi Xu, Anna George and Susan Ellsworth. I have had the pleasure of years of scientific discussion and interchange of ideas about melt structure and dynamics with my co-editors Paul McMillan and Don Dingwell. I again (and again) thank Paul Ribbe for his remarkable efforts as the series editor of the MSA *RiM* volumes, and acknowledge the support of the National Science Foundation for funding our research.

REFERENCES

Abragam A (1961) Principles of Nuclear Magnetism. Clarendon Press, Oxford, 599 p
Angell CA (1985) Strong and fragile liquids. In: KL Ngai and GB Wright (eds) Relaxation in Complex Systems. Office of Naval Research, Washington, DC, p 3-11
Angell CA (1990) Dynamic processes in ionic glasses. Chem Rev 90:523-542
Angell CA, Cheeseman PA, Tamaddon S (1983) Water-like transport property anomalies in liquid silicates investigated at high T and P by computer simulation techniques. Bull Minéral 106:87-97
Bloembergen N, Purcell EM, Pound RV (1948) Relaxation effects in nuclear magnetic resonance. Phys Rev 73:679-712
Boettcher AL, Burnham CW, Windom KE, Bohlen SR (1982) Liquids, glasses, and the melting of silicates to high pressures. J Geol 90:127-138

Brandriss ME, Stebbins JF (1988) Effects of temperature on the structures of silicate liquids: ^{29}Si NMR results. Geochim Cosmochim Acta 52:2659-2670

Bray PJ, and Holupka, E.J. (1985) The potential of NMR techniques for studies of the effects of thermal history on glass structure. J Non-Cryst Solids 71:411-428

Bray PJ, Emerson JF, Lee D, Feller SA, Bain DL, Feil DA (1991) NMR and NQR studies of glass structure. J Non-Cryst Solids 129:240-248

Bray PJ, Gravina SJ, Stallworth PE, Szu SP, Zhong J (1988) NMR studies of the structure of glasses. Exp Tech Phys 36:397-413

Bray PJ, O'Keefe JG (1963) Nuclear magnetic resonance investigations of the structure of alkali borate glasses. Phys Chem Glasses 4:37-46

Brow RK, Phifer CC, Turner GL, Kirkpatrick RJ (1991) Cation effects on ^{31}P MAS NMR chemical shifts of metaphosphate glasses. J Am Ceram Soc 74:1287-1290

Bunker BC, Kirkpatrick RJ, Brow RK, Turner GL, Nelson C (1991) Local structure of alkaline-earth boroaluminate crystals and glasses: II, ^{11}B and ^{27}Al MAS NMR spectroscopy of alkaline-earth boroaluminate glasses. J Am Ceram Soc 74:1430-1438

Bunker BC, Tallant DR, Kirkpatrick RJ, Turner GL (1990) Nuclear magnetic resonance and Raman investigation of sodium borosilicate glass structures. Phys Chem Glass 31:30-40

Burnham CW (1981) The nature of multicomponent aluminosilicate melts. In: DT Rickard, FE Wickman (eds) Chemistry and geochemistry of solutions at high temperatures and pressures. Pergamon Press, New York, p 197-229

Burnham CW, Nekvasil H (1986) Equilibrium properties of granite pegmatite magmas. Am Min 71:239-263

Coté B, Massiot D, Taulelle F, Coutures J (1992a) ^{27}Al NMR spectroscopy of aluminosilicate melts and glasses. Chem Geol 96:367-370

Coté B, Massiot D, Taulelle F, Coutures JP (1992b) ^{27}Al NMR spectroscopy of aluminosilicate melts and glasses. In: LD Pye, WCL Course, HJ Stevens (eds) The Physics of Non-Crystalline Solids. Taylor and Francis, London, p 752-756

Coutures JP, Massiot D, Bessada C, Echegut P, Rifflet JC, Taulelle F (1990) Etude par RMN ^{27}Al d'aluminates liquides dans le domaine 1600-2100°C. C R Acad Sci Paris II 310:1041-1045

Derome AE (1987) Modern NMR Techniques for Chemistry Research. Pergamon Press, New York, 280 p.

Dingwell DB, Webb SL (1989) Structural relaxation in silicate melts and non-Newtonian melt rheology in geologic processes. Phys Chem Minerals 16:508-516

Dingwell DB, Webb SL (1990) Relaxation in silicate melts. Eur J Mineral 2:427-449

Dirken PJ, Jansen JBH, Schuiling RD (1992) Influence of Octahedral Polymerization on Sodium-23 and Aluminum-27 MAS-NMR in Alkali Fluoroaluminates. Am Mineral 77:718-724

Dupree R, Holland D, Mortuza MG (1987) Six-coordinated silicon in glasses. Nature 328:416-417

Dupree R, Holland D, Mortuza MG (1990) A MAS-NMR investigation of lithium silicate glasses and glass ceramics. J Non-Cryst Solids 116:148-160

Dupree R, Holland D, Mortuza MG, Collins JA, Lockyer MWG (1989) Magic angle spinning NMR of alkali phospho-alumino-silicate glasses. J Non-Cryst Solids 112:111-119

Dupree R, Holland D, Williams DS (1986) The structure of binary alkali silicate glasses. J Non-Cryst Solids 81:185-200

Dupree R, Kohn SC, Henderson CMB, Bell AMT (1993) The role of NMR shifts in structural studies of glasses, ceramics and minerals. NATO ASI Ser, Ser C, Nuclear Magnetic Shieldings and Molecular Structure 386:421-434

Dupree R, Kohn SC, Mortuza MG, Holland D (1992) NMR studies of glass structure. In: LD Pye, WC La Course, HJ Stevens (eds) Physics of Non-Crystalline Solids. Taylor and Francis, London, p 718-723

Dye JL, Ellaboudy AS (1991) Solid state NMR of quadrupolar nuclei. In: AI Popov, K Hallenga (eds) Modern NMR Techniques and Their Application in Chemistry. Marcel Dekker, New York, p 217-322

Eckert H (1990) Structural concepts for disordered inorganic solids. Modern NMR approaches and strategies. Ber Bunsenges Phys Chem 94:1062-1085

Ellsworth SE (1994) Borate and borosilicate melts: high temperature MAS NMR and fictive temperature studies. M.S. report, Stanford University, Dept. Mat. Sci. Eng.

Ellsworth SE, Stebbins JF (1994) Borate and borosilicate melts: high T MAS NMR and fictive T studies. EOS, Trans Am Geophys Union 75:714

Emerson JF, Stallworth PE, Bray PJ (1989) High-field ^{29}Si NMR studies of alkali silicate glasses. J Non-Cryst Solids 113:253-259

Engelhardt G, Koller H (1994) ^{29}Si NMR of Inorganic Solids. In: B Blümich (ed) Solid-State NMR II: Inorganic Matter. Springer-Verlag, Berlin, p 1-30

Engelhardt G, Michel D (1987) High-Resolution Solid-State NMR of Silicates and Zeolites. Wiley, New York, 485 p

Engelhardt G, Nofz M, Forkel K, Wihsmann FG, Mägi M, Samosen A, Lippmaa E (1985) Structural studies of calcium aluminosilicate glasses by high resolution solid state ^{29}Si and ^{27}Al magic angle spinning nuclear magnetic resonance. Phys Chem Glass 26:157-165

Farges F, Brown GE, Jr., Calas G, Galoisy L, Waychunas GA (1994) Temperature-induced structural transformations in Ni-bearing silicate glass and melt. Geophys. Res. Lett. 21:1931-1934

Farnan I (1991) NMR spectroscopy in the earth sciences. In: WA Nierenberg (ed) Encyclopedia of Earth System Science. Academic Press, San Diego, CA, p 450-465

Farnan I, Grandinetti PJ, Baltisberger JH, Stebbins JF, Werner U, Eastman M, Pines A (1992) Quantification of the disorder in network modified silicate glasses. Nature 358:31-35

Farnan I, Stebbins JF (1990a) A high temperature ^{29}Si NMR investigation of solid and molten silicates. J Am Chem Soc 112:32-39.

Farnan I, Stebbins JF (1990b) Observation of slow atomic motions close to the glass transition using 2-D ^{29}Si NMR. J Non-Cryst Solids 124:207-215

Farnan I, Stebbins JF (1994) The nature of the glass transition in a silica-rich oxide melt. Science 265:1206-1209

Fenske D, Gerstein BC, Pfeifer H (1992) Influence of thermal motion upon the lineshape in magic-angle-spinning experiments. J Magn Reson 98:469-474

Fiske P, Stebbins JF (1994a) The structural role of Mg in silicate liquids: a high-temperature ^{25}Mg, ^{23}Na, and ^{29}Si NMR study. Am Mineral 79:848-861

Fiske P, Stebbins JF, Farnan I (1994) Bonding and dynamical phenomena in MgO: a high temperature ^{17}O and ^{25}Mg NMR study. Phys Chem Minerals 20:587-593

Fiske PS (1993) The Volumetric properties of Silicate Liquids; Dilatometric and Spectroscopic Studies. PhD dissertation, Stanford University, Stanford, California

Fiske PS, Stebbins JF (1994b) ^{17}O NMR Study of Ti coordination and clustering in silicate glasses. EOS, Trans Am Geophys Union 75:369

Florian P, Massiot D, Poe B, Farnan I, Coutures JP (1995) A time-resolved ^{27}Al NMR study of the cooling process of liquid alumina from 2450°C to crystallisation. Sol St Nucl Magnet Reson (in press)

Fukushima E, Roedder SBW (1981) Experimental Pulse NMR. Addison-Wesley, Reading, MA, 539 p

Gan H, Hess PC (1992) Phosphate speciation in potassium aluminosilicate glasses. Am Mineral 77:495-506

Gan H, Hess PC, Kirkpatrick RJ (1994) Phosphorus and boron speciation in $K_2O-B_2O_3-SiO_2-P_2O_5$ glass. Geochem Cosmochim Acta 58:4663-4648

Gaskell PH (1992) The structure of silicate glasses - new results from neutron scattering studies. In: LD Pye, WCL Course, HJ Stevens (eds) The Physics of Non-Crystalline Solids. Taylor and Francis, London, p 15-21

Gaskell PH, Eckersley MC, Barnes AC, Chieux P (1991) Medium-range order in the cation distribution of a calcium silicate glass. Nature 350:675-677

George AM, Stebbins JF (1994) High temperature ^{23}Na MAS NMR data for albite: comparison to chemical shift models. EOS, Trans Am Geophys Union 75:713

George AM, Stebbins JF (1995) High temperature ^{23}Na NMR data for albite: comparison to chemical shift models. Am Mineral in press

Göbel E, Müller-Warmuth W, Olyschläger H, Dutz H (1979) ^{7}Li NMR spectra, nuclear relaxation, and lithium ion motion in alkali silicate, borate, and phosphate glasses. J Mag Reson 36:371-387

Greaves GN, Gurman SJ, Catlow CRA, Chadwick AV, Houde-Walter S, Henderson CMB, Dobson BR (1991) A structural basis for ionic diffusion in oxide glasses. Phil Mag A 64:1059-1072

Grimmer AR, Blümich B (1994) Introduction to Solid State NMR. In: B Blümich (ed) Solid-State NMR I: Methods. Springer-Verlag, Berlin, p 3-62

Grimmer AR, von Lampe F, Mägi M (1986) Solid-state high-resolution ^{29}Si MAS NMR of silicates with sixfold coordination. Chem Phys Lett 132:549-553

Gupta PK, Lui ML, Bray PJ (1985) Boron coordination in rapidly cooled and in annealed aluminum borosilicate glass fibers. J Am Ceram Soc 68:C-82

Gurman SJ (1990) Bond ordering in silicate glasses: a critique and re-solution. J Non-Cryst Sol 125:151-160

Hallas E, Hähnert M (1985) Solid-state high-resolution ^{27}Al NMR studies of sodium aluminosilicate glasses with different thermal histories. Crystal Res Technol 20:K25-K28

Harris RK (1983) Nuclear magnetic resonance spectroscopy. Pitman, London, 250 p.

Hater W, Müller-Warmuth W, Meier M, Frischat GH (1989) High-resolution solid-state NMR studies of mixed-alkali silicate glasses. J Non-Cryst Solids 113:210-212

Hayakawa S, Yoko T, Sakka S (1995) IR and NMR structural characterization on lead vanadate glasses. J Non-Cryst Solids 183:73-84

Hendrickson JR, Bray PJ (1974) Nuclear magnetic resonance studies of ^{7}Li ionic motion in alkali silicate and borate glasses. J Chem Phys 61:2754-2764

Inagaki Y, Maekawa H, Yokokawa T (1993) Nuclear-magnetic-resonance study of the dynamics of network-glass-forming systems: xNa_2O-$(1-x)B_2O_3$. Phys Rev B 47:674-680

Jäger C, Kunath G, Losso P, Scheler G (1993) Determination of distributions of the quadrupole interaction in amorphous solids by ^{27}Al satellite transition spectroscopy. Sol St Mag Reson 2:73-82

Jain H, Kanert O, Ngai KL (1991) Relationship between nuclear spin relaxation and diffusion in glasses. Defect Diffus Forum 75:163-178

Jakobsen HJ, Skibsted J, Bildsøe H, Nielsen NC (1989) Magic-angle spinning NMR spectra of satellite transitions for quadrupolar nuclei in solids. J Magn Reson 85:173-180

Jeener J, Meier BH, Bachmann P, Ernst RR (1979) Investigation of exchange processes by two-dimensional NMR spectroscopy. J Chem Phys 71:4546

Kanert O, Steinert J, Jain H, Ngai KL (1991) Nuclear spin relaxation and atomic motion in inorganic glasses. J Non-Cryst Solids 131-133:1001-1010

Kanzaki M, Stebbins JF, Xue X (1991) Characterization of the quenched high pressure phases in $CaSiO_3$ system by x-ray diffraction and ^{29}Si NMR. Geophys Res Lett 18:463-466

Kanzaki M, Stebbins JF, Xue X (1992) Characterization of crystalline and amorphous silicates quenched from high pressure by ^{29}Si MAS NMR spectroscopy. In: Y Syono, MH Manghnani (eds) High Pressure Research: Applications to Earth and Planetary Sciences. Terra Scientific Publishing, Tokyo, p 89-100

Kirkpatrick RJ (1988) MAS NMR spectroscopy of minerals and glasses. In: FC Hawthorne (ed) Spectroscopic Methods in Mineralogy and Geology. Rev Mineral 18:341-403

Kirkpatrick RJ, Howell D, Phillips BL, Cong XD, Ito E, Navrotsky A (1991) MAS NMR spectroscopic study of $Mg^{29}SiO_3$ with perovskite structure. Am Mineral 76:673-676

Kirkpatrick RJ, Oestrike R, Weiss CA Jr, Smith KA, Oldfield E (1986) High-resolution ^{27}Al and ^{29}Si NMR spectroscopy of glasses and crystals along the join $CaMgSi_2O_6$ - $CaAl_2SiO_6$. Am Mineral 71:705-711

Kirkpatrick RJ, Sato RK, Haase J, Brow K, Sidebottom D, Green F (1995) High-temperature NMR investigation of the glass transition of Li, Na metaphosphate. Am Ceram Soc Ann Mtg Abs 97:21

Knight CTG, Kirkpatrick RJ, Oldfield E (1990) The connectivity of silicon sites in in silicate glasses as determined by two-dimensional silicon-29 nuclear magnetic resonance spectroscopy. J Non-Cryst Solids 116:140

Kolem H, Kanert O, Schulz H, Guenther B (1990) Design and operation of a variable high-temperature oxygen partial-pressue probe device for solid-state NMR. J Magnet Reson 87:160-165

Koller H, Engelhardt G, Kentgens APM, Sauer J (1994) ^{23}Na NMR spectroscopy of solids: Interpretation of quadrupole interaction parameters and chemical shifts. J Phys Chem 98:1544-1551

Lacey ED (1965) A statistical model of polymerization/depolymerization relationships in silicate melts and glasses. Phys Chem Glasses 6:171-180

Lange RA, Carmichael ISE (1990) Thermodynamic properties of silicate liquids with an emphasis on density, thermal expansion, and compressibility. In: J Nicholls, K Russell (eds) Quantitative Methods in Igneous Petrology. Rev Mineral 24:25-64

Levin EM, Robbins CR, McMurdie HF (1964) Phase Diagrams for Ceramists, Vol 1. American Ceramic Society, Columbus, Ohio, 601 p

Libourel G, Geiger C, Merwin L, Sebald A (1991) High-resolution solid-state ^{29}Si and ^{27}Al MAS NMR spectroscopy of glasses in the system $CaSiO_3$-$MgSiO_3$-Al_2O_3. Chem Geol 96:387-397

Lippmaa E, Mägi M, Samosen A, Engelhardt G (1980) Structural studies of silicates by solid-state high-resolution ^{29}Si NMR spectroscopy. J Am Chem Soc 103:4992-4996

Liu SB, Pines A, Brandriss M, Stebbins JF (1987) Relaxation mechanisms and effects of motion in albite ($NaAlSi_3O_8$) liquid and glass: a high temperature NMR study. Phys Chem Minerals 15:155-162

Liu SB, Stebbins JF, Schneider E, Pines A (1988) Diffusive motion in alkali silicate melts: an NMR study at high temperature. Geochim Cosmochim Acta 52:527-538

MacKenzie KJD, Meinhold RH (1994) ^{25}Mg nuclear magnetic resonance spectroscopy of minerals and related inorganics: a survey study. Am Mineral 79:250-260

Maekawa H (1993) NMR studies of the structure and dynamics of silicate melts. PhD dissertation, Hokkaido University, Japan

Maekawa H, Inagaki Y, Shimokawa S, Nakamura Y, Maekawa T, Yokokawa T (1991a) High-temperature NMR studies of molten oxides and oxide glass. Mater Sci Forum, Molten Salt Chem Technol 73-75:123-129

Maekawa H, Maekawa T, Kawamura K, Yokokawa T (1991b) The structural groups of alkali silicate glasses determined from ^{29}Si MAS-NMR. J Non-Cryst Solids 127:53-64

Marsmann H (1981) Silicon-29 NMR. In: P Diehl, E Fluck, R Kosfeld (eds) NMR Basic Principles and Progress. Springer-Verlag, Berlin, p 65-235

Martin SW (1989) Recent advances in the study of fast ionically conducting glasses using nuclear magnetic resonance techniques. Mat Chem Phys 23:225-265

Massiot D, Bessada C, Coutures JP, Taulelle F (1990a) A quantitative study of ^{27}Al MAS NMR in crystalline YAG. J Magn Res 90:231-242

Massiot D, Florian P, Farnan I, Coutures JP (1994) Monitoring the cooling of liquid alumina from 2500°C by ^{27}Al NMR. 38th Expt NMR Conf, abstr, 315

Massiot D, Taulelle F, Coutures JP (1990b) Structural diagnostic of high temperature liquid phases by ^{27}Al NMR. Colloq Phys 51-C5:425-431

Maxwell R, Eckert H (1994) Chemical equilibria in glass-forming melts: high-temperature ^{31}P and ^{77}Se NMR of the phosphorus-selenium system. J Am Chem Soc 116:682-689

Maxwell R, Eckert H (1995) Molten-state kinetics in glass-forming systems. A high-temperature NMR study of the system phosphorus-selenium. J Phys Chem 99:4768-4778

McMillan P, Akaogi M, Ohtani E, Williams Q, Nieman R, Sato R (1989) Cation disorder in garnets along the $Mg_3Al_2Si_3O_{12}$ - $Mg_4Si_4O_{12}$ join: an infrared, Raman, and NMR study. Phys Chem Minerals 16:428-435

McMillan PF (1994) Water solubility and speciation models. In: MR Carroll, JR Holloway (eds) Volatiles in Magmas. Rev Minral 30:131-156

McMillan PF, Kirkpatrick RJ (1992) Al coordination in magnesium aluminosilicate glasses. Am Mineral 77:898-900

McMillan PF, Wolf GH, Poe BT (1992) Vibrational spectroscopy of silicate liquids and glasses. Chem Geol 96:351-366

Mehring M (1983) High Resolution NMR in Solids. Springer-Verlag, Berlin, 342 p

Merzbacher CI, Sherriff BL, Hartman JS, White WB (1990) A high resolution ^{29}Si and ^{27}Al NMR study of alkaline earth aluminosilicate glasses. J Non-Cryst Solids 124:194-206

Mueller KT, Wu Y, Chmelka BF, Stebbins J, Pines A (1990) High-resolution oxygen-17 NMR of solid silicates. J Am Chem Soc 113:32-38

Murdoch JB, Stebbins JF, Carmichael ISE (1985) High-resolution ^{29}Si NMR study of silicate and aluminosilicate glasses: the effect of network-modifying cations. Am Mineral 70:332-343

Mysen BO (1988) Structure and Properties of Silicate Melts. Elsevier, Amsterdam, 354 p

Mysen BO, Frantz JD (1992) Raman spectroscopy of silicate melts at magmatic temperatures: Na_2O-SiO_2, K_2O-SiO_2 and Li_2O-SiO_2 binary compositions in the temperature range 25-1475°C. Chem Geol 96:321-332

Mysen BO, Frantz JD (1993) Structure of silicate melts at high temperature: In-situ measurements in the system $BaO-SiO_2$ to 1669°C. Am Mineral 78:699-709

Oestrike R, Yang WH, Kirkpatrick R, Hervig RL, Navrotsky A, Montez B (1987) High-resolution ^{23}Na, ^{27}Al, and ^{29}Si NMR spectroscopy of framework aluminosilicate glasses. Geochim Cosmochim Acta 51:2199-2209

Phillips BL, Howell DA, Kirkpatrick RJ, Gasparik T (1992) Investigation of cation order in $MgSiO_3$-rich garnet using ^{29}Si and ^{27}Al MAS NMR spectroscopy. Am Mineral 77:704-712

Phillips BL, Kirkpatrick RJ (1994) Short-range Si-Al order in leucite and analcime: Determination of the configurational entropy from ^{27}Al and variable-temperature ^{29}Si NMR spectroscopy of leucite, its Cs- and Rb-exchanged derivatives, and analcime. Am Mineral 79:1025-1036

Poe BT, McMillan PF, Angell CA, Sato RK (1992a) Al and Si coordination in $SiO_2-Al_2O_3$ glasses and liquids: A study by NMR and IR spectroscopy and MD simulations. Chem Geol 96:333-349

Poe BT, McMillan PF, Coté B, Massiot D, Coutures J (1992b) $SiO_2-Al_2O_3$ liquids: In situ study by high-temperature ^{27}Al NMR spectroscopy and molecular dynamics simulation. J Phys Chem 96:8220-8224

Poe BT, McMillan PF, Coté B, Massiot D, Coutures JP (1993) $MgAl_2O_4$ and $CaAl_2O_4$ liquids: in-stu high temperature ^{27}Al NMR spectroscopy. Science 259:786-788

Poe BT, McMillan PF, Coté B, Massiot D, Coutures JP (1994) Structure and dynamics in calcium aluminate liquids: high-temperature ^{27}Al NMR and Raman spectroscopy. J Am Ceram Soc 77:1832-1838

Prabakar S, Rao KJ, Rao CNR (1991) A MAS NMR investigation of lead phosphosilicate glasses: the nature of the highly deshielded six-coordinated silicon. Mat Res Bull 26:285-294

Ramachandran R, Knight CTG, Kirkpatrick RJ, Oldfield E (1985) A two-dimensional NMR approach to the study of intermolecular scrambling reactions. J Mag Reson 65:136-141

Risbud SH, Kirkpatrick RJ, Taglialavore AP, Montez B (1987) Solid-state NMR evidence of 4-, 5-, and 6-fold aluminum sites in roller-quenched $SiO_2-Al_2O_3$ glasses. J Am Ceram Soc 70:C10-C12

Sanders JKM, Hunter BK (1987) Modern NMR spectroscopy. Oxford University Press, Oxford, 308 p.

Sato RK, McMillan PF, Dennison P, Dupree R (1991a) High resolution ^{27}Al and ^{29}Si MAS NMR investigation of $SiO_2-Al_2O_3$ glasses. J Phys Chem 95:4484

Sato RK, McMillan PF, Dennison P, Dupree R (1991b) A structural investigation of high alumina content glasses in the CaO-Al$_2$O$_3$-SiO$_2$ system via Raman and MAS NMR. Phys Chem Glasses 32:149-154

Schmidt-Rohr K, Spiess HW (1991) Nature of nonexponential loss of correlation above the glass transition investigated by multidimensional NMR. Phys Rev Lett 66:3020-3024

Schramm S, Oldfield E (1984) High-resolution oxygen-17 NMR of solids. J Am Chem Soc 106:2502-2506

Sebald A (1994) MAS and CP/MAS NMR of Less Common Spin-1/2 Nuclei. In: B Blümich (ed) Solid-State NMR II: Inorganic Matter. Springer-Verlag, Berlin, p 91-132

Sen S, George AM, Stebbins JF (1995) Ionic conduction and mixed cation effect in silicate glasses and liquids: ^{23}Na and ^{7}Li NMR spin-lattice relaxation and a multiple-barrier model of percolation. J Non-Crystal Solids 188:54-62

Sen S, Stebbins JF (1994) Phase separation, clustering and intermediate range order in Li$_2$Si$_4$O$_9$ glass: a ^{29}Si MAS NMR spin-lattice relaxation study. Phys Rev 50:822-830

Sen S, Stebbins JF (1995) Structural role of Nd^{3+} and Al^{3+} cations in SiO$_2$ glass: a ^{29}Si MAS NMR spin-lattice relaxation, ^{27}Al NMR and EPR spectroscopic study. J Non-Cryst Solids in press

Sen S, Stebbins JF, Hemming NG, Ghosh B (1994) Coordination environments of B impurities in calcite and aragonite polymorphs: A ^{11}B MAS NMR study. Am Mineral 79:819-825

Shimokawa S, Maekawa H, Yamada E, Maekawa T, Nakamura Y, Yokokawa T (1990) A high temperature (1200°C) probe for NMR experiments and its application to silicate melts. Chem Lett 1990:617-620

Skibsted J, Nielsen NC, Bilsoe H, Jakobsen HJ (1991) Satellite transitions in MAS NMR spectra of quadrupolar nuclei. J Mag Reson 95:88-117

Slichter CP (1990) Principles of Magnetic Resonance. Springer-Verlag, Berlin, 655 p.

Smith KA, Kirkpatrick RJ, Oldfield E, Henderson DM (1983) High-resolution silicon-29 nuclear magnetic resonance spectroscopic study of rock-forming silicates. Am Mineral 68:1206-1215

Spearing DR, Stebbins JF (1989) The ^{29}Si NMR shielding tensor in low quartz. Am Mineral 74:956-959

Stebbins JF (1987) Identification of multiple structural species in silicate glasses by ^{29}Si NMR. Nature 330:465-467

Stebbins JF (1988a) Effects of temperature and composition on silicate glass structure and dynamics: Si-29 NMR results. J Non-Cryst Solids 106:359-369

Stebbins JF (1988b) NMR spectroscopy and dynamic processes in mineralogy and geochemistry. In: FC Hawthorne (ed) Spectroscopic Methods in Mineralogy and Geology. Rev Minerl 18:405-430

Stebbins JF (1991a) Experimental confirmation of five-coordinated silicon in a silicate glass at 1 atmosphere pressure. Nature 351:638-639

Stebbins JF (1991b) Nuclear magnetic resonance at high temperature. Chem Rev 91:1353-1373

Stebbins JF (1995) Nuclear magnetic resonance spectroscopy of silicates and oxides in geochemistry and geophysics. In: TJ Ahrens (ed) Handbook of Physical Constants. American Geophysical Union, Washington, DC, p 303-332

Stebbins JF, Farnan I (1989) NMR spectroscopy in the earth sciences: structure and dynamics. Science 245:257-262

Stebbins JF, Farnan I (1992) The effects of temperature on silicate liquid structure: a multi-nuclear, high temperature NMR study. Science 255:586-589

Stebbins JF, Farnan I, Dando N, Tzeng SY (1992a) Solids and liquids in the NaF-AlF$_3$-Al$_2$O$_3$ system: a high temperature NMR study. J Am Ceram Soc 75:3001-3006

Stebbins JF, Farnan I, Williams EH, Roux J (1989a) Magic angle spinning NMR observation of sodium site exchange in nepheline at 500°C. Phys Chem Minerals 16:763-766

Stebbins JF, Farnan I, Xue X (1992b) The structure and dynamics of alkali silicate liquids: one view from NMR spectroscopy. Chem Geol 96:371-386

Stebbins JF, George AM, Sen S (1994) Alkali cation sites and dynamics in silicate liquids: high temperature NMR spectroscopic studies. EOS, Trans Am Geophys Union 75:714

Stebbins JF, Kanzaki M (1990) Local structure and chemical shifts for six-coordinated silicon in high pressure mantle phases. Science 251:294-298

Stebbins JF, McMillan P (1989) Five- and six- coordinated Si in K$_2$Si$_4$O$_9$ glass quenched from 1.9 GPa and 1200°C. Am Mineral 74:965-968

Stebbins JF, McMillan P (1993) Compositional and temperature effects on five coordinated silicon in ambient pressure silicate glasses. J Non-Cryst Solids 160:116-125

Stebbins JF, Murdoch JB, Schneider E, Carmichael ISE, Pines A (1985) A high temperature nuclear magnetic resonance study of ^{27}Al, ^{23}Na, and ^{29}Si in molten silicates. Nature 314:250-252

Stebbins JF, Schneider E, Murdoch JB, Pines A, Carmichael ISE (1986) A new probe for high-temperature nuclear magnetic resonance spectroscopy with ppm resolution. Rev Sci Inst 57:39-42

Stebbins JF, Sen S (1994) Nucleation and medium range order in silicate liquids: inferences from NMR spectroscopy. Materials Research Society Symposia 321:143-153

Stebbins JF, Sen S, Farnan I (1995a) Silicate species exchange, viscosity, and crystallization in a low-silica melt: In situ high-temperature MAS NMR spectroscopy. Am Mineral 80:861-864

Stebbins JF, Sen S, George AM (1995b) High temperature nuclear magnetic resonance studies of oxide melts. J Non-Cryst Solids in press

Stebbins JF, Spearing DR, Farnan I (1989b) Lack of local structural orientation in oxide glasses quenched during flow: NMR results. J Non-Cryst Solids 110:1-12

Stebbins JF, Sykes D (1990) The structure of $NaAlSi_3O_8$ liquids at high pressure: new constraints from NMR spectroscopy. Am Mineral 75:943-946

Stebbins JF, Xu Z, Vollath D (1995c) Cation exchange rate and mobility in aluminum-doped lithium orthosilicate: high-resolution lithium-6 NMR results. Solid State Ionics 78:L1-L8

Svare I, Borsa F, Torgeson DR, Martin SW (1993) Correlation functions for ionic motion from NMR relaxation and electrical conductivity in the glassy fast-ion conductor $(Li_2S)_{0.56}(SiS_2)_{0.44}$. Phys Rev B 48:9336-9344

Swamy KCK, Chandrasekhar V, Harland JJ, Holmes JM, Day RO, Holmes RR (1990) Pentacoordinate acyclic and cyclic anionic oxysilicates. A ^{29}Si NMR and X-ray structural study. J Am Chem Soc 112:2341-2348

Taulelle F, Coutures JP, Massiot D, Rifflet JP (1989) High and very high temperature NMR. Bull Magn Reson 11:318-320

Timken HKC, Schramm SE, Kirkpatrick RJ, Oldfield E (1987) Solid-state oxygen-17 nuclear magnetic resonance spectroscopic studies of alkaline earth metasilicates. J Phys Chem 91:1054-1058

Tossell JA (1990) Calculation of NMR shieldings and other properties for three and five coordinate Si, three coordinate O and some siloxane and boroxol ring compounds. J Non-Cryst Solids 120:13-19

Tossell JA (1992) Calculation of the ^{29}Si NMR shielding tensor in forsterite. Phys Chem Minerals 19:338-342

Tossell JA, Vaughn DJ (1992) Theoretical Geochemistry: Applications of Quantum Mechanics in the Earth and Mineral Sciences. Oxford University Press, Oxford, 514 p

Turner GL, Chung SE, Oldfield E (1985) Solid-state oxygen-17 nuclear magnetic resonance spectroscopic study of group II oxides. J Mag Res 64:316-324

Turner GL, Kirkpatrick RJ, Risbud SH, Olfield E (1987) Multinuclear magic-angle sample-spinning nuclear magnetic resonance spectroscopic studies of crystalline and amorphous ceramic materials. Am Ceram Soc Bull 66:656-663

Turner GL, Smith KA, Kirkpatrick RJ, Oldfield E (1986) Boron-11 nuclear magnetic resonance spectroscopic study of borate and borosilicate minerals and a borosilicate glass. J Mag Reson 67:544-550

Vermillion KE, Florian P, Grandinetti PJ, Stebbins JF (1995) Structure of network-modified alkali silicate glasses. Am Ceram Soc Ann Mtg Abs 97:23

Weeding TL, deJong BHWS, Veeman WS, Aitken BG (1985) Silicon coordination changes from 4-fold to 6-fold on devitrification of silicon phosphate glass. Nature 318:352-353

Weiden N, Rager H (1985) The chemical shift of the ^{29}Si nuclear magnetic resonance in a synthetic single crystal of Mg_2SiO_4. Z Naturforsch 40A:126-130

Wilson MA (1987) NMR Techniques and Applications in Geochemistry and Soil Chemistry. Pergamon, Oxford, 353 p.

Wind RA (1991) Solid state NMR of spin-1/2 nuclei. In: AI Popov, K Hallenga (eds) Modern NMR Techniques and Their Application in Chemistry. Marcel Dekker, New York, p 125-216

Wolf G, Durben DJ, McMillan P (1990) High-pressure Raman spectroscopic study of sodium tetrasilicate $(Na_2Si_4O_9)$ glass. J Chem Phys 93:2280-2288

Xu S, Pan L, Tian F, Wu X (1988) Temperature-dependent ^{11}B NMR studies of B_2O_3 glass. Act Phys Sin 37:1866-1869

Xu S, Pan L, Wu X (1989) Temperature-dependent ^{11}B NMR studies of $0.5\ Na_2O \cdot B_2O_3$ glass. Sci Rep 20:1533-1535

Xu Z, Stebbins JF (1995) ^6Li and ^7Li NMR chemical shifts, relaxation, and coordination in silicates. Sol St Nucl Magnet Reson in press

Xue X, Stebbins JF (1993) ^{23}Na NMR chemical shifts and the local Na coordination environments in silicate crystals, melts, and glasses. Phys Chem Minerals 20:297-307

Xue X, Stebbins JF, Kanzaki M (1991a) Oxygen-17 NMR of quenched high-pressure crystals and glasses. EOS, Trans Am Geophys Union 72:572

Xue X, Stebbins JF, Kanzaki M (1994) Correlations between O-17 NMR parameters and local structure around oxygen in high-pressure silicates and the structure of silicate melts at high pressure. Am Mineral 79:31-42

Xue X, Stebbins JF, Kanzaki M, McMillan PF, Poe B (1991b) Pressure-induced silicon coordination and tetrahedral structural changes in alkali silicate melts up to 12 GPa: NMR, Raman, and infrared spectroscopy. Am Mineral 76:8-26

Xue X, Stebbins JF, Kanzaki M, Tronnes RG (1989) Silicon coordination and speciation changes in a silicate liquid at high pressures. Science 245:962-964

Yarger J, Diefenbacher J, Poe BT, Wolf GH, McMillan PF (1994) High coordinate Si and Al in sodium aluminosilicate glasses quenched from high pressure. EOS, Trans Am Geophys Union 75:713

Youngman RE, Zwanziger JW (1994) Multiple boron sites in borate glass detected with dynamic angle spinning nuclear magnetic resonance. J Non-Cryst Solids 168:293-297

Zhang Y, Stolper EM (1993) Reaction kinetics, fictive temperature, cooling rates, and glass relaxation. EOS, Trans Am Geophys Union 74:351

Chapter 8

VIBRATIONAL SPECTROSCOPY OF SILICATE LIQUIDS

Paul F. McMillan and George H. Wolf

Department of Chemistry and Biochemistry
Arizona State University
Tempe, Arizona 85287 U.S.A.

INTRODUCTION

Background

A silicate melt phase is an essential component in nearly all igneous processes. In a melt, the viscosity can decrease some 20 orders of magnitude below that of corresponding solids. As a result, timescales for chemical fractionation and heat transport within the planet are dramatically reduced in the presence of a liquid state. Throughout geologic time, such melts have played a central role in determining the chemical and thermal evolution of the Earth. The presence of a molten silicate phase is likely to have played the single most important role in shaping the evolution of the present Earth, through near-global melting in the early Archean (Takahashi, 1990).

Although near-surface melts can be examined directly during volcanic eruptions, magmas formed by complete or partial melting deep within the Earth must be studied indirectly, by field studies of fossil erupted or plutonic rocks, or via seismic observations. For this reason, it is essential to carry out laboratory experiments to observe and understand the properties of natural or model magma compositions, to extrapolate to the behavior of melts within the Earth. There is particular interest in developing an atomic level understanding of the processes which lead to viscous flow, chemical and thermal transport, and densification of silicate melts, under the P-T conditions of the crust and mantle. This level of understanding serves to make an extrapolation from the laboratory results to the behavior of a natural system as reliable as possible.

In such studies, the goal is to develop models of the local, medium-range, and long-range "structure", or relative atomic arrangements in the melt, and to use these to understand the thermodynamic and rheological properties of magma. Many techniques have been applied to investigate the structures of aluminosilicate melts, including all of the classical spectroscopic and diffraction methods employed in physics and chemistry, along with computer simulations of the melt structure. An important consideration is the reliability of the structural inferences drawn from the spectroscopic data. Nearly all techniques require empirical calibration of their spectral features, and some subjectivity can creep into the development of any structural model. It can also prove technically difficult to study the melt directly under conditions appropriate for the Earth's interior. Silicate melts are refractory systems, and in many cases, only recently have the technical advances been made which permit direct investigation of the high temperature liquid. Even now, in situ studies of silicate melt structure under combined high pressure-high temperature conditions are few and far between. In many cases, the problem of dealing with high temperatures has been by-passed by quenching the liquid rapidly to a glass, and carrying out a spectroscopic or structural study on the glassy sample at room temperature. However, this approach presumes that the important structural quantities have been frozen in at the glass transition, during the quench from the liquid state. A further important parameter in melt structural studies is consideration of the timescale over which the spectroscopic measurement itself is made—is it an "instantaneous" or a time-averaged structure which is being probed? These points must be considered and addressed in a systematic way.

Vibrational spectroscopy, primarily via infrared transmission or reflection studies, or Raman scattering, is well known as a classic technique for structure elucidation in chemistry and physics, and have many applications in mineralogy and geochemistry (McMillan, 1984, 1989; McMillan and Hofmeister, 1988). These techniques have been used extensively for investigating the structure of glass (see review by Wong and Angell, 1976). Many vibrational studies of the structure of silicate glasses have been carried out, and the results have often been extrapolated to obtain structural models for the corresponding high temperature melts of geological interest. These studies have been reviewed by Mysen et al. (1982a), McMillan (1984, 1989), and Mysen (1988, 1990a,b). Most spectroscopic studies of silicate glass structure have been carried out for simple binary and ternary compositions. These are not usually directly relevant to rock-forming melts, but serve to establish structural principles which may be extended to the complex multicomponent geological systems. The most powerful studies have often been those in which the results of several spectroscopic and diffraction measurements are integrated to obtain detailed information on the local atomic environments (see the review edited by Hawthorne, 1988), and combined with and bulk property determinations to construct a working model of the glass and melt structure (Navrotsky, 1984; Navrotsky et al., 1985; Mysen, 1988, 1990a: see chapter by Hess in this volume).

The late 1980s and early 1990s represented a new beginning in the field of silicate melt investigation. Much research had been carried out prior to this. However, during this period, new instrumentation and techniques were developed to permit reliable in-situ spectroscopic studies of the high temperature liquids, particularly in the field of nuclear magnetic resonance applied to silicates (see chapter by Stebbins in this volume). The NMR technique is unique in that it readily gives direct information both on the structural sites present in the glass and liquid, and also on the dynamics of exchange reactions or other chemical averaging processes affecting them (Stebbins, 1988; Stebbins et al., 1992). These results led directly to a new vision of the high temperature melt, not as a random mix of entangled silicate polymer species, but as a dynamic, rapidly exchanging chemical environment, with ionic exchange between polymer units occurring on a timescale comparable with those of viscous flow and chemical diffusion. Next, the methods which had been successfully developed to understand the relaxation dynamics of molten salts and polymeric liquids began to be applied to silicate liquids, leading to a direct and intimate link between the calorimetric properties and the rheology of the melt (see chapters by Dingwell and Webb, and Richet and Bottinga, in this volume). In particular, the Adam-Gibbs theory of relaxation allows a quantitative link to be made between the timescale for viscous flow and the configurational entropy of the melt, and leads to a consideration of the "configurational landscape" sampled by the liquid as a function of temperature. One goal of structural studies on melts is to investigate the nature of the local atomic arrangements which contribute to these configurations, and the energetic barriers between them. Finally, quantitative use began to made of Maxwell's treatment of viscoelastic liquids, with a relaxation time related to dissipation of strain induced by some applied chemical, mechanical or thermal stress (see chapters by Webb and Dingwell, and by Moynihan, in this volume). This approach forces a consideration of the relative timescales for the stress dissipation, the applied strain rate, and the time-temperature relations of the microscopic structural relaxation, and leads to a better appreciation of the differences and similarities between silicate liquids and glasses.

Several in-situ high temperature vibrational spectroscopic studies of silicate melts had been carried out prior to this period. These were designed in large part to test the extrapolation of structural models developed from vibrational studies on glasses to the corresponding liquids (Grove and Jellyman, 1955; Markin and Sobolev, 1960; Sweet and White, 1969; Sharma et al., 1978; Shevyakov et al., 1978; Piriou and Arashi, 1980; Seifert et al., 1981; Iwamoto et al., 1984; Kashio et al., 1980, 1982; Domine and Piriou, 1983) (Figs. 1 and 2). Although some changes in the spectra with temperature were noted, it was generally concluded that the glass did provide a satisfactory model for the melt structure, from the observed similarity of the glass and melt spectra. This was a convenient observation, because of the experimental difficulties associated with carrying out in-situ studies under high temperature conditions.

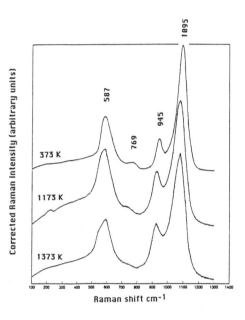

Figure 1. The Raman spectra of glass and melt near sodium disilicate composition ($Na_2O:2SiO_2:0.027 \cdot NaAl_2O_4$), as obtained by Seifert et al. (1981). The liquidus temperature is 850°C. The intensities have been corrected for the (first-order) frequency- and temperature-dependence of the Raman scattering by a function like that in Equation (16). These high quality spectra were taken with a scanning instrument. [Redrawn from data of Seifert et al. (1981), Fig. 4, p. 1883.]

However, as a better understanding of the nature of the glassy and liquid state has been developed, there has been renewed interest in carrying out in-situ vibrational studies at high temperature. It has become clear that the often subtle changes which occur in the vibrational spectra with increasing temperature through the glass transformation range into the supercooled and stable liquid regime can be due to structural relaxation taking place. These changes are often masked by modifications occurring in the vibrational spectrum at high temperatures, due to the intrinsic temperature dependence of the spectral features and the anharmonicity of the sample. Such studies have been facilitated by recent developments in instrumentation for infrared and Raman spectroscopy, along with a better understanding of methods for analyzing the high temperature spectra (Mysen and Frantz, 1992, 1993a,b, 1994a,b; McMillan et al., 1992, 1994; Daniel et al., 1993, 1995a).

"Liquids" versus "glasses"

It is useful at this stage to examine the distinction between "liquids" and "glasses" more closely, in terms of the relaxation processes and timescales involved (Dingwell and

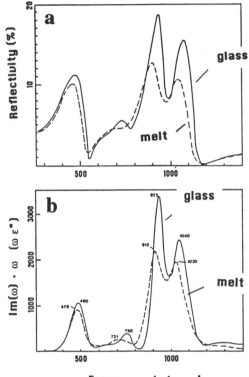

Figure 2. Infrared reflectance (a) and absorption (b) spectra for sodium silicate (Na_2O:1.41 SiO_2) glass (solid line) and melt (at 1243 K) (dotted line), obtained by Domine and Piriou (1983). Note the reduction in reflectivity at high temperature, due to anharmonic effects. The absorption spectrum is the imaginary part of the dielectric function as a function of frequency (ω): $\omega\varepsilon''(\omega)$, obtained from the reflectivity via a Kramers-Kronig relation. The high frequency absorption bands are broader in the high temperature spectrum. [Redrawn from data of Domine and Piriou (1983), Figs. 1 and 2, pp. 126 and 127.]

Webb, 1990: see also chapters by Dingwell and Webb, Moynihan, and Richet and Bottinga in this volume). In general, a structural relaxation time is defined by the time required for the system to dissipate, or relax from, an applied stress (Angell, 1984, 1988; Angell and Torrell, 1983). For example, in a Newtonian regime, the structural relaxation time τ_s associated with a mechanical shear stresses is proportional to the coefficient of shear viscosity, $\tau_s = \eta_s/G_\infty$, where G_∞ is the high frequency shear modulus. (If the stress is applied very rapidly, the behavior becomes non-Newtonian, and brittle failure may occur: see chapter by Dingwell and Webb). It is often useful to consider the applied stress to be a periodic function of time, especially for spectroscopic applications, so that the frequency dependence of the stress-strain relations can be examined relative to the structural relaxation timescale.

When the frequency of the applied stress is lower than that defined by the inverse of the structural relaxation time, $1/\tau$, the system can relax structurally in response to the stress; i.e. the strain energy is dissipated within the timescale of the experiment. This behavior is characteristic of a liquid. Our working definition of "liquid" behavior is usually based on just this shear relaxation behavior, which permits the material to readjust its shape to the form of its container, on a typical human observational timescale. The magnitude of the coefficient for shear viscosity is the usual measure of the "stickiness", or resistance to flow, of the liquid. If the frequency of the applied stress is greater than $1/\tau$, the system response will be *unrelaxed*; i.e. the strain will not be dissipated on the timescale of the experiment. This unrelaxed behavior is used to define a *glass*.

If the frequency of the applied stress is increased during an experiment at constant temperature and pressure, the passage from the relaxed (liquid) regime to the unrelaxed (glass) regime is known as the *glass transition*. Alternatively, if the stress is applied with a fixed frequency, the glass transition can also be observed as a function of temperature. In general, the characteristic time for structural relaxation, is an activated process, and decreases with increasing temperature. The observed *glass transition temperature*, T_g, is thus a function of the characteristic frequency range of the probe experiment used to define the glass transition, relative to the timescale of the relaxation process being investigated.

The "normal" glass transition temperature is usually measured by calorimetric or viscosity measurements (see chapters by Moynihan, Dingwell and Webb, and Richet and Bottinga). These techniques are associated with characteristic measurement timescales on the order of 10^2 to 10^3 s. Because the intrinsic resistance to shear deformation, given by the infinite frequency shear modulus G_∞, of a silicate melt is on the order of 10^{10} Pa (10 GPa) (Wong and Angell, 1976; Dingwell and Webb, 1990), the glass transition as observed by viscosity measurements (often termed T_η: Richet and Bottinga, 1984) is commonly defined to occur when the viscosity attains a value of 10^{12} Pa·s (or 10^{13} poise: 1 poise = 0.1 Pas). This defines a glass transition timescale of $\sim 10^3$ s, for those structural relaxations which are involved in the process of viscous flow. There appears to be a good correlation between the structural relaxation timescale and that of enthalpy relaxation in silicate melts (see Dingwell and Webb, this volume). Although this relation does not necessarily hold for all liquids, especially those which have a highly non-Arrhenian temperature dependence to their relaxation (termed *fragile* liquids: Angell, 1984, 1988, 1991), it means that for silicates, the calorimetric glass transition temperatures ($T_{g, cal}$) are generally equivalent to those defined by viscosity measurements (T_η) (Dingwell and Webb, this volume).

For temperatures lower than T_g, relaxation times are longer than 10^2 to 10^3 s, and structural equilibration is not usually achieved on an experimental timescale. The glass transition occurs when the temperature is sufficiently high that the relaxation time is on the order of this "laboratory" timescale, and the system can reach structural equilibrium during a normal experiment: for example, on the order of a few minutes to several seconds as the run temperature is changed. For silicate systems, typical glass transition temperatures are on the order of several hundred degrees Kelvin (Richet and Bottinga, 1980, 1984; Richet, 1984; Dingwell and Webb, 1990). In in situ vibrational spectroscopic experiments, significant changes associated with structural re-equilibration begin to occur as silicate glasses are heated above their glass transition temperatures, which are recorded in the observed spectra.

The atomic motions in a liquid can be conveniently separated into two classes. In *vibrational* motions, the atoms execute small periodic displacements about their equilibrium positions. In a *relaxational* mode, the nuclei wander from their starting point (Richet and Neuville, 1992). Both types of motion can absorb energy from an incident beam of light, and can be detected and studied by vibrational spectroscopy. The working distinction between glass and liquid is that, in a glass, structural changes are only vibrational in nature, when the structure is probed on a "laboratory" timescale (10^0 to 10^3 s). Above the laboratory glass transition, when the system is allowed to equilibrate for a time on the order of the structural relaxation timescale, these vibrational structural modifications are joined by relaxational effects, in which the supercooled liquid begins to explore additional structural configurations (Richet and Neuville 1992). As the temperature is increased, a greater number of configurations is opened up for exploration, on faster and faster timescales.

As the liquid is quenched, the glass transition, or the passage from the configurational relaxation regime to the purely vibrational excitation regime, is encountered at a temperature (T_g) defined by the quenching rate relative to the rate of structural relaxation. A faster quenching rate corresponds to a higher glass transition temperature, at which configurational exploration is "frozen" in the material. Once a glass has been quenched, it can then subsequently be annealed at a different temperature to T_g, for a time corresponding to the structural relaxation time at the new temperature. If the system does not crystallize, the result is a glass with a newly equilibrated structural configuration, corresponding to the annealing temperature T_f. This temperature is usually known as the "fictive temperature". The fictive temperature is the same as T_g for samples prepared directly from the melt, determined by the quenching rate relative to the relaxation kinetics in the liquid. Glassy samples can be annealed at higher or lower temperatures than their original T_g. In addition, glasses with different values of T_g can be prepared by varying the quench rate. Using the combined techniques of varying the quench rate and annealing in the glassy state, series of glasses with a range of T_g or T_f values can be prepared, corresponding to different states of structural relaxation. The structures of such glass series can then conveniently be examined at room temperature. Any structural changes are due to the configurational dependence of the liquid with temperature. This method can provide a useful complement to in-situ high temperature studies for investigating the supercooled liquid in the glass transformation range.

A few studies of this type have been carried out for silicates, particularly SiO_2 (Mikkelsen and Galeener, 1980; Geissberger and Galeener, 1983; Galeener, 1985). In this case, the relaxation kinetics are slow, and a range of several hundred degrees variation can be obtained in T_g and T_f, sufficient for observable changes to be recorded in the vibrational spectra. However, in general, the accessible temperature range for most silicates is much smaller, usually on the order of a few tens of degrees. Long term annealing of the glass at temperatures in the range of T_g often results in crystallization. In addition, the maximum values of T_f attainable remain well below the liquidus temperatures. For these reasons, it usually becomes necessary to carry out in situ high temperature spectroscopic experiments, in order to obtain a complete picture of structural changes in aluminosilicate liquids at magmatic temperatures. However, the study of series of glasses prepared with different T_g or T_f values should be pursued whenever possible, to provide an insight into the effects of configurational changes on the vibrational spectra, uncomplicated by vibrational effects which may occur at high temperatures.

It is important to distinguish two timescales in in-situ high temperature experiments. These are the "dwell" time, which represents the time for which the system is held at a given temperature before performing the measurement, and the "probe" time, which is the characteristic timescale for the measurement being made. The dwell time is fixed by the structural relaxation time for the liquid at the temperature of interest. If the sample is held at temperature for a shorter time than this, configurational relaxation is not achieved (Dingwell and Webb, 1990). In the case of vibrational spectroscopy, the probe timescale is on the order of 10^{-12} to 10^{-14} s. This is many orders of magnitude shorter than the structural relaxation times of silicate liquids at most temperatures of interest (the vibrational timescale is not approached until temperatures on the order of 3000 to 4000 K are attained for most systems: see chapters by Moynihan, Dingwell and Webb, and Richet and Bottinga), so that these systems never exhibit a relaxed response to the applied stress of the fluctuating radiation field in the vibrational spectroscopic measurement. For this reason, although the silicate melt may well be studied at temperatures well above its liquidus, and be classed as a "liquid" because of its relaxation behavior on the laboratory timescale, it behaves as an unrelaxed "glass" toward the vibrational spectroscopic experiment. By this we mean that the system does not have sufficient time to relax its structural configuration in response to the fluctuating electromagnetic stress field of the applied radiation, but shows unrelaxed vibrational behavior. This is one reason why the vibrational spectra of silicate "glasses" and "melts" are observed to be so similar. As the temperature is increased, and the system studied becomes more fluid, some relaxation may begin to occur. Some of the energy of the applied radiation is dissipated into the system, and diffusional motions occur. This shows up as a broadening in observed vibrational modes, as their lifetime becomes limited by the relaxational modes, and broad low energy absorption or inelastic scattering appears in the spectrum (Rothschild, 1986).

NMR "versus" vibrational spectroscopy

Prior to the early 1980s, most studies of silicate glass structure were carried out using vibrational spectroscopy, primarily via Raman scattering (see reviews by Mysen et al., 1980a, 1982a; McMillan, 1984a, 1989; Mysen, 1988, 1990a). Considerable advances in our understanding of silicate glass structure were made at that time, particularly in the composition dependence of polymerization state, which could be correlated with the thermodynamic properties and rheology of the melts. Around this time, the techniques of high resolution solid state NMR spectroscopy, rendered possible by fast magic angle sample spinning (MAS), began to be applied to the the problem of silicate glass structure (Schramm et al., 1984; Dupree et al., 1984, 1985, 1986; Murdoch et al., 1985; see reviews by Kirkpatrick et al., 1986; Engelhardt and Michel, 1987; Kirkpatrick, 1988; Stebbins, 1988; Dupree et al., 1992, 1993: also Stebbins, this volume). This technique provides a powerful method for detailed investigation of the local site geometry, particularly the coordination number and polymerization state, of the network-forming and network-modifying nuclei which constitute the glass and melt structure: these include ^{29}Si, ^{27}Al, ^{17}O, ^{31}P, ^{10}B, ^{11}B, ^{23}Na, ^{25}Mg, ... The assignment of NMR resonances to structural sites remains empirical, being based mainly on crystal chemical correlations, but the spectra are much simpler to interpret than vibrational data and are highly quantitative in that the strength of the signal is usually simply proportional to the number of sites of a given type (Stebbins, 1987; Maekawa et al., 1991; Xue et al., 1991). In addition, high-temperature NMR methods can be used to examine thermal averaging and chemical exchange processes between sites, through analysis of the NMR lineshape or direct measurement of relaxation times (Stebbins, 1988; Stebbins et al., 1992; Stebbins, this volume). For these reasons, NMR spectroscopy has emerged as a technique of choice for detailed, site-specific studies of local atomic environments in aluminosilicate materials. However, vibrational spectroscopy remains a useful complement to NMR methods for silicate glass and melt studies, for several reasons.

First, based on equation of state data and rheology of magmas, it is known that significant structural changes must occur in silicate systems under high pressure conditions relevant to melts in the crust and mantle (Kushiro, 1976, 1978, 1981, 1983; Kushiro et al., 1976; Stolper et al., 1981; Scarfe et al., 1987; Rigden et al., 1988, 1989; Agee and Walker, 1988; Miller et al., 1991a,b). These changes are most likely associated with Al and Si coordination changes at high pressure (Waff, 1975; Hemley et al., 1986;

Williams and Jeanloz, 1988; Wolf et al., 1990: see chapter by Wolf and McMillan). However, although NMR spectra of glasses quenched from high pressure do reveal the presence of high coordinate Si and Al species (Stebbins and McMillan, 1989, 1993; Xue et al., 1989, 1991; Stebbins and Sykes, 1990; Sykes et al., 1993; Yarger et al., 1994, 1995: see chapters by Stebbins, and Wolf and McMillan), the structural species are not completely retained upon decompression (Hemley et al., 1986; Stolper and Ahrens, 1987; Williams and Jeanloz, 1988, 1989; Wolf et al., 1990; Kubicki et al., 1991). In-situ studies of the high pressure melt structure are necessary. Some pioneering studies in ^{1}H and ^{13}C NMR spectroscopy of organic liquids at high pressure (to several GPa, at room temperature) have recently been carried out (Halvorson et al., 1992; Marzke, et al., 1994). However, these techniques are not yet easily applied to silicate systems, for reasons of resolution and sensitivity. Vibrational spectroscopy remains the technique of choice for in-situ studies of silicate glasses and melts at high pressure.

Next, the study of Fe-bearing glasses and melts by NMR methods is generally precluded by the extreme line broadening effects associated with the presence of paramagnetic species (unpaired electrons). The presence of these species does give rise to strong optical fluorescence, which can limit Raman scattering studies (Mysen et al., 1980b; Virgo and Mysen, 1985; Cooney, 1990; Wang et al., 1993), but infrared spectroscopy can be carried out with little difficulty to study the structures of these glasses and melts.

Finally, use of the NMR technique to distinguish structural sites relies on the observation of characteristic resonances for these sites, which are usually separated in frequency by several tens to thousands of kHz. If the structural environments are interchanged rapidly relative to this timescale through some structural averaging mechanism, the NMR spectrum collapses to a single Lorentzian line. This is the case in ^{29}Si, ^{27}Al and ^{17}O NMR studies of aluminosilicates at temperatures above the glass transformation range, and for nuclei such as ^{23}Na at much lower temperatures (Stebbins et al. 1992; this volume). Analysis of the NMR lineshapes of the sample at high temperature gives information on the kinetics of structural relaxation processes in the liquid, but unambiguous conclusions regarding the structural species present can no longer be drawn. It has been found useful to supplement the NMR studies of rapidly exchanging liquids with ion dynamics simulations, to model the structural species present in the melt (Poe et al., 1992a, 1993a,b) (see chapter by Poole et al., in this volume).

In order to examine the structural nature of the high temperature liquid directly, it is necessary to use a probe technique with a faster response timescale, so that different structural environments are not averaged on the timescale of the experiment. Vibrational spectroscopic techniques have already proven sensitive to small variations in local structure, from studies on silicate glasses, so that they do provide a useful probe of glass or liquid structure. The characteristic timescale for vibrational spectroscopies is determined by that of the light-matter interaction, on the order of 10^{-12} to 10^{-14} s (McMillan, 1984; McMillan and Hofmeister, 1988; McMillan et al., 1992). In particular, the Raman scattering event is essentially "instantaneous", with a timescale determined by the lifetime of the unstable "virtual state" which can be thought to exist as the incident light photon interacts with the sample, on the order of 10^{-14} s.

The principal vibrational techniques used to study silicate liquids include infrared absorption, reflection and emission spectroscopies, and Raman scattering spectroscopy. In infrared absorption, a beam of infrared light is passed through the sample, and the transmission is measured as a function of wavelength. Absorption peaks are observed, which correspond to excitation of the infrared active vibrational modes of the sample. In infrared reflection spectroscopy, the infrared beam is reflected from the sample surface, and the positions of resonances deduced from poles and zeroes in the reflectivity as a function of wavelength (Hadni, 1967; McMillan, 1985; McMillan and Hofmeister, 1988). During infrared emission, dielectric fluctuations due to excited vibrational modes in the heated sample give rise to a spectrum of emitted light, containing the characteristic infrared active vibrational frequencies.

In Raman scattering, molecular vibrations are studied by a light scattering experiment in which the energy of an incident light beam, usually provided by a laser operating in the visible region (400 to 750 nm), is slightly lowered or raised by inelastic interaction with the vibrational modes (McMillan, 1984, 1989; McMillan and Hofmeister, 1988). A related experiment, Brillouin scattering, is also used to study the propagation of sound waves in liquids and glasses (Grimsditch, 1984; Grimsditch et al., 1988; Suito et al., 1992; Wolf et al., 1992; Askarpour et al., 1993; Zha et al., 1994). Raman or Brillouin scattering which involves a shift in the incident light to longer wavelength, or lower energy, is termed *Stokes* scattering, and shifts to shorter wavelength are known as *anti-Stokes* scattering. Purely elastic light scattering, with no change in incident laser wavelength, is known as Rayleigh scattering. Diffuse inelastic scattering, due to some of the incident light energy being dissipated by structural relaxation in the melt, results in some broadening and change in shape of the Rayleigh scattered line, which can be analyzed to give information on these processes (Rothschild, 1987).

There are two principal obstacles to carrying out successful studies of high temperature silicate liquids via infrared or Raman spectroscopy. The first concerns carrying out the experiments, followed by extraction of the vibrational data from the high temperature signal. This has posed formidable technical challenges, but the problems are being diminished as new experimental techniques and methods for analyzing the spectra become available. The second obstacle is more serious, and concerns the reliable assignment of infrared or Raman bands to characteristic vibrations of specific structural units in the glass and melt. Considerable progress has been made in this area for SiO_2 and the binary silicate glasses (see recent reviews by Matson et al., 1983; Mysen et al., 1980a; McMillan, 1984a; Mysen, 1988, 1990a). However, questions still remain, and the detailed interpretation of aluminosilicate glass spectra in general in terms of their structure remains an active area of research. The current state of understanding for the interpretation of infrared and Raman spectra of silicate melts and glasses is discussed in the final section.

THEORETICAL BACKGROUND

Interaction with light: selection rules

To better appreciate the utility of vibrational spectroscopy as a probe of the structure and dynamic properties of silicate liquids, it is necessary to have a basic understanding of the processes which determine the vibrational spectrum, including changes in peak frequency and line width and shape as functions of increasing temperature.

Many basic treatments of molecular vibrations and vibrational spectroscopy remain within the harmonic model as an approximation to introduce the concepts and terminology. However, in high temperature studies such as those described here, it is necessary to go beyond the harmonic approximation, and consider the anharmonic processes which result in changes in the peak positions, absolute and relative intensities, and line shapes with increasing temperature.

The vibrational spectrum of a molecular system is determined by the nature of the atoms present, their relative arrangement, and the forces between them, so that vibrational spectroscopy is a useful technique for structural characterization. In order to do this, characteristic peaks must be assigned to the vibrational modes of structural units present. Although this can not yet be done reliably for most of the spectra of silicate and aluminosilicate glasses and liquids, considerable progress has been made, both in establishing empirical correlations between spectra and structure, and in developing a theoretical understanding of the vibrational modes. There is still uncertainty concerning certain some band assignments, even for quite simple compositions, with resulting ambiguity regarding their implications for the glass and liquid structure. Areas which are the concern of active research, or which deserve further attention, are summarized in the final section of this chapter.

Molecular vibrational frequencies lie in the range 10^{12} to 10^{14} s^{-1}, corresponding to that of light in the infrared region of the electromagnetic spectrum. If the vibration results in a change in the electric dipole moment, the vibrational mode is the seat of a fluctuating electric dipole which can interact with infrared light (Wilson et al., 1955; McMillan, 1985; McMillan and Hofmeister, 1988). If energy is transferred from the light to the vibrational mode of the sample, infrared absorption has occurred, and a series of minima at frequencies corresponding to those of the IR-active vibrations will be observed in a transmission spectrum of the sample.. In an infrared *emission* experiment, the dielectric fluctuations associated with the vibrations of the sample provide a source for electromagnetic radiation, and the sample *emits* a characteristic set of infrared frequencies. The emitted infrared radiation from a heated sample causes the sample to be perceived as "hot" by an observer. Infrared emission spectroscopy can provide a powerful method for studying the vibrational spectrum of a heated sample: however, the signal can be complicated by several overlapping optical effects, and the the spectrum can become surpisingly difficult to analyze (Gervais, 1983).

A quantum mechanical approach is usually taken to gain a more complete understanding of infrared absorption and emission processes (Wilson et al., 1955). If two vibrational states of a molecule are described by vibrational wave functions φ_n and φ_m, where n and m are the vibrational quantum numbers, the probability of a transition (n $\rightarrow$ m) occurring between these states via absorption or emission of infrared light, is proportional to the square of the *transition moment* (M_{mn}^2):

$$M_{mn} = \int_0^\infty \varphi_m^* \hat{\mu} \varphi_n dq \ .$$
(1)

Here q is a generalized vibrational displacement coordinate (i.e. the set of particular atomic displacements involved in the vibrational mode of interest), and φ_m^{*} is the complex conjugate of the wavefunction, which depends upon q. This expression is often written more compactly using the Dirac bracket notation as (McQuarrie, 1973)

$$M_{mn} = \left\langle \varphi_m \left| \hat{\mu} \right| \varphi_n \right\rangle \ .$$
(2)

In these expressions, the *operator* $\hat{\mu}$ is proportional to the classical dipole moment function (Wilson et al., 1955). By expanding $\mu(q)$ in a Taylor's series about the equilibrium geometry, it can be easily shown that M_{mn} is only non-zero when $(d\mu/dq) \neq 0$ (for m $\neq$ n); i.e. there must be a *change in dipole moment* for the vibration to be infrared active. In addition, if the vibrational motion is harmonic, M_{mn} is only non-zero when m = n ± 1, which restricts vibrational transitions to jumps between adjacent levels. These findings constitute the *selection rules* for the infrared absorption or emission process, and determine the appearance of the fundamental absorption or emission bands in an infrared spectrum. If the vibration is anharmonic, the second selection rule is relaxed, and *overtone* and *combination bands* appear in the spectrum, although with lower intensity than the fundamentals (Wilson et al., 1955; McMillan, 1985; McMillan and Hofmeister, 1988).

In Raman scattering, the incident light is in the visible region, and has an electric field which oscillates at a frequency v^*, much greater than the vibrational frequencies, v_i (Long, 1977; McMillan, 1984) Electrons are much lighter than the nuclei, and these respond much more rapidly to the light. The oscillating electric field thus causes a displacement of the electron density relative to the nuclear positions, resulting in an instantaneous *induced dipole moment*, $\bar{\mu}_{ind}$. The magnitude of $\bar{\mu}_{ind}$ depends upon the electric field strength $\bar{E}$ of the incident radiation:

$$\bar{\mu}_{ind} = \alpha\bar{E} \ . \tag{3}$$

The proportionality factor α is known as the molecular *polarizability*. Both $\bar{\mu}_{ind}$ and $\bar{E}$ are vectors, so that α is a second rank tensor. For an isotropic glass or melt, the diagonal terms of α ($\alpha_{xx} = \alpha_{yy} = \alpha_{zz}$) are all equal, as are the off-diagonal elements ($\alpha_{xy} = \alpha_{yz} = \alpha_{zx}$). Because the electric field of the incident light oscillates with frequency v^*, the induced dipole moment is time dependent:

$$\mu_{ind}(t) = \alpha E(t) = \alpha E_o \cos 2\pi v^* t \ . \tag{4}$$

However, the molecular polarizability depends upon the relative positions of the nuclei within the molecule, which oscillate with the vibrational frequency v_i. For small vibrational displacements q_i, α is expanded in a Taylor series about its equilibrium value (Wilson et al., 1955):

$$\alpha = \alpha_o + \left(\frac{d\alpha}{dq_i}\right)_o q_i + \ ... \tag{5}$$

The term in α_o is the static molecular polarizability. Because the vibrational displacements for a given mode, q_i, vary with the vibrational frequency v_i, this results in the following time dependency to α:

$$\alpha(t) = \alpha_o + \left(\frac{d\alpha}{dq_i}\right)_o q_i \cos 2\pi v_i t + ... \ . \tag{6}$$

The expression for the induced dipole moment then becomes:

$$\mu_{ind}(t) = \alpha_o E_o \cos 2\pi v^* t$$
$$+ \tfrac{1}{2}\left(\frac{d\alpha}{dq_i}\right)_o q_i E_o \cos 2\pi(v^* + v_i)t + \tfrac{1}{2}\left(\frac{d\alpha}{dq_i}\right)_o q_i E_o \cos 2\pi(v^* - v_i)t \tag{7}$$

This shows that two new terms appear in the time dependency of the induced dipole, which arise because of the dependence of α on q for the vibrating molecule. These terms result in re-emission of light from the molecule at the shifted frequencies $v^* - v_i$ and $v^* + v_i$, in addition to the incident light frequency v^*. The shifted frequencies correspond to Stokes and anti-Stokes Raman scattering, respectively, and the unshifted light is Rayleigh scattering.

Because of the tensor nature of the polarizability, it is usually necessary to specify the polarization state of the incident and the scattered radiation. For isotropic systems such as glasses and liquids, it is sufficient to distinguish cases where the incident and scattered beam polarizations are parallel or perpendicular to each other. These are often quoted with respect to the polarization state of the radiation exiting the laser, which is usually "vertical" in a typical laboratory orientation. A "vertical-vertical", or VV (equivalent to HH, or "horizontal-horizontal") polarization experiment selects the diagonal elements of the polarizability tensor, or the most symmetric vibrational modes. The perpendicular polarizations VH or HV select the off-diagonal tensor elements, or the least symmetric modes. The value of the depolarization ratio, ρ, defined as the ratio of the perpendicular to the parallel polarized scattered intensity (I_{VH}/I_{VV}) can be used to identify the symmetry type of the vibrational mode, which gives valuable information in structural studies. In addition, detailed measurements of Raman active mode polarizability can be used to investigate the contribution of different vibrational relaxation mechanisms to the scattered linewidth (Rothschild, 1986).

In quantum mechanical terms, the probability of a Raman scattering transition occurring between two vibrational states φ_n and φ_m, is proportional to the square of the magnitude of the transition moment, with

$$M_{mn}^2 = \left\langle \varphi_m \left| \hat{\mu}_{ind} \right| \varphi_n \right\rangle^2.$$ (8)

Expanding the expression for the induced dipole moment μ_{ind} gives the selection rules for Raman scattering. M_{mn} is non-zero when the vibrational mode is associated with a change in molecular polarizability ($d\alpha/dq_i \neq 0$). As in infrared absorption, if the vibration is harmonic, $m = n \pm 1$. This selection rule is relaxed (i.e., overtone and combination bands appear) if anharmonicity is considered. Vibrational excitations in condensed phases are often described in terms of *phonons*, or units of vibrational excitation (Wallace, 1972; Reissland, 1973). A fundamental absorption or inelastic scattering event is a *one-phonon process*. Raman scattering at twice the vibrational frequency due to interaction with the first overtone, allowed because of anharmonic effects, corresponds to *two-phonon* scattering. The fundamental vibration spectrum is also often known as the *first-order* Raman spectrum, and overtone and combination bands constitute the *second order spectrum* (McMillan, 1985; McMillan and Hofmeister, 1988). This becomes important in the analysis of high temperature Raman spectra of silicate liquids (Daniel et al., 1995a).

The vibrational spectroscopic processes can also be usefully described in terms of an energy transfer model, in which the the incident and scattered *light radiation* is considered to be quantized, in addition to the vibrational modes. The incident light photon has energy $E = h\nu*$, where $\nu*$ is the radiation frequency and h is Planck's constant (Loudon, 1983). In infrared absorption, the incident photon has an energy equal to that of the difference between the initial and final vibrational states, $h\nu = E_m - E_n$. During the absorption process, the incident light photon is destroyed, or annihilated, as the vibrational excitation (the phonon) is created. In infrared emission, a photon is created, with a corresponding decrease in the degree of vibrational excitation of the molecule. In a Rayleigh scattering process, one photon with energy $E_0 = h\nu*$ is destroyed, and another with the same energy is created, so that the net number of photons with the energy E_0 of the incident radiation remains the same. In Stokes Raman scattering, an incident photon with energy E_0 is destroyed, and another with energy $(E_0 - \Delta E) = h(\nu* - \nu_i)$ is created, where $\Delta E = h\nu_i$ corresponds to the energy difference between the initial and final vibrational states. A phonon with energy $h\nu_i$ is also formed. In anti-Stokes Raman scattering, the photon with energy E_0 is destroyed, along with the phonon, and a photon with energy $(E_o + \Delta E)$ is created. In second order Raman scattering, the scattering occurs via inelastic interaction with overtone or combination vibrations, for which $\Delta n = \pm 2$ (i.e., two phonons are created or destroyed), and the scattered photon has energy $h(\nu^* \pm 2\nu_i)$.

Frequency shifts with temperature

For harmonic vibrational models, the vibrational frequency is independent of temperature. This is observed *not* to be the case for anharmonic vibrations, which usually show a decrease in mode frequency as the temperature is increased. There are some interesting exceptions to this general behavior in the case of silicate glasses and liquids. For example, the behavior of SiO_2 glass and liquid is discussed below (McMillan et al., 1994). The normal frequency decrease of a vibrational mode can be understood to a first approximation within the *quasi-harmonic* model, in which the effect of temperature enters through the increased volume due to thermal expansion (Wallace, 1972). The mode Grüneisen parameter is defined by

$$\gamma_i = -\frac{\partial \ln \nu_i}{\partial \ln V} = -\frac{V}{\nu_i}\frac{\partial \nu_i}{\partial V} \ . \tag{9}$$

Because the volume at any temperature is defined by the thermal expansion coefficient,

$$\alpha = \frac{1}{V}\left(\frac{\partial V}{\partial T}\right)_P = \left(\frac{\partial \ln V}{\partial T}\right)_P \ , \tag{10}$$

the thermal mode Grüneisen parameter can be written

$$\gamma_{i,T} = -\frac{1}{\alpha}\left(\frac{\partial \ln \nu_i}{\partial T}\right)_P = -\frac{1}{\alpha \nu_i}\left(\frac{\partial \nu_i}{\partial T}\right)_P \ . \tag{11}$$

Vibrational frequency shifts as a function of temperature $(\partial \nu_i / \partial T)_P$ for vibrational modes of silicate crystals, glasses and melts are commonly found to lie in the range 0.01 - 0.1 cm^{-1}/K (McMillan et al., 1992; Mysen and Frantz, 1992, 1993a,b, 1994a,b; Daniel et al., 1995a).

The quasi-harmonic theory does not predict any effect of the temperature on the vibrational linewidth. In addition, vibrational frequency shifts with temperature are often larger than expected from a consideration of thermal expansion alone. To understand this, it is necessary to develop a more sophisticated theory, based on anharmonic interactions between vibrational modes, or phonon-phonon (ϕ-ϕ) interactions (Wallace, 1972; Reissland, 1973; Gervais, 1983). In this approach, the frequency shifts, Δ_i, are given by

$$\Delta_i = \Delta_i^{QH} + \Delta_i^{\phi-\phi} \ . \tag{12}$$

The first contribution to the shift is the quasi-harmonic term, and the second arises from phonon-phonon interaction terms, which are obtained by applying perturbation theory to first- and second-order (Wallace, 1972; Reissland, 1973). Both terms result in linear frequency shifts with increasing temperature, to a very good first approximation, so that any non-linearity in frequency shifts with temperature reveal additional anharmonic effects upon the vibrational frequencies. The anharmonic interactions also result, to a first approximation, in a linear increase in the vibrational linewidth ($2\Gamma_i$ = full width at half maximum: FWHM), so that marked deviations from linearity also reveal an additional relaxation mechanism for the vibrational energy (Gervais, 1983; Gillet et al., 1993a).

Infrared and Raman intensities

It is also important to consider the likely effects of temperature on infrared absorption and Raman scattered intensities, for correct interpretation of the high temperature data. Within the harmonic model, the infrared absorption coefficient for each vibration is given by

$$\kappa = \frac{N\pi}{3c}\left(\frac{d\mu}{dq_i}\right)_o^2 .$$ (13)

q_i is the general vibrational displacement coordinate, weighted by the atomic masses of the atoms participating in the vibration. This expression has no explicit temperature dependence, so that, to first order, we expect infrared absorption intensities measured at the peak maximum to be independent of temperature. This is not observed to be the case. With increasing temperature, anharmonic effects broaden the absorption peak and shift the position of the maximum, usually to lower wavenumber (Gervais, 1983; McMillan, 1984). The same anharmonic processes result in a decrease in reflectivity in infrared reflection studies (Hadni, 1967; Gervais, 1983). This is clearly seen in high temperature reflectivity studies of silicate liquids (Markin and Sobolev, 1960; Sweet and White, 1969). Quantitative studies have indicated, however, that the integrated area under a given absorption band is relatively unaffected by temperature, over a wide temperature range (Sherwood, 1972; Paterson, 1982).

The case for Raman scattering is quite different, in that the peak intensities depend intrinsically upon the temperature, within a harmonic model. The intensity of Raman scattering (I^{Stokes} or $I^{anti-Stokes}$) associated with a given vibration relative to the incident laser intensity I_0, from a collection of N molecules, is described by the *differential cross-section*, $d\sigma_i/d\Omega$, for Raman scattering into a solid angle $d\Omega$. Within the harmonic model, this is given by the following expressions:

Stokes scattering:
$$\frac{d\sigma_i}{d\Omega} = \frac{I^{Stokes}}{NI_o} = \frac{(v*-v_i)^4}{2\varepsilon_o^2 c^4 v_i(1-e^{-\frac{hv_i}{kT}})}\left(\frac{d\alpha}{dq_i}\right)^2$$ (14)

anti - Stokes:
$$\frac{d\sigma_i}{d\Omega} = \frac{I^{anti-Stokes}}{NI_o} = \frac{(v*+v_i)^4}{2\varepsilon_o^2 c^4 v_i(e^{\frac{hv_i}{kT}}-1)}\left(\frac{d\alpha}{dq_i}\right)^2 .$$ (15)

ε_0 is the permittivity of free space. These are the expressions for the temperature- and frequency-dependence of the fundamental Raman lines, which are the only modes allowed within the harmonic model. It is obvious that both the absolute and relative intensities of Raman active modes within a given spectrum depend upon the temperature, and upon the frequency of the vibrational mode (through the terms in $(v*\pm v_i)^4/v_i$), in addition to their intrinsic Raman activity ($d\alpha/dq_i$). These frequency- and temperature-dependent terms can severely distort the spectrum, and must obviously be removed if relative band intensity changes are to be followed as a function of temperature in silicate glass or melt studies. This is usually done by deriving a temperature- and frequency-dependent correction factor from expressions (14) and (15), and applying it to the measured Raman intensity (I_{obs}) observed at temperature T. For the Stokes spectrum, the corrected Raman intensity is given by

$$I^{Stokes}_{corr} = I^{Stokes}_{obs}\frac{v(1-e^{-\frac{hv}{kT}})}{(v*-v)^4} ,$$ (16)

and for the anti-Stokes spectrum,

$$I_{corr}^{anti-Stokes} = I_{obs}^{anti-Stokes} \frac{v(e^{\frac{hv}{kT}} - 1)}{(v*+v)^4} \ . \tag{17}$$

These expressions can also be written in terms of the *mean vibrational occupation number*, or the *mean number of phonons*, defined by

$$\bar{n} = \frac{1}{e^{hv/kT} - 1} \ . \tag{18}$$

It is not usual to apply this correction to the measured Raman intensities in crystalline spectra. In some studies of the Raman spectra of silicate glasses at ambient conditions, corrected spectra have been reported, although this is not usually done in silicate glass studies (e.g. Mysen et al., 1980a, 1982a,b, 1983, 1985; McMillan et al., 1982, 1984b; Matson et al., 1983; Mysen, 1988). In uncorrected spectra, the intensity below approximately 500 to 700 cm^{-1} is artificially enhanced, and there is a maximum in the 50 to 100 cm^{-1} region which does not correspond to a vibrational band (Fig. 3). This maximum results from two opposing effects in the frequency- and temperature-

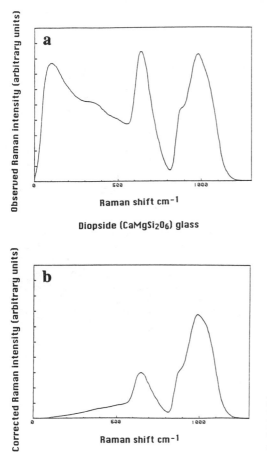

Diopside (CaMgSi$_2$O$_6$) glass

Figure 3. The effect of frequency- and temperature-correction on the observed Raman intensities in the unpolarized room temperature spectrum of CaMgSi$_2$O$_6$ (diopside) glass, using the expression for first-order Raman scattering (Eqn. 16).

dependence of the Raman spectra of glasses. At low frequencies, the glass shows nearly Debye-like behavior, and the frequency spectrum rises smoothly from zero wavenumber, before structure due to the vibrational modes at higher frequency sets in. However, the temperature dependence gives rise to a multiplying factor which is decreasing exponentially at small wavenumber. The two combine to give an apparent maximum in the Raman spectrum, which disappears if the sample is cooled well below room temperature, or if the correction procedure is carried out (Piriou and Alain, 1979).

This low frequency maximum should not be confused with a Raman active vibrational mode of the glass. However, true vibrational modes can and do occur in this region: for example, vitreous SiO_2 has a weak broad band near 90 cm^{-1}, which is only revealed when the intensity correction has been carried out (McMillan et al., 1984). The "spurious" low frequency "band" is distinct from the "*boson peak*", which is currently a topic of intense interest in glass science. This feature has been observed at very low frequency (2 to 15 cm^{-1}) in the inelastic neutron spectra of glassy polymers, and corresponds to vibrational excitations not present for the corresponding crystals (Frick and Richter, 1995). The boson peak is thought to play a role in determining the properties of the system at the glass transition, and perhaps the liquid as well (Angell, 1995). Future vibrational studies of silicate glasses might well focus on the very low frequency region, to study and understand the possible occurrence and potential significance of a boson peak in these systems also.

Because of the obvious importance for realistic comparison of relative band intensities at different temperatures, most of the high temperature Raman spectra of silicate glasses and melts which appear in the literature have been corrected for frequency- and temperature-dependence by an expression like (16) or (17). However, it has recently been realized that this correction is not sufficient to correctly account for the temperature dependence of the Raman scattered intensity (McMillan et al., 1994; Poe et al., 1994; Daniel et al., 1995a,b). The correction factors described above are only appropriate for *first order* Raman scattering; i.e. for the fundamental vibrational modes with $\Delta n = \pm 1$, within the harmonic approximation. The frequency and temperature dependence of *overtone* transitions is different. The overtones from vibrations below 600 cm^{-1} occur throughout the region of the fundamental Raman spectrum of silicate glasses and melts, 0 to 1200 cm^{-1}. For temperatures above a few hundred °C, there is already a non-negligible contribution from such overtone vibrations in the spectrum (McMillan et al., 1994). These appear most obviously as an enhanced intensity in the 600 to 900 cm^{-1} region, in a part of the spectrum which is usually free from strong Raman bands in silicate glass systems. This intensity enhancement can be misleading in silicate melt studies, because this is a spectral region in which vibrational modes characteristic of silicate and aluminate species in high coordination might be expected to appear (Williams and Jeanloz, 1988, 1989; Poe et al., 1994; Daniel et al., 1995b). The enhanced intensity in this region can be identified as a vibrational, rather than a configurational, effect, because it occurs well below the glass transition temperature, and for a system (SiO_2) not expected to undergo any dramatic coordination changes under these conditions (McMillan et al., 1994). For very high temperature studies, above approximately 1200° to 1400°C, the second order (i.e. from overtone vibrations) Raman scattering strongly influences the spectrum, and it must be removed before the first order (fundamental) Raman spectrum can be examined for changes due to structural relaxation. However, because of the different frequency- and temperature-dependence of the second order scattering, application of the first order correction functions, (16) or (17), does not entirely remove the intensity from the second order effects from the spectrum.

For example, the intensity of Stokes second order Raman scattering from overtone vibrations with $\Delta n = 2$ is proportional to

$$I^{Stokes(2)} \propto \frac{(v^* - 2v_i)^4 \bar{n}(\bar{n}+1)}{v_i^2}\left(\frac{d\alpha}{dq_i}\right)^2 \tag{19}$$

(Ganguly and Birman, 1971), so that the appropriate frequency- and temperature-correction factor for the second order scattering becomes of the form

$$I_{corr}^{Stokes(2)} = I_{obs}^{Stokes(2)} \frac{v^2(e^{\frac{hv}{kT}} + e^{-\frac{hv}{kT}})}{(v^* - 2v)^4} . \tag{20}$$

Daniel et al. (1995a) have described a method whereby the second order Raman spectrum in the 0 - 1200 cm^{-1} region can be extracted from a series of high temperature spectra, by first carrying out the "normal" correction via equation (16), then subtracting a reference spectrum from the high temperature spectra and multiplying by a factor derived from the combination of equations (16) and (20). This was carried out for $NaAlSi_3O_8$ and $CaAl_2Si_2O_8$ glasses and liquids, and the form of the second order spectrum remained nearly independent of temperature below T_g, as expected (Daniel et al., 1995a). The second order intensity at temperature T is then reconstructed, using equation (19), and subtracted from the high temperature data. This results in a pure first order Raman spectrum at high temperature, from which configurational effects can be identified and interpreted (Daniel et al., 1995a,b). These high temperature aluminosilicate glass and melt spectra are described in detail in the last section of this Chapter.

Linewidths and lineshapes

The intrinsic width and shape of the resonance observed in infrared absorption or emission for a single molecule is related to the net transition rate (τ) between levels n and m, which can be determined via time-dependent perturbation theory (Pauling and Wilson, 1963; McQuarrie, 1973; Loudon, 1983). The transition probability per unit time, ($\Gamma = 1/\tau$), is given by the lifetime of the excited state with quantum number m, for a given vibrational mode:

$$\Gamma_i \approx \frac{8\pi^2 v_i^2}{3c^2} m\left(\frac{d\mu}{dq_i}\right)_o^2 \tag{21}$$

within the harmonic approximation (Decius and Hexter, 1977). The resulting lineshape is a Lorentzian function, with a width at half height of $2\Gamma_i/c$ wavenumbers (c is the speed of light in cm/s), in the absence of any other mechanism which affects the vibrational relaxation rate. It can be observed that strong infrared absorption or emission lines, for which the dipole moment change (dμ/dr) is large, are also intrinsically short-lived (i.e. $\tau = 1/\Gamma$ is small), and these have a broad linewidth. This is also true for vibrations which occur at high frequency (v): these have a shorter intrinsic lifetime of the excited state than do low frequency vibrations.

Any anharmonicity associated with the vibration results in faster relaxation of the excited vibrational state, due to leakage of the vibrational energy into other coupled atomic motions, which broadens the observed line (Decius and Hexter, 1977). Additional line broadening also occurs in this case, because anharmonicity in the potential energy function causes different transitions between adjacent vibrational levels to have slightly different frequencies, so that the observed absorption or emission peak tends to spread out (Wilson et al., 1955; McMillan, 1984). A similar general result is found for Raman scattering, although the mathematical treatment is more complicated because the interaction involves two photons as well as the molecular vibration, and the time-dependent perturbation must be carried to second order (McQuarrie, 1976; Loudon, 1983).

The peaks observed in infrared absorption and Raman scattering spectra for gas phase molecules and crystalline solids are generally much sharper than the broad bands measured for glasses and liquids (infrared *reflection* bands for condensed phases are very broad, due to the interaction of light with long range Coulomb fields inside the sample). This broadening is a combination of dynamical and structural effects, termed *homogeneous* and *inhomogeneous* broadening effects. To understand the origin of these, it is useful to consider the general changes in the vibrational spectrum as a collection of gas phase molecules is condensed into a liquid, then frozen into a crystal or a glass.

The homogeneous linewidth is determined by the intrinsic lifetime of the vibrational mode, which usually results in a Lorentian lineshape, as described above. For a collection of molecules in the gas phase phase, vibrational relaxation and line broadening occur via Doppler and collision broadening (Steele, 1971; Decius and Hexter, 1977; Person, 1982). The Doppler effect occurs because the molecules are in motion with speed u, so the actual frequency absorbed or emitted is reduced by a quantity u/c. The resulting absorbed or emitted intensity has a Gaussian profile, with a half width determined by the vibrational frequency, the molecular weight, and the sample temperature. *Collision* broadening arises because collisions between gas phase molecules provide an efficient mechanism for transferring vibrational energy into translational and rotational degrees of freedom. This is determined by the gas density, in addition to molecular size, weight and shape.

As the gas density is increased into the fluid phase, the effects of collision broadening become more marked. This results in an increase in the homogeneous linewidth of the vibrational bands, due to more efficient relaxation of excited vibrational states. Any rotational structure present in the gas phase spectrum will also collapse, due to broadening of individual rotational lines, contributing to the overall linewidth. In addition, rotational degrees of freedom begin to couple more strongly with the vibrations, providing a further mechanism for efficient vibrational relaxation, and resulting in additional homogeneous line broadening.

As the sample begins to behave as a dense fluid, translations and rotations of individual molecules are no longer free but begin to be hindered by interactions with neighbouring molecules. Far-infrared absorption or low frequency Raman (Brillouin) scattering can begin to be observed due to inelastic interaction of light with these hindered translations and rotations. The vibrational peaks become further broadened as vibrational excitation energy is lost into these relaxational degrees of freedom, further reducing the lieftime of the excited vibrational state. In addition, density fluctuations begin to give rise to quasi-elastic low frequency Raman scattering, observed as an additional tail on the Rayleigh line.

In this regime, the symmetries of molecular groups can be perturbed by interaction with their neighbours, and the infrared and Raman activities of vibrational modes can be altered. The anharmonicities of individual vibrational modes can change also. Chemical exchange or dissociation reactions may occur, or associated complexes between molecular groups may begin to appear, resulting in the appearance of new features in the vibrational spectra.

In the liquid state, individual vibrating units experience a range of structural and chemical environments, which are no longer averaged on the timescale of the vibrational experiment. These result in a range of geometries, force fields, and effective masses, so that the vibrational lines are broadened due to such "static" structural effects. These effects result in inhomogeneous broadening, which is often associated with a Gaussian profile for the observed lineshape. In addition, vibrational relaxation processes due to anharmonicity, interaction with hindered translations or rotations, or chemical exchange processes, all still occur on the timescale of 10^{-14} to 10^{-9} s. These efficiently relax excited vibrational states on a timescale which approaches that of the vibrational mode itself, so that linewidths and lineshapes can be determined by a combination of inhomogeneous (structural distribution) and homogeneous (vibrational lifetime) contributions. Studying the relative importance of these two contributions will provide valuable information on relaxation processes in silicate melts, in future vibrational spectroscopic studies at high temperature.

In the solid state, a wide range of behavior is encountered. In condensed phases, the rotational and translation degrees of freedom of gas phase molecules, which give rise to much of the structural relaxation in liquids, become true vibrational modes known as *librations*. In crystals, the small range in local geometries and chemical arrangements gives rise to much narrower lines, and broadening is mainly due to anharmonic effects. Because of the long range translational symmetry, vibrational coherence between adjacent vibrating units is observed. This results in the phonon dispersion relations observed for crystalline materials (Bilz and Kress, 1979), and limits those modes observable in infrared and Raman spectra to ones with long wavelength (the $k = 0$ "selection rule") (Decius and Hexter, 1977). In amorphous solids, which can be considered in a first approximation as "frozen" liquids, inhomogeneous linewidths are largely determined by the distribution in local geometries and speciation.

Two principal approaches have been applied in interpreting the structures and vibrational spectra of aluminosilicate glasses and liquids. In the first, the glass is considered to represent a highly disordered crystalline matrix. The inhomogeneous broadening of vibrational bands is considered to result from progressive destruction of the long range crystalline symmetry, so that the entire phonon dispersion relations begin to be observed in the infrared and Raman spectra. This is discussed in detail later for vitreous SiO_2. This approach is probably best suited to those compositions which show little chemical heterogeneity. In the second view, individual structural units within the glass, such as SiO_4 tetrahedra or SiOSi linkages, are considered as "molecules" perturbed by their surroundings, with differing degrees of distortion and coupling between adjacent units. This approach is better adapted to glasses which contain a wide range of structural units, such as silicate glasses of intermediate polymerization. Both interpretations give rise to distributions in local molecular geometry, interatomic force constants, and degree of vibrational coupling between adjacent units, usually resulting in Gaussian shapes for the vibrational bands observed in infrared absorption and Raman scattering. In high temperature silicate liquids, the structural relaxation times do not in general begin to approach the vibrational timescale over the temperature range investigated to date (below 2000 K), so that lineshapes in the liquids remain Gaussian, with a width determined by the range in local structural environments. This may not be true for all vibrations of silicate systems, and the point at which the lineshape and width become determined by dynamic processes is an interesting one to be explored in future studies.

EXPERIMENTAL VIBRATIONAL SPECTROSCOPY
AT HIGH TEMPERATURES

Infrared reflection and emission studies

Until recently, high temperature infrared spectroscopy has been an easier experiment to perform than Raman scattering, and several in situ studies of silicate glasses and liquids at temperatures up to 2000°C have appeared in the literature since the mid 1950s (Grove and Jellyman, 1955; Jellyman and Proctor, 1955; Markin and Sobolev, 1960; Gaskell, 1966a; Sweet and White, 1969; Domine and Piriou, 1983) (Fig. 2). Most of these measurements have been made via infrared reflectivity from the liquid surface, due to the difficulties associated with preparing thin films for absorption studies (powder transmission studies are inappropriate, for obvious problems of reaction between the melt and the mounting medium). In these high temperature IR studies, many types of furnace arrangement have been used to heat and contain the sample.

At high temperature, the sample itself becomes a source of infrared radiation via blackbody emission, and it is important to discriminate against this signal when making measurements (Gervais, 1983). Early infrared studies were carried out using scanning instruments, in which light from the spectrometer source was reflected from the sample, and then separated into its constituent wavelength components via a prism or grating spectrometer system. The blackbody emission from the sample constitutes a continuous radiation source incident on the detector, resulting in a constant background signal voltage (Hadni, 1967). This background can be subtracted if the light incident upon the sample from the spectrometer source is chopped before reaching the sample, resulting in a time-dependent signal. The light reaching the detector when the chopper is closed contains only the blackbody emission of the sample, whereas that obtained with the chopper open also contains the infrared reflection component, so that the pure reflectance spectrum can be obtained by subtraction. This technique was employed in most early IR studies of melts at high temperature.

Modern infrared studies use Fourier transform interferometric (FTIR) techniques. In this experiment, the sample is irradiated with light from the source which covers the entire infrared (and into the visible, depending on the source) region of the electromagnetic spectrum. Before reaching the sample, the light is passed through a Michelson interferometer (see McMillan and Hofmeister, 1988; Ihinger et al., 1994, for a description of the FTIR technique applied to geochemical studies). The beam is split into two components: one which is reflected from a moving mirror and a second, reference, beam which is reflected from a static mirror. The two beams are recombined at the beamsplitter, resulting in a time-dependent interference pattern of infrared light, $I_0(t)$. This interferogram represents the $I(t)$ pattern delivered by the source, over all infrared wavelengths, and is sent to the sample as an incident beam. The incident interferogram serves as a reference for the experiment, and is measured and stored for treatment of the data.

When the incident beam $I_0(t)$ is passed through the sample (or reflected from its surface), it is modified by absorption or reflection of specific infrared wavelengths, by their interaction with the vibrational modes of the sample. This results in a different interferogram, $I_{sample}(t)$. The interferograms from the sample and reference are ratioed, to obtain the interferogram containing the information on the IR-active vibrations. This is then transformed from the time domain into the frequency (v) domain using fast Fourier transform techniques, to obtain the infrared spectrum $I(v)$ of the sample.

The FTIR method is much faster than previous scanning techniques, and results in a much higher signal to noise ratio obtained in much shorter time. This permits examination of much smaller samples. For this reason, micro-beam FTIR instruments have become popular in mineralogical studies (McMillan and Hofmeister, 1988). The FTIR method also results in absolute frequency determinations, because the interferometer is constantly internally calibrated against a reference laser beam, which is used to track and control the moving mirror velocity. Because of these and other desireable features, FTIR spectroscopy has essentially supplanted scanning spectrometry as the technique of choice for infrared studies.

High temperature infrared studies using FTIR instruments do require special care in experiment design, especially if quantitative reflectance spectroscopy is to be carried out. The primary question concerns the blackbody emission from the sample at high temperature. In the case of scanning experiments, this signal can be subtracted by use of a chopped incident beam, as described above. However, in FTIR experiments, one part of the sample emission is sent back along the incident beam optical path, passing through the beamsplitter, and on to the moving mirror. At this point, this generates a time-dependent contribution to the "incident" intensity which is derived from the sample itself, not the infrared source (Gervais, 1983). The combined "incident" beams are then sent back through the sample, to the detector. The problem arises because the heated sample does not emit the same spectral signature as the IR source, which is usually a smooth function of emitted intensity versus wavelength in the infrared region. The emission from the sample contains instead all of the vibrational modes which are to be studied in the experiment. For this reason, the infrared spectrum obtained from the ratioed sample and reference interferograms has an "extra" intensity contribution within the vibrational peaks, due to the thermal emission from the sample (Gervais, 1983). This is a particular problem for IR reflectance measurements, for which it is important to have correct intensity information for analysis of the spectra to obtain the optical constants (Hadni, 1967; Gervais, 1983). The "extra" intensity contribution from sample emission can be eliminated physically in some spectrometer designs, by masking the return beam from the sample in the beamsplitter compartment (Gervais, 1983). This is not always possible, however, and the contribution can simply be measured independently (by obtaining an emission spectrum with the source masked), and subtracted.

In infrared reflectivity studies, the reflectance spectrum must be analyzed to obtain information on the optical constants as a function of wavelength or frequency (Hadni, 1967; Decius and Hexter, 1977; Gervais, 1983; McMillan, 1984; McMillan and Hofmeister, 1988). These constants include the real and imaginary parts of the refractive index (n and k, or n' and n"), and the frequency-dependent complex dielectric constant (ε' and ε'') (Fig. 2). These are conveniently obtained from the reflectivity spectrum by a Kramers-Kronig analysis, which is an application of linear response theory, or an empirical classical oscillator fit (see Hadni, 1967; Gervais, 1983; McMillan, 1985; McMillan and Hofmeister, 1988). Each method has its advantages and drawbacks, and both are usually combined to extract the best set of optical constants for a given material. The infrared reflectance spectrum contains much more information than an IR absorbance or emission spectrum. The low frequency edge of a reflectance band corresponds to the frequency of the "transverse" (TO) optic vibrational mode observed in absorption studies, which gives rise to a pole in the reflectivity function (McMillan, 1985; McMillan and Hofmeister, 1988). The high frequency zero in the reflectivity band occurs at the frequency of the "longitudinal (LO) optic" mode, in which the atomic displacements lie parallel with the propagation direction of the incident beam. This mode can not usually be observed in transmission studies. The imaginary part of the dielectric constant [$\varepsilon''(\nu)$ or $\varepsilon_2(\nu)$] constitutes a dielectric loss function, which reflects the amount of infrared radiation energy lost to the sample through anharmonic processes. The magnitude of this quantity reflects the degree of anharmonicity of the vibrational mode, which is an important parameter in high temperature studies. A high mode anharmonicity results in increased absorption intensity, a decreased reflectivity, and broadening of the absorption band (Markin and Sobolev, 1960; Gervais, 1983; Domine and Piriou, 1983).

Raman scattering

Before the mid-1980s, nearly all Raman instruments in common use were scanning spectrometers. High temperature Raman spectroscopy using these instruments was extremely challenging, in that long scan times were required to obtain useful signal to noise ratios at adequate resolution. This required a high temperature stability of the furnace assembly, and also that the sample composition did not change during the run due to volatility of oxide components. Raman spectroscopy requires sensitive detection of the extremely weak Raman scattered signal, with efficient discrimination against extraneous background radiation. This can occur due to elastic scattering of the incident laser light from sample imperfections (bubbles, fractures, and density fluctuations) and fluorescence from even trace quantities of some impurities, especially iron. For these reasons, Raman spectroscopic studies of geological materials have always posed experimental difficulties, even at ambient conditions. In the high temperature experiments, an additional limiting factor was provided by blackbody emission from the furnace assembly, and from the sample itself. Any heated substance emits thermal radiation, with intensity proportional to the fourth power of the temperature, at any given wavelength. The wavelength (frequency) dependence of the emitted intensity (expressed as the emitted power per unit surface area) is given to a first approximation by Planck's law:

$$I^{blackbody}(\nu, T) = \frac{2\pi \nu^3 h}{c^2} \frac{1}{e^{h\nu/kT} - 1} . \tag{22}$$

For real materials, this expression must be multiplied by a frequency-dependent emission coefficient, $\lambda(\nu, T)$. For temperatures above ~1000 K, the thermal emission spectrum begins to extend into the visible region of the spectrum, and can interfere with the quality of the weak Raman signal. This results in increased background intensity, and even longer scan times are required to improve the signal to noise ratio. For these reasons, only a few experiments were carried out on molten silicates using scanning spectroscopy, usually on simple binary silicate compositions with quite high Raman scattering intensities (Sharma et al., 1978 Piriou and Arashi, 1980; Seifert et al., 1981; Kashio et al., 1982, 1984; Iwamoto et al., 1984).

With the advent of computer control of the experiment and digital data recording, it became possible to signal average, and post-collection data treatment requiring digitization of the spectra became simpler. However, the field of in situ high temperature (and high pressure) Raman spectroscopy was revolutionized by the appearance of multichannel spectrograph instruments, coupled with sensitive diode array or CCD detectors (Hemley et al., 1987; McMillan and Hofmeister, 1988; McMillan et al., 1992; Mysen et al., 1992a). These instruments permit recording the entire spectrum, or a substantial portion of it, in a single detection event, which may only take a fraction of a second. With a diode array, averaging many scans is possible within a few minutes at each temperature point, and the detector sensitivity is high, so that excellent signal to noise ratios can be obtained at extremely high temperatures, even for weakly scattering liquids. With a CCD system, the dynamic range is very large, so high quality spectra are often best obtained in a single acquisition, with a long acquisition time (several tens to hundreds of seconds).

A typical Raman system used for modern high temperature work is a triple monochromator system, incorporating an initial double monochromator stage, which is used to select the spectral range of interest and discriminate against the incident laser line, and a final grating in the spectrograph stage, which disperses the light on to the detector (Hemley et al., 1987). In a diode array detector, the physical size of each diode determines the resolution of the final spectrum, along with the dispersion characteristics of the final spectrometer stage. The dispersed light from the final grating is distributed over the linear array, and the signal intensity read out as a function of wavelength to constitute the spectrum. The signal reaching a diode array detectors is often intensified by placing a fluorescent screen before the detector, which can improve the signal to noise by over an order of magnitude. The charge coupled device (CCD) is also constituted from an array of diodes, but the signal is read out sequentially by shifting the read voltage sequentially across the register. The detector array in a CCD usually forms a two dimensional matrix of pixels, which can be used to construct a Raman image of the sample, and can be extremely useful in spatial filtering extraneous signal, such as the blackbody radiation emitted by the high temperature furnace.

A further improvement in modern Raman spectroscopic techniques for high temperature spectroscopy is provided by micro-Raman instrumentation, coupled with spatial filtering of the recorded signal. The bulk of the extraneous background intensity in high temperature experiments results from blackbody emission from the furnace assembly and the sample surface. With the micro-Raman technique, it is possible to focus inside the high temperature liquid sample. Placing an aperture at an intermediate focus plane before the spectrometer entrance permits only the light from the focus volume (the effective "*Raman scattering cylinder*", with dimensions determined by the optical characteristics of the focusing and detection system) to enter the spectrometer. The spatial filtering can be adjusted to control both the aerial and depth resolution, and the resulting signal to background intensity ratio sampled. The sensitive detection system ensures that the weakest signals can be detected, even when the spatial filter is closed to discriminate against the maximum extraneous background radiation.

These techniques have been combined with the "wire loop" heating method, to easily obtain high quality Raman spectra at high temperatures (up to ~2000 K), with good temperature control, stability and homogeneity over the region sampled. This technique was adapted from a method developed for in situ high temperature optical microscopy and X-ray diffraction (Ohashi and Hadidiacos, 1976), and was applied to micro-Raman studies of silicate melts by Mysen and Frantz (1992; 1993a,b; 1994a,b). The earliest version of the technique involved forming a thermocouple by welding Pt and $Pt_{90}Rh_{10}$ wires, flattening a region near the junction, and drilling a hole (several hundred μm in diameter) through the flattened area at the junction. The powdered sample was placed in this hole, and melted by applying a heating current to the Pt-Rh wire (Mysen and Frantz, 1992). The molten sample is retained in the hole by surface tension. Micro-Raman spectra are then obtained by focusing the laser into the melt (Fig. 4). In this first series of experiments, the temperature was controlled and measured during a separate electrical cycle, using the same drilled thermocouple which served as the sample holder. Excellent thermal stability has now been demonstrated with this technique, with temperatures remaining steady over many hours, to even days, in some experiments.

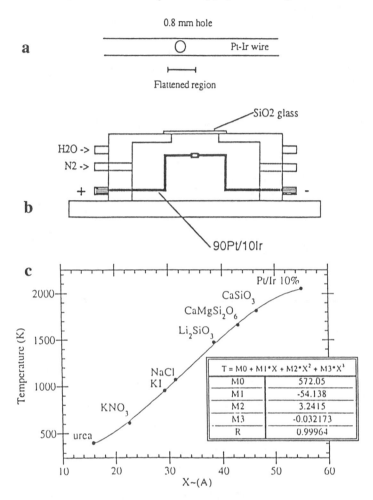

Figure 4. Schematic of the Pt-Ir drilled wire heater used in the high T Raman experiments of McMillan et al. (1994) (a) The wire was mounted in a water-cooled stage body for high temperature microscopy, and the chamber was closed with a silica glass window to avoid damage to the objective, and to retain the inert gas atmosphere (b). In that work, a linear correlation was found between known melting point samples, and the voltage measured across the wire. In the high temperature study of Daniel et al. (1995a), a larger number of melting point standards was investigated, and a third-order polynomial was fit to the T(X) curve, where X was the current through the wire (c).

In a variant of this technique, a drilled thermocouple is not used, but simply a flattened region of refractory wire (Pt, Pt-Ir, Ir, or W) (Richet et al., 1993). In current applications of the wire loop technique, a temperature calibration is established from known melting point samples, using the current through the wire (I), the voltage across the sample area (V), or the power delivered to the sample as the calibration variable (Fig. 4) (Mysen and Frantz, 1993a,b, 1994a,b; McMillan et al., 1994; Poe et al., 1994; Daniel et al., 1995a,b,c). Although it might be expected that the sample temperature should increase linearly with the applied power ($I^2R = IV$), this is not the case experimentally, due to heat losses by radiation and conduction, and changes in the electrical resistance (R) of the wire with temperature. In practice, a good linear correlation is found between I and T, or V and T, over most of the temperature range (Fig. 4). This breaks down at higher temperatures, close to the melting point of the wire, as might be expected. However, temperature calibration curves can be established using polynomial fits (Fig. 4). The results indicate that the temperature is known to within a few °C. It is found that wires with similar dimensions have similar calibration curves, so that a full calibration is not necessary for each new wire. In addition, the calibration remains stable over several heating cycles, so that the same wire can be re-used for successive experiments. In this case, it is important to re-calibrate the wire periodically during the series of runs. Finally, in the case of Ir- (and W-) containing wires, it is important to run the experiment in an inert atmosphere, because the oxides of these metals are volatile, and black IrO_2 becomes deposited on the (expensive) microscope objective in a remarkably short time!

In these high temperature Raman experiments, it has become important to closely examine the procedures for extracting the Raman spectrum from the high temperature data. First, the blackbody radiation emitted by the sample within the scattering cylinder (which can not be eliminated by spatial filtering) must be subtracted. This is simply achieved by running a spectrum with the laser off, to obtain the blackbody background in that spectral region (Figs. 5 and 6). This procedure is much more satisfactory than trying to fit a theoretical background, assuming some form for the black- or grey-body spectrum. We have observed in some high temperature experiments that a thermal background remains after this procedure (McMillan et al., 1994; Daniel et al., 1995a,b) (Fig. 6). This is most likely due to an additional sample heating (by ~20°C) through absorption of the incident laser beam by structural defects (Si-O⁻, Al-O-O⁻ species, etc.) in the high temperature liquid (McMillan et al., 1994; Daniel et al., 1995a,b).

Following the background correction, the spectrum must be normalized for the intrinsic temperature- and wavelength-dependence of the Raman scattered intensity (Figs. 5 and 6). This was discussed in the previous section. As noted there, at sufficiently high temperatures (above ~1500 K), the intensity of *second-order* Raman scattering becomes

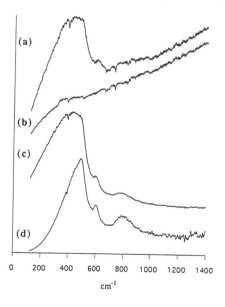

Figure 5. Treatment of high temperature Raman data in the study of SiO_2 glass and supercooled liquid (McMillan et al., 1994), using a diode array instrument. (a) is the total signal measured, including Raman scattering and black body emission, obtained at 1950 K, along with any non-linearities in the instrumental response. The "noise" is a four-pixel pattern associated with the diode array, and occasional sharp negative "peaks" are due to "lazy" diodes. These features are usually removed by carrying out a flat field correction, obtained by broad band illumination of the detector (Wolf et al., 1990). In the high temperature experiments, these are also removed in the following step. (b) represents black body emission from sample, recorded under the same conditions, but with the laser off. (c) is the Raman signal (a) - (b), with black body emission subtracted. This procedure also removes the four-pixel pattern, any non-linear background detector response, and the sharp features (except for one near 400 cm⁻¹!) due to the diode array. (d) Raman signal, corrected for temperature- and frequency-dependence of Raman scattered intensity (according to the first-order harmonic approximation of Eqn. 16).

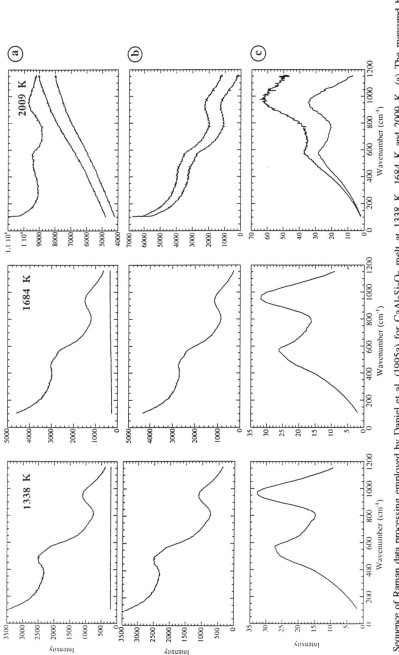

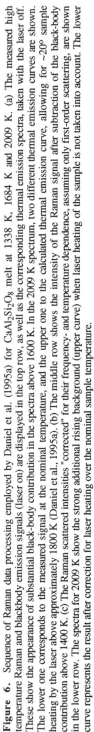

Figure 6. Sequence of Raman data processing employed by Daniel et al. (1995a) for CaAl$_2$Si$_2$O$_8$ melt at 1338 K, 1684 K and 2009 K. (a) The measured high temperature Raman and blackbody emission signals (laser on) are displayed in the top row, as well as the corresponding thermal emission spectra, taken with the laser off. These show the appearance of substantial black-body contribution in the spectra above 1600 K. In the 2009 K spectrum, two different thermal emission curves are shown. The lower one corresponds to the measured signal at the nominal temperature, and the upper one to the calculated thermal emission curve, allowing for ~20° sample heating by the laser above approximately 1800 K (Daniel et al., 1995a). (b) The middle row shows the intensity of the Raman signal after subtraction of the black-body contribution above 1400 K. (c) The Raman scattered intensities "corrected" for their frequency- and temperature dependence, assuming only first-order scattering, are shown in the lower row. The spectra for 2009 K show the strong additional rising background (upper curve) when laser heating of the sample is not taken into account. The lower curve represents the result after correction for laser heating over the nominal sample temperature.

appreciable, and must be accounted for before structural interpretation of the spectra (McMillan et al., 1994; Daniel et al., 1995a,b).

Above ~2000 K, the radiation intensity from blackbody emission becomes so intense that it completely swamps the Raman scattering spectrum, despite spatial filtering and sensitive detection techniques. Studies in this temperature regime will require application of time-resolved Raman techniques, in which the detection system is gated synchronously with a pulsed laser excitation, to achieve a similar result to that achieved by chopping the source beam in scanning infrared spectroscopy.

For in situ experiments under *combined* high P-T conditions, several methods for resistive heating in the diamond anvil cell have been developed. Such techniques have been applied to measurements on germanate melts and glasses up to ~900 K at high pressure (Farber and Williams, 1992), and we have carried out preliminary studies for $Na_2Si_4O_9$ composition to 10 GPa and 770 K (see final section). Higher temperatures will most easily be obtained by CO_2 laser heating methods in the diamond anvil cell. These techniques have been pioneered by Gillet (1993) and Gillet et al. (1993b) for crystalline minerals, who have demonstrated the feasibility of obtaining high quality Raman spectra up to 12 GPa and over 1700 K.

VIBRATIONAL STUDIES OF ALUMINOSILICATE LIQUIDS AND GLASSES

SiO_2

This composition forms the prototype silicate liquid, and knowledge of its structural behavior through the glass transformation range into the stable liquid regime is necessary for developing an understanding of the properties of highly silicic magmas. It is useful to describe its vibrational properties in some detail, as a basis for discussing the spectra of multicomponent silicate and aluminosilicate compositions.

The infrared and Raman spectra of SiO_2 glass and liquid at high temperature have been investigated in several studies. Markin and Sobolev (1960) first studied the infrared reflectivity in the region 870 to 1250 cm^{-1} to 2000°C, and Gaskell (1966a) obtained infrared reflectance spectra above 420 cm^{-1} between room temperature and 900°C, in the glassy regime. Exarhos et al. (1988) reported a high temperature Raman spectrum of SiO_2 liquid at 2250 K, obtained by CO_2 laser heating of a region inside a silica glass sample. McMillan et al. (1994) have used Pt-Ir wire-loop heating methods to investigate the Raman spectrum of SiO_2 glass and supercooled liquid, between room temperature and 1950 K.

Before describing the changes in the spectra at high temperature, it is useful to summarize our current understanding of the structure and vibrational spectra of SiO_2 glass at room temperature. Vitreous SiO_2 consists of a continuous three-dimensional network of corner-sharing SiO_4 tetrahedra (Mozzi and Warren, 1969; Konnert et al., 1982, 1987; Johnson et al., 1983; Elliot, 1991). The total vibrational spectrum for SiO_2 glass has been measured via inelastic neutron scattering spectroscopy by Galeener et al. (1983). (This experiment actually gives $S(\omega)$, which is the vibrational density of states $g(\omega)$ weighted by the inelastic neutron scattering cross-sections of the constituent atoms. However, the neutron scattering lengths of Si and O are comparable (4.1 and 5.8 barns), so that the distinction is not important here). The neutron spectrum shows a dominant band centred near 350 cm^{-1}, with a low frequency shoulder resolved at 100 cm^{-1}. There is a sharp peak at 800 cm^{-1}, and an asymmetric band between 950 cm^{-1} and 1300 cm^{-1} (Galeener et al., 1983). The high frequency feature appears to be split into two

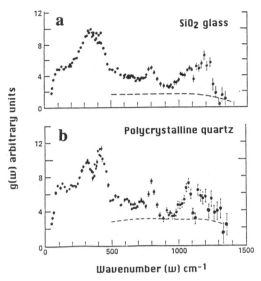

Figure 7. Total vibrational density of states $g(\omega)$ for glassy (top) and polycrystalline (α-quartz) SiO_2, obtained by inelastic neutron scattering (Galeener et al. 1983). Note the similarity between the glass and crystal spectra. The dashed line indicates the approximate extent of two-phonon contributions to the spectrum. [Redrawn from Galeener et al. (1983), Fig. 1, p. 1055.]

components, with maxima near 1050 and 1200 cm^{-1}. There is also broad scattering intensity in the 500 to 800 cm^{-1} region. This vibrational density of states is almost indistinguishable from that of crystalline SiO_2 (Fig. 7).

The vibrational spectrum of SiO_2 glass was initially interpreted by Bell, Dean and co-workers (Dean, 1972; Bell, 1976; Bell and Dean, 1972; Bell et al., 1968, 1970, 1980) in terms of a random network model. This model was physically constructed from distorted, corner-linked, SiO_4 tetrahedral units, and the atomic positions were measured and used in a lattice dynamics calculation, assuming a simple force field model for the bond stretching and angle bending interactions (Bell and Dean, 1966). The highest frequency bands, near 1000 cm^{-1}, were associated with Si-O bond stretching vibrations, in which the Si and O atoms are displaced in opposite directions, approximately to the Si...Si vectors (McMillan, 1984a). The peak near 800 cm^{-1}, along with some of the intensity near 550 cm^{-1}, was interpreted (in the studies of Bell and Dean) as a vibration involving oxygen motion perpendicular to the Si...Si lines, in the planes of their SiOSi linkages. The principal lower frequency band, which dominates the spectrum, was found to be due to a vibration in which the oxygen atoms moved perpendicular to the SiOSi planes.

Later calculations using more extensive force fields have resulted in some modifications to the interpretation of the bands below 800 cm^{-1} (Bates, 1972; Etchepare et al., 1974; McMillan and Hess, 1990). The vibrations near 800 cm^{-1} are now thought to correspond to Si-O stretching vibrations, with a large associated silicon displacement (McMillan, 1984a). The bands below 500 cm^{-1} contain both OSiO and SiOSi bending contributions, and are characterized by a small degree of silicon participation in the motion (McMillan, 1984a; McMillan and Hess, 1990). This general assignment is in agreement with the results of Si and O isotopic substitution experiments (Galeener and Mikkelsen, 1981; Galeener and Geissberger, 1983; Sato and McMillan, 1987).

The Raman spectrum of SiO_2 glass contains two weak, sharp, polarized peaks at 606 and 492 cm^{-1} (Fig. 8). These are not predicted by a random network model, or any based on the structures of low pressure crystalline SiO_2 polymorphs. The assignment of these features has been the subject of much discussion in the literature, and it is currently thought that these peaks correspond to the symmetric oxygen breathing vibrations of three- and four-membered siloxane rings of SiO_4 tetrahedra, respectively, embedded as "defects" within the glass structure (Sharma et al., 1981; Galeener, 1982a,b; O'Keeffe and Gibbs, 1984; McMillan, 1988; Warren et al., 1991; McMillan et al., 1994). The modes are vibrationally decoupled from the rest of the glass network, remaining highly localized within the rings, which explains their narrow width compared to the rest of the glass spectrum. The SiOSi angles within the three-membered rings are constrained to have small values, near 130°, by the Si-O bond length and the tetrahedral OSiO angle (Galeener, 1982a,b). The SiOSi angles for a planar four-membered ring lie near 160° for regular tetrahedra, although this value may be reduced by puckering the ring. It should be noted that these defects correspond to only minor quantities of the small ring species in the glass. Mikkelsen and Galeener (1980) and Geissberger and Galeener (1983) have measured the formation enthalpy of the three-membered ring species in SiO_2 to be approximately 42 kJ/mol. Taking the glass transition temperature to be near 1480 K (Richet and Bottinga, 1984), only approximately 1% of the silicon atoms in the glass are present in three-membered rings.

The Raman spectrum of vitreous SiO_2 is dominated by a strongly polarized band at 430 cm^{-1} (Fig. 8). This band is associated with predominantly oxygen motion, and has been described variously as a symmetric oxygen stretching vibration of the bent SiOSi linkages, as a symmetric bending vibration of the SiOSi linkages, with oxygen motion perpendicular to the Si...Si line, or as a symmetric OSiO angular deformation of the coupled SiO_4 groups (Sen and Thorpe, 1977; Galeener and Geissberger, 1983; Sharma et al., 1984; McMillan and Hess, 1990; McMillan et al., 1994). In fact, all of these designations are largely equivalent. The frequency of such vibrations is known to be dependent upon intertetrahedral angle, so that the large width of the band in vitreous SiO_2 can be understood in terms of the known distribution in SiOSi angles within the structure (Mozzi and Warren, 1969; Dupree and Pettifer, 1986; Konnert et al., 1982, 1987; Devine et al., 1987).

The infrared spectrum of vitreous SiO_2 closely matches the features observed in the Raman spectrum, and also in the total vibrational density of states function, particularly in the high frequency region (Galeener et al., 1983) (Fig. 8). In particular, the infrared absorption spectrum contains a strong peak at 1060 cm^{-1}, at the same position as one of the weak high frequency Raman peaks, and a weaker feature at higher wavenumber which appears as a shoulder in powder transmission measurements: Lippincott et al., 1958; Poe et al., 1992.

If the 1060 cm^{-1} peak were truly common to both the infrared and Raman spectra, this would imply that the glass structure is non-centrosymmetric structure, as might be expected, and that the vibrational mode was *polar* (i.e. associated with a dipole moment change, in both IR and Raman spectra). For a non-centro-symmetric structure, such a polar Raman-active vibration should have two components, corresponding to transverse (TO) and longitudinal (LO) vibrational modes, split in frequency by the long-range Coulombic field in the glass. Such TO-LO splitting gives rise to the broad infrared reflectance bands of crystalline minerals, and is well known to occur in the Raman spectra of non-centrosymmetric crystals, such as α-quartz (She et al., 1971). The possible occurrence of TO-LO splitting in the Raman spectrum of vitreous SiO_2 has been discussed extensively. In the early study of Galeener and Lucovsky (1976), several TO-LO mode pairs were proposed, including the sharp peaks at 490 and 600 cm^{-1} which were later identified as the three- and four-membered ring "defect" bands (see above). In

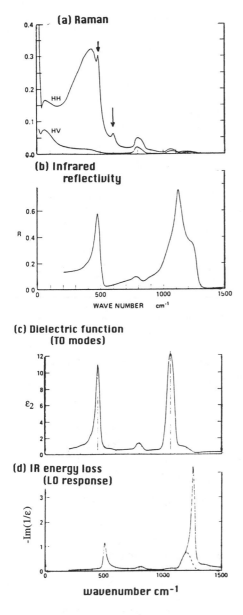

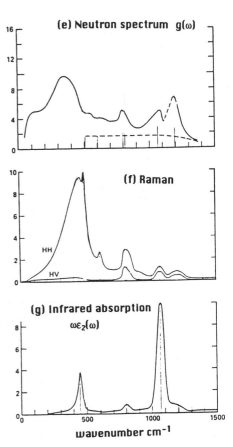

Figure 8. Comparison of the IR, Raman and neutron vibrational spectra of SiO$_2$ glass, redrawn from Galeener et al. (1983). At left are shown the (a) HH and HV polarized room temperature Raman spectra, uncorrected for temperature effects. The "defect peaks" are indicated by arrows on the Raman spectrum. Next, the infrared reflectivity (b) is shown, along with the derived dielectric functions $\varepsilon_2(\omega)$ and $(-\mathrm{Im}(1/\varepsilon)$ [(c) and (d)]. The imaginary dielectric constant curve [$\varepsilon_2(\omega)$] records the response of transverse (TO) vibrational modes propagating in the glass (c), and the energy loss function [$-\mathrm{Im}(1/\varepsilon)$] records the longitudinal LO mode response (d). The dashed line at high frequency in the -Im(1/ε) function indicates where Galeener et al. (1983) thought the maximum of the LO mode response should lie (but see text). At right, the neutron (e), Raman (f) and infrared absorption [$\omega\varepsilon_2(\omega)$], (g) spectra are compared. The Raman data have been corrected by an Equation (16) to remove the first-order temperature and frequency dependence. [Redrawn from Galeener et al. (1983), Figs. 5, 10, and 13, pp. 1059, 1062, 1063.]

the more recent study of Galeener et al. (1983), the principal TO-LO pair was proposed to consist of the high frequency bands at 1060 cm^{-1} and 1200 cm^{-1}. Analysis of the infrared reflectivity spectrum actually showed an LO response (as the energy loss function—Im(1/ε), where ε is the complex dielectric function) with a peak occurring near 1260 cm^{-1}, at higher frequency than the broad Raman band at 1200 cm^{-1}, but Galeener et al. (1983) presented arguments to suggest that the true maximum was coincident with the Raman band position. This assignment is now apparently "accepted" in many discussions of the vibrational spectrum of vitreous SiO$_2$ (Sharma et al., 1984).

If the 1060 and 1200 cm^{-1} Raman bands *did* form a TO-LO pair, they should exhibit *polariton* behavior, in which their frequencies should depend upon the scattering angle θ. This was tested by Denisov et al. (1984), and was found *not* to be the case for the normal Raman spectrum of SiO$_2$ glass. Instead, Denisov et al. (1979, 1984) observed polariton behavior in the *hyper-Raman* spectrum of vitreous SiO$_2$, which has different selection rules, closer to those determining the infrared activity (see McMillan, 1985; McMillan and Hofmeister, 1988). The high frequency polaritons actually found by Denisov et al. (1979) corresponded to one TO mode at 1065 cm^{-1}, which overlaps with the normal Raman band at 1060 cm^{-1}, and a corresponding LO component at 1255 cm^{-1}, the frequency of the maximum in the infrared LO response found by Galeener et al. (1983). A second high frequency TO-LO pair occurred at 1180 cm-1, close to the location of the normal Raman band at 1200 cm^{-1}. These observations demonstrate that the normal Raman modes are *not* polar, so that the bands observed in the Raman spectrum do not in fact correspond to the same vibrations observed in infrared absorption, despite the close correspondence between the IR and Raman frequencies (Denisov et al., 1984). This is a very important observation, because it indicates that, although the infrared and Raman spectra of SiO$_2$ glass may appear to be similar, they are each selecting a *different* subset of the vibrational normal modes within the density of states function. This is also likely to be the case for other, multicomponent, silicate glasses and melts (Gervais, 1988; Parot-Rajaona et al., 1994).

From the vibrational spectra, it is obvious that the high frequency modes are split into two groups, with frequencies centred near 1060 cm^{-1} and 1200 cm^{-1} (in fact, Seifert et al. (1982) and Mysen et al. (1982b) have suggested that the higher frequency band of this pair might even have two components, at 1157-1162 and 1209-1213 cm^{-1}). The presence of two sets of Si-O stretching bands could be associated with the presence of two structure types present in the fully-polymerized glass (Seifert et al., 1982; Mysen et al., 1982a,b). For example, to explain the anomalous properties observed for vitreous silica, Vukcevitch (1972) has proposed a "two-state" model for the SiO$_2$ glass structure, in which there are two distinct populations of SiOSi angles (φ) within the tetrahedral network, with values near 138° (φ$_\alpha$) and 145° (φ$_\beta$). The relative proportion of SiOSi angles with these two values is fixed by the temperature and pressure, and dynamic interchange between φ$_\alpha$ and φ$_\beta$ is possible as an activated process (with activation energies of a few kJ/mol) through cooperative rotations of the interconnected tetrahedra (Vukcevitch, 1972). Sen and Thorpe (1977), Galeener and Sen (1978), and Galeener (1979), have developed a model for the dynamical behavior of tetrahedrally connected AX$_2$ frameworks, including SiO$_2$. In this model, for a framework in which each AXA angle (the TOT angle, with T = Si or Al, in the case of the aluminosilicates) was 90°, the vibrational spectrum would contain a single line for A-X stretching, because the two stretching vibrations from each "leg" of the linkage would be vibrationally decoupled. As the AXA angle is opened out, the A-X stretching vibrations begin to couple with each other, and they separate into a low frequency symmetric combination, and a high frequency antisymmetric stretching combination (Sen and Thorpe, 1977; Galeener and

Sen, 1978; Galeener, 1979). When the AXA angle attains 180°, the vibrations are completely coupled, and are analogous to the symmetric and asymmetric stretching vibrations of a linear triatomic molecule like CO_2 (McMillan, 1985; Ihinger et al., 1994).

This model was applied to the vibrations of the SiO_2 network by Sen and Thorpe (1977), Galeener and Sen (1978), and Galeener (1979). Seifert et al. (1982) proposed that the two high frequency bands near 1060 cm^{-1} and 1200 cm^{-1} might correspond to the asymmetric stretching modes associated with *two* populations of SiOSi groups within the glass structure, in the spirit of the two state model for SiO_2 glass suggested by Vukcevitch (1972). From the wavenumbers of the high frequency modes and also the positions of the proposed low frequency (symmetric stretching) components, Seifert et al. (1982) suggested that the intertetrahedral angles in the two units were approximately 127° and 132°. The model was then extended to a discussion of fully polymerized aluminosilicate glasses, discussed further below (Seifert et al., 1982). In Mysen et al. (1982a), the discussion is worded more generally. Here it is suggested that the Raman spectrum of vitreous SiO_2 is difficult to explain on the basis of a single structure unless concepts such as TO-LO splitting are invoked. This was discussed above. It was proposed that the asymmetry in the observed distribution of SiOSi angles could be explained by the existence of more than one three-dimensional entity in the melt. Further, it was noted that electron microscopic observations of SiO_2 glass (Gaskell, 1975; Gaskell and Mistry, 1979; Bando and Ishizuka, 1979) had led to the suggestion that the glass consists of at least two distinct structures (Mysen et al., 1982a). These statements might be taken to imply that the two structures, corresponding to populations with different SiOSi angles, might constitute separate domains within the glass (Seifert et al., 1982; Mysen et al., 1982a,b, 1983).

The presence of macroscopic domains of this type would constitute a *two-phase* structure for the glass, leading to density-driven phase separation (Wolf et al., 1992; Aasland and McMillan, 1994; Smith et al., 1995). The possibility of such two-phase behavior for SiO_2 glass and liquid has been a topic for considerable recent discussion at high pressure, in the "mixed phase" regime where the silicon coordination is changing from 4- to 6-fold (Wackerle, 1962; Hemley et al., 1986; Williams and Jeanloz, 1988; Schmitt and Ahrens, 1989; Jeanloz, 1990; Willams et al., 1993). The potential occurrence of two phase behavior at lower pressure, in the tetrahedral network regime, is an exciting possibility, and is being actively discussed in relation to proposed phase relations in the analogous framework liquid H_2O (Poole et al., 1992, 1993, 1994, 1995; Stanley et al., 1994; Angell, 1995; Shao and Angell, 1995: see chapters by Poole et al., and Wolf and McMillan, this volume). In the meantime, a *two-state* model for the glass structure (Smith et al., 1995) might certainly be appropriate (Smith et al., 1995), with domains on a nanometric scale corresponding to silica units with the two different average SiOSi angles suggested by Seifert et al (1982) and Mysen et al. (1983).

A qualitatively different explanation has been suggested for the observed high frequency vibrations of SiO_2 glass; that these simply correspond to different types of stretching vibration of the linked SiO_4 tetrahedra within the structure: i.e. the high frequency bands correspond to different vibrational normal modes of the network (McMillan et al., 1982). Different types of Si-O stretching vibration are known to fall into distinct frequency ranges within crystalline SiO_2 polymorphs, for which a two-phase structural model is obviously inappropriate (Bates, 1972; Etchepare et al., 1974; McMillan and Hess, 1990). However, it is not impossible that the interpretations of Seifert et al. (1982) and McMillan et al. (1982) are quite compatible. It is known that SiOSi angles do fall into families within crystalline SiO_2 polymorphs (Vukcevich, 1972), and the different normal modes might be preferentially associated with vibrations of Si-O bonds involved in one or another of these angular populations. This point deserves further study as the vibrational motions in SiO_2 polymorphs become better understood through *ab initio* calculations (Allen and Teter, 1987, 1990; Demkov et al., 1995).

High temperature Raman spectra of SiO_2 glass and liquid are shown in Figure 9. In order to separate between changes in such high temperature spectra due to purely vibrational (anharmonic) effects and those due to structural relaxation, it is important to obtain series of spectra in the glass phase below the glass transition, as well as in the supercooled or stable melt (McMillan et al., 1994).

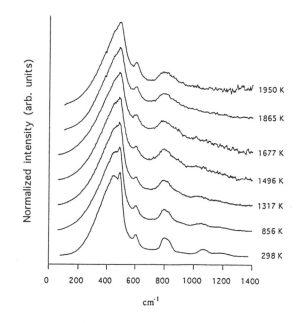

Figure 9. High temperature Raman spectra of SiO_2 glass and super-cooled liquid to 1950 K (McMillan et al., 1994). The data have been reduced to remove the effects of and frequency- and temperature-dependence (first-order Raman scattering only: Eqn. 16), and the spectra have been normalized the the same maximum peak height.

An obvious feature of these spectra is the growth of a broad band which extends under the entire spectrum, with its intensity maximized in the 600 to 800 cm^{-1} region (Fig. 10). This a vibrational effect, rather than changes due to configurational relaxation at high temperature, because it begins well below the glass transformation range ($T_g \sim 1480$ K: Bruckner, 1970; Richet and Bottinga, 1984). The broad scattering signal is due to second order contributions to the Raman spectrum, which arise mainly from sum combinations of the bands below 600 cm^{-1} (McMillan et al., 1994). Sum bands from the 800 cm^{-1} and 1000 to 1300 cm^{-1} bands appear at much higher frequency in the spectrum of vitreous SiO_2 (Stolen and Walrafen, 1976). There is an increased incidence of such two-phonon processes at high temperature. However, the intensity of such sum bands has a different temperature dependence to the first order Raman spectrum, as noted previously (Sherwood, 1972; McMillan et al., 1994; Daniel et al., 1995a).

The "defect" peak near 600 cm^{-1}, due to the symmetric SiOSi bending ("ring-breathing") mode of three-membered siloxane rings in the structure, remains visible to the highest temperatures studied. This peak shows only a small frequency shift with temperature, from 606 cm^{-1} at room temperature to 603 cm^{-1} at 1950 K (McMillan et al., 1994), consistent with a vibration of a species with constrained geometry (Galeener,

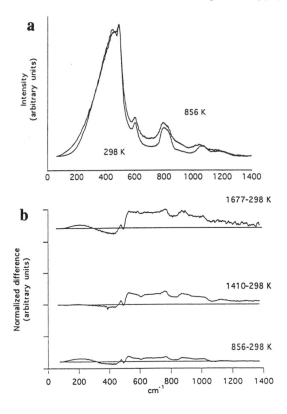

Figure 10. (a) Comparison of normalized, temperature-corrected (first-order Raman scattering only) 856 K and 298 K spectra of SiO_2 in the glassy regime, showing the intensity increase centred near 800 cm^{-1}. This is due to anharmonic effects (second order Raman scattering: McMillan et al., 1994; Daniel et al., 1995a). (b) Difference plots (spectrum at T - 298 K spectrum), showing gradual increase of intensity due to second order Raman spectrum, beginning well below T_g. The second order spectrum is not completely removed by the correction procedure of Equation (16).

1982a). Geissberger and Galeener (1983) measured the intensity of this peak as a function of fictive temperature for a series of quenched or annealed SiO_2 glasses, and calculated an endothermic formation enthalpy for these species (42 kJ/mol). The frequency of the defect peak near 490 cm^{-1}, assigned to four-membered siloxane rings by Sharma et al. (1981), also appears to be quite independent of temperature (McMillan et al., 1994). The formation enthalpy for this species is smaller, approximately 15 kJ/mol (Geissberger and Galeener, 1983). The positive formation energies for these species indicate that the proportion of three- and four-membered rings should increase in the supercooled liquid with increasing temperature. This is apparent in the high temperature spectra for the three-membered ring species, as the 603 to 606 cm^{-1} peak becomes more prominent with increasing temperature. It is difficult to quantify this effect, however, because of broadening in the peak itself, and because of frequency shifts in the strong band at lower wavenumber (McMillan et al., 1994). Using the formation enthalpies determined by Geissberger and Galeener (1983) indicates that, at 2000 K, ~3% of the Si atoms should be present in three-membered rings, and ~10% in four-membered rings (compared with ~1% and ~7%, at 1480 K).

The dominant Raman band, with its maximum near 440 cm^{-1} at room temperature, shows anomalous behavior, in that its frequency *increases* with increasing temperature (Fig. 11: McMillan et al., 1994). This effect is observed to begin well below the glass transition range. In the glass, the frequency shift results from anharmonic vibrational motions. At T_g (~1480 K), the rate of increase in the frequency of this band with

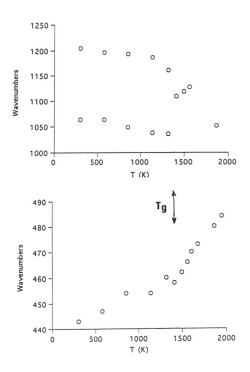

Figure 11. Variation of Raman frequencies of SiO_2 glass and supercooled liquid with temperature (McMillan et al., 1994): top—high frequency bands 1060 and 1200 cm^{-1} at room temperature); bottom—440 cm^{-1} band. This band shows an anomalous frequency *increase* with increasing temperature, due to SiOSi angle narrowing via anharmonic effects. This contributes to a low volume thermal expansion of the glassy state. There is a break in slope at T_g (1480 K) as the anharmonic SiOSi angle narrowing is joined by configurational effects, which also results in narrower angles.

temperature increases by approximately a factor of two, as vibrational effects are joined by structural (configurational) relaxation (McMillan et al., 1994). This behavior can be rationalized in terms of the oxygen displacements within SiOSi linkages, which are not geometrically constrained within small rings in the glass framework. A homogeneous model for the glass structure can be assumed as a first approximation, with an average SiOSi angle ~143° at room temperature (Mozzi and Warren, 1969; Devine et al., 1987). The motion of the two-coordinated oxygen within the bent SiOSi linkages is perpendicular to the Si...Si line (Bates, 1972; Etchepare et al., 1974; Galeener and Sen, 1978; McMillan, 1984a; McMillan and Hess, 1990). This vibration is highly anharmonic, so that increasing its degree of vibrational excitation will move the average oxygen position further from the Si...Si line, and decrease the SiOSi angle (Smyth et al., 1953; Vukcevich, 1972). There is an inverse correlation between the SiOSi bending frequency and SiOSi angle in condensed silicates (Lazarev, 1972; Sen and Thorpe, 1977; Galeener and Sen, 1978; McMillan, 1984a), so that the observed frequency moves to higher values. Above the glass transition, the rate of change of the band frequency increases, indicating that the relaxed supercooled liquid structure contains structures with a smaller average SiOSi angle, decreasing to approximately 137° (estimated from the position of the Raman line) by 2000 K (McMillan et al., 1994). This is consistent with the measurements of Geissberger and Galeener (1983). For silica samples prepared with different fictive temperatures, the position of the dominant Raman line, measured at room temperature, moved to progressively higher wavenumber as T_f increased. This indicates that glassy configurations with smaller average SiOSi angle are frozen at higher temperature. These arguments can also be expressed in terms of the two state angular model of Vukcevich (1972). The shift in the dominant Raman line to higher frequency could indicate an increased proportion of the species with smaller SiOSi angle (ϕ_α) at high temperature. If, in addition, the presence of smaller ring configurations (e.g. four-membered rings) favours smaller SiOSi angles (Sharma et al., 1981; Seifert et al., 1982; Mysen et al., 1982a,b, 1983; Galeener, 1982a,b), the configurations formed above T_g might contain a higher proportion of these smaller ring species.

These angular changes have been used to rationalize the anomalous volume changes observed for SiO_2 glass and melt with increasing temperature (Smyth, 1953; Gaskell, 1966b; Bruckner, 1970a; Vukcevich, 1972), which are also correlated with anomalies in the elastic properties (Bucaro and Dardy, 1974). Such structural effects could be important in determining the rheology of silicic magmas. The volume thermal expansion coefficient is anomalously low over most of the temperature range, much smaller than for quartz or cristobalite, and density maxima are observed at approximately 175 K and 1800 K (Bruckner, 1970). At low temperatures, the anharmonic narrowing of the SiOSi angle tends to displace the silicon atoms towards each other, along the Si...Si line. This occurs because the Si-O bonds are relatively rigid, and the bond length is conserved, and the net effect is one of a structural *densification*. (Smyth, 1953). Such displacements occur via cooperative rotations of adjacent tetrahedra about their Si-O bonds, into free volume present within the low density glass structure. These motions oppose the normal volume expansion due to increased thermal agitation, so that the overall thermal expansion is small, and even exhibits passes through zero just below room temperature (Bruckner, 1970). As the temperature is increased further, the volume expansion effect of increased thermal excitation of the higher frequency Si-O stretching vibrations begin to overcome the contraction due to the decreasing SiOSi angle, and the thermal expansion coefficient increases. However, above the glass transition temperature ($T_g \sim 1480$ K), the supercooled liquid begins to relax to configurations with even smaller SiOSi angles, once more opposing the volume expansion, but this time via a structural relaxation mechanism. Within the supercooled liquid, the thermal expansion coefficient eventually begins to rise once more due to thermal effects, and a second density maximum is observed as α passes through zero.

In the high frequency region, the weak Raman bands at 1064 and 1204 cm^{-1} due to Si-O stretching decrease in frequency with temperature, below T_g (McMillan et al., 1994). This is a consequence of both the anharmonicity of the Si-O bond stretching, and the smaller average SiOSi angle occurring at high temperature (Sen and Thorpe, 1977; Galeener and Sen, 1978). Above T_g, these two bands can no longer be distinguished, and only a very weak, broad feature with its maximum near 1100 cm^{-1} can be observed in some of the high temperature spectra (Fig. 9; McMillan et al., 1994). Although the Raman intensity of these bands is lowered, the Si-O stretching features are neither lost from the vibrational spectrum, nor suffer from extreme broadening above T_g, because the infrared bands remain strong and clearly distinguishable to ~2300 K (Markin and Sobolev, 1960). The room temperature infrared reflection spectrum contains a principal Si-O stretching band with its reflectance maximum at 1120 cm^{-1}, and a shoulder at 1216 cm^{-1} (Gaskell and Johnson, 1976a,b). This shoulder is present in the spectra up to 1273 K (Gaskell, 1966a), but has merged with the principal band for the spectra obtained at 1773 K and 2273 K by Markin and Sobolev (1960). The reflectivity of this band decreases by more than a factor of two between room temperature and 2000°C, indicating considerable anharmonic damping of the Si-O stretching vibrations at high temperature. This could account for the merging of the high frequency features in the IR and Raman spectra, as the two types of normal mode involving Si-O stretching within the glass network become strongly anharmonically coupled (McMillan et al., 1994).

Alkali and alkaline earth silicates

After vitreous silica, the binary and pseudo-binary alkali and alkaline earth silicate glass series SiO_2-MO-M_2O have received most attention. The major structural changes and their effects on the infrared and Raman spectra are now quite well documented and understood, although some important points still remain to be clarified.

As metal oxide component is added to SiO_2, the fully polymerized tetrahedral framework becomes "depolymerized", with "non-bridging oxygens (NBO)" appearing in response to the increased O:Si ratio. This occurs because silicon has a stronger affinity for oxygen (i.e. is a stronger Lux-Flood acid: Lux, 1939; Flood and Forland, 1947; Huheey, 1983) than most other metals, so in nearly all binary silicate systems studied to date, all of the oxygen becomes bound to the silicon. Some exceptions to this behavior are known: for example, Pb-O-Pb linkages have been documented for lead silicate glasses and liquids (Worrell and Henshall, 1978; Piriou and Arashi, 1980), and Al-O-Al linkages do occur in binary (and ternary) aluminosilicate systems (McMillan and Piriou, 1982; McMillan et al., 1982; Putnis and Angel, 1985; Carpenter, 1991a,b; Kubicki and Sykes, 1993). The changing polymerization of the silicate framework is expressed by the successive appearance of tetrahedral silicate units with 1, 2, 3 and 4 NBOs, corresponding to the disilicate (-Si_2O_5), metasilicate (-SiO_3), pyrosilicate (Si_2O_7) and orthosilicate (SiO_4) compositions and structural species, analogous to the behavior of crystalline silicates. These units are now often labelled as Q^n species, using terminology developed within the NMR community to describe aqueous silicate solutions (Engelhardt et al., 1975; Engelhardt and Hoebbel, 1984; Engelhardt and Michel, 1987), where n is the number of *bridging* oxygens (i.e. contained within SiOSi linkages) around the tetrahedral silicon.

The major differences between the glasses or melts and corresponding crystalline species are that (a) there is generally coexistence of several Q^n species at a given composition, with their relative concentrations related by composition-, pressure- and temperature-dependent equilibria, and (b) there is no requirement for long-range symmetrical arrangement of the Q^n species into coherent sheet, chain, or other polymer structures. The presence of these different silicate species, and the compositional dependence of their relative concentrations, was amply demonstrated by Raman spectroscopic studies of silicate glasses (Brawer and White, 1975, 1977; Iwamoto et al., 1975; Virgo et al., 1980; Mysen et al. 1980a, 1982a; Kashio et al., 1980, 1982; Verweij and Konijnendijk, 1976; Furukawa et al., 1981; Matson et al., 1983; McMillan, 1984a,b; Mysen, 1988), and confirmed in subsequent NMR experiments (Schramm et al., 1984; Dupree et al., 1986; Engelhardt and Michel, 1987; Stebbins, 1988; Kirkpatrick, 1988; Maekawa et al., 1991).

The changing polymerization of the silicate framework is expressed in the Raman spectrum by the successive appearance of strong polarized bands near 1100 cm^{-1}, 1000 cm^{-1}, 900 cm^{-1} and 850 cm^{-1}, with decreasing silica content, which can be assigned to the symmetric Si-O stretching vibrations of Q^3, Q^2, Q^1, and Q^0 species, respectively (Figs. 12 and 13) (Brawer and White, 1975; Furukawa et al., 1981; Mysen et al., 1980a; 1982a; Matson et al., 1983; McMillan, 1984a,b; Mysen, 1988). For a given silica content, several Q^n species are present simultaneously, and the symmetric stretching bands are more or less well-resolved, depending on the system. In general, the band resolution in the high frequency region is poorer in Al-free silicate systems containing the small cations Li$^+$ or Mg^{2+} (Brawer and White, 1975; Mysen et al., 1980a; 1982a; Matson et al., 1983; McMillan, 1984a,b). In order to resolve individual components within the

Figure 12. Unpolarized Raman spectra of glasses along the SiO$_2$-CaO and SiO$_2$-CaMgSiO$_4$ joins. Spectra are not corrected for temperature effects. Characteristic high frequency peaks associated with symmetric Si-O stretching vibrations of Q^3, Q^2, Q^1 and Q^0 species are labelled as S, C, D and M (for "sheet", "chain", "dimer" and "monomer" units, respectively). These appear successively as the silica content is reduced (bottom to top). The vibrations for the SiOSi and OSiO deformation vibrations appear in the 500 to 700 cm^{-1} region. [From Mysen et al. (1980a), Figs. 2 and 3, p. 694].

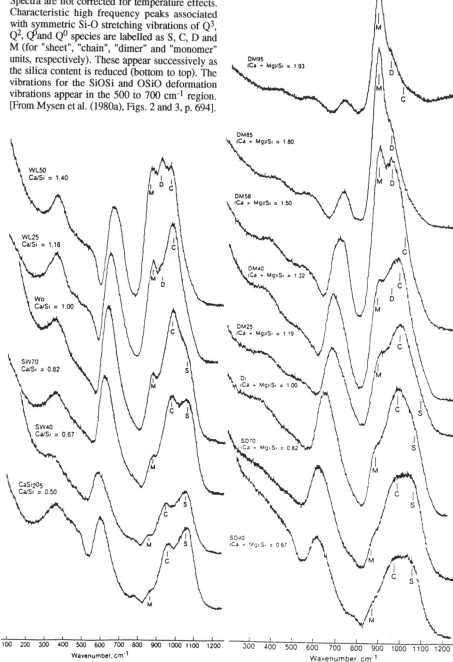

Alkali tetrasilicate glasses

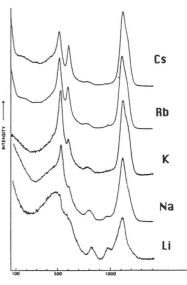

Figure 13. Unpolarized Raman spectra for alkali silicate glasses at the tetrasilicate ($M_2Si_4O_9$) composition (80 mole% SiO_2), from Matson et al. (1983) (uncorrected room temperature spectra). The spectra are dominated by the high frequency band due to Q^3 species, at 1100 cm^{-1}. The high frequency shoulder assigned to $Q^{3'}$ species (Matson et al., 1983), is clearly visible for K-, Rb- and Cs-glasses. [Used by permission of the editor of *Iournal of Non-Crystalline Solids,* from Matson et al. (1983), Fig. 5, p. 331].

Raman shift (cm^{-1})

unresolved profile, curve-fitting schemes are generally used (Mysen et al., 1980a, 1982; McMillan, 1984b). The positions and relative intensities of the principal high frequency band components are now quite well determined for the entire range of silica contents (Mysen et al., 1982a; McMillan, 1984a; Mysen, 1988).

Some questions regarding the high frequency region still remain to be resolved. First, when the Q^2 band first appears as a weak feature at silica contents near 85 mol %, its frequency is low, near 950 cm^{-1} for all alkali silicate systems (Brawer and White, 1975; Furukawa et al., 1981; Matson et al., 1983). Near the metasilicate composition, its relative intensity increases, and its frequency also moves rapidly to a higher value, near 970 cm^{-1}. There is still no good explanation for this behavior, although it must lie in the local structure around Q^2 species in highly polymerized glasses and melts. It is not known how the local structural environment might affect the nature of the vibration and its relative Raman scattering efficiency. Similar behavior is observed for the characteristic peak for the Q^3 species. The frequency of this band is close to 1100 cm^{-1} for all alkali series at high silica content (Brawer and White, 1975; Furukawa et al., 1981; Mysen et al., 1982a; Matson et al., 1983), but it moves to *lower* wavenumber as the metasilicate composition is approached. This band appears to occur near 1050 to 1070 cm^{-1} for all alkaline earth silicate compositions (Mysen et al., 1980a; McMillan, 1984b).

These frequency changes, which may be accompanied by changes in Raman scattering efficiency for the vibration, could pose a problem for quantitative fitting and determination of relative species concentration, especially in alkali silicate compositions. The relative concentrations of Q^n species have been estimated as a function of composition in binary silicate glasses, from relative Raman band intensities, along with mass balance considerations (Mysen et al., 1982a,b; McMillan, 1984b). For glasses along the SiO_2-Na_2O join, the relative abundances of Q^4, Q^3 and Q^2 species obtained by Mysen et al. (1982a) was found not to agree well with species estimations based on NMR spectroscopy (Stebbins, 1987). The effective Raman scattering factors used to estimate the concentrations of anionic silicate species from the intensities of their characteristic bands of each species were modified by Mysen (1990b), to bring the concentration determinations of Raman and NMR techniques into better agreement.

Quantitative Raman spectroscopy has been used to investigate the temperature dependence of silicate speciation reactions in molten alkali silicates (McMillan et al., 1992; Mysen and Frantz 1993a,b,c; 1994a,b). At a given silica content, the co-existing silicate species can be related by a series of disproportionation reactions of the type:

$$2Q^3 \Leftrightarrow Q^4 + Q^2$$
$$2Q^2 \Leftrightarrow Q^3 + Q^1 \tag{23}$$

etc., ...

(Morey and Bowen, 1924; Mysen et al., 1982a; Stebbins, 1988; Mysen and Frantz, 1993a). The equilibrium constant for each reaction is then

$$K_1 = \frac{a_{Q^3}^2}{a_{Q^4} a_{Q^2}} \quad , \tag{24}$$

etc., ...

(Stebbins, 1988; Mysen, 1990b,c; McMillan et al., 1992). The activity coefficients can be set to unity in a first approximation, and the equilibrium rewritten in terms of the mole fractions of the silicate species:

$$K_1 = \frac{X_{Q^3}^2}{X_{Q^4} X_{Q^2}} \quad , \tag{25}$$

(Mysen, 1990b; McMillan et al., 1992; Mysen and Frantz, 1993b,c). Mass balance requirements can be used to further simplify these expressions, to allow the concentrations to be more easily obtained from the spectroscopic data. For example, at the disilicate composition, $X_{Q^4} = X_{Q^2}$, and the equilibrium constant becomes

$$K = \left(\frac{X_{Q^3}}{X_{Q^2}} \right)^2 . \tag{26}$$

This is a useful expression for determination of the equilibrium constant from Raman spectroscopic data, because the characteristic peaks due to Q^3 and Q^2 species are clearly observed in the spectra of alkali disilicate glasses and liquids (Fig. 14). If it is assumed that the ratio of the Raman scattering cross sections for the 950 and 1100 cm-1 peaks assigned to Q^3 and Q^2 species does not change as a function of temperature, the relative intensities of these bands can be used to directly obtain the variation of the equilibrium constant with temperature. This was done by McMillan et al. (1992), who obtained a value of 20.6 ± 5.5 kJ/mol for ΔH^o for the disproportionation reaction (Fig. 15). It was noted that the disproportionation equilibrium did not change below the glass transition temperature ($T_g = 765$ K), as expected (Stebbins, 1988). This is confirmed by the more extensive measurements of Mysen and Frantz (1992, 1993a,b, 1994a).

Using a different approach, in which values for the effective Raman cross-sections ("Raman quantification factors", calibrated by comparison with quantitative NMR studies on alkali silicate systems) of characteristic bands assigned to Q^4, Q^3 and Q^2 units, Mysen and Frantz (1993a,b; 1994a,b) have calculated mole fractions for these species in several series of molten silicates. This then permitted these authors to determine both the absolute value of the equilibrium constant and its variation with temperature (Fig. 15). The results of McMillan et al. (1992) and Mysen and Frantz (1993a, 1994a) can be compared for the $K_2Si_2O_5$ composition. The absolute values of K at any temperature are different between the two studies (by a factor of 4 to 5), both because of differences in detailed band assignments, and because of different assumptions concerning the relative Raman activities of the vibrational modes. However, the values of ΔH^o obtained from the slope of ln K versus 1/T are comparable: Mysen and Frantz (1993a), obtained 14 ± 2 kJ/mol for this quantity, which corresponds to the lower bound estimated by McMillan et al. (1992). In the later study of Mysen and Frantz (1994a), this estimate was revised to 27.7 kJ/mol, which lies close to the *upper* limit of McMillan et al. (1992).

It might be expected that the intrinsic Raman activity of these vibrational modes should vary with temperature (Lines, 1987; Mysen and Frantz, 1994a). McMillan et al. (1992) did suggest that the observed Q^2/Q^3 intensity ratio varied slightly with temperature below T_g, due to such effects. However, Mysen and Frantz (1994a) obtained more points in this temperature region, and found no observable effect of temperature in the glassy regime.

The endothermic ΔH^o values obtained in these studies are consistent with the values obtained from high temperature NMR studies (in the range 10 to 30 kJ/mol, dependent upon composition, Stebbins, 1988; Brandriss and Stebbins, 1988; Stebbins et al., 1992: see also Stebbins, this volume). It is of some interest that this ΔH^o value is also in the same range as that determined for the formation enthalpy of SiO_5 (and also AlO_5!) species, identified as intermediates for O^{2-} exchange between Q^n species in the relaxation process associated with viscous flow in high silica melts (Stebbins, 1991; Stebbins et al., 1992; Poe et al., 1992a,b, 1994), and also for the formation of three-membered siloxane rings in $K_2Si_4O_9$ melt. Could this suggest a common origin for these structural features in silicate melts (see chapter by Wolf and McMillan, for a discussion of similar features in the high pressure spectra)?

There are good reasons for wishing to obtain these equilibrium constants from vibrational spectroscopic data, to complement the high temperature NMR studies. First, the Raman experiments should provide a good independent test of the NMR data. In addition, the NMR spectra can not be followed to very high temperatures, because the Q^n species exchange reactions begin to occur sufficiently rapidly that the NMR signal is averaged to a single Lorentzian line, and the resonances due to individual species can no longer be followed. The faster response timescale of the vibrational spectroscopic technique permits the speciation reaction to be followed to much higher temperature, providing a longer temperature "lever" for the determination of ln K versus 1/T (McMillan et al., 1992). However, there remain ambiguities concerning the assignment of the high frequency bands of silicate glasses and melts, which introduce some uncertainty into the determination of relative species concentrations.

First, the fully-polymerized network of SiO_2 (Q^4 species) gives rise to weak Raman bands near 1060 and 1200 cm^{-1}. As discussed in the previous section, these are weakened and broadened into a single feature near 1100 cm^{-1}, which can only just be distinguished in high temperature Raman spectra of SiO_2, in favourable cases. This would indicate that

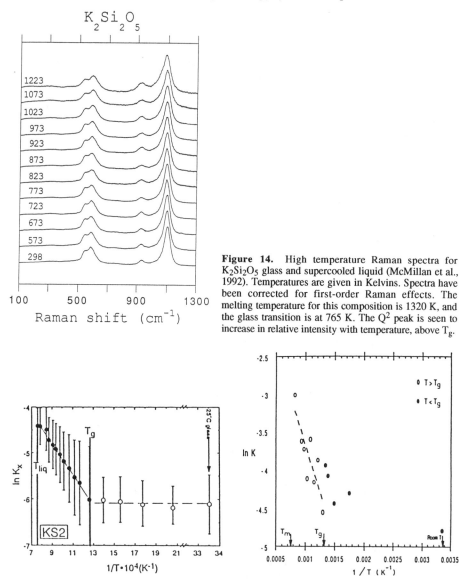

Figure 14. High temperature Raman spectra for $K_2Si_2O_5$ glass and supercooled liquid (McMillan et al., 1992). Temperatures are given in Kelvins. Spectra have been corrected for first-order Raman effects. The melting temperature for this composition is 1320 K, and the glass transition is at 765 K. The Q^2 peak is seen to increase in relative intensity with temperature, above T_g.

Figure 15. Plots of the logarithm of the equilibrium constant for the disproportionation reaction between silicate Q^n species $2Q^3 = Q^4 + Q^5$, determined from the relative Raman intensities of characteristic Si-O stretching peaks in the spectrum of $K_2Si_2O_5$ liquid, as a function of temperature. Plot on the left is taken from Mysen and Frantz (1994a); that on the right from McMillan et al. (1992). The difference in slope between the two plots, and the difference in magnitude of ln K, is due to different assumptions about the assignment of unresolved high frequency bands in the spectra. McMillan et al. (1992) obtained 20.6 ± 5.5 kJ/mol for the reaction, whereas Mysen and Frantz (1994a) obtained 27.7 kJ/mol. McMillan et al. (1992) suggested that there might be an effect of temperature on the relative Raman intensities of the Q^2 and Q^3 bands below T_g (765 K): Mysen and Frantz (1994a) find this not to be the case. [Figure at left used by permission of the editor of *Contributions to Mineralogy and Petrology*, from Mysen and Frantz (1994a), Fig. 12, p. 11].

no high frequency band characteristic of fully polymerized SiO_2 units should be visible in the Raman spectra of high silica melts. However, features are present in this region for alkali silicates at high silica content, and it is not yet known if some of these might represent a modified version of the Q^4 bands of SiO_2, appearing with greater intensity in the binary glasses and melts (Matson et al., 1983; McMillan et al., 1992; Mysen and Frantz, 1993b,c). Additional bands are also expected in this same high frequency region. For example, the asymmetric stretching vibrations associated with Q^3 and Q^2 species should also occur between 1050 and 1150 cm^{-1} (Furukawa et al., 1981; McMillan, 1984a,b), and would give rise to bands of similar intensity to the Q^4 features. Both of these considerations must complicate the quantitative determination of species concentrations from the Raman spectra. Broad features between 1050 and 1150 cm^{-1} do appear in the high frequency spectra of the binary glasses and melts (Furukawa et al., 1981; Matson et al., 1983; McMillan et al., 1992; Mysen and Frantz, 1993b,c), and different authors have chosen to use different band assignments for these. Because of the large width of the these bands, their contribution to the relative intensity can be quite large, and this likely results in the spread of values determined for the equilibrium constant in different studies (Mysen, 1990b; McMillan et al., 1992; Mysen and Frantz, 1993b,c; 1994). A further complication in this spectral region is the occurrence of a shoulder near 1160 cm^{-1} in the spectra of high silica glasses with large alkali cations (K^+, Rb^+, Cs^+) (Matson et al., 1983). This shoulder was assigned by Matson et al. (1983) and Fukumi et al. (1990) to the symmetric stretching vibration of a "modified" $Q^{3'}$ species, presumably distinguished from the "normal" Q^3 unit by the relative arrangement of alkali cations about the non-bridging oxygen. This assignment appears to be borne out by the behavior of this band in high pressure and high temperature studies (Wolf et al., 1990; McMillan et al., 1992; Durben, 1993: see chapter by Wolf and McMillan). This band merges into the "normal" Q^3 band for $K_2Si_4O_9$ glass at high temperature, providing information on the local relaxation of K^+ ions about the non-bridging oxygen (McMillan et al., 1992).

This type of experiment is particularly important in determining the variation in ΔH^o for the Q^3, Q^2 and Q^4 disproportionation reactions as a function of silica content, and of the nature of the charge balancing cation. Mysen and Frantz (1993a) found little dependence of ΔH^o on silica content, for compositions more silica-rich than the disilicate, for Na- and K-bearing silicate liquids. Although the absolute values of ΔH^o are subject to some uncertainty because of the ambiguities in Raman band assignment described above (the error is not likely to exceed a few kJ/mol, which represents the scatter between different experimental determinations), the relative values are probably realistic within a given study. From the data of Mysen and Frantz (1993a, 1994a), the disproportionation reaction is less endothermic, by ~5 to 10 kJ/mol, for Na-silicates compared with K-silicates, and lies near zero for lithium disilicate composition. ΔH^o in fact becomes *exothermic* for higher silica content Li-silicates (Mysen and Frantz, 1993a, 1994a,c). These variations reflect the increased tendency toward unmixing of the silicate melts, in the order Li > Na > K. Mysen and Frantz (1993a, 1994a,b) combined these data with liquidus phase equilibria to extract information on the activity coefficients of the silicate polymer species in the three series of melts. This work was extended to the $BaO-SiO_2$ system by Mysen and Frantz (1993b).

In the infrared reflection spectra of alkali silicate glasses, the strong high frequency reflectance band of SiO_2 glass (the reflectance band maximum occurs near 1150 cm^{-1}, corresponding to the TO resonance at 1060 cm^{-1}, and has a reflectivity close to 70% for pure SiO_2) decreases rapidly in reflectivity with added alkali oxide component, and can no longer be distinguished in the spectra below approximately 75 mol % SiO_2

(Florinskaya, 1959; Sanders et al., 1974) (Fig. 16). The same is true for the high frequency shoulder on this reflectance band. This observation can not be used to conclude that fully-polymerized Q^4 units are no longer present in the glass at lower silica content, because they are observed or required by ^{29}Si and Raman studies on the same glass series (Mysen et al., 1982a; Murdoch et al., 1985; Stebbins, 1987, 1988; Maekawa et al., 1990; Mysen, 1990b; McMillan et al., 1992; Mysen and Frantz, 1992, 1993a,b; 1994a,b). Either the strongly infrared active band due to asymmetric stretching vibrations of the Q^4 units becomes modified due to the increasing depolymerization of the glass network, or it is simply unresolved from the other strong high frequency bands in this region at lower silica content.

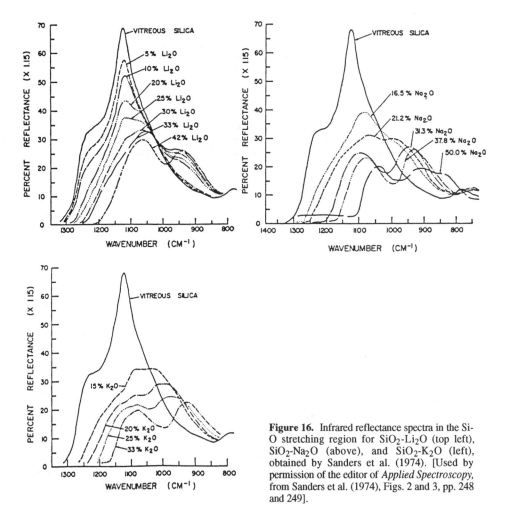

Figure 16. Infrared reflectance spectra in the Si-O stretching region for SiO_2-Li_2O (top left), SiO_2-Na_2O (above), and SiO_2-K_2O (left), obtained by Sanders et al. (1974). [Used by permission of the editor of *Applied Spectroscopy*, from Sanders et al. (1974), Figs. 2 and 3, pp. 248 and 249].

The disappearance of the asymmetric stretching vibrations of the SiO_2 component is accompanied by the growth in intensity of a reflectance band near 1075 cm^{-1}, corresponding to an absorption band with maximum near 1050 cm^{-1} (Sweet and White, 1969; Sanders et al., 1974; Domine and Piriou, 1983). This band shifts to lower wavenumber as the silica content is decreased, and is joined by a lower frequency component (reflectance maximum in the 1000 to 900 cm^{-1} region: the frequency decreases with lower silica content), which begins to grow as the silica content is reduced below approximately 85 mol % (Sweet and White, 1969; Sanders et al., 1974; Takashima and Saito, 1976, 1977; Domine and Piriou, 1983). These two features most likely correspond to asymmetric Si-O stretching vibrations, in silicate polymer units, dominated by Q^3 and Q^2 units, respectively (Gaskell, 1967; Furukawa et al., 1981; Dowty, 1987a,b,c). A weak additional feature appears at lower frequency (800 to 900 cm^{-1}) in the spectra of some glass compositions at lower silica content, due to asymmetric stretching of more depolymerized species (Sanders et al., 1974). A band in this region has been measured in the infrared absorption spectra of orthosilicate glasses, due to the v_3 asymmetric stretching vibration of isolated SiO$_4$ (Q^0) groups (Virgo et al., 1980; Williams et al., 1989). This corresponds to a broad, depolarized band in the Raman spectra of these glasses (McMillan et al., 1981; Piriou and McMillan, 1983; Williams et al., 1989; Cooney et al., 1990). The symmetric stretching vibration for these Q^0 species occurs in the Raman spectrum at slightly lower wavenumber (850 to 870 cm^{-1}), and is highly polarized (Virgo et al., 1980; Mysen et al., 1980a, 1982a; McMillan et al., 1981; Piriou and McMillan, 1983; McMillan, 1984b; Williams et al., 1989).

Although the features due to asymmetric Si-O stretching vibrations in the infrared spectra of binary silicate glasses correspond to depolarized bands at approximately the same wavenumber in the Raman spectra, there is no guarantee that the IR and Raman measurements are in fact sampling the same modes in the vibrational density of states. This was discussed above for the high frequency bands in the IR and Raman spectra of SiO$_2$ glass. The infrared bands are by definition due to polar modes, in which there is a change in net dipole moment during the vibrational motion. If there is an effective inversion centre operating within the glass, over the distance scale sampled during the IR or Raman experiment (one IR wavelength corresponds to ~10^4 nm, and the wavelength of incident blue-green light in a Raman scattering experiment is ~500 nm), the modes sampled in the Raman experiment will *not* be the same as those recorded by infrared spectroscopy, although they may correspond to the same type of Si-O stretching motion. This point, which is relevant for the detailed understanding of the vibrational behavior of silicates at high temperature, has been discussed by several authors (Denisov et al., 1978, 1980, 1984; Gervais et al., 1988; Parot-Rajaona et al., 1994; McMillan et al., 1994). It was considered above for vitreous SiO$_2$, and is discussed further for binary silicate glasses below.

The high frequency IR spectra of silicate melts are not likely to be as useful as the Raman spectra for determining the energetics of the Q speciation reactions, because the IR bands associated with different Q^n species shift with silica content, presumably due to the delocalized nature of the asymmetric stretching vibration. In addition, the absolute reflectivity diminishes rapidly with increasing temperature, due to the increased damping for the anharmonic Si-O stretching modes (Markin and Sobolev, 1960; Gervais, 1983; Domine and Piriou, 1983), and will be associated with a larger absorption coefficient at high temperature. This vibrational effect will interfere strongly with any relative intensity changes due to configurational relaxation in the melt, so that quantitative IR spectroscopy can not be used easily to determine the equilibrium constant for the disproportionation reaction.

On the other hand, the infrared measurements might play a critical role in studying high coordinate species present in silicate melts. It is known from NMR studies that high coordinate (SiO$_5$) species are present in the high temperature melt, formed as an intemediate in the O^{2-} exchange reaction between silicate polymer species (Stebbins, 1991; Stebbins and McMillan, 1993). The concentration of these species is expected to be small at magmatic temperatures, below 0.2% (Stebbins, 1991; Stebbins and McMillan, 1993) (at least, at low pressure: see chapters by Stebbins, and by Wolf and McMillan). No feature has yet been reliably assigned to the symmetric stretching vibration of such species in the Raman spectra of silicate glasses or melts (Dickinson et al., 1990; Wolf et al., 1990; Xue et al., 1991; Mysen and Frantz, 1994a). Even if such a band would be expected to appear (Poe et al., 1992a), it is likely that its intensity would be swamped by the dominant symmetric Si-O stretching vibrations of the tetrahedral silicate units (Mysen and Frantz, 1994a). On the other hand, the infrared absorption spectrum might be expected to contain a feature due to the *asymmetric* stretching vibrations of such species (Poe et al., 1992a). Xue et al. (1991) observed a feature near 900 cm^{-1} in the IR reflectance spectra of Na- and K-silicate glasses prepared at high pressure, which appeared to be correlated with the presence of 5- and 6-coordinated silicate species. Similar bands in the 700 to 900 cm^{-1} range have been observed in in situ high pressure infrared spectra of binary silicate glasses, and assigned to the presence of high coordinate silicon (Williams and Jeanloz, 1988; Kubicki et al., 1992). It is possible that further detailed infrared studies of silicate melts at high temperature will reveal the presence of SiO_5 species, as expected from the NMR work on quenched glasses.

Infrared spectra of binary silicate glasses show a medium-intensity band in the 450-500 cm^{-1} region, which can be assigned to the tetrahedral SiO_4 deformation vibration (δ_{OSiO}), mixed with the symmetric component of the SiOSi bending vibration, for more polymerized silicates (Gaskell, 1967; Lazarev, 1972; Dowty, 1987a). Usually only a single band is observed, which moves to higher frequency as the silica content is decreased (Sweet and White, 1969; Sanders et al., 1974; Dowty, 1987a). Although similar features are observed in the Raman spectrum in this frequency, with the same general assignment, it is likely that the types of vibrational mode sampled are quite different.

First, it appears that the infrared resonances are Lorentzian in form, rather than the Gaussian bands observed in the Raman spectra (Mysen et al., 1982b; Gervais et al., 1988; Mysen, 1990b,c; Parot-Rajaona et al., 1994). This implies that the infrared active vibrations are lifetime-limited, rather than a band shape determined by a distribution of structural units (see second section of this chapter). This observation implies that the polar infrared active modes are coupled through longer range (electrostatic) interactions than the principal modes probed by Raman spectroscopy, which are presumably more localized (Gervais et al., 1990; Parot-Rajaona et al., 1994). This also has implications for the *coherence length* of the infrared active vibrations, which could give valuable additional information on the structural relaxation processes in high temperature melts (Parot-Rajaona et al., 1993). Most of the spectroscopic experiments used in silicate melt studies to date have probed the local structure and the relaxation processes in the *time* domain: in future studies, it will be extremely important to develop techniques to address the *spatial* dependence of relaxation phenomena.

A further interesting phenomenon related to these observations has been described by Gervais et al. (1988). These authors investigated the infrared reflectance spectra of silicate glasses containing sodium, and observed that the absorption band at 460 cm^{-1} due to SiO_4 deformations was asymmetric. This asymmetry was so marked that, in some cases, the absorption coefficient dropped to zero near 330 cm-1, on the low frequency side of the peak, and was enhanced on the high frequency side of the absorption. These authors suggested that this behavior was due to a constructive-destructive interference relationship (i.e. a Fano interaction: Fano, 1961) between the bending vibrations of the silicate glass framework and the coupled displacements of the Na^+ ions within the glass (Gervais et al., 1988). The Na^+ motions give rise to a broad band maximized near 100 cm^{-1}. These vibrations are highly anharmonic, and give rise to a broad, asymmetric absorption band extending to approximately 500 cm^{-1}, under the SiO_4 deformation band. In the analysis of Gervais et al. (1988), the interference results from a vibrational interaction between the near-continuum of highly delocalized, polar Na^+ motion and the more localized tetrahedral bending vibrations. This effect was only observed for glasses containing sodium, not for other alkali or alkaline earth cations (Gervais et al., 1987, 1988; Parot-Rajaona et al., 1992, 1994).

The Raman spectra of silicate glasses and melts in the 500 to 600 cm^{-1} region usually show two bands, whose relative intensities vary with composition, temperature, and pressure (Mysen et al., 1980a, 1982a; Matson et al., 1983; Wolf et al., 1990; Xue et al., 1991; McMillan et al., 1992; Mysen and Frantz, 1994a,b) (Figs. 12, 13 and 14). One band is consistently observed near 600 cm^{-1}, nearly independent of the nature of the alkali cation. Based on its insensitivity to the counter-ion, Matson et al. (1983) suggested that this band might have a similar origin to the 606 cm^{-1} "defect" band of vitreous silica, and assigned it to a symmetric oxygen breathing vibration of three-membered siloxane rings within the glass structure. The frequency of this band was found to be nearly independent of temperature for $K_2Si_4O_9$ glass and melt, further supporting this assignment (McMillan et al., 1992). The intensity of the band is observed to increase with increasing temperature (Fig. 17), giving a formation enthalpy for these three-membered ring species of approximately 19.5 kJ/mol (McMillan et al., 1992). This is smaller than the endothermic three-membered ring defect species in SiO_2 glass (42 kJ/mol: Geissberger and Galeener, 1983), reflecting the relative ease of formation of these structural units in the presence of non-bridging oxygens (see chapter by Wolf and McMillan, for a discussion of the relationship between the formation of three-membered rings and Si^V species).

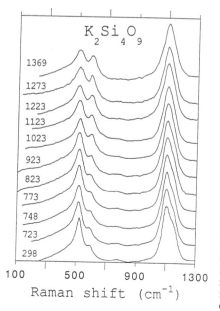

Figure 17. High temperature Raman spectra for $K_2Si_4O_9$ glass and melt (McMillan et al., 1992). Temperatures are given in Kelvins. The melting temperature for this composition is 1043 K, and the glass transition is at 752 K. The 606 cm^{-1} peak, assigned to 3-membered siloxane ring groups in the glass and liquid, increases in relative intensity with temperature above T_g. The formation enthalpy for this species is calculated to be 19.5 kJ/mol.

A similar conclusion was reached by Wolf et al. (1990), who observed that the intensity of the three-membered ring band increased with increasing pressure in $Na_2Si_4O_9$ glass at room temperature. This observation indicates that the ring formation reaction has a negative ΔV relative to larger rings or unclosed chain units with non-bridging oxygens, as might be expected.

The low frequency region of the Raman spectra of binary silicate glasses also contains a band in the 500 to 600 cm^{-1} region, whose frequency depends much more upon the nature of the counter ion, the temperature, and the pressure (Mysen et al., 1980a, 1982a; Matson et al., 1983; Wolf et al., 1990; Xue et al., 1991; McMillan et al., 1992; Mysen and Frantz, 1994a,b). This is most likely due to the SiOSi deformation vibration in depolymerized silicate units, coupled with tetrahedral OSiO bending vibrations, as for the infrared active mode in this region. At higher silica content, a broader feature due to SiOSi linkage vibrations within fully polymerized Q^4 units can also be observed (Matson et al., 1983).

Only a few studies of silicate glasses have investigated very low frequency region (below ~400 cm^{-1}), which contains vibrational modes due to vibrations of the network modifying cations, coupled with complex deformations of the silicate polymer (Exarhos and Risen, 1972). Raman spectra of the heavier alkaline earth silicates, especially Ca^{2+} bearing compositions, often show a broad band in the 200 to 350 cm^{-1} region, which can be identified with the stretching vibration of the metal cation against a (relatively stationary?) oxygen cage (Lazarev, 1972; Brawer and White, 1977; Virgo et al., 1980; Mysen et al., 1980a, 1982a; McMillan, 1984a,b). This is not the case for Raman spectra of Mg-silicates, for which the Mg-O vibrations appear to be strongly coupled with the deformational modes of the silicate framework (McMillan, 1984a,b). This was also suggested to be the case for the far-IR spectra of Ca- and Mg-bearing glasses (Gervais et al., 1987). However, in a more recent study, Hauret et al. (1994) have found well-defined bands which can be attributed to Ca^{2+} and Mg^{2+} cation cage vibrations, at 245 and 340 cm^{-1}, respectively. In the far-IR spectra of sodium silicate glasses, the position of the sodium ion vibration decreases from near 240 cm-1 to approximately 80 cm-1 with increasing degree of polymerization of the silicate framework (Merzbacher and White, 1988). This vibration is highly anharmonic, and gives rise to a broad, asymmetric band extending to above 500 cm^{-1} (Gervais et al., 1988). This corresponds to a coherent displacement of the Na$^+$ ions, coupled through the long range Coulombic field in the glass lattice, giving rise to the polar long wavelength vibration which interferes with the silicate deformation vibrations near 450 cm^{-1}, as discussed above (Gervais et al., 1988). In future studies, it will be important to explore the composition and temperature dependence of these features in silicate glasses and melts, because they are likely to give extremely valuable information on the behavior of alkali and alkaline earth cations both in the melt, and at temperatures below the glass transition, where glassy ionic conductivity might be encountered (Angell, 1990: see chapter by Chakraborty). This would provide an extremely useful complement to NMR studies of the alkali or alkaline earth cation behavior at high temperature (Stebbins, this volume).

Aluminosilicates along the "charge-balanced" SiO_2-$MAlO_2$ or SiO_2-MAl_2O_4 joins

Compositional changes in the Raman spectra of aluminosilicate glasses are not quite as well understood as those for binary silicates. This is largely because the bands are much broader than for the binary systems, presumably due to the perturbing effect of Al on the silicate stretching and bending vibrations, and the greater distribution in structural environments possible, so that it is more difficult to identify individual vibra-

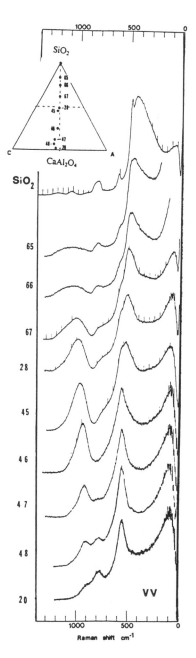

tional band components. From diffraction studies and NMR experiments, the glass framework in the SiO_2-$MAlO_2$ (M = Li, Na, K, Rb, Cs) or SiO_2-MAl_2O_4 (M = Ca, Sr, Ba) systems is well described in terms of a fully-polymerized network of tetrahedral silicate and aluminate units (Taylor and Brown, 1979a,b; Taylor et al., 1980; Murdoch et al., 1985; Engelhardt and Michel, 1987; Oestrike et al., 1987; Merzbacher et al., 1990). This agrees with the general observation that infrared and Raman spectra of glasses along these joins show some similarity to those of their crystalline counterparts (Sharma and Simons, 1981; McMillan et al., 1982; Seifert et al., 1982; Sharma et al., 1983; Matson et al., 1986; Roy, 1987). A ^{27}Al MAS NMR study of glasses along (or at least, very near) the SiO_2-$MgAl_2O_4$ join showed obvious features which could be attributed to the presence of high coordinate (5- and 6-) aluminate species (McMillan and Kirkpatrick, 1992). The Raman spectra of these glasses do not differ substantially (i.e. no obvious new bands are observed, although the Raman bands are less well resolved) from those of samples along the SiO_2-$NaAlO_2$ or SiO_2-$CaAl_2O_4$ joins (Seifert et al., 1982; McMillan et al., 1982), indicating that the effects of high coordinate Al species on the Raman spectra might be subtle (McMillan and Piriou, 1982, 1983; Poe et al., 1992a). This is discussed further below.

The Raman and infrared spectra between 850 and 1250 cm^{-1} contain several unresolved component bands, which are due to stretching vibrations of the tetrahedral aluminosilicate framework. Because the Al-O stretching vibrations of aluminates tend to lie at lower wavenumber (Tarte, 1967; Taylor, 1990; Poe et al., 1992a), these are probably best assigned as Si-O stretching modes, perturbed by the presence of Al coordinating the oxygen (Figs. 18 and 19). When a small amount of aluminate component is added to SiO_2 glass, a polarized band appears at approximately 1140 cm^{-1}, between the 1060 and 1200 cm^{-1} bands of the SiO_2 framework (Fig. 20) (Mysen et al., 1980a;

Figure 18. Parallel (VV) polarized Raman spectra for glasses along the SiO_2-$CaAl_2O_4$ join (McMillan et al., 1982). Glass compositions are shown (in mole %) in the inset (C = CaO; A = Al_2O_3), keyed to sample identification numbers (McMillan et al., 1982). These room temperature spectra have not been normalized for the effects of frequency- or temperature-dependence.

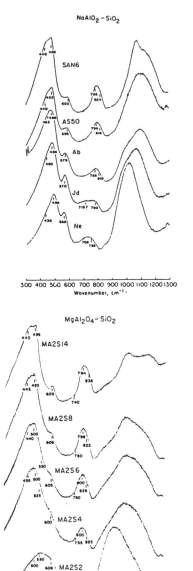

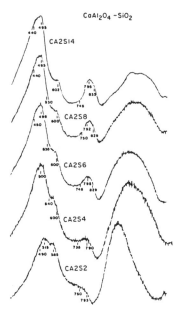

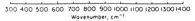

Figure 19. Unpolarized Raman spectra for glasses along the SiO_2-$NaAlO_2$, SiO_2-$CaAl_2O_4$ and SiO_2-$MgAl_2O_4$ joins, obtained by Seifert et al. (1982). These data have been corrected for the temperature dependence of the Raman scattering. The SiO_2 content (in mol %) along the join corresponding to eah glass is: SiO_2-$NaAlO_2$ series: SAN - 87.5; AS50 - 80; Ab - 75; Jd - 67; Ne - 50; SiO_2-$CaAl_2O_4$ series: CA2I4 - 87.5; CA2S8 - 80; CA2S6 - 75; CA2S4 - 67; CA2S2 - 50; CATS - 33; CA6S4 - 25; $CaAl_2O_4$ - 0; SiO_2-$MgAl_2O_4$ series: MA2S14 - 87.5; MA2S8 - 80; MA2S6 - 75; MA2S4 - 67; MA2S2 - 50. [From Seifert et al. (1982), Fig. 3, p. 701].

McMillan et al., 1982; McMillan and Piriou, 1982). McMillan et al. (1982) suggested that this could be assigned to a "symmetric" Si-O stretching vibration of a "Q^4(1 Al)" species (Oestrike et al., 1987; Engelhardt and Michel, 1987): i.e. an SiO_4 tetrahedron in which three oxygens are bridging to other Si atoms, and the fourth is bound to Al (this was given the pictorial symbol: $\equiv$SiOAl in McMillan et al. 1982). Because the Al-O bond is weaker than Si-O, this vibration should behave somewhat like a symmetric Si-O_{NBO} stretching mode, as found in Na- or Mg-silicates. However, the Al-O bond *is* considerably stronger than Na-O or Mg-O, so the vibration will have some of the character of an antisymmetric stretching vibration, as found in the inter-tetrahedral SiOSi linkage. This intermediate character and strong perturbation results in the high frequency of the vibration (1140 cm^{-1}, compared with 1100 cm^{-1} for a "normal" Q^3 stretching mode), and its low intensity and high depolarization ratio (McMillan et al., 1982).

As the silica content is reduced, the intensity of this band increases. It is rapidly joined by a second band with maximum near 1020 cm^{-1} (Fig. 19). This second band is already present as a shoulder on the higher frequency band by 80 mol % SiO_2, and the two are approximately equal in intensity by around 65 mol % SiO_2, for alkali aluminosilicate joins (Mysen et al., 1980a; Sharma and Simons, 1981; Seifert et al., 1982; McMillan et al., 1982). By 50 mol % SiO_2, ($NaAlSiO_4$ or $CaAl_2Si_2O_8$ composition), the 1020 cm^{-1} band is present as a single, quite narrow and symmetric, component. This second band was assigned by McMillan et al. (1982) to the "symmetric" Si-O(Al) stretching vibration of a silicate tetrahedron with two of its oxygens bound to adjacent Si, and two to Al atoms (a Q^4(2 Al) species). Both the "1140 cm^{-1}" and the "1020 cm^{-1}" bands do appear to show a general decrease in their frequency (by 20 to 40 cm^{-1}) with decreasing silica content (Seifert et al., 1982; Mysen et al., 1983), which would reflect the effects of changing next-nearest-neighbours in the aluminosilicate glass framework with composition. Additional components due to silicate tetrahedra surrounded by (3 Al, 1 Si) and (4 Al) were identified in the glass spectra by McMillan et al. (1982), at lower silica content.

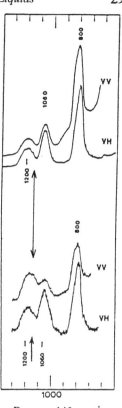

Raman shift, cm^{-1}

Figure 20. High frequency Raman spectra in the Si-O-(Si,Al) stretching region for SiO_2 (top) and a glass along the SiO_2-$KAlSi_3O_8$ join (bottom), showing the appearance of a polarized band near 1140 cm^{-1}, between the bands for SiO_2 at 1060 and 1200 cm^{-1} (McMillan et al., 1982).

The form of the narrow high frequency band for the 50 mol % SiO_2 glasses suggests that these are almost the sole structural species present at this composition, which is in agreement with the results of NMR spectroscopy (Oestrike et al., 1987; Engelhardt and Michel, 1987; Merzbacher et al., 1990). This implies that the glasses at this composition are highly ordered, with a predominance of AlOSi linkages between the Q^4(2 Al) species (Oestrike et al., 1987).

This observation would be consistent with the occurrence of the "aluminium avoidance rule" in the aluminosilicate glasses (Lowenstein, 1954; Murdoch et al., 1985; Oestrike et al., 1987; Engelhardt and Michel, 1987; Merzbacher et al., 1990). McMillan et al. (1982) had originally suggested that many of the intertetrahedral linkages within the glass were AlOAl, based on the assignment of a band at 570 cm^{-1} in the lower frequency region of the spectrum. This would be inconsistent with the ordered glass structure proposed above (Oestrike et al., 1987). This assignment is discussed further below. However, it should be noted that there is nothing intrinsically against the occurrence of AlOAl linkages within the glass or melt. Lowenstein's (1954) arguments were based essentially on the non-occurrence of inter-tetrahedral AlOAl links in minerals with SiO_2 content $\geq$ 50 mol %, for which the most ordered arrangement contains no bonds between aluminate tetrahedra. This does not reflect any inherent "instability" in such linkages: indeed, they are found in the tetrahedral framework of the highly refractory ceramic phase $CaAl_2O_4$ (which has a tridymite-like aluminate framework), and corundum (Al_2O_3, with AlOAl linkages between the weaker Al-O bonds of octahedral groups) melts at 2054°C and is harder than most silicate minerals! In fact, intertetrahedral AlOAl linkages have been described in the crystalline framework of disordered anorthite ($CaAl_2Si_2O_8$) (Carpenter, 1991a,b), and it is highly probable that they do occur in aluminosilicate glasses and melts, even with $\geq$ 50 mol % SiO_2.

Based on the observation that the 1140 cm^{-1} and 1020 cm^{-1} bands appear together, with sub-equal intensities, in the Raman spectra of glasses along the SiO_2-$CaAl_2O_4$ join (Fig. 18), McMillan et al. (1982) suggested that local clustering of structural units into SiO_2-rich and Ca- and Al-rich domains might occur within the glass. This should be more marked than for the SiO_2-$NaAlO_2$ system, in which the 1140 cm^{-1} and 1020 cm^{-1} bands appear sequentially. McMillan et al. (1982) rationalized this by a model based on the local charge balance requirements of alkali and alkaline earth ions, and suggested that the clustering should become more marked in the sequence K < Na < Li < Ca < Mg, based on existing spectroscopic data. A similar observation was made by Sefert et al. (1982), within the context of the structural model developed by those authors. This tendency to clustering was borne out by the results of high temperature solution calorimetry (Navrotsy et al., 1982; Roy and Navrotsky, 1984; Navrotsky, 1985: see chapter by Navrotsky, this volume). Positive deviations from ideality in the heats of mixing were found for glasses along the SiO_2-$MAlO_2$ and SiO_2-MAl_2O_4 joins, and the magnitude of the deviation increased with increasing field strength of the metal cation. This is behavior which might eventually lead to metastable or stable liquid-liquid phase separation, and might provide a driving force for unmixing in aluminosilicate liquids at high silica content (Navrotsy et al., 1982; Roy and Navrotsky, 1984; Navrotsky, 1985: chapters by Navrotsky and by Hess, this volume).

Seifert et al. (1982) and Mysen et al. (1983) developed a different interpretation of the high frequency bands of these "charge balanced" aluminosilicate glasses, described in the discussion of vitreous SiO_2 above, based on a theoretical model for the vibrations of tetrahedral frameworks developed by Sen and Thorpe (1977), Galeener and Sen (1978), and Galeener (1979). This model has recently been extended into a study of the high temperature aluminosilicate liquids by Neuville and Mysen (1995). In the model, the 1140 cm^{-1} and 1020 cm^{-1} bands represent coupled (Si,Al)-O stretching vibrations within distinct aluminosilicate frameworks, distinguished by their average TOT (T = Si,Al) angle. The higher frequency component corresponds to the framework with a larger average inter-tetrahedral angle.

The "two-domain" model developed for SiO_2, with different TOT angular distributions for each "domain", was extended to glasses along the SiO_2-$MAlO_2$ and SiO_2-MAl_2O_4 joins by Seifert et al. (1982), and Mysen et al. (1983, 1985), and has been used to interpret the Raman spectra of melts in the SiO_2-$NaAlO_2$ system (Neuville and Mysen, 1995). The spectral interpretation is important, because the relative intensity changes observed in the two high frequency bands as a function of composition, temperature and pressure might be used to construct models for Al partitioning between the two structural domains, with potential implications for magma behavior (Seifert et al., 1982; Mysen et al., 1983; Mysen, 1988, 1990b).

Systematic infrared transmission and reflectance studies of glasses along these joins have been carried out (Roy, 1987, 1990; Hobert et al., 1985; Merzbacher and White, 1988; Parot-Rajaona et al., 1993, 1994). In the transmission studies, the strong high frequency peak of vitreous SiO_2 near 1100 cm^{-1} moves to lower wavenumber and becomes less intense as aluminate component is added, and a low frequency shoulder near 950 cm^{-1} grows in intensity. These two features are not always well resolved (Roy, 1987, 1990). These vibrations are almost certainly due to asymmetric stretching vibrations within the TOT linkages (Furukawa et al., 1981; Dowty 1987a), but their precise assignment is not yet clear. The nature of these highly polar IR-active modes has been discussed by Parot-Rajaona et al. (1994), who consider the question of vibrational coherence within the disordered glass network. The infrared spectra show a band in the 400-500 cm^{-1} region, assigned to the tetrahedral deformation vibration (δ_{OTO}), as in the Al-free silicates. This always appears as a single peak, which shifts to lower frequency as the silica content is decreased. It occurs at 468 cm^{-1} for pure SiO_2, and at 425 cm^{-1} for $CaAl_2O_4$ glass (Poe et al., 1992a).

As for the case of the binary silicates, the behavior of the Raman spectra in the 450-600 cm^{-1} region is more complicated, and has been studied in greater detail. The vibrations in this region most likely correspond to symmetric bending motions of the oxygen within TOT linkages, coupled with the OTO tetrahedral angular deformation. As aluminate component is added to SiO_2, the dominant Raman band at 430 cm^{-1} moves to higher wavenumber, and becomes slightly narrower (Figs. 18 and 19). A feature near 580 cm^{-1} grows in relative intensity. Over most of the composition range, these two more or less resolved bands are observed. Close to the silica-free end of the join, the two merge to form the single, sharp band of $CaAl_2O_4$ glass near 570 cm^{-1} (Sharma et al., 1979; Mysen et al., 1981; Sharma and Simons, 1981; McMillan et al., 1982; Sharma et al., 1983; Seifert et al., 1982; Mysen et al., 1983; Fleet et al., 1984; Matson and Sharma, 1985; Matson et al., 1986; Poe et al., 1994; Daniel et al., 1995; Neuville and Mysen, 1995).

There has much discussion regarding the assignment of the feature near 580 cm^{-1}. Seifert et al. (1982) and Mysen et al. (1983) considered that this would correspond to the low frequency component of the TOT coupled stretching mode, in the network model of Sen and Thorpe (1977) and Galeener and Sen (1978), corresponding to the aluminosilicate species with smaller TOT angle. McMillan et al. (1982) proposed that it was due to an AlOAl bending vibration due to these species within the glass, because it appeared to grow into the dominant band of silica-free $CaAl_2O_4$ glass. Matson et al. (1986) suggested instead that this 580 cm^{-1} feature was due to three-membered rings of $(Al,Si)O_4$ tetrahedra, analogous to the siloxane rings proposed to account for the 606 cm^{-1} "defect" peak in the spectrum of SiO_2 glass (Galeener, 1982a,b).

This last interpretation now appears to be supported by results of *ab initio* molecular orbital calculations on such three membered ring species (Kubicki and Sykes, 1993), and by the relative insensitivity of this band frequency to increasing temperature (Daniel et al., 1995). The band at lower wavenumber would then correspond to vibrations of TOT linkages with larger average angle (note that, at this point, the interpretation converges with that of Seifert et al., 1982, and Mysen et al., 1983), not constrained within rings. Even this interpretation may be indistinguishable from that of McMillan et al.

(1982). In SiOSi linkages, the angle at the two-coordinate oxygen is relatively unconstrained, unless the tetrahedra are joined into a small ring (McMillan et al., 1994). In contrast, SiOAl and especially AlOAl linkages are associated with the presence of a charge-balancing cation coordinating the bridging oxygen. This must result in an angular constraint through metal...metal repulsion in the oxygen coordination sphere, so that AlOSi and AlOAl angles are smaller, and have a narrower distribution. This is consistent with the narrower Raman bands, at higher frequency, observed for aluminosilicate compositions.

It is likely that the structure of $CaAl_2O_4$ glass and liquid does not consist entirely of three-membered rings of aluminate tetrahedra, although they are most likely present, but of AlOAl linkages all at around the same angle, largely determined by the balance of Al...Al, Ca...Al and Ca...Ca repulsions about the bridging oxygen. It is of some interest that, if this "metal-metal non-bonded repulsion" concept is applied to other glass compositions, it leads to some interesting structural considerations. For example, nepheline ($NaAlSiO_4$) and anorthite ($CaAl_2Si_2O_8$) compositions are usually considered to provide structural analogues, because the silica content and metal oxide:alumina ratio are the same. However, the metal packing schemes should be quite different: 1:1:1 in one case, and 1:2:2 in the other, so that the glass structures and melt relaxation behavior might be different for the two cases.

Recent studies have addressed the structural changes occurring in glasses and melts along the SiO_2-$NaAlSiO_4$ and SiO_2-$CaAl_2Si_2O_8$ joins to magmatic temperatures, using Raman spectroscopy (Seifert et al., 1981; McMillan et al., 1994; Poe et al., 1994; Daniel et al., 1995; Neuville and Mysen, 1995). The changes observed in SiO_2 glass and melt were described above (McMillan et al., 1994). For $NaAlSi_3O_8$ and $CaAl_2Si_2O_8$ compositions, once the spectra were corrected for the frequency- and temperature-dependent effects associated with first-order Raman scattering, the growth of a broad band underneath the entire Raman spectrum, with its intensity maximized between 600-900 cm^{-1}, was observed (Daniel et al., 1995) (Fig. 21). This was an extremely exciting result, because this is the wavenumber range in which stretching vibrations of high coordinate aluminate species (AlO_5 and AlO_6 groups) would be expected to occur (Tarte, 1967; Williams and Jeanloz, 1988, 1989; Taylor, 1990; Poe et al., 1992a). Substantial concentrations of these species had been predicted to be present in high temperature aluminosilicate liquids with similar compositions from ion dynamics simulations, and from ^{27}Al NMR spectroscopy on melts and glasses (Scamehorn and Angell, 1991; Coté et al., 1992a,b; Poe et al., 1992a,b; 1993, 1994; McMillan and Kirkpatrick, 1992). Because these species are expected to play an important role in viscous relaxation in the high temperature melts, their direct observation via vibrational spectroscopy would have been quite important.

However, growth of a similar broad feature had already been reported for pure SiO_2 glass, at temperatures well below the glass transition (McMillan et al. 1994) (Fig. 10). This can not be due to configurational changes, but must be vibrational in nature. The feature in SiO_2 glass was identified as being due to an increase in intensity of the *second-order* Raman spectrum, associated with the increased occurence of two-phonon processes at high-temperature (McMillan et al. 1994). Examination of the data for albite and anorthite indicated that growth of the "additional" intensity in the 600 to 900 cm^{-1} region also began in the glassy regime, well below T_g, and so also was vibrational in nature (Daniel et al., 1995a). This feature is mainly due to the second order Raman spectra of $NaAlSi_3O_8$ and $CaAlSi_2O_8$, arising from combination bands involving the vibrational modes below 600 cm^{-1} (Fig. 22). Its intensity only increases in the "temperature

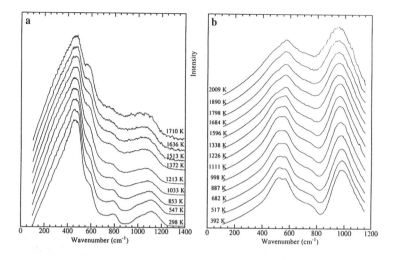

Figure 21. Raman spectra of (a) NaAlSi$_3$O$_8$ glass, supercooled liquid and stable liquid (T$_g$ = 1096 K; T$_m$ = 1380 K) and (b) CaAl$_2$Si$_2$O$_8$ glass and liquid (T$_g$ = 1160 K; T$_m$ = 1830 K) from Daniel et al. (1995a). The spectra in this Figure have been "corrected" for the frequency- and temperature dependence of the first-order Raman scattering only. The growth of the band near 580 cm^{-1} above T$_g$ is apparent in both sets of spectra. A broad feature maximized in the 800 to 900 cm^{-1} region also appears to grow. with increasing temperature.

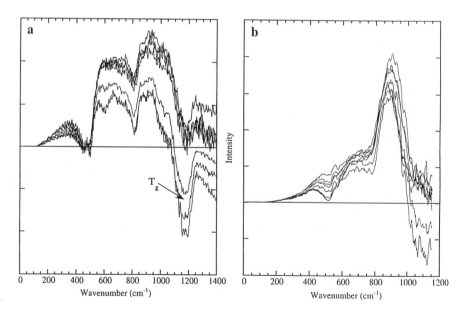

Figure 22. The second order Raman spectra of (a) NaAlSi$_3$O$_8$ and (b) CaAl$_2$Si$_2$O$_8$ glass and melt, obtained via a difference method (Daniel et al., 1995a). Some changes above T$_g$ are observed, due to configurational contributions appearing in the spectra, in addition to the second order Raman scattering.

normalized" spectra, because the temperature dependence of the second order spectrum is different from that for first order Raman scattering, and so is incorrectly accounted for by the standard "normalization" procedure. This was discussed in the second section of this chapter. The procedure developed for evaluating the contribution of the second order Raman spectrum, and applying the correct normalization formalism to the high temperature spectra, is described in detail by Daniel et al. (1995a). Once this is applied, no anomalous increase in intensity in the 600 to 900 cm^{-1} region is observed (Fig. 23).

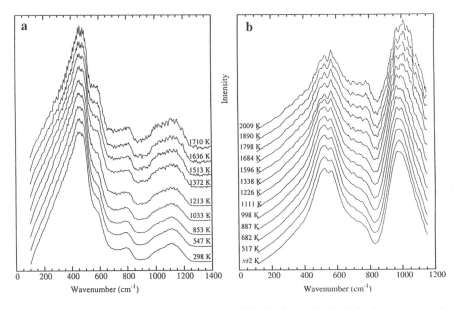

Figure 23. The high temperature Raman spectra of (a) NaAlSi$_3$O$_8$ and (b) CaAl$_2$Si$_2$O$_8$ glass and melt, obtained after temperature normalization, and removing the contribution due to second order Raman scattering (Daniel et al., 1995a).

One obvious change in the high temperature Raman spectra of albite and anorthite liquids concerns the relative band intensities in the 400 to 600 cm^{-1} region (Seifert et al., 1981; Daniel et al., 1995; Neuville and Mysen, 1995) (Figs. 21 and 23). Analogous to the observation already made for SiO$_2$ and the binary silicates, the band near 580 cm^{-1} increases with intensity with increasing temperature above T$_g$. This indicates that the proportion of three-membered aluminosilicate rings, with narrower average TOT angle, increases in the melt. This is consistent with an endothermic formation enthalpy for these species, as already found for three-membered siloxane rings in the Al-free silicates (Galeener and Geissberger, 1983; McMillan et al., 1992, 1994; Mysen and Frantz, 1994a). Analogous to the silicate ring species, it is also likely that these species would have a negative formation volume, so would be expected to be quite abundant in highly polymerized aluminosilicate melts, at high pressure and temperature within the Earth. Since it appears that these species react preferentially with dissolved water in the melt (McMillan, 1994), this could be an important geochemical observation.

In a very recent study, Daniel et al. (1995b) have studied the changes in the Raman spectrum of CaAl$_2$O$_4$ liquid as a function of temperature. In this study, care was taken to correct the spectra for vibrational effects, so that configuration changes could be examined. Above Tg, growth of a broad band in the 700 to 900 cm^{-1} region was observed. These changes were compared with a high pressure study on the same glass. It was concluded that the aluminate units in $CaAl_2O_4$ liquid most likely remained tetrahedral, but that OAl_3 "tricluster" units were formed (Daniel et al., 1995b).

Haplobasaltic and other aluminosilicate compositions

Several studies, primarily those by Brawer and White (1977), Mysen et al. (1981, 1985), Mysen (1990b,c) and Mysen and Frantz (1994b), have investigated the structure of aluminosilicate glasses and liquids for the geochemically relevant compositions between the binary silicate join and the SiO_2-$MAlO_2$ or SiO_2-MAl_2O_4 "charge-balanced" line, which begin to model the compositions of basaltic to andesitic melts. Although the remaining ambiguities in the spectral interpretation of the simpler systems must affect our understanding of the structural changes in these melts, some interesting general observations can be made. The spectra of binary Na- and Li-tetrasilicates show a weak peak due to Q^2 units near 950 cm^{-1}, and a dominant Raman band due to Q^3 (Mysen, 1990c) (Fig. 24). Addition of aluminate component shows the growth of a new band in this region, near 1050 cm^{-1}. This is presumably due to a vibration of an aluminosilicate unit within the glass, which could be the same as that described previously near the same position for glasses along the SiO_2-$MAlO_2$ joins. $K_2Si_4O_9$ glass only exhibits a weak feature near 950 cm^{-1}, indicating the low abundance of Q^2 units for this composition. This peak appears to increase in intensity as the aluminate component is added, which might indicate that the Al-free domains within the glass have lower average polymerization. On the basis of this type of observation, Mysen et al. (1981) and Mysen

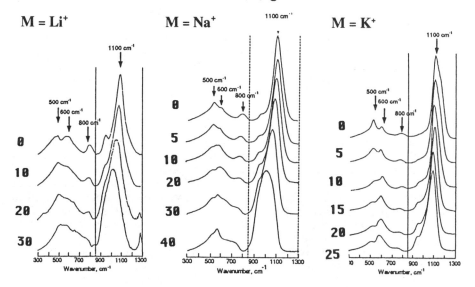

Figure 24. Room temperature Raman spectra of glasses with varying aluminate component, along the joins $M_2Si_4O_9$-$M_2(MAl)_4O_9$ [M = Li$^+$ (left); Na$^+$ (middle); K$^+$ (left)], obtained by Mysen (1990c). The spectra have been corrected for (first-order) temperature effects. The numbers refer to the quantity of $M_2(MAl)_4O_9$ component substituted for $M_2Si_4O_9$ in the glass composition, in mol %. [From Mysen (1990c), Figs. 5, 6 and 7, p. 125 and 126.]

(1990a,b,c) have suggested that Al partitions more strongly into structural units within the glass which have higher polymerization. This parallels a suggestion by McMillan and Piriou (1983) that, for glasses along the binary CaO-Al$_2$O$_3$ join, extended aluminate polymer units with average polymerization near Q3 (sheet structures) might be unstable. This suggestion has important potential consequences for activity-composition relations within the melt, and for crystal-liquid phase relations involving these liquids (Mysen et al., 1985; Mysen, 1988, 1990a,b). Mysen and Frantz (1994b) have noted that the presence of Al affecs the temperature dependence of the Q^2 and Q^3 band intensities in the melt, so that aluminate component must play an important role in determining the energetics of the silicate speciation reactions, as might be expected. It is obvious that much more work needs to be done, both on these multicomponent systems, and on the simpler binary and ternary joins, before a detailed view of melt structure changes can be established from this type of data, but it is certain that these experiments can lead to a deep understanding of the energetics and mixing relations in such aluminosilicate liquids.

Other components

Fe^{2+} and Fe^{3+} are major components of natural melts. There have been several studies of Fe-bearing glasses and melts, often using Raman or IR spectroscopy in conjunction with Mössbauer and X-ray absorption spectroscopies, to begin to ascertain the structural role of iron (Mysen et al., 1980b; Virgo and Mysen, 1985; Cooney and Sharma, 1990; Wang et al., 1993). These studies have also extended to other transition metal-containing glasses and liquids. Titanium is a petrologically important minor element, which can have large anomalous effects on melt structure and properties (Dingwell, 1992; Lange and Navrotsky, 1993). The structural state of titanium has been studied in alkali and alkaline earth titanosilicate glasses (Mysen et al., 1980c), and recently the structural changes in Na$_2$Si$_2$O$_5$-Na$_2$Ti$_2$O$_5$ liquids has been investigated as a function of temperature using Raman spectroscopy (Mysen and Neuville, 1995). It is obvious that this represents a rich field of research for the future. One interesting feature in such studies is that the metal-oxygen bond is quite strong, and vibrates at similar frequencies to Si-O stretching (Mysen et al., 1980c; Williams et al., 1989; Cooney and Sharma, 1990), and there may be competition between the transition metal (or lanthanide) and silicon for the available oxygen: i.e. these cations are strong Lux-Flood acids (Ellison and Hess, 1990).

CONCLUSION

From the studies carried out to date, it is obvious that vibrational spectroscopy is an invaluable technique for probing the atomic level dynamics and configurational changes in aluminosilicate glasses and liquids at magmatic temperatures. The vibrational data are most useful when taken together with results obtained from other spectroscopic or diffraction experiments, and used to construct melt structure models which can be tested against measurements of thermodynamic or transport properties. Although most of the results described here have concerned infrared or Raman spectroscopy carried out at high temperature at ambient pressure, the vibrational techniques are ideally suited to in situ studies at high pressure (see chapter by Wolf and McMillan, in this volume), and a few combined in situ high pressure-high temperature studies are beginning to appear (Farber and Williams, 1992) (see also Fig. 25). This will be necessary to understand the structures and properties of silicate liquids under conditions appropriate to melting in the crust and mantle. Much of the work carried out to date has been concerned with investigating the structural changes in simple binary and ternary silicate and aluminosilicate compositions, which are often not of direct geological relevance. However, the results obtained from these systems can usually be applied almost directly

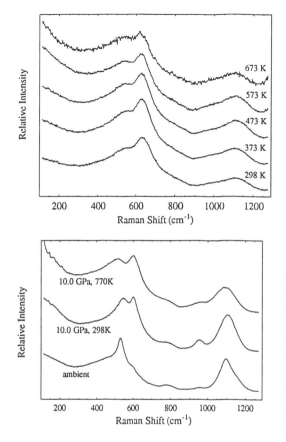

Figure 25. Top: In situ Raman spectrum of Na_2SiO_4 glass as a function of temperature, taken at 10 GPa (S. Robinson, G. Wolf, P. McMillan, unpublished data). Bottom: comparison of the ambient pressure 298 K Raman spectra of Na_2SiO_4 glass decompressed from: (a) 10 GPa and 770 K. (b) 10 GPa and 298 K. (c) an unpressurized sample. At ambient pressure the glass transition temperature is 750 K. This likely decreases ~20° by 10 GPa. Spectral changes due to structural relaxation in the high pressure melt are clearly visible.

to constructing models for the structures and relaxation behavior of geochemically important melts, and studies on multi-component systems are now under way. An important direction for the future of melt studies will be to fully develop our understanding of the structural nature of simple glasses and melts, and how the different structural environments, and apply this knowledge to natural melt systems. Many aspects of this goal are well under way, as emphasized by the contributions in this volume, although much remains to be done. Finally, this discussion has been concerned only with the vibrational properties and structures of dry (i.e. volatile-free) melts. Much information has been obtained on the properties of volatile-containing melts, and the role of species such as H_2O and CO_2 in determining their structural, thermodynamic, and rheological behavior. Vibrational spectroscopy has played an important role in these studies also. The information on such volatile-bearing systems, and their geological implications, has been summarized in a previous *Reviews in Mineralogy* (Volume 30, Carroll and Holloway, 1994).

ACKNOWLEDGMENTS

The authors are indebted to their colleagues, collaborators and students with whom they have studied and debated silicate glass and melt structure, relaxation, and NMR and vibrational spectroscopy over the years. In particular, we thank Isabelle Daniel, Brent Poe and Scott Robinson for their efforts in obtaining high quality Raman spectra at high temperatures, several of which have been used to illustrate this review. We owe a special thanks to Paul Ribbe, Series Editor, without whom there would be no *Reviews in Mineralogy*. PFM also thanks his co-organizers Jonathan Stebbins and Don Dingwell for many stimulating discussions of structure and relaxation processes in melts. Our research on silicate melt studies has been supported by grants from the National Science Foundation, Experimental Geochemistry program.

REFERENCES

Aasland S, McMillan PF (1994) Density-driven liquid-liquid phase separation in the system Al_2O_3-Y_2O_3. Nature 369:633-636

Allen DC, Teter MP (1987) Nonlocal pseudopotentials in molecular-dynamical density-functional theory: application to SiO_2. Phys Rev Lett 59:1136-1139

Allen DC, Teter MP (1990) Local density approximation total energy calculations for silica and titania structure and defects. J Am Ceram Soc 73:3247-3250

Angell CA (1984) Strong and fragile liquids. In: KL Ngai, GB Wright (eds) Relaxations in Complex Systems, p 3-11. ONR and National Technical Information Service, Washington, DC

Angell CA (1988) Structural instability and relaxation in liquid and glassy phases near the fragile liquid limit. J Non-Cryst Solids 102:205-221

Angell CA (1990) Dynamic processes in ionic glasses. Chem Rev 90:523-542

Angell CA (1991) Relaxation in liquids, polymers and plastic crystals-strong/fragile patterns and problems. J Non-Cryst Solids 131/133:13-31

Angell CA (1995) Formation of glasses from liquids and biopolymers. Science 267:1924-1935

Angell CA, Torell LM (1983) Short time structural relaxation processes in liquids: comparison of experimental and computer simulation glass transition on picosecond timescales. J Chem Phys 78:937-945

Bando Y, Ishizuka K (1979) Study of the structure of silica glass by high-resolution electron microscopy. J Non-Cryst Solids 33:375-388

Bates JB (1972) Dynamics of β-quartz structures of vitreous SiO_2 and BeF_2. J Chem Phys 56:1910-1917

Bell RJ, Bird NF, Dean P. (1968) The vibrational spectra of vitreous silica, germania and beryllium fluoride. J Phys Chem 1:299-303

Bell RJ, Carnevale A, Kurkjian CR, Peterson GE (1980) Structure and phonon spectra of SiO_2, B_2O_3 and mixed SiO_2-B_2O_3 glasses. J Non-Cryst Solids 35/36:1185-1190

Bell RJ, Dean P (1966) Properties of vitreous silica: analysis of random network models. Nature 212:1354-1356

Bell RJ, Dean P (1970) Atomic vibrations in vitreous silica. Trans Faraday Soc 50:55-61

Bell RJ, Dean P (1972) Localization of phonons in vitreous silica and related glasses. In: RW Douglas, DA Ellis (ed) 3rd Int'l Conf Phys Non-Cryst Solids, p 443-452. Wiley-Interscience, New York

Bell RJ (1976) Vibrational properties of amorphous solids. Methods Comp Phys 15:215-276

Bilz H, Kress W (1979) Phonon Dispersion Relations in Insulators. Springer-Verlag, New York

Brandriss ME, Stebbins JF (1988) Effects of temperature on the structures of silicate liquids: ^{29}Si NMR results. Geochim Cosmochim Acta 52:2659-2669

Brawer SA, White WB (1975) Raman spectroscopic investigation of the structure of silicate glasses. I. The binary alkali silicates. J Chem Phys 63:2421-2432

Brawer SA, White WB (1977) Raman spectroscopic investigation of the structure of silicate glasses. (II). Soda-alkaline earth-alumina ternary and quaternary glasses. J Non-Cryst Solids 23:261-278

Bruckner R (1970) Properties and structure of vitreous silica. I. J Non-Cryst Solids 5:123-175

Bucaro JA, Dardy HD (1974) High-temperature Brillouin scattering in fused quartz. J Appl Phys 45:5324-5329

Carpenter MA (1991a) Mechanisms and kinetics of Al-Si ordering in anorthite: I. Incommensurate structure and domain coarsening. Am Mineral 76:1110-1119

Carpenter MA (1991b) Mechanisms and kinetics of Al-Si ordering in anorthite: II. Energetics and a Ginzburg-Landau rate law. Am Mineral 76:1120-1133

Carroll M, Holloway JR (eds) (1994) Volatiles in Magmas. Reviews in Mineralogy, Vol 30. Mineralogical Society of America, Washington, DC, 517 p

Cooney TF, Sharma SK (1990) Structure of glasses in the systems Mg_2SiO_4-Fe_2SiO_4, Mn_2SiO_4-Fe_2SiO_4, Mg_2SiO_4-$CaMgSiO_4$, Mn_2SiO_4-$CaMnSiO_4$. J Non-Cryst Solids 122:10-32

Coté B, Massiot D, Taulelle F, Coutures JP (1992a) ^{27}Al NMR spectroscopy of aluminosilicate melts and glasses. Chem Geol 96:367-370

Coté B, Massiot D, Poe BT, McMillan PF, Taulelle F, Coutures JP (1992b) Liquids and glasses: structural differences in the CaO-Al_2O_3 system as evidenced by ^{27}Al NMR spectroscopy. J Phys IV Coll C2:223-226

Daniel I, Gillet Ph, Poe BT, McMillan PF (1995a) In-situ high temperature Raman spectroscopic studies of aluminosilicate liquids. Phys Chem Minerals 22:74-86

Daniel I, Gillet Ph, Poe BT, McMillan PF (1995b) Raman spectroscopic study of structural changes in calcium aluminate ($CaAl_2O_4$) glass at high pressure and high temperature Chem Geol, submitted

Daniel I, Gillet Ph, McMillan PF, Richet P (1995c) An in situ high-temperature structural study of stable and metastable $CaAl_2Si_2O_8$ polymorphs. Mineral Mag 59:25-34

Dean P (1972) The vibrational properties of disordered systems: numerical studies. Rev Mod Phys 44:127-168

Decius JC, Hexter RM (1977) Molecular Vibrations in Crystals. McGraw-Hill, New York

Demkov AA, Ortega J, Sankey OF, Grumbach MP (1995) Electronic structure approach for complex silicas. Phys Rev B 52:1618-1630

Denisov VN, Mavrin BN, Podobedov VB, Sterin KE (1978) Hyper-Raman scattering and longitudinal-transverse splitting of vibrations in fused quartz. Sov Phys Solid State 20:2016-2017

Denisov VN, Mavrin BN, Podobedov VB, Sterin KE (1980) Hyper-Raman scattering by polaritons in fused quartz. JETP Lett 32:316-319

Denisov VN, Mavrin BN, Podobedov VB, Sterin KE, Varshal BG (1984) Law of conservation of momentum and rule of mutual exclusion for vibrational excitations in hyper-Raman and Raman spectra of glasses. J Non-Cryst Solids 64:195-210

Devine RAB, Dupree R, Farnan I, Capponi JJ (1987) Pressure-induced bond-angle variation in amorphous SiO_2. Phys Rev B 35:2560-2562

Dickinson JE, Scarfe CM, McMillan P (1990) Physical properties and structure of $K_2Si_4O_9$ melt quenched from pressures up to 2.4 GPa. J Geophys Res B95:15675-15681

Dingwell DB (1992) Shear viscosity of alkali and alkaline earth titanium silicate liquids. Am Mineral 77:270-274

Dingwell DB, Webb S.L. (1990) Relaxation in silicate melts. Eur J Mineral 2:427-449

Domine F, Piriou B (1983) Study of sodium silicate melts and glass by infrared reflectance spectroscopy. J Non-Cryst Solids 57:125-130

Dowty E (1987a) Vibrational interactions of tetrahedra in silicate glasses and crystals: I. Calculations on ideal silicate-aluminate-germanate structural units. Phys Chem Minerals 14:80-93

Dowty E (1987b) Vibrational interactions of tetrahedra in silicate glasses and crystals: II. Calculations on melilites, pyroxenes, silica polymorphs and feldspars. Phys Chem Minerals 14:122-138

Dowty E (1987c) Vibrational interactions of tetrahedra in silicate glasses and crystals: III. Calculations on simple alkali silicates, thortveitite and rankinite. Phys Chem Minerals 14:542-552

Dupree R, Pettifer RF (1984) Determination of the Si-O-Si bond angle distribution in vitreous silica by magic angle spinning NMR. Nature 308:523-525

Dupree R, Holland D, McMillan PW, Pettifer RF (1984) The structure of soda-silicate glasses: a MAS NMR study. J Non-Cryst Solids 68:399-410

Dupree R, Holland D, Williams DS (1985) A magic angle spinning study of the effect of modifier and intermediate oxides on the local structure in vitreous silicate networks. J Phys 46:C8, 119-123

Dupree R, Holland D, Williams DS (1986) The structure of binary alkali silicate glasses. J Non-Cryst Solids 81:185-200

Durben DJ (1993) Raman Spectroscopic Studies of the High Pressure Behavior of Network Forming Tetrahedral Oxide Glasses. Arizona State Univ, Tempe, AZ

Elliot SR (1991) Medium-range structural order in covalent amorphous solids. Nature 354:445-452

Ellison AJG, Hess PC (1990) Lathanides in silicate glasses: a vibrational spectroscopic study. J Geophys Res B 95:15717-15726

Engelhardt G, Michel D (1987) High-Resolution Solid-State NMR of Silicates and Zeolites. John Wiley & Sons, New York

Etchepare J (1972) Study by Raman spectroscopy of crystalline and glassy diopside. In: RW Douglas, DA Ellis (ed) Amorphous Materials, p 337-346. Wiley-Interscience, New York

Etchepare J, Merian M, Smetankine L (1974) Vibrational normal modes of SiO_2 I α and β quartz. J Chem Phys 60:1873-1876

Exarhos GJ, Risen WM (1972) Cation vibrations in inorganic oxide glasses. Solid state Comm 11:755-758

Exarhos GJ, Frydrych WS, Walrafen GE, Fisher M, Pugh E, Garofalini SH (1988) Vibrational spectra of silica near 2400 K: measurement and molecular dynamics simulation. In: RJH Clark, DA Long (eds) Proc 11th Int'l Conf Raman Spectroscopy, p 503-504. John Wiley & Sons, New York

Fano U (1961) Effects of configuration interaction on intensities and phase shifts. Phys Rev 124:1866-1878

Farber DL, Williams Q (1992) Pressure-induced coordination changes in alkali-germanate melts: an in situ spectroscopic investigation. Science 256:1427-1430

Fleet ME, Herzberg CT, Henderson GS, Crozier ED, Osborne MD, Scarfe CM (1984) Coordination of Fe, Ga and Ge in high pressure glasses by Mössbauer, Raman and X-ray absorption spectroscopy, geological implications. Geochim Cosmochim Acta 48:1455-1466

Flood H, Forland T (1947) The acidic and basic properties of oxides. Acta Chem Scand 1:592-604

Florinskaya VA (1959) Infrared reflection spectra of sodium silicate glasses and their relationship to structure. The Structure of Glass—Proc 3rd All-Union Conf Glassy State, 2:154-168. Consultants Bureau, Leningrad

Frick B, Richter D (1995) The microscopic basis of the glass transition in polymers from neutron scattering studies. Science 267:1939-1945

Fukumi K, Hayakawa J, Komiyama T (1990) Intensity of Raman band in silicate glasses. J Non-Cryst Solids 119:297-302

Furukawa T, Fox KE, White WB (1981) Raman spectroscopic investigation of the structure of silicate glasses. III. Raman intensities and strucutral units in sodium silicate glasses. J Chem Phys 75:3226-3237

Galeener FL (1979) Band limits and vibrational spectra of tetrahedral glasses. Phys Rev B 19:4292-4298

Galeener FL (1982a) Planar rings in glasses. Solid State Comm 44:1037-1040

Galeener FL (1982b) Planar rings in vitreous silica. J Non-Cryst Solids 49:53-62

Galeener FL (1985) Raman and ESR studies of the thermal history of amorphous SiO_2. J Non-Cryst Solids 71:373-386

Galeener FL, Sen N (1978) Theory for the first-order vibrational spectra of disordered solids. Phys Rev B17:1928-1933

Galeener FL, Geissberger AE (1983) Vibrational dynamics in ^{30}Si-substituted vitreous SiO_2. Phys Rev B27, 6199-6204

Galeener FL, Leadbetter AJ, Stringfellow MW (1983) Comparison of the neutron, Raman, infrared vibrational spectra of vitreous SiO_2, GeO_2, BeF_2. Phys Rev B 27:1052-1078

Galeener FL, Mikkelsen JC (1981) Vibrational dynamics in ^{18}O-substituted vitreous SiO_2. Phys Rev B23:5527-5530

Ganguly AK, Berman JL (1967) Theory of lattice Raman scattering in insulators. Phys Rev B162:806-816

Gaskell PH (1966a) Thermal properties of silica. Part 1—Effect of temperature on infra-red reflection spectra of quartz, cristobalite and vitreous silica. Trans Faraday Soc 62:1493-1504

Gaskell PH (1966b) Thermal properties of silica. Part 2—Thermal expansion properties of vitreous silica. Trans Faraday Soc 62:1505-1510

Gaskell PH (1967) The vibrational spectra of silicates. Part I. Phys Chem Glasses 8:69-80

Gaskell PH (1975) Construction of a model for amorphous materials using ordered units. Phil Mag 32:211-229

Gaskell PH, Mistry AB (1979) High-resolution transmission microscopy of small amorphous silica particles. Phil Mag 39:245-257

Geissberger AE, Galeener FL (1983) Raman studies of vitreous SiO_2 versus fictive temperature. Phys Rev B 28:3266-3271

Gervais F (1983) High-temperature infrared reflectivity spectroscopy by scanning interferometry. In: KJ Burton (ed) Infrared and Millimeter Waves. Part 1. Electromagnetic Waves in Matter, 8:279-339. Academic Press, Orlando, FL

Gervais F, Blin A, Massiot D, Coutures JP, Chopinet MH, Naudin F (1987) Infrared reflectivity spectroscopy of silicate glasses. J Non-Cryst Solids 89:384-401

Gervais F, Blin A, Chopinet MH (1988) Fano effet in glasses. Solid State Comm 65:653-655

Gervais F, Lagrange C, Blin A, Aliari M, Hauret G, Coutures JP, leroux M (1990) Comparison of dielectric response deduced from infrared reflectivity and Raman spectra of silicate glasses. J Non-Cryst Solids 119:79-88

Gillet Ph (1993) Stability of magnesite ($MgCO_3$) at mantle pressure and temperature conditions. A Raman spectroscopic study. Am Mineral 78:1328-1331

Gillet Ph, Biellmann C, Reynard B, McMillan P (1993a) Raman spectroscopic studies of carbonates. Part I: High-pressure and high-temperature behavior of calcite, magnesite, dolomite and aragonite. Phys Chem Minerals 20:1-18

Gillet Ph., Fiquet G, Daniel I, Reynard B (1993b) Raman spectroscopy at mantle pressure and temperature conditions. Experimental set-up and the example of $CaTiO_3$ perovskite. Geophys Res Lett 20:1931-1934

Grimsditch M, Bhadra R, Meng Y (1988) Brillouin scattering from amorphous materials at high pressures. Phys Rev B38:7836-7838

Grimsditch M, Bhadra R, Torell LM (1989) Shear waves through the glass-liquid transformation. Phys Rev Lett 62:2616-2619

Grove FJ, Jellyman PE (1955) The infra-red transmission of glass in the range room temperature to 1400°. J Soc Glass Technology 39:3T-15T

Hadni A (1967) Essentials of Modern Physics Applied to the Study of the Infrared. Pergamon Press, Oxford

Halvorson KH, Raffaelle DP, Wolf GH, Marzke RF (1992) Proton NMR chemical shifts in organic liquids measured at high pressure using the diamond anvil cell. In: HD Hochheimer, RD Etters (eds) Frontiers of High Pressure Research, p 217-221. Plenum Press, New York

Hauret G, Vaills Y, Luspin Y, Gervais F, Coté B (1994) Similarities in the behavior of magnesium and calcium in silicate glasses. J Non-Cryst Solids 170:175-181

Hawthorne F (ed) (1988) Spectroscopic Methods in Mineralogy and Geology. Rev Mineral, Vol 18. Mineralogical Society of America, Washington, DC

Hemley RJ, Bell PM, Mao MK (1987) Laser techniques in high pressure geophysics. Science 237: 605-612

Hobert H, Dunken HH, Stephanowitz R, Marx R (1985) Characterization of oxides and glasses by IR spectroscopy. Z Chem 25:358-361

Huheey JE (1983) Inorganic Chemistry. Principles of Structure and Reactivity. 3rd ed. Harper and Row, New York

Ihinger P, Hervig RL, McMillan PF (1994) Analytical methods for volatiles in glasses. In: M Carroll, JR Holloway (eds) Volatiles in Magmas, Rev Mineral 30:67-121

Iwamoto N, Tsunawaki Y, Fuji M, Hattori T (1975) Raman spectra of K_2O-SiO_2 and K_2O-SiO_2-TiO_2 glasses. J Non-Cryst Solids 18:303-306

Iwamoto N, Umesaki N, Dohi K (1984) Structural investigation of $Rb_2O·4SiO_2$ glass and melt by Raman spectroscopy (in Japanese). Yogyo Kyokai Shi 92:201-209

Jeanloz R (1990) Thermodynamics and evolution of the Earth's interior: high pressure melting of silicate perovskite as an example. 1990 Gibbs Symp, Am Math Soc, p 211-226, Washington, DC

Jellyman PE, Procter MA, Procter JP (1955) XVI—Infra-red reflection spectra of glasses. J Soc Glass Technology 39:173t-192t

Johnson AV, Wright AC, Sinclair RN (1983) Neutron scattering from vitreous silica. II. Twin-axis diffraction experiments. J Non-Cryst Solids 58:109-130

Kashio S, Iguchi Y, Goto T, Nishina Y, Fuwa T (1980) Raman spectroscopy study on the structure of silicate slags. Trans Iron Steel Inst Japan 20:251-253

Kashio S, Iguchi Y, Fuwa T, Nishina Y, Goto T (1982) Raman spectroscopic study on the structure of silicate slags (in Japanese). Trans Iron Steel Inst Japan 68:123-129

Kirkpatrick RJ, Dunn T, Schramm S, Smith KA, Oestrike R, Turner F (1986) Magic-angle sample-spinningnuclear magnetic resonance spectroscopy of silicate glasses: a review. In: GE Walrafen, AG Revesz (eds) Structure and Bonding in Noncrystalline Solids, p 303-327. Plenum Press, New York.

Kirkpatrick RJ (1988) MAS NMR spectroscopy of minerals and glasses: In: Spectroscopic Methods in Mineralogy and Geology, FC Hawthorne (ed) Rev Mineral, 18:341-403

Konnert J, D'Antonio P, Huffman M, Navrotsky A (1987) Diffraction studies of a highly metastable form of amorphous silica. J Am Ceram Soc 70:192-196

Konnert J, D'Antonio P, Karle J (1982) Comparison of radial distribution function for silica glass with those for various bonding topologies. J Non-Cryst Solids 53:135-141

Kubicki JD, Sykes D (1993) Molecular orbital calculations of vibrations in three-membered aluminosilicate rings. Phys Chem Minerals 19:381-391

Kushiro I (1976) Changes in viscosity and structure of melts of $NaAlSi_2O_6$ composition at high pressure. J Geophys Res 81:6347-6350

Kushiro I (1978) Viscosity and structural change of albite ($NaAlSi_3O_8$) melt at high pressure. Earth Planet Sci Lett 41:87-90

Kushiro I (1981) Viscosity, density and structure of silicate melts at high pressures, and their petrological applications. In: RB Hargraves (ed) Physics of Magmatic Processes, p 93-120

Kushiro I (1983) Effect of pressure on the diffusivity of network-forming cations in melts of jadeite compositions. Geochim Cosmochim Acta 47:1415-1422

Kushiro I, Yoder HS, Mysen BO (1976) Viscosities of basalt and andesite melts at high pressure. J Geophys Res 81:6351-6357

Lange RA, Navrotsky A (1993) Heat capacities of TiO_2-bearing silicate liquids: evidence for anomalous changes in configurational entropy with temperature. Geochim Cosmochim Acta 57:3001-3012

Lazarev AN (1972) The Vibrational Spectra of Silicates. Consultants Bureau, New York

Lines ME (1987) Absolute Raman intensities in glasses. I. Theory. J Non-Cryst Solids 89:143-162

Loudon R (1983) The Quantum Theory of Light. Clarendon Press, Oxford

Lowenstein W (1954) The distribution of aluminum in the tetrahedra of silicates and aluminates. Am Mineral 32:92-96

Lux H (1939) "Acids" and "bases" in a fused salt bath: the determination of oxygen-ion concentration. Z Electrochem Soc 45:303-309

Markin EP, Sobolev NN (1960) Infrared reflection spectrum of boric anhydride and fused quartz at high temperatures. Optics Spectros 9:309-312

Marzke RF, Raffaelle DP, Halvorson KE, Wolf GH (1994) A ^{1}H NMR study of glycerol at high pressure. J Non-Crysta Solids 172-174:401-407

Matson DW, Sharma SK, Philpotts JA (1983) The structure of high-silica alkali-silicate glasses: a Raman spectroscopic investigation. J Non-Cryst Solids 58:323-352

Matson DW, Sharma SK (1985) Structures of sodium alumino- and gallosilicate glasses and their germanium analogs. Geochim Cosmochim Acta 49:1913-1924

Matson DW, Sharma SK, Philpotts JA (1986) Raman spectra of some tectosilicates and glasses along the orthoclase-anorthite and nepheline-anorthite joins. Am Mineral 71

McMillan P (1984a) Structural studies of silicate glasses and melts-applications and limitations of Raman spectroscopy. Am Mineral 69:622-644

McMillan P (1984b) A Raman spectroscopic study of glasses in the system $CaO-MgO-SiO_2$. Am Mineral 69:645-659

McMillan P (1985) Vibrational spectroscopy in the mineral sciences. In: SW Kieffer, A Navrotsky (eds) Microscopic to Macroscopic: Atomic Environments to Mineral Thermodynamics. Rev Mineral 14:9-63

McMillan P (1988) Vibrational spectroscopy of amorphous SiO_2. In: RAB Devine (ed) The Physics and Technology of Amorphous SiO_2, p 63-70. Plenum, New York

McMillan P (1989) Raman spectroscopy in mineralogy and geochemistry. Ann Rev Earth Planet Sci 17:255-283

McMillan PF (1994) Water solubility and speciation models. In: M Carroll, JR Holloway (eds) Volatiles in Magmas, Rev Mineral 30:131-156

McMillan P, Hofmeister AM (1988) Infrared and Raman spectroscopy. In: F Hawthorne (ed) Spectroscopic methods in Mineralogy and Geology, Rev Mineral 18:99-159

McMillan P, Piriou B, Navrotsky A (1982) A Raman spectroscopic study of glasses along the joins silica-calcium aluminate, silica-sodium aluminate, silica-potassium aluminate. Geochim Cosmochim Acta 46:2021-2037

McMillan PF, Hess AC (1990) Ab initio force field calculations for quartz. Phys Chem Minerals 17:97-107

McMillan PF, Kirkpatrick RJ (1992) Al coordination in magnesium aluminosilicate glasses. Am Mineral 77:898-900

McMillan PF, Piriou B (1982) The structures and vibrational spectra of crystals and glasses in the silica-alumina system. J Non-Cryst Solids 53:279-298

McMillan PF, Piriou B (1983) Raman spectroscopy of calcium aluminate glasses and crystals. J Non-Cryst Solids 55:221-242

McMillan P, Coutures JP, Piriou B (1981) Diffusion Raman d'un verre de monticellite. CR Acad Sci Paris Sér II 292:196-198

McMillan P, Piriou B, Couty R (1984) A Raman study of pressure-densified vitreous silica. J Chem Phys 81:4234-4236

McMillan PF, Wolf GH, Poe BT (1992) Vibrational studies of silicate liquids. Chem Geol 96:351-366

McMillan PF, Poe BT, Gillet Ph, Reynard B (1994) A study of SiO_2 glass and supercooled liquid to 1950 K via high-temperature Raman spectroscopy. Geochim Cosmochim Acta 58:3653-3664

McQuarrie DA (1976) Statistical Mechanics. Harper and Row, New York

Merzbacher CI, White WB (1988) Structure of sodium in aluminosilicate glasses: a far-infrared reflectance spectroscopic study. Am Mineral 73:1089-1094

Merzbacher CI, Sherriff BL, Hartman JS, White WB (1990) A high resolution ^{29}Si and ^{27}Al NMR study of alkaline earth aluminosilicate glasses. J Non-Cryst Solids 124:194-206

Mikkelsen JC, Galeener FL (1980) Thermal equilibration of Raman active defects in vitreous silica. J Non-Cryst Solids 37:71-84

Miller GH, Stolper EM, Ahrens TJ (1991a) The equation of state of a molten komatiite, 1, shock wave compression to 36 GPa. J Geophys Res 96:11831-11848

Miller GH, Stolper EM, Ahrens TJ (1991b) The equation of state of a molten komatiite, 2, application to komatiite petrogenesis and the Hadean mantle. J Geophys Res 96:11849-11856

Morey GW, Bowen NL (1924) The binary system sodium metasilicate-silica. J Phys Chem 28:1167-1179

Mozzi RL, Warren BE (1969) The structure of vitreous silica. J Appl Cryst 2:164-172

Murdoch JB, Stebbins JF, Carmichael ISE (1985) High-resolution ^{29}Si NMR study of silicate and aluminosilicate glasses: the effect of network-modifying cations. Am Mineral 70:332-343

Mysen BO (1988) Structure and Properties of Silicate Melts. 354 p Elsevier, Amsterdam

Mysen BO (1990a) Relationships between silicate melt structure and petrologic processes. Earth Sci Rev 27:281-365

Mysen B (1990b) Effect of pressure, temperature and bulk composition on the structure and species distrubution in depolymerized alkali aluminosilicate melts and quenched melts. J Geophys Res B95:15733-15744

Mysen BO (1990c) Role of Al in depolymerized, peralkaline aluminosilicate melts in the systems Li_2O-Al_2O_3-SiO_2, Na_2O-Al_2O_3-SiO_2, K_2O-Al_2O_3-SiO_2. Am Mineral 75:120-134

Mysen BO, Frantz JD (1992) Raman spectroscopy of silicate melts at magmatic temperatures: Na_2O-SiO_2, K_2O-SiO_2, Li_2O-SiO_2 binary compositions in the temperature range 25-1475°C. Chem Geol 96:321-332

Mysen BO, Frantz JD (1993a) Structure and properties of alkali silicate melts at magmatic temperatures. Eur J Mineral 5:393-407

Mysen BO, Frantz JD (1993b) Structure of silicate melts at high temperature: In-situ measurements in the system BaO-SiO_2 to 1669°C. Am Mineral 78:699-709

Mysen BO, Frantz JD (1994a) Silicate melts at magmatic temperatures: in situ structure determination to 1651°C and effect of temperature and bulk composition on the mixing behavior of structural units. Contrib Mineral Petrol 117:1-14

Mysen BO, Frantz JD (1994b) Structure of haplobasaltic liquids at magmatic temperatures: in situ, high temperature study of melts on the join $Na_2Si_2O_5$-Na_2 $(NaAl)_2O_5$. Geochim Cosmochim Acta 58:1711-1733

Mysen BO, Neuville DR (1995) Effect of temperature and TiO_2 content on the structure of $Na_2Si_2O_5$-$Na_2Ti_2O_5$ liquids and glasses. Geochim Cosmochim Acta 59:325-342

Mysen BO, Virgo D, Scarfe CM (1980a) Relations between the anionic structure and viscosity of silicate melts—a Raman spectroscopic study. Am Mineral 65:690-710

Mysen BO, Seifert F, Virgo D (1980b) Structure and redox equilibria of iron-bearing silicate melts. Am Mineral 65:867-884

Mysen BO, Ryerson FJ, Virgo D (1980c) The influence of TiO_2 on the structure and derivative properties of silicate melts. Am Mineral 65:1150-1165

Mysen BO, Virgo D, Kushiro I (1981) The structural role of aluminum in silicate melts- a Raman spectroscopic study at 1 atmosphere. Am Mineral 66:678-701

Mysen BO, Virgo D, Seifert FA (1982a) The structure of silicate melts: implications for chemical and physical properties of natural magma. Rev Geophys Space Phys 20:353-383

Mysen BO, Virgo D, Seifert FA (1982b) Curve-fitting of Raman spectra of silicate glasses. Am Mineral 67:686-695

Mysen BO, Virgo D, Danckwerth P, Seifert FA, Kushiro I (1983) Influence of pressure on the structure of melts on the joins $NaAlO_2$-SiO_2, $CaAl_2O_4$-SiO_2, $MgAl_2O_4$-SiO_2. Neues Jahrb Mineral Abh 147:281-303

Mysen BO, Virgo D, Seifert FA (1985) Relationships between properties and structure of aluminosilicate melts. Am Mineral 70:88-105

Navrotsky A (1985) Crystal chemical constraints on the thermochemistry of minerals. In: SW Kieffer, A Navrotsky (eds) Microscopic to Macroscopic: Atomic Environments to Mineral Thermodynamics, Rev Mineral 14:225-275

Navrotsky A, Geisinger KL, McMillan P, Gibbs GV (1985) The tetrahedral framework in glasses and melts—Inferences from molecular orbital calculations and implications for structure, thermodynamics, physical properties. Phys Chem Minerals 11:284-298

Navrotsky A, Peraudeau G, McMillan P, Coutures JP (1982) A thermochemical study of glasses and crystals along the joins silica-calcium aluminate and silica-sodium aluminate. Geochim Cosmochim Acta 46:2093-2047

Neuville DR, Mysen BO (1995) Role of aluminium in the silicate network: in situ, high-temperature study of glasses and melts on the join SiO_2-$NaAlO_2$. Geochim Cosmochim Acta, in press

O'Keeffe M, Gibbs GV (1984) Defects in amorphous silica: Ab initio MO calculations. J Chem Phys 81:876-879

Oestrike R, Yang WH, Kirkpatrick RJ, Hervig RL, Navrotsky A, Montez B (1987) High-resolution ^{23}Na, ^{27}Al and ^{29}Si NMR spectroscopy of framework aluminosilicate glasses. Geochim Cosmochim Acta 51:2199-2209

Ohashi Y, Hadidiacos CG (1976) A controllable thermocouple microheater for high-temperature microscopy. Carnegie Inst Wash Yearb 75:828-833

Parot-Rajaona T, Vaills Y, Massiot D, Gervais F (1992) Analysis of lithium aluminosilicate glasses by Raman scattering, infrared reflectivity, and ^{29}Si MAS-NMR spectroscopy. Proc. XVI Int'l. Cong. Glass, Madrid, Oct. 1992. Bol. Soc. Esp. Ceram. Vid., 31-C, vol. 3

Parot-Rajaona T, Coté B, Vaills Y, Gervais F (1993) Degree of coherence of vibrations in silicate glasses. Proc. Eur. Workshop on Glasses and Gels, Montpellier, France. J. Phys. IV Coll C2:227-230

Parot-Rajaona T, Coté B, Bessada C, Massiot D, Gervais F (1994) An attempt to reconcile interpretations of atomic vibrations and ^{29}Si NMR data in glasses. J Non-Cryst Solids 169:1-14

Paterson, M.S. (1982) The determination of hydroxyl by infrared absorption in quartz, silicate glasses and similar materials. Bull Minéral 105:20-29

Pauling L, Wilson EB (1963) Introduction to Quantum Mechanics with Applications to Chemistry. McGraw-Hill, New York. Reprinted 1985 by Dover Publications.

Person WB, Zerbi G (1982) Vibrational Intensities in Infrared and Raman Spectroscopy. Elsevier, Amsterdam

Piriou B, Alain P (1979) Density of states and structural forms related to structural properties of amorphous solids. High Temp High Press Res 11:407-414

Piriou B, Arashi H (1980) Raman and infrared investigations of lead silicate melts. High Temp Sci 13:299-313

Piriou B, McMillan P (1983) The high-frequency vibrational spectra of vitreous and crystalline orthosilicates. Am Mineral 68:426-443

Poe BT, McMillan PF, Sato RK, Angell CA (1992a) Al and Si coordination in SiO_2-Al_2O_3 glasses and liquids: a study by NMR and IR spectroscopy and MD simulations. Chem Geol 96:333-349

Poe BT, McMillan PF, Coté B, Massiot D, Coutures JP (1992b) SiO_2-Al_2O_3 liquids: in situ study by high-temperature ^{27}Al NMR spectroscopy and molecular dynamics simulations. J Phys Chem 96:8220-8224

Poe BT, McMillan PF, Coté B, Massiot D, Coutures JP (1993) The Al environment in $MgAl_2O_4$ and $CaAl_2O_4$ liquids: an in situ study via high temperature NMR. Science 259:786-788

Poe BT, McMillan PF, Coté B, Massiot D, Coutures JP (1994) Structure and dynamics in calcium aluminate liquids: high-temperature ^{27}Al NMR and Raman spectroscopy. J Am Ceram Soc 77:1832-1838

Poole PH, Sciortino F, Essmann U, Stanley HE (1992) Phase behavior of metastable water. Nature 360:324-328

Poole PH, Essmann U, Sciortino F, Stanley HE (1993) Phase diagram for amorphous solid water. Phys Rev E 48:4605-4610

Poole PH, Sciortino F, Grande T, Stanley HE, Angell CA (1994) Effect of hydrogen bonds on the thermodynamic behavior of liquid water. Phys Rev Lett 73:1632-1635

Poole PH, Grande T, Sciortino F, Stanley HE, Angell CA (1995) Amorphous polymorphism. J Comp Mat Sci (submitted)

Putnis A, Angel RJ (1985) Al, Si ordering in cordierite using magic angle spinning NMR II: models of Al, Si order from NMR data. Phys Chem Minerals 12:217-222

Reissland JA (1973) The Physics of Phonons. John Wiley & Sons, New York.

Richet P, Bottinga Y (1980) Heat capacity of liquid silicates: new measurements on $NaAlSi_3O_8$ and $K_2Si_4O_9$. Geochim Cosmochim Acta 44:1535-1541

Richet P, Bottinga Y (1984) Glass transitions and thermodynamic properties of amorphous SiO_2, $NaAlSi_nO_{2n+2}$ and $KAlSi_3O_8$. Geochim Cosmochim Acta 48:453-470

Richet P, Bottinga Y, Téqui C (1984) Heat capacity of sodium silicate liquids. J Am Ceram Soc 67:C6-C8

Richet P, Gillet Ph, Pierre A, Ali Bouhfid M, Daniel I, Fiquet G (1993) Raman spectroscopy, x-ray diffraction, phase relationship determinations with a versatile heating cell for measurements up to 3600 K (or 2700 K in air). J Appl Phys 74:5451-5456

Richet P, Neuville DR (1992) Thermodynamics of silicate melts: configurational properties. In: S Saxena (ed) Thermodynamic Data. Systematics and Estimation, Adv Phys Geochem 10:132-160. Springer-Verlag, New York

Rothschild WG (1986) Band shapes and dynamics in liquids. In: JR Durig (ed) Vibrational Spectra and Structure. A Series of Advances 15:57-109. Elsevier, Amsterdam

Roy BN (1987) Spectroscopic analysis of the structure of silicate glasses along the join $xMAlO_2$-$(1-x)SiO_2$ (M = Li, Na, K, Rb, Cs). J Am Ceram Soc 70:183-192

Roy BN (1990) Infrared spectroscopy of lead and alkaline-earth aluminosilicate glasses. J Am Ceram Soc 73:846-855

Roy BN, Navrotsky A (1984) Thermochemistry of charge-coupled substitutions in silicate glasses: The systems $M_{1/n}{}^{n+}AlO_2$-SiO_2 (M = Li, Na, K, Rb, Cs, Mg, Ca, Sr, Ba, Pb). J Am Ceram Soc 67:606-610

Sanders DM, Person WB, Hench LL (1974) Quantitative analysis of glass structure with the use of infrared reflection spectra. Appl Spectroscopy 28:247-255

Sato RK, McMillan P (1987) Infrared and Raman spectra of the isotopic species of α-quartz. J Phys Chem 91:3494-3498

Scamehorn C, Angell CA (1991) Viscosity-temperature relations and structure in fully-polymerized aluminosilicate melts from ion dynamics simulations. Geochim Cosmochim Acta 55:721-730

Scarfe CM, Mysen BO, Virgo D (1987) Pressure dependence of the viscosity of silicate melts. Geochem Soc Spec Pub No 1:59-67

Schmitt DR, Ahrens TJ (1989) Shock temperatures in silica glass: implications for modes of shock-induced deformation, phase transformation, and melting with pressure. J Geophys Res 94:5851-5872

Schramm SE, deJong BHWS, Parziale VE (1984) ^{29}Si magic angle spinning NMR study of local silicon environments in amorphous and crystalline lithium silicates. J Am Ceram Soc 106:4396-4402

Seifert FA, Mysen BO, Virgo D (1981) Structural similarity of glasses and melts relevant to petrological processes. Geochim Cosmochim Acta 45:1879-1884

Seifert FA, Mysen BO, Virgo D (1982) Three-dimensional network structure of quenched melts (glass) in the systems SiO_2-$NaAlO_2$, SiO_2-$CaAl_2O_4$ and SiO_2-$MgAl_2O_4$. Am Mineral 67:696-717

Sen N, Thorpe MF (1977) Phonons in AX_2 glasses: from molecular to band-like modes. Phys Rev B15:4030-4038

Shao J, Angell CA (1995) Vibrational anharmonicity and the glass transition in strong and fragile vitreous polymorphs. Proc Int'l Conf Glass, Beijing, China, submitted

Sharma SK, Mammone JF, Nicol MF (1981) Raman investigations of ring configurations in vitreous silica. Nature 292:140-141

Sharma SK, Simons B (1981) Raman study of crystalline polymorphs and glasses of spodumene composition quenched from various pressures. Am Mineral 66, 118-126

Sharma SK, Simons B, Yoder HS (1983) Raman study of anorthite calcium Tschermak's pyroxene and gehlenite in crystalline and glassy states. Am Mineral 68:1113-1125

Sharma SK, Virgo D, Mysen BO (1978) Structure of glasses and melts of Na_2O-$xSiO_2$ (x = 1,2,3) composition from Raman spectroscopy. Carnegie Inst Wash Yearb 77:649-652

Sharma SK, Virgo D, Mysen BO (1979) Raman study of the coordination of aluminum in jadeite melts as a function of pressure. Am Mineral 64:779-787

She CY, Masso JD, Edwards, DF (1971) Raman scattering by polarization waves in uniaxial crystals. J Phys Chem Solids 32:1887-1900

Sherwood PMA (1972) Vibrational Spectroscopy of Solids. University PressCambridge

Shevyakov AM, Trofimenko AV, Sizonenko AP, Burkov VP, Zhuravlev GI (1978) Study of the structure and crystallization of litium oxide-silicon dioxide system melts by a high-temperature IR-spectroscopic method (in Russian). Zh Prikl Khim 51:2612-2615

Smith KH, Shero E, Chizmeshya A, Wolf GH (1995) The equation of state of germania glass: two-domain description of the viscoelastic response. J Chem Phys 102:6851-6857

Smyth HT, Skogen HS, Harsell WB (1953) Thermal capacity of vitreous silica. J Am Ceram Soc 36:327-328

Stanley HE, Angell CA, Essmann U, Hemmati M, Poole PH, Sciortino F (1994) Is there a second critical point in liquid water? Physica A 205:122-139

Stebbins JF (1987) Identification of multiple structural species in silicate glasses by ^{29}Si NMR. Nature 330(6147):465-467

Stebbins JF (1988) Effects of temperature and composition on silicate glass structure and dynamics: Si-29 NMR results. J Non-Cryst Solids 106:359-369

Stebbins JF (1991) NMR evidence for five-coordinated silicon in a silicate glass at atmospheric pressure. Nature 351:638-639

Stebbins JF, Farnan I, Xue X (1992) The structure and dynamics of alkali silicate liquids: a view from NMR spectroscopy. Chem Geol 96:371-385

Stebbins JF, McMillan PF (1993) Compositional and temperature effects on five-coordinated silicon in ambient pressure silicate glasses. J Non-Cryst Solids 160:116-125

Steele D (1971) Theory of Vibrational Spectroscopy. W. B. Saunders, Philadelphia

Suito K, Miyoshi M, Sasakuro T, Fujisawa H (1992) Elastic properties of obsidian, vitreous SiO_2, vitreous GeO_2 under high pressure up to 6 GPa. In: Y Syono, MH Manghnani (eds) High-Pressure Research: Application to Earth and Planetary Sciences, p 219-225. American Geophysical Union, Washington, DC/Terra Scientific Publishing Co, Tokyo

Sweet JR, White WB (1969) Study of sodium silicate glasses and liquids by infrared reflectance spectroscopy. Phys Chem Glasses 10:246-251

Sykes D, Poe B, McMillan PF, Luth RW, Sato RK (1993) A spectroscopic investigation of the structure of anhydrous $KAlSi_2O_8$ and $NaAlSi_2O_8$ glasses quenched from high pressure. Geochim Cosmochim Acta 57:3575-3584

Takahashi E (1990) Speculations on the Archean mantle: missing link between komatiite and depleted garnet peridotite. J Geophys Res 95:15941-15954

Takashima H, Saito H (1976) Resolution of infrared reflection spectra in potassium oxide-silicon dioxide glasses (in Japanese). Nagoya Kogyo Gijutsu Shikensho Hokoku 25:246-251

Takashima H, Saito H (1977) Infrared reflection spectra of mixed alkali and alkali-alkaline-earth silicate glasses (in Japanese). Yogyo Kyokai Shi 85:412-418

Tarte P (1967) Infra-red spectra of inorganic aluminates and characteristic vibrational frequencies of AlO_4 tetrahedra and AlO_6 octahedra. Spectrochim Acta 23A:2127-2143

Taylor M, Brown GE (1979a) Structure of mineral glasses. I. The feldspar glasses, $NaAlSi_3O_8$, $KAlSi_3O_8$, $CaAl_2Si_2O_8$. Geochim Cosmochim Acta 43:61-77

Taylor M, Brown GE (1979b) Structure of mineral glasses. II. The SiO_2-$NaAlSiO_4$ join. Geochim Cosmochim Acta 43:1467-1475

Taylor M, Brown GE, Fenn PM (1980) Structure of silicate mineral glasses. III. $NaAlSi_3O_8$ supercooled liquid at 805°C and the effects of thermal history. Geochim Cosmochim Acta 44:109-1119

Taylor WR (1990) Application of infrared spectroscopy to studies of silicate glass structure: examples from the melilite glasses and the systems Na_2O-SiO_2 and Na_2O-Al_2O_3-SiO_2. Proc Indian Acad Sci 99:99-117

Verweij H, Konijnendijk WL (1976) Structural units in K_2O-PbO-SiO_2 glasses by Raman spectroscopy. J Am Ceram Soc 59:517-521

Virgo D, Mysen BO (1985) The structural state of iron in oxidized vs. reduced glasses at 1 atm: A ^{57}Fe Mössbauer study. Phys Chem Minerals 12:65-76

Virgo D, Mysen BO, Kushiro I (1980) Anionic constitution of 1-atmosphere silicate melts: implications for the structure of igneous melts. Science 208:1371-1373

Vukcevich MR (1972) A new interpretation of the anomalous properties of vitreous silica. J Non-Cryst Solids 11:25-63

Wackerle J (1962) Shock-wave compression of quartz. J Appl Phys 33:922-937

Wallace DC (1972) Thermodynamics of Crystals. John Wiley, New York

Wang Z, Cooney TF, Sharma SK (1993) High temperature structural investigation of $Na_2O\cdot0.5Fe_2O_3\cdot$ 3 SiO_2 and $Na_2O\cdot FeO\cdot3$ SiO_2 melts and glasses. Contrib Mineral Petrol 115:112-122

Warren WL, Lenahan PM, Brinker CJ (1991) Experimental evidence for two fundamentally different E' precursors in amorphous silicon dioxide. J Non-Cryst Solids 136:151-162

Williams Q, Jeanloz R (1988) Spectroscopic evidence for pressure-induced coordination changes in silicate glasses and melts. Science 239:902-905

Williams Q, Jeanloz R (1989) Static amorphization of anorthite at 300 K and comparison with diaplectic glass. Nature 338:413-415

Williams Q, McMillan P, Cooney T (1989) Vibrational spectra of orthosilicate glasses. The Mg-Mn join. Phys Chem Minerals 16:352-359

Wilson EB, Decius JC, Cross PC (1955) Molecular Vibrations. McGraw-Hill, New York. Reprinted 1980 by Dover Publications

Wolf GH, Durben DJ, McMillan PF (1990) High pressure Raman spectroscopic study of sodium tetrasilicate ($Na_2Si_4O_9$) glass. J Chem Phys 93:2280-2288

Wolf GH, Wang S, Herbst CA, Durben DJ, Oliver WF, Kang ZC, Halvorson K (1992) Pressure induced collapse of the tetrahedral framework in crystalline and amorphous GeO_2. In: Y Syono, MH Manghnani (eds) High-Pressure Research: Application to Earth and Planetary Sciences, p 503-517. Terra Scientific Publishing Co & Am Geophysical Union, Tokyo/Washington, DC

Wong J, Angell CA. (1976) Glass Structure by Spectroscopy. Marcel Dekker, New York

Worrell CA, Henshall T (1978) Vibrational spectroscopic studies of some lead silicate glasses. J Non-Cryst Solids 29:283-299

Xue X, Stebbins J, Kanzaki M, Poe B, McMillan P (1991) Pressure-induced silicon coordination and tetrahedral structural changes in alkali oxide-silica melts up to 12 GPa: NMR, Raman and infrared spectroscopy. Am Mineral 76:8-26

Zha CS, Hemley RJ, Mao HK, Duffy TS, Meade C (1994) Acoustic velocities and refractive index of SiO_2 glass to 57.5 GPa by Brillouin scattering. Phys Rev B 50:13105-13112

Chapter 9

X-RAY SCATTERING AND X-RAY SPECTROSCOPY
STUDIES OF SILICATE MELTS

Gordon E. Brown, Jr.

*Department of Geological & Environmental Sciences
and Stanford Synchrotron Radiation Laboratory
Stanford University, Stanford, CA 94305 U.S.A.*

François Farges

*Laboratoire de Physique et Mécanique des Géomatériaux
Université Marne-la-Vallée (and UA CNRS 734 and LURE)
2 allée de la Butte Verte, 93166 Noisy le Grand Cedex, France.*

Georges Calas

*Laboratoire de Minéralogie-Cristallographie
UA CNRS 09, Universités Paris 6 et 7 and Institut de Physique du Globe
4, place Jussieu, 75252 PARIS Cedex 05, France.*

INTRODUCTION

Defining the molecular-scale structure and dynamics of silicate melts has been a long-term goal of inorganic geochemists, igneous petrologists, ceramists, glass technologists, and metallurgists because of the desire to understand melt properties at a fundamental level. Achieving this goal has proven to be difficult because silicate melts, unlike crystalline silicates, possess no long-range structural periodicity or symmetry. Thus the standard structural tools of the mineralogist and crystallographer—X-ray and electron diffraction—are of limited use. Silicate melts do possess short-range order, however, which is defined here as structural order within ~1.6 to 3 Å radius of a central atom. Short-range order is usually manifested as cation-oxygen polyhedral units such as tetrahedra and octahedra. In addition, silicate melts may possess varying degrees of medium-range order, depending on bulk composition and temperature. Medium-range order is defined here as that occurring between ~3 and 6 Å radius of a central atom. This radius includes second- and third-neighbor environments around a central ion, and the order could involve a repetitive arrangement of corner-linked polyhedra, such as silicate tetrahedra linked through all four corners to other tetrahedra (i.e. Q^4 units) or six-membered rings of corner-linked silicate tetrahedra. More extended medium-range order in glasses and melts can occur to distances of 6 to 10 Å from a central atom, and could involve a particular network topology. In general, however, structural order (or distance correlations) in glasses and melts diminishes with increasing distance from a reference atom. The distance ranges defined above roughly parallel regions I, II, and III defined by Wright (1988) for short-range order (region I: 1.5 to 2.9 Å, corresponding to the SiO_4 tetrahedron), medium-range order (region II: 2.9 to 5.5 Å, corresponding to the interlinking of SiO_4 tetrahedra), and more extended order (region III: 3.5 to >10Å, corresponding to the network topology) in vitreous SiO_2.

The short- and medium-range structure in silicate melts can be probed using a variety of high-temperature spectroscopic and diffraction methods, many of which are discussed in Volume 18 of *Reviews in Mineralogy* (Hawthorne, 1988). Among the more direct structural probes of silicate glasses and melts are X-ray scattering and X-ray absorption spectroscopy (XAS), both of which provide quantitative measures of interatomic distances and coordination numbers. This chapter discusses these methods and focuses on the structural information they have provided on silicate melts at high temperature. We limit discussion of X-ray studies of silicate glasses under ambient conditions to a few examples because a number of reviews have been published (e.g. Wright, 1974; Wright and Leadbetter, 1976; Calas and Petiau, 1983; Brown et al., 1986; Calas et al., 1987; Brown et al., 1988; Calas et al., 1995). We also limit discussion of X-ray scattering and XAS methods under ambient conditions since they have been reviewed thoroughly (X-ray scattering: e.g. Klug and Alexander, 1954; Warren, 1969; Leadbetter and Wright, 1972; Wright, 1974; Wright and Leadbetter, 1976; Wagner, 1978; Waseda, 1980; Wright et al., 1982; Wright, 1988; Wright, 1993b; Wright, 1994; XAS: e.g. Stern and Heald, 1983; Teo, 1986; Brown et al., 1988; Koningsberger and Prins, 1988; Lytle, 1989).

Structural methods that depend on the interaction of X-ray photons with matter sample a very short time domain, which limits the information they can provide about the atomic motions or relaxation times responsible for melt properties such as diffusion or viscous flow. The time required for an X-ray photon to be absorbed, i.e. to excite an electron to a higher energy level, is about 10^{-16} seconds, compared to about 10^{-12} to 10^{-14} seconds for interatomic vibrations (IR and Raman spectroscopy). Thus XAS takes an instantaneous "snap-shot" of an atom's environment, summing all local atomic arrangements around the absorbing atom during that short time period. Because the time required for measurement of an X-ray absorption spectrum is seconds to tens of minutes, depending on experimental technique, an XAS spectrum sums on the order of 10^{16} to 10^{19} of these "snap-shots". These summations show a range of positions for each atom, which can be large or small depending on the stretching and bending frequencies and amplitudes of individual bonds and linked polyhedral units. In silicate melts, the situation is further complicated because of the diffusional motion of the atoms. Similar time scales apply to elastic X-ray scattering, which involves energy transfer between an X-ray photon and an electron bound to an atom, followed by re-emission of the X-ray photon without loss of energy. Thus an X-ray radial distribution function (defined below) also represents the sum of an extremely large number of structural "snap-shots". Given these limitations, one might conclude that little can be learned about the structure of silicate melts using X-rays because the image will be so smeared. However, because the bonding forces between cations and nearest-neighbor oxygens are relatively strong for many cations, cation-oxygen polyhedra are usually well defined, even in the liquid at high temperatures, unless the cations are large and of low charge. Thus, X-ray methods can provide direct, quantitative, isotropically averaged information on average first- and second-neighbor distances, bond angles, and cation coordination numbers, and they can also provide information on medium-range structural correlations.

This chapter begins with a brief review of historical concepts about silicate glass/melt structure, followed by a discussion of radial distribution functions (rdf's), which are the primary means of displaying structural information on amorphous materials from X-ray (or neutron) scattering experiments. The rdf's of silica and feldspar-composition glasses are used to illustrate the concepts of short- and medium-range structure in glasses and melts and form a basis for discussing silicate melt structure. We then review high-temperature X-ray scattering methods and some of the structural information derived for silicate melts using these methods during the past 20 years. Where appropriate, we compare structural data for glasses under ambient conditions with those for melts. High-temperature XAS methods are discussed next, including several theories which can accurately describe possible anharmonic effects on XAS-derived distances and coordination numbers. A summary of recent high-temperature XAS studies of cation environments in silicate melts is also given. The chapter concludes with discussions of some of the implications of XAFS-derived coordination numbers for divalent transition metal cations in melts for their melt-crystal partitioning and medium-range ordering of cations in silicate melts. Throughout the chapter we specify the formal valence of a cation by Roman numerals in parentheses after the element symbol; cation or anion coordination number is indicated by a superscript number in square brackets before the element symbol.

HISTORICAL PERSPECTIVES

The crystallite and random-network models

One of the first theories of silicate glass structure is attributed to Frankenheim (1835) who postulated that glasses are made up of very small crystals which are referred to as crystallites. He speculated that there is a size distribution of crystallites, with the smaller ones melting earlier than the larger ones, thus "lubricating" the glass and allowing it to flow at temperatures lower than the melting point. This theory was thought to explain the well-known phenomenon known as glass softening. A similar crystallite structure theory was proposed by Lebedev (1921) some 85 years later to explain relatively sharp changes in the properties of silicate glasses at temperatures near the $\alpha \leftrightarrow \beta$ transition of quartz. Randall et al. (1930) carried out one of the first X-ray scattering studies on silicate glasses of geochemical relevance, including the compositions SiO_2, $CaSiO_3$, $NaAlSi_3O_8$, and $KAlSi_3O_8$. They noticed gross similarities between the positions of the scattering maxima of each glass and those of the corresponding crystal. Based on these observations, they concluded that each of these glasses consisted of very small crystallites (10 to 100 Å in diameter) with an atomic arrangement like the crystal of the same composition. Additional support for the crystallite model was provided by Valenkov and Porai-Koshits (1936), who, on the basis of X-ray scattering measurements on silicate glasses, proposed a revised estimate of crystallite dimensions of 7.5 to 25 Å.

Zachariasen (1932, 1933) and Warren (1933) proposed quite a different model for silicate glass/melt structure which disputed the idea of crystallites. This was the random-network model in which silicate tetrahedra are linked through corners (i.e. through bridging oxygens) but display no structural ordering similar to crystals of the same composition at distances beyond about 8 Å. Zachariasen (1932) also envisioned cations such as Na, K, and Ca fitting into the voids created by a random network of corner-linked SiO_4 and AlO_4 tetrahedra and serving to balance valences. Zachariasen (1933) was careful to point out that a continuous random network was not necessary for glass formation in multi-component silicate systems and introduced the concept of network-forming and network-modifying elements, which are discussed below.

Warren et al. (1936) carried out the first Fourier analysis of diffraction data from vitreous SiO_2, concluding that each Si atom is surrounded by four oxygens and that each oxygen is shared by two silicons. However, these observations did not prove or disprove the random-network or crystallite models. The original crystallite model was strongly challenged when Warren and Biscoe (1938) found no measurable small angle X-ray scattering (SAXS) from silica glass. They suggested that the maximum allowable dimension of a hypothetical crystallite could not exceed ~8 Å, based on line-broadening analysis of the glass X-ray scattering data. This dimension is well below the maximum values estimated in the studies by Randall et al. (1930) and Valenkov and Porai-Koshits (1936). Warren and Biscoe further argued that this dimension is of the same order as the unit-cell dimensions of β-cristobalite, concluding that the concept of very small crystals of these dimensions comprising silicate glasses is meaningless. Some years later, an analysis of the lack of SAXS from silica glass by Bienenstock and Bagley (1966) showed that Warren and Biscoe's experimental method was not sensitive enough to detect X-ray scattering in the small-angle region; however, more recent SAXS data on silica glass (e.g. Weinberg, 1962; Renninger and Uhlmann, 1974) are consistent with structural domains no larger than 10 to 12 Å.

The crystallite and random-network models represent two extremes, and for many silicate and oxide glasses, the true structural model probably lies somewhere between these extremes. A modified crystallite model has been proposed (Porai-Koshits, 1958) in which discrete crystallites do not occur. In this model there are spatial fluctuations in the degree of medium-range order in the glass network for simple network glasses like vitreous SiO_2. The more highly ordered regions may have atomic arrangements that approach those of crystals, and these regions are envisioned as being interconnected by less-ordered regions. The total volume fraction of the ordered regions is estimated to be less than 80% (Porai-Koshits, 1985). One could think of these more ordered regions as being precursors to nuclei of crystals; however, such "nuclei" don't grow during the rapid quench of framework liquids, based on diffraction experiments on the glasses. Schematic two-dimensional drawings of the crystallite and random network models of vitreous silica are compared with that of a crystal made up of corner-linked tetrahedra (projected as triangles) in Figure 1. These glass structure models illustrate the suspected importance of ring structures in vitreous SiO_2 and the idea that ordered domains don not extend more than a few tetrahedra. Several more recent glass structure models have been proposed, such as the paracrystal (Phillips, 1982) and strained-crystal (Goodman, 1982) models, which postulate crystal-like domains up to 60 to 80 Å in diameter. However, they have been challenged (see Wright, 1994) because diffraction evidence limits the size of "crystallites" in vitreous SiO_2 to 10 to 12 Å.

The Zachariasen-Warren random-network model and the Porai-Koshits modified crystallite model both have attractive features, particularly when the concept of network-forming and network-modifying cations is included (see, e.g. Rawson, 1967; Nelson et al., 1983). Network formers are those cations that can form polyhedra (typically oxygen

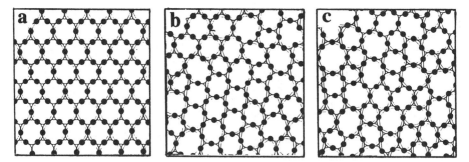

Figure 1. Schematic drawings comparing the crystallite (a) and random-network (b) models of glass structures with that of a crystal (c) made up of corner-linked tetrahedra. The triangles represent projections of the bases of oxygen tetrahedra. (From Wright, 1994.)

tetrahedra or triangles) which link through corners to form a three-dimensional network, whereas network modifiers tend to cause depolymerization of these networks. Table 1 classifies selected cations found in silicate and germanate glasses among these two groups. The basis for this concept was provided by Goldschmidt (1926) who first suggested that certain elements like Si and B with radius ratios (with oxygen as the anion) between 0.2 and 0.4 readily form glasses. Zachariasen extended Goldschmidt's idea to a set of rules for glass formation that bear his name. Zachariasen's rules are closely related to Pauling's rules for stable ionic crystals (Pauling, 1929) and can be stated as follows for A_mO_n compositions (Rawson, 1967):

1. No oxygen atom may be linked to more than two atoms A.
2. The number of oxygen atoms surrounding atoms A must be small.
3. The oxygen polyhedra share corners with each other, not edges or faces.
4. At least three corners of each oxygen polyhedron must be shared.

Dietzel (1942) introduced a parameter related to radius ratio, referred to as field strength, which he defined as the charge on the cation divided by the square of the cation-oxygen distance $[Z_c/d(M-O)^2]$. He classified network formers as cations with field strengths between 1.4 and 2.0, intermediate cations as those with field strengths between 0.5 and 1.0, and network modifiers as those with field strengths between 0.1 and 0.4 (Vogel, 1971). Table 1 provides a listing of network-forming and network-modifying cations important in glasses of geochemical and ceramic interest and includes revised values of their field strengths based on metal-oxygen distances, d(M-O), calculated using effective ionic radii values of Shannon (1976). This table omits the category of intermediate cations proposed by Dietzel and indicates that the original field strength ranges listed by Dietzel should be extended to include the possibility of highly charged cations like Zr(IV), Th(IV), and Mo(VI) acting as network modifiers. The concept of cation field strength will be compared with M-O bond valence later in this chapter.

One of the most important contributions of this early work on silicate glass structure was the suggestion by Zachariasen and others that the same crystal chemical principles that govern the stability of oxide and silicate crystal structures should also be applicable to oxide and silicate glasses. These principles have been utilized in developing plausible structural models of short- and medium-range order in silicate glasses and melts, as discussed later. Additional discussion of early studies of glass structure, especially those carried out in the former Soviet Union, can be found in Bartenev (1970).

Modern studies of glass/melt structure

Since the early days of glass science, there have been numerous structural studies of silicate glasses which have led to refinements of the above models and development of new models for medium-range structure. These include the modified random network model proposed by Greaves et al. (1981) and Greaves (1985), and the densely packed domain model proposed by Gaskell et al. (1991). It is now well recognized that structural data from many different techniques are required to develop models of silicate glasses and melts which can be related to their properties. It is also recognized that deriving a unique structural model for a glass or melt from any experimental method or computer simulation is impossible because of their disordered nature. Nonetheless, these methods have provided valuable insights about local atomic arrangements, and in certain cases, more extended topologies.

X-ray and neutron scattering continue to be among the most important methods for obtaining quantitative structural data on silicate glasses, especially when combined with computer fits of the data to structural models. In fact, the radial distribution functions

Table 1.   Valence (Z), coordination number (CN), ionic radius, mean linear M-O expansion coefficient ($\overline{\alpha}$) ($\times 10^6$ k^{-1}) at 1773 K, mean d(M-O) at 298 K, mean d(M-O) at 1773 K, field strength, and bond valences at 298 K and 1773 K for selected cations occurring in silicate glasses and melts.

Cation	Valence (Z)	CN[1]	Ionic radius (Å)[2]	Mean linear $\overline{\alpha}$ [3]	R = d(M-O)[4] @298 K	d(M-O)[5] @1773 K	Field strength {Z/R²} @ 298 K	Brese & O'Keeffe R₀ value	M-O bond valence @298 K (v.u.)[6]	M-O bond valence @1773 K (v.u.)[7]
Network Modifiers										
Cs	1	8	1.74	32.0	3.10	3.25	0.10	2.42	0.16	0.11
Rb	1	8	1.61	32.0	2.97	3.11	0.11	2.26	0.15	0.10
K	1	8	1.51	32.0	2.87	3.01	0.12	2.13	0.14	0.09
Na	1	6	1.02	24.0	2.38	2.47	0.18	1.80	0.21	0.17
Li	1	4	0.59	16.0	1.95	2.00	0.26	1.466	0.27	0.24
Ba	2	8	1.42	16.0	2.78	2.85	0.26	2.29	0.27	0.22
Sr	2	8	1.26	16.0	2.62	2.68	0.29	2.118	0.26	0.22
Ca	2	6	1.00	12.0	2.36	2.40	0.36	1.967	0.35	0.31
Mg	2	6	0.72	12.0	2.08	2.12	0.46	1.693	0.35	0.32
		5	0.66	10.0	2.02	2.05	0.49		0.41	0.38
		4	0.57	8.0	1.93	1.95	0.53		0.53	0.50
Be	2	4	0.27	8.0	1.63	1.65	0.38	1.381	0.51	0.48
Mn	2	6	0.83	12.0	2.19	2.23	0.42	1.790	0.34	0.30
		5	0.75	10.0	2.11	2.14	0.45		0.42	0.39
		4	0.66	8.0	2.02	2.04	0.49		0.54	0.50
Fe	2	6	0.78	12.0	2.14	2.18	0.44	1.734	0.33	0.30
		5	0.70	10.0	2.06	2.09	0.47		0.41	0.38
		4	0.63	8.0	1.99	2.01	0.51		0.50	0.47
Ni	2	6	0.69	12.0	2.05	2.09	0.48	1.654	0.34	0.31
		5	0.63	10.0	1.99	2.02	0.50		0.40	0.37
		4	0.55	8.0	1.91	1.93	0.55		0.50	0.47
Zn	2	6	0.74	12.0	2.10	2.14	0.45	1.704	0.34	0.31
		5	0.68	10.0	2.04	2.07	0.48		0.40	0.37
		4	0.60	8.0	1.96	1.98	0.52		0.50	0.47
Pb	2	8	1.29	16.0	2.65	2.71	0.28	2.112	0.23	0.20
Al	3	6	0.535	8.0	1.90	1.92	0.83	1.651	0.51	0.48
		5	0.48	6.7	1.84	1.86	0.89		0.60	0.57
Ga	3	6	0.62	8.0	1.98	2.00	0.77	1.730	0.51	0.48
		5	0.55	6.7	1.91	1.93	0.82		0.61	0.58
Fe	3	6	0.645	8.0	2.00	2.02	0.75	1.759	0.52	0.49
		5	0.58	6.7	1.94	1.96	0.80		0.61	0.58
La	3	8	1.16	10.7	2.52	2.56	0.47	2.172	0.39	0.35
Gd	3	8	1.05	10.7	2.41	2.45	0.47	2.065		0.35
		6	0.94	8.0	2.30	2.33	0.57		0.53	0.49
Yb	3	6	0.87	8.0	2.23	2.26	0.60	1.985	0.52	0.48
Ti	4	6	0.605	6.0	1.965	1.98	1.04	1.815	0.67	0.64
		5	0.34	5.0	1.70	1.71	1.38		1.4[8]	1.32
Zr	4	8	0.84	8.0	2.20	2.23	0.83	1.937	0.49	0.46
		6	0.72	6.0	2.08	2.10	0.92		0.68	0.65
Th	4	8	1.05	8.0	2.41	2.44	0.69	2.167	0.52	0.48
		6	0.94	6.0	2.30	2.32	0.76		0.70	0.66
Mo	6	4	0.41	2.7	1.77	1.78	1.92	1.907	1.45	1.42
U	6[9]	2[9]	0.45	1.3	1.82	1.82	1.89	2.075	1.99	1.97
		4	0.89	2.7	2.25	2.26	1.19		0.62	0.61
		4	0.89	6.0	2.25	2.27	0.79	2.112	0.69	0.65

Table 1. Continued

Cation	Valence	CN[1]	Ionic Radius r (Å)[2]	Mean Linear $\overline{\alpha}$ [3]	R = d(M-O) (Å)[4] @298 K	d(M-O) (Å)[5] @1773 K	Field Strength $\{Z/R^2\}$ @ 298 K	Brese & O'Keeffe R_0 value	M-O Bond Valence @298 K (v.u.)[6]	M-O Bond Valence @1773 K (v.u.)[7]
Network Formers										
B	3	4	0.11	5.3	1.47	1.48	1.39	1.371	0.77	0.74
		3	0.01	4.0	1.37	1.38	1.60		1.00	0.98
Al	3	4	0.39	5.3	1.75	1.76	0.98	1.651	0.77	0.74
Fe	3	4	0.49	5.3	1.85	1.86	0.88	1.759	0.78	0.75
Ga	3	4	0.47	5.3	1.83	1.84	0.90	1.730	0.76	0.73
Si	4	4	0.26	4.0	1.62	1.63	1.52	1.624	1.01	0.98
Ti	4	5	0.51	5.0		1.96	1.05	1.815	0.70	0.67
		4	0.42	4.0	1.95[10] 1.78	1.79	1.26		1.10	1.07
Ge	4	4	0.39	4.0	1.75	1.76	1.30	1.748	0.99	0.97
P	5	4	0.17	3.2	1.53	1.54	2.14	1.604	1.22	1.20

[1] Most common or possible coordination number(s) in glasses or melts.

[2] Values from Shannon (1976) unless otherwise specified.

[3] Mean linear expansion coefficients for cation polyhedra based on data from Hazen and Finger (1982) calculated for the temperature range 298 K to 1773 K. These values are larger than those reported in Table 6.3 of Hazen and Finger (1982), which are for the temperature range 298 K to 1273 K.

[4] Sum of cation radius and radius for 3-coordinated oxygen (1.36 Å).

[5] Calculated using $\overline{\alpha}$ values listed in Table above.

[6] Values calculated using Brese and O'Keeffe (1991) bond valence parameters (R_0), d(M-O) values at 298 K listed in Table above, and equation (39).

[7] Values calculated using Brese and O'Keeffe (1991) bond valence parameters (R_0), d(M-O) values at 1773 K listed in Table above, and equation (39).

[8] Values are for the axial bond [d(Ti=O) = 1.70 Å] of ([5]Ti=O)O4 polyhedra from Farges et al. (1995d).

[9] Two values are for equatorial bonds (lower value) [d(U(VI)-O) ≈ 2.22 Å] and axial bonds (higher value) [d(U=O) = 1.70 Å] for (UO2)O4 polyhedra from Farges et al. (1992).

[10] Average values for [5]Ti are for equatorial bonds of ([5]Ti=O)O4 polyhedra from Farges et al. (1995d). Values for the shorter axial titanyl bond (Ti=O) are listed under network modifiers. See [8] above.

derived from these data provide the primary experimental check of molecular dynamic and Monte Carlo simulations of glass and melt structure (see chapter by Poole et al., this volume). Raman, NMR, UV/visible, and X-ray absorption spectroscopies have played major roles in providing new information on glass structure over the past 15 years. Reviews of this literature can be found in the following references: [Raman studies] Mysen (1988), McMillan et al. (1992), McMillan and Wolf (this volume); [NMR studies] Stebbins (1987), Farnan and Stebbins (1990); Stebbins and Farnan (1989, 1992), Farnan et al. (1992), Stebbins (this volume); [UV/visible studies] Calas and Petiau (1983), Keppler (1992), Galoisy and Calas (1993a); [X-ray absorption studies] Greaves (1985), Brown et al. (1986), Calas et al. (1987), Brown et al. (1988), Calas et al. (1995). Structural studies of molten silicates have not received nearly as much attention as silicate glasses during this period because of the difficulty of performing experiments on a liquid held at temperatures of 1000 to 2000 K. Most of these high-temperature studies have utilized X-ray scattering, X-ray absorption spectroscopy, vibrational spectroscopy, or NMR methods. Comparison of melt results with those from the corresponding glasses has shown significant differences in local and medium-range structure for some compositions, so the assumption that the structure of a glass is the same as that of its melt is not always true. One of the next frontiers is to study the structure of silicate melts at high pressures in order to develop a basis for understanding the properties of natural magmas in the earth's interior (see chapter by Wolf and McMillan, this volume). One such study of alkali-germanate melts has been made using Raman spectroscopy to pressures of 2.2 GPa (Farber and Williams, 1992). The sample was held in a resistance-element-heated diamond-anvil high-pressure cell. In situ high-temperature and high-pressure X-ray scattering and XAS studies of silicate melts have not yet been attempted, to our knowledge, but such studies should be greatly facilitated using the new third-generation, high-brightness synchrotron radiation sources in the U.S., Europe, and Japan.

X-RAY SCATTERING STUDIES OF SILICATE GLASSES AND MELTS

Scattered x-ray intensity and radial distribution functions

The theoretical foundation for structural analysis of non-crystalline materials was provided by Debye (1915). A modified version of the Debye scattering equation describes the X-ray intensity scattered by an assemblage of atoms of different types (expressed in electron units) as

$$NI(s) = \sum_i f_i^2(s) + \sum_{i \neq j} \sum f_i(s) f_j(s) \frac{\sin sr_{ij}}{sr_{ij}} + I_{inc}(s) \tag{1}$$

where N is the number of formula units in the scattering volume of the sample, r_{ij} is the distance between atoms i and j, and $f_i(s)$ and $f_j(s)$ are the X-ray scattering factors for atoms i and j. The first summation includes all atoms in the scattering volume and represents X-ray scattering from electrons on the same atom; the second term represents interference scattering from electrons on different atoms. This term is the structurally sensitive part of $I(s)$. The third term represents the incoherent (or Compton) scattering which arises from inelastic X-ray photon-electron scattering. The variable s (referred to as the scattering vector and sometimes designated by Q) is defined as $4\pi\sin\theta/\lambda$, where θ is half the scattering angle and λ is the X-ray wavelength.

The experimentally measured scattered X-ray intensity from a glass or melt, as a function of s, is given by

$$I(s) = I_{obs}(s) \, K \, A(s) \, P(s) \tag{2}$$

where $A(s)$ and $P(s)$ are absorption and polarization corrections and K is a constant which normalizes I_{obs} to one formula unit of the sample. Equations for calculating K using the Krogh-Moe-Norman method (Krogh-Moe, 1956; Norman, 1957) and for absorption and polarization corrections for different sample geometries can be found in Marumo and Okuno (1984) and Waseda (1980). The scattered X-ray intensities from albite composition glass and high albite crystal, as a function of s, are shown in Figure 2. Most silicate glasses and melts have an X-ray intensity function of this general form. Also shown are the smoothly decreasing function with increasing s, which represents the first term in Equation (1) (the sum of scattering from individual atoms) and the smoothly increasing function with increasing s, which represents the third term in Equation (1) (the incoherent or Compton scattering). These two terms must be subtracted from the total scattered intensity, $I(s)$, in order to isolate the structurally sensitive, reduced scattered intensity (or interference function), designated $si_0(s)$. In practice, the first and third terms of Equation (1) can be approximated using analytical expressions for the coherent (Cromer and Mann, 1968) and incoherent (Balyuzi, 1975) atomic scattering, respectively. Comparison of the scattering functions in Figure 2 for crystalline high albite and albite composition glass illustrates the overall similarity of the two.

Inversion of the experimental interference function, $i_0(s)$, using the Fourier integral theorem was first suggested by Zernike and Prins (1927) as a means of obtaining a radial

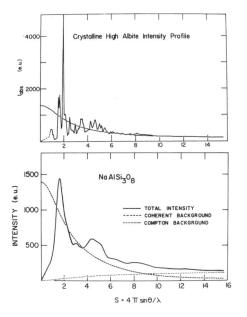

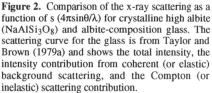

Figure 2. Comparison of the x-ray scattering as a function of s ($4\pi\sin\theta/\lambda$) for crystalline high albite ($NaAlSi_3O_8$) and albite-composition glass. The scattering curve for the glass is from Taylor and Brown (1979a) and shows the total intensity, the intensity contribution from coherent (or elastic) background scattering, and the Compton (or inelastic) scattering contribution.

distribution function for a non-crystalline sample without any a priori knowledge about its atomic arrangement. The *radial distribution function*, or rdf, derived by Fourier transforming $i_0(s)$, can be written as

$$G(r) = 4\pi r^2 \rho(r) = 4\pi r^2 \rho_0 + \frac{2r}{\pi} \int_0^{s_{max}} \frac{s i_0(s)}{g(s)} M(s) \sin(sr) ds \tag{3}$$

where

$$g(s) = \sum_{F.U.} f_i^2(s) \bigg/ \sum_{F.U.} Z_i^2 \tag{4}$$

is a "sharpening" function to correct for the dependence of f_i and f_j on s, and where

$$M(s) = \exp(-\alpha s^2) \tag{5}$$

which is used to reduce the weight of high-angle (high s) data (which have lower signal-to-noise than low-angle data) and to reduce ripples in the Fourier transform due to a finite data range. The value of α commonly used for silicate glasses should be about 0.01 (Wright and Leadbetter, 1976). The sum in the denominator of Equation (4) is over one formula unit (F.U.) of the scattering material and sums the squared number of electrons, Z, on each atom. The term ρ_0 in Equation (3) is the average atomic density of the sample, defined as $\rho_0 = dN/(A \times 10^{24})$, where d is the sample density in gcm^{-3}, N is Avogadro's number, and A is the atomic weight (in g) of one formula unit. G(r), as defined here, has units of electrons^2/Å and represents the probability of finding two atoms separated by a distance r + dr weighted by the product of the number of electrons on each of the two atoms. The variable r in Equation (3) is interatomic distance (in Å).

Another commonly used form of the rdf, sometimes referred to as the difference or differential rdf, is expressed as

$$D(r) = G(r) - 4\pi r^2 \rho_0 \tag{6}$$

The primary difference between $G(r)$ and $D(r)$ is that the former has an r^2 dependence and approaches the parabolic function $4\pi r^2 \rho_0$ at large r values, whereas the latter approaches unity at large values of r. The function $4\pi r^2 \rho_0$ represents a homogeneous distribution of electron density. Deviation of the rdf from $4\pi r^2 \rho_0$ (for the $G(r)$ function) or from unity (for the $D(r)$ function) indicates that the atomic arrangement around a central atom is not random.

$G(r)$ and $D(r)$ functions are not always defined consistently in the literature, with some papers defining them as above and others defining them in the opposite sense or defining them such that they are in units of electrons^2/Å^2. Some papers also multiply the $G(r)$ or $D(r)$ function by r or r^2 to increase the amplitude of the function at higher r. We have defined them here to be consistent with the convention used by Klug and Alexander (1954) and Warren (1969) and adopted by Taylor and Brown (1979a). Wright (1988, 1994) has introduced the $T(r)$, or total correlation, function, which is in units of electrons^2/Å^2. These differences and the attributes of each type of rdf are discussed in more detail in Wright (1994). Modern X-ray rdf studies of glasses often multiply the rdf by a special resolution function, the Lorch modification function (Lorch, 1969), to produce rdf's free of ripples due to truncation of the Fourier transform caused by a finite data range Δs.

Another type of correlation function commonly used is referred to as the *pair correlation function* for two atoms i and j which is defined as (Warren, 1969):

$$P_{ij}(r) = \int_0^{s_{max}} \left[f_i f_j / g^2(s) \right] \exp(-\alpha^2 s^2)(\sin s r_{ij} \sin s r)ds \tag{7}$$

These individual functions can be summed to give a radial distribution function of the following form, which is referred to as the *pair distribution function* (Warren, 1969):

$$\sum_{u.c.} \sum_i \frac{N_{ij}}{r_{ij}} P_{ij}(r) = 2\pi^2 r \rho_e \sum_{u.c.} Z_j + \int_0^{s_{max}} s i(s) \exp(-\alpha^2 s^2)(\sin rs)ds \tag{8}$$

where the first sum is over the "unit cell" or formula unit, N_{ij} is the number of atom neighbors in the ith shell around atom j, ρ_e is the average electron density, and Z_j is the atomic number of atom j. The other terms in Equations (7) and (8) are defined above.

The average number of atoms, N_{ij} located at distances between r_1 and r_2 in a glass or melt can be estimated by integrating the area under the corresponding peak in the $G(r)$ function:

$$N_{ij} = \int_{r_1}^{r_2} 4\pi r^2 \rho(r)dr \tag{9}$$

This number is commonly referred to as the coordination number when it includes only those atoms in the first coordination sphere of a central atom.

$G(r)$, $D(r)$, and $P(r)$ are one-dimensional representations of a three-dimensional structure; thus the implicit assumption is made that the sample is macroscopically isotropic. Also these functions are averages over the irradiated volume of the sample, so they cannot provide direct information about chemical heterogeneities such as phase separation since the small-angle scattering, which would indicate such heterogeneities, is usually not measured or is lost when $i_0(s)$ is extrapolated to zero s, as is typically done. Another limitation of X-ray rdf's is that all pair correlations contribute to them, making their interpretation difficult for multi-component glasses, especially at r > 4 to 5 Å.

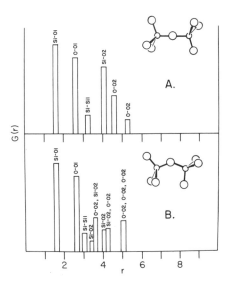

Figure 3. Hypothetical x-ray radial distribution functions, plotted in histogram fashion, for two T_2O_7 molecules (T = Si) with T-O-T angles of 180° (A) and 140° (B). (From Taylor and Brown, 1979a.)

Figure 3 is a simplified example of an rdf intended to help explain its physical meaning. It illustrates in histogram fashion the $G(r)$ functions for two possible arrangements of the Si_2O_7 molecule, which is a small and, arguably, representative piece of silica glass or melt containing two tetrahedra linked through a bridging oxygen. The number appended to the symbol for the jth atom in the figure is defined as follows: 1 for jth atoms on the same SiO_4 tetrahedron and 2 for jth atoms on the adjacent tetrahedron. This notation will be used throughout this chapter. The Si-O1, O-O1, and Si-Si1 distances in the molecule with a T-O-T angle of 180° are 1.6 Å, 2.6 Å, and 3.2 Å, respectively. The Si-Si1 distance and some of the O-O1 distances will have different values for the molecule with a T-O-T angle of 140°. The distances for Si-O2 and O-O2 will also depend on whether the molecule is in a cis or trans configuration. Trans configurations are shown here. This figure was derived by drawing vectors, r_{ij}, between all atom pairs for each molecule. The collection of vectors was plotted on the histogram according to their lengths, and the area of each bar is proportional to twice the number of atom pairs at distance $| r_{ij} |$ multiplied by the number of electrons associated with atoms i and j in each pair.

The area under the first maximum in Figure 3, corresponding to Si-O1 pairs, is $16 \times (14 \times 8) = 1792$ electrons^2 (the integers in parentheses are the atomic numbers of the two atoms, Si and O), whereas that under the third maximum, corresponding to the Si-Si1 pair, is $2 \times (14 \times 14) = 392$ electrons^2. One can imagine how the rdf of a larger portion of the silica glass or melt structure can be derived. In principle, this can be accomplished by adding more tetrahedra that share oxygen corners with the two tetrahedra in the Si_2O_7 molecule and allowing for different Si-O-Si bond angles and different degrees of rotation of each tetrahedron around each Si-O(bridge) bond. The angle of this rotation is referred to as the torsion angle α and is illustrated in Figure 4.

This exercise illustrates several important features about X-ray rdf's. First, the rdf is sensitive to local structure. Second, the interpretation of rdf's becomes complicated after the first few distances. Third, as symmetry is reduced (e.g. in going from the Si_2O_7 molecule with an Si-O-Si angle of 180° to the one with an angle of 140°), the rdf becomes increasingly broadened, with fewer sharp maxima.

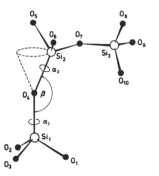

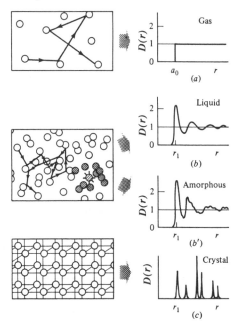

Figure 4 (above). Definition of the torsion angles, α_1 and α_2, for a representative piece of the silica glass struc-ture. β is the Si-O-Si angle. (Modified after Wright et al., 1982.)

Figure 5 (right). Schematic illustrations of the atomic-level structures and pair-distri-bution functions (D(r) versus r in Å) for a crystal, a glass, a melt, and a gas, all with one type of atom. (Modified after Waseda, 1980.)

Figure 5 is a schematic representation of radial distribution functions for a hypothetical crystal, a glass, a liquid, and a gas, all consisting of a single type of atom. It illustrates how the rdf changes with increasing disorder and how diffusional motion of the atoms in the liquid broadens the pair distribution function relative to the glass. In the gaseous state, the atoms have a much longer mean free path than in the liquid, the density is much lower, and the positional correlation is quite small. Thus the rdf becomes a step function, with the minimum distance being defined by the atomic core diameter, a_0, due to the hard-shell repulsive portion of the pair potential.

Quasi-crystalline models of glass/melt structure

The rdf of a multi-component glass or melt becomes quite complicated with increasing distance from a central atom because of the large number of degrees of freedom in atomic arrangements. It is very difficult to interpret an X-ray rdf beyond about 5 Å, even for silica glass (see next section), without the use of a structural model. Such models can be generated in several ways (e.g. from crystal structures of materials with the same composition as the glass or melt, or from computer simulations [see Chapter by Poole et al. (this volume) and later discussion]. They are particularly useful for interpreting the distance region between about 4 and 8 Å in the rdf, which contains information about medium-range structure, such as chains and rings.

One such model, based on the crystal structure approach, was proposed by Taylor (1979) and is used in this chapter. Other quasi-crystalline models have been proposed by Hosemann and Bagchi (1962), Leadbetter and Wright (1972), Konnert et al. (1973), Konnert et al. (1982), Okuno and Marumo (1982), and Yasui et al. (1983), among others. Taylor's approach can be summarized as follows: A sphere of radius r_c is defined about each atom in the asymmetric unit of the model crystal structure, and all interatomic vectors between the central atom and other atoms in the sphere are calculated. A disorder parameter is assigned to each vector based on the experimental X-ray isotropic temperature factors of the two atoms and the length of the vector. Longer interatomic vectors are assigned larger disorder parameters to model the loss of correlation with increasing distance from the central atom. The scattered X-ray intensity is then calculated using the equation

$$i_q(s) = \sum_i \sum_j M_{ij} \left[\exp(-b_{ij}s^2) f_i(s) f_j(s) \sin(sr_{ij}) / sr_{ij} \right] + I_{SA}(s) \tag{10}$$

where the first summation extends over the asymmetric unit of the unit cell of the model structure, the second summation extends over all the atoms within a distance r_c of atom i, and where M_{ij} is a multiplicity factor for r_{ij} given by the multiplicity of atom i in the unit cell divided by the number of formula units per unit cell. In practice, r_c should be chosen about 1 Å larger than the maximum r plotted in the calculated rdf (Taylor, 1979). The term b_{ij} serves to increase the disorder in the model structure to simulate the corresponding glass structure. It is defined as

$$b_{ij} = (B_i + B_j)/(8\pi^2) + (r_{ij}/kr_c)^2 \tag{11}$$

where B_i and B_j are the isotropic temperature factors of atoms i and j, and k is an adjustable constant (= 4 for most glasses). The second term is an arbitrary function to model the loss of correlation between atoms as the distance between them increases. The terms $f_i(s)$ and $f_j(s)$ are the X-ray scattering factors, r_{ij} is the mean distance between atoms i and j calculated from the model structure, and $I_{SA}(s)$ is the small-angle scattering correction for the finite size of the model structure defined as

$$I_{SA}(s) = \sum_i \sum_j f_i(s) f_j(s) 4\pi\rho_0 \left[sr_c \cos(sr_c) - \sin(sr_c) \right] / s^3 \tag{12}$$

Even if a predicted rdf based on a particular structure is found to agree with the experimental rdf, it is not likely to be a unique structure solution. Nonetheless, this approach has been used to provide valuable insights about possible structural units involving pair-correlations to 7 Å and beyond in glass and melt structures.

Figure 6 shows an example of a quasi-crystalline rdf calculated using the method of Taylor (1979) for artificially disordered crystalline high albite. It compares very well with the experimental rdf derived from an X-ray scattering experiment on a sample of polycrystalline high albite. Also shown are the calculated pair correlations for T-O, Na-

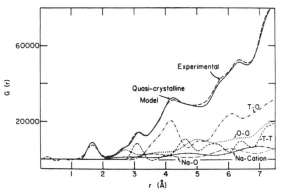

Figure 6. Comparison of the calculated quasi-crystalline rdf of the high albite structure (solid line) with the experimental rdf calculated from x-ray scattering data for polycrystalline high albite (dashed line). Also shown are correlations for various ion pairs in high albite. (From Taylor and Brown, 1979a.)

O, O-O, T-T, and Na-(T, Na), where T represents tetrahedrally coordinated cations (Si, Al). Comparison of the pair correlations for Na-O and Na-(T, Na) with those for T-O, O-O, and T-T illustrates the relative insensitivity of X-ray rdf's to large, weakly bonded monovalent cations which have a range of interatomic distances with oxygens. This example also illustrates how complicated the rdf becomes with increasing distance due to the overlap of different pair correlations.

Radial distribution function of silica glass

The structure of vitreous silica has been the subject of study and speculation since the 1930s. The first modern X-ray scattering study of silica glass using non-photographic methods was that of Mozzi and Warren (1969). They eliminated Compton scattering experimentally by measuring the X-ray fluorescence signal from a metal foil which was excited by X-ray scattering from the glass, and they analyzed their data using the pair distribution function described earlier. Figure 7 shows the pair distribution function they derived from data taken using Rh Kα radiation ($\lambda = 0.6147$ Å) out to an s value of 20 Å^{-1}. A careful analysis of possible configurations of linked SiO$_4$ tetrahedra suggested the following pair correlations (with the peak label given in parentheses): Si-O (B), O-O1 (C), Si-Si1 (D), Si-O2 (E), O-O2 (F), and Si-Si2 (G). The following important conclusions can be derived from the study of Mozzi and Warren: (1) essentially all of the Si atoms are bonded to four oxygens at the corners of a tetrahedron, with an average Si-O distance of 1.62 Å (non-tetrahedral Si, if present, was not detected); (2) essentially all of the oxygens are bonded to two Si atoms; (3) the Si-O-Si angle, β, follows a distribution curve, V(β), (Fig. 8) and varies from 120° to 180°, with a maximum at ~144°; and (4) there appears to be a random orientation about the Si-O bond directions (see Fig. 4). The glass structure developed in the Mozzi-Warren analysis is consistent with that of a random network. No attempt was made by Mozzi and Warren to determine the sizes and distribution of rings of tetrahedra.

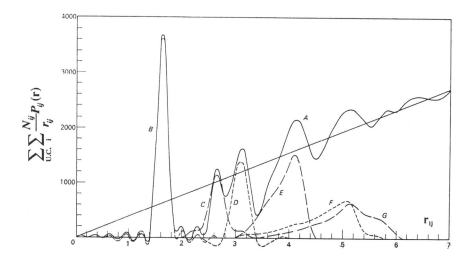

Figure 7. Experimentally derived pair distribution function for SiO$_2$ glass, showing individual pair correlation functions for Si-O1 (peak B), O-O1 (C), Si-Si1 (D), Si-O2 (E), O-O2 (F), and Si-Si2 (G). (Modified after Mozzi and Warren, 1969.)

The small range of Si-O bond lengths, d(Si-O), and the large range of Si-O-Si angles in silica glass can be explained by considering the variation in total energy of the

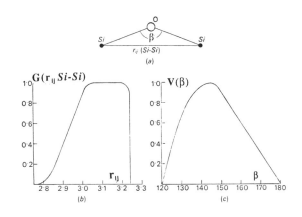

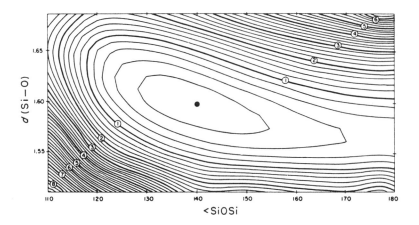

Figure 8. Plots of the distributions of Si-Si1 distances (b) and Si-O-Si angles (c) (designated β) for Si_2O linkages in SiO_2 glass. (Modified after Mozzi and Warren, 1969.)

Figure 9. Potential energy surface for the $H_6Si_2O_7$ molecule plotted as a function of bridging Si-O bond length, d(Si-O), and Si-O-Si angle. The energy minimum (dot) is -2.9×10^6 kJ mol^{-1}. Contours correspond to increments of 2.6 kJ mol^{-1}, relative to this energy minimum. (After Gibbs et al., 1981.)

$H_6Si_2O_7$ molecule, as a function of Si-O distance and Si-O-Si angle, calculated using *ab-initio* quantum mechanical methods with an extended basis set (Gibbs et al., 1981). As can be seen in Figure 9, the potential energy surface consists of a narrow valley, elongated subparallel to the Si-O-Si angle axis. The potential energy contours are very steep along the direction approximately parallel to the d(Si-O) axis. These results indicate that Si-O bond length should be relatively invariant, whereas there is no significant potential energy barrier to Si-O-Si angle changes over the range 120° to 180°. The energy barrier to rotation about the Si-O bridging bonds (the torsion angle α) is also quite low for this molecule (deJong and Brown, 1980a). The potential energy minimum for the disilicate molecule occurs at about 1.60 Å and 140°, which are, within experimental error, the same as those determined by Mozzi and Warren for silica glass. The implications of these and other molecular orbital calculations for glass and melt structure and properties are discussed in Navrotsky et al. (1985) and in deJong and Brown (1980a,b).

Other X-ray studies of the structure of vitreous silica include those by Zarzycki (1957), Cartz (1964), Konnert and Karle (1972), Konnert et al. (1973), Konnert and Karle (1973), DaSilva et al. (1974), Konnert et al. (1982), Morikawa et al. (1982), and others reviewed in Wright (1994). The studies by Konnert and co-workers are particularly noteworthy because of the unique analysis they performed and the controversy these studies have generated (see Wright and Leadbetter (1976) for a discussion). In their 1982 study, Konnert et al. computed quasi-crystalline rdf's (D(r) functions) for quartz, cristobalite, tridymite, and a model of vitreous silica containing 1412 atoms and compared these calculated rdf's with the experimental rdf of silica glass (Fig. 10). A statistical analysis of the correlation coefficients between the quasi-crystalline rdf's and the experimental rdf resulted in correlation coefficients of 0.26, 0.69, 0.82, and 0.91, respectively. Quasi-crystalline rdf models of both cristobalite and tridymite, which contain only six-membered rings of silicate tetrahedra, show strong similarities with the experimental rdf (the earlier study by Cartz (1964) came to a similar conclusion, but the data quality was inferior). However, the Konnert et al. (1982) analysis indicates that the silica glass rdf is best fit by the 1412 atom model, which contains both six- and five-membered rings in the ratio of 2.6:1. These studies also suggest that structurally significant features are present in the rdf of silica glass out to ~20 Å, although this suggestion has been questioned based on a careful evaluation of the data quality (Wright and Leadbetter, 1976). Even if the Konnert et al. (1982) analysis of silica glass is correct to an r value of only 10 Å, it should not be misconstrued as supporting the original crystallite model of glass structure. Instead, it would suggest that the bonding topology in silica glass shows similarities to a tridymite-like bonding topology.

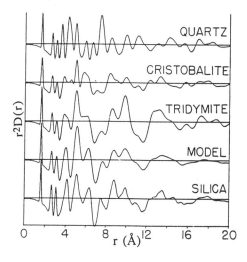

Figure 10. Experimental X-ray rdf ($r^2D(r)$ function) for silica glass (bottom curve) compared with quasi-crystalline model rdf's based on the artificially disordered crystal structures of quartz, cristobalite, tridymite, and a 1412 atom continuous random network model containing six- and five-membered rings in the ratio 2.6:1. (Modifed after Konnert et al., 1982.)

In principle, it should be possible to improve the fit obtained by Konnert et al. (1982) of the experimental rdf of silica glass using reverse Monte Carlo methods. These methods involve movement of the atoms in a cluster to yield a calculated interference function and rdf that agree well with experiment. The reverse Monte Carlo method was applied to vitreous silica by Keen and McGreevy (1990), and the fit with the neutron and X-ray scattering functions is impressive out to $s = 13$ Å^{-1}. However, as pointed out by Wright (1994), their structural model must be rejected because it contains too many silicon atoms which are not four coordinated. This example illustrates the statement made earlier about the non-uniqueness of a glass or melt structure model which fits an experimental rdf. Agreement with diffraction data is necessary, but it is not a sufficient requirement to validate a structural model. However, a structural model which is not consistent with diffraction data must be wrong (Wright, 1994).

Even if excellent fits are achieved using the reverse Monte Carlo approach or a molecular dynamics simulation, it is unlikely that bonding topologies present at low concentrations in glasses can be detected in an X-ray scattering experiment. For example, the presence of small concentrations of planar three- and four-membered rings in vitreous silica at concentrations $\leq 1\%$ would not be detectable in an X-ray scattering experiment. These types of rings were inferred by Galeener (1982) based on the observation of two sharp bands in the Raman spectrum of vitreous silica (D_1 at 495 $\AA^{-1}$ and D_2 at 606 $\AA^{-1}$), which he assigned to ring stretches of planar four- and three-membered rings of silicate tetrahedra, respectively. It is also quite unlikely that an X-ray scattering experiment could detect "defect levels" of five- and six-coordinated Si, as have been observed at low concentrations in several Na- or K-containing silicate glasses prepared by rapid isobaric quenching at high pressures (Xue et al., 1989, 1991; Stebbins and McMillan, 1989, 1993). Raman scattering and NMR spectroscopy are typically more sensitive to these dilute levels of particular structural units in a glass.

A recently published X-ray scattering study of vitreous silica (Poulsen et al., 1995) utilized very high energy X-rays (95 keV) from a high-brightness synchrotron X-ray source (HASYLAB in Hamburg, Germany). The advantages of hard X-rays for scattering studies of amorphous materials include (1) an enlarged s range (from 0.8 to 32 $\AA^{-1}$ in this experiment), thus better direct space resolution than experiments with softer X-rays; (2) considerably smaller absorption corrections; and (3) smaller backgrounds. The results of this study are consistent with the earlier study of Mozzi and Warren (1969), but the errors are smaller. The maximum of the Si-O-Si angle distribution was found to be 147°, which is closer to the original Mozzi-Warren value (144°) than to the value (152°) derived by Da Silva et al. (1974) from a reanalysis of the Mozzi-Warren data.

The first X-ray scattering experiment on silica glass at high pressures (up to 42 GPa) was performed by Meade et al. (1992) using a very small glass sample in a diamond cell and very intense synchrotron X-rays at the National Synchrotron Light Source, Brookhaven National Laboratory, USA. The use of a very small (10-μm diameter) X-ray beam was required to avoid strong Compton scattering from the diamonds. They found clear evidence in their radial distribution functions for a significant change in short- and medium-range order with increasing pressure, which involves an increase in average Si-O bond length (Fig. 11). These changes signal an increase in the average coordination of Si as the glass becomes densified (see chapter by Wolf and McMillan, this volume).

Radial distribution functions of framework aluminosilicate glasses

The X-ray rdf's of feldspar-composition glasses ($NaAlSi_3O_8$ (AB), $KAlSi_3O_8$ (OR), $CaAl_2Si_2O_8$ (AN)) were determined by Taylor and Brown (1979a), and those for AB and AN glasses are shown in Figure 12. Also shown are quasi-crystalline rdf's, based on the albite and anorthite crystal structures, calculated using the method of Taylor (1979). The most important conclusion reached in this study is that the medium-range structures of AB and OR glasses are inconsistent with the four-membered tetrahedral ring topology of crystalline feldspars (see inset feldspar structure in Fig. 12), whereas that of AN glass is similar to this topology. Instead, the rdf's of AB and OR glasses were found to be more consistent with a six-membered ring topology similar to that in crystalline tridymite (Fig.

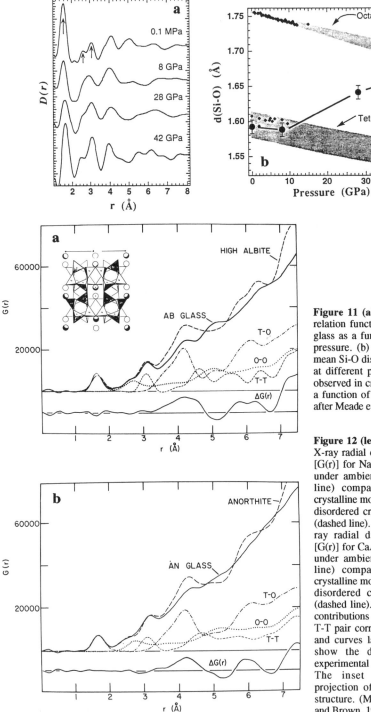

Figure 11 (above). (a) Pair correlation functions, D(r), for SiO_2 glass as a function of increasing pressure. (b) Comparison of the mean Si-O distances in SiO_2 glass at different pressures with those observed in crystalline silicates as a function of pressure. (Modified after Meade et al., 1992.)

Figure 12 (left). (a) Experimental X-ray radial distribution function [G(r)] for $NaAlSi_3O_8$ (AB) glass under ambient conditions (solid line) compared with a quasi-crystalline model rdf of artificially disordered crystalline high albite (dashed line). (b) Experimental X-ray radial distribution function [G(r)] for $CaAl_2Si_2O_8$ (AN) glass under ambient conditions (solid line) compared with a quasi-crystalline model rdf of artificially disordered crystalline anorthite (dashed line). Also shown are the contributions from T-O, O-O, and T-T pair correlations (T = Si,Al) and curves labeled $\Delta G(r)$ which show the difference between experimental and calculated rdf's. The inset shows an (010) projection of the albite feldspar structure. (Modified after Taylor and Brown, 1979a.)

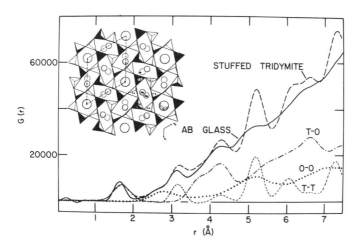

Figure 13. Experimental x-ray radial distribution function [G(r)] for NaAlSi₃O₈ (AB) glass under ambient conditions (solid line) compared with a quasi-crystalline model rdf of artificially disordered crystalline nepheline (dashed line), a stuffed derivative of tridymite. Also shown are the contributions from T-O, O-O, and T-T pair correlations (T = Si, Al). The inset shows an (001) projection of the nepheline crystal structure. (Modified after Taylor and Brown, 1979a.)

13). The key feature in the AB glass rdf which led to this conclusion is the local maximum at ~5.1 Å (Fig. 12), which was interpreted as being due to T-T2 pair correlations (where T represents the tetrahedrally coordinated cations Si or Al) typical of a six-membered ring of tetrahedra. At the r-value of about 4.6 Å, where T-T2 pair correlations typical of four-membered rings of tetrahedra would be predicted, the rdf's of both AB and OR glasses show a minimum. For AN glass, there is a maximum at about 4.6 Å and a minimum at 5.1 Å (Fig. 12), which is consistent with the presence of four-membered rings of tetrahedra, similar to the ring topology of the anorthite crystal structure. These two types of ring topologies are shown in Figure 14. The large circles in this figure represent alkali cations. Figure 13 shows the stuffed tridymite-like topology, which is consistent with the rdf's of alkali feldspar composition glasses under ambient conditions, as well as a comparison of the experimental G(r) function for AB glass and the quasi-crystalline G(r) calculated assuming a stuffed-tridymite-like bonding topology.

The T-O distances in the feldspar glasses were found to increase from 1.63 Å in AB and OR glasses, with a Si:Al ratio of 3:1, to 1.66 Å in AN glass, with a Si:Al ratio of 1:1.

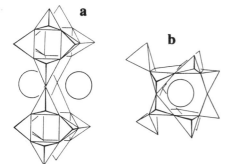

Figure 14. Schematic drawings of the ring topologies of the feldspar (a) and nepheline (b) crystal structures. The triangles represent Si- or Al-containing TO₄ tetrahedra and the circles represent alkali cations. (From Taylor and Brown, 1979a.)

These values are slightly longer than that in silica glass, as expected, because of the partial substitution of the larger Al(III) for Si(IV). The T-O-T bond angles are 146° for AB and OR glasses and 143° for AN glass. Finally, silica, AB, and OR glasses have more medium-range order than AN glass, as indicated by the loss of correlation above about 7.2 Å in the G(r) functions for silica, AB, and OR glasses compared with about 6 Å in AN glass. Using the short-range order parameter, ζ, suggested by Waseda and Suito (1977) (see later discussion of alkali silicate melts), ζ values of 4.4 and 3.6 are obtained for AB-OR and AN glasses, respectively. These values are lower than the ζ-value for silica glass (6.2) reported by Waseda and Suito, suggesting that the feldspar-composition glasses have less medium-range order than silica glass.

The measured densities of AB (2.38 g cm^{-3}) and OR (2.37 g cm^{-3}) glasses are significantly different from those of high albite (2.62 g cm^{-3}) and sanidine (2.57 gcm^{-3}), whereas the measured density of AN glass (2.69 g cm^{-3}) is much closer to that of crystalline anorthite (2.76 g cm^{-3}). The densities of AB and OR glasses were calculated based on hypothetical nepheline-like and kalsilite-like structures, respectively, both of which are stuffed-tridymite derivatives. The densities predicted in this fashion are 2.40 gcm^{-3} and 2.31 gcm^{-3}, respectively, which are close to those observed. These arguments were used to support the lack of feldspar-like bonding topology in AB and OR glasses and the presence of a stuffed-tridymite-like topology (Taylor and Brown, 1979a). This study was extended to jadeite- (NaAlSi$_2$O$_6$ (JD)) and nepheline-composition (NaAlSiO$_4$ (NE)) glasses (Taylor and Brown, 1979b), both of which were found to have rdf's similar to those for AB glass (Fig. 15). The densities of these two glasses are predicted better with the stuffed-tridymite model than ones based on the jadeite or carnegeite crystal structure (Fig. 16), although not as well as for silica or AB glass. The rdf's of AB, JD, and NE glasses, as well as a rhyolite-composition glass (Hochella and Brown, 1984) and AB and JD glasses quenched from 10 kb pressure (Hochella and Brown, 1985), are similar to that of silica glass, which was found to have a bonding topology similar to tridymite (see earlier discussion of the rdf of silica glass) (Fig. 15).

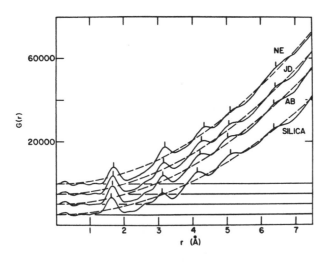

Figure 15. Comparison of G(r) functions for silica, AB, jadeite- (JD), and nepheline-composition (NE) glasses. The dashed lines are the function $4\pi r^2 \rho_0$, which represents a homogeneous distribution of electron density. (From Taylor and Brown, 1979b.)

The effect of water on the short- and medium-range structures of glasses in the albite-anorthite-quartz system was studied by Okuno et al. (1987) using X-ray scattering methods. They found small but significant differences in the radial distribution functions for correlations beyond 3.5 Å, and attributed these differences to the network-modifying effects of water. The difference was largest for the pair of AB and AB-H_2O glasses and smallest for AN-Q and AN-Q-H_2O glasses, where Q represents the SiO_2 component. No evidence was found in this study for a change in the average coordination number of Al caused by dissolution of water in AB melt.

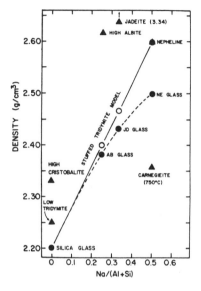

Figure 16. The observed densities of SiO_2, $NaAlSi_3O_8$ (AB), $NaAlSi_2O_6$ (JD), and $NaAlSiO_4$ (NE) glasses (solid circles) compared with densities calculated assuming stuffed-tridymite models (based on the nepheline unit cell) of the glasses (open circles) plotted as a function of Na/(Al+Si). Also shown are the densities of various related crystalline materials (solid triangles). (From Taylor and Brown, 1979b.)

Henderson et al. (1984) carried out an X-ray scattering study of $NaFeSi_3O_8$ and $KFeSi_3O_8$ glasses, hoping that the replacement of Al by the heavier Fe, with greater scattering power, would enhance resolution of some of the structural features of these glasses relative to the study of Taylor and Brown (1979a). Using quasi-crystalline models, they found that $KFeSi_3O_8$ is composed predominantly of six-membered rings, similar to the finding of Taylor and Brown for $KAlSi_3O_8$ glass. For $NaFeSi_3O_8$ glass, however, they found that the experimental rdf was closer to a quasi-crystalline model representing a mixture of nepheline and high albite bonding topologies, i.e. containing both six- and four-membered rings of tetrahedra. The Fe(III)-O distance could not be resolved from the Si-O distance, giving an average d(T-O) value of 1.70 Å. The XAFS spectroscopy study by Brown et al. (1978) found Fe(III) to be four-coordinated in glasses of these compositions.

Experimental approaches to high-temperature X-ray scattering studies of silicate melts

A number of sample geometries and furnace designs have been used for in situ high-temperature X-ray scattering studies of silicate melts; the ones discussed here were used to collect the data presented in the next section. The study by Taylor et al. (1980) used a cylindrical glass sample mounted vertically (in the $\chi = 0°$ position) on a four-circle diffractometer fitted with a Mo X-ray tube. An alumina-coated Pt-10% Rh resistance heater, modified after the design of Brown et al. (1973), was used to heat the sample without interfering with the incident or scattered X-ray beams. Sample temperature was monitored using a Pt-10% Rh thermocouple mounted just above the molten sample and out of the X-ray beam. The θ-2θ step-scanning method was used for data collection, an incident-beam graphite monochromator was used to produce Mo$K\alpha$ wavelengths, and the scattered X-rays were detected using a scintillation detector coupled with a pulse-height analyzer. Temperatures in excess of 1300 K for extended time periods can be reached with this design. However, this technique works only for high viscosity melts, such as those of feldspar composition, because the sample will change shape if the melt viscosity is too low. Changes in sample shape during data collection would create serious difficulties because the scattering volume would change and exact absorption corrections would be impossible. Also with this design, there is no enclosure around the sample in which to control oxygen partial pressure, so melt samples with cations that can be oxidized or reduced cannot be studied using this simple apparatus.

The high-temperature X-ray scattering studies of silicate melts by Waseda and co-workers utilized a flat sample in an alumina sample holder oriented in the horizontal position in a θ-θ diffractometer. In this type of diffractometer, the sample remains stationary, and the X-ray tube (Mo for these experiments) and scintillation detector are scanned symmetrically about the sample. The sample was heated by means of a W resistance heater in the form of a flat strip, on which the sample holder was mounted. In addition, a W multi-wire resistance heater was mounted above the sample to increase the maximum temperature attainable and to minimize temperature gradients. Sample temperature was monitored using a W-5% Re/W-26% Re thermocouple touching the back of the sample holder-W heater assembly. This design allows temperatures in excess of 2500 K to be reached and maintained stably for several hours, and highly absorbing samples can be studied because of the reflection mode of data collection. The heater and sample are located inside a vacuum chamber which can be evacuated to 10^{-6} torr; thus, oxidizing and reducing atmospheres in contact with a melt sample can be controlled. Details about this apparatus and its operation are given in Waseda et al. (1975). Several of the high-temperature melt studies by Marumo and co-workers utilized a similar heater design and diffractometer (Marumo and Okuno, 1984).

The high-temperature studies of Nukui and co-workers and some of those by Marumo and co-workers utilized an oxygen-propane gas-flame heater after the designs of Nukui et al. (1972) and Miyata et al. (1979), respectively. The samples were cut prisms of a crystalline material mounted in the $\chi = 180°$ (top) position of a four-circle diffractometer that utilized a Mo X-ray tube. These micro gas-flame heaters are capable of producing a sample temperature in excess of 2500K and can be positioned to melt only the bottom of the prism, which typically assumes a well-defined hemispherical shape. Sample temperatures are measured using an optical pyrometer. The main advantages of this type of heater are the small amount of sample required and the high temperatures that can be attained. The main disadvantages are the fact that it cannot be used with samples having high vapor pressures, because the material will evaporate rapidly, and it is not suitable for samples with large absorption coefficients because the X-ray beam must pass through the sample. Even for relatively refractory glass samples, it is possible that heating to very high temperatures will cause element loss and a change in bulk composition.

A specially designed diffractometer and sample heater for X-ray scattering studies of high-temperature liquids at a synchrotron radiation source have been described by Marumo et al. (1989) and installed on an X-ray beam line at the Photon Factory in Tsukuba, Japan. This facility was used to study melts in the $CaMgSi_2O_6$-$CaTiAl_2O_6$ system at temperatures as high as 1873K (Marumo et al., 1990).

A number of other types of sample heaters and X-ray diffractometers have also been designed and built for high-temperature studies of liquids, including a Pt wire loop and the furnace assemblies described by Levy et al. (1966) and Mitchell et al. (1976), but these have not been used in X-ray scattering studies of silicate melts to our knowledge. Irrespective of the type of furnace used in studies of silicate melts, it is important to check the compositions of quenched samples after they have been melted because of the possibility of loss of volatile and non-volatile elements at very high temperatures, as well as the addition of components from sample holders (e.g. alumina or Pt). Unfortunately, this is rarely done.

It is possible to measure the X-ray scattering from a liquid sample using an energy dispersive detector, but to the best of our knowledge, no experiments using such a detector have been carried out on silicate melts. Such studies can sample a very large s range, thus are capable of providing very high resolution in the radial distribution function. The reader interested in energy dispersive methods is referred to the energy dispersive X-ray study of liquid Hg by Prober and Schultz (1975) and to the work of Egami (e.g. Aur et al., 1982).

X-ray scattering results for silicate melts

SiO_2 and Al_2O_3 melts. Waseda and Toguri (1977, 1990) and Nukui et al. (1978) determined the rdf's of molten silica at temperatures of 1750°C and 1850°C, respectively. These studies found that the SiO_4 structural unit persists in these melts and that the random network is a valid model for the melt structure. Interatomic distances and coordination numbers (in parentheses) from the most recent study are: Si-O: 1.62 ± 0.01 Å (3.8 ± 0.2 oxygens); O-O1: 2.65 ± 0.01 Å (5.6 ± 0.2 oxygens); Si-Si1: 3.12 ± 0.1 Å (3.9 ± 0.2 silicons). These values are very similar to those determined by Mozzi and Warren (1969) in their study of vitreous silica under ambient conditions, which emphasizes the fact that Si-O bonds are strong, undergoing little if any thermal expansion, and that the average bonding topology of the silica glass random network is not strongly affected at these temperatures. Clearly there will be bond breakage, atom hopping (diffusion), and structural relaxation (viscous flow) in molten silica at these high temperatures, but these changes are not observed in the experimental time frame of an X-ray scattering experiment.

Alumina melt at ~2370 K has been studied by Nukui et al. (1976) and Waseda and Toguri (personal communication, 1988) using high-temperature X-ray scattering as well as high-temperature NMR methods (see chapter by Stebbins, this volume). Interatomic distances and coordination numbers (in parentheses) from the X-ray scattering study of Waseda and Toguri are: Al-O1: 2.02 Å (5.5 oxygens); O-O1 and Al-Al1: 2.82 Å (3.2 atoms); Al-Al1: 3.41 Å (5.3 atoms). It is possible that a low-r shoulder of the 2.82 Å feature located at ~2.6 Å in the rdf of Waseda and Toguri is real and could correspond to O-O1 distances only, whereas the 2.82 Å feature could be due primarily to Al-Al1 correlations. Nukui and co-workers compared the experimental rdf [D(r) function] of Al_2O_3 melt with those calculated assuming bonding topologies like in α-Al_2O_3 and γ-Al_2O_3. The quasi-crystalline rdf for the γ-Al_2O_3 topology fits the experimental rdf significantly better than the one for α-Al_2O_3, although there is some mismatch in the 2.7 to 4 Å region between the experimental rdf and the rdf calculated for γ-Al_2O_3-like topology. Nukui and co-workers argued that the γ-Al_2O_3 bonding topology is more likely in the melt at high temperatures because of the observation by Thompson and Whittemore (1968) that alumina melted at very high temperatures crystallizes into γ-Al_2O_3, whereas when melted at lower temperatures, it crystallizes into α-Al_2O_3. Consideration of the crystal structures of α-Al_2O_3 (Ishizawa et al., 1980) and γ-Al_2O_3 (Zhou and Snyder, 1991) shows that the O-O1 and Al-Al1 distances overlap in the 2.5 to

3.0 Å region, which is consistent with the assignments of the pair correlations in the melt made above. Furthermore, the high-temperature structural study of α-Al$_2$O$_3$ (Ishizawa et al., 1980) showed that the two independent Al-O distances increased from 1.971 and 1.852 Å at 300 K to 2.024 and 1.880 Å at 2170 K. The observed average Al-O distance in Al$_2$O$_3$ at 2370 K is consistent with the longer of these two values.

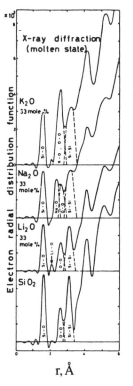

Figure 17. Experimental xrf's [G(r)] for melts of these compositions (in mol %):
33 Li$_2$O   67 SiO$_2$,
33 Na$_2$O   67 SiO$_2$,
33 K$_2$O   67 SiO$_2$,
100 SiO$_2$
at temperatures of 1150°, 1000°, 1100°, and 1750°C, respectively (see Table 2 for details). (From Waseda and Suito, 1977.)

Alkali silicate melts. The X-ray rdf's of binary alkali silicate melts in the systems Li$_2$O-SiO$_2$, Na$_2$O-SiO$_2$, and K$_2$O-SiO$_2$ were determined by Waseda and Suito (1977) and are shown in Figure 17. The interatomic distances (r_{ij}), coordination numbers (N_{ij}), and root mean square atomic displacements ($(\Delta r_{ij})^{1/2}$) for molten samples with 33 mol % M$_2$O - 67 mol % SiO$_2$ and for molten SiO$_2$ are given in Table 2 and are compared with similar data for the corresponding glasses. In general, the short-range order ·in the M$_2$O-SiO$_2$ melts examined by Waseda and Suito is similar to that in the corresponding glasses. Waseda and Suito (1977) introduced a measure of short-range order which they used to compare the various M$_2$O-SiO$_2$ melts. This dimensionless parameter is defined as $\zeta = r_s/r_1$, where r_s is the distance beyond which the G(r) function is experimentally indistinguishable from the $4\pi r^2\rho_0$ function (which represents a homogeneous distribution of electron density), and r_1 is the shortest observed pair correlation (Si-OI = 1.62 Å). The value of ζ for SiO$_2$ melt is 6.2, whereas those for the M$_2$O-SiO$_2$ melts are between 5.5 and 5.2. This change in ζ, as the first 10 mol % M$_2$O is added to SiO$_2$ melt, signals a significant break-up of the random network of tetrahedra, as expected. From 20 to 40 mol % M$_2$O, ζ is essentially constant, whereas from 40 to 60% M$_2$O, ζ decreases slightly. Waseda and Suito pointed out the similarity between the trends for ζ vs. mol % M$_2$O and the activation energy for viscous flow vs. mol % M$_2$O.

It is interesting to note that the cation-oxygen distances in the melts appear to be about the same as those in the corresponding glasses (Table 2), suggesting little bond expansion with increasing temperature. This is somewhat surprising in light of measured bond expansions as a function of temperature for crystalline alkali-containing silicates, which are relatively high. The mean [6]Li-O, [6]Na-O, and [6]K-O linear expansion coefficients over the temperature range 300 to 1300 K are 20×10^{-6} K^{-1}, $\sim17 \times 10^{-6}$ K^{-1}, and 21×10^{-6} K^{-1}, respectively (Hazen and Finger, 1982), and the calculated [6]Li-O, [6]Na-O, and [6]K-O bond lengths at 1300 K are 2.26, 2.86, and 3.05 Å, respectively. Thus the expected increases in the average Li-O, Na-O, and K-O

Table 2. Structural data for molten and glassy binary alkali silicates (Waseda and Toguri, 1977a), molten alkaline earth silicates (Waseda and Toguri, 1990), glassy alkaline earth silicates (Waseda and Toguri, 1977b), and molten 2FeO-SiO_2 (Waseda and Toguri, 1978).

Sample and temperature (°C)	Pair Corre- lation	X-ray (molten state) [*]			X-ray (glassy state) [*]		
		r_{ij} (Å)	N_{ij} (atoms)	$(\Delta r_{ij})^{1/2}$ (Å)	r_{ij} (Å)	N_{ij} (atoms)	$(\Delta r_{ij})^{1/2}$ (Å)
SiO_2							
T=1750	Si-O	1.62	3.8	0.096	1.62	3.9	0.087
	O-O	2.65	5.6	0.124	2.65	5.5	0.102
	Si-Si	3.12	3.9	0.187	3.11	3.9	0.141
$0.33Li_2O$-$0.67SiO_2$							
T=1150	Si-O	1.61	3.8	0.117	1.62	3.7	0.090
	Li-O	2.08	4.1	0.131	2.07	3.8	0.095
	O-O	2.66	5.5	0.195	2.65	5.6	0.101
	Si-Si	3.13	3.8	0.260	3.13	3.8	0.143
$0.33Na_2O$-$0.67SiO_2$							
T=1000	Si-O	1.62	4.1	0.095	1.62	4.0	0.086
	Na-O	2.36	5.9	0.151	2.36	5.8	0.101
	O-O	2.66	5.6	0.202	2.65	5.2	0.112
	Si-Si	3.20	3.8	0.279	3.21	3.6	0.146
$0.33K_2O$-$0.67SiO_2$							
T=1100	Si-O	1.62	3.9	0.124	1.62	3.8	0.086
	K-O	2.66	13.0[**]	0.182[**]	2.65	13.2[**]	0.120[**]
	O-O	2.66			2.65		
	Si-Si	3.23	3.7	0.257	3.23	3.5	0.154
MgO-SiO_2							
T=1700	Si-O	1.62	3.9	0.109	1.63	3.7	0.096
	Mg-O	2.12	4.3	0.151	2.14	4.6	0.108
	O-O	2.65	5.4	0.215	2.65	5.7	0.151
	Si-Si	3.16	3.3	0.282	3.15	3.4	0.213
CaO-SiO_2							
T=1600	Si-O	1.61	3.9	0.127	1.63	3.8	0.109
	Ca-O	2.35	5.9	0.171	2.43	5.9	0.125
	O-O	2.67	5.2	0.206	2.66	5.5	0.183
	Si-Si	3.20	3.1	0.264	3.23	3.4	0.199
2FeO-SiO_2 [†]							
T=1400	Si-O	1.62	3.9	0.147	--	--	--
	Fe-O	2.05	3.9	0.214	--	--	--
	O-O	--	--	--	--	--	--
	Si-Si	3.27	3.1	0.302	--	--	--

[*] Errors for r_{ij}, N_{ij}, and $(\Delta r_{ij})^{1/2}$ are ±0.01Å, ±0.3 atoms, and ±0.005 Å, respectively.

[**] The pair-correlations for K-O and O-O overlap, so coordination numbers and peak widths could not be determined independently.

[†] O-O pair correlation overlaps with Fe-Fe correlations, thus d(O-O) could not be determined independently.

distances over this temperature range are about 0.04, 0.05, and 0.06 Å, respectively. Here we are assuming that the mechanism of thermal expansion and the degree of anharmonicity in glasses and melts are similar to those in crystals of similar compositions.

It is not clear why there are no significant differences between the Li-O, Na-O, and K-O bond lengths observed by Waseda and Suito (1977) in the alkali silicate melts at 1300 K and the corresponding glasses at 300 K. These M-O bonds are the weakest in the glass, melt, and crystal structures, so they would be expected to expand more than other M-O bonds. Because of the weakness of M-O bonds relative to Si-O bonds, the M-O bonds are likely to relax to their unexpanded distances during the quench, unless the interlinked SiO_4 tetrahedra to which the alkali cations are bonded relax in such a way so as to prevent this. Structural relaxation of the tetrahedral portion of the melt and glass structure during heating or quenching should involve changes in the Si-O-Si (β) angle and the α_1 and α_2 torsion angles, but not in Si-O distances (see Figs. 4 and 9), whereas relaxation of the MO_x polyhedra should involve changes in M-O bond lengths as well as in M-O-Si, M-O-M, and O-M-O angles. Thus a difference in M-O bond length is expected between the glass under ambient conditions and the glass or melt at high temperature. In addition, it is likely that the coordination environments of alkali cations are less affected by changes in the tetrahedral topology in melts and glasses of low tetrahedral polymerization than in those with a high degree of tetrahedral polymerization. A likely explanation for the observed similarity of alkali cation-oxygen distances in glasses at 300 K and the melts at 1300 K is that the high-temperature rdf studies of these melts were not sufficiently sensitive to longer M-O pair-correlations which overlap with O-O and Si-Si pair-correlations (Fig. 17), thus they were not clearly observed. These longer M-O bonds, if present, should undergo greater thermal expansion than the shorter ones. The X-ray rdf studies of alkali silicate glasses at 300 K by Yasui et al. (1983, 1994) and Imaoka et al. (1983) developed quasi-crystalline models for these glasses which support the presence of a relatively broad range of M-O distances for each types of cation. These studies also found a significant dependence of average Si-O-Si angle on the type of alkali cation, suggesting that their size exerts a strong influence on geometric details of tetrahedral linkages.

As the concentration of M_2O in these melts increases above 50 mol %, Waseda and Suito (1977) found that the number of second-neighbor Si atoms around a central Si decreased from 4 to 3, which they interpreted as indicating a reduction in tetrahedral polymerization and an increase in the randomness of arrangement of SiO_4 units. These high-temperature data are relatively insensitive to possible changes in tetrahedral polymerization or Q speciation, which is known to occur when smaller amounts of M_2O are added to SiO_2 melt from high-temperature ^{29}Si NMR studies (Stebbins et al., 1992). However, higher resolution X-ray rdf studies of alkali disilicate (33 mol % M_2O) and alkali metasilicate (50 mol % M_2O) glasses (Yasui et al., 1983; Imaoka et al., 1983; Yasui et al., 1994) at 300 K provide evidence for Q^2 (single chains of tetrahedra) and Q^3 (sheets of tetrahedra) groups, respectively. The coordination numbers of Li, Na, and K were reported to be 4, 6, and 7, respectively, in all the M_2O-SiO_2 melt compositions studied (10-60 mol % M_2O) (Waseda and Suito, 1977), which are similar to those reported for the glasses (Yasui et al., 1983; Imaoka et al., 1983). However, as pointed out by Yasui et al. (1983), it is difficult to define precisely the alkali cation coordination number because of the relatively continuous nature of the M-O distance distribution.

In this and the following discussion of results of high-temperature X-ray scattering studies of silicate melts, it should be remembered that a high-temperature X-ray rdf of a silicate melt represents the static average of many different structural configurations around each cation (see discussion in the introduction of this chapter) and that in rapidly relaxing melts such as the alkali silicate compositions, alkali cations are very likely to be diffusing rapidly during the X-ray scattering measurements. Thus it is difficult to derive a precise description of the local structural environments of alkali cations in silicate melts from high-temperature X-ray rdf's.

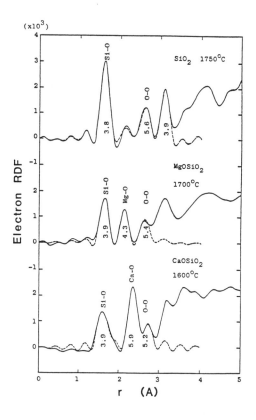

Figure 18. Experimental x-ray rdf's of SiO_2 (1750°C), MgO-SiO_2 (1700°C), and CaO-SiO_2 (1600°C) melts. The numbers under the peaks represent the number of nearest-neighbor atoms (see Table 2 for details). (From Waseda and Toguri, 1990.)

Alkaline-earth-silicate melts. The X-ray rdf's of $MgSiO_3$ (EN) and $CaSiO_3$ (WO) melts determined by Waseda and Toguri (1978, 1990) are shown in Figure 18 and the distances are summarized in Table 2. The numbers quoted are from the more recent study. Si in these melts was found to be coordinated by four oxygens. As the mol % MO was increased in these melts beyond the equimolar compositions, the coordination number of the Si-Si pair was found to decrease from 4 to 3, which was interpreted as indicating a decrease in tetrahedral polymerization, similar to the finding for alkali-silicate melts. The coordination numbers of Mg and Ca were reported to be about four and six, respectively. The observed average Ca-O distance (2.35 Å) is consistent with the sum of radii (2.36 Å) with Ca coordinated by six oxygens (Shannon, 1976), without accounting for possible thermal expansion (which would add about 0.06 Å; Hazen and Finger, 1982). However, the observed Mg-O distance (2.12 Å) is significantly longer than would be expected for [4]Mg (sum of [4]Mg and [3]O radii values gives 1.93 Å (Table 1), and thermal expansion would add another 0.04 Å). The reported Mg-O distance is the same as that expected for [6]Mg, including thermal expansion (distance calculated from the sum of radii + thermal expansion ~2.13 Å). Thus the coordination number reported for Mg in EN melt by Waseda and Toguri (1991) is not consistent with the reported Mg-O distance. There is some overlap between the Ca-O1 and O-O1 pair correlations in WO melt, whereas the Mg-O1 pair correlation is cleanly separated from both the Si-O1 and O-O1 pair correlations in EN melt (Fig. 18). This would make the coordination number of Ca somewhat less certain than that of Mg. A temperature-dependent study of WO melt from 1600 to 1780°C showed that the peaks in the interference function and the pair correlations in the rdf's broaden slightly, but that the short- and medium-range structures do not change. However, as expected, there are increases in the root mean square displacements of the atoms $[(\Delta r_{ij})^{1/2}]$ in the Si-O1, M-O1, O-O1, and Si-Si1 pairs with increasing temperature.

The coordination environments of Mg and Ca in these melts differ from those observed in separate X-ray scattering studies of EN (Yin et al., 1983) and WO (Yin et al., 1986) glasses at 300 K. In these glasses, Mg was reported to be six-coordinated, with four short Mg-O distances of 2.08 Å and two longer Mg-O distances of 2.50 Å, and Ca was reported to be six-coordinated with an average Ca-O distance of 2.43 Å (four Ca-O pairs at 2.34 Å and two pairs at 2.54 Å). The two longer Mg-O and Ca-O distances reported are estimates based on fitting a pigeonite-like quasi-crystalline model to the EN and WO glass data, so it is not clear how much weight should be placed on these conclusions since they are model dependent. Yin et al. (1983, 1986) also calculated X-ray scattering functions for both glasses using the results of molecular dynamics simulations (Matsui et al., 1982) and found reasonable but not perfect agreement. Another MD simulation of $MgSiO_3$ glass at 300 K (Kubicki and Lasaga, 1991) fits the X-ray data of Yin et al. (1983) more poorly, thus should be viewed with caution. The results of Yin et al. (1983, 1986) are consistent with two different structural models, which emphasizes the need for caution in interpreting X-ray scattering data from glasses. It is not clear how the difference in reported Mg coordination environment in EN melt and glass can be reconciled; however, the reported average Mg-O bond lengths in the melt tend to favor a Mg coordination between five and six rather than between four and five for this composition. On the other hand, the Mg-O pair correlation in the rdf of EN melt (Waseda and Toguri, 1990) is cleanly separated from other pair correlations and has an integrated area consistent with [4]Mg. The two longer Mg-O distances reported by Yin et al. (1983) are located beneath the rdf feature for O-O1 pair correlations, so their existence cannot be confirmed. One possible explanation for these observations is that the local environment of Mg consists of four oxygens, and that the two more distant oxygens reported by Yin et al. are not really part of the coordination sphere of Mg in EN glass, assuming they are present. We favor this explanation in light of EXAFS and UV/visible evidence that Fe(II) and Ni(II) are dominantly four and five coordinated in similar glasses and melts, respectively, as discussed below. Mg and Ni(II) are very similar in size and would be expected to adopt similar coordination environments in silicate glasses and melts. In addition, the coordination chemistries of Mg and Fe(II) are similar in crystalline silicates, so they would also be expected to have similar coordination environments in silicate glasses and melts. This important issue will be revisited in a later discussion of the partitioning of cations between silicate melts and coexisting crystals.

Yin et al (1986) compared the short-range order parameter, ζ (discussed above), for EN and WO glasses with those of silica glass and the alkali silicate glasses studied by Waseda and Suito (1977). ζ-values of about 4.4 and 5.0, respectively, were observed for these two glasses. Yin et al. also observed a near-linear trend between the ζ-parameter and a coulombic force, $2z/r^2_{M-O}$, for these glasses, with short-range order decreasing as the coulombic force between modifier-cation and oxygen increases. These concepts should also be applicable to silicate melts of the same compositions.

Mg- and Ca-containing silicate glasses of pyrope (PY) and grossular (GR) compositions also have been investigated by X-ray rdf methods (Okuno and Marumo, 1993). The Si and Al cations in these glasses were reported to be four-coordinated by oxygens, with T-O distances of 1.68 and 1.70 Å, respectively for PY and GR glasses. The average Mg-O distance in PY glass is 2.1 Å, and the reported Mg coordination number is 4.8. In GR glass, the average Ca-O distance is 2.4 Å, and the reported Ca coordination number is about 7. The short-range structure of PY glass was found to be very similar to that of EN glass out to a radial distance of 6 Å, whereas that of GR glass showed significant differences relative to WO glass, especially beyond about 3 Å.

Iron-silicate melts. The structures of melts in the systems $FeO\text{-}SiO_2$ and $FeO\text{-}Fe_2O_3\text{-}SiO_2$ were studied by Waseda and Toguri (1978) and Waseda et al. (1980) as a function of SiO_2 content, temperature, and partial oxygen pressure, and the results are shown in Figures 19, 20, and 21 and reported in Table 2. The compositions studied ranged from 0 to 44 mol % SiO_2 in the ferrous iron-containing system and from 20 to 35 wt % SiO_2 in the ferric iron-containing system. In all of the iron-bearing melts studied by Waseda and co-workers, Si is four-coordinated by oxygens. The Si-Si distance was found to decrease with increasing SiO_2 content in both systems, and the number of second-neighbor Si atoms was found to increase. This was interpreted as indicating an increase in polymerization of SiO_4 tetrahedra with increasing SiO_2 content, as shown schematically in Figure 21. Beyond about 30 wt % SiO_2, these trends flatten out, suggesting that changes in polymerization cease with increasing SiO_2 content. Also, Waseda and co-workers observed that the Si-Si distance in this region is about 3.18 Å, which they

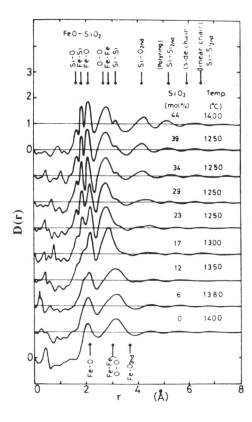

Figure 19. Experimental x-ray D(r) functions of melts in the system $FeO\text{-}SiO_2$. The D(r) function for fayalite-composition melt is the curve labeled 34 mol %. The partial pressure of oxygen during these experiments was held constant at 10^{-11} atm. (Modified after Waseda and Toguri, 1978.)

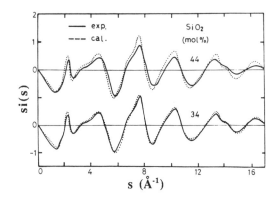

Figure 20. Comparison of the experimental (solid line) and calculated (dotted line) si(s) functions (reduced intensity functions) for molten fayalite (labeled 34 mol % SiO_2) and for a melt of composition $54FeO$-$44SiO_2$. (Modified after Waseda and Toguri, 1978.)

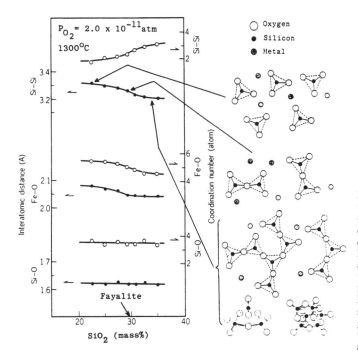

Figure 21. Plots of the variation of distances (solid circles) and coordination numbers (open circles) for ion pairs in the FeO-Fe_2O_3-SiO_2 system as a function of the mass % SiO_2. Data shown are for melts without Fe(III). Drawings on the right show the possible anion groups at 3 values of mass % SiO_2. (From Waseda and Toguri, 1990.)

pointed out is closer to the value expected for silicate rings than chains. In both systems, they found that the Fe-O distance gradually decreases (from 2.08 to 2.02 Å) with increasing SiO_2 content, which they attributed to a change from octahedral to tetrahedral coordination of Fe. Fayalite composition melt was predicted to have dominantly four-coordinated Fe(II) based on the high-temperature rdf, whereas molten FeO has dominantly six-coordinated Fe(II) and a NaCl-like structure. Temperature was found to cause a small amount of thermal expansion of Fe-O, Fe-Si, and Si-Si distances, whereas changes in the partial pressure of oxygen from 10^{-11} to 10^{-7} atm was found to have a negligible effect on interatomic distances and coordination numbers. A quasi-crystalline model generated for fayalite composition melt (Fig. 20), consisting of a random packing of SiO_4 and $Fe(II)O_4$ tetrahedra, gave a satisfactory fit of the X-ray interference function for molten Fe_2SiO_4 but not for a melt of composition $0.56\ FeO$-$0.44\ SiO_2$. One troubling result from this impressive study is the very short Fe-Si correlation (1.79 Å) reported by Waseda and Toguri, which yields an unrealistically small Si-O-Fe angle of 55°. We have no explanation for this confusing result except to point out that it is not consistent with high-temperature EXAFS data for Fe_2SiO_4 melt (Jackson et al., 1993), which are discussed later in this chapter.

Feldspar-composition melts. The rdf of a supercooled albite-composition (AB) melt at 805°C ($T_g \approx 765$°C) was determined by Taylor et al. (1980). The scattering function and rdf are very similar to those observed for the same composition glass under ambient conditions (Taylor and Brown, 1979a). They concluded that the supercooled melt has a bonding topology more similar to that of nepheline, a stuffed-tridymite derivative with six-membered rings of tetrahedra, than high albite, with four-membered rings. During the course of their high-temperature X-ray experiments, the glass devitrified into high albite. Based on the changes in scattering functions as a function of time, Taylor et al. concluded that T-O bonds were broken and reformed in the melt during crystallization, resulting in a different network topology. This conclusion helps rationalize the observed kinetic barrier to melting and crystallization of anhydrous $NaAlSi_3O_8$.

An X-ray scattering study of albite and anorthite (AN) composition melts (Okuno and Marumo, 1982; Marumo and Okuno, 1984) at temperatures of 1200 and 1600°C, respectively, came to essentially the same conclusions as Taylor and Brown (1979a) in their study of AB and AN glasses under ambient conditions. Okuno and Marumo concluded that the bonding topology of AB melt is similar to that of stuffed tridymite, whereas that of AN melt is similar to the bonding topology of feldspar. T-O and T-T distances determined for AB melt are 1.64 and 3.14 Å, respectively, and those for AN melt are 1.68 and 3.12 Å, respectively. The T-O-T angles are 146.4° (AB) and 136.4° (AN). Quasi-crystalline models of anorthite, based on a structure refinement at 1430°C (reported in Smith, 1974), and low tridymite under ambient conditions, resulted in T-O and T-T distances and T-O-T angles which are reasonably close to those observed. Those derived from a quasi-crystalline model of monalbite at 1060°C (data from Winter et al., 1979) did not fit the data as well.

Molecular dynamics simulations of melts and glasses in the system $NaAlSiO_4$-SiO_2 have been carried out by Scamehorn and Angell (1991) and Stein and Spera (1995) (see chapter by Poole et al., this volume, for further discussion). The study by Scamehorn and Angell suggests that T-O distances increase and T-O-T angles decrease in these melts (T = Si, Al) as temperature increases. This prediction is consistent with the slight increase in T-O distance (1.66 to 1.68 Å) and decrease in T-O-T angle (146° to 136°) in $CaAl_2Si_2O_8$ glass at 24°C (Taylor and Brown, 1979a) versus $CaAl_2Si_2O_8$ melt at 1600°C (Okuno and Marumo, 1982). No significant differences were observed in these parameters in $NaAlSi_3O_8$ glass at 24°C (Taylor and Brown, 1979a), $NaAlSi_3O_8$ supercooled melt at 805°C (Taylor et al., 1980), and $NaAlSi_3O_8$ melt at 1200°C (Okuno and Marumo, 1982). These different behaviors of $CaAl_2Si_2O_8$ and $NaAlSi_3O_8$ melts are consistent with the suggestion by Scamehorn and Angell that the fragility of the former is greater due to its higher Al content and the greater violation of the Al-avoidance principle (Lowenstein, 1954) involving the formation of Al-O-Al linkages. Stein and Spera (1995) obtained similar results and also found a strong inverse correlation between T-O distance and T-O-T angle as observed in crystalline framework silicates (Brown et al., 1969) and as predicted by molecular orbital calculations on $H_6T_2O_7$ clusters (Gibbs et al., 1981). However, a comparison of the X-ray rdf's calculated by Stein and Spera (1995) using their MD results for NaAlSi$_3$O$_8$, NaAlSi$_2$O$_6$, and NaAlSiO$_4$ melts at 3000 K with the experimentally measured rdf's of NaAlSi$_3$O$_8$ and NaAlSi$_2$O$_6$ glasses quenched from 10 kbar pressure (Hochella and Brown, 1985) and of NaAlSiO$_4$ glass (Aoki et al., 1986) give poor agreement beyond the first coordination sphere around T cations. These relatively poor comparisons between MD simulations and experiment are caused in part by the simple two-body potentials used in the simulations, the sensitivity of angle distributions to the potential used, and the large differences in temperature of the MD simulations and the experiments to which they are being compared. Wright (1993a) discusses the need for quantitative checks of the quality of MD simulations of glass/melt structure by comparison with X-ray scattering data.

X-RAY ABSORPTION SPECTROSCOPY

Basic principles

X-ray absorption spectroscopy (XAS) differs from X-ray scattering in several important respects. First, XAS is element specific whereas X-ray scattering is not. This attribute of XAS greatly simplifies interpretation of rdf's derived by Fourier transforming Extended X-ray Absorption Fine Structure (EXAFS) data because only pair correlations involving the absorbing element are present. In the case of rdf's derived from X-ray scattering, all possible pair correlations are present, often creating serious overlap problems (see e.g. X-ray rdf for 33K$_2$O-67SiO$_2$ melt in Fig. 6). Another important difference is that XAS provides only short- and medium-range structural information due to the relatively short mean free path of electrons in solids. X-rays have much greater penetrating power than electrons, so X-ray scattering provides information on both short-range and distant correlations, limited in the case of non-crystalline materials mainly by the increase in positional disorder with increasing distance from a reference atom (as reflected by an increase in the Debye-Waller factor). A third difference is the fact that interatomic distances derived from XAS must be corrected for a phase shift of the photoelectron wave as it interacts with the potentials of the absorbing atom and the surrounding backscattering atoms. X-rays undergo no such phase shift. Examination of Equations (13) and (15) (see below) shows that they contain a phase-shift term [$\Sigma \phi(k, R)$] in the sine function describing the oscillatory nature of the photoelectron waves. The magnitude of the phase-shift correction is of the order of 0.3 to 0.5 Å, depending on the types of atoms in the atom pair and the interatomic distance. The phase-shift correction in XAS analysis must be determined by calibration using XAFS spectra from crystalline model compounds for which the distance of the same atom pair is accurately known or by calculation from theory. Both approaches are straightforward but result in data analysis that is more involved and time consuming than for X-ray scattering data.

XAS and X-ray scattering also differ in their sensitivity to positional and thermal disorder of neighboring atoms around a central atom, with XAS being much more sensitive to these effects and requiring special data analysis procedures when these types of disorder become important. It is well known (Eisenberger and Brown, 1979; Crozier et al., 1988; Crozier, 1995) that both positional (or static) and thermal disorder around an absorbing atom can create a non-Gaussian distribution of interatomic distances. If this effect is large, interatomic distances and coordination numbers can be reduced significantly relative to their correct values. Errors of this type can be important in EXAFS analyses of silicate glasses and melts (where the additional effect of anharmonic vibrations between atoms becomes significant) if the analysis is carried out using a Gaussian or harmonic approximation of disorder effects (see later discussion). The effects of thermal and static disorder on EXAFS are discussed in some detail below, including appropriate theories that take these effects into consideration.

A current experimental limitation of XAS for high-temperature studies of silicate melts is the difficulty of probing the local environments of elements with atomic numbers (Z) less than that of K or Ca, which includes the geochemically important elements Na, Mg, Al, and Si. The low energy of K-absorption edges of these elements (< 2 keV) necessitates the use of a vacuum system to prevent absorption of these X-rays by gas molecules. High-temperature X-ray scattering studies of silicate melts, while less sensitive to low-Z than high-Z elements, does not require in-vacuum measurements to be able to determine Na-O or Si-O distances. However, it is only a matter of time and effort before high-temperature XAS studies of light atoms in silicate melts will become relatively routine. In-vacuum furnaces have been developed and are commonly used for annealing solid samples up to temperatures of 1000°C prior to surface characterization studies (e.g. LEED, photoemission, surface EXAFS, and X-ray standing wave spectroscopy), but they have not been optimized for XAS studies of low Z elements in silicate melts. Furthermore, suitable sample containers must be developed which do not interfere with the soft incident X-rays or the fluorescent X-rays emitted from the sample. One of the main problems that will be encountered in high-vacuum XAS studies of silicate melts is loss of volatile elements, like Na, which will be exacerbated in a UHV system. Constraints on high-pressure XAS studies of low-Z elements in silicate glasses are particularly severe using the diamond cell because of the very strong absorption of soft X-rays by the diamonds and gasket material. At present, such studies are limited to elements with Z > 27 (Co) due to the high absorption of X-ray with energies lower than about 7.5 keV. Combined high-pressure-high-temperature XAS studies of silicate melts are even more challenging and have not yet been carried out, to our knowledge.

The Ti-K X-ray Absorption Fine Structure (XAFS) spectrum of the cation-excess crystalline spinel $Ni_{2.6}Ti_{0.7}O_4$ is shown in Figure 22 as an example of the general form of an X-ray absorption spectrum as a function of incident X-ray energy. The acronym XAFS is now commonly used to indicate the X-ray absorption spectrum of an element at one of its absorption edges (e.g. K, L_{III}, etc.). It is convenient to divide an XAFS spectrum into three energy regions in which different types of electronic processes dominate. These are (1) the pre-edge region (5 to 20 eV below an absorption edge), (2) the X-ray absorption near-edge structure, or XANES, region (a few eV below to about 100 eV above the absorption edge), and (3) the extended X-ray absorption fine structure, or EXAFS, region (from 100 eV to as much a 1000 eV above an absorption edge). These regions are designated for the Ti K-edge spectrum in Figure 22. It is now possible to calculate a full XAFS spectrum, using *ab initio* multiple-scattering codes (FEFF 6: Fig. 22). Independent structural information can be obtained from the three regions of an

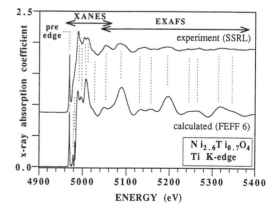

Figure 22. Top curve: Experimentally observed Ti K-edge x-ray absorption fine structure (XAFS) spectrum of a cation-excess spinel $Ni_{2.4}Ti_{0.7}O_4$ showing the three spectral regions referred to as the pre-edge, XANES, and EXAFS. Also shown is the XAFS spectrum for [4]Ti in this spinel calculated using the *ab-initio* multiple-scattering code FEFF6.

XAFS spectrum, as reviewed below. More complete discussions of XAS can be found in a number of references, including Stern and Heald, 1983; Teo, 1986; Brown et al., 1988; Koningsberger and Prins, 1988; and Lytle, 1989.

Pre-edge region. Pre-edge features are sometimes present in an XAFS spectrum (particularly for the *K*-absorption edges of metal cations (M) from Ca to Cu) (Fig. 22). They are commonly attributed to bound-state electronic transitions from M1s energy levels to M3d/O2p molecular orbital levels (e.g. Grunes, 1983; Wong et al., 1984; Bianconi et al., 1985a,b; Waychunas et al., 1983; Waychunas, 1987; Brown et al., 1988; Uozumi et al., 1992; Galoisy and Calas, 1993a; Farges et al., 1995d). A 1s→3d transition is forbidden by dipole selection rules but becomes allowed when p-d orbital mixing occurs such as when M is located in a site without a center of symmetry (such as a tetrahedron, a trigonal bipyramid, or a distorted octahedron). The height and position of the pre-edge feature are direct functions of the degree of p-d mixing (and oxidation state) and are, therefore, sensitive to M-coordination and M site distortion. These features, when present, are unaffected by anharmonicity, as verified in a number of Ti, Fe, Ni, and Mo model compounds studied as a function of temperature (Fig. 23). We will make use of the height and position of pre-edge features of Ti and Ni in silicate melts to assess their local coordination environments.

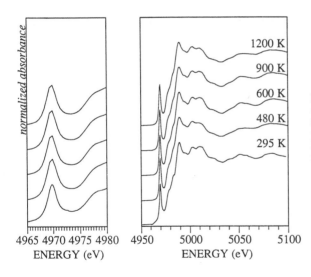

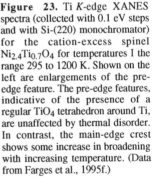

Figure 23. Ti *K*-edge XANES spectra (collected with 0.1 eV steps and with Si-(220) monochromator) for the cation-excess spinel $Ni_{2.4}Ti_{0.7}O_4$ for temperatures I the range 295 to 1200 K. Shown on the left are enlargements of the pre-edge feature. The pre-edge features, indicative of the presence of a regular TiO_4 tetrahedron around Ti, are unaffected by thermal disorder. In contrast, the main-edge crest shows some increase in broadening with increasing temperature. (Data from Farges et al., 1995f.)

As previously mentioned, it is now possible to use *ab initio* multiple scattering codes (see XANES section below) (see also Fig. 22) to calculate pre-edge features and to verify some trends observed empirically for model compounds, glasses, and melts (Bianconi et al., 1985b; Davoli et al., 1992a,b; Paris et al., 1995; Farges et al., 1995d). A bound-state transition (such as the 1s→3d transition primarily responsible for the most intense pre-edge features of first-row transition metals) is modeled in this formalism as a very large number of photoelectron scattering events of high order along the same localized paths, e.g. the Ti→O→Ti→O→Ti path of a TiO_4 tetrahedron (referred to as a path of order 4, because the full path consists of 4 individual ones).

Use of the pre-edge region to obtain information on the coordination environment of first-row transition metals can be quite rapid, even for very dilute species, but quantitative interpretations require careful analysis of the pre-edge features of well-characterized model compounds, which cover a range of coordination environments of the metal of interest. It is also essential that the data are normalized properly so that the pre-edge is normalized to the edge jump. Without proper data normalization, it is ill-advised to compare pre-edge features from different studies.

XANES region. The XANES region corresponds to low values of the photo-electron wave number, k (below 2 Å^{-1}), which is defined as k = $[0.262(E-E_0)]^{1/2}$, where E is the kinetic energy of the photoelectron and E_0 is the photoelectron energy at k = 0. Some authors have attributed XANES features to a variety of electronic transitions (such as 1s→4p) and/or other electronic effects (such as shake-down effects, etc.: Bair and Goddard, 1980; Greegor et al., 1984; Hawthorne et al., 1991; Li et al., 1994, 1995). However, there is now general agreement that these features are attributable to single- and multiple-scattering of the ejected photoelectron wave from the central absorbing atom among its nearest and next-nearest neighbors (e.g. Kutzler et al., 1980; Bianconi et al., 1985a; Natoli and Benfatto, 1986; Davoli, 1990; Filipponi et al., 1991; Davoli et al., 1992a,b; Rehr et al., 1992; Tyson et al., 1992; Paris et al., 1995). Traditionally, the XANES region is thought to be dominated by multiple-scattering events, whereas the EXAFS region is dominated by single-scattering events. These ideas are generally true, and several computer codes now exist which can perform high-level single- and multiple-scattering calculations and accurately predict XANES and EXAFS spectra (Natoli and Benfatto, 1986; GNXAS: Filipponi et al., 1991; FEFF6: Rehr et al., 1992; Zabinsky et al., 1995). These codes utilize a variety of potentials (e.g. Dirac-Hara, Xα, Hedin-Lundqvist), but the Hedin-Lundqvist potentials are generally recognized as the most accurate ones currently available for multiple-scattering calculations of XAFS. However, there is still some uncertainty about their use for insulators and about the degree of overlap of muffin-tin radii. Such overlap is necessary to help reduce the effects of discontinuities in regions between muffin-tin spheres around atoms and to give a better molecular potential for materials with highly inhomogeneous electron density distributions, such as oxide compounds, including glassy and molten silicates (Farges et al., 1995d,e,f). However, multiple scattering can be an important effect in the EXAFS region, particularly when collinear or near-collinear configurations of atoms occur around the absorbing element. Such configurations are very common when the symmetry of the compound is high (e.g. in fcc structures). In general, multiple-scattering is usually limited to the XANES region when atoms are in more disordered environments, such as those found in triclinic crystal structures and in silicate glasses and melts.

The Ti *K*-XANES spectrum of β-Ba_2TiO_4 calculated by Farges et al. (1995d) using *ab initio* multiple-scattering theory (FEFF6: Rehr et al., 1992; Zabinsky et al., 1995), is compared with the experimental spectrum in Figure 24. XANES calculations of multiple-scattering paths of order 2 made using a TiO_4 tetrahedron embedded in a representative portion of the β-Ba_2TiO_4 crystal structure reproduce the gross shape of the main absorption edge, but they don't reproduce the fine structure or the pre-edge feature. Similar calculations with paths up to order 8 do reproduce the pre-edge feature well, but still don't give the observed fine structure on the main edge. Inclusion of eight second-neighbor Ba atoms around the central TiO_4 tetrahedron is necessary to reproduce the main features of the observed Ti *K*-XANES spectrum of β-Ba_2TiO_4. The recent use of such multiple-scattering calculations to generate XANES spectra for possible short and medium-range environments of cations in silicate glasses (see Farges et al., 1995d,e,f) has added a powerful new tool for studying the structure of glasses and melts.

The effects of positional and thermal disorder on the XANES region are significantly less than in the EXAFS region (see discussion below). This is true, in part,

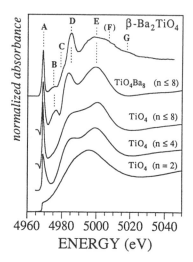

Figure 24. Comparison of the experimental Ti *K*-XANES spectrum of β-Ba$_2$$^{[4]}TiO_4$ with theoretical spectra from FEFF6 multiple-scattering calculations on TiO$_4$ and TiO$_4$Ba$_8$ clusters. In the small cluster calculations, the TiO$_4$ unit was embedded in the larger cluster, which is representative of the medium-range environment around Ti in this crystal structure. Calculations for the TiO$_4$ units were done for n = 2, n ≤ 4, and n ≤ 8, which represent the order of scattering paths considered. The calculation for the TiO$_4$Ba$_8$ cluster was for n ≤ 8. (Modified after Farges et al., 1995d.)

because anharmonicity is a direct function of the photoelectron wave number k, and the XANES region corresponds to low values of k (below 2 Å^{-1}). We will make use of this difference in later discussions of XAS analysis of silicate melts. In these compounds, the first EXAFS oscillation occurs in the XANES energy region and often arises from M-O first neighbors (Brown et al., 1988). Its maximum is a direct indication of the average M-O bond length, even if anharmonicity is present. However, *ab initio* XANES calculations of the type discussed above are still needed to justify this model because some aspects of this protocol are not entirely understood. Also, it is a good idea to test the structural interpretation of the XANES region using structural information from other parts of the XAFS spectrum, such as pre-edge features (when present) or the EXAFS oscillations, as discussed below.

EXAFS region. Extended X-ray absorption fine structure is due to constructive and destructive interference between the outgoing and backscattered photoelectron ejected when the energy of the incident X-rays matches that required to excite an electron from a deep core level. Because the EXAFS signal is damped with increasing energy or k value, EXAFS analysis of dilute metal species in silicate melts or glasses is generally limited to elements at concentrations >20 to 30 ppm using fluorescence detection methods (see later discussion). This decrease in signal results in much longer data collection times than for pre-edge or XANES spectra of an element. However, whenever possible, EXAFS data should be collected along with pre-edge and XANES data so that comparisons can be made and quantitative measures of interatomic distances and coordination numbers can be derived.

EXAFS can be modeled accurately, in the absence of multiple-scattering of the photoelectron among different atoms in the vicinity of the absorber, by using the single-scattering formalism, where the modulations of the normalized absorption coefficient χ(k) are given by (Crozier et al., 1988):

$$\chi(k) = S_0^2 rf \sum_j \frac{N_j |F_{cw}|(k,R)}{k} \int_0^\infty \frac{g(R_j)}{R_j^2} e^{-2R_j/\lambda} \sin\left(2kR_j + \sum \phi(k,R)\right) dR \qquad (13)$$

where S_0^2 is the amplitude reduction factor and rf is the reduction factor for the total central atom loss. Also, for every shell of neighboring atoms j, N_j is the number of backscattering atoms; $|F_{cw} (k, R)|$ is the effective, curved-wave backscattering amplitude; R_j is the average distance between the central and backscattering atoms; $g(R_j)$ is the (partial) radial distribution function of the neighboring distances around the absorbing element; λ is the photoelectrons mean free path; and $\Sigma\phi (k, R)$ is the sum of the phase-shift functions (central and backscattering phase-shifts). This formalism is valid for any experimental XAFS data. However, it can be greatly simplified by the use of some approximations, as explained below.

The harmonic approximation. When atomic vibrations are harmonic or when the distribution of interatomic distances is symmetrical, a Gaussian pair-distribution function can be used to represent $g(R_j)$ in Equation (13) and can be defined for the jth shell of neighboring atoms as:

$$g(R_j) = \frac{1}{\sigma_j \sqrt{2\pi}} \exp[-(R_j - \overline{R}_j)^2 / 2\sigma_j^2] \tag{14}$$

where σ_j^2 expresses the mean-square variation of R_j from the average value of R_j (designated $\overline{R}_j$). In the Gaussian (or harmonic) approximation, Equation (13) can be rewritten as:

$$\chi (k) \approx S_0^2 \, rf \sum_j \frac{N_j |F_{cw}(k,R)|}{k \, R_j^2} \, e^{-(2R_j / \lambda + 2 \, k^2 \sigma_j^2)} \, \sin (2kR_j + \sum \phi(k,R)) \tag{15}$$

Equation (15) should only be used to model structural environments of atoms in which Equation (14) correctly describes the pair distribution being probed. This is often the case in crystalline silicates under ambient conditions where there is not much static disorder (or radial distortion) of the metal sites being probed by XAFS spectroscopy. Also, as will be shown below, the harmonic approximation often holds for highly charged cations with low coordination numbers in silicate glasses and melts. However, for cations with low formal charge (+1 or +2) and high coordination numbers in glasses, which often results in high static disorder around the cation and/or increasing anharmonic vibrations with increasing temperature, the harmonic approximation may fail. EXAFS analysis of cation environments in these situations using the harmonic approximation often results in bond lengths and coordination numbers which are too small.

This effect can be illustrated by an example based on the EXAFS spectrum of two hypothetical NiO_5 polyhedra (one regular and one distorted) calculated using the FEFF6 multiple-scattering code. The calculation for the regular NiO_5 polyhedron, with all five oxygen neighbors at the same distance from Ni (d(Ni-O) = 2.0 ± 0.01 Å), results in an EXAFS spectrum with a single frequency (Fig. 25, middle). However, when the five oxygens are at a range of distances (d(Ni-O) = 2.0 ± 0.05 Å), the simulated EXAFS spectrum shows more than one frequency and results in a phase difference between the two models, which is noticeable at higher k values. In addition, the maximum of the Fourier transform is shifted to lower R (Fig. 25, right), despite the fact that the average distance is the same. Also note that the XANES region is not affected by this polyhedral distortion (Fig. 25, left). Fitting of the EXAFS spectrum produced by the distorted NiO_5 polyhedron using the harmonic approximation would fail to reproduce the asymmetric tail on the high-R side of the Fourier transform (Fig. 25, right), resulting in a Ni-O distance that is too short (by ~0.05 Å) and a coordination number that is significantly less than 5.

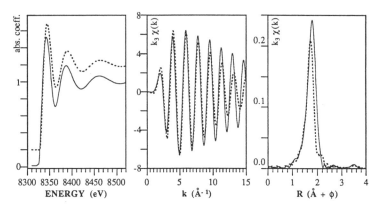

Figure 25. Comparison of normalized XANES spectra (left), background-subtracted EXAFS spectra (middle) (both XANES and EXAFS were calculated using FEFF6 for a hypothetical NiO_5 unit with d(Ni-O) = 2.0 Å), and the Fourier transform of the EXAFS spectrum (right) based on a regular NiO_5 polyhedron (solid curves) and a distorted NiO_5 polyhedron (dashed curves). For the regular polyhedron, the individual Ni-O distances are distributed by ±0.01 Å (which represents a normal Debye-Waller factor) around the average d(Ni-O) value of 2.0 Å, whereas for the distorted polyhedron, they are distributed by ± 0.05 Å around the average value of 2.0 Å. The Fourier transforms are uncorrected for phase-shifts, so d(Ni-O) is about 0.4 Å shorter than the properly fit value.

Anharmonicity. The derivation of accurate EXAFS structural parameters from Equation (15) becomes more challenging when the $g(R_j)$ function is not symmetrical due to static (or positional) disorder and/or when the interatomic potential of a given atom pair permits significant anharmonic vibrations (e.g. when it is relatively shallow and asymmetric). The former condition is often the case in silicate glasses at ambient temperature and in crystalline materials with highly distorted metal sites, whereas the latter can be quite important in silicate melts at high temperature, especially for weakly bonded cations.

As shown by the NiO_5 example above, significant positional disorder, resulting in an asymmetric distribution of pair correlations, causes the harmonic EXAFS approximation (Eqn. 15) to underestimate interatomic distance and coordination number. Increased anharmonicity with increasing temperature can cause a similar problem (Stern et al., 1991). There are several ways to model these effects and to obtain accurate distances and coordination numbers, including methods based on effective pair potentials, cumulant expansions, and an empirical model that relates anharmonicity to bond valence. We have used all of these methods in analyzing high-temperature XAFS data for silicate melts and discuss them below.

Effective pair-potential method. Thermally averaged EXAFS data collected at high temperature can be expressed as

$$\langle \chi(k) \rangle = \int g(R_j)_{asym} \chi(k, R(z)) dR \tag{16}$$

where $\chi(k,R(z))$ is the distance-dependent EXAFS model and $g(R_j)_{asym}$ is an asymmetric distance distribution function. The variable z denotes the atomic displacements relative to the average pair distance R and is designated R(z). The choice of $g(R_j)_{asym}$ determines the quality of the fit to the high-temperature EXAFS data and can be related to the interatomic pair potential function using Boltzmann statistics in a manner analogous to relating the X-ray radial distribution function, G(r), to pair potentials by Born-Green or

Percus-Yevick type equations (see Waseda, 1980). An expression for $g(R_j)_{asym}$ can be written as (Teo, 1986; Mustre de Leon et al., 1991):

$$g(R_j)_{asym} = \frac{\dfrac{U(R_j)}{k_B T}}{\displaystyle\int_0^\infty \dfrac{U(R_j)}{k_B T} dR} \tag{17}$$

where $U(R_j)$ is the interatomic pair potential, k_B is Boltzmann's constant, and T is the temperature (in K). Various types of effective-pair potentials have been used (Madelung, Morse, Born-Mayer, Lennard-Jones, etc.; see Teo, 1986; Jackson et al., 1991; Mustre de Leon et al., 1991; Jackson et al., 1993; Crozier, 1995). For example, in the high-temperature XAFS study of fayalite-composition liquid, Jackson et al. (1991) found that a Lennard-Jones-like potential of the following form best fit the [6]Fe(III)-O bond-length thermal expansion data of andradite:

$$U(R_{ij}) = a\left\{\left[\frac{R_0}{R}\right]^x - 2\left[\frac{R_0}{R}\right]^y\right\} \tag{18}$$

where a is the depth of the potential well, x and y are the coefficients representing the "repulsive" and "attractive" parts of the potential, respectively, R is the Fe-O distance at high temperature, and R_0 is the Fe-O distance corresponding to the bottom of the potential well (typically chosen as the Fe-O distance at room temperature or the lowest temperature at which the Fe-O distance has been measured). The a, x, and y parameters were determined by a non-linear least-squares fit to the temperature-dependent Fe K-EXAFS data for andradite garnet, and the derived values are x = 5.0, y = 3.5, and a = 165.4 kJ/mol for the [6]Fe(III)-O bond. This potential function is plotted in Figure 26, together with the resulting distribution functions, $g(R_j)_{asym}$, for different temperatures. The values derived for magnesiowüstite are x = 5.6, y = 2.8, and a = 120.6 kJ/mol for the[6]Fe(II)-O bond. The deeper potential well for the [6]Fe(III)-O bond, relative to the [6]Fe(II)-O bond, is consistent with the higher strength of the former and the fact that it is less affected by anharmonicity than the latter. The potential function for andradite accurately reproduces the known thermal expansion of the Fe-O bond and that for (Mg,Fe)O gives a reasonable fit to the bond expansion data for magnesiowüstite. The refined potential for the [6]Fe(III)-O bond in andradite reflects the contribution of each shell of atoms around Fe, as opposed to a single atomic pair, thus should be good starting points for molecular dynamics simulations of ferrosilicates containing Fe(III) in six-fold coordination.

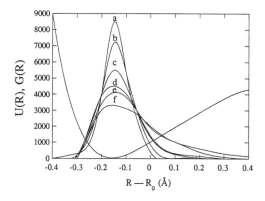

Figure 26. Potential energy function (U(Rij) (concave upward curve) and derived asymmetric pair distri-bution functions $g(R_j)_{asym}$ (concave downward curves) for the [6]Fe^{3+}-O bond in andradite (Ca$_3$Fe$_2$Si$_3$O$_{12}$) as a function of interatomic distance R at temperature T relative to the Fe-O distance R_0 at 300K (horizontal axis in Å) and temperature (in K). The temperatures for the various $g(R_j)_{asym}$ functions are (a) 90 K, (b) 300 K, (c) 415 K, (d) 670 K, (e) 870 K, (f) 1070 K, and (g) 1573 K. (After Jackson et al., 1991.)

Cumulant expansion method. In the cumulant expansion method, a statistical approach is used. The partial radial distribution function $g(R_j)$ is expressed as a sum of cumulants, C_n, of increasing order (Crozier et al., 1988):

$$\int_0^\infty \frac{g(R_j)}{R_j^2} \, e^{-2R_j/\lambda} \, e^{i\,2k\,(R_j - \overline{R_j})} \, dR \equiv \exp\left[\sum_{n=0}^\infty \frac{2\,i\,k^n}{n!} \, C_n\right] \quad (19)$$

where R_j and k are the average distance (including anharmonic effects) and the photoelectron wave number, respectively. Solving Equation (19) results in the curved-wave XAFS function, Equation (15), but with an additional multiplicative amplitude term, δA, and an additive phase-shift term, $\delta\phi$, defined as follows (Crozier, 1995):

$$\delta A = \exp\left[\sum_{n=0}^\infty \frac{-1^n}{2n!} \, (2k)^{2n} \, C_{2n}\right] \quad (20)$$

$$\delta\phi = \sum_{n=0}^\infty \frac{-1^n}{2n+1!} \, (2k)^{2n+1} \, C_{2n+1} + 4\,k\,\sigma^2 \frac{(1 + R/\lambda)}{R} \quad (21)$$

The harmonic and periodic components of the EXAFS formalism are given for $n = 1$ ($\delta\phi = 2\,k\,R_j$) and $n = 2$ ($\delta A = e^{-2\,k^2\,\sigma_j^2}$), respectively. A perturbation in the harmonic shape of the $g(R_j)$ function is created when several additional cumulants (C_3, C_4, ...) are included as:

$$\delta A = e^{\frac{2}{3}C_4 k_4} \quad (22)$$

$$\delta\phi = -\frac{4}{3}C_3 k^3 - 4k\sigma^2\left(\frac{1}{R} + \frac{1}{\lambda}\right) \quad (23)$$

Application of the cumulant model gives excellent results in structures that show moderate anharmonicity as a function of temperature, such as most oxide compounds. Figure 27 from Farges et al. (1995d) shows the results of fitting high-temperature EXAFS data for rutile (r-TiO$_2$) using harmonic and cumulant expansion models, compared with results from high-temperature X-ray diffraction structure refinements from Meagher and Lager (1978).

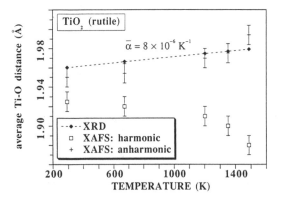

Figure 27. Variation of the EXAFS-derived average Ti-O distance in rutile (r-TiO$_2$) as a function of temperature. Open squares represent results from fitting using a harmonic model; black diamonds represent results from fitting using cumulant expansion (up to order 3); crosses represent x-ray diffraction-determined average Ti-O distances as a function of temperature (x-ray data from Meagher and Lager, 1978). (From Farges et al., 1995d.)

Farges-Brown empirical model. Farges and Brown (1995b) have developed a model to correct for anharmonicity based on high-temperature XAFS data collected for crystalline model compounds for which bond length expansion as a function of temperature has been measured using X-ray diffraction. We previously mentioned that when anharmonicity is not negligible (due to positional or thermal disorder, or both), an additional component ($\delta\phi$) appears in the phase term (Eqn. 23). A reduced phase term ($\Sigma\phi\,(k, R) + \delta\phi$) can be defined as:

$$\left(\sum \phi(k,R) + \delta\phi\right) = (\text{total phase term}) - 2kR \tag{24}$$

This quantity can be evaluated from experimental spectra collected on model compounds for which the crystal structure is known as a function of temperature (which is needed to calculate the quantity 2kR). For example, we calculated the reduced phase term $\delta\phi$ for the Ni-O pair in NiO from the measured variation in Ni-O distance over the temperature range 300 to 1300 K from high-temperature X-ray diffraction studies (Bobrovskii et al., 1973) (Fig. 28). The temperature- and k-dependence of the experimental reduced phase term at temperature T can be modeled reasonably well using the empirical equation:

$$\left(\sum \phi(k,R) + \delta\phi\right) \approx \sum \phi(k,R)\left(1 - \overline{\alpha}\,\Delta T\,R_0(k - k_0)\right) \tag{25}$$

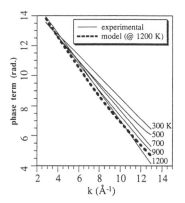

Figure 28. Plot of the reduced phase term ($\Sigma\phi + \delta\phi$) versus the photoelectron wave number k for NiO at 300, 500, 700, 900, and 1200 K. The 1200 K spectrum can be fit reasonably well using Equation (25). (From Farges and Brown, 1995b.)

where $\overline{\alpha}$ is the average linear thermal expansion for the bond studied (in K^{-1}); $\Delta T = T - 273$ K; R_0 is the average M-O bond length at room temperature, and k_0 is the starting value for the photoelectron wave number used during the modeling of the XAFS signal (usually 3 to 4 $Å^{-1}$). From Equation (25), one can derive an expression for the anharmonic contribution to the phase term (Fig. 28):

$$\delta\phi \approx -\sum \phi(k,R)\,\overline{\alpha}\,\Delta T\,R_0(k - k_0) \tag{26}$$

This model takes advantage of the simple variation of $\delta\phi$ with k, due to the quasi-linear variation of $\Sigma\phi\,(k,\,R)$ with k. Also, the $\delta\phi$ term does not have units and is, therefore, homogeneous to the periodic term of Equation (13).

An example of the application of this empirical approach is shown in Figure 29 for the Ni *K*-EXAFS of Ni-olivine (γ-Ni$_2$SiO$_4$). The cumulant expansion theory and this empirical model converge to similar solutions, and both agree, within error, with the X-ray diffraction values as a function of temperature. Without these corrections, anharmonic effects cause significant underestimation of the [6]Ni-O bond length at 1300 K (-0.10 Å).

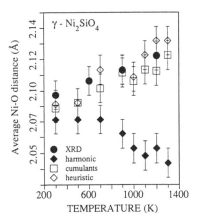

Figure 29. Variation of the average Ni-O distance in γ-Ni_2SiO_4 as a function of temperature: the black circles represent x-ray diffraction values (data from Lager and Meagher, 1978); the black diamonds represent XAFS data analyzed assuming an harmonic model for the Ni-O pair correlation; the open squares represent XAFS data analyzed assuming the cumulant expansion anharmonic model for the Ni-O pair correlation; the open diamonds represent XAFS data analyzed assuming the Farges-Brown heuristic anharmonic model. (From Farges and Brown, 1995b.)

Relationship of anharmonicity to bond thermal expansion coefficients. Linear thermal expansion coefficients are defined as (Hazen and Finger, 1982):

$$\overline{\alpha} = \frac{1}{R}\frac{\partial R}{\partial T} \approx \frac{1}{R_0}\frac{R - R_0}{T - T_0} \tag{27}$$

where R and R_0 are the average bond lengths at the temperatures T and T_0, respectively. Linear thermal expansion coefficients for metal-oxygen (M-O) bonds can be predicted reasonably well using the Hazen and Finger (1982) equation:

$$\frac{\overline{\alpha}\, s\, Z_a\, Z_c}{N} = 4 \times 10^{-6} \tag{28}$$

where s is an ionicity factor, Z_c and Z_a are, respectively, the formal charge for the cation and the anion involved in the bond considered, and N is the coordination number of the cation. In oxide compounds or melts, s and Z_a are 0.5 and 2, respectively, so Equation (28) can be simplified to :

$$\overline{\alpha}\,\frac{Z_c}{N} = 4 \times 10^{-6} \tag{29}$$

In the cumulant expansion theory, the changing behavior in anharmonicity (C_3) around any element M can be related to differences in the M-O linear thermal expansion coefficients $\overline{\alpha}$ (Stern et al., 1991) as follows:

$$\overline{\alpha} = \frac{C_3}{2\sigma^2\, T\, R_0} \tag{30}$$

in which T is the actual temperature (in K) of the compound, R_0 is the distance observed at room temperature, and σ^2 is a Debye-Waller type factor (mean square variation of R). Therefore, $\overline{\alpha}$ can be determined from the direct measurement of R as a function of temperature, or it can be calculated from refined values of C_3, the most important anharmonic parameter.

Relationship of anharmonicity to Pauling bond valence. Linear thermal expansion coefficients can be correlated with the Pauling bond valence, v ($= Z_c/N$; in valence units, v.u., where N is the cation coordination number), so Equation (30) simplifies to:

$$C_3 = 4 \times 10^{-6}\,\frac{\sigma^2\, T\, R_0}{v} \tag{31}$$

Similarly, the empirical model of Farges and Brown (1995b) can be rewritten as a function of ν :

$$\delta\phi \approx -4\times 10^{-6} \sum \phi(k,R) \, \nu \, \Delta T \, R_0 \, (k-k_0) \qquad (32)$$

When used in Equation (15), these formalisms properly account for anharmonicity with no additional fitting parameters other than those used in the harmonic XAFS model (N, R, and σ^2: see also Frenkel and Rehr, 1993). Equations (31) and (32) also indicate that anharmonicity is related to the average bond valence of the pair correlation probed. This is reasonable because stronger bonds should be less subject to anharmonic effects at high temperature than weaker bonds.

Prediction of anharmonicity in crystals and melts

The above formalisms can be used to predict anharmonicity for a given bond. Examination of Equations (31) and (32) also suggests that the anharmonic correction for two different bonds should be similar if their average bond valences are the same (Fig. 30). For example, $^{[6]}$Fe(III)-O bonds should show anharmonicity similar to that for $^{[4]}$Fe(II)-O bonds, because the classical Pauling bond valence is the same in both cases (0.5 v.u.). This model thus provides a theoretical justification for the use of Fe(III)-containing model-compounds (e.g. andradite garnet) to correct for anharmonicity around $^{[4]}$Fe(II) in molten fayalite (Jackson et al., 1991, 1993). Average distances for strong bonds derived in the harmonic XAFS approximation, such as those around $^{[4]}$Ge(IV) or $^{[4]}$Ti(IV), need no anharmonic correction at 1500 K (ΔR is 0 Å). In contrast, $^{[6]}$Ni(II)-O bonds in NiO need to be corrected by ~+0.10 Å at 1500 K.

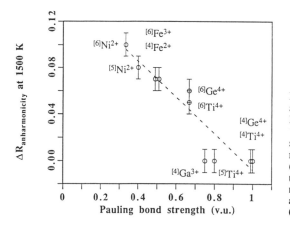

Figure 30. Summary of the variation of anharmonic corrections for a given Pauling bond strength (calculated at 1500 K). Data from various literature studies of model compounds. Anharmonicity is referred to here as the ΔR (in Å) needed to correct the harmonic fit to the values derived using other techniques (mostly high-temperature X-ray and neutron diffraction experiments). (From Farges, 1994, with pemission.)

Finally, Equations (31) and (32) indicate that the medium-range structure around an absorber should exert little influence on changes in M-O bond length (where M is the absorbing cation) as a function of temperature. The similar anharmonic behavior observed for NiO and γ-Ni$_2$SiO$_4$, which have very different medium-range structures (locally undistorted Ni sites in NiO versus distorted Ni sites in γ-Ni$_2$SiO$_4$ and quite different second- and third-neighbor configurations), is in agreement with this suggestion. Therefore, this anharmonic formalism should also be useful in correcting XAFS data for anharmonic effects in aperiodic structures, such as silicate glasses and melts.

Model-independent measure of the effective pair-potentials and their g(R) functions

In molten lead, Stern et al. (1991) showed that, for moderately anharmonic systems, anharmonicity can reasonably be considered as a simple diatomic oscillator associated with third and fourth order perturbations:

$$V(R) = \frac{1}{2} f R^2 - g R^3 - h R^4 - \dots \tag{33}$$

$$\text{with :} \quad f = \frac{k_B T}{\sigma^2} \tag{34}$$

$$g = \frac{C_3}{T^2} \tag{35}$$

$$h = \frac{9 g^2}{2 f} \tag{36}$$

However, for such systems, the third-order perturbation is often sufficient to account for anharmonicity. Similarly, for systems with little anharmonicity, the third- and fourth-order perturbations can be neglected. The corresponding g(R) calculated from integrating Equation (17) is given by:

$$g(R) = \frac{1 + \gamma (R - R_0)^3}{\sqrt{2 \pi \sigma^2}} \exp\left[\frac{-(R - R_0)^2}{2\sigma^2} \right] \tag{37}$$

$$\text{where} \quad \gamma \approx \left[\frac{g}{k_B T} \right] \tag{38}$$

As for V(R), this anharmonic g(R) function is similar to the harmonic g(R) function, but with an R-dependent amplitude term, varying as a function of $(R-R_0)^3$, the mean cubic variation of R. Figure 31 shows the shape of the g(R) function for the Ge-O pair correlation as measured for the two polymorphs of GeO_2 (rutile and quartz modifications). The values for g(R) functions are refined from high-temperature EXAFS data, using Equations (37) and (38). The potential for the quartz modification best fits the experimental XAFS data for molten GeO_2. The difference in the g(R) maximum between GeO_2 (quartz) and GeO_2 (rutile) is due to the presence of four- and six-coordinated Ge in these compounds, respectively. Increasing temperature tends to decrease the magnitude

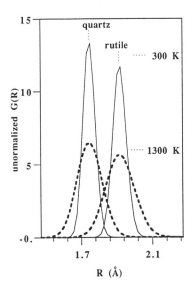

Figure 31. Plot of the g(R) function calculated using Equations (37) and (38) for high temperature XAFS data taken on the rutile and quartz modifications of GeO_2 at the Ge *K*-edge. (Data from Farges et al., 1995g.)

of the g(R) function. The g(R) function for GeO_2 (rutile) is slightly asymmetric at high temperature (above 900 K), whereas that for GeO_2 (quartz) is close to the harmonic shape, even at 1200 K. This difference is related to the higher bond valence of the Ge-O bond in GeO_2 (quartz) compared to that for GeO_2 (rutile) (1.0 and 0.67 v.u., respectively).

High-temperature XAS experimental methods

Several types of furnace assemblies have been used to collect high-temperature XAFS data on melts. Their geometries depend on whether data are collected in transmission or fluorescence modes. Both types are described below.

Transmission mode. In this data collection mode, the sample is mounted vertically. X-ray absorption by the sample is measured by monitoring X-ray flux before and after the X-rays pass through the sample, using two ionization chambers (Fig. 32). Using this experimental design, one can collect XAFS data on samples in which the cation of interest is relatively concentrated (> a few wt %). However, due to the low viscosity of some melts, it is not always possible to maintain the sample in a vertical position, and the sample thickness may vary during a scan, which makes spectral analysis difficult or impossible. Several "tricks" have been used to prevent the sample from flowing. One involves grinding the sample and mixing it with a non-corrosive, relatively X-ray transparent material such as boron nitride (Stern et al., 1991). However, for a number of molten oxides, BN reacts quite readily to form new compounds. Carbon powders have also been used to hold powdered samples in vertical sample holders and to minimize interaction with melts (Filipponi and di Cicco, 1994). This technique is a good way of studying reduced melts, such as molten metals at ultra-high temperatures (~2800 K and above). However, for oxide melts containing cations such as Fe(II) or Fe(III), the use of a carbon binder may result in reduction of the cation. Even in materials that are not easily reduced, the use of C or BN binders may present a problem, as was observed in a high-temperature study of GeO_2, where Ge(IV) was found to transform into Ge metal at a temperature of a few hundred degrees (Filipponi, 1994; Itié, 1994, pers. comm.).

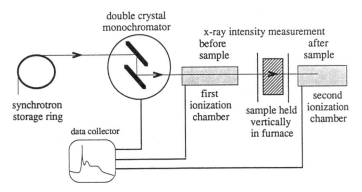

Figure 32. Schematic illustration of XAFS data collection in the transmission mode.

Pt-grids have also being used to hold melt samples vertically for XAS studies (Seifert et al., 1993). The powdered sample is pressed into the grid before heating, and because of the high surface tension of most silicate melts in contact with Pt, the liquid phase may be maintained vertically on the heating grid. However, the X-ray absorption measured is that of the melt sample plus the grid. Therefore, corrections must be made to the XAFS data prior to analysis.

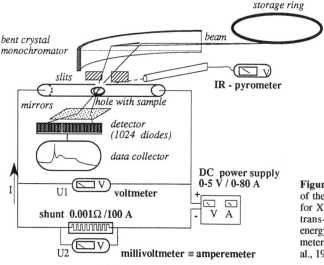

Figure 33. Schematic drawing of the heating-wire furnace used for XAFS data collection in the trans-mission mode with the energy-dispersive XAFS spectro-meter at LURE. (From Farges et al., 1995a.)

Another approach is to work on very small melt samples mounted in a small hole in a Pt wire heater and to reduce the size of the X-ray beam so as to avoid interference by the Pt (Fig. 33) (Richet et al., 1993). This approach has been tested successfully on the energy-dispersive EXAFS spectrometer at LURE (XAS 10, Orsay, France) (Farges et al., 1995a). In these experiments, a 70-mm long $Pt_{0.9}Rh_{0.1}$ wire was used to heat a sample to 2000 K during XAFS data collection in the energy-dispersive mode (the nominal melting point of this alloy is 2070 K). Even higher temperatures (up to ~3500 K) can be attained using Ir, Mo, Ta, or W wires under vacuum conditions (Richet et al., 1993; Daniel et al., 1995). A 500-µm hole was drilled in a flattened section of the $Pt_{0.9}Rh_{0.1}$ wire (1 cm long). Heating was provided by a stabilized power supply (10 V, 80 A). Prior to data collection, a powdered sample is packed in the hole. Special care is needed to produce a sample of uniform thickness to achieve an absorption-edge jump of between 1µ and 3µ, where µ is the absorption coefficient. In the experiments at LURE, the hole containing the sample was positioned where the X-rays are focused by a bent crystal [usually Si-(111) or Si-(311)]. The transmitted X-ray intensity was monitored using an array of 1024 photodiodes located behind the heating cell. To calculate the absorption coefficient, the incident X-ray intensity is measured using the same photodiode array but in the absence of sample. The main advantages of this technique are the very small amount of sample needed, very fast data collection times (average of 3 sec/scan) which allow large numbers of scans to be collected at different temperatures over short time periods (e.g. 50 scans between 295 and 2000 K can be collected in ~10 minutes), small temperature gradients (± 5 K at 2000 K), the ability to study structural changes in a time-resolved mode, and very high reproducibilities in the EXAFS-derived structural parameters (beam fluctuations are insignificant because of the rapidity of data collection). There are some disadvantages of this method, however. For example, sample temperatures are difficult to measure accurately using a thermocouple because it creates a heat sink; optical pyrometers can be used to eliminate this problem. Also dilute samples cannot be studied easily in transmission; however, a recent study succeeded in obtaining EXAFS spectra of a diopside composition silicate glass and melt containing as little as 2 wt % Ni using this device (Farges et al., 1995b).

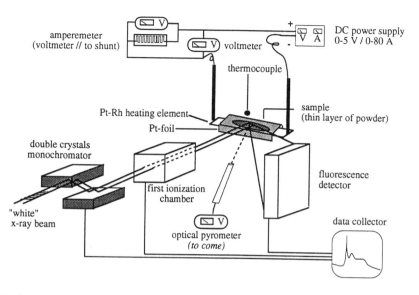

Figure 34. Schematic drawing of a high-temperature cell used for XAFS data collection in the fluorescence mode at SSRL. (From Farges et al., 1995d.)

Fluorescence mode. The use of fluorescence detection makes XAFS data collection on melts much easier than with the transmission methods discussed above because the X-ray beam does not pass through the sample and the sample geometry can be simplified (Fig. 34). An inexpensive, efficient ion-chamber detector has been developed for fluorescence XAS (Lytle et al., 1984), and it can be fitted to a high-temperature furnace assembly quite easily. Most of the high-temperature XAFS data on silicate melts have been collected in this mode (Waychunas et al., 1987, 1988, 1989; Jackson et al., 1991, 1993; Farges et al., 1994; Farges and Brown, 1995a,b; Farges et al., 1995c,e,f). The sample is held at 45° to the X-ray beam and is mounted horizontally, thus minimizing problems with possible changes in sample shape during data collection. Another advantage of this data collection mode is the fact that high-temperature XAFS studies can be made on samples with very dilute concentrations of the element of interest. For example, data were recently taken on Th at a concentration of 100 ppm in silicate melts at temperatures up to 1400 K (Farges and Brown, 1995a). Because of the relatively low solid-angle of data collection using current furnace designs, most of the fluorescence emitted by the melt sample is not collected by the detector. However, this is not a serious problem since the fluorescence detection mode is quite sensitive to dilute amounts of the absorbing element (as low as 10 to 20 ppm). The major disadvantage of the fluorescence detection mode is the possibility of self-absorption effects. When the element studied is too concentrated (above ~10 wt %), the edge features and EXAFS oscillations will be reduced in amplitude as the element concentration increases. However, corrections for self-absorption can be made, if needed, using existing *ab initio* codes (e.g. ATOMS from the University of Washington which is based on McMaster X-ray cross sections of the elements) or, if possible, by collecting an XAFS spectrum for a compound simultaneously in the transmission and fluorescence modes, then comparing the two. If the transmission spectrum has higher EXAFS amplitude than the fluorescence spectrum, the fluorescence spectrum can be rescaled in amplitude. The fluorescence mode can be used to collect XAFS data for elements at concentrations ranging from major to trace levels, although its use at high concentrations must be made with self-absorption effects in mind. Self absorption affects amplitudes, thus EXAFS-derived coordination numbers, but has little effect on phases, thus on EXAFS-derived distances. Implementation of the heating wire technique in the fluorescence detection mode using a "quick-EXAFS" facility (with a continuously scanning monochromator instead of a step-scanning monochromator) on third generation synchrotron sources (APS, Chicago; ESRF, Grenoble) should dramatically improve the quality of high-temperature XAFS spectra in the near future.

High-temperature XAS results for cations in silicate melts

There have been a limited number of high-temperature XAS studies of specific elements in silicate and germanate melts, including highly charged cations that are network-formers [Ti(IV), Ge(IV)], highly charged cations that act as network modifiers [Zr(IV), Mo(VI), Th(IV), and U(VI)], and divalent cations that can act as network formers or modifiers [Fe(II), Ni(II), and Zn(II)]. Selected results are presented below. There have been no high-temperature XAS studies of trivalent cations in silicate melts.

Highly charged cations

Titanium (IV). Interest in the coordination chemistry of Ti in silicate glasses and melts at high temperature has been stimulated by observations that (Na, K)-titanosilicate glasses display positive anomalies in their heat capacities just above the glass transition temperature, T_g (Richet and Bottinga, 1985; Lange and Navrotsky, 1993; Bouhifd, 1995; also see chapters by Navrotsky and Webb and Dingwell, this volume). Lange and Navrotsky (1993) suggested that this anomaly might be related to a change in Ti coordination. Paris et al. (1993) and Seifert et al. (1993) studied the local environment of Ti in a Rb-tetrasilicate melt at 1700°C using XANES spectroscopy and found no clear evidence for a Ti-coordination change in the melt. However, this composition should not have a large heat capacity anomaly (Bouhifd, 1995), therefore may not undergo a Ti coordination change near T_g if the anomaly is caused by this structural change. In order to test the hypothesis that the heat capacity anomaly above T_g is caused by a change in Ti coordination, Farges et al. (1995e,f) carried out a comprehensive high-temperature XAS study of Ti(IV) in nine Na-, K-, and Ca-titanosilicate glass/melts between 293 to 1650 K with TiO_2 concentrations ranging from 2.7 to 30.5 wt %. Four of these compositions show large Cp anomalies. Figure 35 shows examples of XANES spectra for K_2O-TiO_2-$2SiO_2$ (KTS2) glass and melt at temperatures from 295 to 1560K. The height and position of the pre-edge feature are consistent with dominantly [5]Ti in the glass and melt. [5]Ti occurs in a number of crystalline oxides and silicates as well as in silicate glasses (Yarker et al., 1986; Henderson and Fleet, 1995).

In silicate glasses and melts with high Ti-contents (>10 mol %), Farges and co-workers (1995e,f) found square pyramidal ([5]Ti=O)O_4 to be the dominant Ti species, with one short axial Ti=O bond (~1.7 Å) and four longer Ti-O bonds (~1.95 Å), based on anharmonic analysis of EXAFS data as well as detailed comparisons between theoretical and experimental pre-edge and XANES spectra. For example, *ab initio* multiple-scattering calculations (using FEFF6) for a ([5]Ti=O)$O_4Si_4K_8$ cluster, in which Ti is located in a square pyramid, simulates the experimental spectrum reasonably well (Fig. 36). Multiple-scattering calculations for TiO_4 tetrahedra do not match the observed spectra well. The constancy of the pre-edge feature and XANES spectrum and anharmonic analysis of the EXAFS (Fig. 37) indicate that the first coordination shell of Ti does not change with increasing temperature, including the temperature range just above T_g (~885 K for this glass), where anomalous heat capacities have been measured

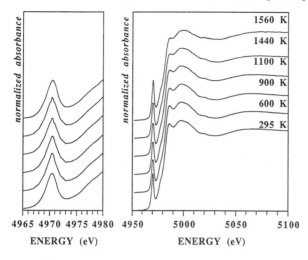

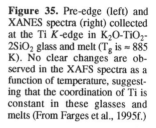

Figure 35. Pre-edge (left) and XANES spectra (right) collected at the Ti *K*-edge in K_2O-TiO_2-$2SiO_2$ glass and melt (T_g is ≈ 885 K). No clear changes are observed in the XAFS spectra as a function of temperature, suggesting that the coordination of Ti is constant in these glasses and melts (From Farges et al., 1995f.)

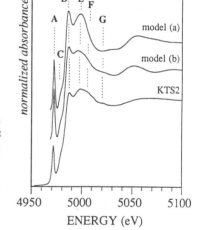

Figure 36. Comparison of calculated spectra (using FEFF6) for a $(^{[5]}Ti=O)O_4Si_4K_8$ cluster (models (a) and (b)) with that measured for KTS2 glass (K_2O-TiO_2-$2SiO_2$) (bottom). Model (a): spectrum predicted for this cluster with four Ti-K distances at 3.2 Å and four Ti-K distances at 3.5 Å. Model (b): spectrum predicted with Ti-K distance distributions disordered by randomly increasing or decreasing them by ±0.2 Å around the mean value of 3.4 Å. (From Farges et al., 1995e.)

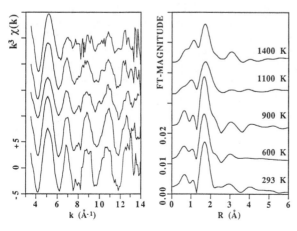

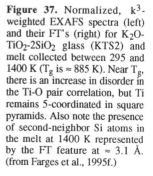

Figure 37. Normalized, k^3-weighted EXAFS spectra (left) and their FT's (right) for K_2O-TiO_2-$2SiO_2$ glass (KTS2) and melt collected between 295 and 1400 K (T_g is ≈ 885 K). Near T_g, there is an increase in disorder in the Ti-O pair correlation, but Ti remains 5-coordinated in square pyramids. Also note the presence of second-neighbor Si atoms in the melt at 1400 K represented by the FT feature at ≈ 3.1 Å. (from Farges et al., 1995f.)

on this same glass (Bouhifd, 1995). $(^{[5]}Ti=O)O_4$ square pyramids are connected to the tetrahedral framework with Ti-Si distances of ~3.2 ± 0.1 Å (Ti-O-Si angles of ~130°). Theoretical analysis of the Ti K-edge XANES for the melts suggests the disappearance of K second neighbors around Ti above T_g.

The results of the calculations discussed above for $(^{[5]}Ti=O)O_4$ polyhedra do not match those reported by Paris et al. (1995) in their theoretical modeling of the Ti-K-XANES spectra of several Ti-bearing silicate glasses, which are similar to that of a Rb-Ti tetrasilicate melt collected at 1700 K (Paris et al., 1993; Seifert et al., 1993). They interpreted the glass spectrum as indicating a mixture of $^{[5]}Ti$ (as trigonal bipyramids) and $^{[6]}Ti$. The spectrum of these glasses (unspecified composition; taken from Dingwell et al., 1994) is very similar to those collected for Ti-bearing sodium and potassium disilicate and tetrasilicate glasses by Marumo et al. (1990) and Farges et al. (1995f) (Fig. 35). The calculated XANES spectrum reported by Paris and coworkers for square-pyramidal TiO_5 units shows a very small pre-edge feature, which is inconsistent with the experimental spectra collected for the glasses (where a rather intense pre-edge is observed, as in the melt; see Fig. 35). In contrast, their calculated XANES spectrum for the TiO_5 trigonal bipyramid shows an intense pre-edge feature, which appears to be a closer match to the XANES spectra of the glasses and melts. Farges et al. (1995e,f) pointed out that the XANES spectrum calculated by Paris and coworkers (1995) for square pyramid TiO_5 units (particularly the pre-edge feature) does not match the spectra of a variety of model compounds containing these Ti polyhedra (as in $Ba_2TiSi_2O_8$ and $KNaTiO_3$: Farges et al., 1995d). Also, the suggestion by Paris et al. (1993) and Dingwell et al. (1994) that Ti occurs in trigonal bipyramids in titanosilicate glass/melts is not consistent with neutron scattering, Raman, and XAFS spectroscopy measurements on similar glasses and melts, which indicate that $(^{[5]}Ti=O)O_4$ square pyramids are the major Ti species (see Yarker et al., 1986; Brown et al., 1988 and references therein). More recent calculations of the XANES spectrum for $^{[5]}Ti$ in square pyramid units (Farges et al., 1995d,e,f) match the observed Ti-XANES spectra of these glasses, melts, and model compounds reasonably well and contain a strong pre-edge feature which is characteristic of the Ti square pyramidal environment, a conclusion that is again supported by a recent re-investigation of the structure of the potassium titanium disilicate glass by neutron scattering methods (Cormier et al., 1995).

In the most dilute Ti-glass studied by Farges et al. (1995e) (K-metasilicate with 2.7 wt % TiO_2), $^{[4]}Ti$ was detected, and its local environment was found to be relatively well ordered in the glass (especially the 2nd-neighbor alkalis). This order decreases dramatically above T_g, as indicated by the large change in the integrated intensity of the main absorption edge features with increasing temperature (Fig. 38). The pre-edge spectra of this melt at different temperatures are typical of a mixture of major amounts of $^{[4]}Ti$ and $^{[5]}Ti$, with $^{[4]}Ti$ being dominant. The average Ti coordination does not change during the glass-to-melt transition. However, the two K-metasilicate glasses (before and after quenching) do not have the same pre-edge features (Fig. 38), suggesting that $^{[5]}Ti$ is favored in very rapidly quenched liquids. The slowly quenched sample was kept at 450 K during data collection to avoid hydration of this highly hygroscopic glass by atmospheric moisture; however if hydration had occurred, it should favor the presence of $^{[6]}Ti$, the most stable Ti-coordination in hydroxides. No evidence for $^{[6]}Ti$ in this sample was found. The main absorption edge features of the glasses and the melt are markedly different. Based on *ab initio* XANES calculations, these changes have been attributed by Farges et al. (1995f) to the breakdown of alkali-containing percolation domains around Ti. The percolation model is discussed in more detail in a later section on models of medium-range order in silicate melts.

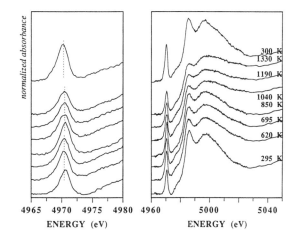

Figure 38. Pre-edge (left) and XANES spectra (right) collected at the Ti *K*-edge for potassium metasilicate glass and melt containing 1.8 wt % Ti (sample KS1 of Farges et al., 1995d.) The spectrum of the original glass sample is shown at the bottom of the figure, the high-temperature glasses and melts are in the middle, and the slowly quenched liquid is shown at the top. (From Farges et al., 1995f.)

The presence of titanyl moieties in silicate glasses and melts has been interpreted as indicating that Ti acts simultaneously as network modifier and former because Ti is surrounded by both non-bridging (O of the axial Ti=O bond) and bridging oxygens (O of the Ti-O bonds). The ($[5]$Ti=O)O_4 polyhedra are believed to promote local structural heterogeneities referred to as percolation domains, involving interfaces between polymerized and less polymerized regions. Configurational changes in these proposed percolation domains are thought to be responsible for anomalous heat capacity variations observed in some of these systems above T_g (Farges et al., 1995f).

Because the coordination environment of Ti in silicate glasses is similar to that of the melts, the pressure-induced coordination change around Ti in densified silicate glasses observed using XANES spectroscopy (Paris et al., 1994) is likely to be representative of a pressure-induced coordination change in the melt.

An EXAFS study of Ti(IV) in glasses along the $CaMgSi_2O_6$ (Di) - $CaTiAl_2O_6$ (Tp) composition join under ambient conditions (Marumo et al., 1990) reported dominantly [4]Ti in the glasses. Even though they did not carry out an anharmonic data analysis, the Ti-O distances should be relatively unaffected by anharmonicity because of the high Ti-O bond valence (Farges and Brown, 1995b; also see discussion of anharmonicity and bond valence above). In addition, they did not consider the possibility of [5]Ti. In fact, examination of their pre-edge features suggests that [5]Ti is the dominant species and that [4]Ti and [6]Ti may also be present. This suggestion is consistent with the Ti-O distances they obtained [1.96 Å (Tp10), 1.93 Å (Tp20), 1.93 Å (Tp30), 1.89 Å (Tp40), and 1.88 Å (Tp50)], which suggest that Ti in these glasses is dominantly five-coordinated, based on the average [4]Ti-O (1.84 Å), [5]Ti-O (1.91 Å), and [6]Ti-O (1.96 Å) distances reported by Farges et al. (1995 d,e,f). In Tp10, [6]Ti is probably the majority coordination state, based on the observed Ti-O distance. Marumo et al. (1990) noted an increase in [4]Ti with increasing Ti content in their glasses. Another interpretation for this result is an increase in the ratio of five-:six-coordinated Ti with increasing Ti content. If ($[5]$Ti=O)O_4 square pyramids occur in these glasses, with one short Ti=O and four long Ti-O bonds, this trend could be explained by noting that [5]Ti increases at the expense of [6]Ti as Al replaces Si. It should be energetically favorable for [4]Al to bond to oxygens of both long and short Ti-O bonds of ($[5]$Ti=O)O_4 groups, whereas [4]Si should bond only to oxygens in the long Ti-O bonds. Thus, glasses with low Tp contents should favor [6]Ti.

Other relevant EXAFS studies of Ti in silicate glasses include those by Greegor et al. (1983), Dingwell et al. (1994), Paris et al. (1994), and Paris et al. (1995). The study by Greegor and co-workers reported that [6]Ti is favored at low TiO_2 contents (< 0.05 wt %) along the TiO_2-SiO_2 composition join, whereas [4]Ti is favored at TiO_2 contents ranging from 2.0 to ~8.0 %. Above ~9 wt % TiO_2, the sixfold/fourfold ratio was found to increase appreciably, and at ~15 wt % TiO_2, rutile was found to have crystallized. Greegor et al. (1983) did not consider the possibility of [5]Ti in their glasses, however, the pre-edge features of these glasses do not indicate significant amounts of [5]Ti. Dingwell et al. (1994) carried out an XAFS study of alkali- and alkaline-earth-titanium tetrasilicate and metasilicate glasses. They found that the pre-peak heights for the alkali-bearing glasses increased in the order Li<Na<K<Rb<Cs, whereas it remained constant for the Ca- and Sr-bearing glasses. They interpreted these results as indicating a decrease in the average Ti coordination number with increasing size of the alkali ion and suggested that Ti is present in these glasses in three different site geometries: tetrahedra, trigonal bipyramids, and distorted octahedra. The conclusion that [5]Ti occupies trigonal bipyramids in these glasses (instead of square pyramids) is inconsistent with the neutron scattering and EXAFS data collected for similar glasses (Yarker et al., 1986; Farges et al., 1995e).

Germanium (IV). The local environment of this tetravalent ion has been studied in two types of melts using high-temperature XAS: GeO_2 (Farges et al., 1995g), $SrGeO_3$ and $CaGeO_3$ (Andrault et al., 1995), and in $Bi_xGe_yO_z$ melt using high-temperature X-ray scattering methods (Takeda et al., 1993a,b). In all of these melts, Ge is mostly four-coordinated like in the corresponding glasses at ambient pressure (Lorch, 1969; Konnert et al., 1973; Okuno et al., 1986; Itié et al., 1989). No clear evidence for [5]Ge and/or [6]Ge was found, unlike Si in silicate glasses and melts which is present in five- and six-coordinated forms in small amounts (Xue et al., 1991; McMillan et al., 1992; Poe et al., 1992; Stebbins and McMillan, 1993). Minor amounts (< 5%) of a specific coordination state of a cation in the presence of a majority state are usually not detectable using XAFS spectroscopy.

The XANES spectrum for GeO_2 (rutile) (Fig. 39) is characterized by an intense absorption edge (feature A) which is typical of [6]Ge. GeO_2 (quartz) and melt have a less intense absorption edge, suggesting the presence of mostly [4]Ge in these two forms. Also the position of feature B for the melt sample indicates an average Ge-O distance of ~1.75 ± 0.01 Å, which is the same as in GeO_2 (quartz). This distance is consistent with the

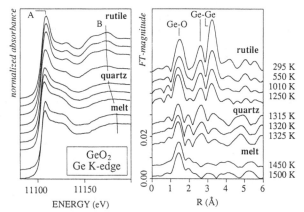

Figure 39. High-temperature Ge *K*-edge XANES spectra of the polymorphs of GeO_2: GeO_2 (rutile) (four top spectra, collected between 295 and 1250 K); GeO_2 (quartz) (3 middle spectra; 1315 to 1325 K); and GeO_2 melt (two bottom spectra: 1450 to 1500 K). (Left): Ge *K*-edge XANES spectra; (Right): FT of the EXAFS. (From Farges et al., 1995a.)

presence of major amounts of $^{[4]}$Ge in the melt. The FT shows two major contributions (distances are uncorrected for backscattering phase-shifts): one near 1.4 Å, arising from Ge-O correlations centered around 1.89 Å for GeO_2 (rutile) and 1.75 to 1.76 Å for GeO_2 (quartz) and melt; the second is related to second-neighbor Ge-Ge correlations. Note the disappearance of the Ge-Ge pairs near 3 Å in the FT (~3.3 Å when corrected for Ge-Ge backscattering phase-shift) during the GeO_2 (quartz)-to-melt transformation.

The Ge K-edge XAFS results for GeO_2 melt are consistent with those from EXAFS studies of Bi_2O_3-GeO_2 melts (Ohmote and Waseda, 1994) and a combined EXAFS and anomalous X-ray scattering study of glasses in the GeO_2-P_2O_5 system (Shimizugawa et al., 1994). They are also consistent with results from the XAS and X-ray rdf study of $NaAlGe_3O_8$ and $NaGaGe_3O_8$ glasses (Okuno et al., 1984) and the XAS studies of GeO_2 glasses (Lapeyre et al., 1983; Okuno et al., 1986b). These studies report Ge-O distances ranging from 1.73 to 1.78 Å, which are consistent with $^{[4]}$Ge.

Zirconium (IV). The coordination environment of Zr(IV) has been investigated in $Na_2Si_2O_5$ (NS2) and $Na_2Si_3O_7$ (NS3) glasses and melts between 293 and 1550 K containing between 0.1 and 10 wt % Zr (Farges and Brown, 1995a). The Zr-K edges of the melts studied are typical of $^{[6]}$Zr (Fig. 40), with average Zr-O distances of ~2.07 to 2.10 Å. The Zr sites in these melts are relatively regular, based on comparison of Zr XANES spectra of the melts with those of model compounds. Evidence for Zr-Si correlations in the melts was also detected, as in Zr glasses (Zr-Si distances of ~3.6 ± 0.1 Å: Farges et al., 1991a; Farges, 1995). Thus, there is no detectable coordination change around Zr during the melting of these glasses. However, there does appear to be a coordination change from 6 to 8 prior to the crystallization of zircon from Zr-containing $NaAlSi_3O_8$ glass (Farges et al., 1991a). This change is thought to be a precursor to the formation of zircon-like structural units in the melt. TEM examination of these glasses found no evidence for small nuclei at the 20 to 30 Å scale (F. Farges, unpublished).

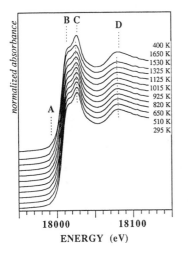

Figure 40. Zr K-edge XANES spectra of sodium-trisilicate glass and melt containing 3 wt % Zr for temperatures between 295 and 1650 K (the 400 K spectrum at top of figure is that collected in the glass quenched in the XAFS furnace) (From Farges and Brown, 1995a.)

Ab initio XANES calculations (cf. Farges, 1995) have suggested that feature A in the XANES spectrum of Zr in $Na_2Si_3O_7$ (NS3) glasses and melts (Fig. 40) is an electronic transition; its shape is indicative of the presence of relatively regular ZrO_6 octahedra. Feature B (Fig. 40) is due primarily to single scattering from second-neighbor Si, whereas feature C is due to multiple scattering involving first-neighbor O around Zr.

The height difference between features B and C is indicative of the presence of corner-shared SiO_4 tetrahedra connected to ZrO_6 polyhedra. Feature D in the Zr *K*-XANES spectrum of these glasses and melts is the maximum of the first EXAFS oscillation, the position of which indicates an average Zr-O distance of ~2.07 ± 0.01 Å. This distance is consistent with the presence of regular ZrO_6 octahedra. Features C and D, which are related to the Zr-O pair correlation, are shifted to slightly lower energies, suggesting that the average Zr-O distance increases by ~0.02 Å with increasing temperature. This change in the average Zr-O distance is due to the thermal expansion of the Zr-O bond. *Ab initio* Zr *K*-XANES calculations also support a regular coordination environment of Zr in the glasses and melts studied by Farges et al. (1991a) and Farges (1995), in agreement with anharmonic modeling of the Zr *K*-edge EXAFS.

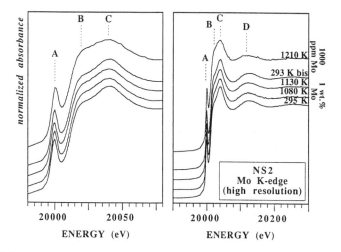

Figure 41. High-resolution Mo *K*-edge XANES spectra (left) and a blow-up of the pre-edge region (right) of Na-disilicate glass/melts (NS2) containing 1000 ppm Mo (top spectrum) and 1 wt % Mo (four bottom spectra, respectively. From top to bottom: glass sample as quenched in the XAFS furnace (293 K bis); melts at temperatures of 1130 and 1080 K; and the original glass sample (295 K). (From Farges and Brown, 1995a.)

Molybdenum (VI). Mo *K*-edge XAFS spectra have been collected for sodium di- and trisilicate glasses and melts between 295 and 1500 K with Mo contents between 1 and 3 wt % (Farges and Brown, 1995a). The Mo *K*-edge spectra of the glasses and melts are characterized by a strong pre-edge feature (Fig. 41), which is attributed to $1s{\rightarrow}4d$ electronic transitions, as observed in a number of model compounds containing MoO_4 tetrahedra. This feature is allowed because Mo(VI) is located in $Mo(VI)O_4^{2-}$ tetrahedra. The main edge features (B and C) are due to single-and multiple scattering within these tetrahedra. The first EXAFS oscillation (feature D) is typical of a Mo-O distance of   ~1.77 ± 0.02 Å, which is consistent with the presence of molybdate tetrahedra in these glasses and melts. No evidence for second neighbors around Mo was found. Application of the bond-valence model (Farges et al., 1991b; and discussed below) to these molybdate tetrahedra suggests that the oxygens around Mo(VI) cannot bond to Si. A Mo-Si correlation, if present, should be detectable in silicate melts using XAFS spectroscopy, based on detection of Zr-Si (Farges and Brown, 1995a), Fe-Si (Jackson et al., 1993), Ni-Si (Farges et al., 1995c), and Ti-Si correlations (Farges et al., 1995f) in melts of similar compositions. In contrast, molybdate tetrahedra are likely to bond to several network modifiers which are difficult to detect by XAFS spectroscopy because their distributions are usually too disordered, thus reducing their XAFS signal. Current knowledge of the coordination chemistry of Mo(VI) in silicate glasses and melts also suggests that Mo(VI) does not undergo a significant coordination change above T_g.

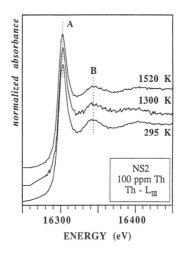

Figure 42. Th L_{III}-XANES spectra of a sodium disilicate glass and melt (NS2) (temperatures up to 1520 K) containing 100 ppm Th. (From Farges and Brown, 1995a.)

Thorium (IV). The coordination chemistry of Th at minor to trace-level concentrations (3 wt %—100 ppm) has been investigated using high-temperature XAS in sodium disilicate and trisilicate glasses and melts, at temperatures up to 1550 K (Farges and Brown, 1995a). In silicate glasses of other compositions, Th is dominantly six- and eight-coordinated up to the maximum Th solubility (Veal et al., 1986; Farges, 1991). At very low Th concentrations (100 ppm), a larger fraction of the Th is eight-coordinated. Only weak evidence for second neighbors has been detected in XAFS studies of most Th-bearing glasses, leaving some uncertainty about how ThO_6 and ThO_8 units are connected to the tetrahedral framework. The glass-to-melt transition in the sodium disilicate composition is not accompanied by any significant change in the positions or intensities of Th L_{III}-XANES features (Fig. 42), suggesting that the local environment around Th in this melt is similar to that in the glass at ambient temperature and pressure. A very similar local topology around Th was observed in the sodium disilicate and sodium trisilicate melts studied, which were found to contain ThO_6 and ThO_8 polyhedra. Because U(IV) has a coordination chemistry similar to that of Th(IV) (Farges, 1991; Farges et al., 1992), it is likely that U(IV) can also be present as a mixture of $U(IV)O_6$ and $U(IV)O_8$ units in silicate melts.

Uranium (VI). XAFS spectroscopy of U(VI) (at the U-L_{III} edge) in silicate glasses suggests that U(VI) occurs as uranyl moieties, with two $U(VI)$-O_{axial} distances at 1.78 ± 0.03Å and 4 to 5 $U(VI)$-$O_{equatorial}$ distances at ~2.25 ± 0.03Å (Petiau et al., 1984; Petit-Maire et al., 1986; Veal et al., 1987; Farges et al., 1992). UV/visible spectra of these glasses supports this conclusion (Calas, 1979; Farges et al., 1992). Sodium disilicate melts (containing between 0.5 and 3 wt % U) were investigated using XAS at temperatures up to 1550 K (Farges and Brown, 1995a). U L_{III}-XANES spectra from this study are shown in Figure 43. Features A, B, and C are attributed to multiple-scattering around the uranyl moieties. No significant changes were detected in the position and intensity of these features with increasing temperature. Only a slight broadening of feature C was detected above T_g (around 900 K).

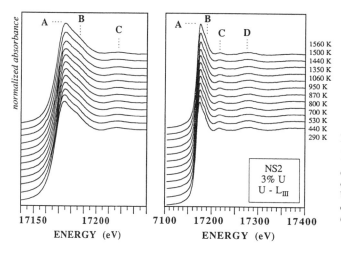

Figure 43. U L_{III} XANES spectra (right) and blow-ups of the main absorption edge (left) of a sodium-disilicate glass/melt (NS2) between 290 and 1560 K containing 3 wt % U(VI). (After Farges and Brown, 1995a.)

Divalent cations

Iron (II). Ferrous iron was the first cation to be studied in molten silicates using high-temperature XAFS spectroscopy (Waychunas et al., 1987, 1988, 1989; Jackson et al., 1991, 1993). The glass and melt compositions examined in these studies included $M_2FeSi_3O_8$ (M = Li, Na, K, Ca) and Fe_2SiO_4. Special precautions were taken to control the oxygen partial pressure during the glass preparation and in situ high-temperature XAFS measurements. Furthermore, Mössbauer spectroscopy on the glasses quenched after the high-temperature measurements showed no evidence of Fe(III) or Fe(0). The main finding in these studies - that Fe(II) is present dominantly as FeO_4 tetrahedra in melts covering a range of compositions and degrees of tetrahedral polymerization - has generated some controversy. Optical spectra collected for a variety of Fe(II)-bearing glasses have been interpreted as being consistent with ferrous iron in distorted six-coordinated oxygen polyhedra (Boon and Fyfe, 1972; Fox et al., 1982; Keppler, 1992; Keppler and Rubie, 1993). However, Mössbauer spectroscopy has suggested the presence of a lower coordination around Fe(II) in silicate glasses quenched from the melts studied by Waychunas et al. (1988) and Jackson (1991), with isomer shifts consistent with those of four- and five-coordinated Fe(II) in well-characterized crystalline model compounds. Furthermore, the 1s→3d pre-edge features of the glasses and melts studied by Waychunas et al. (1988), Jackson (1991), Jackson et al. (1991); Henderson et al. (1991), and Jackson et al. (1993) have heights and positions similar to those observed for hercynite spinel at 1073 K and Fe(II)-bearing leucite, both of which have Fe(II) in tetrahedral coordination. In contrast, the pre-edge feature of Fe(II) in magnesiowüstite at a series of temperatures (190 to 673 K) was found to be only 1/3 as high and is consistent with Fe(II) in a relatively undistorted octahedral site. A survey of Fe(II) in octahedral sites with different degrees of distortion in crystalline silicates and oxides (Waychunas et al., 1988) showed that the Mössbauer isomer shift values and pre-edge heights of the most distorted Fe(II) octahedra are significantly larger and smaller, respectively, than those observed for Fe(II) in the glasses and melts, indicating that ferrous ions in the glasses and melts are in sites with lower coordination numbers. Clearly, the coordination of Fe(II) in the glasses reflects the local structure at T_g, which may be different from that of the melts. As discussed below, the coordination of Fe(II) in these melts is dominantly four-coordinated, whereas in the quenched glasses, Fe(II) is dominantly five-coordinated.

Keppler (1992) dismissed the EXAFS evidence for Fe(II) in tetrahedral coordination in the silicate glasses and melts studied by Waychunas et al. (1988) because he felt these data should be affected in the same way as those for Fe(II) in crystalline fayalite (Waychunas et al., 1986), which were found to give an average Fe-O distance which is too short (2.12 Å versus 2.168 Å from the X-ray crystal structure refinement of Birle et al., 1968), and a coordination number which is too small (4.3 oxygens versus 6 oxygens). Waychunas et al. (1986) were careful to point out this discrepancy and the possibility that their EXAFS analysis, in the harmonic approximation, was not able to detect the longer Fe-O bonds due to positional disorder effects or to spectral interferences. We now know that significant positional disorder, as is true for the M1 and M2 octahedral sites of fayalite, can lead to incorrectly short Fe-O distances and small coordination numbers for Fe and other cations unless these effects, collectively referred to as anharmonicity, are accounted for explicitly. Application of cumulant expansions, which were discussed earlier in this chapter, to a reanalysis of Fe K-EXAFS spectra of crystalline fayalite collected up to k = 13 Å^{-1} yielded an average Fe-O distance of 2.167 Å (Jackson et al., 1995), which is the same as that observed by X-ray diffraction. Similar anharmonic corrections applied to the Fe K-EXAFS spectra of the glasses studied by Waychunas et al. (1988) indicate that the Fe-O distances originally reported for the glasses are ~0.04 Å too short, whereas those for the melts at 1200 K are ~0.08 Å too short. However, this correction does not change the conclusions of that study, i.e. that Fe(II) is dominantly four-coordinated in the melts. This result is consistent with pre-edge data for iron in a variety of Fe(II)-bearing molten silicates; pre-edge features are relatively insensitive to anharmonic effects, so they provide an independent check of coordination geometry.

The interpretation of optical absorption spectra of Fe(II) in silicate glasses must be reconsidered in light of XAFS studies of Fe(II) in silicate glasses and melts. The strong band at ~9000 cm^{-1} and the weaker band at ~5000 cm^{-1} (Jackson, 1991; Keppler, 1992) are classically interpreted as arising from $^{[6]}$Fe(II) and $^{[4]}$Fe(II), respectively (e.g. Calas and Petiau, 1983). Keppler (1992) argues that the weaker 5000 cm^{-1} band is most likely due to Fe(II) in a distorted octahedral site. In order to help resolve this issue, Jackson et al. (1995) carried out a magnetic circular dichroism (MCD) study of Fe(II) in a glass of composition $Na_{1.46}Ca_{0.24}Fe_{1.08}Si_{2.97}O_8$ which was shown to contain only Fe(II) by Mössbauer spectroscopy. The optical spectrum of this glass (designated NC6) displays two bands (9200 and 5320 cm^{-1}) which were thought to represent $^{[6]}$Fe(II) and $^{[4]}$Fe(II), respectively. The MCD spectra as a function of temperature and magnetic field strength contain three features at ~8500 cm^{-1}, ~6700 cm^{-1}, and ~4500 cm^{-1}. Analysis of possible crystal-field transitions and consideration of magnetic saturation effects and polarization dependences suggest that these bands are due to three distinct ferrous sites in this glass (Jackson et al., 1995). The lowest energy band (4500 cm^{-1}) is consistent with Fe(II) in a four-coordinated site, whereas the two higher energy bands are consistent with Fe(II) in two five-coordinated sites. No clear evidence for $^{[6]}$Fe(II) was found in the MCD analysis. These results are in good agreement with those from anharmonic analysis of the Fe K-edge EXAFS, which yielded an Fe(II)-O distance of 2.03 Å, a coordination number of 4.7 oxygens, and a Debye-Waller factor of 0.0081 Å^2. EXAFS analysis in the harmonic approximation yielded a distance of 1.99 Å, a coordination number of 4.6 oxygens, and a Debye-Waller factor of 0.0096 Å^2. The Fe-O distance predicted for $^{[6]}$Fe(II), $^{[5]}$Fe(II), and $^{[4]}$Fe(II) from Shannon (1976) radii values (assuming three-coordinated oxygen) are 2.14, 2.06, and 1.99 Å, respectively, at ambient temperature and pressure (see Table 1). Thus the observed Fe-O distance of 2.03 Å is consistent with a 50:50 mixture of $^{[4]}$Fe(II) and $^{[5]}$Fe(II) in NC6 glass. Taken together, the anharmonic XAFS results, XANES pre-peak heights, Mössbauer isomer shifts, and MCD analysis of

NC6 glass all point to a mixture of four- and five-coordinated Fe(II).

EXAFS analysis of Fe(II) in Fe_2SiO_4 melt also yielded strong evidence for dominantly four-coordinated ferrous iron (Jackson et al., 1993), as was also found by Waseda and Toguri (1978) in high-temperature X-ray scattering experiments on fayalite liquids (see earlier discussion). Anharmonicity was analyzed using the effective pair-potential method discussed earlier, and the Fe-O bond-length correction was +0.10 Å at 1575 K, yielding an Fe(II)-O distance of 1.98 ± 0.02 Å, compared with ~2.22 Å in crystalline fayalite at the melting temperature (1478 K) (Jackson et al., 1993). The average $[4]$Fe(II)-O distance observed in Fe_2SiO_4 melt at 1575 K by Waseda and Toguri is slightly longer (2.04 Å), and the Fe(II) coordination number they report (~4) is slightly lower than that predicted using the Fe-O distances reported in Table 1. The change in Fe-K pre-edge feature and edge position during melting is shown in Figure 44, as is the predicted change in medium-range structure around a typical oxygen in fayalite versus Fe_2SiO_4 melt. Although the high-temperature EXAFS data were noisy, a possible second-neighbor feature at ~2.3 Å (uncorrected for phase shift) in the Fourier transform of the Fe K-EXAFS spectrum of the melt is consistent with an Fe and/or Si backscatterer. No evidence was found for a second-neighbor Si at about 1.8 Å, as reported by Waseda and Toguri (1978) in their X-ray scattering study of Fe_2SiO_4 melt (see Fig. 19). Based on their XAFS analysis, Jackson et al. (1993) proposed a network-like structure for Fe_2SiO_4 melt, with a local arrangement like that in crystalline phenakite (Be_2SiO_4) and willemite (Zn_2SiO_4), where each oxygen of an SiO_4 tetrahedron is shared with two $Fe(II)O_4$ tetrahedra. This structural model was used to estimate a melt density of 3.86 gcm^{-3}, which compares well with the measured value of 3.75 gcm^{-3}. If instead, the Fe_2SiO_4 melt structure is assumed to be similar to that of crystalline fayalite, the predicted melt density is 4.26 gcm^{-3}.

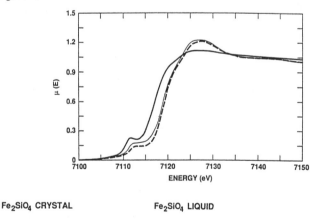

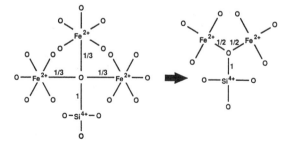

Fe$_2$SiO$_4$ CRYSTAL   Fe$_2$SiO$_4$ LIQUID

Figure 44. (Top) Fe K-XANES spectrum of Fe_2SiO_4 melt at 1575 K (heavy solid line) compared with the spectrum of crystalline Fe_2SiO_4 before (dashed line) and after (light solid line) melting. (Bottom) Schematic illustrations of the medium-range environment of Fe(II) in Fe_2SiO_4 crystal and melt. The numbers by the bonds to the central oxygen in each model are the Pauling bond valences. (Modified after Jackson et al., 1993.)

Anharmonic EXAFS analysis of Fe_2SiO_4 glass (reported in Jackson et al., 1993) showed that the average Fe(II)-O bond length (2.02 ± 0.02 Å) is slightly longer than in the melt (1.98 ± 0.02 Å), indicating a larger amount of [5]Fe(II) in the glass (see Table 1). This result suggests that caution must be applied in inferring coordination numbers for Fe and other transition metals in silicate melts from spectroscopic data on these metals in silicate glasses (cf. Keppler and Rubie, 1993). For example, Cooney and Sharma (1990) interpreted Raman spectra of Fe_2SiO_4 glass as indicating a coordination number for Fe(II) lower than 6. However, as indicated earlier, the local coordination environment of an element in a silicate glass reflects the local structure at T_g.

Nickel (II). Four- and five-coordinated Ni(II) were found in silicate glasses of a range of compositions using optical and EXAFS spectroscopies, with [5]Ni dominating in most of the glasses, especially the sodic ones (Galoisy and Calas, 1991, 1992, 1993a,b). As for Fe(II), these Ni K-edge XAFS results have been questioned based on crystal field reasoning and the possibility that static disorder effects may cause an underestimation of the XAFS-derived average Ni-O distance (Keppler and Rubie, 1993). However, Ni pre-edge studies (Galoisy and Calas, 1993a), *ab initio* theoretical Ni K-XANES calculations (Farges and Brown, 1995b), and anharmonic modeling of the EXAFS oscillations (Farges et al., 1994) have shown that disorder effects are relatively small for Ni in these glasses. Four- and five-coordinated Ni species were also detected in a variety of silicate melt compositions (Farges et al., 1994; Farges and Brown, 1995b; Farges et al., 1995b,c) at temperatures up to 2000 K and for Ni-concentrations ranging from 1.5 to ~20 wt %. Figure 45 shows representative Ni K-XANES and EXAFS spectra for $Na_2Si_2O_5$ glass (at 293 K) and melt (at 1175 K) containing 2 wt % NiO from the study by Farges et al. (1994). [4]Ni was found to be the major Ni-species in most of the melts examined; [6]Ni, if present in these melts, is too dilute to be detected (< than 10 % of the total Ni atoms present in the melt). Ni-Si correlations are clearly seen in the Fourier transform of the glass EXAFS spectrum (Fig. 45) and are present, but weaker, in the melt. These correlations indicate that Ni is closely associated with the silicate tetrahedral network in these materials.

Most of the Ni in the $Na_2Si_2O_5$ glass studied by Galoisy and Calas (1993a) is present in five-coordinated sites, with minor amounts in four-coordinated sites. The melt of this composition, however, shows major amounts of [4]Ni, as indicated by analysis of the pre-edge (feature A in Fig. 45, left), which is much higher in the melt than in the glass), the XANES (features B and C in Figure 45 (left) are shifted to higher energies in the melt, suggesting a lower average Ni-O distance in the melt), and the anharmonic analysis of the EXAFS signal (Farges et al., 1994).

Figure 45. Ni K-edge XANES spectra (left), EXAFS spectra (middle), and the FT of the EXAFS spectra (right) for 2 wt % NiO in a sodium disilicate glass (295 K) and melt (1175 K). (From Farges and Brown, 1995b.)

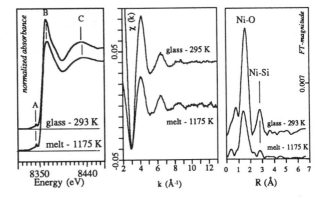

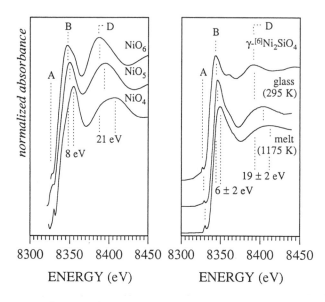

Figure 46. (Left) Ni *K*-edge XANES spectra calculated using the FEFF6 code for a variety of 4-, 5-, and 6-coordinated NiO_n polyhedra found in Ni-oxide compounds ($^{[4]}NiCr_2O_4$, $K^{[5]}NiPO_4$ and γ-$^{[6]}Ni_2SiO_4$). (Right) Experimental Ni *K*-edge XANES spectra collected for γ-$^{[6]}Ni_2SiO_4$ and a Na-Ni disilicate glass (295 K) and melt (1175 K). (From Farges and Brown, 1995b.)

    Ab initio Ni *K*-XANES calculations have verified the presence of $^{[4]}Ni$ in $Na_2Si_2O_5$ melt, as shown in Figure 46 (Farges and Brown, 1995a). The XANES region is a sensitive probe of Ni coordination, because features related to the Ni-O pair correlation (features A, B, and D in Fig. 46) are shifted to higher energies with decreasing Ni coordination (by 8 and 21 eV for features B and D, respectively, when Ni coordination decreases from 6 to 4). Similar shifts are observed for features B and D in the melt. When compared to Ni-olivine, these shifts suggest that the average Ni-coordination is ~4 in $Na_2Si_2O_5$ melt, whereas it is ~5 in $Na_2Si_2O_5$ glass. A mixture of ~50:50 $^{[4]}Ni$ and $^{[6]}Ni$ in the glass sample would also result in an average Ni coordination of 5, but this would result in significantly broadened XANES features, which are not observed.

    Zinc (II). Zinc in crystalline hemimorphite ($Zn_4Si_2O_7(OH)_2 \cdot H_2O$) is six-coordinated by oxygens, but its coordination environment in the anhydrous melt or glass produced by melting hemimorphite is unknown. Farges et al. (1995b) have investigated the coordination environment of Zn in an anhydrous melt and glass of "hemimorphite" composition, at temperatures up to 2000 K. Zn *K*-edge XANES spectra are shown in Figure 47. Note that the first EXAFS oscillation (feature A) is shifted toward lower energies in the crystal with increasing temperature, suggesting a lengthening of the Zn-O bond (thermal expansion). In the melt, feature A is shifted toward higher energies, consistent with the presence of dominantly $^{[4]}Zn$. An intermediate Zn coordination environment (average Zn coordination number ≈ 5) is observed in the glass. Thus the mineral-to-melt transition is accompanied by a decrease in the Zn coordination number from 6 to ~4, as observed for ferrous iron in the melting of fayalite (Jackson et al., 1991, 1993). It is possible that the local environment around an average oxygen in this melt·is like that in the willemite structure (see Fig. 44) as has been suggested for Fe_2SiO_4 melt (Jackson et al., 1993). The coordination number of Zn in anhydrous "hemimorphite"-composition melt is consistent with that observed in X-ray rdf's of K_2O-ZnO-SiO_2 glasses by Musinu and Piccaluga (1994), who reported average Zn-O distances of ~1.96 Å and Zn coordination numbers of 3.8 to 4.7 (see Table 1).

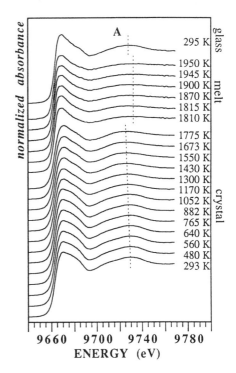

Figure 47. Normalized Zn *K*-edge XANES spectra for hemimorphite crystal (293 to 1775 K), anhydrous melt (1810 to 1950 K), and anhydrous glass (295 K after melting—top spectrum). (From Farges et al., 1995b.)

Coordination chemistry of cations in silicate melts: an XAFS perspective

Network-former and modifier roles of cations in silicate melts. Dietzel's original classification of cations in silicate glasses as network formers, intermediate cations, and network modifiers based on field strengths (Table 1) is also a simple and valuable concept for silicate melts, although we prefer to simplify the classification to include just network formers and modifiers. XAFS results for cations in silicate glasses and melts are generally consistent with this classification, but reveal several new points which help define the structural role of certain cations in silicate glasses and melts. For example, the discovery of $(^{[5]}Ti=O)O_4$ square pyramids in silicate melts, coupled with bond-valence arguments discussed below, led to the proposal that Ti(IV) can act simultaneously as network former and network modifier, with oxygen of the short Ti=O axial bond (bond valence = 1.4 v.u.) forming bonds only to low field strength cations, and the four oxygens of the longer Ti-O equatorial bonds (bond valence = 0.8 v.u.) linking to silicate tetrahedra. Examination of Table 1 also suggests that U(VI) in the uranyl form in silicate melts and glasses may play a structural role similar to Ti(IV), with oxygens of the two short U=O axial bonds (bond valence = 2.0 v.u.) incapable of forming bonds with Si, and oxygens of the four to five equatorial U-O bonds (bond valence = 0.6 v.u.) forming bonds with Si. This is indeed the case for U(VI) in some silicate glasses (Farges et al., 1992). It is relatively easy to classify cations with low field strength (e.g. Na and K) and cations with high field strengths (e.g. Si and B) as networks modifiers and formers, respectively. However, it is often less clear how to categorize cations of intermediate field strength [e.g. Fe(II) and Ni(II)] in these two categories, as pointed out by Henderson et al. (1991), without direct measurements of their first-neighbor coordination numbers and medium-range environments in silicate glasses and melts.

Hess (1991) proposed the use of bond valence to help understand the structural role of cations in glasses and melts and suggested that network modifiers have bond valences of less than 0.75 valence units. He also argued that even if divalent cations like Mg and Fe form MgO_4 and FeO_4 tetrahedral units in silicate melts, they should not be classified as network formers, since the substitution of Mg or Fe(II) for a normal network former, such as Si or Al, should disrupt the network and change its properties. While we generally agree with this argument, we point out that the structural model proposed for Fe_2SiO_4 melt (Jackson et al., 1993), with two $Fe(II)O_4$ tetrahedra sharing each oxygen of an SiO_4 tetrahedron (Fig. 44), and possibly extending in a three-dimensional fashion with a phenakite-like bonding topology, should result in physical properties which differ from those of a depolymerized silicate liquid. In fact, the viscosity of Fe_2SiO_4 melt represents a local maximum along the $FeO-SiO_2$ join (Shiraishi et al., 1978) possibly due to this proposed bonding topology. Nonetheless, the [4]Fe(II)-O and [4]Mg-O bonds are weaker than [4]Si-O and [4]Al-O bonds in silicate melts, and their presence will cause physical properties different from those of an aluminosilicate framework melt (see Henderson et al., 1991). The idea that four-coordinated divalent cations, with bond valences of ~0.5 v.u., are part of a melt network, as proposed for Fe(II) by Waychunas et al. (1988) and Henderson et al. (1991) and for Ni(II) by Galoisy and Calas (1992), simply indicates that these polyhedra share bridging oxygens with adjacent SiO_4 tetrahedra. Consideration of bond valence rather than field strength also indicates that oxygens bonded to highly charged cations like Mo(VI), which occur as tetrahedral MoO_4 units (Mo-O bond valence = 1.4), cannot form bonds with Si, thus act as network modifiers. Therefore, we suggest modification of Hess' (1991) idea that network modifiers should form bonds to oxygens with bond valences < 0.75 v.u.. Instead, network formers should form bonds having bond valences of 1 ± 0.25 v.u., which implies that cations with M-O bond valences < 0.75 or > 1.25 should act as network modifiers. The higher limit provides a more stringent definition of network modifier in that a cation-oxygen bond with bond valence > 1.25 cannot share oxygens with SiO_4 tetrahedra. In contrast, cation polyhedra with M-O bond valences < 0.75 can share oxygens with SiO_4 tetrahedra, thus can be embedded in the tetrahedral network.

The coordination environment of Mg in silicate glasses and melts.

The observation that divalent Fe and Ni are dominantly four coordinated in the silicate melts studied by XAFS spectroscopy has important implications for the coordination chemistry of Mg in silicate melts. Because Mg is essentially the same size as Ni(II) (Table 1), it would also be expected to have a similar coordination environment in silicate melts. There is some preliminary XAFS evidence for this hypothesis from the study by Henderson et al. (1992) which reports mean Mg-O distances of 1.94, 1.92, 1.98, 1.97, and 1.98 Å for glasses of anorthite-forsterite, K-Mg leucite, diopside, anorthite-diopside, and anorthite-diopside-quartz compositions, respectively. Anharmonic effects at 298 K are expected to add $\le$ 0.03 Å to these distances (see discussion above of anharmonic corrections of Fe(II)-O distances in glasses at 298 K). Thus these values are consistent with a mixture of [4]Mg and [5]Mg (see Table 1). A 50:50 mixture of [4]Mg and [5]Mg in a glass at 298 K is predicted to have an average Mg-O distance of 1.98 Å. The Mg-O distance reported for a "basalt"-composition glass by Henderson et al. (1992) is 2.07 Å, which is consistent with dominantly [6]Mg in this glass. There is also indirect evidence from the XAFS study by Marumo et al. (1990) that Mg occurs in four-coordinated sites in glasses of composition $CaMgSi_2O_6$ - $CaTiAl_2O_6$. In addition, as discussed earlier, in their X-ray rdf study of $MgSiO_3$ melt, Waseda and Toguri (1977) assigned a coordination number of 4 to Mg. However, one problem with the hypothesis that Mg is four-coordinated in silicate glasses and melts is a potentially contradictory result provided by a recent high-temperature NMR study of ^{25}Mg in Na-Mg, K-Mg, and Ca-Mg silicate melts (Fiske and Stebbins,

1994). They found evidence that Mg is six-coordinated in $Ca_{0.29}Mg_{0.14}Si_{0.57}O_{1.57}$ melt at 1400°C, with an Mg-O bond length of 2.08 Å, estimated on the basis of chemical shifts. They also suggested that Mg in $Na_{0.28}Mg_{0.18}Si_{0.54}O_{1.54}$ melt at 1400°C, with an estimated Mg-O bond length of 2.00 Å, has [5]Mg, a mixture of [4]Mg and [6]Mg, or a mixture of all three Mg coordination states. The average [4]Mg-O bond length in akermanite is 1.92 Å (Kimata and Ii, 1981), and the [4]Mg-O and [5]Mg-O bond lengths estimated on the basis of Shannon (1976) radii values (Table 1) are 1.93 and 2.02 Å, respectively, at 298 K. The anticipated thermal expansion of an [4]Mg-O bond at 1400°C is about 0.03 Å (Hazen and Finger, 1982); thus the estimated Mg-O bond length (~2.00 Å) for the Na-Mg silicate melt studied by Fiske and Stebbins is consistent with a 75:25 mixture of [5]Mg and [4]Mg (see Table 1). Their suggestion that [6]Mg is the dominant form of Mg in Ca-Mg silicate melt can be rationalized by considering the higher field strength of Ca relative to Na (Table 1) or the higher bond valence of the Ca-O bond relative to the Na-O bond. These factors would favor a higher coordination number for Mg in the Ca-Mg silicate melt as pointed out by Fiske and Stebbins. The study of Fe(II) in Ca-bearing silicate glasses and melts (Jackson, 1991; Jackson et al., 1995) also found that Fe(II) has a higher average coordination number when Ca is more abundant that Na. Thus the coordination chemistries of Mg, Fe(II), and Ni(II) appear to be similar in silicate melts, as expected.

Coordination changes above T_g. Examination of the information on coordination chemistry of metal ions in silicate melts derived from high-temperature XAFS spectroscopy suggests that highly charged cations do not show major changes in their coordination environment during the glass-to-melt transition and at temperatures above T_g (Ti, Zr, Mo, Th, U, Ge). These elements have relatively low coordination numbers (4 for Ge and Mo; 4 to 5 for Ti; 6 for Zr, Th, and U) in most silicate glasses and melts studied up to their maximum solubility. However, higher coordinated species (five- and six-coordinated Ge, six-coordinated Mo and Ti, eight-coordinated Zr, Th, and U) could also be present in small concentrations (<5 atom %) in the silicate melts studied, but may not be detected by XAFS spectroscopy.

Metal ions with lower charge (divalent Fe, Ni, and Zn) also exhibit lower coordination numbers in melts (mostly 4) than in crystals. However, Fe(II) and Ni(II) have higher average coordination numbers in glasses (five-coordinated on average) than in melts of the same composition. These results indicate that the coordination states of lower field strength cations like Fe(II), Ni(II), and Zn(II) (see Table 1) in silicate glasses may not be representative of their coordination states in silicate melts of the same compositions. Therefore, in situ high-temperature-high-pressure studies of melts are required to understand the coordination chemistry of divalent transition elements under conditions representative of the Earth's upper mantle. In contrast, the local structure around highly charged cations studied by XAFS spectroscopy in melts is similar to that in their glassy counterparts.

Implications of XAFS-derived coordination numbers for Fe(II) and Ni(II) for cation partitioning between silicate melts and crystals. There has been a great deal of interest in the partitioning behavior of first-row transition metal cations between silicate melts and coexisting igneous minerals since the introduction of crystal field theory to geochemistry (Williams, 1959; Burns and Fyfe, 1964). This well-known theory provides a reasonable framework for rationalizing the order of uptake of these cations into crystals. Burns (1993) summarizes many of the ideas and observations that have been used to build a structural view of transition metal ion behavior in silicate melts, but he cautions that much of the reasoning to date is based on observations of these ions in silicate glasses under ambient conditions, not melts under high-temperature conditions. The implicit assumption made is that the coordination environments of these ions are the same in glasses and melts of the same composition. Williams (1959) first suggested that the partitioning of a cation from melt to crystal usually involves an increase in average coordination number and a decrease in average interatomic distance. He further speculated that transition metal ions which are particularly stabilized in octahedral coordination will gain stability in leaving sites of irregular coordination in the melt and entering more regular octahedral sites in crystals. This reasoning is generally correct (except for the suggestion that the M-O distance is generally smaller in the crystal), but its validity depends on knowledge of average transition metal ion coordination sites in silicate melts.

Spectroscopic studies (primarily UV/visible spectroscopy and XAS) of transition metal ions in silicate glasses suggest that Cr(III), V(III), and Mn(III) are predominantly octahedrally coordinated (Keppler, 1992), whereas Fe(III) occurs mainly in tetrahedral coordination and is a network former (Brown et al., 1978; Fleet et al., 1984). Mn(II) is thought to occupy either a distribution of octahedral and tetrahedral sites or distorted sites lacking a center of symmetry (Nelson and White, 1980; Kohn et al., 1990; Keppler, 1992); however, in glasses of orthosilicate composition, they occupy dominantly tetrahedral sites (Cooney and Sharma, 1990; Jackson, 1991). Co(II) is thought to occupy dominantly tetrahedral sites in silicate and borosilicate glasses studied to date (Nelson and White, 1986; Corrias et al., 1986; Keppler, 1992). The coordination geometries of Fe(II) and Ni(II) in silicate glasses are controversial, with some workers favoring distorted octahedral coordination for Fe(II) and Ni(II) (e.g. Keppler, 1992) and others favoring four- and five-coordinated sites (Cooney and Sharma, 1990; Jackson et al., 1991, 1993; Galoisy and Calas, 1991, 1993a). One of the few spectroscopic studies of a transition metal in a molten material made prior to 1990 was the UV/visible study of Goldman and Berg (1980), which suggested that Fe(II) is dominantly six-coordinated in a borosilicate glass at 24°C and also in the corresponding melt at 1260°C. Based on these and other spectroscopic observations (see Calas and Petiau, 1983 for additional spectroscopic data on these and other first-row transition metal ions in silicate glasses), it appears that the local coordination environments of transition metal ions in glasses are often similar to those in crystals of similar compositions, with a few important exceptions. Predictions of melt-crystal partitioning behavior of these ions, based on crystal field stabilization energies (CFSE) derived from UV/visible spectroscopic studies of silicate glasses, appear to be valid only for Ni(II) and Cr(III), which have significantly higher values of CFSE in crystals than in glasses relative to the other ions (Calas and Petiau, 1983). Both Calas and Petiau (1983) and Keppler (1992) present indirect evidence that melt-crystal partition coefficients for transition metal ions depend on melt polymerization, with these ions partitioning preferentially into crystals as melts become more polymerized. An important question remains, however. Are the coordination environments of transition metal ions in silicate glasses the same as those in silicate melts?

High-temperature XAS studies of Fe(II) and Ni(II) in silicate melts of several compositions (Fe_2SiO_4, $NaSi_2O_5$, and $M_2(Fe,Ni)Si_3O_8$, where M = Li, Na, K, Rb) have shown that they are dominantly four-coordinated (Waychunas et al., 1987, 1988, 1989; Jackson et al., 1991, 1993; Farges et al., 1994, 1995c). These results provide some important constraints on melt structure and on the structural reasoning used in explaining the partitioning behavior of these ions. For olivine-composition melts, assumptions commonly made about melt structure are (1) that these melts have both octahedral (Oct) and tetrahedral (Tet) sites, by analogy with the distorted hexagonal closest-packing of oxygens in the olivine structure; (2) that olivine-composition melts have a larger Oct/Tet ratio than more compositionally complex silicate melts; and (3) that Fe(II) in such melts occupies Oct sites (see, e.g. Boon and Fyfe, 1972; Roeder, 1974; Irvine and Kushiro, 1976; Takahashi, 1978; Burns, 1993). Assumption (3) is also commonly made for Ni(II) in olivine-composition melts (e.g. Irvine and Kushiro, 1976). The data of Takahashi (1978) show the order of melt→crystal partitioning to be Ni > Co > Fe > Mn for olivine crystals relative to coexisting $(Mg_{0.5}Fe_{0.5})_2SiO_4$ - $K_2O \cdot 4SiO_2$ melts, which is the order of their octahedral site preference energies (OSPE). These observations are similar to the predictions of Roeder (1974) and suggest that the cations occupy fewer octahedral sites in the melt than in the coexisting crystal. Comparison of OSPE for Fe(II) (~0.17 eV) (Burns, 1993) with kT (~0.14 eV at 1575 K, where k is the Boltzmann constant) indicates that Fe(II) should not show strong preference for octahedral sites in the crystal or melt at this temperature. However, the studies by Jackson et al. (1991, 1993) show that Fe(II) occupies mainly four-coordinated sites in Fe_2SiO_4 melt. Ni(II) has a much larger OSPE (~0.9 eV) (Burns, 1993), and thus it should be stabilized in octahedral sites in the melt (at 1575 K), if they exist. Ni(II) is dominantly four-coordinated in olivine-composition melt (Farges et al., 1995b) and in $Na_2Si_2O_5$ melt (Farges et al., 1994, 1995c; Farges and Brown, 1995b), and it is dominantly tetrahedral and pentahedral in more SiO_2-rich glasses ($CaNiSi_2O_6$ and $(Na,K)_2NiSi_3O_8$) (Galoisy and Calas, 1991, 1992, 1993a). Thus the existing high-temperature structural data for Fe(II) and Ni(II) in silicate melts do not support the general assumption that these cations are six-coordinated in the melt (cf. Boon and Fyfe, 1972; Goldman and Berg, 1980; Mysen, 1988; Keppler, 1992). There is also evidence that Mg is dominantly four-coordinated in silicate glasses of various compositions (Henderson et al., 1992) and is four- to five-coordinated rather than six-coordinated in (Na,K)-bearing silicate melts (Fiske and Stebbins, 1993). Thus the coordination chemistry of Mg appears to be consistent with that of Fe(II) and Ni(II) in similar melts.

In more polymerized silicate melts of composition $M_2Fe(II)Si_3O_8$, Fe(II) is four- to five-coordinated, with no detectable Fe(II) in six-coordinated sites (Waychunas et al., 1988). The longer Fe(II)-O distances in these melts (~2.03 Å—which includes an anharmonic correction of 0.09 Å applied to the distances reported by Waychunas et al., 1988) relative to Fe_2SiO_4 melt (1.98 ± 0.02 Å—which is corrected for anharmonic effects) may help explain the observation of Calas and Petiau (1983) that the CFSE of transition metal ions is lower in more polymerized melts. Henderson et al. (1991) obtained similar results for Fe(II) in highly polymerized $MFe(II)_{0.5}Si_{2.5}O_6$ glasses (M = K, Rb, and Cs) with leucite stoichiometry, assuming a similar correction of their data for anharmonicity.

Assuming that natural ultramafic melts also contain [4]Fe(II), a pressure-induced [4]Fe to [6]Fe coordination change in these melts is possible (Waychunas et al., 1988; Waff, 1975), with attendant changes in the density and, therefore, buoyancy of ultramafic melts in the earth's low velocity zone. Based on the observation of dominantly [4]Mg in silicate glasses of "mafic" compostions (Henderson et al., 1992) and of a mixture of [4]Mg and [5]Mg in a (Na,K)-silicate melt, one could also speculate that a similar coordination change for Mg in ultramafic silicate melts is possible with increasing pressure. A recent in situ Raman spectroscopic study of alkali-germanate melts at pressures up to 2.2 GPa (Farber and Williams, 1992) found evidence for this type of coordination change for Ge, which behaves like Si in silicates. Similarly, Itié et al. (1989) found evidence for an increase in the coordination of Ge in GeO_2 glass from 4 to 6 with increasing pressure. Simultaneous high-temperature, high-pressure Fe- and Ni-XAFS studies of Fe_2GeO_4 and Ni_2GeO_4 liquids are required to test this hypothesis for Fe(II) and Ni(II) in compositionally simple liquids.

MODELS OF MEDIUM-RANGE ORDER IN SILICATE GLASSES AND MELTS

Bond valence models of medium-range order in silicate melts

As mentioned earlier, crystal chemical principles formed the basis for the early models of glass structure and the rules governing glass formation. Farges et al. (1991) employed modern extensions of these principles, namely a combination of Pauling's electrostatic valence principle (Pauling, 1929) and the empirical bond valence-bond length model developed by Brown and Shannon (1973), to propose plausible structural models of the medium-range environment around cations in silicate glasses and melts containing Zr(IV). This environment, as explained in the introduction, extends from the first atomic shell around a cation up to 6 to 10 Å. These types of empirical models have since been extended to Th(IV) (Farges, 1991), U(IV) and U(VI) (Farges et al., 1992), Ti(IV) (Farges et al., 1995e,f), and Ni(II) (Galoisy and Calas, 1993a; Farges et al, 1994) using the bond valence (s)-bond length (R) relationship of Brown and Altermatt (1985) and Brese and O'Keeffe (1991), which is defined as

$$s = \exp\left[\frac{(R_0 - R)}{0.37}\right] \tag{39}$$

R is the observed M-O distance and R_0 is the bond valence parameter of M, which is tabulated by Brown and Altermatt (1985) and Brese and O'Keeffe (1991) for most cations in different valence states. The R_0 values for an element vary as a function of the cation's valence but not as a function of coordination number in these tables. This formulation might seem to result in bond valences that could differ significantly from the original Pauling definition (Z_c/N), which explicitly accounts for both cation valence (Z_c) and cation coordination number (N). However, variations in bond valence expected as cation coordination number varies are reflected in the observed M-O bond length in the Brown-Shannon model. The variation of M-O bond length as a function of cation coordination number is well documented for most M-O bonds (Shannon and Prewitt, 1969). Brese-O'Keeffe values of R_0 for selected cations are listed in Table 1.

The bond-valence model should be useful in developing constraints for medium-range structures around any cation, in principle, but it has the greatest utility for highly charged cations that form bonds of high bond valence because such cations place the greatest constraints on possible atomic arrangements. An example of the application of this model to the medium-range environment around Zr is taken from Farges et al. (1991). Bond valence-bond length correlations for each M-O bond type present in the Zr-containing glasses studied by Farges et al. (1991) are shown in Figure 48, including Si-O, Al-O, Zr-O, and Na-O. Their classical Pauling bond valences are $s([4]Si\text{-}O) = 1.0$; $s([4]Al\text{-}O) = 0.75$; $s([6]Zr\text{-}O) = 0.67$; $s([8]Zr\text{-}O) = 0.5$; $s([6]Na\text{-}O) = 0.167$ valence units, v.u.). The bond-valence-bond length relationship (Eqn. 39) indicates that shorter bonds should have higher bond valence. Because of the lack of periodicity in glasses and melts and the freedom this provides M-O bonds to adjust their length locally, depending on local bonding requirements, M-O distances have a range of values, the range varying inversely with the bond valence for a given M-O bond. We combined this idea with the basic requirement that Pauling's second rule (Pauling, 1929) should be valid in silicate glasses and melts, i.e. the sum of M-O bond valences to an oxygen should equal 2 v.u. or be near this value (within ±0.1 v.u.) (Farges et al., 1991).

In developing possible models for the local and medium-range environments of Zr in silicate glasses, we have made use of this well-known principle and have also accounted for the possibility of variations in individual cation-oxygen distances in a glass

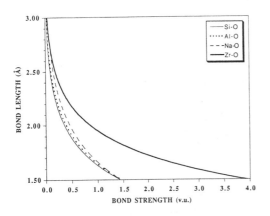

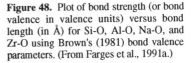

Figure 48. Plot of bond strength (or bond valence in valence units) versus bond length (in Å) for Si-O, Al-O, Na-O, and Zr-O using Brown's (1981) bond valence parameters. (From Farges et al., 1991a.)

or melt structure. Consideration of numerous, well-refined sodium aluminosilicate crystal structures shows that individual Si-O bond lengths may vary between about 1.54 and 1.70 Å, with bond valences ranging between 1.28 and 0.82 v.u., and that individual Al-O bond lengths can vary between about 1.63 and 1.82Å, with bond valences between 1.04 and 0.63 v.u. (Brown et al., 1969; Brown and Gibbs, 1969; 1970). Similarly, Na-O bond lengths may vary from about 2.2 to ~3.4 Å with bond valences between 0.29 and <0.02 v.u. The expected coordination number of Na in aluminosilicate glasses and melts is six (McKeown et al., 1985; Xue et al., 1991). Individual Zr-O bond lengths show less variation than Na-O bond lengths in crystal structures, as expected, with values between about 2.03 and 2.11 Å for six-coordinated Zr and between 2.13 and 2.35 Å for eight-coordinated Zr, corresponding to bond valences between 0.78 and 0.30 v.u. Adjustments of Si-O, Al-O, and Na-O bond lengths within these limits for a given molecular model (see Figs. 49 and 50), such that the total bond valence to each type of oxygen is near 2.0 v.u., can be made to test the validity of different models with d[Zr-O] fixed at the EXAFS-derived value of 2.07 Å.

This modeling indicates that [6]Zr(IV) cannot bond directly to a bridging oxygen in the albite-composition melt network at the observed [6]Zr-O distance (2.07 Å) without significantly lengthening {Si,Al}-O bonds and disrupting the tetrahedral network. For example, if [6]Zr bonded directly to Si-O-Si or Si-O-Al linkages (Fig. 49a), the Si-O and Al-O bonds would be required to lengthen beyond their observed maximum value in aluminosilicates in order to result in a bond valence sum near 2.0. This is unlikely. It would be possible for [6]Zr to bond in part to Al-O-Al linkages. Although such linkages are less likely than Si-O-Si and Si-O-Al linkages in sodium aluminosilicate glasses, they may be stabilized by [6]Zr. It is unlikely, however, that a major portion of the Zr in the glasses studied bonds preferentially to bridging oxygens in Al-O-Al linkages because of the observed decrease in Zr solubility in peraluminous melts relative to peralkaline melts (Watson, 1979). An alternative configuration could have [6]Zr bonding to an oxygen in an Si-O(nonbridging) bond without other charge-balancing cations (Fig. 49a); however, this would require the Si-O bond to shorten to its minimum observed value, which is also unlikely. A more likely structural configuration is the linkage of [6]Zr to several non-bridging oxygens as shown in Figure 49b. In this model, the bond valence sums at the three types of oxygens shown are within 0.1 v.u. of satisfying Pauling's second rule and the EXAFS results of Farges et al. (1991). In contrast, [8]Zr is predicted to bond preferentially to bridging oxygens (Fig. 50b) rather than non-bridging oxygens (Fig. 50a) on the basis of similar bond-valence bond-length reasoning. However, better local charge balance can be achieved when [8]Zr bonds to Al-O-Si linkages rather than to Si-O-Si linkages. In the latter case, Si-O bonds would be required to lengthen to their maximum values, whereas in the former case much less change in length from average observed values is required. The thermodynamic activity of ZrO_2 would be predicted to be highest in a melt with local environments like those in Figures 49a and 50a and lowest in a melt where Zr is bonded stably to nonbridging (Fig. 49b) or bridging oxygens (Fig. 50b).

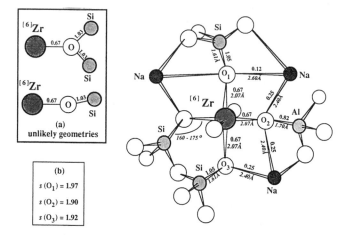

Figure 49. Bond length-bond valence models for ZrO_6 polyhedra in a silicate glass/melt. (a) examples of unlikely geometries with over- and under-bonded oxygens; (b) bond valence (s) sums to the three non-equivalent oxygens for a possible environment around ZrO_6 satisfying the Zr K-EXAFS results of the study by Farges et al. (1991a) and Pauling's second rule. Distances and Brown-Altermatt bond valences are shown for some of the bonds. (From Farges et al., 1991a.)

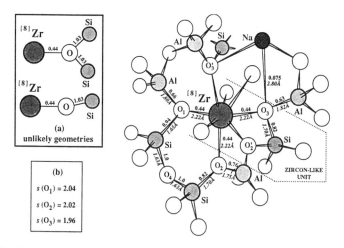

Figure 50. Bond length-bond valence models for ZrO_8 polyhedra in a silicate glass/melt. (a) examples of unlikely geometries with over- and under-bonded oxygens; (b) bond valence (s) sums to the three non-equivalent oxygens for a possible environment around ZrO_6 satisfying the Zr K-EXAFS results of the study by Farges et al. (1991a) and Pauling's second rule. Distances and Brown-Altermatt bond valences are shown for some of the bonds. Some medium-range environments are similar to that observed in zircon (outlined by dotted line) and may be responsible for an increase in $ZrSiO_4$ activity in highly-polymerized melts. (From Farges et al., 1991a.)

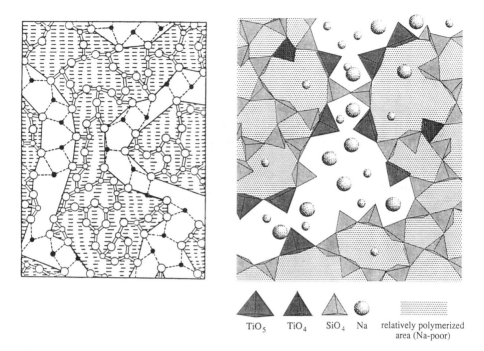

TiO$_5$ TiO$_4$ SiO$_4$ Na relatively polymerized area (Na-poor)

Figure 51 (left). Modified random network model for a "two-dimensional" oxide glass. Small open circles represent network-forming cations (C$_F$), small filled circles represent network-modifying cations (C$_M$), and large open circles represent oxygens (O). The boundaries are through the C$_F$-O (non-bridging) bonds and are intended to highlight percolation channels where C$_M$ cations are concentrated. (From Greaves, 1985.)

Figure 52 (right). Model of NTS2 glass/melt containing (mol %) 50SiO$_2$·25TiO$_2$·25Na$_2$O, showing the percolation domain where ([5]Ti = O)O$_4$ polyhedra are concentrated, with the titanyl oxygen forming bonds to Na (fuzzy circles) in the percolation domain and oxygens of the four longer Ti-O bonds forming bonds to SiO$_4$ tetrahedra. Also shown are SiO$_4$ and TiO$_4$ polyhedra. (From Farges et al., 1995e.)

Modified random network model and percolation domains in silicate glasses and melts

A modification of the Zachariasen-Warren random network model has been proposed by Greaves et al. (1981) and Greaves (1985) for Na$_2$O-SiO$_2$ glasses which involves two interlacing "sublattices", i.e. network regions made up of network formers and inter-network regions made up of network modifiers. A "2-dimensional" representation of this model is shown in Figure 51. This "modified random network" (MRN) model has several appealing features. First, it does not violate any of the well-known principles of glass formation (see earlier discussion of historical perspectives). Second, the MRN proposes "percolation channels" which could facilitate the diffusion of network modifier cations. Greaves (1985) estimated that in three dimensions the network-modifier regions would extend to percolation channels when their volume fraction exceeds 16%. Correlations have been noted between the mol % alkali oxide and the correlation factor for ionic diffusion (e.g. in Na$_2$O-GeO$_2$ glasses; Friebele et al., 1979), which is consistent with the percolation threshold concept suggested by Greaves. Further discussion of the relationship between ionic conductivity and structure in silicate glasses can be found in Greaves et al. (1991), Greaves (1992), and Vessal et al. (1992). The diffusion properties of many binary alkali silicate glasses are predicted using the MRN. Third, the MRN is consistent with independent reasoning using the valence-bond model discussed above that led to the proposal of percolation channels or domains in Na and K titanosilicate glasses by Farges et al. (1995f) (Fig. 52). Alkali cations are shown concentrated in these domains whose boundaries are defined by the presence of ([5]Ti=O)O$_4$ polyhedra. Due to the high bond valence of the short Ti=O bonds (1.4 v.u.), the titanyl oxygens cannot be shared with SiO$_4$ tetrahedra because the resulting bond valence sum (~2.4 v.u.) would greatly exceed 2. However, the titanyl oxygen can be shared with several alkali cations which form bonds of low bond valence ($\leq$ 0.2 v.u.). In contrast, the longer Ti-O bonds in ([5]Ti=O)O$_4$ polyhedra have bond valences of 0.8 v.u., thus the oxygens in these bonds can be shared with SiO$_4$ tetrahedra. This reasoning led to the model shown in Figure 52, with the titanyl oxygens pointing in toward the alkali-rich percolation domains and the longer Ti-O bonds pointing toward the alkali-poor polymerized region. Farges et al. (1995f) suggested that the breakdown of these percolation domains above T_g helps explain the anomalous heat capacities above T_g in these glasses. No changes were found in Ti coordination number above T_g in these glasses, so this was eliminated as a cause of the Cp anomalies. Finally, the MRN model is consistent with the results of neutron scattering experiments by Gaskell et al. (1991, 1992) which found evidence for "clustering" of Ca in silicate glasses. The existence of percolation channels in binary alkali silicate glasses has not been verified by small angle X-ray or neutron scattering, to our knowledge, although it is possible that the electron density contrast between the network former and network modifier domains is too small or their dimensions are too small to be detected. However, there is some structural evidence for the MRN model from XAFS spectroscopy, X-ray and neutron scattering, and anomalous X-ray scattering as discussed below.

Experimental evidence consistent with the concept of percolation domains, in which alkalis or alkaline earths are concentrated in silicate glasses, has been provided by XAFS spectroscopy and X-ray and neutron scattering. These methods provide information about possible C_M - C_M correlations, where C_M represents a network-modifying cation. XAFS spectroscopy is generally limited to correlations of less than 4 to 5 Å, whereas X-ray and neutron scattering, which are not limited by the short mean free path of electrons, can provide quantitative information on more distant correlations. Results on C_M - C_M correlations in glasses from scattering methods are reviewed by Gaskell (1991a,b). One way to enhance the scattering signal from such correlations is to study glasses containing a heavy metal as a major component. In this case, the scattering frequency responsible for the C_M–C_M pair correlation in the X-ray or neutron rdf will be stronger than other pair correlations. This approach was used by Hanson and Egami (1986) in an X-ray scattering study of Cs-silicate glasses. They found a strong pair-correlation in the rdf at 4.1 Å, which they assigned to Cs-Cs1, and they observed additional features at 6.9 and 8.0 Å, which they interpreted as being due to extensive Cs clustering in the glass.

Another approach to the problem of detecting domains of network-modifying cations compares the neutron scattering from two different isotopes of an element having significantly different neutron scattering lengths. A pioneering neutron scattering study of this type was carried out by Yarker et al. (1986) using Ti isotopically substituted K$_2$O-TiO$_2$-SiO$_2$ glass. They found Ti-Ti correlations at 3.4 Å in the neutron rdf. A random distribution of Ti atoms would give a correlation length of 6.1 Å, so the shorter observed distance was interpreted as indicating preferential clustering of Ti in this glass. Eckersley et al. (1988), Gaskell et al. (1991) and Gaskell et al. (1992) performed similar experiments on CaO-SiO$_2$ and NiO-2CaO-3SiO$_2$ glasses using the "double difference method" which is explained in Gaskell et al. (1991) and Wright et al. (1991). The first of these glasses was prepared with two different Ca isotopes and the second with two different Ni isotopes. In both studies, evidence for cation clustering was found. For example, in CaO-SiO$_2$ glass, Ca-Ca correlations at 3.8 Å and 6.4 Å were detected. The shorter of these distances is consistent with edge-shared CaO$_6$ octahedra, whereas the longer Ca-Ca distance was interpreted as evidence for planar arrangements of edge-shared CaO$_6$ octahedra. Anomalous X-ray scattering can also provide useful information on longer-range C$_M$ - C$_M$ pair correlations and takes advantage of the anomalous variations of the absorption coefficient in the vicinity of an X-ray absorption edge of an element (Warburton et al., 1987). This method has been used recently to study the clustering of Y cations in Y$_2$O$_3$-Al$_2$O$_3$-SiO$_2$ glass (Nukui et al., 1992) and of Sr cations in SrO-SiO$_2$ glasses (Creux et al., 1995). Verification of similar network-modifier clustering in silicate melts using the neutron "double-difference" or anomalous X-ray scattering methods has not been attempted, to our knowledge.

Changes in medium-range order caused by nucleation in silicate glasses and melts

When the medium-range structure in a silicate glass or melt becomes enough like that of a crystalline phase, nucleation and crystal growth may occur given a sufficient energy drive such as undercooling. Such nucleation has been observed in an ex situ fashion by XAFS spectroscopy, including the studies of Dumas and Petiau (1986), Dumas et al. (1985), and Petiau and Calas (1985), which detected significant changes in second neighbors around Ti, Zn, and Zr in silicate glasses of various compositions prior to the development of an X-ray diffraction pattern. In situ high-temperature Ni *K*-edge XAFS experiments on a Ni-containing sodium-disilicate glass/melt near T$_g$ (Farges and Brown, 1995b) have detected the nucleation of NiO crystals (i.e. appearance of strong Ni-Ni pairs in the EXAFS spectra) from the glass ~50 K below T$_g$. In addition, the formation of nuclei of Ni-diopside crystals was observed in situ by XAFS spectroscopy near 1100 K in a diopside-composition glass containing 2 wt % Ni (Farges et al., 1995 a). These in situ observations may be indirect evidence for the presence of densely-packed units around Ni in these glasses at room temperature as suggested by Gaskell et al. (1992). Such domains may favor the nucleation of NiO just below T$_g$ where structural reorganization can occur. If they are present, however, they would not be observable by XAFS spectroscopy because the contributions of longer-range Ni-Ni correlations are too weak and disordered. Neutron scattering on silicate melts is necessary to determine if densely packed domains of network modifiers are present in melts.

SUMMARY OF AVERAGE M-O DISTANCES AND CATION COORDINATION NUMBERS IN SILICATE GLASSES AND MELTS

A selective synthesis of the M-O distances and M coordination numbers derived from X-ray and neutron scattering and XAFS studies of glasses and melts reviewed in this chapter is presented in Table 3.

Table 3 is on the next ten pages. Its caption follows:

Table 3. Selected cation-oxygen distances, coordination numbers (CN) and angles in silicate melts and glasses from x-ray and neutron scattering and EXAFS spectroscopy.

Table 3.
Network-Forming Cations (in silica, germania, and alumina glasses and melts)

Composition	T (K)	Method	Bond	Cation CN	d(T-O) (Å)	T-O-T (°)	Reference
SiO$_2$	298	x-ray rdf	Si-O	4[1]	1.62	120-180[2]	Mozzi & Warren (1969)
	298	neutron rdf		4[1]	1.60	<144>	Lorch (1969)
	298	Si-EXAFS		4[1]	1.6	145[3]	Greaves et al. (1981)
	298	x-ray rdf		4[1]	1.595	~140[3]	Konnert & Karle (1973)
	298	x-ray rdf		4[1]	1.626	149[3]	Wright (1994)
	298	x-ray rdf		4[1]	1.612	142[3]	Poulsen et al. (1995)
	298	neutron rdf		4[1]	1.608	147	Wright (1994)
	2023	x-ray rdf		4[1]	1.62	146[3]	Waseda & Toguri (1977)
	2123	x-ray rdf		4[1]	1.60	149[3]	Nukui et al. (1978)
	298, 10^{-4}GPa	x-ray rdf		4[1]	1.59	134[3]	Meade et al. (1992)
	298, 28 GPa	x-ray rdf		4-6[4]	1.64	150[3]	Meade et al. (1992)
	298, 42 GPa	x-ray rdf		4-6[4]	1.66	<150	Meade et al. (1992)
GeO$_2$	298	x-ray rdf	Ge-O	4[1]	1.70	125-152	Zarzycki (1957)
	298	neutron rdf		4[1]	1.72	--	Lorch (1969)
	298	neutron rdf		4[1]	1.72	132	Wright et al. (1982)
	298	x-ray rdf		4[1]	1.74	132	Wright et al. (1982)
	298	Ge-EXAFS		4[1]	1.734	130[3]	Lapeyre et al. (1983)
	80	Ge-EXAFS		4[1]	1.732	130	Okuno et al. (1986b)
	298	Ge-EXAFS		4[1]	1.736		Okuno et al. (1986b)
	1473	x-ray rdf		4[1]	1.73		Zarzycki (1957)
	1500	Ge-EXAFS		4[1]	1.75	137-180	Farges et al. (1995a)
	298, 10^{-4}GPa	Ge-EXAFS		4[1]	1.75	--	Itié et al. (1989)
	298, 10 GPa	Ge-EXAFS		4-6	1.80	--	Itié et al. (1989)
	298, 18 GPa	Ge-EXAFS		6	1.84	--	Itié et al. (1989)
Al$_2$O$_3$	2373	x-ray rdf	Al-O	6[1,7]	1.85	104[3]	Nukui et al. (1976)
	2363	x-ray rdf		5.5	2.02	--	Waseda & Toguri (pers com)
93.9(SiO$_2$)-6.1(Al$_2$O$_3$)[4]	298	x-ray rdf	T-O[5]	4.3[8]	1.62	--	Morikawa et al. (1982)
33 (SiO$_2$)- 67(Al$_2$O$_3$)[4]	298	x-ray rdf	T-O[5]	4.7[8]	1.81	--	Morikawa et al. (1982)
			T-O[5,6]	4	1.74	--	
			Al-O[6]	6	1.91	--	

Network-Forming Cations (in alkali- and alkaline-earth silicate glasses and melts)

Composition	T (K)	Method	Bond	Cation CN	d(T-O) (Å)	T-O-T (°)	Reference
33 Li$_2$O-67 SiO$_2$	298	x-ray rdf	Si-O	4[1]	1.61	--	Waseda & Suito (1977)
	298	neutron rdf		4[1]	1.62	--	Waseda & Suito (1977)
	1423	x-ray rdf		4[1]	1.61	--	Waseda & Suito (1977)
Li$_2$SiO$_3$	298	x-ray rdf	Si-O	~4[9]	~1.6[9]	135	Yasui et al. (1994)
Li$_2$Si$_2$O$_5$	298	neutron rdf	Si-O	4[1]	1.62	--	Misawa et al. (1980)
33 Na$_2$O-67 SiO$_2$	298	x-ray rdf	Si-O	4[1]	1.62	--	Waseda & Suito (1977)
	298	neutron rdf		4[1]	1.61	--	Waseda & Suito (1977)
	1273	x-ray rdf		4[1]	1.62	--	Waseda & Suito (1977)
Na$_2$SiO$_3$	298	x-ray rdf	Si-O	~4[9]	~1.6[9]	143	Yasui et al. (1994)
Na$_2$Si$_2$O$_5$	298	neutron rdf	Si-O	4[1]	1.63	--	Misawa et al. (1980)
	298	Si-EXAFS		4[1]	1.6	~140[3]	Greaves et al. (1981)
33 K$_2$O-67 SiO$_2$	298	x-ray rdf	Si-O	4[1]	1.62	--	Waseda & Suito (1977)
	298	neutron rdf		4[1]	1.61	--	Waseda & Suito (1977)
	1373	x-ray rdf		4[1]	1.62	--	Waseda & Suito (1977)
K$_2$SiO$_3$	298	x-ray rdf	Si-O	~4[9]	~1.6[9]	158	Yasui et al. (1994)
K$_2$TiSi$_2$O$_7$	298	neutron rdf	Si-O	4[1]	~1.62	--	Yarker et al. (1986)
Cs$_2$SiO$_3$	298	x-ray rdf	Si-O	~4[9]	~1.6[9]	171	Yasui et al. (1994)
44 MgO-56 SiO$_2$	298	x-ray rdf	Si-O	4[1]	1.63	150[3]	Waseda & Toguri (1977)
	1973	x-ray rdf		4[1]	1.63	152[3]	Waseda & Toguri (1977)
MgSiO$_3$	298	x-ray rdf	Si-O	4[1]	1.63	139[3]	Yin et al. (1983)
	1993	x-ray rdf		4[1]	1.62	154[3]	Waseda & Toguri (1990)
45 CaO-55 SiO$_2$	298	x-ray rdf	Si-O	4[1]	1.63	164[3]	Waseda & Toguri (1977)
	1873	x-ray rdf		4[1]	1.63	171[3]	Waseda & Toguri (1977)
CaSiO$_3$	298	x-ray rdf	Si-O	4[1]	1.64	--	Yin et al. (1986)
	1873	x-ray rdf		4[1]	1.61	171[3]	Waseda & Toguri (1990)
Mg$_3$Al$_2$Si$_2$O$_{12}$	298	x-ray rdf	T-O[5]	4[1]	1.68	--	Okuno & Marumo (1993)
Ca$_3$Al$_2$Si$_2$O$_{12}$	298	x-ray rdf	T-O[5]	4[1]	1.70	--	Okuno & Marumo (1993)
Fe$_2$SiO$_4$	1573	x-ray rdf	Si-O	4[1]	1.62	--	Waseda & Toguri (1978)

Network-Forming Cations (in feldspar-composition glasses and melts)

Composition	T (K)	Method	Bond	Cation CN	d(T-O) (Å)	T-O-T (°)	Reference
$NaAlSi_3O_8$ (AB)	298	x-ray rdf	T-O[5]	4[1]	1.63	146	Taylor & Brown (1979a)
	1078				1.63	146	Taylor et al. (1980)
	1473				1.64	146	Okuno & Marumo (1982)
	298	Al-EXAFS	Al-O	4[1]	1.76	--	Brown et al. (1982); McKeown et al. (1985b)
$KAlSi_3O_8$	298	x-ray rdf	T-O[5]	4[1]	1.63	146	Taylor & Brown (1979a)
$CaAl_2Si_2O_8$ (AN)	298	x-ray rdf	T-O[5]	4[1]	1.66	143	Taylor & Brown (1979a)
	1873	x-ray rdf			1.68	136	Okuno & Marumo (1982)
$NaAlSi_2O_6$	298	x-ray rdf	T-O[5]	4[1]	1.62	--	Taylor & Brown (1979b)
	298	Al-EXAFS	Al-O	4[1]	1.78	--	Brown et al. (1982); McKeown et al. (1985b)
$NaAlSiO_4$	298	x-ray rdf	T-O[5]	4[1]	1.62	--	Taylor & Brown (1979b)
$AB-H_2O$	298	x-ray rdf	T-O[5]	4[1]	1.63	148[3]	Okuno et al. (1987)
$AN-Q-H_2O$[10]	298	x-ray rdf	T-O[5]	4[1]	1.62	149[3]	Okuno et al. (1987)
$NaFeSi_3O_8$	298	Fe-EXAFS	Fe(III)-O	4[1]	1.87	--	Brown et al. (1978)
	298	x-ray rdf	T-O[11]	4[1]	1.70	140	Henderson et al. (1984)
$KFeSi_3O_8$	298	x-ray rdf	T-O[11]	4[1]	1.70	152	Fleet et al. (1984)
$NaGaSiO_4$	298[12]	Ga-EXAFS	Ga-O	4[1]	1.83	--	Fleet et al. (1984)
	298[13]	Ga-EXAFS	Ga-O	4[1]	1.83	--	Fleet et al. (1984)
$NaGaSi_3O_8$	298	Ga-EXAFS	Ga-O	4[1]	1.88	--	Okuno et al. (1984)
	298	x-ray rdf	T-O[14]	4[1]	1.74	128	Okuno et al. (1984)
$NaAlGe_3O_8$	298	Ge-EXAFS	Ge-O	4[1]	1.76	--	Okuno et al. (1984)
	298	x-ray rdf	T-O[15]	4[1]	1.76	128	Okuno et al. (1984)
$NaGaGe_3O_8$	298	Ge-EXAFS	Ge-O	4[1]	1.75	--	Okuno et al. (1984)
	298	Ga-EXAFS	Ga-O	4[1]	1.88	--	Okuno et al. (1984)
	298	x-ray rdf	T-O[16]	5[1]	1.78	124	Okuno et al. (1984)
$KGaGe_3O_8$	298	x-ray rdf	T-O[16]	5[1]	1.78	126	Okuno et al. (1986a)
$Na_{0.5}Ca_{0.5}$ $Ga_{1.5}Ge_{2.5}O_8$	298	x-ray rdf	T-O[16]	5[1]	1.82	120	Okuno et al. (1990)
	298	Ge-EXAFS	Ge-O	4[1]	1.76	--	Okuno et al. (1990)
	298	Ga-EXAFS	Ga-O	4[1]	1.83	--	Okuno et al. (1990)
$CaGa_2Ge_2O_8$	298	x-ray rdf	T-O[16]	5[1]	1.83	125	Okuno et al. (1990)
	298	Ge-EXAFS	Ge-O	4[1]	1.76	--	Okuno et al. (1990)
	298	Ga-EXAFS	Ga-O	4[1]	1.85	--	Okuno et al. (1990)

Network-Modifying Cations (alkali cations in silicate glasses and melts)

Composition	T (K)	Method	Bond	Cation CN	d(M-O) (Å)	Reference
33 Li$_2$O-67 SiO$_2$	298	x-ray rdf	Li-O	4[1]	2.07	Waseda & Suito (1977)
	298	neutron rdf		4[1]	2.07	Waseda & Suito (1977)
	1423	x-ray rdf		4[1]	2.08	Waseda & Suito (1977)
33 Na$_2$O-67 SiO$_2$	298	x-ray rdf	Na-O	6[1]	2.36	Waseda & Suito (1977)
	298	neutron rdf		6[1]	2.34	Waseda & Suito (1977)
	1273	x-ray rdf		6[1]	2.36	Waseda & Suito (1977)
Na$_2$Si$_2$O$_5$	298	Na-EXAFS	Na-O	5[1]	2.3	Greaves et al. (1981)
	298	Na-EXAFS	Na-O	6[1]	2.61	McKeown et al. (1984a)
NaAlSi$_3$O$_8$	1473	x-ray rdf	Na-O	8[17]	2.35[17]	Okuno & Marumo (1982)
NaAlSiO$_4$	298	Na-EXAFS	Na-O	5[1]	2.62	McKeown et al. (1984a)
Na$_{0.75}$Al$_{1.2}$Si$_{2.9}$O$_8$	298	Na-EXAFS	Na-O	6[1]	2.61	McKeown et al. (1984a)
Na$_2$CaSi$_5$O$_{12}$	298	Na-EXAFS	Na-O	6[1]	2.58	Greaves et al. (1981)
33 K$_2$O-67 SiO$_2$	298	x-ray rdf	K-O	≈8[1]	≈2.65	Waseda & Suito (1977)
	298	neutron rdf		≈8[1]	≈2.65	Waseda & Suito (1977)
	1373	x-ray rdf		≈8[1]	≈2.66	Waseda & Suito (1977)
K$_2$Si$_2$O$_5$	298	K-EXAFS	K-O	≈7[18]	2.8	Greaves (1992)
KCsSi$_2$O$_5$	298	K-EXAFS	K-O	≈7[18]	2.7	Greaves (1992)
KAlSi$_3$O$_8$	298	K-EXAFS	K-O	8.9	3.00	Jackson et al. (1987)
AB$_{30}$OR$_{70}$[19]	298	K-EXAFS	K-O	9.6	3.03	Jackson et al. (1987)
AB$_{50}$OR$_{50}$[19]	298	K-EXAFS	K-O	10.4	3.06	Jackson et al. (1987)
AB$_{70}$OR$_{30}$[19]	298	K-EXAFS	K-O	9.5	3.02	Jackson et al. (1987)
KCsSi$_2$O$_5$	298	Cs-EXAFS	Cs-O	8[18]	3.0	Greaves (1992)
Cs$_2$Si$_2$O$_5$	298	Cs-EXAFS	Cs-O	8[18]	3.1	Greaves (1992)
Cs$_{0.8}$Na$_{1.2}$Si$_3$O$_7$	298	x-ray rdf	Cs-O	8[18]	3.1	Hanson & Egami (1986)

Network-Modifying Cations (alkaline-earth cations in silicate glasses and melts)

Composition	T (K)	Method	Bond	Cation CN	d(M-O) (Å)	Reference
MgSiO$_3$	298	x-ray rdf	Mg-O	4	2.08	Yin et al. (1983)
				2[20]	2.50[20]	
	1993	x-ray rdf		4[1]	2.12	Waseda & Toguri (1990)
44 MgO-56 SiO$_2$	298	x-ray rdf	Mg-O	4[1]	2.14	Waseda & Toguri (1977)
	1973	x-ray rdf		4[1]	2.16	Waseda &Toguri (1977)
Mg$_3$Al$_2$Si$_3$O$_{12}$	298	x-ray rdf	Mg-O	5[1]	2.1	Okuno & Marumo (1993)
CaMgSi$_2$O$_6$	298	Mg-EXAFS	Mg-O	4[31]	1.98	Henderson et al. (1992)
CaAl$_2$Si$_2$O$_8$ - CaMgSi$_2$O$_6$	298	Mg-EXAFS	Mg-O	4[31]	1.97	Henderson et al. (1992)
CaAl$_2$Si$_2$O$_8$-Mg$_2$SiO$_4$	298		Mg-O	4[31]	1.94	Henderson et al. (1992)
K$_2$MgSi$_5$O$_{12}$	298	Mg-EXAFS	Mg-O	4[31]	1.92	Henderson et al. (1992)
"Basalt"	298	Mg-EXAFS	Mg-O	6[30]	2.07	Henderson et al. (1992)
45 CaO-55 SiO$_2$	298	x-ray rdf	Ca-O	6[1]	2.43	Waseda & Toguri (1977)
	1873	x-ray rdf		6[1]	2.41	Waseda & Toguri (1977)
CaSiO$_3$	298	x-ray rdf	Ca-O	4	2.34	Yin et al. (1986)
				2[20]	2.54[20]	
	1873	x-ray rdf		6[1]	2.35	Waseda & Toguri (1990)
CaAl$_2$Si$_2$O$_8$	298	Ca-EXAFS	Ca-O	7[1]	2.64	Binsted et al. (1985)
	298	Ca-EXAFS		6[1]	2.44	Combes et al. (1991)
	1873	x-ray rdf		8[22]	2.45[22]	Okuno & Marumo (1982)
CaMgSi$_2$O$_6$	298	Ca-EXAFS	Ca-O	8[1]	2.63	Binsted et al. (1985)
	298	Ca-EXAFS		6[1]	2.38	Combes et al. (1991)
CaAl$_2$SiO$_6$	298	Ca-EXAFS	Ca-O	6[1]	2.45	Combes et al. (1991)
CaTiSi$_2$O$_5$	298	Ca-EXAFS	Ca-O	6[1]	2.44	Combes et al. (1991)
"Rhyolite"[23]	298	Ca-EXAFS	Ca-O	6[1]	2.49	Combes et al. (1991)
Ca$_3$Mg$_4$Al$_2$Si$_7$O$_{24}$[24]	298	Ca-EXAFS	Ca-O	8[25]	2.47[25]	Hardwick et al. (1985)
Ca$_3$Al$_2$Si$_3$O$_{12}$	298	x-ray rdf	Ca-O	7[1]	2.4	Okuno & Marumo (1993)

[1] Coordination number was rounded to the nearest integer which is within the reported error range (e.g., 4.3±0.3 is rounded to 4).

[2] Angle distribution reported by the authors.

[3] T-O-T angle calculated from reported T-O and T-T distances.

[4] Numbers before parentheses are mol % of the oxide.

[5] T represents Si and Al.

[6] Distances reported are from model fits of the si(s) functions. The model used was that of 2:1 mullite.

[7] An ^{27}Al NMR study of Al_2O_3 melt (Poe et al., 1992) indicates that the average coordination of Al is [5]Al.

[8] More recent ^{27}Al NMR studies of glasses in this system (Risbud et al., 1987; Sato et al., 1991) show that they contain significant quantities of [5]Al as well as [4]Al and [6]Al. ^{27}Al NMR studies of melts in this system (Poe et al., 1992) also indicate significant quantities of [5]Al.

[9] Estimated by visual inspection of rdf curves.

[10] AN-Q-H$_2$O represents a glass composition in the system anorthite-quartz-water.

[11] T represents Si and Fe(III).

[12] Glass quenched at 0.01 GPa.

[13] Glass quenched isobarically at 2.5 Gpa.

[14] T represents Ga and Si.

[15] T represents Al and Ge.

[16] T represents Ga and Ge.

[17] Distance and coordination number derived by fitting quasi-crystalline model to si(s) data for albite-composition melt (Okuno and Marumo, 1982).

[18] Coordination number was estimated by the authors of this chapter by comparison of observed K-O and Cs-O distance with those predicted using Shannon (1976) radius values for different K and Cs coordination numbers.

[19] AB represents the composition NaAlSi$_3$O$_8$ and OR the composition KAlSi$_3$O$_8$. The subscripts represent mol %.

[20] Yin et al. (1983) reported these two more distant oxygens at 2.50 Å around Mg based on fits to a "pigeonite-like" quasi-crystalline model.

[21] Coordination number estimated by comparison of observed Mg-O distance (Henderson et al., 1992) with Mg-O distances predicted for different Mg coordination numbers using Shannon (1976) radius values (see Table 1).

[22] Distance and coordination number derived by fitting quasi-crystalline model to si(s) data for anorthite-composition melt (Okuno and Marumo, 1982).

[23] The composition of this simplified "rhyolite" glass is Na$_{1.2}$K$_{1.2}$Ca$_{0.3}$Mg$_{0.1}$Fe$_{0.5}$Al$_{3.0}$Si$_{12.6}$O$_{32}$.

[24] Represents a simplified basalt composition.

[25] Values are from coordination model E of Hardwick et al. (1985).

Network-Modifying Cations (first-row transition metal cations in silicate glasses and melts)

Composition	T (K)	Method	Bond	Cation CN	d(M-O) (Å)	Reference
Na₂TiSi₂O₇	298	Ti-EXAFS	Ti(IV)-O	5[26]	1.70, 1.94[27]	Farges et al. (1995e)
	1300	Ti-EXAFS		5[26]	1.71, 1.94[27]	Farges et al. (1995f)
K₂TiSi₂O₇	298	neutron rdf[28]	Ti(IV)-O	5[1]	1.65, 1.95[27]	Yarker et al. (1986)
	298	Ti-EXAFS		5[29]	1.64, 1.93[27]	Yarker et al. (1986)
	298	Ti-EXAFS		5[26]	1.65, 1.95[27]	Farges et al. (1995e)
	1440	Ti-EXAFS		5[26]	1.78, 1.96[27]	Farges et al. (1995f)
GKW[30]	298	Ti-EXAFS	Ti(IV)-O	5[26]	1.78, 1.96[27]	Farges et al. (1995e)
	1550	Ti-EXAFS		5[26]	1.81, 1.95[27]	Farges et al. (1995f)
Mn₂SiO₄	298	Mn-EXAFS	Mn(II)-O	4-5[31]	2.12[32]	Jackson (1991)
Mn₁.₅Fe₀.₅SiO₄	298	Mn-EXAFS	Mn(II)-O	4-5[31]	2.13[32]	Jackson (1991)
	298	Fe-EXAFS	Fe(II)-O	4-5[31]	2.04[33]	Jackson (1991)
MnFeSiO₄	298	Mn-EXAFS	Mn(II)-O	5-6[31]	2.18[32]	Jackson (1991)
	298	Fe-EXAFS	Fe(II)-O	4-5[31]	2.04[33]	Jackson (1991)
Mn₀.₅Fe₁.₅SiO₄	298	Mn-EXAFS	Mn(II)-O	5-6[31]	2.15[32]	Jackson (1991)
	298	Fe-EXAFS	Fe(II)-O	4-5[31]	2.04[33]	Jackson (1991)
Fe₂SiO₄	1573	x-ray rdf	Fe(II)-O	4[1]	2.04	Waseda & Toguri (1977)
	298	Fe-EXAFS		5[31]	2.06[33]	Jackson (1991)
	1575	Fe-EXAFS		4[1]	1.98	Jackson et al. (1993)
FeMgSiO₄	298	Fe-EXAFS	Fe(II)-O	4-5[31]	2.04[33]	Jackson (1991)
FeCaSiO₄	1525	Fe-EXAFS	Fe(II)-O	4[1]	1.95	Jackson et al. (1991)
Fe₃Al₂Si₃O₁₂	1585	Fe-EXAFS	Fe(II)-O	6[1]	2.10	Jackson et al. (1991)
Na₁.₉Fe₀.₈Si₃.₁O₈	298	Fe-EXAFS	Fe(II)-O	4-5[31]	2.06[34]	Waychunas et al. (1988)
	1123	Fe-EXAFS		4[1]	2.02[35]	Waychunas et al. (1988)
K₁.₇Fe₀.₇Si₃.₂O₈	298	Fe-EXAFS	Fe(II)-O	4-5[31]	2.04[34]	Waychunas et al. (1988)
	1173	Fe-EXAFS		4[1]	2.02[35]	Waychunas et al. (1988)
Na₁.₉Ni₀.₅Si₁.₉O₄.₇₅	298	Ni-EXAFS	Ni(II)-O	5[31]	2.00	Galoisy & Calas (1993a); Farges et al. (1994)
	1175	Ni-EXAFS		4-5[31]	1.95	Farges et al. (1994); Farges & Brown (1995b)
K₀.₆₈Zn₀.₀₈SiO₂.₇₆	298	x-ray rdf	Zn-O	4.7	1.96	Musinu & Piccaluga (1994)

Network-Modifying Cations (other cations in silicate glasses and melts)

Composition	T (K)	Method	Bond	Cation CN	d(M-O) (Å)	Reference
NaAlSi$_3$O$_8$(AB) + 2000 ppm Yb	298	Yb-EXAFS	Yb(III)-O	5[31]	2.11	Ponader & Brown (1989)
Na$_{1.7}$AlSi$_7$O$_{17}$ (PR) + 2000 ppm Yb	298	Yb-EXAFS	Yb(III)-O	6[31]	2.21	Ponader & Brown (1989)
Na$_2$Si$_3$O$_7$ (TS) +2000 ppm Yb	298	Yb-EXAFS	Yb(III)-O	6[31]	2.20	Ponader & Brown (1989)
MgCaSiO$_4$ (MO) + 3 wt % Th	298	Th-EXAFS	Th(IV)-O	8[31]	2.41	Farges (1991)
MgCaSi$_2$O$_6$ (DI) + 3 wt % Th	298	Th-EXAFS	Th(IV)-O	8[31]	2.41	Farges (1991)
Silicate Glasses[38] + 3 wt % Th	298	Th-EXAFS	Th(IV)-O	6-8[31]	2.35-2.40	Farges (1991)
Na$_{0.87}$Al$_{0.96}$Si$_{3.07}$O$_8$+ 1 wt.% Th	298	Th-EXAFS	Th(IV)-O	6[31]	2.33	Farges (1991)
Na$_2$Si$_2$O$_5$ (DS) +100 ppm -3 wt % Zr	1550	Th-EXAFS	Th(IV)-O	6[31]	2.35-2.40	Farges & Brown (1995a)
Na$_2$Si$_3$O$_7$ (TS) + 100 ppm -3 wt % Zr	1550	Th-EXAFS	Th(IV)-O	6-8[31]	2.35-2.40	Farges & Brown (1995a)
Na$_{0.87}$Al$_{0.96}$Si$_{3.07}$O$_8$+ 1 wt % U(VI)	298	U-EXAFS	U(VI)-O	6-7[39]	1.82,2.23[40]	Farges et al. (1992)
Na$_{1.7}$AlSi$_7$O$_{17}$ (PR) + 2000 ppm U(V,VI)	298	U-EXAFS	U(V,VI)-O	6-7[39]	1.78,2.22[40]	Farges et al. (1992)
Silicate Glasses[41] + 0.2-3.3 wt % U(V)	298	U-EXAFS	U(V)-O	6[31]	2.19-2.24	Farges et al. (1992)
Silicate Glasses[42] + 0.2-3.3 wt % U(IV)	298	U-EXAFS	U(IV)-O	6[31]	2.26-2.29	Farges et al. (1992)
Na$_2$Si$_2$O$_5$ (DS) + 0.5 -3 wt % U(VI)	1550	U-EXAFS	U(VI)-O	6-7[39]	1.82,2.23[40]	Farges & Brown (1995a)

Network-Modifying Cations (other cations in silicate glasses and melts)

Composition	T (K)	Method	Bond	Cation CN	d(M-O) (Å)	Reference
$Y_{0.75}AlSi_{1.2}O_2$	298	X-ray rdf[36]	Y(III)-O	5.6	2.37	Nukui et al. (1992)
$NaAlSi_3O_8$ (AB)[37] + 2000 ppm Zr	298	Zr-EXAFS	Zr(IV)-O	6-7	2.10	Farges et al. (1991a)
$Na_{1.7}AlSi_7O_{17}$ (PR) + 2000 ppm Zr	298	Zr-EXAFS	Zr(IV)-O	6[1]	2.07	Farges et al. (1991a)
$Na_2Si_2O_5$ (DS) + 0.1 to 10 wt % Zr	1550	Zr-EXAFS	Zr(IV)-O	6[31]	2.07-2.10	Farges & Brown (1995a)
$Na_2Si_3O_7$ (TS) + 0.1 to 10 wt % Zr	1550	Zr-EXAFS	Zr(IV)-O	6[31]	2.07-2.10	Farges & Brown (1995a)
$NaAlSi_3O_8$ (AB) + 2000 ppm Mo	298	Mo-EXAFS	Mo(VI)-O	4[31]	1.73	Farges et al. (1991b)
$Na_{1.7}AlSi_7O_{17}$ (PR) + 2000 ppm Mo	298	Mo-EXAFS	Mo(VI)-O	4[31]	1.75	Farges et al. (1991b)
$Na_2Si_2O_5$ (DS) + 1 to 3 wt % Mo	1210	Mo-EXAFS	Mo(VI)-O	4[31]	1.77	Farges & Brown (1995a)
$Na_2Si_3O_7$ (AB) +2000 ppm Mo	298	Mo-EXAFS	Mo(VI)-O	4[31]	1.75	Farges et al. (1991b)
$Na_2Si_3O_7$ (TS) + 1 to 3 wt % Mo	1210	Mo-EXAFS	Mo(VI)-O	4[31]	1.77	Farges & Brown (1995a)
$NaAlSi_3O_8$ (AB) + 2000 ppm La	298	La-EXAFS	La(III)-O	9[31]	2.59	Ponader & Brown (1989)
$Na_{1.7}AlSi_7O_{17}$ (PR) +2000 ppm La	298	La-EXAFS	La(III)-O	7[31]	2.43	Ponader & Brown (1989)
$Na_2Si_3O_7$ (TS) + 2000 ppm La	298	La-EXAFS	La(III)-O	7[31]	2.42	Ponader & Brown (1989)
$NaAlSi_3O_8$ (AB)+ 2000 ppm Gd	298	Gd-EXAFS		6[31]	2.30	Ponader & Brown (1989)
$Na_{1.7}AlSi_7O_{17}$ (PR) + 2000 ppm Gd	298	Gd-EXAFS	Gd(III)-O	7[31]	2.36	Ponader & Brown (1989)
$Na_2Si_3O_7$ (TS) +2000 ppm Gd	298	Gd-EXAFS	Gd(III)-O	8[31]	2.43	Ponader & Brown (1989)

[26] Coordination numbers were derived from both EXAFS analysis and the height and position of the pre-edge feature of the Ti-K absorption edge. The dominant Ti coordination polyhedron in these glasses and melts is the ([5]Ti=O)O$_4$ square pyramid.

[27] The two distances reported for these glasses and melts are for the one short axial Ti=O (titanyl) bond and the four longer equatorial Ti-O bonds of the ([5]Ti=O)O$_4$ square pyramid.

[28] The neutron rdf was derived using difference techniques on glasses isotopically substituted with [46]Ti and [48]Ti.

[29] The Ti coordination numbers were fixed (with one oxygen and four oxygens for the short axial bond and the longer equatorial bonds of the ([5]Ti=O)O$_4$ square pyramid, respectively, during least squares refinement of the Ti-O distances.

[30] GKW stands for the composition Ca$_{0.11}$Ti$_{0.05}$Al$_{0.15}$Si$_{0.76}$O$_{1.96}$.

[31] Coordination number was estimated on the basis of EXAFS-derived Fe(II)-O distance (see Table 1) after correction of distance for anharmonic effects. The coordination numbers derived from the EXAFS amplitudes are lower because of self absorption effects in the fluorescence EXAFS measurements for these concentrated Fe and Mn compounds.

[32] The Mn(II)-O distance reported by Jackson (1991) for the glasses at 298 K were corrected for anharmonic effects by adding 0.04 Å to the reported values. The magnitude of this correction is consistent with that determined for Fe(II) in NC6 glass [see section on Fe(II)].

[33] The Fe(II)-O distance reported by Jackson (1991) for the glasses at 298 K were corrected for anharmonic effects by adding 0.04 Å to the reported values. The magnitude of this correction is consistent with that determined for Fe(II) in NC6 glass [see section on Fe(II)].

[34] The Fe(II)-O distance reported by Waychunas et al. (1988) for the glasses at 298 K were corrected for anharmonic effects by adding 0.04 Å to the reported values. The magnitude of this correction is consistent with that determined for Fe(II) in NC6 glass [see section on Fe(II)].

[35] The Fe(II)-O distance reported by Waychunas et al. (1988) for the melts at 1100-1200 K were corrected for anharmonic effects by adding 0.08 Å to the reported values. The magnitude of this correction is consistent with that determined for Fe(II) in Fe$_2$SiO$_4$ melt at a similar temperature [see section on Fe(II)].

[36] The x-ray rdf was a partial rdf based on anomalous x-ray scattering near and far from the Y K-absorption edge.

[37] The abbreviations AB, PR, DS, TS refer to the albite, peralkaline, disilicate, and trisilicate glass compositions studied by Ponader and Brown (1989), Farges et al. (1991a,b), and Farges and Brown (1995a).

[38] The Th-silicate glasses studied by Farges (1991) containing 3 wt.% Th (besides MO and DI compositions) ranged in composition from CaAl$_2$Si$_2$O$_8$ and NaAlSi$_3$O$_8$ to olivine basalt, tholeite basalt, dacite, and rhyolite compositions.

[39] U coordination consists of two closest uranyl oxygens and 4-5 more distant equatorial oxygens.

[40] Two distances reported are for the two short axial U=O bonds and the 4-5 longer axial U-O bonds.

[41] The U-bearing glasses referred to here include the AB and TS compositions as well as a calc-alkaline rhyolite composition (RH).

[42] The U-bearing glasses referred to here include the DS, TS, AN (anorthite), and DI (diopside) compositions.

CONCLUSIONS AND FUTURE PROSPECTS

X-ray scattering studies of silicate glasses over the past 60 years have led to the most widely used conceptual models of glass/melt structure, including the random network and modified crystallite models. These models provide a useful framework within which X-ray scattering studies of glasses and melts can be compared. One major assumption made in formulating these models is that silicate glasses and melts have local atomic arrangements that are dictated by the same crystal chemical principles as the local atomic arrangements in silicate crystals. This assumption has proven to be true in general, although there are exceptions in silicate glasses and melts to the coordination numbers usually adopted by cations in crystals.

In situ, high-temperature X-ray scattering studies of silicate melts have provided unique information on interatomic distances and coordination numbers (including first- and more distant neighbors) of cations in melts, as well as information on network topologies. Conventional X-ray or neutron rdf's contain contributions from all possible pair correlations in a glass or melt, thus are difficult to analyze quantitatively above about 4 Å without some type of structural model. This is true for compositionally simple melts like SiO_2 and is particularly true for compositionally complex melts of geochemical interest. Structural models can be generated by assuming that the local and medium-range bonding topologies in a crystal are like those in a melt of similar composition, then tested against the experimental rdf and interference function. Such quasi-crystalline models have proven to be very useful in interpreting rdf's of silicate glasses and melts, but they are not unique. Structural models of silicate glasses and melts can also be generated using molecular dynamics or Monte Carlo simulations, but they must be compared with experimental rdf's and X-ray interference functions to verify their utility.

The use of very intense synchrotron X-ray sources to study silicate glass and melt structure has just begun (Leadbetter, 1994) and should lead to substantial improvements in data quality in X-ray scattering experiments on silicate glasses and melts as already demonstrated in a few studies. In particular, the ability to collect X-ray scattering data to very high s values should lead to substantial improvements in distance resolution in experimental X-ray rdf's. Synchrotron X-ray sources open several new avenues of investigation of glass and melt structure, including studies of silicate melts at high pressures and time-resolved high-temperature studies of transient phenomena such as nucleation and growth of crystals and structural fluctuations that result in phase separation. The ability to collimate a synchrotron X-ray beam to a very small size and still have sufficient flux to carry out a scattering experiment will benefit high pressure studies in particular. In addition, the ability to tune the wavelength of synchrotron X-rays permits anomalous X-ray scattering on silicate glasses and melts. In such experiments, X-ray scattering data are collected at an X-ray energy near the absorption edge of a selected element and at an energy away from that edge. Subtraction of the two data sets and Fourier transformation provides pair correlations involving the element of interest. This information is similar to that provided by "double-difference" neutron scattering studies on isotopically substituted glasses or melts. Such experiments are particularly useful in probing medium-range structure, and "double-difference" neutron scattering studies have already led to the concept of densely packed domains of network modifiers in silicate glasses.

Although only about 20 years have elapsed since the first X-ray absorption spectroscopy experiment at a synchrotron radiation laboratory, a great deal has been learned about the local coordination environments of cations in silicate glasses and melts using XAS. Over the past decade, high-temperature XAS studies of cations in silicate melts have contributed unique information on their local environments and have shown that cation environments in silicate glasses are not always the same as those in melts of the same composition. XAS has several major attributes relative to conventional X-ray scattering methods. It can probe the structural environment of a wide range of elements, including those that are "spectroscopically silent" using other techniques. XAS, particularly in the fluorescence detection mode, can be used to study specific elements in glasses and melts at very dilute concentrations (≤ 100 ppm). X-ray scattering and many other spectroscopic methods do not have this sensitivity. Another significant attribute is the fact that a radial distribution function derived by Fourier transforming a normalized and background-subtracted EXAFS spectrum contains only pair correlations involving the element of interest. This attribute greatly simplifies interpretation of the rdf relative to one derived from conventional X-ray scattering. However, one limitation of XAS relative to X-ray scattering is the short distance range accessible due to the relatively short mean free path of electrons in the EXAFS energy region. Thus XAS is a short-range structural probe which rarely "sees" atoms at a distance of more than ~4 Å from a central absorber in a glass or melt. Nonetheless, information from XAS experiments can be combined with crystal chemical principles, such as bond valence-bond length relationships and Pauling's rules, to develop plausible models of medium-range structure for silicate glasses and melts. The modified random network model was derived from a consideration of medium-range structural information from XAS experiments on binary alkali silicate glasses.

Structural information on glasses and melts derived from XAS studies must be extracted with care, because both positional and thermal disorder can result in interatomic distances and coordination numbers that are artificially small. Over the past five years, several methods have been developed to take these "anharmonic" effects into account, and accurate structural data can now be derived from XAFS studies of cations in glasses and melts, even at temperatures above 2000 K.

X-ray absorption near edge structure (XANES) spectra provide structural information on cations in silicate glasses and melts complementary to that provided by EXAFS. XANES spectra are also sensitive to the oxidation state of a cation as well as the symmetry of its first coordination sphere. XANES spectra also have significantly more amplitude than EXAFS spectra, particularly for cations at very dilute concentrations. Until quite recently, XANES spectra have been used mainly in the "fingerprint" mode because of the lack of a sufficiently accurate theory and accessible computer code that could calculate XANES spectra which match experimental spectra. This situation has changed dramatically within the last year, and it is now possible to calculate accurate XANES spectra for most cations in oxides and silicates. XANES spectra are particularly sensitive to the local and medium-range environments of a cation because they are produced by multiple scattering of the photoelectron among the atoms in the extended environment around the absorbing cation. Also, XANES is not as sensitive to anharmonic effects as EXAFS, so no anharmonic corrections are required in XANES studies of glasses and melts. This multiple-scattering theory of XANES has been used to develop structural models of both the short- and medium-range environments of cations in silicate glasses and melts.

Third-generation synchrotron X-ray sources should extend the sensitivity of XAS studies to even lower element concentrations than currently detectable with existing sources. Conventional fluorescence XAFS methods can detect ≥ 20 to 30 ppm currently, using wiggler-magnet beamlines on second-generation synchrotron radiation sources, but with third-generation sources, this limit will likely be lowered to ≤ 1 ppm. Higher fluxes will certainly help to improve the signal-to-noise ratio of EXAFS spectra collected at high k-values (12 to 20 Å^{-1}), where the intensity of EXAFS from cations in silicate glasses and melts is usually too low to be useful with current synchrotron X-ray sources. This k region is critical for better defining the medium-range structure of silicate glass/melts. Therefore, some improvement in the definition of medium-range structure in these materials based on EXAFS measurements can be expected in the next few years. These new synchrotron X-ray sources will also permit time-resolved studies of transient phenomena in silicate melts. XAS data on glasses and melts can now be collected quite rapidly (a few seconds per scan) using the dispersive XAS method, but data must be collected in transmission, which ultimately limits the concentration of an element to greater than several thousand ppm. "Quick-EXAFS" can be used in the fluorescence mode, thus can detect much lower concentrations and may make it possible to carry out time-resolved studies of trace elements in melts on third-generation synchrotron radiation sources.

Another area which is certain to develop over the next few years is the study of structural environments of low Z elements (Na, Mg, Al, Si) in silicate melts using XAS. Such studies must be done under vacuum because of the strong attenuation of soft X-rays by air. Vacuum-compatible furnaces and fluorescence detectors have been in use for some time for solid surface studies, so much of the technology exists for high-vacuum melt studies. What is required is an appropriate sample holder which will not react with silicate melts and will not interfere with or absorb too much of the incident X-ray beam or fluorescent X-ray signal.

A final comment concerns the need to combine structural information on silicate melts from a number of spectroscopic and scattering methods as well as modern computational modeling in order to develop the most accurate structure-property models. Because of the complexity of silicate melts, it is unlikely that any single structural method will lead to a robust model which can be used for accurate predictions of melt properties ranging from density to viscosity.

ACKNOWLEDGMENTS

Gordon Brown thanks the U.S. National Science Foundation for long-term support of X-ray scattering and X-ray absorption spectroscopy studies of silicate glasses and melts at Stanford University, and acknowledges NSF Grant EAR-9305028 for support of recent work, including the preparation of this chapter and XAFS studies of silicate melts over the past two years. The collaborations between Brown, François Farges, and Georges Calas involving XAS studies of silicate melts began five years ago and have been greatly facilitated by a travel grant from the NSF-CNRS International Program (Grant INT-9116008) which allowed us to travel between Stanford (and SSRL) and the University of Paris (and LURE) for joint studies. Brown also wishes to acknowledge the long-term collaboration of Glenn Waychunas (Stanford University) on XAS studies of silicate glasses and melts, as well as fruitful collaborations with a number of former graduate students and post-doctoral fellows (Mark Taylor, Bernard deJong, Phil Fenn, Keith Keefer, Michael Hochella, David McKeown, Carl Ponader, Heather Boek, Bill Jackson, and Jean-Marie Combes) and with Jose Mustre de Leon (U. Mexico). Farges wishes to acknowledge collaborations with M. Madon, P. Richet, D. Andrault, G. Fiquet, Jean-Paul Itié, Alain Polian and former graduate students P.E. Petit and C. Remy. We thank Paul McMillan, Jonathan Stebbins, and Glenn Waychunas for very helpful reviews of this chapter. Finally, we thank the scientific staffs of SSRL and LURE for their technical support during our XAS studies. SSRL is supported by the U.S. DOE and NIH.

REFERENCES

Andrault D, Itié J-P, Farges F (1995) High temperature structural study of germanates, perovskites and wollastonites. Eur J Mineral (submitted)

Aoki N, Yambe S, Inoue H, Hasegawa H, Yasui I (1986) An X-ray diffraction study of the structure of Na_2O, Al_2O_3, $2SiO_2$ glass. Phys Chem Glasses 27:124-127

Aur S, Egami T, Berkowitz AE, Walter JL (1982) Atomic structure of amorphous particles produced by spark erosion. Phys Rev B26:6355-6361

Bair RA, Goddard WA (1980) Ab initio studies of the X-ray absorption edge in copper complexes. 1. Atomic Cu^{2+} and $Cu(II)Cl_2$. Phys Rev B22:2767-2776

Balyuzi HHM (1975) Analytical approximations to incoherently scattered X-ray intensities. Acta Cryst A31:600-602

Bartenev GM (1970) The Structure and Properties of Inorganic Glasses. 246 p. Wolters-Noordhoff, Groningen, The Netherlands

Bianconi A, Fritsch E, Calas G, Petiau J (1985a) X-ray absorption near-edge structure of 3-d transition elements in tetrahedral coordination. The effects of bond-length variation. Phys Rev B32:4292-4295

Bianconi A, Garcia J, Marcelli A, Benfatto M, Natoli CR, Davoli I (1985b) Probing higher order correlation functions in liquids by XANES (X-ray Absorption Near Edge Structure). J Physique C9 46:101-106

Bienenstock A, Bagley BG (1966) Calculation of upper bounds to the small-angle scattering from crystallite models of amorphous materials. J Appl Phys 37:4840-4847

Binstead N, Greaves GN, Henderson CMB (1985) An EXAFS study of glassy and crystalline phases of compositions $CaAl_2Si_2O_8$ (anorthite) and $CaMgSi_2O_6$ (diopside). Contrib Mineral Petrol 89:103-109

Birle JD, Gibbs GV, Moore PB, Smith JV (1968) Crystal structures of natural olivines. Am Mineral 53:807-824

Bobrovskii AB, Kartmazov N, Finkel VA (1973) (no title) Izvestia Akademia Nauk. SSSR (Neorganic Materials) 9:1075-1076

Boon JA, Fyfe WS (1972) The coordination number of ferrous ions in silicate glasses. Chem Geol 10:287-298

Bouhifd M-A (1995) Propriétés thermodynamiques de minéraux et liquides d'intérêt géophysique. Thèse de l'Université Denis Diderot, Paris (Ph.D. dissertation, in French)

Brese NE, O'Keeffe M (1991) Bond-valence parameters for solids. Acta Cryst B47: 192-197

Brown GE Jr, Calas G, Waychunas G, Petiau J (1988) X-ray absorption spectroscopy and its applications in mineralogy and geochemistry. In: Spectroscopic Methods in Mineralogy and Geology, FC Hawthorne (ed) Rev Mineral, Vol. 18, p 431-512. Mineral Soc Am, Washington, DC

Brown GE Jr, Dikmen FD, Waychunas GA (1983) Total electron yield K-XANES and EXAFS investigation of aluminum in amorphous and crystalline alumino-silicates. Stanford Synchrotron Radiation Lab Report 83/01:148-149

Brown GE Jr, Gibbs GV (1969) Oxygen coordination and the Si-O bond. Am Mineral 54:1528-1539

Brown GE Jr, Gibbs GV (1970) Stereochemistry and ordering in the tetrahedral portion of silicates. Am Mineral 55:1587-1607

Brown GE Jr, Gibbs GV, Ribbe PH (1969) The nature and variation in length of the Si-O and Al-O bonds in framework silicates. Am Mineral 54:1044-1061

Brown GE Jr, Keefer KD, Fenn PM (1978) Extended X-ray absorption fine structure (EXAFS) study of iron-bearing silicate glasses. Prog Abstr Ann Mtg Geol Soc Am, p 373 (abstract)

Brown GE Jr, Sueno S, Prewitt CT (1973) A new single-crystal heater for the precession camera and four-circle diffractometer. Am Mineral 58:698-704

Brown GE Jr, Waychunas GA, Ponader CW, Jackson WE, McKeown DA (1986) EXAFS and NEXAFS studies of cation environments in oxide glasses. J Physique C8 47:C8661-C8668

Brown ID (1981) The bond valence method: An empirical approach to chemical structure and bonding. In: Structure and Bonding in Crystals, Volume II, M O'Keeffe, A Navrotsky (eds) p 1-30. Academic Press, New York

Brown ID, Altermatt D (1985) Bond-valence parameters obtained from a systematic analysis of the inorganic crystal structure database. Acta Cryst B41:244-247

Brown ID, Shannon RD (1973) Empirical bond-strength-bond-length curves for oxides. Acta Cryst A29:266-282

Burns RG (1993) Mineralogical Applications of Crystal Field Theory, 2nd Edition, p 312-323, Cambridge University Press, Cambridge, UK

Burns RG, Fyfe WS (1964) Site preference energy and selective uptake of transition metal ions during magmatic crystallization. Science 144:1001-1003

Calas G (1979) Etude expérimentale du comportement de l'uranium dans les magmas: états d'oxydation et coordinance. Geochim Cosmochim Acta 43:1521-1531

Calas G, Brown GE Jr, Farges F, Galoisy L, Itié J-P, Polian A (1995) Cations in glasses under ambient and non-ambient conditions. Nucl Instr Meth Phys Res B (in press)

Calas G, Brown GE Jr, Waychunas GA, Petiau J (1987) X-ray absorption spectroscopic studies of silicate glasses and minerals. Phys Chem Minerals 15:19-29

Calas G, Petiau J (1983) Structure of oxide glasses. Spectroscopic studies of local order and crystallochemistry. Geochemical implications. Bull Minéral 106:33-55

Cartz L (1964) An X-ray diffraction study of the structure of silica glass. Zeit Krist 120:241-269

Combes J-M, Brown GE Jr, Waychunas GA (1991) X-ray absorption study of the local Ca environment in silicate glasses. In SS Hasnain (ed) XAFS VI, Sixth Int'l Conf on X-ray Absorption Fine Structure, p 312-314, Ellis Horwood, Chichester, UK

Cooney TF, Sharma SK (1990) Structure of glasses in the systems Mg_2SiO_4-Fe_2SiO_4, Mn_2SiO_4-Fe_2SiO_4, Mg_2SiO_4-$CaMgSiO_4$, and Mn_2SiO_4-$CaMnSiO_4$. J Non-Crystal Solids 122:10-32

Cormier L, Galoisy L, Calas G, Gaskell PH (1995) Medium-range order studies of oxide glasses by EXAFS and neutron diffraction. Ber Deutsch Mineral Gesell 1:49 (abstr)

Corrias A, Magini M, de Moraes M, Sedda AF, Musinu A, Paschina G, Piccaluga G (1986) X-ray diffraction investigation of Co(II) ions in borosilicate glasses. J Chem Phys 84:5769-5774

Creux S, Bouchet B, Gaskell PH (1995) Anomalous X-ray scattering study of medium-range order in SrO-SiO_2 and SrO-Al_2O_3-$4SiO_2$ glasses. J Non-Crystal Solids (in press).

Cromer DT, Mann JB (1968) X-ray scattering factors computed from numerical Hartree-Fock wave functions. Acta Cryst A2:321-324

Crozier ED (1995) Impact of the asymmetric pair distribution function in the analysis of XAFS. Physica B 208-209:330-333

Crozier ED, Rehr JJ, Ingalls R (1988) Amorphous and liquid systems. In: X-ray Absorption. Principles, Applications, Techniques of EXAFS, SEXAFS and XANES, Chemical Analysis, Vol. 92, DC Koningsberger, R Prins (eds) p 373-442. John Wiley & Sons, New York

Daniel I, Gillet P, Poe BT, McMillan PF (1995) In situ high-temperature Raman spectroscopic studies of aluminosilicate liquids. Phys Chem Minerals 22:74-86

DaSilva JRG, Pinatti DG, Anderson CE, Rudee ML (1974) A refinement of the structure of vitreous silica. Phil Mag 31:713-717

Davoli I (1990) Principles and recent developments of XANES spectroscopy. In: Absorption Spectroscopy in Mineralogy, A Mottana, F Burragato (eds) p 206-226, Elsevier, Amsterdam

Davoli I, Paris E, Benfatto M, Seifert F (1992a) XAS study on densified SiO_2 glasses. In: The Physics of Non-Crystalline Solids, LD Pye, WC LaCourse, HJ Stevens (eds) p 484-488. Taylor & Francis, London

Davoli I, Paris E, Stizza S, Benfatto M, Fanfoni M, Gargano A, Bianconi A, Seifert F (1992b) Structure of densified vitreous silica: silicon and oxygen XANES spectra and multiple scattering calculations. Phys Chem Minerals 19:171-175

Debye P (1915) Zerstreuung von Röntgenstrahlen. Ann Physik 46:809-823

deJong BHWS, Brown GE Jr (1980a) Polymerization of silicate and aluminate tetrahedra in glasses, melts and aqueous solutions: I. Electronic structures of $H_6Si_2O_7$, $H_6Al_2O_7^{1-}$ and $H_6Al_2O_7^{2-}$. Geochim Cosmochim Acta 44:491-511

deJong BHWS, Brown GE Jr (1980b) Polymerization of silicate and aluminate tetrahedra in glasses, melts and aqueous solutions: II. The network modifying effects of Mg^{2+}, K^+, Na^+, Li^+, H^+, OH^-, F^-, Cl^-, H_2O, and CO_2 in silicate melts. Geochim Cosmochim Acta 44:1627-1642

Dietzel A (1942) Die Kationenfeldstärken und ihre Beziehungen zu Entglasungs-vorgängen, zur Verbind-ungsbildung und zu den Schmelzpunkten von Silikaten. Zeit Electrochemie 48:9-23

Dingwell DB, Paris E, Seifert F, Mottana A, Romano C (1994) X-ray absorption study of Ti-bearing silicate glasses. Phys Chem Minerals 21:501-509

Dumas T, Petiau J (1986) EXAFS study of titanium and zinc environments during nucleation in a cordierite glass. J Non-Crystal Solids 81:201-220

Dumas T, Ramos A, Gandais M, Petiau J (1985) Role of zirconium in nucleation and crystallization of a (SiO_2, Al_2O_3, MgO, ZnO) glass. Mat Sci Letters 4:129-132

Eckersley MC, Gaskell PH, Barnes AC, Chieux P (1988) Structural ordering in a calcium silicate glass. Nature 335:525-527

Eisenberger P, Brown GS (1979) The study of disordered systems by EXAFS: limitations. Solid State Comm 29:481-484

Farber DL, Williams Q (1992) Pressure-induced coordination changes in alkali-germanate melts: an in situ spectroscopic investigation. Science 256:1427-1430

Farges F (1991) Structural environment around Th^{4+} in silicate glasses. Implications for the geochemistry of incompatible Me^{4+} elements. Geochim Cosmochim Acta 55:3303-3319

Farges F (1994) Ordre local autour d'éléments fortements chargés dans des silicates amorphes: métamictes, vitreux et fondus. Habilitation à diriger les recherches, Université Marne-la-Vallée, Noisy le Grand, France (in French)

Farges F (1995) Does Zr-F complexation occur in magmas? Chem Geol (in press)

Farges F, Andrault D, Itié J-P (1995b) Crystal chemistry of Ni and Zn in molten phosphates and silicates using high temperature XAFS up to 2000 K. Ber Deutsch Mineral Gesell 1:62 (abstract)

Farges F, Brown GE Jr (1995a) Coordination chemistry of trace levels of Zr, Mo, Th, and U in synthetic silicate melts and natural glasses—An in situ, high-temperature XAFS spectroscopy study. EOS Trans Am Geophys Union (in press)

Farges F, Brown GE Jr (1995b) An empirical model for the anharmonic analysis of high-temperature XAFS spectra of oxide compounds with applications to the coordination environment of Ni in Ni-olivine and Ni-Na-disilicate glass and melt. Chem Geol (in press)

Farges F, Brown GE Jr, Calas G, Galoisy L, Waychunas GA (1994) Structural transformation in Ni-bearing $Na_2Si_2O_5$ glass and melt. Geophys Res Letters 21:1931-1934

Farges F, Brown GE Jr, Calas G, Galoisy L, Waychunas GA (1995c) Local structure change around 2 wt % of Ni in a $Na_2Si_2O_5$ melt. Physica B 208-209:381-382

Farges F, Brown GE Jr, Navrotsky A, Gan H, Rehr JJ (1995e) Coordination chemistry of Ti(IV) in silicate glasses and melts. II. Glasses under ambient conditions. Geochim Cosmochim Acta (submitted)

Farges F, Brown GE Jr, Navrotsky A, Gan H, Rehr JJ (1995f) Coordination chemistry of Ti(IV) in silicate glasses and melts. III. Glasses and melts between 293 and 1650 K. Geochim Cosmochim Acta (submitted)

Farges F, Brown GE Jr, Rehr JJ (1995d) Coordination chemistry of Ti(IV) in silicate glasses and melts. I. XAFS study of Ti coordination in oxide model compounds. Geochim Cosmochim Acta (submitted)

Farges F, Fiquet F, Andrault D, Itié J-P (1995g) In situ high temperature XAFS and anharmonicity. Physica B 208-209:269-272

Farges F, Itié J-P, Fiquet G, Andrault D (1995a) A new device for XAFS data collection up to 2000 K (or 3700 K under vacuum). Nuc Instr Meth Phys Res B 97:155-161

Farges F, Ponader CW, Brown GE Jr (1991a) Structural environments of incompatible elements in silicate glass/melt systems: I. Zr at trace levels. Geochim Cosmochim Acta 55:1563-1574

Farges F, Ponader CW, Brown GE Jr (1991b) EXAFS study of the structural environments of trace levels of Zr^{4+}, Mo^{6+} and $U^{6+}/U^{5+}/U^{4+}$ in silicate glass/melts systems. In SS Hasnain (ed) XAFS VI, Sixth Int'l Conf on X-ray Absorption Fine Structure, p 309-311, Ellis Horwood, Chichester, UK

Farges F, Ponader CW, Calas G, Brown GE Jr (1992) Local environment around incompatible elements in silicate glass/melt systems. II: U(VI), U(V) and U(IV). Geochim Cosmochim Acta 56:4205-4220

Farnan I, Grandinetti PJ, Baltisberger JH, Stebbins JF, Werner U, Eastman MA, Pines A (1992) Quantification of the disorder in network-modified silicate glasses. Nature 358:31-35.

Farnan I, Stebbins JF (1990) A high temperature ^{29}Si NMR investigation of solid and molten silicates. J Am Ceram Soc 112:32-39

Filipponi A, diCicco A (1994) Development of an oven for X-ray absorption measurements under extremely high temperature conditions. Nucl Instr Meth Phys Res B 93:302-310

Filipponi A, diCicco A, Tyson TA, Natoli CR (1991) *Ab initio* modeling of X-ray absorption spectra. Solid State Comm 78:265-275

Fiske PS, Stebbins JF (1994) The structural role of Mg in silicate liquids: A high-temperature [25]Mg, ^{23}Na, and ^{29}Si NMR study. Am Mineral 79:848-861

Fleet ME, Herzberg CT, Henderson GS, Crozier ED, Osborne MD, Scarfe CM (1984) Coordination of Fe, Ga and Ge in high pressure glasses by Mössbauer, Raman, and X-ray absorption spectroscopy, and geological implications. Geochim Cosmochim Acta 48:1455-1466

Fox KE, Furukawa T, White WB (1982) Transition metal ions in silicate melts. Part 2. Iron in sodium silicate glasses. Phys Chem Glasses 23:169-178

Frankenheim ML (1835) Die Lehre von der Cohäsion, Breslau, p 389. (as quoted in Wright, 1994)

Frenkel AI, Rehr JJ (1993) Thermal expansion and X-ray absorption fine-structure cumulants. Phys Rev B48:585-588

Friebele EJ, Griscom DL, Stapelbroek M, Weeks RA (1979) Fundamental defect centers in glass: The peroxy radical in irradiated, high-purity, fused silica. Phys Rev Letters 42:1346-1349

Galeener FL (1982) Planar rings in glasses. Solid State Comm 44:1037-1040

Galoisy L, Calas G (1991) Spectroscopic evidence for five-coordinated Ni in $CaNiSi_2O_6$ glass. Am Mineral 76:1777-1780

Galoisy L, Calas G (1992) Network-forming Ni in silicate glasses. Am Mineral 77:677-680

Galoisy L, Calas G (1993a). Structural environment of nickel in silicate glass/melt systems. Part I. Spectroscopic determination of coordination states. Geochim Cosmochim Acta 57:3613-3626

Galoisy L, Calas G (1993b) Structural environment of Ni in silicate glass/melt systems: Part 2. Geochemical implications. Geochim Cosmochim Acta 57:3627-3633

Gaskell PH (1991a) Models for the structure of amorphous solids. Glass Sci Tech 5:175-278

Gaskell PH (1991b) The structure of silicate glasses and crystals - towards a convergence of views. In: Trans Am Cryst Assoc 27:95-112, AC Wright (ed) Polycrystal Book Service, P.O. Box 3439, Dayton, OH

Gaskell PH, Eckersley MC, Barnes AC, Chieux P (1991) Medium range order in the cation distribution of a calcium silicate glass. Nature 350:675-677

Gaskell PH, Zhao J, Calas G, Galoisy L (1992) The structure of mixed cation oxide glasses. In: The Physics of Non-Crystalline Solids, LD Pye, WC LaCourse, HJ Stevens (eds) p 53-58, Taylor & Francis, London

Gibbs GV, Meagher EP, Newton MD, Swanson DK (1981) A comparison of experimental and theoretical bond length and angle variations for minerals, inorganic solids, and molecules. In: Structure and Bonding in Crystals I, M O'Keeffe, A Navrotsky (eds) p 195-225. Academic Press, New York

Goldman DS, Berg JI (1980) Spectral study of ferrous iron in Ca-Al-borosilicate glass at room temperature and melt temperatures. J Non-Crystal Solids 38/39:183-188

Goldschmidt VM (1926) Geochemische Verteilungsgesetze der Elemente. Skrifter Norske Videnskaps Akademy (Oslo), I. Math-naturwiss Kl. No. 8:7-156

Goodman CHL (1982) Constitutional diagrams and glass formation. In: The Structure of Non-Crystalline Materials, PH Gaskell, JM Parker, EA Davis (eds) p 151-163, Taylor & Francis, London

Greaves GN (1985) EXAFS and the structure of glass. J Non-Crystal Solids 71:203-217

Greaves GN (1992) Glass structure and ionic transport. In: The Physics of Non-Crystalline Solids, LD Pye, WC LaCourse, HJ Stevens (eds) p 453-459, Taylor & Francis, London

Greaves GN, Fontaine A, Lagarde P, Raoux D, Gurman SJ (1981) Local structure of silicate glasses. Nature 293:611-616

Greaves GN, Gurman SJ, Catlow CRA, Chadwick A., Houde-Walter S, Henderson CMB, Dobson BR (1991) A structural basis for ionic diffusion in oxide glasses. Phil Mag A64:1059-1072

Greegor R.B, Lytle FW, Ewing RC, Haaker RF (1984) Ti-site geometry in metamict-, annealed, and synthetic complex Ti-Nb-Ta oxides by X-ray absorption spectroscopy. Nucl Instr Meth Phys Res B 1:587-591

Greegor RB, Lytle FW, Sandstrom DR, Wong J, Schultz P (1983) Investigation of TiO_2-SiO_2 glasses by X-ray absorption spectroscopy. J Non-Crystal Solids 55:27-43

Grunes LA (1983) Study of the K edges of 3d transition metals in pure and oxide form by X-ray absorption spectroscopy. Phys Rev B27:2111-2131

Hanson CD, Egami T (1986) Distribution of Cs^+ ions in single and mixed alkali glasses from energy dispersive X-ray diffraction. J Non-Crystal Solids 87:171-184

Hardwick A, Whittaker EJW, Diakun GP (1985) An extended X-ray absorption fine structure (EXAFS) study of the calcium site in a model basaltic glass, $Ca_3Mg_4Al_2Si_7O_{24}$. Mineral Mag 49:25-29

Hawthorne FC (ed) (1988) Spectroscopic Methods in Mineralogy and Geology, Rev Mineral 18, 698 p, Mineral Soc Am, Washington, DC

Hawthorne FC, Groat LA, Raudsepp M, Ball NA, Kimata M, Spike FD, Gaba R, Halden NM, Lumpkin GR, Ewing RC, Greegor RB, Lytle FW, Ercit TS, Rossman GR, Wicks FJ, Ramik RA Sherriff BL, Fleet ME, McCammon C (1991) Alpha-decay damage in titanite. Am Mineral 76:370-396

Hazen RM, Finger LW (1982) Comparative Crystal Chemistry. Temperature, Pressure, Composition and the Variation of Crystal Structure. 231 p. John Wiley & Sons, New York

Henderson CMB, Charnock JM, Helz GR, Kohn SC, Pattrick RAD, Vaughan DJ (1991) EXAFS in earch sciences research. In: SS Hasnain (ed) X-ray Absorption Fine Structure, p. 573-578. Ellis Horwood Ltd., Chichester, UK

Henderson CMB, Charnock JM, van der Laan G, Schreyer W (1992) X-ray absorption spectroscopy of Mg in minerals and glasses. Synchrotron Radiation: Appendix to Daresbury Annual Report 1991/92, 77

Henderson GS, Fleet ME (1995) The structure of Ti-silicate glasses by micro-Raman spectroscopy. Can Mineral 33:399-408

Henderson GS, Fleet ME, Bancroft GM (1984) An X-ray scattering study of vitreous $KFeSi_3O_8$ and $NaFeSi_3O_8$ and reinvestigation of vitreous SiO_2 using quasi-crystalline modelling. J Non-Crystal Solids 68:333-349

Hess PC (1991) The role of high field strength cations in silicate melts. In: Physical Chemistry of Magmas, LL Perchuk, I Kushiro (eds) p 152-191, Springer-Verlag, New York

Hochella MF Jr, Brown GE Jr (1984) The structures of albite and jadeite composition glasses quenched from high pressure. Geochim Cosmochim Acta 49:1137-1142

Hochella MF Jr, Brown GE Jr (1985) Structure and composition of rhyolitic composition melts. Geochim Cosmochim Acta 48: 2631-2640

Hosemann R, Bagchi H (1962) Direct Analysis of Difffraction by Matter. 120 p. North-Holland Pub Co, Amsterdam

Imaoka M, Hasegawa H, Yasui, I. (1983) X-ray diffraction study of the structure of silicate glasses. Part 2. Alklai disilicate glasses. Phys Chem Glasses 24:72-78

Irvine TN, Kushiro I (1976) Partitioning of Ni and Mg between olivine and silicate liquids. Carnegie Inst Washington Yrbk 75:668-675

Ishizawa N, Miyata T, Minato I, Marumo F, Iwai S (1980) A structural investigation of α-Al_2O_3 at 2170 K. Acta Cryst B36:228-230

Itié JP, Polian A, Calas G, Petiau J, Fonatine A, Tolentino H (1989) Pressure-induced coordination changes in crystalline and vitreous GeO_2. Phys Rev Letters 63:398-401

Jackson WE (1991) Spectroscopic Studies of Ferrous Iron in Silicate Liquids, Glasses, and Crystals. Ph.D. dissertation, Stanford University, Stanford, CA, 162 p

Jackson WE, Brown GE Jr, Ponader CW (1987) X-ray absorption study of the potassium coordination environment in glasses from the $NaAlSi_3O_8$-$KAlSi_3O_8$ binary - structural implications for the mixed-alkali effect. J Non-Crystal Solids 93:311-322

Jackson WE, Waychunas GA, Brown GE Jr, Mustre de Leon J, Conradson S, Combes J-M (1993) High-temperature XAS study of Fe_2SiO_4: evidence for reduced coordination of ferrous iron in the liquid. Science 262:229-233

Jackson WE, Waychunas GA, Brown GE Jr, Mustre de Leon J, Conradson S, Combes J-M (1991) In situ high-temperature X-ray absorption study of ferrous iron in orthosilicate crystals and liquids. In: SS Hasnain (ed) X-ray Absorption Fine Structure, p 298-301. Ellis Horwood Ltd., Chichester, UK

Jackson WE, Yeager M, Mabrouk PA, Waychunas GA, Farges F, Solomon EI, Brown GE Jr (1995) Spectroscopic study of ferrous iron in amorphous silicates: implications for the structure of silicate melts of geochemical interest. J Am Chem Soc (submitted)

Keen DA, McGreevy RL (1990) Structural modelling of glasses using reverse Monte Carlo simulation. Nature 344:423-425

Keppler H (1992) Crystal field spectra and geochemistry of transition metal ions in silicate melts and glasses. Am Mineral 77:62-75

Keppler H (1993) Influence of fluorine on the enrichment of high field strength trace elements in granitic rocks. Contrib Mineral Petrol 114:479-488

Keppler H, Rubie DC (1993) Pressure-induced coordination changes of transition-metal ions in silicate melts. Nature 364:54-56

Kimata M, Ii N (1981) The crystal structure of synthetic akermanite, $Ca_2MgSi_2O_7$. N Jahrb Mineral Monat:1-10

Klug HP, Alexander LE (1954) X-ray Diffraction Procedures for Polycrystalline and Amorphous Materials. 716 p. John Wiley & Sons, New York

Kohn SC, Charnock JM, Henderson CMB, Greaves GN (1990) The structural environments of trace elements in dry and hydrous silicate glasses. A manganese and strontium K-edge X-ray absorption spectroscopic study. Contrib Mineral Petrol 105: 359-368

Koningsberger DC, Prins R, Eds. (1988) X-ray Absorption: Principles, Applications, Techniques of EXAFS, SEXAFS, and XANES. 673 p. John Wiley & Sons, New York

Konnert JH, Karle J (1972) Tridymite-like structure in silica glass. Nature (Phys Sci) 236:92-93

Konnert JH, Karle J (1973) The computation of radial distribution functions for glassy materials. Acta Cryst A29:702-710

Konnert JH, D'Antonio P, Karle J (1982) Comparison of radial distribution functions for silica glass with those for various bonding topologies: use of correlation functions. J Non-Crystal Solids 53:135-141

Konnert JH, Karle J, Ferguson G (1973) Crystalline ordering in silica and germania glasses. Science 181:177-179

Krogh-Moe J (1956) A method for converting experimental X-ray intensities to an absolute scale. Acta Cryst 9:951-953

Kubicki JD, Lasaga AC (1991) Molecular dynamics simulations of pressure and temperature effects on $MgSiO_3$ and Mg_2SiO_4 melts and glasses. Phys Chem Minerals 17:661-673

Kutzler FW, Natoli CR, Misemer DK, Doniach S, Hodgson KO (1980) Use of one elecron theory for the interpretation of near edge structure in K-shell X-ray absorption spectra of transition metal complexes. J Chem Phys 73:3274-3288

Lager GA, Meagher EP (1978) High-temperature structural study of six olivines. Am Mineral 63:365-377

Lange AR, Navrotsky A (1993) Heat capacities of TiO_2-bearing silicate liquids: evidence for anomalous changes in configurational entropy with temperature. Geochim Cosmochim Acta 51:2931-2946

Lapeyre C, Petiau J, Calas G, Gautier F, Gombert J (1983) Ordre local autour du germanium dans les verres du systeme SiO_2-GeO_2-B_2O_3-Na_2O: etude par spectrometrie d'absorption X. Bull Minéral 106:77-85

Leadbetter AJ (1994) The role of large facilities in understanding silicate glasses. J Non-Crystal Solids 179:116-124

Leadbetter AJ, Wright AC (1972) Diffraction studies of glass structure. I. Theory and quasi-crystalline models. J Non-Crystal Solids 7:23-36

Lebedev AA (1921) Trudy Gos, Otical Institute 2:1 (as quoted in Bartenev, 1970)

Levy HA, Danford MD, Narten AH (1966) Data collection and evaluation with an X-ray diffractometer designed for the study of liquid structure. ORNL-3960, Oak Ridge National Laboratory, Oak Ridge, TN

Li D, Bancroft GM, Kasrai M, Fleet ME, Feng XH, Tan KH (1994) High-resolution Si and P *K*- and *L*-edge XANES spectra of crystalline SiP_2O_7 and amorphous SiO_2-P_2O_5. Am Mineral 79:785-788

Li D, Bancroft GM, Fleet ME, Feng XH (1995) Silicon *K*-edges XANES spectra of silicate minerals. Phys Chem Minerals 22:115-122

Lorch E (1969) Neutron diffraction by germania, silica and radiation-damaged silica glasses. J Phys C (Solid State Phys) 2:229-237

Lowenstein W (1954) The distribution of aluminum in the tetrahedra of silicates and aluminates. Am Mineral 39:92-96

Lytle FW (1989) Experimental X-ray absorption spectroscopy. In: Applications of Synchrotron Radiation, H Winick et al. (eds) p 135-223. Gordon and Breach, New York

Lytle FW, Greegor RB, Sandstrom DR, Marques DR, Wong J, Spiro CL, Huffman GP, Huggins FE (1984) Measurement of soft X-ray absorption spectra with a fluorescence ion chamber detector. Nucl Instr Meth Phys Res 226:542-548

Marumo F, Morikawa H, Shimizugawa Y, Tokonami M, Miyake M, Ohsumi K, Sasaki S (1989) Diffractometer for synchrotron radiation structural studies of high temperature melts. Rev Sci Instr 60:2421-2424

Marumo F, Okuno M (1984) X-ray structural studies of molten silicates: anorthite and albite melts. In: Materials Science of the Earth's Interior, I Sunagawa (ed) p 25-38. Terra Scientific, Tokyo

Marumo F, Tabira Y, Mabuchi T, Morikawa H (1990) Coordinations of transition metals in amorphous silicates. In: Dynamic Processes of Material Transport and Transformation in the Earth's Interior, F Marumo (ed) p 53-65. Terra Scientific, Tokyo

Matsui Y, Kawamura K, Syono Y (1982) Molecular dynamics calculations applied to silicate systems: Molten and vitreous $MgSiO_3$ and Mg_2SiO_4 under low and high pressures. In: High Pressure Research in Geophysics, S Akimoto, M. Manghani (eds) p 511-524, Reidel, Tokyo

McKeown DA, Waychunas GA, Brown GE Jr (1985a) EXAFS and XANES study of the local coordination environment of sodium in a series of silica-rich glasses and selected minerals within the Na_2O-Al_2O_3-SiO_2 system. J Non-Crystal Solids 74:325-348

McKeown DA, Waychunas GA, Brown GE Jr (1985b) EXAFS and XANES study of the local coordination environment of aluminum in a series of silica-rich glasses and selected minerals within the Na_2O-Al_2O_3-SiO_2 system. J Non-Crystal Solids 74:349-371

McMillan PF, Wolf GH, Poe BT (1992) Vibrational spectroscopy of silicate liquids and glasses. Chem Geol 96:351-366

Meade C, Hemley RJ, Mao HK (1992) High pressure X-ray diffraction of SiO_2 glass. Phys Rev Letters 69:1387-1390

Meagher EP, Lager GA (1978) Polyhedral thermal expansion in the TiO_2 polymorphs: Refinement of the crystal structures of rutile and brookite at high temperature. Can Mineral 17:77-85

Mitchell EWJ, Poncet PFJ, Stewart RJ (1976) The ion pair distribution functions in molten rubidium chloride. Phil Mag 31:721-732

Misawa M, Price DL, Suzuki K (1980) The short-range structure of alkali disilicate glasses by pulsed neutron totral scattering. J Non-Crystal Solids 37:85-97

Miyata T, Ishizawa N, Minato I, Iwai S. (1979) Gas-flame heating equipment providing temperatures up to 2,600K for the four-circle diffractometer. J Appl Cryst 12:303-305

Morikawa H, Miwa S, Miyake M, Marumo F, Sata T (1982) Structural analysis of SiO_2-Al_2O_3 glasses. J Am Ceram Soc 65:78-81

Mozzi R, Warren BE (1969) The structure of vitreous silica. J Appl Cryst 2:164-172

Musinu A, Piccaluga G (1994) X-ray diffraction studies of multicomponent oxide glasses. J Non-Crystal Solids 177:81-90

Mustre de Leon J, Conradson SD, Bishop AR, Raistric ID, Batistic I, Jackson WE, Brown GE Jr, Waychunas GA (1991) XAFS analysis in the anharmonic limit: Applications to Hi-T_c superconductors and ferrosilicates. In: XAFS VI, Sixth International Conference on X-ray Absorption Fine Structure, SS Hasnain (ed) p 54-57. Ellis Horwood Ltd. Publishers, Chichester, UK

Mysen BO (1988) Structure and Properties of Silicate Melts. 354 p. Elsevier, Amstedam

Navrotsky A, Geisinger KL, McMillan PF, Gibbs GV (1985) The tetrahedral framework in glasses and melts- Influences from molecular orbital calculations and implications for structure, thermodynamics, and physical properties. Phys Chem Minerals 11:284-298

Natoli CR, Benfatto M (1986) A unifying scheme of interpretation of X-ray absorption spectra based on multiple scattering theory. J Physique 47 C8:11-23

Nelson C, Furukawa T, White WB (1983) Transition metal ions in glasses: network modifiers or quasi-molecular complexes? Mat Res Soc Bull 18:959-966

Nelson C, White WB (1980) Transition metal ions in silicate melts: I. Manganese in sodium silicate melts. Geochim Cosmochim Acta 44:887-893

Nelson C, White WB (1986) Transition metal ions in silicate melts: IV. Cobalt in sodium silicate and related glasses. J Mat Sci 1:130-138

Norman N (1957) A Fourier transform method for normalizing intensities. Acta Cryst 10:370-373

Nukui A, Iwai S, Tagai H (1972) Gas flame heating equipment providing up to 2300°C for an X-ray diffractometer. Rev Sci Instr 42:1299-1301

Nukui A, Shimizugawa U, Inoue S, Ozawa H, Uno R, Oosumi K, Makishima A (1992) A structural study of Y_2O_3-Al_2O_3-SiO_2 glass employing partial RDFs obtained by anomalous scattering. J Non-Crystal Solids 150:376-379

Nukui A, Tagai H, Morikawa H, Iwai S (1976) Structural conformation and solidification of molten alumina. J Am Ceram Soc 59:534-536

Nukui A, Tagai H, Morikawa H, Iwai S (1978) Structural study of molten silica by an X-ray radial distribution analysis. J Am Ceram Soc 61:174-176

Okuno M, Marumo F (1982) The structures of anorthite and albite melts. Mineral J 11:180-196

Okuno M, Marumo F (1993) The structure analysis of pyrope ($Mg_3Al_2Si_3O_{12}$) and grossular ($Ca_3Al_2Si_3O_{12}$) glasses by X-ray diffraction method. Mineral J 16:407-415

Okuno M, Marumo F, Morikawa H (1986a) The structure analyses of $KAlGe_3O_8$ and $KGaGe_3O_8$ glasses by X-ray diffraction and EXAFS measurements. Rept Res Lab Engin Materials, Tokyo Inst Technol 11:1-9

Okuno M, Marumo F, Morikawa H, Nakashima S, Iiyama JT (1987) Structures of hydrated glasses in the albite-anorthite-quartz system. Mineral J 13:434-442

Okuno M, Marumo F, Sakamaki T, Hosoya S, Miyake M (1984) The structure analysis of $NaGaSi_3O_8$, $NaAlGe_3O_8$ and $NaGaGe_3O_8$ glasses by X-ray diffraction and EXAFS measurements. Mineral J (Japan) 12:101-121

Okuno M, Yin CD, Morikawa H, Marumo F, Oyanagi H (1986b) A high resolution EXAFS and near edge study of GeO_2 glass. J Non-Crystal Solids 87:312-320

Omote K, Waseda Y (1994) A structural study of the molten Bi_2O_3-GeO_2 system by the EXAFS method. J Non-Crystal Solids 176:116-126

Paris E, Dingwell DB, Romano C, Seifert FA (1993) X-ray absorption study of Ti in silicate melts. EOS Trans Am Geophys Union 74:347 (abstract)

Paris E, Dingwell DB, Seifert FA, Mottana A, Romano C (1994) Pressure-induced coordination change of Ti in silicate glass: a XANES study. Phys Chem Minerals 21: 510-515

Paris E, Romano C, Wu Z (1995) Application of multiple scattering calculations to the study of local geometry of titanium in silicate glasses of geological interest. Physica B 208-209:351-353

Pauling L (1929) The principles determining the structure of complex ionic crystals. J Am Chem Soc 51:1010-1026

Petiau J, Calas G (1985) EXAFS and edge structures: application to nucleation in oxide glasses. J Physique 47 C8:949-953.

Petiau J, Calas G, Dumas T, Heron AM (1984) EXAFS and edge studies of transition elements in silicate glasses. In: EXAFS and Near Edge Structure III, KO Hodgson, B Hedman, JE Penner-Hahn (eds) p 291-296, Springer-Verlag, New York

Petit-Maire D, Petiau J, Calas G, Jacquet-Francillon N (1986) Local structure around actinides in borosilicate glasses. J Physique 47 C8:849-852

Philips JC (1982) Spectroscopic and morphological structure of tetrahedral oxide glasses. Solid State Phys 37:93-171

Poe BT, McMillan PF, Angell CA, Sato RK (1992) Al and Si coordination in SiO_2-Al_2O_3 glasses and liquids; a study by NMR and IR spectroscopy and MD simulations. Chem Geol 96:333-349

Poe BT, McMillan PF, Coté B, Massiot D, Coutures J-P (1992) SiO_2-Al_2O_3 liquids: In situ study by high-temperature ^{27}Al NMR spectroscopy and molecular dynamics simulation. J Phys Chem 96:8220-8224

Ponader CW, Brown GE Jr (1989) Rare earth elements in silicate glass/melt systems: I. Effects of composition on the coordination environments of La, Gd, and Yb. Geochim Cosmochim Acta 53:2893-2903

Porai-Koshits EA (1958) The Structure of Glass, p 25, Consultants Bureau, New York

Poulsen HF, Neuefeind J, Neumann H-B, Schneider JR, Zeidler MD (1995) Amorphous silica studied by high energy X-ray diffraction. J Non-Crystal Solids 188:63-74

Prober JM, Schultz JM (1975) Liquid-structure analysis by energy-scanning X-ray diffraction: mercury. J Appl Cryst 8:405-414

Randall JT, Rooksby HP, Cooper BS (1930) X-ray diffraction and the structure of vitreous solids-I. Zeit Krist 75:196-214

Rawson H (1967) Inorganic Glass-Forming Systems. 267 p. Academic Press, New York

Rehr JJ, Zabinsky ZI, Albers RC (1992) High-order multiple scattering calculations of X-ray-absorption fine structure. Phys Rev Letters 69:3397-4000

Renninger AL, Uhlmann DR (1974)Small angle X-ray scattering from glassy SiO₂. J Non-Crystal Solids 16:325-327

Richet P, Bottinga Y (1985) Heat capacity of aluminium-free liquid silicates. Geochim Cosmochim Acta 49:471-486

Richet P, Gillet P, Pierre A, Bouhifd A, Daniel I, Fiquet G (1993) Raman spectroscopy, X-ray diffraction, and phase relationship determinations with a versatile heating cell for measurements up to 3600 K (or 2700 K in air). J Appl Phys 74:5451-5456

Risbud SH, Kirkpatrick RJ, Taglialavore AP, Montez B (1987) Solid-state NMR evidence of 4-, 5-, and 6-fold aluminum sites in roller-quenched SiO₂-Al₂O₃ glasses. J Am Ceram Soc 70:C10-C12

Roeder PL (1974) Activity of iron and olivine solubility in basaltic liquids. Earth Planet Sci Letters 23:397-410

Sato RK, McMIllan PF, Dennison P, Dupree R (1991) High-resolution ²⁷Al and ²⁹Si MAS NMR investigation of SiO₂-Al₂O₃ glasses. J Phys Chem 95:4483-4489

Scamehorn CA, Angell CA (1991) Viscosity-temperature relations and structure in fully polymerized aluminosilicate melts from ion dynamics simulations. Geochim Cosmochim Acta 55:721-730

Seifert F, Paris E, Dingwell DB, Davoli I, Mottana A (1993) In situ X-ray absorption spectroscopy to 1500 °C: cell design, operation, and first results. Terra Abstracts 1:366Shannon RD (1976) Revised effective ionic radii and systematic studies of interatomic distances in halides and chalcogenides. Acta Cryst A32:751-767

Shannon RD, Prewitt CT (1969) Effective ionic radii in oxides and fluorides. Acta Cryst B25:925-945

Shimizugawa Y, Marumo F, Nukui A, Ohsumi K (1994) Anomalous scattering and EXAFS study of GeO₂-P₂O₅ glass system. J Non-Crystal Solids 176:76-84

Shiraishi Y, Ikeda K, Tamura A, Saito T (1978) On the viscosity and density of the molten FeO-SiO₂ system. Trans Japan Inst Metals 19:264-274

Smith JV (1974) Feldspar Minerals 1. Crystal Structure and Physical Properties. 627 p. Springer-Verlag, New York

Stebbins JF (1987) Identification of multiple structural species in silicate glasses by ²⁹Si NMR. Nature 330:465-467

Stebbins JF, Farnan I (1989) Nuclear magnetic resonance specroscopy in the earth sciences: structure and dynamics. Science 245:257-263

Stebbins JF, Farnan I (1992) Effects of high temperature on silicate liquid structure: a multinuclear NMR study. Science 255: 856-589

Stebbins JF, Farnan I, Xue X (1992) The structure and dynamics of alkali silicate liquids: A view from NMR spectroscopy. Chem Geol 96:371-385

Stebbins JF, McMillan PF (1989) Five- and six-coordinated Si in K₂Si₄O₉ glass quenched from 1.9 GPa and 1200 °C. Am Mineral 74:965-968

Stebbins JF, McMillan PF (1993) Compositional and temperature effects on five- coordinated silicon in ambient pressure silicate glasses. J Non-Crystal Solids 160: 116-125

Stein DJ, Spera FJ (1995) Molecular dynamics simulations of liquids and glasses in the system NaAlSiO₄-SiO₂: methodology and melt structures. Am Mineral 80:417-431

Stern EA, Heald SM (1983) Basic principles and applications of EXAFS. In: Handbook on Synchrotron Radiation 1b955-1014. EE Koch (ed) North Holland, New York

Stern EA, Livins P, Zhang Z (1991) Thermal vibration and melting from a local perspective. Phys Rev B43:8850-8860

Takahashi E (1978) Partitioning of Ni²⁺, Co²⁺, Fe²⁺, Mn²⁺, and Mg²⁺ between olivine and silicate melts: compositional dependence of partition coefficient. Geochim Cosmochim Acta 42:1829-1844

Takeda S, Petkov VG, Sugiyama K, Waseda Y (1993a) Energy-dispersive X-ray diffraction (EDXD) facility for determining structure of high temperature melts with a stationary specimen goniometer. Mat Trans, Japan Inst Metal 34:410-414

Takeda S, Sugiyama K, Waseda Y (1993b) Structural analysis of molten bismuth germanate compounds. Japan J Appl Phys Part 1 32:5633-5336

Taylor M (1979) A quasi-crystalline model for interpretation of radial distribution functions: theory and experimental confirmation. J Appl Cryst 12:442-449

Taylor M, Brown GE Jr (1979a) Structure of mineral glasses I: The feldspar glasses NaAlSi₃O₈, KAlSi₃O₈, CaAl₂Si₂O₈. Geochim Cosmochim Acta 43:61-75

Taylor M, Brown GE Jr (1979b) Structure of mineral glasses II: The SiO₂-NaAlSiO₄ join. Geochim Cosmochim Acta 43:1467-1473

Taylor M, Brown GE Jr, Fenn PM (1980) Structure of mineral glasses III: NaAlSi₃O₈ supercooled liquid at 805°C and the effects of thermal history. Geochim Cosmochim Acta 44:109-117

Teo B (1986) EXAFS: Basic Principles and Data Analysis. Inorg Chem Concepts 9, 349 p. Springer-Verlag, Berlin

Thompson VS, Whittemore OJ (1968) Structural changes on reheating plasma-sprayed alumina. Am Ceram

Soc Bull 47:637-641

Tyson TA, Hodgson KO, Natoli CR, Benfatto M (1992) General multiple-scattering scheme for the computation and interpretation of X-ray-absorption fine structure in atomic clusters with applications to SF_6, $GeCl_4$, and Br_2 molecules. Phys Rev B46:5997-6019

Uozumi T, Okada K, Kotani A, Durmeyer O, Kappler, JP, Beaurepaire E, Parlebas JC (1992) Experimental and theoretical investigations of the pre-peaks at the Ti K-edge absorption spectra in TiO_2. Europhys Letters 18:85-90

Valenkov N, Porai-Koshits E (1936) X-ray investigation of the glassy state. Zeit Krist 95:195-229

Veal BW, Mundy JN, Lam DJ (1987) Actinides in silicate glasses. In: Handbook on the Physics and Chemistry of the Actinides, AJ Freeman, GH Lander (eds) p 271-312. Elsevier, Amsterdam

Vessal B, Greaves GN, Marten PT, Chadwick AV, Mole R, Houde-Walter S (1992) Cation microsegregation and ionic mobility in mixed alkali glasses. Nature 356:504-506

Vogel W (1971) Structure and Crystallization of Glasses (Ed. Leipzig). 246 p. Pergamon Press, New York

Waff HS (1975) Pressure-induced coordination changes in magmatic liquids. Geophys Res Letters 2:193-196

Wagner CNJ (1978) Direct methods for the determination of atomic-scale structure of amorphous solids (X-ray, electrons, and neutron scattering). J Non-Crystal Solids 31:1-40

Warburton WK, Ludwig KF Jr, Wilson L, Bienenstock A (1987) Differential anomalous X-ray scattering technique for determination of liquid and amorphous structures. In: Phase Transitions in Condensed Systems—Experiments and Theory. GS Cargill III, F Spaepen, NT King (eds) p 211-225. Mat Res Soc, Pittsburg, PA

Warren BE (1933) X-ray diffraction of vitreous silica. Zeit Krist 86:349-358

Warren BE (1969) X-ray Diffraction. 381 p. Addison-Wesley, New York

Warren BE, Biscoe J (1938) The structure of silica glass by X-ray diffraction studies. J Am Ceram Soc 21:49-54

Warren BE, Krutter H, Morningstar O (1936) Fourier analysis of X-ray patterns of vitreous SiO_2 and B_2O_3. J Am Ceram Soc 19:202-206

Waseda Y (1980) The Structure of Non-Crystalline Materials. 326 p. McGraw-Hill, New York

Waseda Y, Hirata K, Ohtani M (1975) High-temperature thermal expansion of platinum, tantalum, molydenum, and tungsten measured by X-ray diffraction. High Temp- High Press 7:221-226

Waseda Y, Shiraishi Y, Toguri JM (1980) The structure of molten $FeO-Fe_2O_3-SiO_2$ system by X-ray diffraction. Trans Japan Inst Metals 21:51-62

Waseda Y, Suito H (1977) The structure of molten alkali metal silicates. Trans Iron Steel Inst Japan 17:82-91

Waseda Y, Toguri JM (1977) The structure of molten binary silicate systems $CaO-SiO_2$ and $MgO-SiO_2$. Metal Trans B 8B:563-568

Waseda Y, Toguri JM (1978) The structure of the molten $FeO-SiO_2$ system. Metal Trans B 9B:595-601

Waseda Y, Toguri JM (1990) Structure of silicate melts determined by X-ray diffraction. In: Dynamic Processes of Material Transport and Transformation in the Earth's Interior, F Marumo (ed) p 37-51. Terra Scientific, Tokyo

Watson EB (1979) Zircon saturation in felsic liquids: experimental results and applications to trace element geochemistry. Contrib Mineral Petrol 70:407-419

Waychunas GA (1987) Synchrotron radiation XANES spectroscopy of Ti in minerals. Effects of Ti bonding distances, Ti valence, and site geometry on absorption edge structure. Am Mineral 72:89-101

Waychunas GA, Apted MJ, Brown GE Jr (1983) X-ray K-edge absorption spectra of Fe minerals and model compounds: I. near-edge structure. Phys Chem Minerals 10:1-9

Waychunas GA, Brown GE Jr, Apted MJ (1986) X-ray K-edge absorption spectra of Fe minerals and model compounds: II. EXAFS. Phys Chem Minerals 13:31-47

Waychunas GA, Brown GE Jr, Jackson WE, Ponader CW (1989) In situ high temperature X-ray absorption study of iron in alkalisilicate melts and glasses. Physica B 158:67-68

Waychunas GA, Brown GE Jr, Ponader CW, Jackson WE (1987) High temperature X-ray absorption study of iron sites in crystalline, glassy and molten silicates and oxides. Stanford Synchrotron Radiation Lab Report 87/01:139-141

Waychunas GA, Brown GE Jr, Ponader CW, Jackson WE (1988). Evidence from X-ray absorption for network forming Fe^{2+} in molten silicates. Nature 332:251-253

Weinberg DL (1962) Surface effects in small-angle X-ray scattering. J Appl Phys 33: 1012-1013

Williams RJP (1959) Deposition of trace elements in basic magmas. Nature 184:44

Winter JK, Okamura FP, Ghose S (1979) A high-temperature structural study of high albite, monalbite, and the analbite→monalbite phase transition. Am Mineral 64:409-423

Wong J, Lytle FW, Messmer RP, Maylotte DH (1984) K-edge absorption spectra of selected vanadium compounds. Phys Rev B30:5596-5610

Wright AC (1974) The structure of amorphous solids by X-ray and neutron diffraction. In: Advances in

Structure Research by Diffraction Methods 5:1-84. W Hoppe, R Mason (eds) Pergamon Press, New York

Wright AC (1988) Neutron and X-ray amorphography. J Non-Crystal Solids 106:1-16

Wright AC (1993a) The comparison of molecular dynamics simulations with diffraction experiments. J Non-Crystal Solids 159:264-268

Wright AC (1993b) Neutron and X-ray amorphography. In: Experimental Techniques of Glass Science, Ceram Trans, CJ Simmons, OH El-Bayoumi (eds) p 205-314. Am Ceram Soc, Westerville, OH

Wright AC (1994) Neutron scattering from vitreous silica V. The structure of vitreous silica: What have we learned from 60 years of diffraction studies? J Non-Crystal Solids 179:84-115

Wright AC, Clare AG, Bachra B, Sinclair RN, Hannon AC, Vessal, B (1991) Neutron diffraction studies of silicate glasses. In: Trans Am Cryst Assoc 27, AC Wright (ed) p 239-254, Polycrystal Book Service, P.O. Box 3439, Dayton, OH

Wright AC, Etherington G, Desa JAE, Sinclair RN, Connell GAN, Mikkelsen JC Jr (1982) Neutron amorphography. J Non-Crystal Solids 49:63-102

Wright AC, Leadbetter AJ (1976) Diffraction studies of glass structure. Phys Chem Glasses 17:122-145

Xue X, Stebbins JF, Kanzaki, M, McMillan PF, Poe B (1991) Pressure-induced silicon coordination and tetrahedral structural changes in alkali oxide-silica melts up to 12 GPa: NMR, Raman, and infrared spectroscopy. Am Mineral 76:8-26

Xue X, Stebbins JF, Kanzaki M, Tronnes RG (1989) Silicon coordination and speciation changes in a silicate liquid at high pressures. Science 245:962-964

Yarker CA, Johnson PAV, Wright AC, Wong J, Greegor RB, Lytle FW, Sinclair RN (1986) Neutron diffraction and EXAFS evidence for TiO_5 units in vitreous $K_2O \cdot TiO_2 \cdot 2SiO_2$. J Non-Crystal Solids 79:117-136

Yasui I, Akasaka Y, Inoue H (1994) Re-examination of detailed structure of alkali silicate glasses based on two types of diffraction data. J Non-Crystal Solids 177: 91-96

Yasui I, Hasegawa H, Imaoka M (1983) X-ray diffraction study of the structure of silicate glasses. Part 1. Alkali metasilicate glasses. Phys Chem Glasses 24:65-71

Yin CD, Okuno M, Morikawa H, Marumo F (1983) Structure analysis of $MgSiO_3$ glass. J Non-Crystal Solids 55:131-141

Yin CD, Okuno M, Morikawa H, Marumo F, Yamanaka T (1986) Structural analysis of $CaSiO_3$ glass by X-ray diffraction and Raman spectroscopy. J Non-Crystal Solids 80:167-184

Zabinsky SI, Rehr JJ, Ankudinov A, Albers RC, Eller MJ (1995) Multiple scattering calculations of X-ray absorption spectra. Phys Rev B52:2995-3006

Zachariasen WH (1932) The atomic arrangement in glass. J Am Chem Soc 54:3841-3851

Zachariasen WH (1933) The vitreous state. J Chem Phys 3:162-163

Zarzycki, J (1957) Bond angle of Si-O-Si in vitreous silica and Ge-O-Ge in vitreous or liquid germanium oxide. Verres Refract 11:3-8

Zernike F, Prins JA (1927) The bending of X-rays in liquids as an effect of molecular arrangement. Zeit Physik 41:184-194

Zhou R-S, Snyder RL (1991) Structures and transformation mechanisms of the η, χ and θ transition aluminas. Acta Cryst B47:617-630

Chapter 10

DIFFUSION IN SILICATE MELTS

Sumit Chakraborty

Mineralogisch-Petrographisches Institut
Universität zu Köln
Zülpicher Straße 49b
50674 Köln, Germany

INTRODUCTION

Diffusion rates of various species in silicate melts had already begun determining the course of events in the early solar system (e.g. flash melting) before the birth of our planet and continue to influence processes that profoundly affect our lives today (e.g. explosive volcanism). Geochemists have been studying diffusion in silicate melts with at least three objectives of (1) calculating mass transport rates and time scales due to diffusion, (2) understanding the mechanism of this most elementary and fundamental mode of transport and (3) understanding and exploiting the relationship between the process of diffusion and other transport/equilibrium properties to systematize data, check for consistency and reduce experimental effort. This chapter is intended to be an introduction to the various concepts and approaches used in the study of diffusion in melts. In keeping with the objectives of this volume, the interdisciplinary nature of diffusion studies in silicate melts is highlighted. Diffusion studies are "interdisciplinary" on two different levels. Studies in other fields (e.g. materials science) help us to obtain data and understand the process of diffusion in melts of geochemical interest. Within the area of geosciences, studies of diffusion provide data and help us to understand various other properties—both transport and equilibrium, in silicate melts in general. Side by side with these unifying features, there are important boundaries within the field of study of diffusion in silicate melts. Most notably, diffusion in viscous granitic melts is quite different from diffusion in fluid basaltic melts—not only in the magnitudes of diffusion rates but in the very nature of the process itself. This distinction will be considered at several places in this chapter.

We begin this chapter with a historical perspective to set up the present context for the study of diffusion in melts in the geosciences. Following this, an outline of the topics covered in various sections is provided for the reader interested in only certain aspects of diffusion and some conventions are set up. Broadly, the chapter covers four topics: (1) macroscopic, thermodynamic approach to diffusion, (2) relationship between diffusion and various other features of silicate melts, (3) microscopic, statistical approach to diffusion with a view to understanding the significance of some of the macroscopic observables, and (4) empirical aspects of diffusion where experimental procedures and trends in data and examples of applications are considered.

HISTORICAL BACKGROUND AND PRESENT CONTEXT

It is hardly ever recognized that one of the earliest observations of the process of diffusion in the condensed state was actually the result of what would be considered a geochemical investigation today. When the 19th century Scottish botanist Robert Brown lowered his microscope on a fluid inclusion—a drop of water trapped in a crystal of quartz, he observed the constant but random movement of suspended particles in this fluid inclusion.

The kind of motion, of course, came to be known as Brownian motion, and its explanation by Albert Einstein in 1905 founded the microscopic basis of diffusion (as a by-product, it led to the first determination of the Avogadro's number and the mass of an atom!).

Diffusion in silicate melts had already been studied by N.L. Bowen as early as 1921, before the birth of most of the concepts we shall be discussing in this chapter. Following this early start, however, there was a lull in the study of diffusion in melts until the 1970s. Several factors contributed to this inactivity—the realization from the early studies of Bowen that diffusion in melts is in general ineffective as a means of large scale transport of matter, the inability to measure concentration gradients accurately in the absence of microbeam techniques and a focus of research interests in characterizing *systems* through the use of phase equilibria rather than the *processes* leading to a given state of a system, were some of these. The development of the electron microprobe in the 1960s into a routine geochemical analytical tool and questions raised about the applicability of equilibrium relations in trace element modelling led to a renewed interest in the study of diffusion. On the one hand, studies of natural glasses using the electron microprobe revealed the existence of concentration gradients and hence disequilibrium in melts (e.g. Sato, 1975). On the other hand, as quantitative studies of partitioning of trace elements became common, the necessity of diffusion data was felt by both experimentalists ("How long does one have to do an experiment to obtain equilibrium partitioning?") and modellers ("On what spatial and temporal scales can equilibrium partitioning of trace elements be expected in nature?"). This led to a flurry of experimental activity (e.g. Watson, 1979; Hofmann and Margaritz, 1977; Jambon and Semet, 1978) such that by 1980 it was considered necessary to write review articles on diffusion in silicate melts (Hofmann, 1980; Jambon, 1983).

If the decade of 1970s and early 1980s saw a rapid growth in the body of diffusion data, the remarkable features of the late 1980s were related to advances in computer technology. As computers became desktop entities, they influenced the study of diffusion in at least three distinct ways. First, they allowed geoscientists a closer, quantitative look at many processes through numerical modelling and brought to light the need for diffusion data—for minor, trace, as well as major elements. Secondly, modelling on completely different scales allowed structures of melts and diffusion mechanisms to be explored using the methods of statistical and to a lesser extent, quantum mechanics. Thirdly, improvements and accessibility of numerical techniques allowed retrieval of diffusion data from a variety of experimental procedures. Parallel with these developments came major breakthroughs in spatially resolved analytical capabilities. Tools like the ion microprobe, PIXE, infrared microscopy and micro-Raman spectroscopy now allow measurements of an ever expanding range of chemical species with increasing accuracy and spatial resolution. At the same time, improvements in spectroscopy (NMR, Raman, infrared) that allow in-situ observation of melts at high temperatures have largely advanced our understanding of melt structure and dynamics. This renewed ability to measure and understand has led to a more recent outpouring of both diffusion data and models. Since understanding mechanisms is often a major objective, the level of accuracy being sought in these new measurements are much higher than in the earlier studies. The lack of interest in diffusion for a long time owed a lot to the recognition that diffusion is a very inefficient mode of mass transfer in melts. Curiously, the recent revival of interest is largely for the same reason! With the emerging interest in processes, it has become increasingly apparent that the sluggishness of diffusion may make it the rate determining step for many geological processes. Thus, knowing diffusion rates may often be equivalent to knowing the rates of geological processes themselves.

While geochemists have been interested primarily in diffusion rates in silicate liquids, an interest in the diffusion rates in silicate glasses as well as liquids has existed for a long time in materials sciences. For example, metallurgists studied diffusion rates in silicate slags produced in blast furnaces (e.g. Sugawara et al., 1977). Understanding diffusion behavior in silicate glasses and liquids helped to explain various empirical observations in the glass and ceramics industry and led to optimization of many industrial procedures. Questions like "Why does doping a melt with a small amount of transition metal cation make it easier to obtain bubble free glass?" were answered only when the complex diffusion behavior of oxygen in these melts was understood. Chemical engineers have been concerned with corrosion rates of various glasses, amorphous electrochemical sensors and batteries and the process of strengthenning glasses by diffusing large cations into near surface regions. Optical scientists have studied the diffusion rates of various dopants that impart specific optical characteristics to the glasses. For example, ionic transport in amorphous thin films is the basis of electrochromic devices—an example of which may be the "dazzle-free" rear view mirrors of automobiles. The recent interest in the use of electrical characteristics of amorphous materials in various electronic devices has considerably advanced our understanding of ionic transport (diffusion or conduction) in these materials. More recently, answers to environmental and cultural issues like why some stained glass windows in the Gothic Cathedrals of Southern France are degrading depend also in part on understanding diffusion in silicate glasses. It is an objective of this chapter to draw upon the large body of knowledge acquired over the years in these fields to further our understanding of geochemical materials and processes.

SCOPE OF THE CHAPTER AND SOME CONVENTIONS

The overall approach in this chapter will be to introduce individual concepts used in the study of diffusion in silicate melts and use specific examples and possible applications to illustrate them, whenever practical. The wealth of information that may be gleaned from the non-geochemical literature will be emphasized. Studies of simple systems in a rigorous manner offer us insight into the behavior of silicate melts which may then guide us in our attempts to understand and measure properties in complex geological systems. For each aspect, the basic ideas will be introduced and the reader will be directed to appropriate research articles, reviews or books for a detailed exposure. This is the only practical manner to cover reasonable ground within the framework of a chapter in a volume such as this.

An introduction to general diffusion theory may be found in a number of well known books and monographs written mostly for solids or aqueous liquids and solutions (e.g. Shewmon, 1963; Jost, 1960; Borg and Dienes, 1988; Kirkaldy and Young, 1987, Cussler 1976, 1984; Lasaga, 1995). In spite of many similarities, silicate melts differ in many important aspects from both of these classes of materials, an outline of some of the salient differences between silicate melts and other fluids is provided in the next section. Although Kirkaldy and Young (1987) and Lasaga (1995) contain some discussion of diffusion in silicate melts, it was felt necessary to highlight how the general theory of diffusion adapts to some of the unique aspects of silicate melts. In doing this, we will use the above works as well as the brief introduction to diffusion theory provided in the *Reviews in Mineralogy* (Volume 30) chapter by Watson (1994) as a basis. Some of the peculiarities of multicomponent diffusion have not been discussed in previous reviews and will therefore be covered in some detail here. A second aspect that has not been covered in any detail in the earlier reviews is the role of correlations. Diffusion in silicate melts is dominated by correlations—both between the successive jumps of an individual particle (i.e. the motion is not truly random) as well as between the motion of different particles during chemical diffusion (diffusive coupling). We will try to see why such correlations arise and how they influence macroscopic transport rates, their measurement and interpretation.

The four topics of this chapter are organized into nine sections. Much of the geochemical diffusion literature is based on a macroscopic, phenomenological, thermodynamic approach. Accordingly, the first three sections of this chapter will be developed following this approach. We begin this chapter with a discussion of diffusion concepts in the simplest case of one diffusing species in a system. The physics of such a process will be discussed and some mathematical concepts are introduced along with the pertinent terminology. Section II introduces the next level of complication by bringing in a second diffusing species. The question: "During the simultaneous diffusion of two species, how does the motion of each influence the diffusion rate of the other?" is addressed. This is the simplest possible case of chemical diffusion which is necessary for modelling processes involving concentration gradients. Finally, in Section III the most general case of a multicomponent system with n diffusing components will be discussed. Such systems have many peculiarities not found in the simpler systems; these will be pointed out. Parts of this section are somewhat mathematical, the reader interested in the general aspects only may skip these parts. A discussion of how diffusion relates to other physical properties, parameters and melt structure will follow in Section IV. Relationships and connections between thermodynamic mixing, spectroscopic features and diffusion parameters are pointed out to close this part of the chapter using thermodynamic approaches to study diffusion.

The next two sections will focus on more practical aspects of diffusion studies. Often, the lack of measured diffusion data necessitates estimation. A number of available empirical methods for making such estimates will be discussed along with their relative advantages and limitations in Section V; Section VI will deal with experimental methods for obtaining diffusion data. This will bring us to an appropriate point where the reasons for some of the experimental observations as well as theoretical interrelationships between properties can be explored using microscopic approaches. Section VII, of necessity somewhat more mathematical like parts of Section III, will cover selected microscopic aspects of diffusion in silicate melts, including computer simulations. It is written in a manner such that it is possible to skip this section and continue on with Section VIII, dealing with diffusion data in silicate melts. It is *not* the intention to review all available data on diffusion in silicate melts, the reader is referred to related literature instead. General trends are discussed and the influences of composition, temperature and pressure on diffusion rates are considered. In doing so, it is illustrated how these dependencies provide information about the mechanisms of diffusion in silicate melts. Section IX focusses on applications. Once again, instead of describing specific applications, general features are highlighted. Some guiding points, intended to be stepping stones for the reader to the practical area of problem solving using the tools discussed in the previous sections, concludes this last section of the chapter.

Although attempts have been made to present material in a logical sequence, it was impossible to avoid cross referencing back and forth between sections to avoid repetition. Such references are always to the name (or a keyword from the name) of the section written in *italics*, e.g. (see *Microscopics* for further detail). To avoid irritation, the reader may ignore most of the cross-references in a first quick reading and only pay attention to these in subsequent, more careful readings.

Silicate "melts" for the purpose of this chapter include both liquids and glasses. This definition is unlikely to satisfy all readers—the concept of a "melt" existing below the melting temperature (as a glass does) may be disturbing to many, but does serve the purpose of providing a convenient means of describing condensed medium lacking long range order. So we will use the convention in this chapter, without in any way advocating general use of such terminology. The terms "liquid" and "glass", as used in this chapter, refer to amorphous matter above and below the glass transition temperature, respectively.

The glass transition temperature itself depends on the time scale of observation (see the chapter by Dingwell, this volume, for a discussion as well as definition of glass transition)—in the general context, the time scale of interest would be that corresponding to low frequency macroscopic mechanical and calorimetric probes.

We will use the terms "network formers" and "network modifiers" frequently in this chapter as another convenient and very general shorthand, recognizing that these are by no means absolute designation for an element—the same element may be present as both in a melt, may play different roles in different melt compositions and may change its role with intensive variables (P, T, fO_2) in the same melt.

SILICATE MELTS VERSUS AQUEOUS SOLUTIONS

Because natural magmas are liquids, it is a frequent temptation to use concepts developed for diffusion in aqueous solutions (e.g. Cussler, 1984) to deal with diffusion in silicate melts. There are similarities but the concepts are not always directly transferrable. Let us point out some important differences between the two kinds of materials at the outset here, from the point of view of how diffusion is treated:

(1) Aqueous electrolyte solutions are often completely dissociated, silicate melts may be similar only at unreasonably high temperatures and only for some compositions. Thus, coupling of diffusion between different species is much more important in silicate melts than in aqueous solutions.

(2) At the other extreme, in polymer solutions the chains and other polymer units always remain intact. On the other hand in silicate melts the polymeric units are transient, having definite lifetimes. Further, the nature of the units change with temperature in the same melt (e.g. see the chapter by Stebbins in this volume).

(3) Aqueous solutions are frequently dilute, silicate melts are almost never so. Thus, in the former, diffusion of atoms against a constant background of water is often a good approximation; diffusion of atoms against a constant background of silica is hardly ever realistic (except perhaps for self diffusion of trace elements, see below).

(4) Even when speciation changes occur in aqueous solutions, these may often be well characterized. In contrast, although the formalism for treating homogeneous reactions is analogous, methods to identify these in silicate melts are not well developed.

(5) Viscosities of silicate melts and aqueous solutions differ by many orders of magnitude.

WHAT IS DIFFUSION?

A definition of what is meant by diffusion must precede any discussion about the topic. Often, the process is defined in terms of random jumps of particles of matter (atoms, ions, molecules) from one site to another within a medium. Such a definition clearly presumes the existence of sites, i.e. locations within a melt, characterized by a certain geometry and local structure, where an atom may prefer to reside. Preference in this case means local potential energies are low at these locations and that atoms or ions spend considerably more time sitting or vibrating around these sites relative to the time of flight from one such site to another. As we shall see, although such a definition is ideal for studying diffusion in solid phases, there are circumstances in melts where definition of such a "site" may become problematic. Pre-empting such conflicts, we define the process of diffusion following Onsager (1945) in his classic work on liquid diffusion: *"Diffusion is a relative motion of its* (the liquid's) *different constituents"*. This definition has the advantage of being purely descriptive without any implications about the underlying mechanisms.

I. DIFFUSION OF ONE KIND OF PARTICLE IN A MELT

Fick's first law and some of its consequences

To measure the motion of atoms (or any other particles) one defines the physical quantity flux (more rigorously, *flux density*), J which is a vector characterized by a magnitude and a direction. A flux, J_i, gives the number of atoms of a particular kind, i, crossing a unit area perpendicular to the direction of the vector in unit time. Fick, in 1855, empirically observed that in a diffusing system the flux is proportional to the local gradient in concentration. This observation has since been verified experimentally in numerous systems and the microscopic basis of its validity has been found for many circumstances. Mathematically Fick's observation is stated as

$$J_i = -D\nabla C_i, \tag{1}$$

the commonly known statement of Fick's first law of diffusion, where the symbol ∇ is mathematical shorthand for 'gradient'. One recognizes the parameter D in this equation as the constant of proportionality, or, the diffusion coefficient. We will return to more precise definitions of different kinds of diffusion coefficients later in this chapter. The negative sign is a matter of convention which ensures that for positive diffusion coefficients, flux is positive in the direction of decreasing concentration, i.e. things move downhill! We will find later that in complex systems this may not always hold. As the above equation constitutes the basis of diffusion studies for most practical purposes, it is worth taking some time to examine different aspects of this relationship in detail.

First of all, with the definition of flux that we have used above, the units of concentrations are determined—the number of particles crossing a particular area should be proportional to the gradient in number of particles per unit volume. What if we want to use some other kind of concentration scale? We may, for example, introduce other concentration scales like weight percent, mole fractions and so on as long as we use consistent units of flux and in particular, bring in the appropriate multiplying factors to take care of conversions. One very important kind of scale is to use concentrations in terms of mass per unit volume (density) in conjunction with mass fluxes—these come in useful when one needs to use conservation laws, as we shall soon find out. Often, approximations of various kinds are made because the density of a particular silicate melt is not well known. For practical usage, one needs to be aware of these approximations and the uncertainties they may propagate. Whatever we do, the dimensions of the diffusion coefficient should turn out to be $L^2 S^{-1}$, i.e. D should have units of cm^2/sec or, in the S.I. system, $m^2 s^{-1}$. It is important to be careful about this point because if the units are not consistently chosen in a calculation, then the multiplicative constants which should have been used for conversion of units would get absorbed in the diffusion constant (in the simplest possible kind of mess!). When results from two studies are compared, for example, diffusion coefficients would be found to differ by an arbitrary factor. More dangerously, when diffusion coefficients which have some conversion factor buried in them are used to calculate fluxes or time scales using concentration gradients of a different kind, one could obtain grossly incorrect results!

In stating Fick's law above, we used the term 'local gradient'. What does this mean? Consider the solid line ($t = t_1$) in Figure 1 where the concentration gradient changes from point to point. The gradient at point A is different from that at B. 'Local gradient' means in this case 'gradient at a given place at a given time'. Thus, different gradients at A and B imply that the fluxes at A and B will be different (assuming D is the same). Concentrations at A will increase and those at B will decrease such that after some amount of diffusion one obtains a distribution like that shown by the dashed line ($t = t_2$) in Figure 1. The flux at A will now be determined by the gradient at A at this point of time, which would clearly be less than what it was earlier (because the gradient is smaller now). A similar situation holds for the point B. The process will end when the concentration differences are erased. This example serves to highlight a number of important characteristics of the process of diffusion. Firstly, we have seen that the process slows down as it approaches equilibrium—a relatively flat gradient is adjusted much more slowly than the initial steep gradient. Conversely, a sharp change in composition is rounded off very quickly. Secondly, constant diffusion coefficients do not imply constant rate of change of composition—concentration gradients play an important role. The direction of diffusion flux is also determined by the local gradient (imagine a wavy profile instead of the smooth one of Fig. 1). And finally, since the gradient involved is a local gradient (in space and time), Fick's first law is by no means a "steady state" or a "time independent" law (in the sense that it is valid only at steady state), as some text books would like to imply.

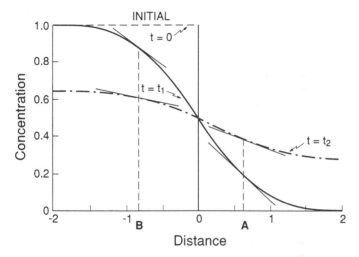

Figure 1. The concept of local gradients in space and time: a schematic illustration of how diffusion proceeds in a typical diffusion couple. Two melts having initial concentrations 1 and 0 of a species on an arbitrary scale are placed in contact along a plane whose trace is along x = 0 (vertical solid line) at time t = 0. The concentration profile resulting from diffusion that would be observed at a later time t = t_1 is shown by the solid curved line. The diffusion at each point in the system is driven by the concentration gradient present locally at that point, as illustrated for two points A and B. At a later time t_2, the gradients at the same points would have changed (dot-dashed line) and further diffusion would follow the gradients at A and B at this point of time. Note that the gradients, and hence diffusive flux, decrease with time at all points.

Finally, why use the ∇ symbol instead of the more familiar dC/dX? There are two reasons. The first one is purely mathematical—if we want to relate a scalar without any direction, such as concentration, to a vector such as flux we need to use the 'grad' operation. But more practically, d/dX is simply the one dimensional version of ∇. In geosciences, we need to study diffusion in 1-, 2-, or 3-dimensions and in various geometries (e.g. sphere, slab, cylinder). ∇ is a catch-all operator that describes gradient in all of these cases. For each specific case, it takes on different forms. We shall illustrate this with examples after we have obtained Fick's second law.

Fick's second law

Although Fick's first law relates flux to concentration gradient at a given place and time, it does not explicitly tell us how the concentration at a point evolves with time. This is obtained from the so called Fick's second law of diffusion. This, in fact, is not an independent law but rather the application of the continuity relation to Fick's first law—that is how it was developed by Fick. Let us look in detail at what is meant by this. Within a given volume of melt, concentration of any kind of particle, i, may change because of two possible reasons: either because i is internally produced or destroyed within this volume, or because of the flux, J_{total} which carries i in or out of the volume. J_{total} may consist of two parts—flux of i in or out of the volume due to diffusion, or that due to advection. Now in the absence of any production or destruction of i within the volume (often equivalent to saying that there is no chemical reaction within the volume which produces or absorbs i) and convection, the only kind of flux is that due to diffusion. Thus, one can bring in some form of conservation law e.g. the conservation of mass to say that the change in concentration of i within the volume is equal to the diffusive flux into the volume minus the diffusive flux out of the volume. This statement, couched in the language of mathematics, is Fick's second law of diffusion and reads

$$\partial C/\partial t = -\nabla \cdot J. \tag{2}$$

One can apply this to Equation (1) to obtain

$$\partial C/\partial t = \nabla \cdot (D\nabla C) \tag{3}$$

or, if D is the same (constant) at every point within the melt,

$$\partial C/\partial t = D\nabla^2 C. \tag{4}$$

To obtain Equation (2) above, one needs to use some amount of vector calculus—but the logic, as we saw, is straightforward. The special case for one dimensions is illustrated fairly easily and the reader may wish to try this out, Watson (1994) provides the solution. At any rate, Equations (2)-(4) contain the variable time explicitly and thus describe the time evolution of concentration. To obtain concentrations at any point in the melt at any time, C(x,t), one needs in addition some information about conditions prevailing at the boundaries of the volume under consideration and the initial concentration distribution within the melt. Given these information, one can solve the differential equations above to obtain the concentration distribution at any time. The methods for doing so, and numerous solutions for many common cases, may be found in the books by Crank (1975) and Carlsaw and Jaeger (1959). When a solution to the above equations cannot be expressed as a formula (analytical solution), one can still use numerical methods to obtain a numerical answer to a given problem.

Once again, before proceeding further, let us pause for a moment to ponder about some implications of Fick's second law. First of all, the logical arguments developed above should have made it clear why units of mass per unit volume are often very useful in discussing diffusion problems. To present the same arguments in terms of, say, mole fractions, would be more complicated—especially if diffusion across a phase boundary is involved so that mole fractions in one phase (say, Mg in olivine) has to be related to mole fractions in another phase (say, Mg in a melt). You may wish to try it out. However, as long as consistency is maintained, mole fractions would have been a perfectly valid set of concentration variables and would have led to similar results (with some extra multiplicative factors to take care of units).

Secondly, this is a suitable time to illustrate the benefits of using the ∇ symbol instead of the more familiar $\partial/\partial X$. For example, we already know that in one dimension, Fick's first law is written as $J_i = -D(\partial C_i/\partial X)$, and the second law with constant diffusion coefficients as $\partial C_i/\partial t = D(\partial^2 C_i/\partial X^2)$, instead of the general form we have seen before. We could use such equations to study diffusion out of a tabular crystal into a melt, for example. For a sphere where diffusion is radial, the first law would be written as $J_i = -D(\partial C_i/\partial r)$, where r is the radius. This is very similar in appearance to the one dimensional case with 'r' replacing 'x'. However, the second law for a sphere takes on a rather different appearance. For constant diffusion coefficient and only radial diffusion, $\partial C_i/\partial t = D[\partial^2 C_i/\partial r^2 + (2/r)(\partial C_i/\partial r)]$. This equation would be useful when instead of a tabular crystal we have diffusion from a spherical melt inclusion out into the enclosing crystal. When diffusion occurs in directions other than radial as well, the equations become formidable indeed with extra terms added to the above. There are other forms for different geometries, Crank (1975) considers the most commonly encountered ones. Thus, we see the benefits of using the ∇ symbol—it takes care of all of these situations and is therefore a true shorthand. To perform actual numerical calculations one would of course need to "unfold" the ∇ as we have done above for the two examples, and work with the full-blown form.

Lastly, it is worth noting the fundamental difference between the first and the second laws. The first law is a constitutive equation which relates concentration to flux. This particular chosen form is empirical, and it is possible to use other functional forms to express this relationship. The thermodynamic formulation due to Onsager discussed later in this chapter (Eqn. 9) is an example of such an alternative form. The second law is simply a conservation law in which a particular constitutive equation is used to relate the conserved property and its flux. The constitutive equation used in this case is the First law, but the conservation law could equally well be applied to any other form of constitutive relation (e.g. the Onsager equation discussed later).

Reference frames and units

As we speak of motion and flux, it is clear that the flux should be defined with respect to a reference frame. In the simplest case, we are used to defining motion with respect to ourselves—the laboratory frame of reference in technical jargon. However, for a number of reasons, it is not always easy to measure diffusion rates with respect to this reference frame. If we recall the definition we have used for the process of diffusion, it becomes apparent that a natural choice for the study of diffusion is to use some part of the melt itself as the reference frame. In crystalline silicates, since diffusion rates of silicon and oxygen are very much slower than that of other cations, this can be achieved quite easily by using the fixed silicate lattice framework as a reference frame in which ions jump from site to site. This is the so called lattice fixed frame, which often coincides with the laboratory frame. The same concept can, in principle, be used for glasses—where the fixed "lattice" is an aperiodic silicate network, which shows no long range order. The diffusing atoms can be thought of as hopping in this aperiodic lattice to effect transfer of mass. However, in the true liquid state, the atoms that define the lattice i.e. silicon, oxygen and perhaps other network formers may begin to move (diffuse) themselves within the time scales of observation of diffusion[1]. In this case, one can still define the reference frame in terms of one of these kinds of atoms (e.g. a frame that moves with the O atoms), but that would no longer be the same as the laboratory frame (e.g. Onsager and Fuoss, 1932, p. 2759). The O atoms would now be "drifting" with respect to the laboratory frame as diffusion

---

[1] Note that this is one major difference between diffusion in crystalline silicates and melts. This difference in diffusion behavior of Si between melts and crystals is not related to the difference in diffusion rates -- it is relatively slow (compared to other cations in the same material) in both. In solid silicates, constraints of stoichiometry on the structure defined in terms of the Si-O lattice do not allow a continuous variation of Si content keeping the structure the same. This makes designing Si diffusion experiments in solids difficult, other than by imposing isotopic gradients. On the other hand, continual variation of Si in melts is rather the rule because there are no constraints of stoichiometry or long range periodic structure. Thus, Si atoms can "move" in a melt -- a feature that results in fundamental differences in the way diffusion occurs in amorphous vs. crystalline matter, as we shall see later.

proceeds. Obviously, the atom (or species) which is chosen to define the frame will have zero flux. This however does not mean that the species has not moved with respect to, say, the walls of the sample container. A common variant of this for liquids is to choose the solvent, or the most abundant species, as being fixed. Such solvent fixed reference frames are often useful for understanding diffusion mechanisms of the dilute constituents—the physical picture being that of the dilute atoms "swimming" in a sea of the solvent. Generally, the flux calculated on the basis of one of these "component fixed reference frame" will not be the same as the flux calculated on the "laboratory reference frame". One more reason to pay attention to the conditions under which data were obtained before comparing or using numerical values.

Often, it is difficult to know a priori whether a species is going to be mobile or not on the time scale of the diffusion experiment. Thus, for studies in liquids, frequently this approach of choosing the laboratory or a component reference frame is abandoned altogether in favour of using one of the so called "average" reference frames. For example, one can choose to define motion with respect to the center of mass of a system (barycentric frame)—a reference frame that has many practical advantages under certain circumstances. Similarly, one defines volume fixed or molar reference frames. The general equation for the definition of such an average reference frame is $\sum J_i Q_i = 0$, where Q_i is the property being used to fix the reference frame.

The subtleties of all these choices have given rise to a rich and controversial literature. The reader interested in details of this aspect are referred to Brady (1975a) for an eloquent presentation in the geochemical literature and Miller et al. (1985) for a somewhat more involved discussion. For the lion-hearted, the recommended authoritative but highly technical text is de Groot and Mazur (1984). From a practical standpoint of those who use diffusion data, however, the important points to note are:

(1) Flux rates in different reference frames are different for the same system and it is important to have consistency in comparing or combining systems.

(2) Conversion of results from one reference frame to another are always possible, and are usually, simple. The general relationship between flux in a frame 1, J^1, and in another frame 2, J^2, is given by

$$J^1 = J^2 + vC$$

where v is the local velocity of frame 2 with respect to frame 1 and C is concentration. Note that the above is a vector sum where each component of the vectors J^1, J^2 and v need to be added. This leads to a generalized version of Fick's law,

$$J^1_i = -D\nabla C_i + v^1 C_i \tag{5}$$

which takes care of a seeming contradiction that was developing—if the flux is the product of diffusion coefficient and local concentration gradient, how can it change depending on the choice of reference frame? The answer lies in the fact that Fick's first law, as stated earlier, implicitly assumed that the reference frame was static (v^1 was zero). If it moves with a local velocity v (recall local gradients?), then this motion needs to be added to the simple gradient term to obtain the flux according to this particular frame of reference. This kind of equation also takes care of diffusion under any kind of "background drift," e.g. convection plus diffusion. One can think of a physical example in the spreading of a blob of ink in a flowing river. Local concentration gradients generate diffusive flux and cause the blob to spread out while the background of flowing water makes the entire system move and provides an additional component to the flux measured by an observer standing on the bank.

(3) The question of units again, for the last time. Note that once the reference frame is defined, the units of flux have been defined. And if the units of flux are defined, the units of concentration have been defined. Thus, one needs to pay attention to consistency, again.

Before leaving this discussion on reference frames, we may add as an aside, a note on glass transitions. It has been discussed at length in this volume that the temperature of transition from a glassy to a liquid state in a silicate melt depends on the kind of observation in question (the experimental time scale). From the viewpoint of diffusion, one can consider the following—as long as the long-range disordered Si-O network remains fixed on a laboratory reference frame $(v^1 = 0$ in Eqn. 5), any other diffusing atom i in the system moves within a frozen, glassy background. Temperatures at which the Si-O network begins to move measurably $(v^1 \neq 0)$ with respect to the laboratory reference frame within the duration of observation (diffusion experiment), diffusion of the species i will occur within a dynamic, liquid-like medium which provides a drifting background. Depending on the kind of reference frame being used, one may observe a change (increase or decrease) of flux at this point, as seen from Equation (5). Note that this temperature is not unique for a system and depends on (1) the duration of the diffusion experiment and (2) the resolution with which motion of the Si-O network can be measured. We return to these topics in detail in *Diffusion, glass transition and relaxation*.

Limitations of Fick's law and non-Fickian diffusion

Fick's first law as stated at the beginning has already been modified to take into account the variability of reference frames. It can be further extended to take care of simultaneous diffusion of multiple species in a multicomponent melt.

Within the framework of diffusion of one kind of particle itself, however, there are still some limitations and qualifications to Fick's law which need to be recognized. First of all, Fick's law is a continuum law, i.e. it treats the medium in which diffusion takes place as continuous. The continuum approach breaks down at short spatial scales when individual atomic jumps are being observed (e.g. using some spectroscopic methods and in computer simulations) and Fick's first law would be a poor description. Indeed, the kinds of argument developed above in connection with Fick's second law using flux balance in a unit volume illustrate that on scales where a "volume" and "concentration" cannot be defined, Fick's laws become difficult to apply. A "scale" is defined by a unit of length and a unit of time (the effect of a slow diffusion process for a long time is the same as the effect of a fast diffusion process for a short time!). Thus, Fick's laws may be inapplicable on very short time scales as well. These are cases when one has to resort to statistical methods, discussed later in the chapter.

Fick's law may also be inapplicable in some cases where diffusion occurs in a medium whose structure is changing (e.g. Crank, 1975). Such cases may occur in melts near the glass transition region (see *Diffusion, glass transition and relaxation* below; also see Crank (1975) for some mathematical methods to treat these).

Note that Fick's law works for any particle, as long as the "linear regime" holds. This goes back to the fact that Fick's first law is empirical, i.e. it was based on a set of experimental observations which were carried out under conditions close to equilibrium. However, it has since been shown that under conditions far removed from equilibrium (e.g. under turbulent convection, exceedingly steep concentration gradients etc.) fluxes may not necessarily be merely proportional to concentration gradients and may be more complex functions. To account for such situations other forms of equation (e.g. the telegraphic equation, Kirwan and Kump, 1987) have been used. We will not deal with such complications in this chapter.

Different kinds of diffusion coefficients— Self and tracer diffusion coefficients

So far, we have consciously left the nature of the diffusion coefficient somewhat vague. A microscopic statistical treatment of diffusion shows that random motion of atoms (as opposed to a directed *flux*) occur in a medium whether or not a concentration gradient (or any other driving force) is present. This issue is taken up in some detail in the section on *microscopics* and a preliminary account has been given by Watson (1994). Indeed, such random motion is the well known Brownian motion that we mentioned in the section on *historical background* earlier. Thus, in a homogeneous phase if one can follow the displacement of an atom through time the diffusion coefficient that governs this displacement would be called the self-diffusion coefficient. The diffusion of Si atoms due to random motion in a SiO_2 melt would be an example of self diffusion. The displacement of a particular Si atom in a melt of SiO_2 is however difficult to follow without marking the atom in some way. This is often achieved by using a specific isotope of the element concerned, say, ^{29}Si in this case. The diffusion of this ^{29}Si may then be followed using a number of different methods (see *Experimental methods* below). And at this point, differences in opinion regarding definition of diffusion coefficients between experts come into play. According to some authors, the diffusion of this ^{29}Si would be governed by a coefficient called the tracer diffusion coefficient. Others prefer to preserve the term tracer diffusion coefficient for atoms that are only present in dilute quantities (e.g. small amounts of Na impurity in a SiO_2 melt) and define the diffusion coefficient for ^{29}Si in the current example also as a self-diffusion coefficient. Thus, the terms tracer- and self-diffusion coefficients have been used interchangeably to some extent in the literature. In some cases, the distinction between the two may be critical. It may therefore be a good practice to define the diffusion coefficient (or make the meaning clear by context) for every usage, irrespective of personal preference of terminology. For practical purposes, it is important to note a few points:

(1) The diffusion coefficient of an atom is different from that of a marked species of the same element by a factor called the correlation factor, f, which lies for most practical cases between 0.1 and 1. Thus, $D_A = f \cdot D^*_A$ where D^*_A is the diffusivity of the marked species of the atom A. Such a relation exists irrespective of the concentration of A. In terms of the microscopic mechanisms of diffusion by jumping from one site to another, the factor f is related to correlation between one jump and the next and hence its name. See the sections on *Microscopics* and *Diffusion and electrical properties* for more on this topic.

(2) These diffusion coefficients can in principle be (and in practice are) functions of composition—nothing suggests that they are independent of composition, as implied in some discussions in the literature.

(3) The most significant point to note from a practical standpoint is that these diffusion coefficients (tracer- or self-diffusion coefficients, whichever way they are defined) describe diffusion rates in the *absence of any driving forces e.g. concentration gradients in a thermodynamically non-ideal system*. These can be very different (but are not necessarily so) from diffusion rates of the same atoms in the presence of a concentration gradient. Thus, diffusion coefficients for use in solving a particular problem should be chosen carefully—this point is treated in detail in Sections II and III of this chapter.

II. DIFFUSION OF TWO COMPONENTS IN A MELT

Chemical diffusion coefficients and diffusive coupling

So far, we have considered only the flux of one particle in a medium. The presence of a concentration gradient, however, implies that as the relative concentration of one component decreases the concentration of at least one other component must increase. This is a direct consequence of the closure problem of concentrations, i.e. the relative concentration at each point in a system must add up to a constant (typically 100 or 1). In other words, it is impossible to have a concentration gradient of only one component in a single phase system. As flux is proportional to concentration gradient according to Fick's law, all components for which concentration gradients exist (a minimum of two components) must diffuse. In the process of such diffusion, the components may "interfere" with each other such that the diffusion rate of a given component may be increased or decreased with respect to its self- or tracer-diffusion coefficient in the same melt, as described above. In this section we restrict our attention to a binary system, i.e. a system where the decrease of one component along a concentration gradient is compensated by the increase of concentration of only *one* other component in the opposite direction. In such a case the diffusion process is one in which the two components simply "exchange places" with each other. We will study in some detail the similarities and dissimilarities between this kind of "exchange" diffusion and the "random motion" diffusion of an element discussed in the previous section. Focussing on the binary system allows many concepts to be explicitly illustrated. Following this, extension to a more general multicomponent system is essentially a step on to somewhat more involved mathematical treatments with only very few new additional concepts.

In discussing the differences between the processes of tracer-diffusion and "exchange diffusion", the perceptive reader will have already observed that while we spoke in terms of diffusion of particles like *elements or ions* for tracer diffusion, we have switched to a terminology involving the diffusion of *components* for the exchange diffusion process. The reason for this is that a concentration gradient may be defined for only a component in a thermodynamic sense (an unit that can be independently added or removed from a system) and not for elements or ions. This may be made clear by using a specific example—the system Na_2O-SiO_2. The constraint of local electrical neutrality prevents one from adding or removing only one kind of atom (Na, Si or O) to the melt while maintaining a single phase system. What can be added or removed are oxide components (Na_2O or SiO_2) or any other neutral combination of cations and anions. Accordingly, these are also the quantities for which concentration gradients and exchange diffusion can be described. Of course, it is possible to deal with diffusion in terms of ions while explicitly using electrical charge balance equations at every step to ensure that this condition is not violated. However, it is much more convenient to deal in terms of components which may be independently added or removed without imposing any additional constraints. More significantly, it may be shown that if fluxes of cations are of interest then using oxide components in a simple diffusion equation is equivalent to the use of individual ions and a explicit treatment of charge balance. This brings us to a very important point which is necessary to remember when dealing with diffusion in silicate melts and cannot be overemphasized: for describing any physical process involving motion of particles one needs to use tracer- or self- diffusion coeffcients, for describing any chemical process involving mass transport leading to a change in composition one needs to use chemical diffusion coefficients. These two are different physical quantities, although under some circumstances they may be numerically very close to each other. When referring to diffusion coefficients, it is therefore very important to have a clear concept as to which one of these quantities is being dealt with.

Some characteristics of chemical diffusion coefficients

The next important characteristic of exchange diffusion treated in this manner is that it is a macroscopic treatment. Instead of explicitly treating the mechanism of charge balance and the motion of individual atoms or ions, we are dealing with hypothetical components, which may or may not have any physical existence. In other words, use of oxide components does not imply that we postulate "oxides" to be present in the melts in any form. Alternately, any other neutral set of components may be chosen in a thermodynamically consistent manner to describe melt compositions and the choice would be equally valid.

This leads us to one of the most important aspects of macroscopic treatment of exchange diffusion—it is a construct to describe the evolution of concentrations and mass transfer in melts. Since there are numerous alternate ways of choosing components in a system, a particular choice of components does not necessarily imply anything about the actual diffusion mechanism or structural units present in the melt. In this regard, this kind of treatment has all the advantages as well as disadvantages of the thermodynamic method. As in thermodynamics, the initial and final states of the system are described without any knowledge whatsoever of the intermediate steps involved. For example, in our model Na_2O-SiO_2 system, any change in concentration along a gradient may be described in terms of a flux of Na_2O. This is irrespective of how complicated the actual process is—it may involve several intermediate steps involving complicated and variable combinations of all three atoms, Na, Si and O. This is an advantage in that mass transport calculations may be done without any knowledge of the actual diffusion mechanisms. On the other hand, the disadvantage is that knowing diffusion rates may not tell us much about actual mechanisms. This is a particular problem when extrapolations need to be made—a situation encountered all too often in geosciences. Other aspects of the macroscopic thermodynamic tradition are inherited by this method of treating diffusion as well. In any system, possibilities are almost endless for a permissible choice of components. However, it is well known in the realm of thermodynamic modelling that the choice of a component similar to a real physical entity often simplifies the models and provides insights into mechanisms that are not afforded by other choices. The same applies to macroscopic treatments of exchange diffusion. If possible, one should attempt to choose components which are similar to the actual diffusing species—this makes subsequent mathematical manipulations simpler (indeed, sometimes these are the only cases where a mathematical treatment is possible!), provides insight into diffusion mechanisms and helps to understand the behavior of a system. A detailed discussion of components in the context of diffusion is given by Brady (1975b).

A brief comment with regard to the choice of reference frames in the context of the various choices of components is in place here. Note that we have been implicitly speaking in this section in terms of a laboratory reference frame, i.e. the viewpoint of an observer standing outside the system and observing it. If one chooses one of the components themselves as the reference frame (e.g. Na_2O), then of course by definition one would observe the flux of only one component, SiO_2—in our system. This should highlight how important it is to define reference frames when discussing fluxes. To emphasize the aspect, let us elaborate this example to a more specific extreme. Consider a case where an observer viewing a melt in a crucible is capable of seeing the motion of individual ions/atoms and sees that only the Na and O atoms are moving around while Si atoms remain fixed. If this state is then described with respect to a reference frame fixed to the Na_2O component, then the flux of Na_2O will be zero by definition. This of course, does not imply that the Na_2O component did not "move", in a conventional sense. Hence, before making any conclusions about diffusion mechanisms from diffusion data, the fluxes must be referred to appropriate reference frames—otherwise misleading conclusions like the one in the above example may be obtained.

Relationship between chemical and tracer diffusion coefficients

Having discussed the nature of the exchange diffusion process, it is now worthwhile to see how diffusion coefficients which describe this process are related to the tracer- or self-diffusion coefficients discussed in Section-I. To do this, one usually makes use of a quantity called mobility. Mobility is defined as the limiting velocity of an atom (ion) under the action of an unit force acting on it in a given medium, everything else remaining fixed. One makes the assumption that the mobility of an atom remains unchanged whether a concentration (chemical potential) gradient is present or not. Then, in the presence of a concentration gradient, the diffusion rate of each kind of atom is calculated assuming they do not interfere. In practice, there are physical constraints which force them to interact. In the classical derivations for metallic alloys in a volume fixed frame (e.g. Darken, 1948), this constraint is that volume changes are resisted. For an ideal system, imposing this constraint on the diffusion rates obtained using various mobilities leads to

$$D = X_1 D_2^* + X_2 D_1^* \tag{6}$$

where D_i^* is the self diffusion coefficient of i at a given composition and D is the chemical diffusion coefficient, X_i is the mole fraction of i (see Allnatt and Lidiard, 1993, for a rigorous derivation of the relationship between mobility and tracer diffusion coefficients). A more stringent constraint in systems involving ions is that local build-up of electrical charges are resisted, i.e. local neutrality is preserved. This implies, that if one ion tends to move faster than another then the local electrical field that develops will generate a force on it to slow it down and prevent local charge buildup. Again, the independent flux rates that are obtained using mobilities are related to each other subject to the condition that the net flux of charges is zero. The equation thus obtained is then compared with the simple Fick's law statement of exchange diffusion to obtain a relationship between the tracer-diffusion coefficients and the exchange or chemical diffusion coefficients. The equation has the form

$$D = \frac{D_1^* D_2^*}{X_1 D_1^* + X_2 D_2^*}. \tag{7}$$

The mathematical derivation is given in Barrer et al. (1963), Manning (1968), and Brady (1975) but may also be carried out as an exercise following the steps outlined above. The expression given above is for a thermodynamically ideal mixing behavior. For non-ideal solutions, an additional *thermodynamic factor* ϕ appears in Equations (6) and (7), and one obtains forms like

$$D = \left[\frac{D_1^* D_2^*}{X_1 D_1^* + X_2 D_2^*} \right] \phi. \tag{8}$$

We will return to thermodynamic factors below, but there are several points to note about these equations in general. First, the form of the Darken and Manning equations are different. The long range and higher magnitude of Coulomb forces in ionic systems suggest that they would be the more dominant restorative forces; hence if only one effect can be addressed, equations derived assuming local charge neutrality are likely to be more appropriate for silicate systems. This is supported by experimental data in simple systems where Equation (7) was shown to be valid (Meier et al., 1990). In the perfect situation of course both effects should be taken into account, see *Multicomponent diffusion* below for more on this. The essential physical difference between the two kinds of equations is that in the Darken equation, the exchange rate is dominated by the fastest moving species (larger D*) whereas in the Manning equation the slower moving species has a more determinative influence on the net exchange process. However, for a dilute component i (X_i very small in Eqns. 6 and 7), both forms of equation predict that the ideal part of the chemical diffusion coefficient D is essentially equal to the tracer diffusion coefficient of the dilute component, D_i^*. This is a very important characteristic of chemical diffusion that holds even in complex multicomponent systems (see below). The significant implication is that tracer diffusion coefficients, corrected for the thermodynamic effect, may be used to calculate mass transport rates of trace elements without any knowledge of the diffusion rates or activities of the other components in any melt. The quantity ($D_i^* \phi$) has often been termed the *intrinsic diffusion coefficient* in the literature, although the same term has also been used sometimes to describe completely unrelated quantities. Lesher (1994) has used these intrinsic diffusion coefficients with thermodynamic activities measured from Soret diffusion experiments to successfully predict mass transport behavior of dilute constituents, Sr and Nd in complex diabase-felsic gneiss diffusion couples.

Another difference of some practical consequence between the Darken and the Manning equations is that the two forms of equations lead to different kinds of compositional dependence of exchange or chemical diffusion rates. With regard to applications, the calculated exchange diffusion rates using the same values of tracer- or self-diffusion coefficients are very similar for the two models for many cases—particularly if all tracer diffusion coefficients are similar in magnitude. Thus, although one kind of model may be theoretically more justified for various reasons, from a practical standpoint it is often difficult to distinguish between the two. In diffusion systems where tracer diffusion coefficients of different species differ by orders of magnitude the differences are, however, significant. In silicate melt systems such situations may arise quite commonly during the mixing of major elements because tracer diffusion rates of alkali atoms, and sometimes those of other network modifiers as well, can differ from those of Si and other network formers by several orders of magnitude. In these cases, use of an accurate model is necessary to predict chemical diffusion rates properly from tracer diffusion data.

Thermodynamic factor

The thermodynamic factor that we have neglected in the preceding discussion is the term ϕ of the form

$$\phi = 1 + \frac{\delta \ln \gamma_i}{\delta \ln X_i}$$

where γ_i is the activity coefficient of component i in the melt. The physical nature of the term may be understood as follows: when mixing of the two components in a binary system results in a net reduction of the enthalpy of the system, then there is an increased tendency to mix. In this case, the thermodynamic factor ϕ is greater than unity and the diffusion rate is enhanced by this factor. When mixing leads to an increase of enthalpy, the thermodynamic factor works to retard diffusion rates and the value of the multiplicative factor ϕ is less than unity. To get an idea about the magnitude of this thermodynamic factor we can make a simple calculation. For a binary melt in which thermodynamic mixing between components 1 and 2 may be described by a simple mixture model

$$R T \ln \gamma_1 = W_{12} X_2^2$$

where W_{12} is an interaction parameter it can be shown (Chakraborty and Ganguly, 1992) that the thermodynamic factor is given by

$$\phi = 1 - \frac{2W_{12}[X_1(1-X_1)]}{RT}.$$

If we now arbitrarily take numerical values of 8000 J/mol for the interaction parameter and calculate φ at 1000°C for an intermediate (half-and-half mixture of components 1 and 2) melt composition we obtain φ = 0.62. This suggests that the thermodynamic factor may be a fairly moderate correction term to the ideal assumption in many circumstances. Note that the form of the equations also show that φ becomes smaller with increasing temperature—this is a common feature which is observed even when more complex activity-composition models are chosen. There are, however, notable exceptions to this rule. Much larger interaction parameters on the order of 40,000 J/mol have been observed in silicate melts (e.g. see Berman and Brown, 1984). But more importantly, when compositions approach a solvus the thermodynamic factor becomes large and may modify diffusion coefficients by orders of magnitudes (e.g. see Cussler, 1984, Brady and McCallister, 1983, although the examples are not from silicate melt systems).

A second point that is interesting to note about φ is that the *gradient* in compositional space of the activity coefficients γ and not the coefficients themselves are involved. The relationships between tracer and chemical diffusion coefficients discussed here become much more complex for multicomponent systems but this dependence on gradients rather than on absolute values remains unchanged in many of the complex multicomponent models as well (e.g. Lasaga, 1979a,b; Cooper, 1965). This fact can be exploited in many cases where activity coefficients are poorly known. For example, when small concentration gradients are involved it is reasonable to neglect the gradient of activity coefficients to a first approximation in many cases, even for strongly nonideal melts (where the presence of a solvus can be precluded). This allows chemical and tracer diffusion coefficients to be related in many complex melts even when activity coefficients are unknown, an example of which may be found in Chakraborty et al. (1995b).

Thermodynamic formulation of diffusion

In describing diffusion above we have been relying on the empirical Fick's law, i.e. flux of an atom is proportional to the concentration gradient. However, at equilibrium between two phases it is the chemical potential of a component that should be equal and not the concentrations in the coexisting phases. Indeed, element partitioning between phases at equilibrium (i.e. different concentrations of an element in two phases at equilibrium with each other) is a key concept in geochemistry. Thus, a little consideration tells us that the flux of material should be dictated by gradients in chemical potentials and not concentrations, such that on reaching equilibrium gradients in chemical potentials disappear. This was recognized and developed into a theory by Onsager in his classic treatise on diffusion (Onsager, 1931b, 1945; Onsager and Fuoss, 1932). Thus in the same spirit as Fick's law, one can state that flux of a component is proportional to its chemical potential gradient instead of the concentration gradient, i.e.

$$J_i = -L_i \frac{\partial \mu_i}{\partial X} \tag{9}$$

where L_i is called the *kinetic coefficient* or the *Onsager coefficient* to distinguish it from the Fickian diffusion coefficient, D_i defined by Equation (1). In the geological literature such equations have been used by metamorphic petrologists to understand the development of metasomatic zones (e.g. Fisher, 1973; Joesten, 1977 etc.). Using the definition of chemical potential, $\mu_i = \mu^0 + RT \ln a_i$, where a_i is the activity (given by the product of concentration, C_i and the activity coefficient, γ_i) in Equation (9) we find

$$J_i = -L_i \frac{\partial (\mu^0 + RT \ln a_i)}{\partial X} = -L_i \frac{\partial (\mu^0 + RT \ln C_i + RT \ln \gamma_i)}{\partial X}$$

For an ideal solution, $\gamma_i = 1$, the above expression reduces to

$$J_i = -\left[\frac{L_i RT}{C_i}\right]\frac{\partial C_i}{\partial X}.$$

Comparing this with the standard Fick's law it is easy to see that in a thermodynamically ideal system, D_i is related to the kinetic coefficients by

$$D_i = \frac{L_i RT}{C_i}. \qquad (10)$$

This expression highlights a number of properties of the kinetic coefficients. From the point of view of a theoretician, the Onsager coefficients are superior to the standard Fickian coefficients because Onsager (1945) related these to fundamental microscopic motion of atoms in a statistical mechanical sense (see the section on *microscopics*). From a practical point of view, however, there are a number of difficulties in using these. Firstly, the Onsager coefficients often depend more strongly on composition than the standard Fickian coefficient. This makes them somewhat inconvenient to use on geochemical materials where compositional variability is the rule. Secondly, and more critically, these coefficients relate flux to chemical potential gradients which cannot be directly measured. Chemical potentials can only be calculated from measured compositions but activity-composition relations in complex geological melts are poorly known—characterizing the free-energy surface of complex geological material continues to be a daunting task. This lack of knowledge of component activities is a crippling disadvantage that has restricted the use of the theoretically preferable Onsager coefficients in the solution of practical problems. We will see later however that in complex multicomponent systems considering these coefficients becomes unavoidable and in the face of frustration various empirical schemes have been devised.

III. DIFFUSION IN A MULTICOMPONENT MELT

Fick-Onsager relations

In a multicomponent system it is possible to have concentration gradients of more than one component at a time which results in simultaneous, multicomponent flux of atoms. Generalization of the concepts developed above for diffusion in a binary system tells us that in such a situation the flux of the different components will interfere with each other. This happens because the various components will try to diffuse at different rates consistent with their individual mobilities but will be constrained by a number of factors, e.g. considerations of local volume and/or charge conservation. In liquids, additionally, a diffusing particle will try to choose a path to minimize the frictional resistance or viscous drag on itself (e.g. Batchelor, 1976; Beenaker and Mazur, 1983). It is clear that the interaction of various mobilities with these different kinds of restorative forces will lead to rather complex and variable diffusion behavior—the possibilities multiply rapidly as the number of components increase. Accordingly, a general mechanistic theory of multicomponent diffusion in liquids is expected to be rather complex and does not exist. However, Onsager (1945) was able to suggest a phenomenological extension of Fick's law for multicomponent solutions. Seeing that the diffusion of each component will be influenced by the diffusion of others, he postulated that the flux of a component will depend not only on its own concentration gradient as in Fick's law but on the concentration gradient of all components in the system. And, as long as one is not too far removed from equilibrium (the driving forces are not too large), he made the reasonable assumption that this dependence will be linear, i.e. the flux of a component will be a linear combination of the concentration gradients of all components in the system. This is the well known multicomponent extension of Fick's law due to Onsager, stated as

$$J_1 = -D_{11}\frac{\partial C_1}{\partial X} - D_{12}\frac{\partial C_2}{\partial X} - \ldots -D_{1n}\frac{\partial C_n}{\partial X}. \tag{11}$$

The constants used in the linear combination above are the various diffusion coefficients, D_{ij}. Each of these D_{ij} characterize the influence of a component j on the flux of a component i. Since it is possible to write one such flux equation for each of (n-1) components[1], one obtains a (n-1)×(n-1) **D** matrix for a system with n-neutral components (with n+1 elements—n cations plus oxygen). The influence of a component i on the diffusion rate of j is usually not the same as the influence of j on i, i.e. $D_{ij} \neq D_{ji}$. Unlike in aqueous solutions (e.g. Cussler, 1976, 1984), the diffusive coupling is usually significant (i.e. $D_{ij} \sim D_{ii}$) in silicate melt systems (e.g. see Chakraborty et al., 1995a, Kress and Ghiorso, 1993 and references therein).

Up to this point, there is nothing special about this form of the multicomponent flux equation—it is just one of many mathematical possibilities to describe diffusive coupling. However, Onsager (1945) was able to relate it to the principle of microscopic reversibility in statistical mechanics developed earlier by him (Onsager 1931a,b). All the reasoning presented above for chemical potential gradients being a better variable for characterizing the driving force of chemical diffusion in a binary system apply equally to multicomponent systems. This led Onsager (1945) to derive a parallel to the binary thermodynamic flux law (Eqn. 9) for multicomponent systems as well

$$J_i = -\sum_j L_{ij}\frac{\partial \mu_j}{\partial X} \tag{12}$$

Just like in the binary system, one can introduce the definitions of chemical potential to obtain a relationship between the **D** and the **L** matrix (containing L_{ij} terms for all components from (n-1) equations like Eqn. 12 above). The algebra can be a little cumbersome, but leads to the very simple looking matrix equation

$$\mathbf{LG} = \mathbf{D} \tag{13}$$

where the matrices **L** and **D** have already been defined and the matrix **G** is a matrix of thermodynamic mixing properties, whose elements, G_{ij} are given by

$$G_{ij} = \frac{\partial (\mu_i - \mu_n)}{\partial C_j}$$

The exact form of the elements depend on the choice of the mixing model. The important point to note is that Equation (13) is valid for thermodynamically ideal as well as non-ideal solutions and even for ideal solutions the matrices **D** and **L** are not the same.

Some important characteristics of the L and D matrices

The matrix **L** has a number of very important characteristics as a consequence of the laws of irreversible thermodynamics. Most importantly, the principle of microscopic reversibility demands that the matrix be a symmetric (i.e. all $L_{ij} = L_{ji}$, note the difference with **D** here) and a positive definite (i.e. all eigenvalues are positive) matrix. In the one case where enough data were available in silicate melt systems, the Onsager symmetry relations

---

[1] Following considerable discussion in the literature (see de Groot and Mazur, 1984 for details) after Onsager's work, it has been shown that in order to satisfy the microscopic basis which Onsager provided, the fluxes must be independent. Since in a n-component system with a defined reference frame, knowing the fluxes of (n-1) components automatically fixes the flux of the nth component, one has to arbitrarily eliminate one of the components as dependent. Thus, in a Fick-Onsager treatment of a n-component system one always works with only (n-1) forces and fluxes, the remaining one being determined by difference.

were found to be valid (Spera and Trial, 1993; Chakraborty, 1994). At local equilibrium[1], the **G** matrix has to be positive definite as well (Prigogine and Defay, 1954). This implies that the **D** matrix, which is a product of the two, must also be a positive definite matrix—a very important mathematical characteristic for practical applications, as we will see below. Some other important characteristics of the **L**-coefficients are that (1) they are *local* coefficients, like the Ds, and (2) they can be functions of T, P, etc. but are independent of forces, e.g. gradients of concentration, chemical potential or temperature.

The most powerful aspect of the **L** matrix formulation is however the fact that it provides a direct link to microscopic motion of individual particles (atoms/ions). This is shown in the section on *microscopics* later in this chapter. At this point it is important again to recall the differences between diffusion of particles and diffusion of components. The coefficients that relate to the atomic jumps (say, L') are not the same as the L coefficients defined above for flux of components. However, the two sets are related and knowing any one set the other can be calculated very easily. Equations for such conversions may be found in Allnatt and Lidiard (1993) or, specifically for silicate melt systems in Petuskey and Schmalzried (1980).

This direct link to individual microscopic jumps, the strong experimental support from studies in a variety of systems, and some of the resulting mathematical simplicity of this form of the multicomponent extension are the features that make it so much more attractive and well used compared to various other possible mathematical functions that may be considered (see Cussler, 1976, 1984, for some other approaches, which may sometimes be useful in very special cases).

The fact that the **D** matrix is positive definite means the various elements of a **D** matrix are related to each other according to specific mathematical rules. These relationships have been derived in a number of places (e.g. Gupta and Cooper, 1971; Kirkaldy and Young, 1987) and are found to be useful in two different ways in the study of multicomponent diffusion. First, the retrieval of **D** matrices from experimentally measured concentration profiles is tricky and may often lead to spurious numbers. Checking whether best-fit **D** matrices to experimental data satisfy the required mathematical conditions is one means of establishing their veracity (e.g. Chakraborty et al., 1995a). The second use of the relationships is more general and in mathematical terms relates to the fact that a positive definite matrix is always amenable to a similarity transform. This in turn implies that coupled diffusion equations may always be decoupled by diagonalizing the **D** matrix (Toor, 1964; Cullinan, 1965). The procedure for such decoupling may be found in a number of works (e.g. Lasaga, 1979a; Kirkaldy and Young, 1987). In non-mathematical terms, this means that a multicomponent diffusion problem may always be reduced to a set of simple, uncoupled binary diffusion problems which can be solved using the methods and solutions

---

[1] The role of local equilibrium in diffusion needs to be clarified at this point. The use of a diffusing component made up of more than one species (e.g. an oxide component AO is made up of a cation, A and oxygen, O) implicitly assumes that within a small incremental volume the reaction of formation A+O = AO attains equilibrium with the local conditions very quickly in comparison to the time scale of diffusion. Further, in a multicomponent system it is also implied that the progress of a reaction A+O = AO in this elemental volume is not affected by the progress of another reaction B+O = BO. In our discussion of Fick's second law as well as chemical diffusion in a binary system, we have already invoked the concept of these small, incremental volumes. For multicomponent systems, unless the diffusion system is mentally discretized into such incremental volumes, one cannot use the Onsager relations to describe vectorial quantities like fluxes of atoms (see de Groot and Mazur, 1984 for details). Thus thermal (see the section on *microscopics*) and chemical equilibrium in such incremental volumes is a basic requirement for the use of Fick-Onsager relations.

widely available for simple diffusion equations (e.g. Crank, 1975; Carslaw and Jaeger, 1959). A very significant conclusion follows from this—a diffusion problem in a multicomponent system is exactly the same as a binary diffusion problem and is no more difficult in any way; in all cases where a binary problem can be solved, a multicomponent problem can be solved as well! The only exception to this is when the boundary conditions for a problem are not homogeneous, e.g. a given boundary of the system of interest is insulating for some components but not for others. However, in most such cases it is usually possible to use mathematical tricks to overcome the handicap. The main difficulty in dealing with multicomponent diffusion therefore is not in the application, but in the experimental determination of all the multicomponent diffusion coefficients. We will return to this issue in the section on *Experimental Methods*.

The use of **D** matrices implies that diffusion fluxes may be treated using vector algebra, and this allows constraints to be placed on allowable forms of compositional evolution during diffusion and the magnitudes of various diffusion coefficients. These have been discussed most notably by Cooper (1974) and Gupta and Cooper (1971).

Domain of validity of Fick-Onsager relation

The multicomponent diffusion equation stated above can only be valid as long as the linear approximation underlying it is valid. Under situations far removed from equilibrium (e.g. diffusion of gases in a flame) the approach cannot be used, but fortunately such situations are not common in viscous geological melts. There are, however, at least two possible situations where the Fick-Onsager equations are of limited use for silicate melts without modifications. The first one of these is the case where diffusion occurs in the presence of a biasing force field (e.g. an externally driven mechanical, electrical or gravitational field). The influence of the gravitational field has attracted some attention in the recent literature (e.g. Vitagliano et al., 1992; Liang et al., 1994). The basic idea is that if diffusion of different components at different rates leads to a density reversal within the diffusion zone then the liquid may become unstable with respect to convection and convective overturn will disrupt the diffusion zone. In this case diffusion itself may still proceed according to the Fick-Onsager equations—only the resulting diffusion layers are unstable.

A second more common and potentially serious violation may occur where a component in the melt occurs in more than one species (e.g. by forming different complexes). In that case, diffusion will proceed in parallel with homogeneous internal reactions between different species. The flux of a component will then no longer be determined by concentration gradients alone but also by the stoichiometry and rates of various internal reactions. Mathematically, this means that fluxes become non-linear functions of composition—a condition which Onsager (1945) had already clearly recognized as unsuitable for treatment using his formulation. It has now been shown experimentally that Fick-Onsager equations are invalid in multicomponent aqueous electrolyte solutions where such speciation is well known (e.g. Spallek et al., 1990). In geological melts, a particularly problematic case may be the diffusion of components in water bearing multicomponent melts. Diffusion of water itself is governed by equations which may be non-linear due to the occurrence of internal reactions involving the interconversion of free water molecules to hydroxyls and vice versa. Such nonlinearity has been predicted (e.g. Wasserburg, 1988; Chekhmir et al., 1988), measured and systematized (Zhang et al., 1991a,b; Jambon et al., 1992). It is also well known that the presence of water enhances diffusion rates of other components in the melt (see review by Watson, 1994). The combination of these two features suggests that chemical diffusion in multicomponent water bearing melts may be difficult to describe within the framework of the simple Fick-Onsager relations without taking the non-linear effects of species interconversion into account. On purely theoretical grounds, the same argument may apply to diffusion of oxygen as well because oxygen can exist in silicate melts as bridging oxygen, non-bridging oxygen and molecularly dissolved oxygen (see *Diffusion and melt structure* below as well as Zhang et al., 1991b). However, in spite of these, oxygen diffusion rates can often be described by an effective diffusion coefficient which is not strongly dependent on composition (e.g. Young et al., 1994b). This would suggest that in many cases (but not all) Fick-Onsager relations would provide a good description of the macroscopic diffusion processes although the link to microscopic aspects may be obscured (since the oxygen diffusion rates are only "effective;" see *Diffusion and melt structure* and *microscopics* for further discussion on this). There is a sufficient body of experimental data available now which attest to this general validity of the Fick-Onsager relations (see Chakraborty et al., 1995a,b; Kress and Ghiorso, 1995 and the references in these works) as well as occasional inability to describe features developed during diffusion in silicate melts (e.g. Kress and Ghiorso, 1993). To my knowledge, studies aimed at specifically testing the effects of independently measurable speciation on multicomponent diffusion have not yet been performed in silicate melts, although some attempts have been made to relate *assumed* speciation to diffusion rates (see *Diffusion and melt structure* below).

Models relating tracer to chemical diffusion coefficients in multicomponent systems

It is intuitively expected that analogous to the binary systems, tracer- or self-diffusion coefficients should be related to chemical diffusion coefficients in multicomponent systems as well. However, the various available relations for doing so are much more complex and are derived based on different assumptions. The fact that differences between models are often very small means that diffusion coefficients must be precisely known and preferably measured on the same material for proper comparison. Such studies are now just beginning to be undertaken (LaTourette et al., 1995; Richter et al., 1994, etc.). In the absence of an unequivocal choice, I will simply present the logic behind the derivation of some of the more commonly used models. Because of the uncertainties involved, I refrain from reproducing the cumbersome algebraic expressions relating tracer and chemical diffusion here, the interested reader is urged to obtain them from the original works cited below.

The factors that modify the random walk mobility[1] of an individual ion during chemical diffusion in a silicate melt include:

(1) constraints of local electroneutrality: if an ion moves much faster or slower than others, there will be a local charge build up. To avoid this, there must be coupling with at least one other ion. The movement of the ions must be slowed down/accelerated in a manner that charge imbalance is compensated. This leads to a Manning type of equation discussed in the binary system.

(2) constraints of thermodynamic activity: as in the binary system, the movement of ions will be optimized to reduce the free energy of mixing.

(3) constraints of speciation: if an ion can be part of different mobile units within the melt, the stoichiometry of the speciation reactions control the net mobility of the ion. In the simplest case, at least oxygen is part of all components in the melt.

In the commonly used volume fixed reference frame for liquids, one has additionally:

(4) constraints of local volume conservation: the movement of different ions are optimized such that volume changes are compensated. This is the basis of the classical Darken equations developed for metallic alloys.

---

[1] The relationship between tracer diffusion coefficients and mobilities already involves the correlation factor, f. This should be taken into consideration in very precise work.

There is no available model for silicate melts which takes care of all of these factors. Okongwu et al. (1973) developed a model for glasses assuming (1) to be dominant and therefore accounting for this factor only. Cooper (1965) developed a model which takes care of all of the factors excepting (3). However, he had to make the assumptions that (a) the volume of a melt is primarily determined by the volume of the anions and (b) the chemical potential of the anions (i.e. fO_2 in the usual case) is constant all through the diffusion system at all times. Lasaga (1979 a) provide models for silicates, which when applied to melts take care of (1) and (2). Lasaga (1979b) developed a model for aqueous solutions which accounts explicitly for (1), (2) and (3). Finally, Lasaga (1995) illustrates how to combine these approaches for an ideal silicate melt, i.e. taking care of all factors excepting (2). It is possible to extend his derivation to obtain a relationship that accounts for all of the factors, but the resulting expression is not available in published form. It is an algebraic exercise that the reader may wish to try out.

In spite of the unavailability of a complete model, it is apparent from the above studies as well as numerous results in aqueous systems that certain generalizations may hold. Just like in the binary system, the chemical diffusion rate of a dilute component is largely determined by its tracer diffusivity (modified by the thermodynamic factor). Continuing the analogy with binary systems, it is the gradients in activity coefficients with respect to composition, rather than the absolute values of the activity coefficients themselves that are important in relating chemical and tracer diffusivities. Lastly, in an interesting derivation Lasaga (1995) delineates the circumstances under which Darken-type equations (accounting mainly for (4)) and Manning type equations (accounting mainly for (1)) are valid, respectively. With the assumption that melt molar volume is determined by the volume of anions, he shows that when oxygen mobilities are low (e.g. in a glass or a viscous fluid), Manning type equations are applicable. This is reasonable, being analogous to the case in solid oxides where cations hop in a fixed oxygen lattice. On the other hand, when oxygen mobilities are high, Darken type equations are applicable. Again, this is a reasonable result because in this case volume changes due to the motion of large oxygen ions would be the dominating feature and the Darken equations optimize ionic motion to counter volume change. Finally, in spite of the obvious shortcomings, Baker (1992a) and Chakraborty et al. (1995b) were able to use even the simple Lasaga (1979a) model with moderate success to describe multicomponent diffusion in silicate melts. This suggests that some of the more complete models, even if incomplete in some respects, may turn out to be adequate for practical purposes.

Effective binary diffusion coefficients and multicomponent D matrices

Almost all studies of chemical diffusion in natural melt compositions have treated the process in terms of a single effective binary diffusion coefficient (EBDC) so far. The concept, developed by Cooper (1968), treats the diffusing component as a "solute" and the remaining matrix as a "solvent."

$$J_i = -D_i(EB)\frac{\partial C_i}{\partial X} \tag{14}$$

where $D_i(EB)$ is the effective binary diffusion coefficient of component i. The effects of diffusive coupling are absorbed in this single diffusion coefficient and consequently, these coefficients are very strong functions of composition and more notably, of concentration gradients as well. Comparing Equation (14) with the Fick-Onsager relation (Eqn. 11) it may be shown that (Cooper, 1968; Chakraborty and Ganguly, 1992) the relationship between the effective binary diffusion coefficients and the Fick-Onsager coefficients, D_{ij} is given by

$$D_i(EB) = D_{ii} + \sum_{j=1}^{j=(n-1)} D_{ij, j \neq i} \frac{\partial C_j}{\partial C_i}. \tag{15}$$

This relationship helps to clarify a number of characteristics of EBDCs. First, ability to describe a diffusion process by an effective binary diffusion coefficient does not imply that diffusive coupling is weak (i.e. D_{ij} are negligible). Second, the expression shows how the EBDC depends on concentration differences, ∂C_j. This means that measured diffusion coefficients are not transferable unless the concentration gradients to be modelled are similar to those used in the measurements. From a physical standpoint, this implies that the coefficients are not true transport coefficients because they depend not only on state parameters like P and T but also on forces e.g. concentration (chemical potential) gradients. Thus, unlike true binary interdiffusion coefficients which are the same for *both* components, EBDC for each component in a multicomponent system are typically different. Diffusion coefficients obtained by fitting a binary model to measured profiles in multicomponent silicate melts have been found (Chakraborty et al., 1995a) to agree with $D_i(EB)$ calculated using Equation (15). A single, constant EBDC could be used to fit the measured data but the errors were much higher than in the measured D_{ij} from the same data. This is an expected result—using a single, constant EBDC to fit a concentration profile means terms involving the local and variable ∂C_j and ∂C_i along a concentration profile in Equation (15) are replaced by mean ΔC_j and ΔC_i, so that this error gets incorporated into the EBDC retrieved from the profiles. Some potential pitfalls in the use and determination of EBDC which can be avoided by careful measurements have also been outlined in Chakraborty et al. (1995a).

It has been shown by Cooper (1968) that all multicomponent diffusion problems can be reduced to effective binary ones provided (1) the system is infinite with respect to diffusion, i.e. diffusion does not reach the boundaries of the system, like in many diffusion couple experiments, and (2) there are no inflections in the concentration profiles e.g. in the case of uphill diffusion (see below). Chakraborty and Ganguly (1992) suggest a method of incorporating the thermodynamic factor into this approach in the spirit of an effective binary treatment. Lastly, note that Zhang et al. (1989) used an effective binary diffusion coefficient to treat the problem of diffusion of major elements during crystal dissolution which is different from the quantity described here. Their approach incorporates additional constants related to equilibrium partitioning into the definition of the diffusion coefficients.

Diffusive coupling and uphill diffusion

One of the main qualitative differences between diffusion in simple binary systems and multicomponent systems is that it is possible to observe uphill diffusion—diffusion of a component against its own concentration gradient. The effect arises from diffusive coupling and may be anticipated from the Fick-Onsager equation (Eqn. 11). If the non-diagonal terms of the **D** matrix, D_{ij}, $_{i \neq j}$, as well as the concentration gradients of various components j are large, then they may overwhelm the flux due to the gradient in compositional space of the diffusing component, i, itself, i.e.

$$\sum_{j, j \neq i} D_{ij} \frac{\partial C_j}{\partial X} > D_{ii} \frac{\partial C_i}{\partial X}$$

Depending on the nature of the other concentration gradients and the signs of various D_{ij}, diffusion of a component i may then progress against its own concentration gradient[1].

---

[1] It is possible to have diffusion of a component against its own chemical potential gradient as well. Although this seems counterintuitive, note that this is theoretically consistent with the Onsager equation (Eqn. 12) where high values of the cross terms L_{ij} can lead to diffusion of a component against its own

A consequence of this is that initial concentration differences of a component become larger with diffusion and inflections appear in a concentration profile. An example from experimentally measured data is shown in Figure 2, where profiles of the components SiO_2 and Al_2O_3 show inflections due to uphill diffusion and the initial differences in concentration of these components have been magnified in the diffusion zone. As diffusion proceeds, these inflections migrate inwards (away from the interface). Once diffusion reaches the ends of the diffusion couple, the shape of the profiles will change and the inflections will begin to decay such that at the end of the mixing process all compositional gradients are erased. Another manifestation of the same phenomenon which is less easily observed is that the diffusion of a component may be *enhanced* due to coupling. This can occur if the signs of the various D_{ij} and concentration gradients are suitably disposed to each other. As long as the system remains "infinite" with respect to diffusion, however, not only can such effects be rationalized—they can even be predicted if the **D** matrix is known.

Solutions to differential equations representing multicomponent extensions of Fick's second law (which is obtained by applying the continuity relation to the Fick-Onsager equations, just like for binary systems) describe the concentration of different components as a function of time at different points within the system. The important feature of these solutions is that in a given system (i.e. given the values of the diffusion coefficients, D_{ij}) they can always be expressed in the form $C_i = \Delta C_i \cdot f(x,t,\theta)$ where $x \equiv$ distance, $t \equiv$ time and $\theta \equiv$ a variable that describes direction in composition space. An explicit example may clarify this latter variable. Let the two independent components A and B of a ternary system ABC be plotted along the x- and y-axes of a Cartesian plot. If we have diffusion between two endmembers, α and β, containing concentrations a_α, b_α and a_β, b_β of the components A and B, respectively, then α and β will plot as two points with co-ordinates (a_α, b_α) and (a_β, b_β). The angle that the line joining these two points makes with the x-axis is defined as θ. Now, any concentration profile is a line describing compositions as a function of distance at a *fixed* time, t. Therefore, whether inflections (maxima or minima) due to uphill diffusion occur along a profile is asked mathematically by posing the question—does (dC_i/dX) equal zero at any point along the profile? From the form of the solution given above, it is seen that at fixed time the values of (dC_i/dX) are a function of the variable θ only. This means, for a given **D** matrix (which depends on composition, temperature etc.), it is the nature of the initial concentration difference that determines whether uphill diffusion occurs. It is possible to obtain uphill diffusion of different components, even with the same **D** matrix, simply by changing the initial gradients. Figure 2, showing data from Chakraborty et al. (1995b), is an example of such a situation where diffusion according to the same **D** matrix (i.e. same average composition and temperature) results in uphill diffusion of different components (Al_2O_3 and SiO_2, respectively). Thompson and Morral (1986) provide equations which allow one to calculate values of θ for which uphill diffusion of different components for a given **D** matrix will occur; Chakraborty et al. (1995b) illustrate the method with silicate melts. It is intuitively expected that in a physical sense, uphill diffusion occurs because a component is "dragged along" by others against its own concentration gradient. Such a coupling may be either thermodynamic (i.e. the activity of a component depends on the concentration of other components) or kinetic (i.e. the

---

chemical potential gradient. Mechanistically, an explanation for this phenomenon may be provided as follows: Chemical potential gradients drive the flux of a component as long as it is *free* to move i.e. not under the influence of any other forces. In the presence of forces arising from diffusive coupling, chemical potential gradients are no longer the only driving forces and it is possible for a component to locally move up its own chemical potential gradient if that maximizes the overall entropy. At the end of the diffusion process of course all chemical potentials must be equal at all points of the system. This phenomenon has been observed in real systems e.g. metals (see Dayananda, 1985); it is difficult to assess the importance of the process in silicate melts because of the poor knowledge of thermodynamic mixing properties.

motion of constituent atoms of a component are coupled to atoms present in other components). Whatever the reason, information about such coupling is contained in the structure of the **D** matrix—it is the objective of the next section to show how.

Before concluding this section on diffusive coupling, we should mention another effect which has been found to be useful for some applications recently. It follows from the Fick-Onsager equation that it is possible to arrange the different concentration gradients in a manner such that they "balance out" and the net flux across a plane is zero for a component, even though concentration gradients of this component exists. A plane across which such a combination of gradients exists has been defined (e.g. see Dayananda, 1985) as the *zero flux plane* (ZFP) and initial compositional differences for which such planes develop during diffusion have been defined as *zero flux directions* (ZFD) in compositional space (e.g. a ternary diagram for a 3-component system). Dayananda and coworkers have developed and used these concepts for various purposes in studies of multicomponent diffusion in metallic alloys (summarized in Dayananda, 1985) and Thompson and Morral (1986) provide equations to calculate the location of these planes and directions for various

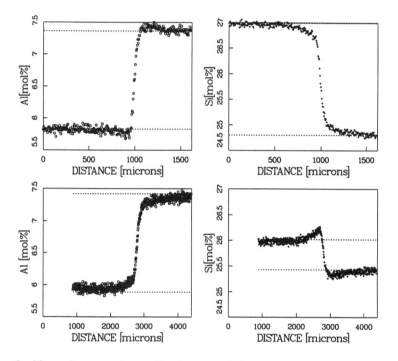

Figure 2. Measured concentration profiles from two diffusion couples with very similar mean compositions but different initial concentration gradients in the system K_2O-Al_2O_3-SiO_2 at 1600°C (from Chakraborty et al., 1995b). Horizontal dotted line show approximate initial compositions. Profiles from one couple (top row) show uphill diffusion of Al_2O_3 and normal diffusion of SiO_2 whereas the data from the other couple (bottom row) shows just the opposite -- uphill diffusion of SiO_2 and normal diffusion of Al_2O_3. Best fits to the two datasets yield the same diffusion coefficients.

components from measured **D** matrices. Since their development is based only on the phenomenological equations, the results may be directly used for other kinds of systems as well—an example of such use in silicate melt systems is described below in the section on *Diffusion and thermodynamic mixing properties*.

Diffusion paths

For ternary systems, much insight about diffusion mechanisms may be obtained by plotting compositions along a diffusion profile in composition space—this may either be a cartesian xy-plot or a ternary diagram. In a xy-plot, the concentrations of any two components in the system are plotted along the x- and the y-axes. Such plots provide a pictorial view of how compositions evolve during diffusion. The significance of these plots may be appreciated by noting that the distance axis of a typical concentration profile plot (e.g. Fig. 2) may be equated to a normalized time scale. If the two initial, endmember compositions plot as two points then the composition to be expected due to complete mixing of these components must lie on the line joining these two endmember compositions. In an infinite diffusion couple, the average composition is attained immediately at the interface. On moving further away from the interface on each side along the concentration profile, one finds compositions that are less evolved towards the average composition until one reaches the initial undisturbed composition at some distance from the interface. This situation holds as long as diffusion has not reached the boundaries of the system (i.e. the system is infinite with respect to diffusion). Thus, a plot of all of these compositions shows how the composition at *any one point of space* evolves with time— hence the name *diffusion path* for plots of such compositional arrays. A most notable feature of these diffusion paths is that they are time invariant, as long as the system remains infinite. In practical terms, this means they describe the compositional evolution in the early stages of mixing (until one endmember composition is exhausted) in any system— whatever the spatial and temporal scales involved. Even in the presence of convection, for example, the actual chemical mixing has to follow such diffusion paths.

Let us now consider how such plots provide information about speciation or mechanisms of diffusion. In the simplest case, the components A and B merely exchange places during diffusion. Clearly, the compositions would evolve along a straight line at 45% (1:1 slope) to the axes. Conversely, when a plotted diffusion path follows such a trend, one can infer that the diffusion mechanism can be described by a simple one to one exchange of A and B with no intermediate species involved. A system where diffusion may be closely approximated by such a mechanism is the exchange of B_2O_3 and SiO_2 between boron bearing and boron free haplogranitic melts (Fig. 3a). More typically, however, diffusion involves several homogeneous exchange reactions[1] within the melt which proceed at different rates. In these cases, compositions evolve away from the line joining the two endmember compositions at the beginning (Fig. 3b). However, mass balance dictates that compositions cannot keep changing arbitrarily away from the mean composition. Therefore at some point (determined by both the rate as well as the stoichiometry of the homogeneous reactions) the diffusion path finally turns around to reach the average composition that should be reached on mixing. Analysis of these S-shaped paths, following the same logic as for the simple case above (but more involved mathematics) provides information about the stoichiometry and rates of the homogeneous reactions in the melt that are responsible for mass transport. Detailed treatment of the mathematical approach is beyond the scope of this chapter, but we note some significant points here. Such analysis is common in materials science and in the geochemical literature it may be found in Chakraborty et al. (1995b) where the analysis has been performed with a step by step description and graphical illutrations for melts in the system K_2O-Al_2O_3-SiO_2.

---

[1] The term "reaction" is used here in the sense that any compositional change may be written formally as a chemical exchange, see Brady (1975b) for a clear exposition of this topic.

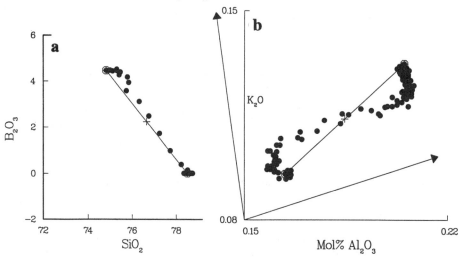

Figure 3. Illustration of different kinds of diffusion paths in two systems. (a) Path due to *almost* binary exchange of SiO_2-B_2O_3 between a haplogranite melt and a haplogranite + 5 wt % B_2O_3 melt at 1200°C, 24 hours (data from Chakraborty et al., 1993). Solid circles: measured composition, Solid line: 1:1 mixing line, cross in circle: initial compositions, plus symbol: mean composition. (b) S-shaped diffusion path due to multicomponent mixing in the system K_2O-Al_2O_3-SiO_2 shown in Al_2O_3-K_2O compositional space. Solid circles: measured composition, Plus symbol: mean composition, cross in circles: initial compositions. Lines with arrows show the direction of the two eigenvectors obtained by fitting the data, the eigenvector close to vertical corresponds to the larger eigenvalue (from Chakraborty et al., 1995b).

We have already referred to the possibility of decoupling multicomponent diffusion equations using eigenvalues and mentioned that the structure of the **D** matrix contains information about the nature of diffusive coupling. The specific attribute of the matrix that contains this information is the eigenvector. Nowadays, mathematical software packages routinely come with the capability of calculating eigenvalues and eigenvectors and these are therefore no longer the computational nuisance that they used to be, at least for diffusion problems which do not involve large matrices. What is more difficult to obtain is a physical insight into these quantities. A very simple, but somewhat abstract, example borrowed from Davis (1973) may help in this regard. Each column of a (2×2) matrix may be taken to contain the coordinates of a point on a xy-plane. If these points are plotted (points labelled 1 and 2 in Fig. 4), then it is possible to draw an ellipse which goes through these two points and has as its center the origin of the coordinate system. The two eigenvalues of the matrix would then give the lengths of the long and the short axis of the ellipse, respectively. The eigenvectors would give the slopes of these axes. In terms of diffusion, then, the eigenvectors describe the direction in composition plane (i.e. stoichiometry) of the fastest and slowest compositional change in this ternary (2×2) system. The eigenvalues give the rates—the relative lengths of the two axes of the ellipse are directly related to the relative speeds of compositional change along the two axes. To those with a geological background it is probably clear by now that we are basically describing a "diffusion indicatrix"—just like an optical indicatrix or a strain ellipse in structural geology. The eigenvalues and eigenvectors are simply defining the principal axes of diffusion in the composition plane. The advantage of the mathematical analysis is that it can be generalized to any number of components and one is not restricted to the graphical limitations of dealing with 3 or at most 4 components. Thus, it is possible to obtain information about diffusive coupling and speciation during diffusion for systems containing any number of components—as long as the **D** matrix can be measured.

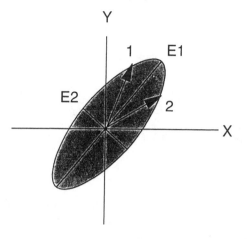

Figure 4. Geometric explanation of the significance of eigenvalues and eigenvectors. If the coordinates of the points labelled 1 and 2 are written in columns next to each other, they define a (2 × 2) matrix. The direction of the axes (E1 and E2) of the ellipse drawn through the points 1 and 2 (with center at the origin) are the two eigenvectors. The lengths of the two axes of the ellipse are proportional to the eigenvalues.

The vector algebraic treatment of Cooper and coworkers (Cooper, 1974; Gupta and Cooper, 1971) allow them to define some characteristics of diffusion as direct consequences of properties of vectors. We will illustrate one of these as an example here. If diffusion occurs by various homogeneous reactions within the melt, we would expect intuitively that the compositional change due to the fastest one of these reactions would dominate at the beginning of diffusion. Then from the preceding discussions it follows that the diffusion path near the endmember compositions would parallel the direction of the major eigenvector (i.e. eigenvector with the largest eigenvalue). This is exactly the result which was derived by Gupta and Cooper (1971) using vector analysis. They were able to arrive at a number of such generalizations, many of which are less intuitive. A summary of their results as well as those obtained by other workers along these lines is provided by Kirkaldy and Young (1987). Some of these results are useful for the evaluation of experimental data (e.g. see Chakraborty et al., 1995b). In general, they help in simplifying the analysis of multicomponent diffusion. As a case in point, our example here tells us that simple chemical analysis of compositions near the least altered part of a diffusion couple suffice to know the nature of the fastest homogeneous reaction in the melt. This information can be obtained without any complex analysis, irrespective of how complex the system is!

IV. RELATION BETWEEN DIFFUSION AND OTHER PROPERTIES

Diffusion, relaxation and glass transition

The glass transition and some related concepts. The transition between a glassy and a liquid state in a silicate melt occurs in numerous processes in nature, industry and laboratory. In order to understand the process of diffusion as well as to systematize and apply diffusion data obtained from various sources, it is necessary to understand how the process of glass transition can affect diffusion rates. The nature of glass transition, in any material, is a matter of intensive research and debate. However, the Adam-Gibbs theory of glass transition based on configurational entropy has been found to be very successful in explaining a number of characteristics of silicate melts (see the chapter by Richet and Bottinga in this volume for a detailed exposition). In brief, the theory asserts that the difference between the liquid and the glassy states is that configurational changes are allowed in the liquid but not in the glass, i.e. the configurational entropy does not change after crossing the glass transition temperature during cooling (measurements on a glass will yield the residual configurational entropy of the liquid frozen in at the glass transition). In other words, in the liquid state the atoms are dynamic enough to allow configurational changes to occur during the time that the melt is being observed. By "change in position of atoms" we are refering to atomic displacements on spatial and temporal scales larger than the typical displacements due to vibration about a mean position (which contribute to the vibrational entropy of the melt).

To connect glass transition to diffusion, the concepts of "strong" and "fragile" liquids, developed by C.A. Angell (e.g. Angell, 1985) are also found to be useful. In brief, strong liquids are those that show relatively small changes in properties (and by inference, in structure) on crossing the glass transition temperature. Fragile liquids, on the other hand, show larger changes in properties on crossing the glass transition temperature and are characterized by deviations from an Arrhenian dependence of viscosity (progressively higher activation energies or energy barriers as one goes to lower temperatures). As energy barriers to diffusive processes depend primarily on local structure (see sections on *Structure and diffusion* and *microscopics*), we would expect diffusion behavior in fragile melts to change significantly at the glass transition while diffusion rates in strong melts may vary smoothly across the glass transition. For silicates, silica rich melts show "strong" behavior (e.g. SiO_2, albite etc.) while silica poor melts (e.g. diopside, anorthite) sometimes show "fragile" behavior. This is one more reason to treat diffusion in silica rich melts on a different footing compared to the process in silica poor melts. The main point to note is that the phenomenon of glass transition, in its various manifestations, is expected to affect diffusion rates in fragile liquids much more than in strong liquids.

On a microscopic scale, motion may be vibrational, rotational or translational. Of these, the latter two in combination with bond rearrangements are mainly responsible for configurational changes. It is reasonable to anticipate that rotation and some bond rearrangements may be energetically more economic than translational motion of a particle across unfavorable structural environments. Consequently, it is possible that in many melt compositions configurational changes due to non-translational motions dominate just above the glass transition—we will see how this may explain some observed characteristics of diffusion in silicate melts.

In considering how silicate liquids fall out of equilibrium during cooling, it is conceivable that various subsystems in a multicomponent melt freeze at different temperatures. Particularly, since a melt is thought to consist of different kinds of Si-O units (see the chapter by Hess, this volume and the section on *Melt structure and diffusion*), it is realistic to expect that various units have different dynamics which do not all freeze at the same time. General theoretical models and some experimental data on organic liquids supporting such dynamic heterogeneity among domains have also been presented (Donth, 1982; Cicerone et al., 1995). In case the various components of such dynamic heterogeneity can be identified and measured in silicate liquids, Zhang (1994) provides some models to treat them. However, it is instructive to note that for fast cooling rates typical of glass formation and moderate energy barriers (tens of kcals), differences in freezing temperatures for different processes would lie within a few degrees, and the processes of structural reorganization in a liquid, within the framework of the Adam-Gibbs theory, are highly cooperative in nature so that freezing of one dynamically heterogenous domain could be rapidly followed by freezing of other cooperating units. Thus, the success of the "fictive temperature" (see chapter by Dingwell, this volume) concept in explaining many aspects of the behavior of silicate melts suggests that the transition, at least for network formers, occurs over a narrow range around the fictive temperature (though perhaps not exactly *at* that temperature) for a given cooling rate. This is the picture we will have in mind when trying to relate diffusion to glass transition. If dynamic heterogeneity is indeed critical, the situation will become more complex but the general features we discuss below would still be valid.

Lastly, we recall that the temperature at which the properties of a glass are frozen (i.e. the "fictive temperature") depends on the cooling rate—for faster cooling the glass freezes at higher temperatures. A consequence of this is that one can cross the glass transition by annealing a glass at low temperatures for long times or at high temperatures for short times.

With this view of the glass transtion, we consider three topics in this section: (1) in what ways can typical experiments to measure diffusion coefficients be affected by the glass transition, (2) how can the above picture of glass transition help us explain some aspects of diffusion data in silicate melts, and (3) a recent model that provides an unified treatment of ionic transport in silicate melts above and below the glass transition temperature and provides some insight into the mechanisms of ionic transport in silicate melts.

Glass transition and measurement of diffusion coefficients. A typical experimental process to measure diffusion coefficients involves the following steps:

(a) preparation of the starting glass by melting followed by a quench. Depending on the conditions of melting and the cooling rate following that, a structure corresponding to a particular temperature (T_f) on the cooling path is frozen. Note that this applies to all glasses used for an experiment, even if the glass is natural.

(b) working of the glass e.g. cutting, polishing, crushing etc. which may or may not affect the structure of the glass.

(c) heating up to the diffusion anneal conditions

(d) anneal at a particular temperature, T_a, for a length of time, t_a.

(e) quench of the melt to a glass again, which is analyzed chemically to provide diffusion rates.

First of all, note that depending on the conditions of the experiment, the liquid-glass transition may be crossed at steps (a), (c) and (e) from different directions and at different rates. It is also possible that the transition occurs during the anneal, i.e. at step (d). Depending on the details of the total thermal history of the entire process, various possibilities may arise which are illustrated schematically in Figure 5:

(1) $T_f > T_a$—glass transition occurs only once in this process, during the formation of the starting material. This is the classic case of diffusion measurements in a glass (Fig. 5, path I). The measured diffusion data pertains to a glass whose structure is frozen at T_f. Therefore, the role of preparation methods is critical and diffusion rates in various starting materials (different T_f) *may* be found to be different; avoiding crystallization is an important consideration sometimes.

(2) $T_f < T_a$—glass transition is crossed thrice in this process, the frozen structure re-equilibrates to the new conditions corresponding to T_a during the anneal and the diffusion data pertains to diffusion in the liquid state (Fig. 5, path IV). During the last quench from the annealing temperature, a structure is frozen in which in the general case corresponds to a temperature different from both T_a and T_f. Of course, depending on the material, it may be possible to take measures to freeze in the structure corresponding to T_a or T_f, if required. The state of the starting material is *relatively* unimportant in this case.

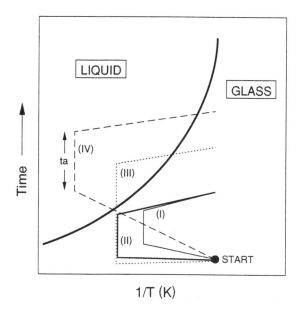

Figure 5. Relationship between thermal history during diffusion anneal and glass transition. Bold line separates the fields of liquid and glass in time-temperature space. All experiments start with a glass and are annealed at a given temperature for a time period, t_a. The paths labelled (I)-(IV) describe the four possibilities: **(I)** complete anneal in glassy state **(II)** short anneal at a higher temperature than (I), still in the glassy state through the entire duration **(III)** anneal at same temperature as (II) but long enough such that the transition from glass to liquid occurs during the anneal and **(IV)** anneal at temperatures high enough that the entire process takes place in a liquid. Note that during cooling the paths cross the glass transition at different temperatures (t_f), so that the structural state of the glass at the end may be the same as at the start for paths (I) and (II) but not for (III) and (IV). In particular, note the difference between paths (II) and (III) corresponding to anneals at the same temperature.

(3) T_f similar to, but slightly less than T_a—this is the most complicated case, dealing with diffusion in a supercooled liquid. Depending on the exact heating and cooling rates at the various steps and the duration of the diffusion anneal, t_a, one may obtain different behaviors. Here, t_a plays a crucial role. For short t_a, the available time may not be enough for the structure frozen in at T_f to change and one would obtain diffusion data in a glass (akin to (1) above, Fig. 5, path II). For very long t_a the structure would change and equilibrate to the new conditions at T_a—the diffusion data pertains to that in a liquid (like (ii) above). At intermediate times, diffusion would occur in a continuously changing structure during the anneal—these are the results which are most difficult to interpret (Fig. 5, path III). Thus diffusion coefficients in this case *may be* (more likely for fragile liquids) functions of annealing time. Note that this kind of time dependence of diffusion coefficient is completely different from the kind discussed below in the section on *Microscopics*. Some mathematical methods to treat these special cases are discussed in Crank (1975).

The features discussed above may become more complex if there are multiple relaxation mechanisms at work or if structure/compositions change significantly during the process (e.g. addition of water to dry glasses for diffusion anneals after step b, high pressure experiments on glasses synthesized at low pressures etc.). For applications where diffusion data are used to model geological processes occuring near the glass transition temperature (analogous to case (iii) above) one should be cautious, particularly if fragile melts are involved. Unlike most other cases, geological processes involving melts may occur on time scales shorter (e.g. cooling of a volcanic bomb, degassing of a lava flow) or longer (mixing in a magma chamber) than t_a.

Effect of glass transition on diffusion data. In the last section we saw how temperature-time histories of the annealing process can affect structure and thereby potentially influence diffusion rates in a melt. In this section we focus on the fact that different constituents of silicate melts move on very different time scales. With respect to glass transition or configurational changes, we can delineate three situations such that during the duration of observation: (1) no configurational changes occur, (2) only configurational changes involving non-translational motions occur, and (3) configurational changes involving all kinds of motion occur. The sequence is approximately one of increasing energy requirement so that one may expect to go from (1) to (3) with increasing temperature with the glass transition occuring between (1) and (2). Also, note the similarity of this division in many respects with that in the previous section.

(1) *The time scale of restructuring of network formers is much greater than the time scale on which the network modifers jump, t_a is short enough that no change of configuration occurs.* In this case, the diffusion of the network modifying atoms occurs essentially in a fixed, aperiodic (no long range order) lattice of the network formers. The situation describes diffusion in a glass or a very viscous liquid and is the kind of system treated in detail in the sections on *microscopics* and *Melt structure and diffusion.* A graphic analog may be the case of a spreading blob of ink in a static pool of water. Size effects may be important—ions of same charge but different sizes may have different diffusivities. Diffusion of alkali atoms in a dry granitic melt may correspond to this situation at reasonable geological temperatures.

(2) *Time scale of restructuring of network formers is much greater than the time scale on which the network modifers jump as before, however, t_a is long enough for perceptible rearrangements of the network to take place.* This is a complicated case in some ways analogous to diffusion in a supercooled liquid discussed above—diffusion takes place in a matrix that is changing with time. An example of this illustrates the situation and we present below the reasoning of Dingwell (1990) in a slightly different form. Consider the case of diffusion of Na in a Na_2O-$3SiO_2$ melt. Experiments are performed for a given time, t_a, to measure the diffusion coefficient at various temperatures and at low temperatures the situation corresponds to (1) above. As temperature increases, more energy is available to the system until a point is reached where t_a is long enough for perceptible structural reorganization to take place. At this temperature, diffusion begins to occur in a matrix whose structure is changing with time. This is manifested in the diffusion data by a change in the activation energy of Na diffusion—showing clearly that the energy barriers to diffusion have changed. At this temperature, diffusivity of network formers are still much too slow which indicates that the configurational changes are taking place by non-translational motions. However, that the configurations do change at this temperature over a time t_a is verified by the fact that calorimetry done on the glass at the same conditions shows the peaks in heat capacity characteristic of glass transition. Note that the alkali-silicate melt of interest here is a fragile melt—so the fact that glass transition affects structure and diffusion rates would have been expected based on our earlier discussion. In other, stronger melts, crossing the glass transition may not necessarily cause such change in diffusion behavior.

(3) *Time scale of jumps by the network modifers is comparable to the time scale of configurational restructuring of the network formers by translational motion.* This is the kind of situation which may be expected in melts with low viscosity e.g. at high temperatures. Returning to our analog of a spreading blob of ink in water, this case would

correspond to the situation where the pool of water has waves going back and forth through it (or flowing in a given direction, if diffusion in the presence of a driving force is being considered). The spreading of the ink blob would now be enhanced because it not only diffuses itself as before[1] but is also carried along as the background matrix moves in waves. In fact, the waves may be the major contribution to the spreading of the ink blob. The effect may be understood from Equation (5) in our discussion of reference frames earlier. If diffusion is described in the laboratory reference frame, then in a glass (case 1) the velocity v of the silicate lattice is zero whereas in this case it has a finite value. Thus, the flux observed is enhanced—it is the result of a "purely" diffusive term $(-D\ \partial C/\partial X)$ added to the velocity v of the silicate lattice. Since network formers dominate in a melt, it is possible that the "ride along" effect may determine the diffusive behavior of the modifiers. Thus, in silicate melts where diffusion rates of network modifiers and formers are *similar* (within 1 to 2 orders of magnitude), the activation energies of diffusion are also similar. Moreover, diffusion rates of ions of different sizes (with the same charge) are very similar to one another. It follows that no network modifier can diffuse slower than the network formers defining the matrix. Note that we have spoken about translation of network formers by local structural reorganization whereas for transport of network modifiers we have used the concept of "jumps"—various later discussions, particularly under *Diffusion and melt structure* and *Diffusion data in silicate melts* show why. An explanation of the diffusion enhancement of network modifiers with increasing temperature on the basis of a quantitative model for a simple system is now discussed below.

Unified model for ionic transport above and below the glass transition tem-perature—Caillot et al. (1994). A long-lasting discussion in the glass and ceramics literature has concerned the role of free volumes versus activated processes in controlling transport rates. Recently, Caillot et al. (1994) have presented a model that incorporates *both* of these approaches and succeeds in explaining many features of transport in non crystalline silicates—in particular the diffusion and electrical conduction behavior over a large range of temperature. Although their model is developed in fairly general terms, the illustrative examples were all for alkali disilicate melts. Behavior similar to that predicted by their model has also been observed in slag melts in the system CaO-Al_2O_3-SiO_2 (Fig. 6) and in melts along the albite-anorthite join (Behrens, 1992). To what extent their model is applicable to more complex melts remains to be seen and should be the topic of future investigations. Here we present the general concepts of the model. The essence of the model is that it considers ionic motion below the glass transition temperature to be due to a thermally activated transport process described by an Arrhenius relation. At higher temperatures, it considers a second mechanism of transport due to a free volume displacement mechanism which is superposed on the thermally activated mode of trans-port. Microscopically, this means that at high temperatures ionic transport occurs either because of thermally activated jumps across an energy barrier or because the energy barrier vanishes following a local reorganization of surrounding atoms. This combined modes of transport leads to an equation for temperature dependence of transport rates of the form:

$$D \;=\; A \exp\left(-\frac{B}{R\,(T\text{-}T_2)} - \frac{C}{R\,T}\right).$$

Here, A, B and C are material dependent parameters which can take on different meanings depending on the details of the microscopic mechanism being envisaged. T_2 is a temperature towards which transport rates decrease strongly, it is often equated to the ideal glass transition temperature, T_0 (the glass transition temperature for structural relaxation at

---

[1]   The rate would be somewhat different, since the environment of the ink particles are moving and hence the energy barriers for diffusion would be slightly different compared to that in the static matrix.

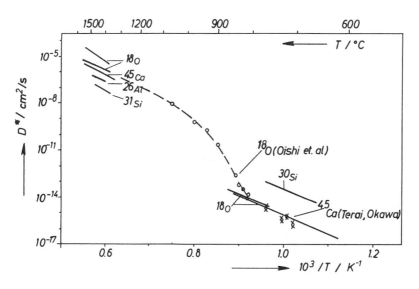

Figure 6. Diffusion data from various sources in the system $CaO-Al_2O_3-SiO_2$ for glassy as well as liquid states as a function of temperature (reproduced from Frischat, 1975).

very slow cooling rates). An equation of the form (16) explains some observed deviations from Arrhenius law and non-linear (in an Arrhenius plot) high temperature increases in ionic transport rates in some silicate melts (Fig. 7). Some examples are presented by Caillot et al. (1994) to illustrate how different combinations of the values of the parameters may lead to the extremes of highly non-linear or practically Arrhenian behavior over a large temperature range. For example, it is clearly apparent that for melts with C » B, transport rates would be continuous and Arrhenian across the glass transition temperature (strong melts?). Fitting transport data (electrical conductivity or diffusivity) to the model allows the independent contributions of the different modes of transport to be evaluated. It should be mentioned (as Caillot et al. (1994) point out that themselves) that the form of dependence of transport rates given by Equation (16) has been proposed by a number of authors before this. However, the particular strength of the Caillot et al. (1994) model is that it is testable–it relates the parameters A, B, C and T_2 to specific physical quantities which may be independently measured or constrained. For example, the activation energy term (involving C in Eqn. 16) is treated in the same way as in models of atomic transport involving Frenkel defects (atom jumping to the next site via an interstitial) in crystalline solids. This is consistent with other available models of ionic transport in glasses, see the section on *Diffusion and melt structure* below. The term with B, on the other hand, is a measure of the probability of the diffusing atom accessing a new site due to local restructuring. B can be expressed either in terms of parameters from the free-volume theory (e.g. Caillot et al., 1994) or in terms of thermodynamic parameters obtained from the configurational entropy based theory of glass relaxation (Adam and Gibbs, 1965). It should be interesting to try and see if thermodynamic parameters calculated according to the configurational entropy theory can successfully predict values of B that describe measured ionic transport rates in amorphous silicates (Richet, 1984, carried out an analogous exercise for viscous transport). Finally, the parameter T_2 is often related to the so called ideal glass transition temperature which means it will be lower than the T_g typically measured in calorimetric experiment. It can however, be estimated from a structural relaxation time-temperature curve (see chapter by Dingwell, this volume).

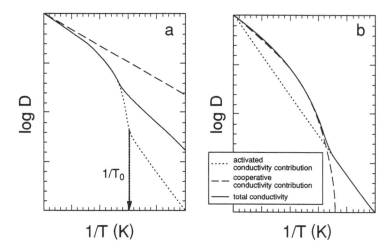

Figure 7. The ionic transport model of Caillot et al. (1994). (a) Depending on various combinations of values of A, B and C in the model (Eqn. 16) it is possible to obtain strictly Arrhenian (dashed line), weakly non-Arrhenian (solid line) or strongly non-Arrhenian (dotted line) behavior across the glass transition. For non-Arrhenian behavior, transport rates decrease rapidly as temperatures approach a value T_o. It is likely that strong melts show behavior similar to the dashed line whereas progressively weaker melts behave like the solid and the dotted lines, respectively. (b) Fitting Equation (16) to overall transport data (diffusivity/conductivity -- showed by solid line) can allow the contributions from the two separate modes of transport -- activated (short dashed line) and cooperative (long dashed line) to be distinguished. This may be particularly useful for nearly Arrhenian behavior where the contributions may not be as graphically apparent. Redrawn after Caillot et al. (1994).

Diffusion and melt structure

Approaching the question of relating melt structure to diffusion, it is interesting to recall W. H. Zachariasen's response to a (one of many, at the time!) critique of his classic work "The atomic arrangement in glass". It was questioned why he had studied glass formation from the point of view of crystal structure rather than liquid structure—glasses are, after all, frozen liquids. Zachariasen's response: "it is apparent from my paper that I avoided 'explaining' the structure of glass in terms of the conditions in the melt, rather, I made correlations to the crystalline state. It is, in my opinion, no real progress to present a theory for the structure of glass in terms of the equally unknown structure of liquids" (abstracted from Cooper, 1982). The same view may be taken today in trying to correlate diffusion to melt structure. Much of what we know today about the influence of structure on diffusion rates has been learned from studies of crystalline solids with a periodic lattice. If we are to develop an understanding of the relationship between structure and diffusion in silicate liquids, a potentially rewarding approach to the problem seems to be to ask the progressively difficult questions: (1) How is the process of diffusion affected if the medium is changed from a periodic, long-range ordered lattice to an aperiodic one? In other words, how is diffusion in a glass different from that in a crystal? (2) If the medium is dynamic, i.e. the structural units constituting the material are themselves in motion, and perhaps even changing their identity with time, how is diffusion affected? In other words, how is the role of structure on diffusion in a liquid different from that in a glass?

Before embarking on a discussion of the role of structure, it is necessary to define what one means by the structure of an amorphous material. Structure can be defined on different spatial scales. For example, short range structure relates to the immediate environment of an atom, e.g. whether Al in a melt is 4-, 5-, or 6-coordinated. Medium range structure relates to the question of the presence of chemical species, e.g. are there chemical species involving Si-O chains in the melt? Long range structure includes everything from small network islands to the periodicity of a crystalline lattice, approximately features on a scale larger than ~10 to 20 Å. Short range structural data is provided by techniques such as EXAFS, medium as well as short range structural information is obtained from various vibrational (e.g. Raman, Infrared) and nuclear (e.g. NMR) spectroscopic techniques. Long-range structure is classically obtained using diffraction techniques. Direct information about long to medium range features in amorphous materials are difficult to obtain, leading to much of the uncertainty in our knowledge of the structure of glasses—some information comes from measurements of radial distribution functions from X-ray and neutron diffraction studies of glasses. On the other hand, numerous spectroscopic data attest to the occurrence of short- to medium-range structure in glasses and recent in situ measurements show that many of these features persist in high temperature liquids.

Relation between structure and diffusion in a glass. In answering the first question regarding the relationship between structure and transport, let us recall that within the framework of transition state theory (manifested by Arrhenian temperature dependence of diffusion rates), the height of the energy barrier for a diffusive jump is largely determined by the local structure. There is ample evidence in support of this from various computer simulations as well as from microscopic theories (see Allnatt and Lidiard, 1993, for a detailed discussion). This suggests that much of the formalism developed for discussing atomic transport in solids can be adopted for considering diffusion of network modifiers in melts with fixed (or almost fixed) lattice, even if it is aperiodic. Thus, probabilities of atomic jumps and activation energies at a given site may be calculated and interpreted using methods developed for solids. Experimentally, this implies that spectroscopic methods which provide information on local structure can tell us something about diffusion activation energies. On the other hand, frequencies of attempted jumps depend more strongly on the overall structure including long-range order and therefore will be expected to be significantly different between crystalline and glassy materials. In fact, it has been recently shown (Tsekov and Ruckenstein, 1994) that the very lack of periodicity in the distribution of potential energy barriers may account for enhanced diffusion rates in amorphous solids compared to their crystalline counterparts; we shall see below how a specific model (the site-mismatch model) leads to this feature. The preceding discussion *does not* imply that macroscopically observed activation energies for diffusion in crystalline and the corresponding amorphous material will be the same or even similar.

Because of these similarities in local jumps, atomic transport of network modifiers in the glassy state has followed the same lines of development as defect assisted diffusion in polar solids. Most of the work in this area has been done in connection with electrical conduction studies, but dc conduction in silicate glasses is dominated by the motion of network modifiers and thus provides direct insight into diffusion mechanisms as well. In brief, transport of ions is thought to be due to thermally activated hopping of a certain concentration, c, of mobile ions. These "mobile ions" are ions that occupy particularly convenient positions for diffusion to take place and the concentration, c, of these mobile ions may themselves be thermally activated. In complex geological melts with various kinds of network forming and network modifying ions, one can obviously have an array of different kinds of sites which allow different degrees of "mobility" for a particular kind of ion. However, as shown in the section on *microscopics*, the resulting transport process can still be described in terms of a single average thermally activated process—obviously at the price of losing information on microscopic mechanisms.

Given the possibility of thermally activated jumps and the "mobile" ions, two main kinds of arguments have been made classically to calculate energy barriers to ionic motion and in the most recent theories the two trends seem to be converging. In the following, we present short outlines of the basic ideas of the different approaches, also see the chapter by Moynihan in this volume.

Anderson and Stuart (1954) model. This is one of the oldest ideas forwarded to explain ionic migration in glasses. The number of sites where jumps can be made is considered fixed and the energy barrier is considered to be composed of two parts—an electrostatic part dealing with the electrical work of moving a charged particle and a mechanical part which deals with work done to move through intervening space posing steric hindrance, e.g. regions of high concentrations of bridging oxygen.

Weak electrolyte model (Ravaine and Souquet, 1977; Ingram et al., 1980). The second kind of model which has been around for a while treats a glass as a weak electrolyte. In essence, these models assume that there are at least two kinds of sites and ionic transport rates are determined by the number of ions thermally populating the higher energy site. In a formal way, Ravaine and Souquet (1977) described this process by writing homogeneous dissociation reactions of the form $Na_2O \Leftrightarrow ONa^- + Na^+$ taking place when a Na_2O enters a melt, resulting in the two kinds of Na. Experimental evidence for the existence of two kinds of sites with one kind contributing more significantly to transport comes from optical basicity studies (e.g. Duffy et al., 1993), at least for alkali cations in simple silicate melts.

Martin and Angell (1986) have shown that these two kinds of models are equivalent to some extent and for simple melts (e.g. alkali-silicates) can help explain a number of characteristics of transport in these melts. They discuss the models and the extent of their equivalence with several well illustrated examples, and the interested reader is referred to that work for details.

Modified random-network (MRN) transport model (Greaves et al., 1991). This model views the structure of a glass as consisting of local regions of network formers separated by "channels" where non-bridging oxygen and network modifiers are concentrated. Diffusion is easier along these "channels" and hence, higher the concentration of network modifiers (and hence non-bridging oxygens), faster the transport rates. Beyond this point however, the detailed model as presented by Greaves et al. (1991) requires changing the population of bridging and non-bridging oxygens to effect transport. This amounts to changing configurations in the glassy state, which would be inconsistent with the Adam-Gibbs theory of glass relaxation. Further, the modified random network model of structure is very closely similar to the phenomenological "cluster-tissue" model of glass structure developed by Ingram (e.g. Ingram, 1987). The latter has been related to the site mismatch model described below and explains all of the transport features that the MRN model does without any conflict with the Adam-Gibbs theory. Thus, in spite of the shortcoming in the transport model of Greaves et al. (1991), the structural model based on EXAFS and NMR data (Greaves, 1985) for MRN may be realistic and we use this to schematically illustrate the differences between transport in a glass and a liquid (Fig. 8). At the level of abstraction of Figure 8, it can also be taken to represent the "cluster-tissue" model of Ingram (1987)—the two important points are that (a) there are structural pathways or "channels" for easy diffusion of network modifiers and (b) the "cluster" regions of high concentration of network formers are still amorphous, very unlike older "microcrystalline" models.

Jump relaxation model (Funke, 1991). This model considers the effects of the local electrostatic imbalances that result from the hop of a charged particle. In essence, the model assumes that immediately after an ion has jumped the surrounding atmosphere "remembers" where the ion *had* been and tries to get the ion back to the old position. A jump is only "successful" when after a while the ion atmosphere is rearranged to suit the environment at the new position. Within this rearrangement time period, there is a strong possibility that the ion will jump back and annul its displacement from the first jump. Although this model accounts well for the strong correlation between jumps (see *Microscopics* and *Diffusion and electrical properties*) and shows that macroscopic diffusion rates would be smaller than that expected from microscopic jump rates, it has difficulties explaining the compositional dependence of transport rates in melts.

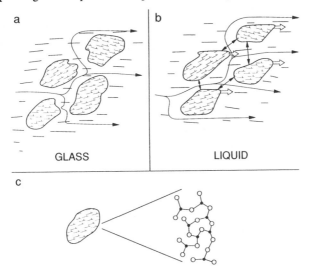

Figure 8. Schematic illustration of difference in diffusion behavior in glasses vs. liquids in relation to structure. The structure of melts are taken to consist of orientationally dissimilar amorphous regions of high concentrations of network formers (shown by islands with crosses in different directions) with intervening regions (dashes) of high concentrations of network modifiers and non-bridging oxygens. Whether the "network former islands" are interconnected or isolated depends on the composition of the melt. This representation is consistent with the modified random network (MRN) model of Greaves (1985) as well as the phenomenological cluster-tissue model of Ingram (1987) and is drawn based on figures in these works. (a) Diffusion in a glass: Arrows show easy pathways for network modifiers through regions of high concentration of non-bridging oxygens ("tissue"), the network formers present in the islands ("clusters") are essentially immobile. (b) Diffusion in a liquid: Long lines with arrows show pathways for network modifiers as in (a). Additionally, however, short double arrowed lines show that the islands ("clusters") may interact with each other (configurational changes) as well as move themselves (double lines with arrowheads). Thus diffusion of network modifiers take place in a changing matrix. The extent of non-Arrhenian behavior (Fig. 7) probably relates to the relative rates of these three processes. (c) A schematic enlargement of the island ("cluster") regions in (a) and (b) highlighting the amorphous nature of these regions of concentration of network formers and bridging oxygens. For clarity, Si atoms (solid circles) are shown with only three bonds.

Site mismatch model (Bunde et al, 1991; Ingram et al, 1991). This model brings together features from both of the classical models discussed above and shows how a distribution of sites rather than their exact geometric disposition can explain many features of ionic transport in melts. Since the model shows promise to explain many aspects of ionic transport in glasses, we treat it in some detail here. The model is based on the structural observation (from EXAFS, for example) that in simple alkali silicate glasses, alkali atoms of different kinds create their own short range environment. In other words, M-O bond lengths are characteristic of the alkali atom, M, and not of the overall composition of the glass. The glass structure has to adjust itself to accommodate an alkali atom in its preferred coordination polyhedra, rather than an atom fitting into the available polyhedral space. It follows that an alkali atom will be able to move out of its current site and into a new one only if the new site has been "prepared" for its arrival. To show how it works, Ingram et al. (1991) describe a *site memory* effect. According to this, a site recently visited by a cation A retains a predisposition to accept either the same or another A atom for a certain time τ following the jump. Such a recently vacated site may be considered analogous to a vacancy in a crystal and has been denoted as an $\bar{A}$ site by Ingram et al. (1991). Other cation sites in the melt, not "pre-treated" like this, are defined as $\hat{C}$ sites—these are analogous to interstitial sites in a crystal, which are not particularly inviting for an atom to hop into. After a time τ following a jump, an $\bar{A}$ site relaxes to a $\hat{C}$ site (this is a local relaxation which has nothing to do with structural relaxation). Clearly, the longer τ is, the longer is the lifetime of "good sites" into which jumps may occur and the higher the transport rate. Note that jumps into an $\bar{A}$ site as well as a $\hat{C}$ site are allowed, only the jump probabilities are different—jump of an A atom into an $\bar{A}$ site is more likely. Thus, the model incorporates features from most of the earlier models—it has the multiple site feature of the Weak Electrolyte model, the memory effect of the Jump Relaxation model and the steric hindrance aspect of the Anderson-Stuart model. In other words, both the local structure as well as long range Coulombic interactions are necessary to describe ionic transport. The authors do point out however, that distortions due to a size effect are not the only aspect that make a site unfavorable, there are "chemical" mismatches (probably related to bond rearrangements) which make the site recently vacated by one atom unfavorable for a different atom to jump into for a short period following the departure.

At this point, we have a partial answer to our original question about similarities and differences in diffusion mechanisms between glasses and crystals. Concepts like vacancies and interstitials and their role in diffusion can be adapted from crystals, but a major difference is that atoms in melts do not have to sit in "good" or "defective" sites—they are always in "good" sites by definition. It is the empty sites that become "good" or "bad" *with time,* and this ability to vary in character is directly related to the fact that a glass has no long range order and is hence more adaptable to local variations. Secondly, as the definition of the sites are determined by the diffusion process itself and the model relies on a distribution of sites rather than any specific geometry, it should be applicable to complex geologic glasses as well. Thirdly, the model should be valid even in liquids as long as the glass structure relaxes on a time scale much longer than the average residence time of cations in their sites (e.g. alkali diffusion in viscous melts at relatively low temperatures).

Before moving onto liquids, it is instructive to see how the model of transport explains two well known features of diffusion in glasses—the enhancement of diffusion rates with increasing alkali content in simple alkali-silicate melts and the reduction of diffusion rates in glasses with two alkalis (or other network modifiers), the so called mixed alkali effect.

The enhancement is classically related to the increase in the number of non-bridging oxygens as the alkali:silica ratio of a melt increases. According to the Anderson-Stuart model, reduced distance between non-bridging oxygen sites (due to their higher concentration) leads to a reduction in the electrostatic work term and hence enhances transport rates (by reducing the barrier height). According to the weak electrolyte model, the increased concentration of non-bridging oxygens increases the population of the high energy sites conducive to transport and hence results in increased transport rates. In keeping with the spirit of compromise, the site mismatch model incorporates both of these aspects. According to it, at low alkali contents there are only a few $\bar{A}$ sites (even if τ is long) so that the macroscopic diffusion is dominated by the low probability jumps from A sites to $\hat{C}$ sites. With increasing alkali content, the number of $\bar{A}$ sites increase, A-$\bar{A}$ pairs come closer to each other and diffusion is dominated by A- $\overline{A}$ jumps which are easier (implying lower activation energy and hence more jumps).

The mixed alkali effect follows as a natural consequence of the site mismatch model. It has simply to do with the fact that a site just vacated by an A atom is not suitable right away for a B atom to jump into. Thus, A- $\overline{A}$ jumps are preferred over jumps from an A site to either a $\hat{C}$ site or a site just vacated by a B atom. Similar arguments hold for B atoms. At small contents of B, there are still enough A- $\overline{A}$ pairs to form an interconnected pathway leading to only slight reduction of transport rates. However, with roughly equal amounts of A and B there is no interconnected "easy" path for either A or B atoms—they obstruct each other and to diffuse, both are forced to jump frequently into uncomfortable sites (either $\hat{C}$ or a site just vacated by the other species) leading to dramatic reduction in transport rates. There are two important consequences of explaining the mixed alkali effect in this manner. First, since the explanation does not depend on specific geometric arguments it can be extended to the liquid state as well. Second, as temperatures increase, the mismatch between sites become smaller and also the readjustment time τ becomes shorter. This means more sites become accessible to a given species of atom and the mixed alkali effect is reduced, consistent with observation. Macroscopically, this accessibility to more sites would be described as an increase in configurational entropy with temperature leading to a reduction and eventual disappearance of the mixed alkali effect—exactly what has been shown by Neuville and Richet (1991). Thus, the site mismatch model succeeds in providing a nice transition from microscopic structural to macroscopic thermodynamic descriptions. Of course, as temperatures are increased much beyond the glass transition additional mechanisms for diffusion may become available as well (see below and the model presented under *Glass transition, relaxation and diffusion*).

The ideas of transport of network modifying atoms along easy "channels" in a silicate melt receive support from measured correlation factors (see section on *Microscopics* and *Diffusion and electrical properties*), from studies of pressure dependence of diffusion rates (see *Diffusion data in silicate melts—pressure dependence of network modifiers*) and from considerations based on percolation theory. According to the latter, one can calculate that in a melt composition given by $[(M_2O)_x(SiO_2)_{1-x}]$ where M is an alkali atom, the channels of non-bridging oxygen should become interconnected (cross the percolation threshold) when x reaches 16% (Zallen, 1983). It is found that addition of alkalis to pure silica glass reduces activation energy for diffusive transport quite drastically until x equals about 10 to 20%; on further addition after this the activation energy remains essentially constant (e.g. see Frischat, 1975, for data on sodium-silicates). In other words, initial addition of network modifiers open up new channels and pathways but once the channels are interconnected, further addition of alkalis do not make diffusion any easier.

Relation between structure and diffusion in a liquid. So far we have concerned ourselves with diffusion in a static structure i.e. in a glass or a liquid where the network formers move at rates much slower than that of the network modifiers. What are the possible consequences of having mobile network formers i.e. how can structure of a liquid affect diffusion rates when the network formers and modifiers move on comparable time scales? One obvious consequence is that the local structure and hence the energy barriers may change. This is manifested as non-linear temperature dependences of transport rates (diffusion, viscosity) in Arrhenius plots—well known for viscosity of fragile liquids (see chapter by Dingwell, this volume) and observed in some cases for diffusivity (see Fig. 6). Additional possibilities for diffusion may become available to network modifier ions *during* the course of dynamic restructuring of the network; enhancement of transport rates resulting from such additional pathways have been discussed above (see *Glass transition, relaxation and diffusion*).

In this section let us take a look at how the diffusion rates of the network formers themselves may be affected due to dynamic reorganization. So far, we have held the approximate view that a glass is made up of small orientationally dissimilar regions in which bridging oxygens and network formers are concentrated and these are separated from each other by channels along which non-bridging oxygens and network modifiers are concentrated (Fig. 8). The identity and position of these "network regions" were considered to remain unchanged through the process of diffusion. In a high temperature liquid, these "network regions" may themselves move (changing the geometry of the "channels") or react with each other, changing the identity of the units themselves. In order to follow the motion of a constituent of the network (a network former), therefore, one has to follow the motion of each kind of unit as well as monitor the interconversion between units. Monitoring the interconversion means keeping track of the abundance of different units at any given instance (i.e. calculating the equilibrium distribution due to various homogeneous speciation reactions) as well as tracking the rate of interconversion of units (i.e. calculating the kinetics of different homogeneous reactions). For a constituent like oxygen, which may be an integral part of the network as a bridging oxygen, reside outside of the network as dissolved molecular oxygen (O_2) or play the intermediate role of a non-bridging oxygen, the situation is even more complex. The essence of the mathematical problem is that what was a diffusion problem has now become a problem of diffusion plus reaction (homogeneous in this case). A treatment of the problem, specifically dealing with various forms of oxygen and water, has been given by Zhang et al. (1991 a,b). Their model was eminently successful in describing water diffusion in silicate melts, see Watson (1994) for a review. They also proposed treatment of oxygen diffusion data for cases where only one homogeneous reaction is present (e.g. interconversion of bridging, non-bridging oxygen and molecular oxygen, assuming there are only one kind of each). Models dealing with diffusion of different kinds of polyanions that may *not* interconvert during diffusion (either complete equilibrium or disequilibrium speciation) have been discussed in the fitting of experimental results by Oishi et al. (1982), Nanba and Oishi (1983), Kress and Ghiorso (1993) etc. However, since the process of diffusion being studied in these works was chemical diffusion in multicomponent systems, the number of parameters that could be varied were relatively large, allowing only very general conclusions to be made.

Recently Young et al. (1994b), expanding on an earlier effort by Kieffer and Borchardt (1987), developed a model which treats both the diffusion of individual units as well as simultaneous reactions which condense or split the various units. They were able to account for the equilibrium as well as kinetic behavior of these homogeneous reactions. Even in these complex models some assumptions about the nature of the network units had to be made. With a further restrictive assumption that the motion of the different units are not correlated to each other (random walk for each unit), they were able to simulate experimental tracer diffusion profiles for melts in simple binary systems like CoO-SiO_2, CaO-SiO_2 and PbO-SiO_2. In their terminology, the *single, average* diffusion coefficient obtained by fitting observed concentration profiles to a model is called the effective diffusion coefficient, D_{eff}. They found that the compositional dependence of D_{eff} is an outcome of a distribution of effective drift velocities of the diffusing species which results from their temporary incorporation into polyanions of different sizes. A similar conclusion was reached by Zhang et al. (1991a) in discussing the compositional dependence of water diffusion rates in silicate melts. Further, Young et al. (1994b) found that exchange of atoms between units played an important role in determining diffusion behavior in the system CoO-SiO_2 but not in PbO-SiO_2 or CaO-SiO_2. As a result, concentration profiles in the system CoO-SiO_2 could not be described by standard solutions to Fick's law for the given boundary conditions of the experiments. The role of speciation was clearly illustrated in this study by showing that the diffusion behavior of oxygen tracer atoms in the melt depended on whether the tracer isotope was introduced in the melt from CoO or SiO_2

during melting. This study thus helps to illustrate some of the complex effects that may appear during diffusion in silicate liquids because of the dynamic variations of the structure of the liquid itself with time.

Needless to say, in realistic geological compositions with various kinds of network formers and modifiers the potential for variations are unlimited—this is a direction of investigation that definitely deserves future attention. We should recognize two important aspects of the relationship between structure and diffusion in silicate liquids from the preceding discussion of the behavior in simple systems. First, the problem has some attributes of the classical chicken and egg syndrome—does structure (i.e. speciation) determine mobility or does mobility determine structure? Second, the kinetic as well as equilibrium behavior of internal speciation reactions play a key role in the transport of atomic species, particularly for network formers. This means that information obtained from structural probes of a direct (e.g. spectroscopy) as well as indirect (e.g. thermodynamic models) nature need to be combined with kinetic considerations (e.g. lifetime of a species in a melt) before they can be related reliably to diffusion behavior. However, if it can be established (e.g. through in situ spectroscopy on the right time scale at various temperatures) that kinetics of speciation plays a subordinate role in a particular transport process by being either too fast or too slow, then structural probes of short to medium range structure can provide us information about activation energies and diffusion pathways. In this regard, the observations of Shimizu and Kushiro (1991) are noteworthy—they found no simple relationship between diffusion rates (of Si, O, Mg and Ca) in melts on the diopside-jadeite join and speciation in terms of NBO/T ratios determined from Raman spectra on quenched glasses. However, the data for diffusion rates of network former Si at intermediate compositions were found to be well described by a model involving configurational entropy of mixing of diopside and jadeite endmembers. This observation only serves to further highlight the cooperative nature of diffusion in complex liquids. For the purposes of practical usage, we reiterate the conclusions of Shimizu and Kushiro (1991) that the relationship between structure and diffusion is complex—beyond the general conclusion that increased polymerization reduces transport rates, it is difficult to make quantitative predictions without using complex models and computer simulations.

Diffusion and electrical properties

Ionic motion dominates electrical transport in many kinds of silicate glasses. Therefore, studying electrical transport is an important alternative approach to understanding diffusion behavior. The increased sensitivity of electrical measurements in comparison to concentration measurements and the opportunities to change polarities and frequencies allow access to properties and phenomena which are beyond the reach of normal diffusion measurements. The two simplest examples are at the high and low temperature extremes— electrical conductivity can be measured to very low temperatures where conventional diffusion measurements are not feasible anymore; electrical conductivity can be measured in situ in high temperature melts whereas classical diffusion measurements invariably study quench products. Moreover, studies of frequency dependent a.c. conductivity and polarization relaxation provide insights into the nature of the jump process during ionic migration. Details of some electrical properties of silicate melts may be found in the chapter by Moynihan in this volume. In this section, after providing the basic relationship between electrical conductivity and diffusion, we will discuss two examples of how electrical measurements may help us gain a better understanding of diffusion in silicate melts: (1) the measurement of Haven ratios, and (2) measurement of electrical transference numbers. In addition, much of the discussion in the section on *Melt structure and diffusion* is based on information obtained from studies of electrical properties.

The basic relationship between electrical conductivity (d.c.) and diffusion coefficients is given by the Nernst-Einstein equation. If one accepts that the electrical conductivity of the melt is due to the motion of one kind of ion, then this equation takes the form:

$$D_i^* = \frac{\sigma k T}{q^2 N} H_R$$

where D_i^* is the self diffusion coefficient, σ is the d.c. or low frequency conductivity, $k \equiv$ Boltzmann constant, $T \equiv$ absolute temperature, $q \equiv ze$ where $z \equiv$ valence, $e \equiv$ elementary charge, $N \equiv$ number of charge carriers and H_R is the Haven ratio. The equation above may be taken to be the definition of Haven ratio (although a more rigorous definition can be given using irreversible thermodynamics, see Kahnt et al., 1988), we will return to its significance in a moment. For now, we note that H_R usually has a value between 0.1 and 1 so that even without knowing anything about its magnitude, it is possible to get a reasonable (within an order of magnitude) estimate of diffusion rates if conductivity is measured, and vice versa. Secondly, as may be intuitively clear, if there is more than one kind of ion responsible for electrical conduction one has to take into account partial conductivities due to individual ions alongwith the coupling between the flows of different kinds of ions. In that case, the general form of the Nernst-Einstein equation is retained although it becomes much more complicated algebraically with additional parameters coming in, for a detailed discussion of these aspects see Allnatt and Lidiard (1993).

The Haven ratio (H_R) is related to the correlation factor, f of the diffusion process (see the section on *Microscopics*). It is a measure of the correlation between successive atomic jumps. One calculates the correlation factor from geometrical and statistical considerations for different jump mechanisms and different kinds of lattices. Comparison of the calculated and observed correlation factors (or, Haven ratios) then provides information about the nature of the diffusion mechanism operating. Haven ratios have been measured using two kinds of experimental approaches: (i) the isotope effect method, where differences in transport of two isotopes of the same element are utilized and (ii) the so called Chemla experiment, where the electrical conduction and diffusion coefficient are simultaneously measured by observing the diffusion of an isotope in the presence of an electrical field. Chemla experiments have been performed on alkali-silicate glasses (e.g. Kahnt et al., 1988; Laborde et al., 1989; Heinemann and Frischat, 1993) as well as on rhyolitic glass (Stanton et al., 1990). Detailed analysis of the low Haven ratio obtained in sodium-disilicate melts (~0.25) show that a model of highly correlated, collinear jumps is most likely. This, in turn, points to a diffusion mechanism where motion occurs along percolation channels, with forward and backward jumps frequently cancelling each other. See also the sections on *Diffusion and melt structure* and *Microscopics* for related discussion.

As pointed out above, if more than one kind of ion is responsible for ionic conduction in a melt, then partial ionic conductivities have to be determined. These are obtained through the measurement of electrical transference numbers. The electrical transference numbers are simply ratios which measure what fraction of the total ionic current is carried by a particular kind of ion. They can be measured using galvanic cells with liquid junctions (i.e. a liquid diffusion couple). Transference numbers are then determined from the measured voltage and independently known thermodynamic properties—see Petuskey and Schmalzried (1980) and Takada and Schmalzried (1983) for details. Transference number measurements can be particularly useful in diffusion studies because they can be related to the Onsager kinetic coefficients, L_{ij}, which are connected directly to microscopic motion (see section on *Microscopics*). The appropriate theoretical relationships for simple systems have been derived by Schmalzried and coworkers and may be found in the works cited above.

Diffusion and viscosity

The difference between diffusion and viscous flow was clearly stated by Onsager (1945) as "Viscous flow is a relative motion of adjacent portions of a liquid. Diffusion is a relative motion of its different constituents." Therefore, although the diffusive motion of a particular species in a multicomponent liquid may be the rate determining step for viscous flow, the two processes are not the same. A rigorous connection between diffusion and viscous flow can only be established by taking into account the correlated motion of all species in a melt. This is to be expected from the configurational theory of viscous flow as well (see chapter by Richet and Bottinga in this volume), which emphasizes cooperative motion. However, a number of observations suggest that to a very close approximation viscous flow in many melts may be related to transport rates of network formers like silicon and oxygen, i.e. the larger part of the energy barrier to viscous flow is due to the transport of these species:

(1) diffusion rates of network formers are always slower (often by orders of magnitude) than those of network modifiers (with the exception of oxygen in some cases)

(2) configurational changes in a melt are affected most readily by changing a non-bridging oxygen to a bridging oxygen and vice versa

(3) creating and breaking a Si-O bond is among the most energetically demanding processes in a silicate melt, and

(4) addition of very small amounts of certain components (most notably water, see Watson, 1994, for a review; also boron in some cases, see Chakraborty et al., 1993) alter the rates of viscous flow as well as diffusion of some network formers quite dramatically, but similarly.

Some recent work suggests that transport of the other significant network-former in geologic melts, Al, may be more, or at least equally, closely related to viscous flow (Baker, 1992a; McMillan et al., 1994; cf. *Pressure dependence of diffusion coefficients* below).

If the standpoint is taken that thermally activated jumps control diffusion as well as viscous flow, then the Eyring model based on the absolute reaction rate theory, when applied to *diffusion rates of the rate determining species*, should yield viscosities of melts to a good approximation. For simple alkali silicate melts, high temperature NMR studies support such a standpoint (see chapter by Stebbins, this volume). Clearly, where transport mechanisms other than thermally activated jumps may become important (see the section on *Glass transition, relaxation and diffusion*), e.g. at high temperatures for some fragile liquids, the model may not work as well. However, if there are multiple "sites" and jump mechanisms, their combined effect may often be treated in terms of a single average diffusion coefficient (see the section on *Microscopics*). With these qualifications, the Eyring model is stated as

$$D = \frac{kT}{\eta \lambda}$$

where D is the self-diffusion coefficient of the rate determining species, $\eta \equiv$ viscosity, $\lambda \equiv$ the jump distance, $k \equiv$ Boltzmann constant and $T \equiv$ absolute temperature. A major uncertainty is the lack of knowledge of λ in silicate melts. Nevertheless, taking λ to be the distance between two oxygen atoms touching each other, the above equation has been used with remarkable success in a variety of silicate melts to relate diffusivity of oxygen or silicon and viscosity (e.g. Oishi et al., 1975; Yinnon and Cooper, 1980; Dunn 1982; Shimizu and Kushiro, 1984; 1991; Dingwell, 1990; Baker, 1992b; Chakraborty et al., 1993; Rubie et al., 1993). In melts where oxygen and silicon diffusivities are somewhat dissimilar (e.g. in jadeite), transport rates of oxygen are found to be more closely related to viscosity. In spite of the success of the model and uncertainties in λ, it is interesting to observe that activation energies of diffusion are in most cases slightly lower than the activation energies of viscous flow. This is consistent with the discussion above—for viscous flow, work in addition to the jump of the rate determining species has to be expended.

At this point it is worth discussing an alternative model that relates viscosity to diffusion rates—the Stokes-Einstein model. The shortcomings of this formulation in describing experimentally observed behavior of silicate melts have been clearly pointed out in previous reviews (Hofmann, 1980; Dunn, 1986), but the model continues to be used in the literature. Thus, we will repeat some of the reasoning already presented in the earlier works and outline conditions where the equation may be used. The Stokes-Einstein model is stated in its simplest form (which is used in most applications to silicate melts) as

$$D = \frac{kT}{6\pi r \eta}$$

where r is the radius of the diffusing particle, taken to be a sphere. The model is derived for uncharged particles and for large spheres moving through a fluid of relatively low viscosity, neither of which is a very good approximation for silicate melts. The second approximation may hold for oxygen in some cases, but certainly not for silicon or various other cations. In fact, at first sight the Stokes-Einstein and the Eyring model appear very similar—both predict that diffusivity is inversely related to viscosity. However, there are important differences between the two models. First, unlike the Eyring model involving λ, the Stokes-Einstein model relates diffusivity and viscosity through the radius of the diffusing atom, which is a well known quantity. This means that in all examples above where the Eyring model was found to work well in relating diffusivity of oxygen to viscosity, the Stokes-Einstein model would yield a diffusivity about 10 times (3π) too high. Secondly, the manner in which the Stokes-Einstein model is derived does not require it to relate diffusivity of any particular species (unlike the Eyring model) to viscosity—it should be valid for any particle moving through the fluid. This means one should be able to take the diffusion coefficient and ionic radius of any element and relate it to viscosity. That such an approach fails miserably has already been shown (Watson, 1979; Hofmann, 1980). It may be emphasized by pointing out that two elements like Ca and Na with similar ionic radii but very different diffusion rates (noting that alkalis diffuse much faster than other elements) will predict very different viscosities for the same melt! The situation worsens if pressure dependence of viscosity is predicted from pressure dependence of network modifiers—even the sign of the dependence is predicted opposite (Angell et al., 1982; Watson, 1979). It is however fair to point out that various adaptations of the Stokes-Einstein equation for charged, non-spherical particles exist in the polymer literature. Such expressions are usually much more complex than the one quoted above, requiring the knowledge of various other parameters and have therefore not been used in studies involving silicate melts. The Stokes-Einstein relation, as stated above, should predict viscosities well with diffusion data for various noble gases (uncharged spherical particles) in basaltic liquids (low viscosity fluid), but such a comparison has not been made to my knowledge. Analogously, the equation seems to work well for fluid melts at extremely high temperatures such as those typically seen in computer simulations (Scamehorn and Angell, 1991).

The success of the Eyring model to relate diffusivity and viscosity is a very useful feature from the point of view of many geological applications. On the one hand, for modelling processes in the deep mantle, viscosities of melts under high pressure are required. While measuring viscosities under very high pressures poses considerable experimental difficulties, diffusion measurements have been performed (e.g. Rubie et al., 1993). This allows viscosities of these melts under high pressure to be estimated using the

Eyring equation. On the other hand, measurement of diffusion rates at low temperatures are difficult, but techniques are available for measuring viscosities under these conditions (e.g. Dingwell et al., 1992).

As an example, Chakraborty et al. (1993) used the Eyring relation and the Arrhenian temperature dependence of viscosity in boron-bearing granitic melts to argue that extrapolation of diffusion data measured at high temperatures should be reliable. Further, combination of diffusion and viscosity data allowed them to demonstrate that addition of small amounts of boron affect both transport processes—a feature which was shown to be of likely importance in the elucidation of emplacement mechanisms of peraluminous granites.

Diffusion and thermodynamic mixing

Chemical diffusion is transport of matter in response to chemical potential gradients. It is therefore expected that characterization of chemical diffusion requires knowledge of chemical potentials or thermodynamic activities of components in a silicate melt. We have already seen how diffusion rates are influenced by thermodynamic activity through the thermodynamic factor, ϕ, in a binary system and its generalized version, the **G** matrix in a multicomponent system. However, since diffusion coefficients are often easier to measure than activity-composition relations in complex multicomponent melts, it is a worthwhile pursuit to try to constrain thermodynamic mixing behavior from diffusion data. Either exact theoretical relationships or trends observed empirically in diffusion couple experiments may be used. Different methods based on both approaches have been suggested in the literature.

Use of exact relationships.

Dilute consituents in multicomponent systems. Recall that chemical diffusion of a dilute constituent is controlled essentially by its own tracer diffusivity (irrespective of the kind of model relating chemical and tracer diffusion coefficients) multiplied by the thermodynamic factor, even in a multicomponent system. This provides a simple way to determine activity-composition relationships, as suggested and illustrated by Lesher (1994). One measures tracer diffusion coefficients independently in two different melts using experiments involving no chemical gradients. Then one performs chemical diffusion experiments using the same two melts and retrieves the chemical diffusion coefficients from observed profiles. The ratio of these two quantities (the chemical and the tracer diffusion coefficient, provided their compositional dependence can be well characterized) in this case gives the thermodynamic factor, ϕ which can yield activity coefficients, γ_i. Because the quantity one obtains is actually the dependence of γ_i on composition, X_i, the method is particularly useful if activity-composition relations are independently known in one endmember melt composition. As retrieval of chemical diffusion coefficients usually follow an effective binary approach, it is important to ensure that the concentration profiles are suitable for such retrieval (see discussion on *effective binary diffusion coefficients* earlier). Note that the method is not as useful for major elements because the chemical diffusion rates of these are controlled not only by their own tracer diffusion coefficients and the thermodynamic factor but also all other tracer diffusion coefficients in the system and there are considerable uncertainties in the models relating these quantities, as discussed earlier.

Major components in multicomponent systems. We have already pointed out that in spite of the many advantages of knowing the Onsager kinetic coefficients, L_{ij}, it is very difficult to measure them. Notwithstanding, it is possible to use the fact that in any system

$L_{ij} = L_{ji}$, i.e. the **L** matrix is always symmetric, to constrain thermodynamic mixing properties. The basis of the approach is Equation (13) which may be rewritten as

$$\mathbf{L} = \mathbf{DG^{-1}}.\tag{17}$$

The elements of the **G** matrix are given by

$$G_{ij} \quad = \quad \frac{\partial}{\partial X_j} \left[\, \mu_i - \mu_n \right]$$

where n denotes the dependent component and X_j is the mole fraction of the jth component. For any chosen thermodynamic model, the G_{ij} can be written in terms of interaction parameters, W_{ij}—the exact expression depending on the chosen model. If only some of these W_{ij} are known, then it is possible to use measured **D** matrices to constrain the values of the unknown parameters. One simply multiplies **D** (whose numerical values are known) by $\mathbf{G^{-1}}$ (written symbolically) and requires the unknown W_{ij} values to be such that the product is a symmetric matrix, as **L** must be. Another way of stating this is to say that the W_{ij} must satisfy the set of relationships (from Onsager, 1945; his Eqn. 17b)

$$\sum_j \frac{\partial \mu_i}{\partial C_j} D_{jk} \quad = \quad \sum_j \frac{\partial \mu_k}{\partial C_j} D_{ji} \; .$$

The approach has been illustrated for silicate melts and used very successfully for solid silicates (Chakraborty, 1994; Chakraborty and Ganguly, 1994). It is assumed that the kind of thermodynamic model to be used is known a priori and only some parameters are missing. The nature of the model and the number of missing parameters determine how well they can be determined. Use of different models will of course yield different values for the same interaction parameters.

A number of variants on this basic approach are also possible. For example, when the thermodynamic parameters are known along with some tracer diffusivities, one can use a model relating tracer and chemical diffusion coefficients in combination with Equation (17) to constrain the unknown diffusion coefficients. The robustness of this approach depends on how well the model relating the two kinds of diffusion coefficients is known. Alternately, when data on a system come from various sources one can use Equation (17) to check for consistency. Finally, the most promising application of the approach is perhaps not in the determination of primary thermodynamic properties but in constraining or exploring possible dependencies on temperature and pressure. Well constrained thermodynamic data are usually available over a limited temperature and pressure range whereas diffusion data can be obtained over a considerably larger range. Since Equation (17) must be satisfied at all conditions, it may be used to characterize possible variations in mixing properties with temperature and pressure (i.e. excess entropy and volume terms)—see Chakraborty (1994) for details. Whatever the goal, it should be remembered that Equation (17) is an exact relationship, the uncertainties and inaccuracies resulting from the method come from the various mixing and diffusion models that may be involved.

Use of empirical observations. The two methods described below are both based on observations that during diffusive mixing of two liquids, activities of some components seem to evolve along well-defined paths. Since the theoretical basis for these observations are not yet clearly defined, the limits of validity of these approaches remain to be determined—either experimentally or from an exact theoretical justification.

Models based on continuous partitioning of elements. Richter (1993) suggested a method to determine activities of fast diffusing components in a melt which is based in principle on the two liquid partition model of Watson (1982). Watson (1982) had observed that in a diffusion couple (liquid-liquid or crystal-liquid), fast diffusing components tend to attain a distribution "necessary to maintain constant chemical potentials of network-modifiers in a zone over which changes in melt structure occur". Richter (1993) has quantified this statement and derived expressions, subject to certain assumptions, which yield acitivities of the fast diffusing component from measured concentrations along a diffusion profile. The model of Richter (1993) also identifies the time scales after which such equilibrium distribution will be attained. While the general trend predicted by Watson (1982) has since been vindicated in a number of studies (see section on *Empirical Methods*), how well the model of Richter (1993) succeeds in quantitative prediction of activity-composition relationships in different systems remains to be experimentally tested.

Zero flux planes (ZFP). It has been empirically observed in a variety of systems that in composition space (e.g. a ternary diagram for a three component system), the line joining the composition of the endmembers of a diffusion couple and the composition at the ZFP is a isoactivity contour. As long as this relationship holds, it provides a method to determine iso-activity compositions in multicomponent systems because zero flux directions (ZFD) may be calculated from measured **D** matrices. Development of the method and examples of calculated iso-activity directions for ternary silicate melts may be found in Chakraborty (1994). Although the relationship has been found to be approximately valid in all systems where tests have been possible so far, the theoretical basis for this observation is not clear. Therefore, the extent of applicability of the method needs to be determined through experiments in different systems where independent data are available.

Relation of diffusion to spectroscopic data

Spectroscopic methods relate to diffusion studies in two distinct ways. They can be used as a tool to observe diffusion itself or alternately, they perform the more classical role of structural probes and provide information which can be used to understand diffusion. In this section we are concerned with the latter aspect, we postpone discussion of direct diffusion studies to the later section on *Experimental methods*. The major strength of spectroscopic methods is that they probe local interactions (in both space and time), which makes them ideally suited to provide information about energy barriers and pathways for diffusion. Secondly, recent advances allow in situ measurements to be made (NMR, Raman, infrared, EXAFS—each of which are discussed in different chapters in this volume) which allows melt structures to be observed at the conditions where diffusion occurs. However, two disadvantages inherent in these methods are (1) each kind of spectroscopy has a characteristic frequency i.e. it can only observe features over a certain time scale, and (2) different spectroscopic techniques are sensitive to different species, i.e. any given spectroscopic tool cannot observe "everything" in a melt. Thus, not all features of a dynamic melt will be recorded by every spectroscopic method—an important point to remember when interpreting in situ spectroscopic data.

Simple illustrative examples may make these two aspects clear. Vibrational spectroscopy (e.g. Raman, Infrared) provide snapshots of a melt on time scales of 10^{-13} seconds or less. Say, in situ Raman spectra of a granitic melt at 1100°C are collected and they show a certain kind of anionic speciation (distribution of bridging and non-bridging oxygens among various kinds of Si-O units). Now, a reasonable diffusion coefficient of 1×10^{-13} m²/sec for a Na atom in these conditions implies that an atom resides at a site for about 10^{-7} seconds between jumps (on a scale of tens of Å). This means it is possible to change the local environment of a given Na atom considerably by processes occuring over longer time scales that would not be visible to Raman spectroscopy. Similarly, spectra on quenched glasses can tell us the ratio of bridging to non-bridging oxygens in a melt. However, they cannot necessarily distinguish between different kinds of non-bridging oxygen (recall the "optimized" sites in the site-mismatch model of Bunde et al. (1991) in the discussion on *Diffusion and melt structure*), which can play a key role in the transport process. We can draw the important conclusion that relating data from any one kind of spectrum to a diffusion process may not always be possible, and a lack of correlation should even be expected in many cases. Thus, the failure of Shimizu and Kushiro (1991) to relate their diffusion data quantitatively to anionic speciation observed from Raman spectroscopy is not entirely surprising.

On the other hand, when results from a number of different spectroscopic techniques are combined, it may often be possible to make qualitative predictions about diffusion pathways. An example is provided by the combined Raman and NMR study of Gan and Hess (1992). Based on their spectroscopic data, they concluded that phosphate speciation in silica rich glasses depend strongly on the alkali:aluminum ratio. In Al-rich glasses, P forms complexes with Al whereas in K-rich glasses it combines with K. Further, they concluded that P forms its own network system and preferred to stay outside of the network formed by Si. These conclusions are supported from diffusion and viscosity measurements on liquids of similar compositions. Diffusion couple experiments have been carried out using diffusion couples made of P-free/P-bearing pairs of peraluminous, subaluminous and peralkaline compositions (Chakraborty and Dingwell, 1993 and unpublished data). In each case, silicon diffuses uphill away from P-rich compositions. In the peraluminous couple, Al diffuses uphill to accumulate in the P-rich side whereas in the peralkaline couple the alkalis (Na and K) do the same. This behavior remains unchanged over a range of temperature of 300°C. These kinds of coupled diffusion are consistent with the speciation model of Gan and Hess (1992). However, there are additional details in the diffusion data which are not predictable from the spectroscopic observations. For example, in the peralkaline couple Al also shows weak uphill diffusion—accumulating away from the P-rich side. A preliminary quantitative treatment of the diffusion process in terms of homogeneous reactions (see above under *Multicomponent diffusion*) seems to indicate that all components are coupled to each other in all melt compositions studied—only the degree of coupling varies. At any rate, the example serves to illustrate how diffusion paths and uphill diffusion in liquids might have been predicted from spectroscopic evidence on glasses in this case, although only qualitatively.

V. EMPIRICAL METHODS

The complex dependence of diffusion rates on structure, speciation and other properties of simple silicate melts indicate that prediction of diffusivities in complex geological melts is likely to remain a difficult problem for some time to come. To overcome some of this handicap, several schemes have been developed over the years to predict diffusion rates in melts based on empirical observations. In this section I provide a brief introduction to some of these approaches. The topic is covered in two parts: (1) methods to predict elemental (tracer) diffusion rates, and (2) methods to predict chemical or component diffusion rates.

Methods to predict elemental diffusion rates

Size-charge correlations. Some simple correlations have been observed for diffusion rates of cations in silicates (melts as well as solids) in general. On the whole, it is a good first order estimate to say that diffusion rates increase with smaller ionic size and charge, the effect of ionic charge being stronger. Plots of log D against the parameter log $(Z^2 r)$ where Z $\equiv$ charge and r $\equiv$ ionic radius often show good negative correlation for diffusion data at a consant temperature in a given matrix (e.g. data for basalts at 1300°C have been illustrated by Hofmann, 1980). It should be emphasized that this is a first order estimate and exceptions are not rare (see Hofmann, 1980, for examples).

Compensation law. An empirical relation that has often been used successfully is the so called *Compensation law*, suggested by Winchell (1969) and Winchell and Norman (1969) for silicates. It is important to point out at the outset that the "law" is a strictly empirical observation based on available data at the time and is not based on any theory. Since its appearance in the literature, numerous attempts have been made to rationalize it with varying degrees of success. In spite of these, a theoretical basis continues to be absent and thus the term "law" is a misnomer. After stating the empirical observation, we will discuss some of these attempts at rationalization.

Using data from solid, glassy as well as liquid silicates, Winchell (1969) was able to show that the activation energy, Q and $\log_{10} D_0$ show a positive linear correlation [where Q and D_0 are parameters of an Arrhenius equation, $D = D_0 \exp (-Q/RT)$]. There was considerable scatter in the correlation, which was improved when only data from amorphous silicates were considered. The term "compensation" comes from the fact that as a consequence of the correlation, the two quantities Q and D_0 compensate each other to some extent so that at some critical temperature, T_c, all species have the same diffusion coefficient. In a subsequent study, Hart (1981) considered the topic of compensation in detail and concluded that different linear correlations should be used for basalts and obsidian. Henderson et al. (1985), added a third, composition-independent compensation law for diffusion of alkali atoms. If the highly non-linear effect of water on cation diffusion rates is considered (see Watson, 1994, for a review), one obtains yet another compensation law and consideration of tracer and chemical diffusivities of oxygen yield two more (Wendlandt, 1991). The important point to recognize is that all of these refer to *rough* correlations and not *exact* predictions. This has two consequences:

(i) newly measured diffusion data can be easily compared to pre-existing ones on a suitable compensation plot ($\log_{10} D_0$ vs. Q). Much of subsequently measured data have been found to follow the linear correlations mentioned above, even when non-cationic species like noble gases and anions (like F^-) or chemical diffusion coefficients were considered. Therefore, for any data that does not follow the trend it is worth re-considering experimental accuracy or alternately, anomalous diffusion mechanisms (e.g. water diffusion, see Zhang et al., 1991a).

(ii) since diffusion coefficients are extremely sensitive to temperature, calculation of the critical temperature T_c from the approximate correlation of Q and $\log_{10} D_0$ is highly uncertain. This has been illustrated by Hofmann (1980) who shows that one can easily obtain T_c values that differ by more than 400°C by simply choosing different subsets of a dataset that appears to follow the compensation law as a whole. Given these uncertainties and the lack of any strong theoretical significance, major efforts to determine T_c do not seem to be very meaningful. Similarly, it is possible to develop various schemes to calculate diffusion coefficients utilizing T_c and estimated activation energies from size-charge correlations of the kind mentioned above. It follows from the above discussion that such predictions are likely to be only approximate, under the best circumstances.

In their original work itself, Winchell and Norman (1969) noted that a correlation between Q and $\log_{10} D_0$ is to be expected from the absolute reaction rate theory (Glasstone et al., 1941), if diffusion occurs by elementary jump processes. The basic reasoning is simple—an atom that requires less energy to move (low activation energy) will jump more often (higher jump frequency). From the various discussions in this chapter it should be clear that atomic migration in silicate melts is far more complicated than an elementary jump process—even in the simplest case there are at least a distribution of jump lengths rather than a single well defined one. However, it can be shown (Dosdale and Brook, 1983, Isard, 1992) that if the observed diffusion process is a series-parallel combination of a number of elementary thermally activated jump processes, then compensation law behavior (i.e. linear correlation of Q and $\log_{10} D_o$) would be expected for each class of material. Based on our current understanding of ionic diffusion in silicate melts, this seems to be a plausible explanation for the validity of compensation law. If this is indeed the underlying reason for compensation behavior, then Dosdale and Brook (1983) further suggest that well controlled measurements of diffusion rates of different species in the same material may allow one to separate the contributions of the individual jump processes to the overall transport rate.

Methods to predict chemical diffusion rates

Chemical diffusion coefficients in complex systems depend on various tracer diffusion coefficients and thermodynamic mixing properties. Since it is uncommon to have all of these data available for the variety of geological melt compositions, empirical observations that allow one to predict diffusion behavior are very valuable. We discuss below two models that provide some guidance.

Transient two-liquid partitioning (Watson, 1982). Watson (1976) had studied element partitioning between immiscible acidic and basic liquids and concluded that certain elements (like alkalies) prefer to be in more polymerized acidic melts while others (like P) partition preferentially into the basic melts. In a subsequent work, Watson (1982) studied the contamination of basalt by continental crust in a series of experiments. He found that diffusion of fast diffusing components like alkalies may be described as a two stage process. In the first stage, the alkalies diffuse rapidly such that they attain an equilibrium distribution with melts of different degrees of polymerization within a diffusion zone. In other words, they tend to be enriched in more acidic melts, even if this involves uphill diffusion against their own concentration gradients. This is the so called *transient two liquid partitioning*. In the second stage, concentration of *all* components in the diffusion zone evolve relatively slowly governed by the diffusion rates of the network former components like SiO_2.

This behavior allows a number of predictions to be made about diffusion in complex multicomponent melts:

(1) concentration distributions during the initial stages of mixing may be predicted from a knowledge of equilibrium partitioning between the endmembers (at least for the fast diffusing components)

(2) the rate of diffusion during the latter stages of mixing for *all* components may be predicted from a knowledge of the diffusion rates of SiO_2.

(3) whether uphill diffusion of a fast diffusing component will occur can be predicted if the two liquid partitioning behavior is known.

These features have helped understand many features of magma mixing, assimilation and selective contamination of melts in natural systems. Concentration gradients that develop around crystals dissolving in a melt can also be explained qualitatively using these concepts. Watson and Baker (1991) provide a detailed review of these topics. Recently, quantitative treatment of multicomponent diffusion data has shown that homogeneous reactions which occur in a melt during diffusion are consistent with this picture of initial rapid diffusion followed by a phase of slow, coupled diffusion of the alkalies (Chakraborty et al., 1995b). Such treatment allows one to separately obtain the rates of the two processes. The behavior can also be rationalized from a microscopic standpoint. The initial phase of rapid diffusion would relate to appropriately charge compensated motion of alkali ions in a fixed (or, nearly fixed) lattice formed by the network formers, analogous to diffusion in a glass. The latter stage may be related to diffusion dominated by configurational changes in the network itself, where the alkali atoms "ride along" with network units undergoing translational motion.

Modified effective binary model (Zhang, 1993). Recently, Zhang (1993) has quantified some aspects of this transient two liquid partitioning approach and has developed a modified effective binary model. The essence of the model lies in the assumption that *all, or at least most,* of the diffusion coupling observed in a multicomponent system is due to thermodynamic effects. In other words, flux of a component is simply proportional to its activity gradients and the proportionality constant is a diffusion coefficient which is termed the *intrinsic effective binary diffusion coefficient* by Zhang (1993). It should be noted that this coefficient is different from *intrinsic diffusion coefficients* defined in the literature (e.g. Carman, 1968) and discussed above as well. The assumption above implies that the influence of the cross terms of the L matrix, L_{ij}, are negligible. Even with this assumption, it is difficult to use activity gradients in complex melts because activities are not measurable. To overcome this problem, Zhang (1993) makes a further assumption that activities of all components are linear functions of the $(SiO_2+Al_2O_3)$ content of the melt. With these two assumptions he was able to show that the model was capable of describing concentration profiles and directions of diffusion around crystals dissolving by a diffusion controlled process in various silicate melts in laboratory experimental charges. In a subsequent study, van der Laan et al. (1994) have shown that diffusion behavior in rhyolite-andesite and rhyolite-rhyolite melt couples can be well described by the model.

It is apparent, as recognized by Zhang (1993) as well, that the model cannot be universally applicable. For example, he points out that the model cannot be used for diffusion in non-infinite reservoirs; nor is it applicable where uphill diffusion of network former species (Si, Al, P etc.) occur. Such uphill diffusion of network formers are quite common in diffusion experiments (e.g. Sugawara et al., 1977; Chekhmir and Epelbaum, 1991; Chakraborty et al., 1995a,b; Chakraborty and Dingwell, 1993) and pose a major limitation for the use of the model. In order to use the model to predict whether uphill diffusion of a component will occur, one has to determine a priori whether the possibility of uphill diffusion of network formers may be eliminated so that the model may be applicable in the first place. From our discussions of macroscopic as well as microscopic behavior it should be apparent that dynamic correlations between diffusing species leading to significant magnitudes of L_{ij} should be expected in many silicate melts. Lastly, the assumption that activities of various components are linear functions of $(SiO_2+Al_2O_3)$ content of melts is clearly a simplification that cannot be generally valid in all complex multicomponent systems. Therefore, although the model can be useful under some circumstances it should be used with caution and should not be taken to be a generally valid one. It is an empirical approximation which does not substitute a general representation using full **D** or **L** matrices, when possible. If used judiciously, however, the model does have the merit of removing the compositional dependencies arising from concentration dependence of mixing properties from the diffusion process (the intrinsic effective binary diffusion coefficients are much weaker functions of composition compared to the standard effective binary diffusion coefficients).

VI. EXPERIMENTAL METHODS

Experimental methods for the determination of diffusion coefficients have been discussed in detail in a number of recent reviews. Watson (1994) provides a brief introduction and reviews by Baker (1993) and Ryerson (1987) deal exclusively and in detail with experimental methods for the determination of diffusion coefficients. Thus, only a general survey of the experimental approach will be provided here by following the broad steps taken by an experimentalist to measure diffusion coefficients in a classical study. Along the way, a few specifics not covered in earlier works will be introduced where appropriate. Some experimental techniques recently developed in the materials science community are discussed. The worker interested in making hands-on measurements is strongly urged to look at the reviews cited above for more details. Spectroscopic and other non-traditional methods of measuring diffusion rates are treated in some detail.

Macroscopic measurements

Problems of handling melts and differences between experiments on "granites" vs "basalts". From the point of view of an experimentalist or someone wishing to critically evaluate experimental data, it is extremely important to distinguish between viscous (usually silica rich, "granitic") and fluid (usually silica poor, "basaltic") melts. The vast majority of melts studied in the glass industry typically belong to the second category. We have seen that the theoretical concepts required to understand diffusion in these two kinds of materials are also different to some extent and have been dealt with at several places in this chapter (*Diffusion and melt structure*, *Diffusion, relaxation and glass transition* and *Microscopics*). A distinction is necessary because the experimental problems in dealing with these two different kinds of melts are completely different and the strategy of experimental design needs to be adapted accordingly. For the more viscous, silica-rich melts the problem is producing homogeneous, bubble-free starting melt compositions. The high viscosity does not allow classical procedures for melt production e.g. pouring on a plate, drawing into a capillary etc. to be used effectively. Homogenizing the melts (structurally and compositionally) is often a problem, epitomized in the considerable amount of literature that has build up around the production and characterization of albite glass (see e.g. Knoche et al., 1992, for discussions). The problem has been typically overcome by using natural rhyolitic glass (e.g. Harrison and Watson, 1983), producing glasses under high pressure or by fining at high temperatures (e.g. Chakraborty et al., 1993). See Baker (1993) for some other recommended alternatives. A second disadvantage of this kind of material is that diffusion rates are slow and so it is necessary to perform relatively long experiments to induce measurable diffusion. In some cases, for diffusion studies in glasses this means that crystallization may intervene. The advantages, however, are that convection is almost never a problem and localized bubbles may often remain practically stationary through the duration of even a long diffusion anneal (e.g. see Chakraborty et al., 1995a).

On the other hand, for the less viscous, silica poor melts the problems are just the opposite. They may often be easily produced (sometimes crystallization is a problem, particularly for Fe-bearing melts) and homogeneity of starting material is rarely a problem. The melt may easily be drawn into capillaries or poured into desired forms. However, the low viscosity makes buoyancy-driven convection, effects due to surface tension and movement of bubbles critical problems in many studies. In such fluid melts, bubbles are a particular problem because even if great care is taken to prepare glasses without bubbles, during remelting for the diffusion anneal dissolved molecular oxygen and other gases often exsolve to form bubbles (the "reboil" phenomenon) which can then grow and migrate. Once formed, bubbles are usually extremely mobile in such melts. Preparation of glasses under vacuum have helped to minimize this effect. Buoyancy-driven convection is usually reduced by starting with an assembly where the denser melt is placed at the bottom. The effects of buoyancy-driven convection may be suppressed by using capillaries, but with increasing length to width ratio the effects due to surface tension become significant and may lead to destabilization of a diffusion couple. Thus there is a trade-off region in the dimensions of a diffusion cell for which both convection due to buoyancy as well as surface tension effects may be suppressed and meaningful diffusion data can be obtained. However, there are situations where it is not possible to avoid both of these effects simultaneously using conventional setups—diffusion in alkali silicate melts above 1000°C being a case in point (Meier and Frischat, 1993). Among geologically interesting samples, diffusion measurement in some alkaline mafic and ultramafic melts may be affected by these problems. Thus, over the years several elaborate and/or imaginative solutions to specific problems have been found. For example, microgravity experiments have been performed in space (e.g. Meier and Frischat, 1990; Meier et al., 1990) or under reduced-gravity conditions on earth using centrifuges (e.g. Meier and Frischat, 1993). Kieffer et al. (1986) and Young et al. (1994a), rather than struggle against bubble formation, developed an ingenous design for a sample capsule which allows an escape path for bubbles without affecting diffusion! In the latter study involving diffusion of oxygen they had to improvise on the method of Kieffer to allow bubbles formed inside the melt to escape without contaminating the melt with externally derived oxygen. Once again, see Baker (1993) as well as Kress and Ghiorso (1995) for various strategies on handling low viscosity melts. Relatively fast diffusion rates in low viscosity melts call for short annealing times which may lead to problems at the other extreme sometimes—(1) it may be necessary to correct for heatup times (see below), and (2) having a reservoir that is large enough to be considered infinite may be difficult.

Besides the factors mentioned above, diffusion path plays an important role in determining the stability of diffusion interfaces in all kinds of melts. Intermediate compositions that develop during diffusion may cause instability if they lie below the solidus (crystallization may occur), a particular problem for regions of the phase diagram with steep liquidus slopes (Chakraborty, unpublished data, for obvious reasons!). At some point in the diffusion zone the density of melts of intermediate compositions that form may be greater than melts that lie below leading to convective turnover in an initially stable system. In a multicomponent system, the nature of diffusive coupling of fluxes may be such that some interfaces may be inherently unstable (see Vittagliano et al., 1992, and Schmalzried, 1995, for a discussion of the theory; Liang et al. (1994) provide some practical examples in silicate melt systems).

The diffusion anneal. The general approach in a classical experiment designed to measure diffusion coefficients is to bring together a *reservoir* and a *source*. The reservoir is the medium in which diffusion rates are to be measured. The species or component whose diffusion rate is to be measured is enriched in the source and diffuses from the source to the reservoir. In a proper self-diffusion experiment the reservoir and the source have the same chemical compositions and the only difference between them is that the source is enriched in some isotope of the diffusing species (e.g. Shimizu and Kushiro, 1984) which is used as a tracer. In chemical or exchange diffusion experiments, the source and reservoir switch roles for different components—the source for one component is the reservoir for another (e.g. in a Fe-Mg exchange experiment a diffusion couple is prepared where the Fe-rich side is the source for Fe and the reservoir for Mg). For more details, see Baker (1993).

The point of time where the source and the reservoir are brought together to the annealing conditions (P, T, etc.) is the initial time, t = 0. Ideally, this should occur instantaneously, i.e. no diffusion should have occured along the way. In practice, this is often not possible due to instrumental limitations, particularly for high pressure experiments (e.g. see Rubie et al., 1993). In that case there are two options available: (1) to perform "zero time" experiments where a run is quenched immediately on reaching annealing conditions. The amount of diffusion is measured and subsequently subtracted from subsequent runs of longer durations. This method is practical when the approach to all annealing conditions in a given study always follows the same path (temperature, time etc.). Usually however, this is not the case (e.g. it takes longer to heat to higher temperatures). Then the only alternative is to (2) take the non-isothermal nature of the anneal into account and use iterative methods to reduce the data (e.g. see Shewmon, 1963; Braedt and Frischat, 1988). The number of variable parameters increase and so do the uncertainties in the measurement. Overall, this approach is not commendable but sometimes unavoidable. Both of these methods may be useful where the objective is to measure concentration dependence at a given temperature but are problematic when measurement of temperature dependence is the goal.

Once the run conditions are attained, the samples are annealed for a given time, t. The main problem in studying diffusion in liquids is to suppress convection and ensure that the only mode of mass transport is diffusion. Considerable effort has been spent in the literature to ensure this and to demonstrate lack of convection a posteriori. In addition to the discussion above, see the reviews by Baker (1993) and Watson (1994) as well as Kress and Ghiorso (1993) and Chakraborty et al. (1995a) for some additional observations that may help to demonstrate lack of convection during a diffusion experiment.

One of the experimental challenges sometimes is to maintain the annealing conditions constant for a reasonable length of time e.g. for experiments at extreme conditions of pressure and temperature (e.g. Rubie et al., 1993). Another point worth noting here is the need for the control and measurement of *absolute* temperatures during such anneals. The reason for this is that a constant error in temperature translates to *different* errors at different temperatures on an Arrhenius plot and may result in the determination of incorrect activation energies. Hence, it is worthwhile to calibrate thermocouples against known melting temperatures and determine hotspots of furnaces carefully—the latter a particularly important issue in high pressure experiments. For experiments involving melts, minimizing thermal gradients are critical to avoid Soret diffusion or convection. Alternately, of course, precisely such gradients can be exploited to measure Soret diffusion, see Lesher and Walker (1991) for details. Some works which discuss thermal gradients in solid media high pressure apparatus in the context of diffusion measurements include Baker (1993), Elphick et al. (1985) and Chakraborty and Ganguly (1992).

After the anneal—measurement of concentration profiles. Once the diffusion anneal is completed, the samples are typically quenched for measuring the amount of diffusion that has taken place. This introduces two sources of uncertainty right away—it is assumed that the composition and species distribution remains unaltered on quenching and that the diffusion penetration lengths measured in glass at room temperature provide information about penetration lengths at higher temperatures. The first of these issues has been discussed at length in Zhang et al. (1991a), the second issue has been addressed in Sugawara et al. (1977), Chakraborty et al. (1995a) etc. For both of these problems, there are no general answers—each case needs to be judged on its own. Hence, we leave the topic by simply guiding the reader to some examples. A further problem arises in high pressure experiments due to deformation and distortion of the experimental charge. Sample containers have been designed to overcome this problem by Watson and coworkers (e.g. Watson, 1979; Ayers et al., 1992; see details in Baker, 1993). Concentrations in the quenched sample are then measured and a variety of techniques are available for doing so. Many methods rely on measuring bulk gain or loss of a component from the reservoir but the general opinion seems to be that methods which specifically measure concentration gradients are superior. A number of methods have been used beginning with the classic measurement of Bowen (1921) using refractive indices. Serial sectioning and analysing successive sections, either by wet chemistry or by counting radioactive tracer concentrations is another classical approach. But the most commonly used methods nowadays are microbeam techniques like the electron microprobe and the ion microprobe (either by depth sputtering or by step scanning), proton- and xray-microprobes have also been used (e.g. Baker, 1990). Nuclear reaction techniques have gained popularity in recent years. See Ryerson (1987) for details of these methods and their relative merits and demerits. A powerful tool that has appeared recently (since the review by Ryerson) is the infrared microscope which allows spatially resolved analysis of water speciation in silicate glasses. This has served to resolve long lasting questions regarding the nature of water diffusion in silicate melts (see Zhang et al., 1991, for a detailed description of the method; Watson, 1994, for a summary of the results). Finally, following the lead provided by Bowen (1921), some studies have measured variation of a property which is a function of composition, rather than composition itself, in the diffusion zone. A recent example of this are studies of water diffusion carried out by Chekhmir and coworkers (1991, for a summary in English) who measured mobility of hydrogen bearing species in albite and other melts using redox sensing color pigments distributed in the melt as "traps".

We conclude this discussion on analytical methods used in macroscopic diffusion studies with one comment on the popular microbeam techniques. Any microbeam technique analyzes the chemical composition of a small volume of the sample surrounding the beam. The finite size of the excitation volume results in a smearing out of compositional gradients measured by step scanning a microbeam. The extent of smearing depends on analytical conditions and the nature of the sample. Such smearing out affects the retrieved diffusion ceofficients from measured concentration profiles because the observed penetration depth is larger than the true penetration depth. The effect is moderate for small beam sizes in an electron microprobe, crystalline material and very short profiles (Ganguly et al., 1988, showed that for profiles larger than 15 μm the effect was negligible). However, for the larger defocussed beams and less dense glassy material, this effect has been found to be much more pronounced (Chakraborty et al., 1995a) and profiles upto 150 μm in length were found to be affected. Such spatial averaging effects in analyses of glasses using infrared spectroscopy (Zhang et al., 1991) and ion microprobe (Chakraborty et al., 1993) have also been considered. The effect may become critically important when the diffusing species is a trace constituent on one side of the diffusion couple and a major element on the other side (e.g. Zr diffusion from solid zircon into a granitic melt—Harrison and Watson, 1983). A consideration of these effects reduces one out of the many sources of inaccuracy in diffusion data and considering the simplicity of the methods (Ganguly et al., 1988, provide very simple methods for assessing such effects), they should be a part of studies involving microbeam analytical techniques for the measurement of diffusion profiles.

Fitting a model to measured profiles—tracer and binary diffusion, constant diffusion coefficients. Once a concentration profile is measured, it is typically fit to a model equation in which the diffusion coefficient is one of the, or ideally the only, parameter which is unknown. In the literature such model equations have almost invariably been analytical solutions to the so called Fick's second law of diffusion. Such solutions are valid for a particular geometry of the diffusion medium (e.g. slab, sphere etc.) and configuration of the concentration gradient (e.g. radial, longitudinal etc.). This is why it is necessary to obtain and maintain a specific geometry during the diffusion anneal (avoid deformation, flow of melts etc.). Two commonly accessible compendium of solutions to the diffusion equation are the works by Crank (1975) and Carslaw and Jaeger (1959) where usable results for most common experimental geometries may be found—the latter work is for heat flow, but the solutions are usable for diffusion studies with no or very little modification. Usually one needs to measure some quantities (concentrations at certain points of the sample before the diffusion experiment, the duration of diffusion anneal, some aspects of the dimension of the diffusion reservoir in some cases) in addition to the diffusion profile. Uncertainties in all these measurements get propagated to the final uncertainty in measured diffusion coefficients. The solutions themselves are typically non-linear involving exponential functions or an error-function (which is nothing but a sum of many exponentials). Many schemes have been developed to linearize these equation to make data reduction easier. Practical guides to using these equations for measuring diffusion coefficients for common experimental geometries and linearizing the solutions have been provided in the reviews by Ryerson (1987), Baker (1993) and Watson (1994) and will not be repeated here. A critical consideration usually is whether the diffusion medium (*reservoir*, as defined above) is finite or infinite with respect to diffusion of a species (i.e. whether diffusion reaches the "ends" of a sample or not). In general, solutions for infinite media are simpler in form and an important objective in experimental design is to achieve this condition by having a large enough reservoir. In some cases this is difficult to achieve, common examples being (1) diffusion of very fast diffusing species (e.g. H_2O and other hydrogen bearing species) (2) simultaneous diffusion of multiple species one of which diffuses rapidly and the other very slowly (e.g. alkali atoms and silicon in some melts) and (3) high pressure experiments where constraints of space are imposed by the available size of the pressure cell. A solution for such cases (finite diffusion couples) which is not easily available but may be derived from relationships given in Crank (1975), is provided below:

$$C(t,x) = \frac{C_1 + C_2}{2} + \frac{2(C_1 - C_2)}{\pi} \sum_{n=1}^{\infty} \frac{1}{n} \exp\left[\frac{-Dn^2\pi^2 t}{l^2}\right] \cos\left[\frac{n\pi x}{l}\right] \sin\left[\frac{n\pi x_i}{l}\right]$$

Here, $C(t,x)$ concentration at time t and distance x (i.e. the measured profile after a given experimental duration, t), C_1, $C_2 \equiv$ initial concentrations in the two end endmembers of the diffusion couple, $D \equiv$ diffusion coefficient, $l \equiv$ total length of diffusion couple (often this is the length of the sample container), $x_i \equiv$ location of interface. Unlike most of the solutions for infinite systems, this cannot be linearized and one would have to perform non-linear least squares fits to obtain diffusion coefficients. It is possible to retrieve more parameters than just D from the above equations e.g. one can retrieve one or both of the starting compositions, C_2 and C_1, as a check on the fitting procedure. However, a word of caution is in order here—the function above is periodic in the length of the cell, l. This means if l is left as a refinable parameter, it is possible to obtain "wavy" best fit profiles which may then lead to incorrect interpretations.

Even if none of the available analytical solutions are suitable for a particular experimental setup, diffusion coefficients can always be retrieved by numerical fitting as long as a solution exists in principle. The wide availability of user-friendly mathematical software packages allow such tasks to be performed nowadays without any specialized skills.

Fitting a model to measured profiles: chemical diffusion, non-constant diffusion coefficients. The problem of non-constant diffusion coefficients has been discussed in the literature for a long time. The classic example of such cases appears when diffusion rates depend on composition: during a diffusion experiment as diffusion alters the composition at a point the diffusion rate itself changes as well. This manifests itself in asymmetric final measured profiles with diffusion coefficients that depends on distance (because composition changes with distance along a profile). The standard method to retrieve these diffusion coefficients from the measured profiles is the so called Boltzmann-Matano method, which is widely discussed in texts on diffusion as well as all the reviews mentioned earlier. The method assumes that there is no volume change on mixing of the various components. This may be a good approximation for many solids and for liquid diffusion couples with small concentration differences, but in many cases may be a poor approximation e.g. in diffusion couples involving dry and volatile bearing melts as end members. Also, the method requires the location of the interface position quite accurately. To avoid such problems, alternatives have been suggested by Sauer and Freise (1962) and Wagner (1969). In the special case of no volume change their expression reduces to the typical Boltzmann-Matano formulation and hence may be considered to be more general. A disadvantage is that the methods call for a knowledge of the molar volumes as a function of composition which is sometimes not available for multicomponent silicate melts although recent work has gone a long way in alleviating this problem, at least for volatile-free melts (Lange and Carmichael, 1990; Knoche et al., 1995). The formula for calculating D according to this method is

$$D(N_2^*) = \frac{(N_2^+ - N_2^-)V_m(N_2^*)}{2t\left(\dfrac{\partial N_2}{\partial x}\right)_{x=x^*}}\left[(1-Y^*)\int_{-\infty}^{x^*}\frac{Y}{V_m}dx + Y^*\int_{x^*}^{\infty}\frac{1-Y}{V_m}dx\right]$$

where N_i is the mole fraction of component i in units of mol per unit volume, (N_i/V_m), Y is an auxiliary composition variable of the form given by

$$Y = \frac{N_2 - N_2^-}{N_2^+ - N_2^-}$$

and the superscripts +, - and * on a variable refer to the values at the two end members of the diffusion couple and the composition at which the diffusion coefficient is to be determined, respectively. x^* is the distance at which $N = N_2^*$. For an example of application of this relation in earth sciences, the reader may consult Misener (1974). In these methods to determine compositionally dependent diffusion coefficients where calculation of areas under curves or local slopes are important, often the calculations are performed on spline or other high order polynomial fits to the profiles rather than the raw data itself. This step is necessary to avoid artefacts due to local irregularities caused by analytical scatter, unless the analytical data is of very high quality. In finite systems with non-constant diffusion coefficients, almost in all cases numerical solutions are required.

One kind of experiment that has been widely used by Watson and coworkers (e.g. Harrison and Watson, 1983) is the use of diffusion profiles developed in a melt during the dissolution of a refractory mineral (see Watson, 1994, for a review). For the poorly soluble accessory minerals used in these studies it is assumed that the movement of the interface due to dissolution of the minerals is negligible and there are no concentration gradients developed due to diffusion in the minerals. Such an assumption has been shown to be justified in the particular cases studied. A further assumption in these studies was that diffusivity was considered to be constant. However, it is also possible to retrieve diffusion coefficients (Appel, 1968) from more general cases where (1) the movement of the interface is significant, (2) the diffusion coefficient is a function of composition, and (3) there is a discontinuity in composition at the interface.

This method is particularly suitable for trace elements because a very small amount of dissolution of the crystal (in which the element of interest is a major consituent, e.g. Zr, Th and P from dissolution of zircon, monazite and apatite, respectively, Harrison and Watson, 1983, 1984) provides enough material for measurable diffusion profiles to develop but at the same time the movement of the mineral-melt boundary due to dissolution can be neglected. Moreover, if the element of interest is a trace constituent in the melt (within the Henry's law regime) then the boundary conditions for diffusion become considerably simplified. The method is not as amenable for studies of major element diffusion because the crystal-melt boundary cannot be taken to remain constant anymore, the boundary conditions involving complex mineral-melt equilibria become complicated and effects due to multicomponent coupling during diffusion become significant. Efforts to overcome some of these difficulties have been made using empirical approaches in recent times (e.g. Zhang et al., 1989; Chekhmir and Epel'baum, 1991), see the section on *empirical methods* for more details.

Fitting a model to measured profiles—multicomponent diffusion. As pointed out earlier, the main problem with multicomponent diffusion is the experimental determination of these coefficients, not their use in mass transfer problems. The origin of the problem is simply that the number of diffusion coefficients required to describe diffusion in a system multiplies approximately as the square of the number of components. Thus, it is necessary to choose components judiciously for a study and to exploit mathematical dependencies between diffusion coefficients to minimize experimental effort. Additionally, the model equations to be fit to retrieve diffusion coefficients do not lend themselves easily to statistical fitting procedures. This means that typically more data than the bare minimum mathematical requirement are necessary to actually obtain the diffusion coefficients unambiguously.

In general, to obtain diffusion coefficients in a melt system with (n+1) elements (n neutral components) one has to determine $(n-1)^2$ diffusion coefficients. For this, one tries to set up diffusion couple experiments fulfilling the following conditions:

(a) The compositional difference between the endmembers are small enough that the compositional dependence of diffusion coefficients may be neglected (to make the models simpler) but at the same time large enough that the differences may be measured well enough to yield good compositional profiles which allow detection of subtle effects due to multicomponent coupling. The nature of the observed profiles allow *a posteriori* determination of whether compositional dependence of diffusion coefficients are significant (see Chakraborty et al., 1995a,b for details).

(b) Several endmembers are chosen such that the mean composition of each diffusion couple is the same. Each diffusion couple experiment yields (n-1) independent compositional profiles. (n-1) of such diffusion couple experiments, all done at the same conditions, then yield $(n-1)^2$ independent profiles from which the $(n-1)^2$ diffusion coefficients may be determined. This means doing two experiments for a ternary system, four for a quaternary, and so on. In practice, to minimize errors, performing more than the minimum required number of experiments is recommended. Proper choice of initial compositional gradients between endmembers also allow the errors in determining diffusion coefficients to be reduced. This entire problem has been discussed mathematically by Trial and Spera (1994) who also suggest strategies for fitting multicomponent diffusion models to data.

It is apparent from this discussion that any approach that allows one to reduce the number of experiments required to determine the multicomponent diffusion coefficients would be a welcome addition to the repertoire of the experimentalist. The fact that theoretically the $(n-1)^2$ diffusion coefficients are not all independent of each other (e.g. see de Groot and Mazur, 1984) further encourages one to search for mathematical relationships that allow the experimental effort to be reduced. For example, we have referred above to effects that may arise from multicomponent coupling. These may include the occurrence of uphill diffusion, the occurrence of zero flux planes etc. The occurrence of such special features in the diffusion profiles is related to the various diffusion coefficients through very specific mathematical relations (e.g. see Thompson and Morral, 1986). These may be exploited systematically to reduce the number of experiments required to determine all diffusion coefficients in a multicomponent system. An example may be found in Chakraborty et al. (1995a) where a method suggested by Schut and Cooper (1982) was modified slightly to obtain all four diffusion coefficients in a ternary system from only one diffusion couple experiment. Data were collected over a range of temperature and composition and tests using additional experiments showed the method to be reliable in that there was no difference between diffusion coefficients obtained using the new method and those obtained using the more traditional approach outlined above.

Another theoretically elegant approach uses the relationship between tracer and chemical diffusion coefficients to reduce the number of experiments required. The elegance of the method comes from the fact that instead of $(n-1)^2$ diffusion coefficients one determines only the n tracer diffusion coefficients at the same composition and a single experiment yields tracer diffusion coefficients, chemical diffusion coefficients as well as the compositional dependence of the chemical diffusion coefficients. To do this one uses a theoretical model to relate tracer- to chemical diffusion coefficients. Observed profiles are then fitted directly using numerical techniques to a solution which incorporates this model. The critical point is that a well-founded model should be available and that in the most general case the thermodynamic mixing behavior in the sysem should be well known. The method has been used very successfully for crystalline silicates (Loomis et al., 1985; Chakraborty and Ganguly, 1991, 1994; and verified through independent tracer diffusion experiments by Chakraborty and Rubie, 1995). However, for the silicate melts the lack of a comprehensive model (see above) as well as paucity of information on mixing behavior of components makes such applications somewhat difficult at this time. Nevertheless, Chakraborty et al. (1993) applied the method successfully in a binary diffusion problem and Chakraborty et al. (1995b) obtained reasonable estimates of tracer diffusion coefficients in a ternary system using the method. The latter work also discussed the potential merits and shortcomings of the approach, the method itself is outlined in detail in Loomis et al. (1985) and Chakraborty and Ganguly (1991). One distinct advantage arises from the fact that the tracer diffusion coefficients depend less strongly on composition than the chemical diffusion coefficients (see discussion above), which allows a wider range of compositions to be modelled in terms of constant, compositionally independent diffusion coefficients.

Spectroscopic methods

In contrast to macroscopic tools which observe spatially and temporally averaged quantities, spectroscopic methods have the potential to provide a direct window into time- and space- resolved microscopic motion which leads to macroscopic diffusion. All spectroscopic methods work on the same principle—an incoming radiation interacts with the sample (or some specified part of it, e.g. a specific nucleus), gets modified in some way and the outcoming radiation provides information on the interaction which has taken place. This can be used in the simplest case as a "fingerprinting" technique whereby the outcoming radiation is used to measure the concentration of some species in the sample as a function of space *after* the diffusion anneal. In these cases (e.g. the application of infrared spectroscopy by Zhang et al. (1991a) mentioned above) spectroscopy is exactly analogous to conventional analytical tools and provides data on macroscopic diffusion rates. In this section we will focus on in situ methods which yield information on microscopic motion.

Spectroscopic methods observe either the motion of atoms directly, or some consequence of it. Each spectroscopic method operates over a given frequency range. For diffusion studies, this means a specific method can observe diffusion only when the diffusion rates lie within certain ranges. Two techniques that have already been used to study dynamics in silicate melt systems under *in situ* conditions are NMR and infrared spectroscopy.

Nuclear spectroscopic techniques. Nucleii interact with their surroundings primarily through magnetic dipolar and electric quadrupolar interactions. These interactions, classified as hyperfine interactions, occur because of magnetic fields and electric field gradients at the nucleii generated by the environment of atoms/ions. Under favorable circumstances it is possible to use these interactions through several nuclear spectroscopic techniques (NMR, Mössbauer) to observe specific nucleii as they move.

Nuclear Magnetic Resonance (NMR). Detailed discussions of studies of dynamic processes by NMR have been provided by Stebbins (1988, and also this volume), in the following we present only a brief introduction to applications of NMR to diffusion studies in melts. The raw data from NMR used for diffusion measurements are spectral linewidths or relaxation times. The basic information (pertinent to diffusion) provided by a NMR experiment is the motional correlation time, τ_c, which for translational diffusion can be interpreted as the mean residence times of an atom on a site. The diffusion coefficient is related to τ_c via the Einstein-Smoluchowski equation, i.e. $D = a^2/6\tau_c$ where a is the distance between sites of individual atomic jumps. To obtain accurate values of D (better than an order of magnitude) from the observable quantities in NMR spectra, one often needs to have a structural model to relate these observables to τ_c. This proves to be difficult in many cases for amorphous material and sets a limit on the attainable accuracy.

The simplest application of NMR to diffusion studies is through the comparison of spectral shapes measured under different conditions. The general effect of atomic motion (on the NMR time scale) is to cause NMR lines to decrease in width as line broadening influences average to zero. In viscous silicate melts, local atomic motion leading to flow as well as configurational changes may be sampled by NMR. In a glass one observes different NMR lines due to Si and O in different environments. With increasing temperature, in situ measurements show that these lines gradually merge to a single peak—a consequence of motional averaging and chemical exchange. The motional line narrowing in glasses can be related to activation energies of diffusion in principle. In practice, the relation seems to be sensitive to the uses of different models and to the choice of different functions to describe line shapes (e.g. Lorentzian vs. Gaussian), so that unique activation energies are difficult to obtain (see Frischat, 1975, for a detailed discussion). However, the line narrowing and ultimate merging due to chemical exchange near the glass transition region is a direct look at diffusive motion on a microscopic scale. This technique has been developed and widely used for silicate melts by Stebbins and coworkers (e.g. Liu et al., 1987; Farnan and Stebbins, 1990, 1994; see Stebbins this volume, for a review). Of particular interest is the use of 2-D NMR spectra (e.g. Farnan and Stebbins, 1990; see the chapter by Stebbins for an account) which effectively separate the total atomic motion from those that result in a net displacement and thus provide a direct measure of the correlation between successive atomic jumps (see sections on *Microscopics* and *Diffusion and electrical properties*).

A second feature of NMR spectra which can be used for dynamic measurement results from a more complicated effect. Here, rather than sample the system as is using a certain radiation, one actually perturbs the state of the system by exciting certain nucleii and then monitors the time evolution of the nuclear magnetization. Basically, one excites a nuclear spin system to a high energy state using a radio-frequency pulse and monitors how this state decays. There are various variants on this theme which come under the category of "nuclear spin relaxation" (see Stebbins, 1988, for details). Although in principle this can provide dynamical information over a wide temperature range, interpretation of the results is often difficult (e.g. the spin-lattice relaxation time (T_1) needs to be isolated from other relaxation effects). Finally, to relate the spin-lattice relaxation time (T_1) or the spectral density of local magnetic dipole field ($J(\omega)$) obtained from NMR experiments to diffusion (ensemble average of the molecular motion of the nuclear magnetic dipoles), one needs to have a model of microscopic motion, which is often difficult to obtain for complex materials. However, for a simple non-correlated, Markovian (memoryless) random walk diffusion in a non-metal with a single jump frequency, it can be shown that

$$\left(\frac{1}{T_1}\right) = \langle\Delta\omega_b^2\rangle \, J(\omega) \quad \sim \quad \frac{\langle\Delta\omega_b^2\rangle}{\omega_b} \; \frac{\omega_b \tau}{1+\omega_b^2\tau^2}$$

if $J(\omega)$ is a Lorentzian. Here, ω_0 is the Larmor frequency. Having obtained τ one proceeds to obtain the diffusion coefficient, D, using the relation $D = a^2/6\tau$. Two corrections need to be made for comparing this D to macroscopic tracer diffusion coefficients: (1) for isotopic mass, which is usually negligible for heavier atoms, and (2) for the temporal correlation function (see the section on *Microscopics*).

The difficulties notwithstanding, the method has been applied to study diffusion in silicate melts. Franke and Heitjans (1992) studied spin-lattice relaxation using ^{7}Li-NMR in both the glassy and crystalline (ß-spodumene) forms of $Li_2O\cdot Al_2O_3\cdot 4SiO_2$. This study shows pronounced diffusion-induced peaks in plots of spin-lattice relaxation rate ($1/T_1$) vs. reciprocal absolute temperature ($1/T$ in kelvin) for both the solid as well as the glassy material. The asymmetry of this peak in both kinds of material indicates that the standard diffusion model of non-correlated, Markovian random walk is not strictly valid for these materials. The data show further that (a) the coupling between the nuclear spin system (^{7}Li) and the lattice system is smaller for the glass and (b) the jump rate of Li^+ is faster and the activation energy is smaller for the glass.

Liu et al. (1987) studied diffusion of Na and Si in albite glass and found Na activation energies in good agreement with macroscopically measured ones. Fiske and Stebbins (1993) studied spin-lattice relaxation times for ^{25}Mg and ^{23}Na in melts of the composition $(Na_2O)_{0.28}$-$(MgO)_{0.18}$-$(SiO_2)_{0.54}$ between 1150° and 1360°C. They obtained apparent activation energies for ^{25}Mg-relaxation of 119 kJ/mol and for ^{23}Na-relaxation of 85 kJ/mol. They speculated that the higher activation energy of Na-relaxation compared to that expected from diffusion data in sodium-silicate melts of similar degree of polymerization could have resulted from a mixed-cation effect between Na and Mg in these melts, akin to the well known mixed alkali effect (see the section on *Diffusion and melt structure*). It is also possible that the differences observed in activation energy simply relate to the ambiguity in relating relaxation times to diffusion rates; direct measurements of diffusion rates in the Na-Mg melts studied by Fiske and Stebbins (1993) should resolve the issue.

Finally, Heitjans et al. (1991) have studied diffusion in glassy silicates using the relatively new method of ß-NMR (ß -radiation-detected NMR). The method allowed them to observe the effect of local disorder on diffusion as well as anomalous diffusion behavior (see section on *Microscopics*) in silicate glasses.

The main disadvantages of NMR as a tool for studying diffusion are:

(1) Its low sensitivity which requires the sample to have a high concentration of NMR sensitive atoms. Roughly 10^{18} to 10^{20} nucleii must be present in the sample to obtain high quality results (Stebbins, 1988), i.e. either large amount of material or highly isotopically enriched systems are required,

(2) paramagnetic impurities need to be avoided, and

(3) like any other spectroscopic method, NMR detects motion over a range of time scales only. This means only diffusion coefficients that fall within a certain

window are measurable using NMR. However, this is offset to a large extent by the variety of NMR measurements that are possible, see below.

In comparison, there are many reasons for using NMR in diffusion studies:

(1) A major advantage of NMR is that the mobile atom in a multicomponent system is readily identified. So NMR is very powerful in identifying if the same atomic species diffuses by more than one process, or to distinguish between localized vs. long-range motion.

(2) A large range of diffusion coefficients may be measured using different kinds of NMR information (i.e. various relaxation modes), so that it provides the advantage of allowing measurement of D over many orders of magnitude using a single technique.

(3) Unlike Mössbauer and some other nuclear spectroscopic techniques, NMR sensitive nucleii are spread all over the periodic table.

Finally, modern pulsed field gradient methods (pfg-techniques) allow diffusion coefficients to be measured directly without recourse to any structural model and are also less affected by paramagnetic impurities. These advances show promises to remove the major barrier in the way of getting accurate diffusion data from NMR methods (see Chadwick, 1990, for more details) over a significant range of temperatures.

Vibrational spectroscopic methods. In general, because of the short time scale on which vibrational spectroscopy probes a sample it is not suitable for diffusion studies. However, a non-nuclear spectroscopic technique that has been used for *in-situ* observation of dynamics in silicate melts is infrared spectroscopy. Keppler and Bagdassarov (1993) observed the changes in the infrared absorption peaks of a rhyolite glass at temperatures up to 1300°C using a heating stage, a FTIR spectrometer and the focussing optics of an IR microscope. An interesting prediction of this study was that at high temperatures OH groups might contribute significantly to the mobility of H_2O in silicate melts, which would be in contrast to the behavior observed so far at lower temperatures (e.g. see Zhang et al., 1991a, and Watson, 1994, for details).

Other methods

A number of electrochemical techniques have also been used to obtain diffusion rates of different melt constituents. Haskin and coworkers (e.g. Semkow et al., 1982; Colson et al., 1990) have used linear sweep voltametry or cyclic voltametry to obtain information about diffusion rates of various metals (Ni, Eu, Mn, Cr, In) in diopsidic melts; Takahashi and Miura (1980) have used the same technique to obtain diffusion rates of Zn, Ni, Co and Fe^{3+} in simple silicate melts. Chronopotentiometry has been used to measure oxygen diffusion rates in melts in the system $CaO-MgO-SiO_2$ (Semkow and Haskin, 1985). All of these techniques have to assume diffusion with reaction (oxidation/reduction) only at the electrode. This may be reasonable for a dilute tracer e.g. in a Ni diffusion experiment. However, the assumption may not be taken to be valid a priori for an abundant network former like oxygen in a liquid (see section on *Diffusion and melt structure*). Further, the method assumes there is only *one* kind of polymerized oxygen (characterized by one activity) and there is only one kind of polymerization-depolymerization reaction. As diffusion coefficients depend quite strongly on these assumptions, these need to be justified for individual systems.

As discussed in the section on *Diffusion and electrical properties*, measurements of electrical properties can provide much insight into rates and mechanisms of ionic motion in silicate melts. See the chapter by Moynihan for some details of electrical measurements as well as Martin and Angell (1986) for a discussion of dc and ac conductivity measurements in glasses.

Additionally, diffusion coefficients have been estimated from rates of mineral growth or dissolution in a number of studies as a by product. The variety of processes from which such estimates can be made are numerous and the data are often subject to various assumptions. We do not deal with this topic in detail in this chapter.

Uncertainties in diffusion data

It is considered useful to include a discussion on the uncertainties of diffusion data because uncertainties which are considered unacceptable in many physical measurements are often the norm in diffusion studies of silicates. The situation has improved considerably in recent years with improvement in techniques of sample preparation, experimentation and analysis. In addition, improved computational capability now allows many models with realistic boundary and initial conditions to be fitted to diffusion data instead of unnecessarily simplified ones (e.g. see Tarento, 1989, for an example). It is now possible to quantitatively estimate the errors from various sources and propagate these errors through the statistical fitting procedure to obtain estimates of uncertainties on diffusion coefficients from propagated errors. Such estimates often range to within 10% (e.g. Chakraborty et al., 1995a) and for a given study are vindicated by the reproducibility of data from various experiments. However, one major source of uncertainty in diffusion studies of refractory materials arises from the nature of the starting material itself. Since measurements in a given study are usually performed on a single kind of starting material (prepared or obtained from one source) for a given composition, the estimates of errors based on propagated uncertainties are in effect, minimum uncertainties in the knowledge of diffusion coefficients. Different methods applied on the same material sometimes yield differing results. The variance in diffusivity in various starting materials of the same composition has hardly been explored for diffusion studies in silicate melts. This is likely to be less of a problem for liquids where the structure is equilibrated. However, even in these, small differences in composition may affect diffusion rates of some species quite dramatically (e.g. see Watson, 1994, for effect of water and other volatiles on diffusion rates of various cations; see Chakraborty et al., 1993, for the effect of small amounts of B on diffusion rates of Si). As such, for the purposes of geological applications (almost invariably involving some extrapolation in composition) it is prudent to expect uncertainties in our knowledge of diffusion coefficients to be higher than that reported in individual studies.

Studies of structural and physical properties of melts have shown that melts of the same composition synthesized using different methods (e.g. sol-gel, melting of oxide mix or high pressure amorphization) or even simply obtained by melting at different temperatures (e.g. Martin and Angell, 1986) can have different characteristics. Melts quenched from different pressures have been shown to have different structural features in a number of studies (e.g. see the chapter by Wolf and McMillan in this volume). All of these have two implications regarding uncertainties of diffusion data. First, from the point of view of understanding diffusion processes in silicate melts it would be interesting to study diffusion in melts of the same composition but different structural states. From the point of view of applications, diffusion data obtained on glasses or at temperatures only slightly above the glass transition temperature are thus prone to even higher uncertainties than diffusion rates in liquids.

VII. MICROSCOPIC ASPECTS OF DIFFUSION IN SILICATE MELTS

Apart from the basic objective of understanding mass transport on a microscopic scale there are a number of practical reasons for studying the atomistic basis of diffusion. The classical motivation for studying mechanisms of diffusion stems from the recognition that extrapolations are inevitable in diffusion studies of geological systems. Experimentally measured diffusion coefficients have to be used in mass transport problems under conditions (temperature, pressure, composition, etc. and most significantly, time scales) which are often far removed from those of the measurements. An understanding of microscopic mechanisms provides a basis for extrapolations and often helps to delineate conditions up to which extrapolations may be valid (i.e. the same mechanisms as those of the laboratory measurements operate). In recent times, however, a more utilatarian purpose needs to be served by linking microscopic and macroscopic processes. Some data may now be obtained directly from observation of microscopic processes through the use of various spectroscopic techniques or methods of computer simulations. In order to relate these data (e.g. atomic jump distances or frequencies) to macroscopic observables such as net mass transport rates, we require a theoretical framework.

Statistical mechanics provides the theoretical framework for relating discrete, microscopic steps made by atoms or clusters of atoms (molecules, polymeric chains, etc.) to continuum, macroscopic description of processes. The means for doing so is to take averages over numerous such microscopic steps in a given volume and over a given time period. The actual task of taking these averages for even the simplest "real" systems is mathematically demanding. However, recognizing the fact that macroscopic properties are averages already tells us something very important—as in all averaging processes, some information is lost and the macroscopic variables cannot provide us complete details of the underlying microscopic mechanisms. Therefore, even studies of the simplest model systems are often very instructive and may bring to light phenomena whose existence we do not suspect from a purely macroscopic description of a process.

In what follows, I treat some selected topics chosen mainly to illustrate through some examples (1) how one approaches the problem of relating microscopic and macroscopic quantities and (2) how this may help us to understand different aspects of macroscopically observed diffusion behavior. Most notably absent in this selection is any treatment of transition state theory, which plays a key role in our understanding of diffusion in silicate melts. However, excellent reviews of the topic are available in Volume 8 of this series (Lasaga and Kirkpatrick, 1981) and specific applications of the theory to diffusion problems are treated in detail by Lasaga (1995) and Allnatt and Lidiard (1993). The discussion below is based to a large extent on derivations and results provided by Allnatt and Lidiard (1993) and Philibert (1991).

Random walk in amorphous medium

The simplest averaging process is the classic "random walk of a drunken sailor problem". The problem is illustrated in practically every textbook and review on diffusion processes (e.g. see the article by Watson, 1994, in this series) and will not be repeated here for the sake of brevity. In essence, one calculates the mean squared displacement (X^2) after n jumps (in time τ) of a particle hopping randomly from node to node in a periodic lattice. The calculation leads to the well known Einstein formula for diffusion that relates X^2 to the parameter $D\tau$, where D is the diffusion coefficient. There are two assumptions inherent in the standard derivation: (1) the jumps take place in a periodic lattice i.e. the jump distance from one site to another is constant in a given direction, and (2) the likelihood of jumping in any direction is always the same i.e. successive jumps are uncorrelated. Both of these conditions are likely to be violated in a silicate melt. Let us explore the consequences of that on the macroscopic diffusion coefficients.

Consider the case where a continuous distribution of jump lengths is available, as is to be expected in an amorphous material. For a single projected jump length, λ, it can be shown that the probability, $W(X,n)$, of a particular atom being displaced a distance X along the projection direction after n jumps is given by

$$W(X, n) = \frac{1}{\sqrt{2\pi n \lambda^2}} \exp\left(-\frac{x^2}{2n\lambda^2}\right).$$

We note that the above expression has the form of a Gaussian distribution with variance $n\lambda^2$, which is simply the mean of the squared displacement, denoted by $\langle X^2 \rangle$. Now, let us consider the case where there are a sequence of jumps n_1 of length λ_1 and n_2 of length λ_2. A very important result from probability theory states that the sum of two Gaussian distributions is itself a Gaussian. Moreover, the variance of the resulting Gaussian is simply the sum of the variances of the component Gaussians. Thus, the mean squared displacement due to the two kinds of jumps occuring together is given by $n_1\lambda_1^2 + n_2\lambda_2^2$. This process can be extended to any arbitrary number of jump lengths. The mean squared displacement, $\langle X_T^2 \rangle$, due to all of these jumps together is given by

$$\langle X_T^2 \rangle = n_1 \lambda_1^2 + n_2 \lambda_2^2 + \ldots\ldots + n_n \lambda_n^2.$$

One can then simply use the Einstein relation relating mean squared displacement to diffusion coefficient, D, to obtain the diffusivity resulting from a ccombination of these various types of jumps

$$D = \frac{\langle X_T^2 \rangle}{2\tau} = \frac{1}{2}\left[\frac{n_1}{\tau}\lambda_1^2 + \frac{n_2}{\tau}\lambda_2^2 + \ldots\ldots + \frac{n_n}{\tau}\lambda_n^2\right]$$

or in terms of frequencies, $\Gamma_i = (n_i/\tau)$

$$2D = (\Gamma_1\lambda_1^2 + \Gamma_2\lambda_2^2 + \ldots + \Gamma_n\lambda_n^2).$$

This is often decomposed into partial diffusion coefficients D_i such that $D_i = (1/2)\Gamma_i\lambda_i^2$ and one obtains

$$D = \sum_i D_i.$$

Generalizing this to a continuous distribution of jump lengths, if we denote the frequency of jumps of length λ to $\lambda+d\lambda$ by $\Gamma(\lambda)d\lambda$, we obtain

$$D = \frac{1}{2}\int_0^\infty \Gamma(\lambda)\lambda^2 d\lambda.$$

This is the result applicable to amorphous structures and serves to illustrate a number of points. First, it tells us why the form of equations obtained from simple derivations considering only one jump length work so incredibly well in describing processes in amorphous systems where we know that in the absence of a lattice with translational symmetry a number of jump lengths must be possible. It is easy to see how the above may be reduced to diffusion with a single average jump length and diffusion coefficient D. Second, the equations above illustrate how the process of averaging erases detail— knowing only the single diffusion coefficient D it is impossible to obtain information about the different microscopic jump lengths, λ_i and jump frequencies, Γ_i. Thirdly, if a spectroscopic method is sensitive to only a particular (or range of) frequency or jump length, it will yield a partial diffusion coefficient, D_i which may or may not be the same as the bulk, macroscopic diffusion coefficient, D. Finally, we have obtained an explicit relation between frequency of jumps and diffusion coefficients. Macroscopic measurements of electrical and mechanical properties which yield frequency dependent results may now be related to diffusion coefficients through such treatments. We also note in passing that Isard (1992) has shown that similar averaging is applicable for activation energies as well i.e. a single activation energy will describe the macroscopic motion of ions (diffusion, d.c. or a.c. conductivity) even if there are a distribution of activation energies at individual sites governing the motion of individual ions.

Correlation factors

Let us now consider the second approximation that goes into the standard derivation of the Einstein equation mentioned above --namely, that there is no correlation between successive jumps. This means that at any given instant a particle can hop in all directions with a likelihood given by pure random walk calculations— its motion is not dependent on its previous history (i.e. preceding jump), nor is it dependent on the motion of other atoms in the system. In a real system these conditions may be violated and correlations may arise in a number of ways. The simplest argument, borrowing from concepts of vacancy diffusion in crystalline solids, goes as follows. If an atom jumps to a neighbouring site, it leaves a "vacant" site behind. Unless substantial configurational changes accompany the jump, then the previous site is by definition now an empty site where the atom may reside and hence the probability of the atom jumping back to that site is higher than jumping in any other direction. This is clearly an argument based on availablility of space. Similar arguments based on steric factors and the concept of local relaxation have been used to develop the *site mismatch* model of ionic transport in glasses (see *Diffusion and melt structure*). Some additional factors that may make jumps to a given configuration more likely than others include Coulombic repulsions, correlated hops of groups of particles (coupled diffusion) or lattice relaxation effects (other than the ones already incorporated in the "site mismatch" model). From the various discussions in this chapter, it should be clear that more than one of these factors are likely to be involved in causing deviations from random walk diffusion in a silicate melt. In fact, experimental evidence of correlated motion in melts come from Chemla experiments, NMR measurements (discussed before) and the fact that electrical conductivity is a function of frequency (see Moynihan, this volume). The correlation effects are embodied in a parameter called the *correlation factor*, usually denoted by f, which multiplies the random walk diffusion coefficient, D to yield macroscopically observable diffusion coefficients, D^*, i.e. $D^* = f \cdot D$. Typically, f is a number between 0.1 and 1. This means that (1) measuring f is difficult and requires very precise measurements of diffusion coefficients and (2) if measured, f can provide information about microscopic mechanisms of diffusion. Classical methods to measure f include using different isotopes of the same element and the so called Chemla experiment (see *Diffusion and electrical properties*).

In cases where the correlation factor has been determined, it can provide considerable insight into diffusion mechanisms. For example, correlation factors for transport of alkalis have been measured in alkali-silicate glasses (e.g. Kahnt et al., 1988; Laborde et al., 1989; Heinemann and Frischat, 1993) and the values have been found to be low in general (<0.3). This indicates that the atoms spend a lot of time jumping back and forth between two sites with no net displacement. Such values of correlation factors can be expected, for example, when diffusion occurs along preferred channels, as discussed earlier in the section on *Structure and diffusion*. The important points to note here are that (a) such low values of correlation factors have been observed in glasses as well as liquids, indicating that similar diffusion mechanisms persist in the liquid at least up to some temperatures and

(b) diffusion coefficients obtained using any method that directly observes atomic jumps may be considerably faster (by a factor 1/f) than those measured using conventional macroscopic methods (see *Experimental methods—NMR* for an example).

At this point it is also worth noting briefly that traditionally diffusion coefficients are taken to be independent of time scales, unless the melt structure itself changes during the duration of diffusion (see section on *Relaxation and diffusion*). However, as methods to observe atomic motion in real time are becoming common (e.g. through light or neutron scattering experiments or NMR) this assumption may no longer be implicitly true in all cases. In the theory of diffusion of colloids and aqueous solutions, a distinction between short-term and long-term diffusion coefficients have been made for a long time (e.g. Beenaker and Mazur, 1983). Some recent molecular-dynamics studies have shown that the short term diffusion coefficient can be considerably larger than the classical long term diffusion coefficient in a liquid where diffusion shows strong dynamic correlations (Dong et al., 1990). The phenomenon of so called anomalous diffusion—processes in which mean squared displacements depend on some power of time rather than t itself (see Bouchaud and Georges, 1990, for a review) also leads to time scale dependent diffusivity. It has been suggested that anomalous diffusion can explain some high frequency conductivity data in phosphate glasses and points to the role of strong interactions during diffusion (Sidebottom et al., 1995), indications for anomalous diffusion in silicate glasses have been found from ß-NMR studies (Heitjans et al., 1991). At this stage, it is not clear to what extent such phenomena may affect geologically interesting silicate melts. Therefore, we mention these features here to simply make the reader aware of the possibility of time scale dependence of diffusion coefficients.

Relationship of L-matrix to microscopic motion

We have mentioned earlier that the L-coefficients (Onsager kinetic coefficients) provide a useful description of the diffusion process because they are firmly based on microscopic theory. Their measurement allows a direct insight into microscopic details of diffusion, unlike any other single macroscopic physical quantity. The actual derivation of the relationship between microscopic parameters and the L-coefficients is an elaborate mathematical exercise and we refer the interested reader to Kreuzer (1981) and Allnatt and Lidiard (1993). We will provide below a logical development followed by an explicit statement of the relationship for a somewhat simplified case. This will allow us to explore the kinds of information about microscopic processes contained in the L-matrix.

The simplified model we have in mind is that of diffusion occuring by hopping from site to site in an aperiodic array. In other words, we are dealing with diffusion in a glass or perhaps a set of fast diffusing network modifiers (e.g. alkalis) in a viscous liquid. This implies that (1) the irregular framework in the melt does not change with time, i.e. "sites" can be defined, say, in terms of immobile network formers and (2) the mean time of stay at a site is much longer than both the time of flight between sites and the mean time of vibration about any such site. That these are reasonable assumptions can be easily seen by noting that even very high diffusion rates of about 10^{-10} m²/sec correspond to a mean time of stay of about 10^{-10} sec, about 1000 times longer than typical vibrational time scales. Since these atomic displacements occur infrequently relative to the period of vibrations, it is also reasonable to assume that local thermal equilibrium will be re-established between jumps. Therefore, it is allowable to use equilibrium statistical thermodynamics[1] to calculate

---

[1]   At this point we note that if transport in the melt occurs by a mechanism similar to the site mismatch model mentioned earlier, the treatment as given here is not strictly applicable. Even then, it does serve the purpose of elucidating the microscopic nature of the L-coefficients.

the frequency of the fluctuations in atomic positions and momenta required for a displacement to occur. Indeed, the justification of these assumptions are provided by molecular dynamic simulations of silicate melts (see the chapter by Poole et al. in this volume).

Since time of stay at a "site" is much longer compared to the time of vibration, one can average out the vibrational motions and assume that the atomic jumps, when they occur, do so instantaneously. This situation may then be described by specifying states of the melt $\alpha, \beta, \gamma, ...$ by the way the atoms are distributed over the various lattice sites. These are the same configurations that one averages over to get the configurational entropy. An atomic jump then corresponds to a transition from one such state α to a different state β. We can then introduce the probability, p_α, that any given system in the ensemble is in the microscopic state α. We also denote the frequency of transition from state α to state β by $w_{\beta\alpha}$ with the assumption that this frequency depends only on the initial and the final states i.e. it is independent of the history of the system. Strictly speaking, this assumption is not applicable to silicate melts, as we saw above. But it does serve to obtain simple results which illustrate general features. The rate at which systems are leaving the state α is given by a summation of all $(w_{\beta\alpha}p_\alpha)$ for $\beta \neq \alpha$. Similarly, the rate of all arrivals to state α is given by a sum over all $w_{\alpha\beta}p_\beta$. We can then describe the time evolution of the occupation probability of any state α, $p_\alpha(t)$, by taking the sum of these two

$$\frac{dp_\alpha}{dt} = -\sum_{\beta \neq \alpha} w_{\beta\alpha}p_\alpha + \sum_{\beta \neq \alpha} w_{\alpha\beta}p_\beta$$

When similar equations for the occupation of each of the states are written out, they may be collected together in the form of a matrix equation which is known as the Master equation of statistical mechanics. Formal integration of the Master equation yields a matrix whose elements are of the form [exp -Pt]$_{p\sigma}$—these give the probability that the system will be in state p at time t given that it was in the state σ at time zero. These conditional probabilities are at the heart of the derivation of practically all significant results.

To go from the specific atomic configurations to a continuum representation, one carries out averages over regions that are small in a macroscopic sense but large enough to contain many lattice points. Then, one obtains the macroscopic transport coefficients for an isothermal system of volume V containing atomic species i,j, etc. in numbers N_i, N_j, etc.

These coefficients are the equilibrium ensemble averages of functions of the individual displacements, $r_{\beta\alpha}$, of all the atoms in all possible transitions $\alpha \to \beta$ undergone by the system. This is why in the macroscopic development of thermodynamic description for diffusion earlier we had added stipulations of local equilibrium over small volumes. This is the step where most of the complicated mathematics appears in the form of Fourier transforms and involved statistics. After going through the math, the final expressions relating the displacement of each particle for each transition from one state to another is still rather complicated. However, we are interested collectively in the motion of atoms of a particular kind (e.g. chemical element or isotope) and not in the motion of each individual particle. Therefore, we can sum over all atoms of a particular kind within the given volume and obtain the macroscopic transport coefficients, L_{ij}. The diagonal coefficients are given by

$$L_{ii} = \frac{1}{6VkT} \left[\sum_m \frac{\langle (\Delta r_i^{(m)})^2 \rangle}{t} + \sum_{m,n;\, m \neq n} \frac{\langle \Delta r_i^{(m)} \cdot \Delta r_i^{(n)} \rangle}{t} \right]$$

where the (vector) displacement of atom m of type i in time t is denoted by $\Delta r_i^{(m)}$. By the nature of thermodynamic ensemble average, each $\langle \Delta r_i^{(m)} \rangle^2$ (with $\langle \, \rangle$ denoting average of the quantity enclosed, as usual) is the same for every atom m of species i so that one can simply drop the suffix m. The displacements of atoms of another chemical species, j are of course different from that of species i. Similarly, the corresponding expression for L_{ij} (i≠j) is

$$L_{ij} = \frac{1}{6VkT} \left[\sum_{m,n} \frac{\langle \Delta r_i^{(m)}. \Delta r_j^{(n)} \rangle}{t} \right]$$

in which $\langle \Delta r_i^{(m)}. \Delta r_j^{(n)} \rangle$ is again the same for all pairs of different atoms. These two relations show us how the macroscopically measurable L_{ij} coefficients may be related to microscopic displacements of atoms and the correlations between their movemements. For example, the microscopic displacements are typical quantities calculated in computer simulations—the equations above show how the macroscopic L coefficients may be directly obtained from such simulations. Allnatt and Lidiard (1993) further show that in the dilute limit of an equilibrium system, the diagonal coefficients, L_{ii}, consist of two terms: a term involving the tracer diffusion coefficient and a term which is the sum of two atom correlations. The correlations in this case are between the motions of atoms of the same species. The off-diagonal coefficients, L_{ij}, on the other hand are made up of only two atom correlation terms (between atoms of species i and j), as indicated by the relationship above. This microscopic view of the system helps to emphasize a very important point regarding the L-coefficients. As a species i becomes dilute, the average separation between two i atoms increase and any mutual correlation between the movement of two i atoms vanishes. In other words, L_{ii} becomes related to the mean-squared-displacement only i.e. the tracer diffusion coefficient. On the other hand, the same does not apply for L_{ij}—as i becomes dilute, it is still surrounded by many j atoms at any given time. So the effects of (i-j) correlations do not vanish and the L_{ij} term does not become negligible. What does happen, however, is that the ratio L_{ij}/L_{ii} becomes independent of composition! Note that in this entire discussion the L-matrix under consideration is that which relates to the flux of the individual elemental species and not the oxide (or other neutral) components.

Computer models—molecular dynamics, Monte Carlo, etc.

Computer modelling of silicate melts is the topic of an independent chapter in this volume and the use of molecular dynamics simulations for diffusion studies in silicate melts has been recently reviewed in detail by Kubicki and Lasaga (1991), also see Stein and Spera (1995). Therefore, we sketch below only a very broad outline of some of the ways in which computer simulations are used to study diffusion in silicate melts.

Computer simulations related to diffusion are broadly of two different kinds—those which are purely statistical in nature and those concerned with details of mechanisms. In both approaches, one uses a "cell" which is a collection of particles, usually represented by an array in the computer memory. Since these cells are of necessity restricted in size, one has to resort to tricks to avoid the effects of "boundaries" from affecting the simulations. This is accomplished typically by choosing so called periodic boundary conditions which simply means that the number of particles in the cell are kept constant. One accomplishes this by ensuring that when a particle exits the cell on one side during a simulation, it enters back from the opposite side. Thus the cell of limited size is used to simulate an infinite system without boundaries and therefore obtain properties of bulk matter as opposed to those of surfaces and interfaces. A consequence of this is that most computer simulations are carried out at constant volume, energy and the number of particles rather than at constant pressure and temperature. This fact should be taken into account in comparing results of computer simulations to experimental data, although the differences are often expected to be small.

The first kind of simulation is typified by Monte Carlo methods where the objectives are to calculate statistical parameters like mean-squared displacement, jump frequencies and, most notably, correlation factors. The process is ideally suited for treating diffusion of a dilute component (the classic example being a vacancy in a solid) in a lattice (which may be aperiodic) and in essence is carried out as follows. One chooses a set of lattice sites (say, $20 \times 20 \times 20$ or more) and an array in the computer memory stores the occupancy of each lattice point. The dilute component is introduced randomly at some of these points. To start the simulation, one chooses an atom at random, and checks if this is the component of interest. If not, one tries again. On finding the component of interest (e.g. a vacancy or a hole), one calculates a jump frequency for that particle (vacancies are "pseudo-particles") according to some rules (determined by the physics of the problem) and compares the calculated frequency with a random number. Both the random number and the frequency are normalized to vary between 0 and 1. If the frequency is larger than the random number, the jump is considered successful. If not, one starts with the choice of a new particle again. This is continued long enough to accumulate statistically meaningful parameters, which amounts to something on the order of 10^6 moves. At the end of the simulation, statistical mechanical equations for calculating ensemble averages of a system of particles are used to obtain the various physical quantities of interest from the recorded displacements of the particles.

The second kind of approach to computer simulation is usually more system specific. The goals of these kinds of studies are to observe mechanisms of diffusion, to perform exact calculations on "ideal materials" created in the computer or to simulate an experimental process to allow better interpretation of macroscopic observations, e.g. NMR spectra. The first of these is by far the more common, at least in the geological literature. Such modelling is usually carried out by means of molecular dynamic simulations, although a variant of the Monte Carlo method, quite different from the one described above, has also been used. For these methods, the systems are typically "annealed" i.e. allowed to attain an equilibrium configuration before starting the actual diffusion simulations.

Molecular dynamics is primarily based on classical mechanics—one calculates the paths of all the particles in the "cell" using equations of motion of classical dynamics. The key question in these simulations is the choice of suitable interatomic potentials. The potentials may be defined empirically or through ab initio quantum mechanical calculations on small clusters. There is a huge literature on the topic, see Poole et al. (this volume) and Kubicki and Lasaga (1991) for details. Given a form of the potential, one calculates the force on each atom at each time step from the gradient of this potential and then based on this calculated force the particle is displaced using Newton's laws of motion. This is carried out for each particle in the cell for each time step and once again, periodic boundary conditions are usually maintained. Apart from the choice of potentials, an important factor is the choice of the time step—an example from Philibert (1991) illustrates the point. The time step should be less than 1/10 the vibrational period of the system i.e. around 10^{-14} to 10^{-15} seconds. Thus, a compromise between the size of the "cell" (the number of interactions to be calculated) and the number of iterations (to give statistically meaningful averages) must be found. For example, if in a cell containing 100 particles it takes 1 second in real time to calculate the 10^4 interactions then computer times become prohibitively long for more than 10^5 iterations. Thus, no process can be studied for longer than $10^5 \times 10^{-14} = 1$ ns! This highlights one of the main problems of studying diffusion in silicate melts using molecular dynamic methods—to capture a statistically significant portion of a disordered medium (as opposed to crystals with translational symmetry), one needs a relatively large cell size. The question of how cell size affects results of simulations of silicate melts has been specifically addressed by Rustad et al. (1990). In large cells, the number of interactions become prohibitively large and diffusion can be followed for only short time periods. Therefore, most studies of silicate melts have been carried out with relatively fluid melts, at very high temperatures (around 2000 K or more) and often under high pressures. This means that in order to compare results of simulations to macroscopic measurements large extrapolations are necessary in most cases. It is even conceivable that the mechanisms of diffusion observed in the high temperature simulations are different from those observed by macroscopic measurements at lower temperatures. However, the situation is rapidly changing due to improvements in both hardware and software of computers and the conditions of computer calculations are approaching those of laboratory measurements. Nevertheless, some very exciting contributions to the understanding of diffusion mechanisms in liquid silicates have already come from such simulations, see for example the section on *Pressure dependence of diffusion coefficients*. An important conclusion that we obtain from molecular dynamic and other forms of computer simulations is that the energetics of diffusion processes are controlled largely by the local structure and bonding around the diffusing atom. This is useful as a conceptual tool when one tries to construct models relating melt structure to diffusion.

VIII. DIFFUSION DATA IN SILICATE MELTS

As mentioned in the introduction, reviewing available diffusion data is not a goal of this work. The objective of this section, after guiding the reader to some sources of diffusion data in silicate melts, is to familiarize him or her to general trends and to provide an idea of some numerical magnitudes.

Several excellent compilations of data are available in the glass literature—notably Freer (1980, 1981), Frischat (1975), Bansal and Doremus (1986) and Mazurin et al. (1983, 1987). Reviews of diffusion in melts in the geochemical literature may be found in Hofmann (1980), Dunn (1986), Jambon (1983), Watson and Baker (1991) and Watson (1994). All of these works summarize available data at the time of writing alongwith some discussion of diffusion theory, experimental techniques and some applications of diffusion data. Two reviews concerned specifically with experimental measurements of diffusion rates are those of Ryerson (1978) and Baker (1993). Finally, specific aspects of diffusion in silicate melts are covered in the review articles by Kubicki and Lasaga (1991) (molecular dynamics simulations), Lesher and Walker (1991) (thermal diffusion in silicate melts) and Watson (1994) (role of volatiles in diffusion in silicate melts). For mathematical methods of problem solving and an introduction to relevant numerical techniques, the reader is referred to texts like Crank (1975), Carlsaw and Jaeger (1959) or Ghez (1988).

It should be apparent by now that the relative mobility of network formers and network modifiers controls the essential nature of diffusion in a melt and determines how atoms migrate, how to treat the process theoretically (the physics as well as the mathematical forms can change) and how to measure these rates experimentally. Therefore, it is better to make a distinction between the cases where diffusion rates[1] of network formers and modifiers are widely different and the instances where these two diffusivities are relatively close to each other (within an order of magnitude or two). Mobility of network formers control viscosity, which is an easily observable and well measured macroscopic property of melts. Thus, relating diffusion rates of network modifiers to viscosity of melts is a good way to systematize data. This brings us to the first rule of thumb with regard to diffusion data—melts with similar viscosity often (but not always) behave similarly with regard to diffusion. In terms of other familiar concepts, this means depolymerized[2] and polymerized melts behave differently with regard to diffusion. In geological terms, these amount to saying that there are important differences in diffusion behavior between silica-rich "granitic" melts and silica-poor "basaltic" melts. This is a feature we have highlighted

---

[1] In this discussion we always refer to tracer or self-diffusivities that relate directly to the mobility of a particle rather than to chemical diffusivities of components, unless otherwise specified.

[2] Degree of polymerization refers to a ratio indicating the number of bridging oxygen atoms per tetrahedrally coordinated network-former atom, not to any permanent geometric features in the melt.

several times during our discussions of theoretical as well as experimental aspects earlier.

Effect of composition on diffusion

It follows from our first rule of thumb that features which tend to increase viscosity usually reduce diffusivity of most species, if there is a strong effect at all. Thus, low silica content, high alkali contents and high temperature—all tend to enhance diffusion rates, although exceptions exist, as we shall see below. Superimposed on this general pattern, detailed chemistry plays a very important role in determining the absolute values of diffusivity. An important difference in compositional dependence of diffusivities in silicate melts from that in many other kinds of media is that compositional dependence does not always increase with decreasing temperature.

Along with the distinction between polymerized and depolymerized melts, a good way to survey diffusion data is to further subdivide elements into the following categories: (1) alkali ions (Na, K, etc.), (2) network modifiers other than alkalies, mostly divalent cations (Ca, Mg, etc.), (3) network former cations (Si, Al plus other cations in smaller concentrations like Ti, B, P), and (4) oxygen and other anions. We also note that many elements have the ability to fulfill more than one of these roles in a melt, their diffusion behavior is accordingly complicated. The most notable examples are Al and Fe (as Fe^{2+} and Fe^{3+}). Following these, we have a brief discussion of (5) compositional effects in chemical diffusion.

(1) Diffusion of alkali ions. These have the simplest diffusion behavior among all species. Diffusion rates of alkali ions are in general much faster than that of all other non-volatile species, consistent with the thinking that cationic charge plays a more important role than size in determining diffusion rates. The compositional dependence for these species is very weak to non-existent (at least at the high temperatures of most measurements in liquids), changing by only about a factor of four in going from basalts to obsidian (Henderson et al, 1985). The interesting aspect is that the *slightly higher diffusion rates are found in the more silica rich melts*, providing one small exception to the trend described above. A typical diffusion rate at 1300°C is about 10^{-9} m^2/sec. In low temperatures liquids and glasses, of course, one finds the strong compositional influence of the well known mixed-alkali effect where diffusion rates at intermediate compositions are slower than for either alkali endmember.

(2) Diffusion of network modifiers other than alkalies. Non-alkali atoms show two kinds of compositional dependence. In low viscosity ($<10^5$ Pa s) melts at high temperatures, the diffusion rates of different cations are similar (within two orders of magnitude) to each other and they decrease with increasing silica content. A typical diffusivity is 10^{-11} m^2/sec at 1300°C. This situation applies to most mafic melts. In high viscosity melts at low temperatures ($>10^7$ Pa s), the diffusivities still remain similar to each other but the compositional dependence changes to one where diffusion rates stay constant or increase very slightly with increasing silica content. This behavior is similar to that of alkalies. A typical diffusion rate is 10^{-14} m^2/sec at 1000°C in melts with viscosities of about 10^7 Pa s. This is the kind of behavior expected in granitic melts. The difference in composition dependence of diffusivity in melts of different viscosities have been rationalized using concepts of structural relaxation of silicate melts (Dingwell, 1990). At low temperatures a mixed-cation effect analogous to the mixed-alkali effect is seen as well. Detailed studies show other more subtle compositional effects, some of these are discussed under *pressure dependence of diffusion data.* An important gap in diffusion data in this category is the absence of systematic studies of the effect of oxygen fugacity on diffusion rates of transition metals as a function of composition.

(3) Diffusion of network forming cations. Among the network former cations, the largest amount of data is available for Si; additional values have been calculated from viscosities using the Eyring relation (the viscosities themselves being either measured or calculated using algorithms like those of Shaw, 1972, or Bottinga and Weill, 1972). Directly measured diffusion data for Al is conspicuous in its rarity (being limited to one study, Henderson et al., 1961) due to the absence of suitable isotopic tracers. This major gap in our knowledge of diffusion rates in silicate melts has provided much room for speculation in models to understand transport processes. This is one area where NMR measurements or techniques that do not require special isotopes (see *Experimental methods*) can make substantial contributions. Diffusion rates in "analog melts" containing Ge, Ga and B have been measured in attempts to complement the existing database and improve our understanding of the role of network formers (e.g. Kushiro, 1983; Baker, 1992a).

As discussed in detail already, diffusion rates of network formers scale directly to viscosity and are the slowest diffusing species in a silicate melt under all circumstances. Apart from this aspect, the diffusion behavior of network formers are complex and depend on many factors, as may be anticipated from our earlier discussion on *Diffusion and melt structure*. The difference between diffusivity of network formers and modifiers is smaller for high temperature, low viscosity melts. The diffusion rates of network formers are strongly affected by presence of small amounts of volatiles (see Watson, 1994); Si diffusion rates are affected by the presence of other network formers (e.g. Al, P, B—see, for example, Harrison and Watson, 1984; Chakraborty et al., 1993; Chakraborty and Dingwell, 1993). Consistent with size-charge arguments, P diffusion rates in viscous, dry granitic melts (Harrison and Watson, 1984) are among the slowest known in silicate melts ($\sim 10^{-15}$ m^2/sec at 1300°C, which is comparable to Fe-Mg diffusion rates in *crystalline* olivine at the same conditions!). Relative abundances of different network formers and modifiers are often an important control on diffusion rates, the most important among these being the alkali:aluminum ratio. When such dependence exists, diffusion rates in the alkali-rich endmembers are usually the fastest and a diffusivity minimum may occur (inferred from viscosity and chemical diffusion data in Baker, 1990) along the alkali-aluminum join at around the 1:1 composition. Effects analogous to a mixed-cation effect are seen for network formers as well, but in an *opposite* sense. When diffusion rates across a join such as diopside-jadeite are measured, one finds a *maxima* in Si diffusivity at an intermediate composition which can be rationalized by considering configurational entropy changes on mixing (Shimizu and Kushiro, 1991). It remains to be determined whether this behavior is caused by the structural effect of varying degrees of polymerization across the join or the chemical effect of replacing Si by Al, although arguments presented by Shimizu and Kushiro (1991) would seem to favour the latter.

Because the diffusion rates of network formers depend on all these parameters and scale with viscosity which varies over many orders of magnitude with melt composition and temperature, it is not very meaningful to state typical diffusion rates. As two examples, we find diffusion rates for Si of about 10^{-12} m^2/sec in a basalt and 10^{-14} m^2/sec in a dry obsidian at 1300°C.

(4) Diffusion of anions. Diffusion studies of anions are dominated by studies of oxygen diffusion. The positive feature that has emerged from the many studies of oxygen diffusion in geological melts is that oxygen tracer diffusion rates can be related to viscosities through the Eyring equation. It should be noted, however, that oxygen tracer diffusion rates in silica-rich granitic melts have not been measured and it is not clear whether the Eyring equation works equally well for these melts. Aside from this aspect, in spite of the numerous studies (in geological as well as ceramics literature) and the potential importance of oxygen diffusion in determining viscous behavior the process remains one of

the less well understood aspects of transport in silicate melts. Complications arise because of a number of reasons:

(a) Oxygen in silicate melts occurs in at least three different forms—bridging oxygen ions, non-bridging oxygen ions and dissolved molecular oxygen. Thus, diffusion of oxygen should be treated as a multispecies diffusion problem (see Zhang et al., 1991b), which has not been done so far for geological materials. Moreover, all oxygen ions of a kind (bridging or non-bridging) are not located at energetically equivalent positions in a melt and therefore will behave differently during diffusion—in the ideal case speciation between these should also be taken into account. The study of Young et al. (1994b) discussed earlier illustrates how complicated the process can be even in a simple binary melt and Beerkens and de Waal (1990) illustrate how such speciation may affect net mass transfer rates.

(b) Depending on kinetics of speciation reactions, transport of oxygen may be dominated by one or the other kind of species (e.g. ionic or molecular oxygen) leading to very different transport rates and activation energies in similar melt compositions.

(c) Depending on the details of the experiment, the measured diffusion coefficients can be either tracer or chemical diffusion coefficients. Because ambient atmosphere, oxygen fugacity and electrochemical forces can induce oxygen transport, evaluating oxygen diffusion experiments is not always straightforward.

(d) Significant differences have been found in the chemical and tracer diffusion behavior of oxygen so that it is necessary to separate, independently explain and relate the two kinds of data.

There is a large literature on oxygen diffusion in silicate melts. The review by Dunn (1986) discusses data available until that time. We provide here a selection of articles which can guide the reader to current literature on the topic: Shimizu and Kushiro (1991, 1984); Semkow and Haskin (1985); Wendlandt (1991); Canil and Muehlenbachs (1990); Kalen et al. (1991) and Beerkens and de Waal (1990). For purposes of estimating oxygen tracer diffusion rates in the absence of measured data, the best approach at present seems to be to use the Eyring equation. Estimating chemical diffusion rates is more complicated because of likely compositional dependencies but the two available models are those of Dunn and Scarfe (1986) and the oxygen chemical diffusion compensation law of Wendlandt (1991).

Data on other anions are limited to halides and have been discussed by Watson (1994).

(5) Chemical diffusion coefficients. As chemical diffusion rates depend completely on the nature of the endmembers involved, it is not possible to provide any general picture. Given a situation where measured data do not exist, one can either use one of the various models relating tracer and chemical diffusion coefficients to calculate the latter (in the worse case, after estimating the former!) or use one of the empirical approaches outlined above. The guiding principle to bear in mind is that chemical diffusion rate of a component is some kind of a weighted average of the tracer diffusion rates of constituent atoms/ions, with the slower species having more "say" in the process if diffusion of charged particles are involved. The thermodynamic factors can, however, truly complicate things in some cases. In dealing with any situation involving estimation of chemical diffusion rates in the absence of measured data, it is worth remembering that

(a) uphill diffusion may occur, so a priori use of effective binary diffusion coefficients may be misleading,

(b) chemical diffusion coefficients are typically strong functions of composition; much more so than the tracer- or self-diffusion coefficients. Thus, using compositionally independent tracer diffusivites to calculate chemical diffusivities as a function of composition is often a good compromise and

(c) to calculate mixing time scales one needs to take diffusive coupling into account. In view of the strong coupling of chemical diffusion of all components (even including alkalis, when complete mixing is to be studied rather than initial contamination rates), one should use diffusion rates of Si or other network formers rather than tracer diffusion rates of the species themselves. In concrete terms this means to calculate mixing time scales of Mg between two melts, using tracer diffusivity of Si may be better than using the tracer diffusivity of Mg itself!

Temperature dependence of diffusion coefficients

Almost all diffusion data obtained on silicate melts in the geological literature so far show Arrhenian temperature dependence, i.e. they obey $D = D_0 \exp(-Q/RT)$ where Q is the activation energy and D_0 is the pre-exponential factor (e.g. Fig. 9). This, combined with the fact that compensation laws are obeyed for silicate melts, allows very reasonable predictions of temperature dependence of diffusion coefficients to be made. A consequence of the compensation law is that the activation energies for slower diffusing species are higher. Thus, activation energies for diffusion in silicate melts range from about 100 kJ/mol (25 kcal/mol) for alkali atoms, through about 125 to 210 kJ/mol (30 to 50 kcal/mol) for non-alkali network modifiers to about 300 kJ/mol (70 kcal/mol) or more for network formers. Some exceptions occur for silicon diffusivity in very fragile melts at high temperatures (e.g. in diopside, Shimizu and Kushiro, 1991) and also for oxygen in some cases. The lower activation energies (~165 kJ/mol or 40 kcal/mol) obtained for Si in diopside is consistent with the idea that energy barriers decrease at high temperatures in fragile liquids. The highest known activation energy in a silicate melt is perhaps that of P diffusion in a dry granitic melt (Harrison and Watson, 1984): 600 kJ/mol (145 kcal/mol)! As a consequence of the kind of dependence that exists between chemical and tracer diffusion coefficients, the former are also Arrhenian for melts studied so far, for both chemical diffusion coefficients and for EBDC's over limited temperature ranges.

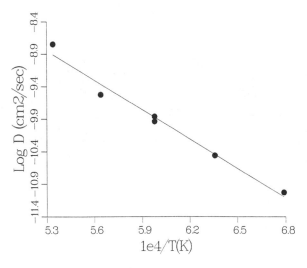

Figure 9. Example of an Arrhenian temperature dependence of diffusion coefficients in a silicate melt. The data are for chemical diffusion in a haplogranite-(haplogranite + 5 wt % B_2O_3) system (Reproduced from Chakraborty et al., 1993).

Non-Arrhenian temperature dependence is, however, well known in many simple silicate melts (see previous discussion and Frischat, 1975, for some examples), particularly when data across the glass transition are available. Some cases of non-Arrhenian behavior in geologically interesting material have been reported. While it is not clear if the dataset of Rálková (1965) is reliable (see Hofmann, 1980), the more recent measurements of Behrens (1992) clearly show non-Arrhenian temperature dependence of Na diffusion in melts along the albite-anorthite join. There may be two reasons for not observing such behavior in geologically interesting material more frequently. Firstly, it is possible that the degree of non-Arrhenian behavior is negligibly small for geologically interesting melts, most of which are fairly "strong". In this connection, it should prove interesting to measure activation energies for Si diffusion in diopside at lower temperatures—these could be higher and closer to more "normal" values of ~300 kJ/mol (70 kcal/mol). Secondly, there is the trivial reason that careful measurements across the glass transition region are rare in geological materials. Note that non-Arrhenian behavior becomes strongly apparent only near the ideal glass transition temperature (which is lower than the T_g measured calorimetrically), if the Caillot et al. (1994) model is correct. Only careful future studies can clarify whether the absence of non-Arrhenian behavior in geological melts is real or simply an artefact of paucity of data. In the meantime, one should recognize the possibility of non-Arrhenian behavior when extrapolating data and be particularly careful when dealing with fragile melts (e.g. some peralkaline basalts).

An interesting aspect which provides important insights into diffusion mechanisms in silicate melts is that activation energies in network formers are often on the order of 400 kJ/mol (100 kcal/mol)—about the energy required to break a single Si-O bond. This value of the activation energy hints at the fact that Si is not transported by breaking completely free from a tetrahedron (4 Si-O bonds). Rather, it breaks the bond to only one of the oxygen ions and then somehow changes its configuration, while still coupled with the remaining oxygen ions in the tetrahedra, to reach its new position. Figure 10 shows a simulation of Si diffusion based on molecular dynamic calculations (Kubicki and Lasaga, 1993) which shows exactly how this may take place in a partially depolymerized melt. Note, for later reference, that the intermediate state of the Si atom is 3-fold coordinated in this case.

Pressure dependence of diffusion coefficients

The effect of pressure on diffusion rates in silicate melts is an active field of research at present. The role of high pressure diffusion data as a substitute for viscosity measurements has already been alluded to. A recent exciting development has been the use of high pressure diffusion data, in combination with computer simulations, to elucidate diffusion mechanisms in silicate melts. Pressure dependence of diffusion rates is usually described by an equation similar to the Arrhenius equation, $D(P,T) = D(1 \text{ bar},T)\exp(-P\Delta V/RT)$ where ΔV is the activation volume, a parameter that provides a measure of pressure dependence. The pressure and temperature dependencies can be combined into a single equation which is often convenient for practical purposes. The reader should note that the parameter ΔV is not a volume of any species but a volume change (related to the difference between the volumes of the activated complex and the ground state in the framework of transition state theory). Therefore ΔV can be positive, negative or zero without any difficulty—a negative ΔV simply implying that the activated complex is denser than the ground state. A positive value of ΔV implies diffusion rates decrease with increasing pressure. Pressure dependence of diffusion rates of network formers can be qualitatively different from those of network modifiers and are treated separately below.

Pressure dependence for network formers. The first hints that the diffusion rates of network formers in silicate melts may show anomalous pressure dependence came from the studies of Kushiro (1976) and Woodcock et al. (1976). The former measured viscosity in jadeite melt as a function of pressure and the latter carried out molecular dynamic simulations of amorphous SiO_2 under pressure. Subsequently, Angell et al (1982)

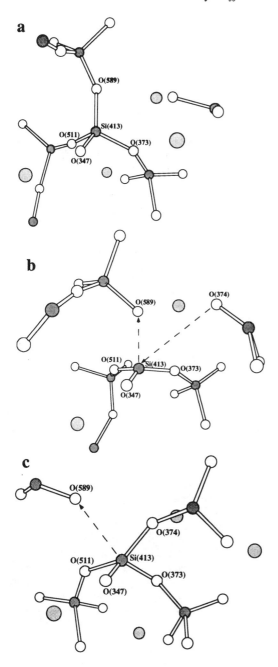

Figure 10. Picture showing how a Si ion may diffuse, as obtained from molecular dynamic simulations (Kubicki and Lasaga, 1993) for melts in the system $MgSiO_3$-Mg_2SiO_4. The motion of the Si(413) atom is followed. In **a** the SiO_4 tetrahedron is shown before diffusion begins. An oxygen ion, O(589) breaks from the tetrahedron in **b**, and a near trigonal planar SiO_3 complex is formed. A new Si-O bond is formed to O(374) in **c**, and the SiO_4 is stable and relatively stationary for the next 40 ps. Note that although a Si-O bond is broken, only the Si diffuses in this case and not the O (reproduced from Kubicki and Lasaga, 1993).

predicted the anomalous diffusion behavior from computer simulations and Shimizu and Kushiro (1984) demonstrated the presence of at least part of it by directly measuring negative activation volumes for oxygen diffusivity in jadeite melt upto pressures of 40 kbar. The complete prediction, from computer simulations (Angell et al., 1982, 1983) is that diffusion rates of network formers would go through a maximum as pressure increases i.e. at some high pressure activation volumes would change sign and become positive. The lowest predicted pressure for the transition, for simple silicates that could be simulated, is about 200 kbar, which is still beyond the experimental range.

Detailed studies of pressure dependence of Si and O along the join diopside-jadeite carried out by Shimizu and Kushiro (1991) reveal some interesting features. Diffusion rates of the two species are practically indistinguishable in diopside whereas oxygen diffuses slightly faster in jadeite (within an order of magnitude, still). The interesting feature, however, is that pressure enhances diffusion rates in jadeite but reduces the diffusivities in diopside. The crossover in pressure dependence occurs at a composition of about $Di_{30}Jd_{70}$. The differences between diopside and jadeite are in degree of polymerization, in fragility and in Al content—the question is, what causes the difference in pressure dependence of diffusivity? Detailed analysis by Shimizu and Kushiro (1991) seemed to suggest that details of polymerization, as measured by Raman spectroscopy, do not explain the observed diffusion behavior very well. Further insight comes from studies on some simple systems where the effects of individual variables may be isolated to some extent.

Rubie et al. (1993) and Poe et al. (1994) have carried out oxygen and silicon tracer diffusion experiments in $Na_2Si_4O_9$ melts at pressures upto 15 GPa in a multi-anvil apparatus. They observe negative activation volumes for both species over the entire pressure range ($\sim$ -3 cm^3/mol), but the pressure dependence is much less than that observed in jadeite. The degree of polymerization of $Na_2Si_4O_9$ is more like that of albite (i.e. more polymerized than jadeite), but the melt contains no Al. This led Poe et al. (1994 and pers. commun.) to conclude that the presence of Al may play an important role in the negative pressure dependence, in addition to degree of polymerization. The differences in activation volumes could also be related, at least to some extent, to the differences in temperatures between the experiments of Shimizu and Kushiro (1984, 1991) and the measurements done in the multianvil apparatus.

Computer simulations and spectroscopic studies have been remarkably successful in not only predicting but also explaining the unexpected behavior. The essence of the explanation of anomalous pressure dependence is that under high pressures, silicon can form five and six coordinated species in an amorphous material. Such highly coordinated species have been predicted from computer simulations (Brawer, 1985; Angell et al., 1982, 1983; Kubicki and Lasaga, 1988[1]) and observed spectroscopically in glasses quenched from high pressures (e.g. Xue et al, 1991). The general idea to explain the pressure dependence is that during initial compression silicon begins to form 5-coordinated species which can act as intermediates to help the diffusion process. As pressure increases, the possibilities for 5-coordination are exhausted and compression occurs by other mechanisms—diffusion begins to slow down with pressure at this stage and the activation volume becomes positive. The 5 coordinated species in this case has a true minimum energy configuration i.e. there is an equilibrium abundance of this species in the melts at high pressures (and also at atmospheric pressures in some compositions). It is therefore *not* a transition state complex for the diffusion process, although it is an intermediate step (see Kubicki and Lasaga, 1991 for more detail). Further, the details of the mechanism are different depending on the availability and abundance of non-bridging oxygens.

Figure 11 from McMillan et al. (1994) shows a schematic picture (based on suggestions from the computer simulations mentioned earlier) of how 5 coordinated silicon might help the diffusion process in a fully polymerized melt, see Poole et al. (this volume) for more details. Note the similarity and differences of this process with that illustrated earlier (Fig. 10). In the computer simulation of Figure 10, showing the effect of

---

[1]   The role of 5-coordinated Si in pure SiO_2 melt has also been challenged, however, based on simulations carried out with different intermolecular potentials and cell sizes. See Rustad et al. (1990) for details.

temperature in a partially depolymerized melt, one forms *3-fold coordinated* silicon as an intermediate to aid diffusion; in the schematic model to illustrate the effect of pressure in a fully polymerized melt, one forms *5-fold coordinated* silicon as an intermediate. In both cases (i.e. with increasing temperature as well as pressure) Si diffusivity is enhanced, although by formation of *different* defect states (Kubicki and Lasaga, 1991).

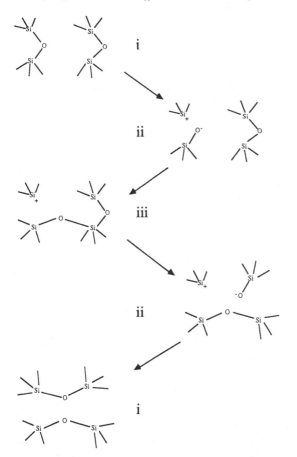

Figure 11. Schematic view of how oxygen diffuses through the intermediary of a five coordinated Si by bond breaking and reforming in a simple silicate (reproduced from McMillan et al., 1994).

The problem that remains is that none of the experimental studies have observed the actual maxima in diffusivity of Si or oxygen i.e. the changeover to positive activation volumes. Considering the uncertainties in computer calculations and the difficulties in high pressure experiments beyond 15 GPa, it does not seem reasonable to pursue an experimental effort to observe the predicted maxima for the simple melts that were simulated. However, Poe (pers. comm.) noted that if it is five coordinated species that are the key players in the anomaly then it is much easier to form five and six coordinated Al species, as verified from NMR spectroscopy and molecular dynamic calculations (e.g. Poe et al., 1994). Secondly, as we saw earlier, the indications from high pressure data were that Al content of the melt plays an important role in the anomalous pressure dependence of network formers. Thirdly, Angell et al. (1982) had predicted that the diffusivity maxima would be expected to shift to lower pressures with decreasing temperature. From these considerations it should be expected that the maxima would be observed most readily in low temperature measurements done on relatively polymerized melts containing Al.

A search of the literature reveals that indeed such a change has been observed in chemical diffusion experiments done on melts on the join albite-anorthite in the glass transformation region (Weinberg, 1987). Weinberg (1987) found that the diffusion rates of all components as well as activation energies in the system Ab-An show a maximum at about 15 kbar for experiments done between 750° to 860°C (Fig. 12). Data from the joins Jd-An and Jd-Ab show no such maxima for measurements done at the same conditions. The fact that he observes maxima in diffusivity of all components is not entirely surprising because of the low temperatures of his experiments (more coupling between cations and oxygen, also see discussion on *network modifiers* below) and his use of effective binary diffusion coefficients (which contain contributions from all components, see *effective binary diffusion coefficient* above). Preliminary tracer diffusion data on $Na_3AlSi_7O_{17}$ for Si and O also show a diffusivity maxima at about 8 GPa (Poe, pers. comm.).

Finally, there is the question of applicability of these considerations to realistic geologic melt compositions. Baker (1990) has obtained negative activation volumes for the diffusion of the SiO_2 component in experiments with dacite-rhyolite diffusion couples. Additionally, viscosity data show the negative activation volumes to be a common feature of polymerized silicate melts (see Dingwell, this volume, for details). Pressure dependence of diffusivities of other network-formers (Ti, Zr) also yield negative activation volumes in some basaltic liquids (preliminary data from Shimizu and Kushiro, 1991).

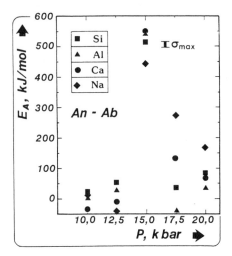

Figure 12. Activation energy obtained from effective binary diffusion coefficients of various oxide components in the system Ab-An as a function of pressure (reproduced from Weinberg, 1987).

Pressure dependence for network modifiers. In contrast to the complexities of the network formers, the diffusivity of network modifiers show relatively simpler behavior with pressure. Diffusion rates of network modifiers decrease with pressure and activation volumes are positive. The magnitude of the activation volumes are often close to the molar volumes of the diffusing ion (e.g. Watson, 1979). This is consistent with the notion of network modifiers diffusing along "channels" (regions of enhanced concentration of non-bridging oxygen in a melt) which close up with increasing pressure and thus reduce mobility. This picture of diffusion has been arrived at from interpretations of experimental data (Watson, 1979; Shimizu and Kushiro, 1991, etc.) as well as from computer simulations (Angell et al., 1982; Kubicki and Lasaga, 1991) and is consistent with current notions about the structure of melts (see earlier discussion).

However, two complicating trends need to be mentioned before we leave the topic. First, pressure dependence of Mg diffusivity along the diopside-jadeite join does show a reversal—activation volumes become negative for jadeite rich compositions at some point between Di_{20} and Di_{60}. Activation volumes for Ca diffusion remain positive over the entire join. Further, the composition dependence of diffusion rates are much stronger for Mg compared to Ca at the jadeite rich end. These observations led Shimizu and Kushiro (1991) to suggest that in some melts there are important distinctions between the diffusion behavior of Ca and Mg and a model where all network modifiers simply diffuse along "channels" is too simpleminded. This is reminiscent of the kind of site specificity seen earlier in the *site mismatch model* for diffusion in glasses, although the detailed process in these high temperature liquids must be different because of the motion of the network formers. A reversal in pressure dependence of diffusion rates of network modifiers has also been predicted from computer simulations (Angell et al., 1982, 1983), but such effects of increased coordination of network modifiers are expected to manifest themselves only at much higher pressures than are currently accessible.

The second complicating feature in the general trend is   actually to be expected—*chemical diffusion* rates, as opposed to *tracer diffusion rates*, of network modifiers in polymerized silicate melts do show negative activation volumes. In the only study available so far that systematically explored chemical diffusivities as a function of pressure, Baker (1990) observed that diffusion rates of all non-alkali network modifiers increase with pressure. The increase was clearly related to the negative activation volumes of silicon and oxygen diffusion and is simply a consequence of the strong diffusive coupling during chemical diffusion in silicate melts.

IX. APPLICATIONS

The range of natural processes affected by diffusion in silicate melts extend far beyond terrestrial scales—both in space and in time. Temporally, the influence of diffusion in silicate melts played a role in processes ranging from flash melting events in the early solar nebula (e.g. Greenwood and Hess, 1995) to recent events within our own experience like the eruption of Mt. St. Helens (e.g. Toramaru, 1995). Spatially, diffusion in silicate melts affects processes deep within the earth (mantle metasomatism), on the surface of the earth (crystallization, assimilation and degassing in lava flows and lakes) as well as far beyond the earth (e.g. assimilation of lunar mare basalts, Finnila et al., 1994). Consequently, it is not possible to discuss, or indeed, foresee, all possible applications of diffusion data. In keeping with the approach in the rest of this chapter, we will therefore provide some examples where applications of diffusion data in silicate melts have been discussed. As mentioned at the beginning of the chapter, the connecting thread in most of these applications is a simple thought—diffusion can be a very inefficient mode of mass transfer. So, when a process in which diffusion is involved does go to completion, it allows us to set a limit on the minimum times that must have been available for the process to take place. On the other hand, there is a good chance that any process which depends on diffusion may be frozen without going to completion. These frozen features can provide us important clues in our attempts to identify and understand processes that we cannot directly observe—provided, we understand the process of diffusion itself.

Understanding diffusion not only helps us in the identification of rate determining processes, but also in the determination of the actual rates and paths of the processes themselves, once diffusion coefficients are known. Some processes where diffusion in silicate melts is involved, in addition to those already mentioned, are: crystal growth (e.g. Lasaga, 1982; Ghiorso, 1987) and dissolution (e.g. Harrison and Watson, 1983, 1984; Scarfe and Brearley, 1987; Zhang et al., 1989), element fractionation between minerals and melts (Hofmann, 1980; Albarede and Bottinga, 1972; Hanson, 1978; Prinzhofer and

Allègre, 1985; Bedard, 1989; Sawyer, 1991; Hofmann and Hart, 1978; Qin, 1992, etc.), kinetics of partial melting (Brearley and Rubie, 1990; Rubie and Brearley, 1990), development of convective boundary layers (e.g. Trial and Spera, 1988) and mixing and contamination of magma bodies (e.g. Watson, 1982; Baker, 1990, 1991; Rosing et al., 1989; Chakraborty et al, 1995b).

The discussion above relates to *direct* applications of diffusion data. As we have seen, diffusion can be related to various equilibrium and transport properties of silicate melts. Many processes in silicate melts are indirectly influenced by diffusion rates—in particular the relative *differences* in diffusion rates of the constituents of a melt. Thus, the relationship between diffusion and various processes like viscous flow, electrical conduction and thermodynamic mixing can be exploited to improve our knowledge of all of these processes. The connection between diffusion and melt structure, although complex, can be utilized to improve our understanding of both. An example where such interrelationships between properties were exploited to understand some aspects of emplacement of peraluminous granite plutons may be found in Chakraborty et al. (1993). Such a listing can continue, but we will simply let the reader use imagination in finding other applications of the study of diffusion. In this pursuit, one problem often faced by newcomers to the field is purely conceptual—how to take a geological problem, formulate it as a diffusion problem and solve it using the proper form of diffusion data and equations? An important aspect of such modelling is to make assumptions that retain the essential physical aspects of a process but simplify the mathematical treatment. Space prohibits delving into this topic, but the reader is encouraged to consult Ghez (1988) for a very readable introduction. In the search for geological situations where diffusion in melts may play an important role, we remind the reader of some general characteristics which we have learned from our discussions of theory and experiments in this chapter:

(1) Diffusion is too inefficient to cause transformations on a global scale but it can profoundly affect the course of events by limiting rates of chemical transformation on local scales.

(2) Chemical mixing of two liquids can take place only by diffusion. Convection can bring individual parcels of liquid into increasingly intimate contact by thinning, stretching out and folding layers, but it cannot erase compositional differences. It can drastically reduce time scales of mixing, but the actual mixing has to occur by diffusion. The mixing path (in composition space) is therefore always that governed by diffusion rates of different components in the liquid.

(3) Diffusion dominates transport processes when convection is suppressed for some reason. Convection is suppressed in melts of high viscosity (e.g. granites) and in thin capillaries (e.g. thin films of melts in grain boundaries during early stages of partial melting)—these are the likely situations for diffusion to dominate.

Armed with these guidelines, it is hoped that the reader will succeed in extending the list of applications given above. The highly correlated (back and forth jumps) random walk through this chapter would be considered a success if it results in a mean squared displacement towards better understanding of the process of diffusion in silicate melts.

ACKNOWLEDGMENTS

I take this opportunity to thank Don Dingwell for nurturing my interest in silicate melts and for being patient during hours of "healthy debate and lively discussion". The scientific as well as editorial support of Jonathan Stebbins made working on this chapter a very pleasant experience. Brent Poe generously shared his latest experimental results (some of which are unpublished) and thoughts; discussions with him improved my understanding of many aspects of melt structure and spectroscopy. Ruth Knoche read earlier versions of the chapter and her liberal criticisms may spare the readership of some truly incomprehensible logic, convoluted text and a flawed model of melt structure. Comments from an anonymous reviewer, Jonathan Stebbins and Jibamitra Ganguly improved the chapter materially and stylistically. Prof. G.H. Frischat, Dr. J.D. Kubicki and Dr. W. Weinberg helped by providing copies of figures from their works.

REFERENCES

Adam G, Gibbs JH (1965) On the temperature dependence of cooperative relaxation properties in glass-forming liquids. J Chem Phys 43:139-146

Albarede F, Bottinga Y (1972) Kinetic disequilibrium in trace element partitioning between phenocrysts and host lava. Geochim Cosmochim Acta 36:141-156

Allnatt AR, Lidiard, AB (1993) Atomic transport in solids. Cambridge University Press, Cambridge

Angell CA (1985) Strong and fragile liquids. In: Ngai K and Wright GB (eds) Relaxation in complex systems. National Technical Information Service, Springfield, VA, p 3-11

Angell CA, Cheeseman PA, Tamaddon S (1982) Pressure enhancement of ion mobilities in liquid silicates from computer simulation studies to 800 kilobars. Science 218:885-887

Angell CA, Cheeseman P, Tamaddon S (1983) Water-like transport property anomalies in liquid silicates investigated at high T and P by computer simulation techniques. Bull Minéral 106:87-97

Anderson OL, Stuart PA (1954) Calculation of activation energy of ionic conductivity in silica glasses by classical methods. J Am Ceram Soc 37:573-580

Appell M (1968) Solution for Fick's 2nd law with variable diffusivity in a multi-phase system. Scripta Metall 2:217-222

Ayers JC, Brenan JM, Watson EB, Wark DA, Minarik WG (1992) A new capsule technique for hydrothermal experiments using the piston-cylinder apparatus. Am Mineral 77:1080-1086

Baker DR (1990) Chemical interdiffusion of dacite and rhyolite: Anhydrous measurements at 1 atm and 10 kbar, application of transition state theory, and diffusion in zoned magma chambers. Contrib Mineral Petrol 104:407-423

Baker DR (1991) Interdiffusion of hydrous dacitic and rhyolitic melts and the efficacy of rhyolite contamination of dacitic enclaves. Contrib Mineral Petrol 106:462-473

Baker DR (1992a) Tracer diffusion of network formers and multicomponent diffusion in dacitic and rhyolitic melts. Geochim Cosmochim Acta 56:617-631

Baker DR (1992b) Estimation of diffusion coefficients during interdiffusion of geologic melts: Application of transition state theory. Chem Geol 98:11-21

Baker DR (1992c) Diffusion of 3^+ and 4^+ network-forming cations in albite melt: Implications for viscous transport. Trans Am Geophys Union EOS 73:599

Baker DR (1993) Measurement of diffusion at high temperatures and pressures in silicate systems. In: Luth RW (ed) Experiments at high pressure and applications to the earth's mantle. Short Course Handbook 21:305-355, Mineral Assoc Canada

Bansal NP, Doremus RH (1986) Handbook of Glass Properties. Academic Press, Orlando, Florida

Barrer RM, Bartholomew RF, Rees LVC (1963) Ion exchange in porous crystals. Part II- The relationship between self- and exchange-diffusion coefficients. J Phys Chem Solids 24:309-317

Batchelor GK (1976) Brownian diffusion of particles with hydrodynamic interaction. J Fluid Mech 74:1-29

Bédard JH (1989) Disequilibrium mantle melting. Earth Planet Sci Lett 91:359-366

Beenaker CWJ, Mazur P (1983) Self-diffusion of spheres in a concentrated suspension. Physica 120A:388-410

Beerkens RGC, de Waal H (1990) Mechanism of oxygen diffusion in glass melts containing variable-valence ions. J Am Ceram Soc 73:1857-1861

Behrens H (1992) Na and Ca tracer diffusion in plagioclase glasses and supercooled melts. Chem Geol 96:267-275

Berman RG, Brown TH (1984) A thermodynamic model for multicomponent melts, with application to the system $CaO-Al_2O_3-SiO_2$. Geochim Cosmochim Acta 48:661-678

Bowen, NL (1921) Diffusion in silicate melts. J Geol 29:295-317

Borg RJ, Dienes GJ (1988) An introduction to solid state diffusion. Academic Press, Boston

Bottinga Y, Weill DF (1972) The viscosity of magmatic silicate liquids: A model for calculation. Am J Sci 272:438-475

Bouchaud J-P, Georges A (1990) Anomalous diffusion in disordered media: Statistical mechanisms, models and physical applications. Phys Rep 195:127-29

Brady JB (1975a) Reference frames and diffusion coefficients. Am J Sci 275:954-983

Brady JB (1975b) Chemical components and diffusion. Am J. Sci 275:1073-1088

Brady JB, McCallister RH (1983) Diffusion data for clinopyroxenes from homogenization and self-diffusion experiments. Am Mineral 68:95-105

Braedt M, Frischat, GH (1988) Sodium self diffusion in glasses and melts of the system Na_2O-Rb_2O-SiO_2. Phys Chem Glasses 29:214-218

Brawer, S (1985) Relaxation in Viscous Liquids and Glasses. The American Ceramic Society, Columbus, Ohio

Brearley AJ, Rubie DC (1990) Effects of H_2O on the disequilibrium breakdown of muscovite + quartz. J Petrol 31:925-956

Bunde A, Maass P, Ingram MD (1991) Diffusion limited percolation: A model for transport in ionic glasses. Ber Bunsenges Phys Chem 95:977-983

Caillot E, Duclot MJ, Souquet JL, Levy M, Baucke FGK, Werner RD (1994) A unified model for ionic transport in alkali disilicates below and above the glass transition. Phys Chem Glasses 35:22-27

Canil D, Muehlenbachs K (1990) Oxygen diffusion in an Fe-rich basalt melt. Geochim Cosmochim Acta 54:2947-2951

Carman PC (1968) Intrinsic mobilities and independent fluxes in multicomponent isothermal diffusion. I. Simple Darken systems. J Phys Chem 72:1707-1712

Carslaw HS, Jaeger JC (1959) Conduction of Heat in Solids. 2nd Edn. Clarendon Press, Oxford

Chakraborty S (1994) Relationships between thermodynamic mixing and diffusive transport in multicomponent solutions: Some constraints and potential applications. J Phys Chem 98:4923-4926

Chakraborty S, Dingwell DB (1993) Multicomponent diffusion and viscous flow in melts in the system Na_2O-K_2O-Al_2O_3-SiO_2-P_2O_5: The influence of the alkali/aluminum ratio. Trans Am Geophys Union EOS 74:617-618

Chakraborty S, Ganguly J (1992) Cation diffusion in aluminosilicate garnets: Experimental determination in spessartine-almandine diffusion couples, evaluation of effective binary diffusion coefficients, and applications. Contrib Mineral Petrol 111:74-86

Chakraborty S, Ganguly J (1994) A method to constrain thermodynamic mixing properties and diffusion data in multicomponent solutions. In: Bocquet JL, Limoge Y (eds) Reactive phase formation at interfaces and diffusion processes. Materials Science Forum 155-156:279-284, Trans Tech Publishers, Aedermannsdorf, Switzerland

Chakraborty S, Rubie DC (1995) Mg tracer diffusion in aluminosilicate garnets at 750-850°C, 1 atm and 1300°C, 8.5 GPa. Contrib Mineral Petrol, in press

Chakraborty S, Dingwell DB, Chaussidon M (1993) Chemical diffusivity of boron in melts of haplogranitic composition. Geochim Cosmochim Acta 57:1741-1751

Chakraborty S, Dingwell DB, Rubie DC (1995a) Multicomponent diffusion in ternary silicate melts in the system K_2O-Al_2O_3-SiO_2: I. Experimental measurements. Geochim Cosmochim Acta 59:255-264

Chakraborty S, Dingwell DB, Rubie DC (1995b) Multicomponent diffusion in ternary silicate melts in the system K_2O-Al_2O_3-SiO_2: II. Mechanisms, systematics, and geological applications. Geochim Cosmochim Acta 59:265-277

Chekhmir AS, Epel'baum MB and Simakin AG (1988) Transport of water in magmatic melts. Geokhimiya 10:303-305 (in Russian)

Chekhmir AS and Epel'baum MB (1991) Diffusion in magmatic melts: New study. In: Perchuk LL, Kushiro I (eds) Physical chemistry of magmas, Advances in Physical Geochemistry 9:99-119, Springer Verlag, New York

Chadwick AV (1990) Nuclear magnetic resonance methods of studying mass transport in solids. J Chem Soc Faraday Trans 86:1157-1165

Cicerone MT, Blackburn FR, Ediger MD (1995) How do molecules move near T_g? Molecular rotation of six probes in *o*-terphenyl across 14 decades in time. J Chem Phys 102:471-479

Colson RO, Haskin LA, Crane D (1990) Electrochemistry of cations in diopsidic melt: Determining diffusion rates and redox potentials from voltammetric curves. Geochim Cosmochim Acta 54:3353-3367

Cooper AR Jr (1965) Model for multi-component diffusion. Phys Chem Glasses 6:55-61

Cooper AR Jr (1968) The use and limitation of the concept of an effective binary diffusion coefficient for multicomponent diffusion. In: Wachtman JB Jr, Franklin AD (eds) Mass Transport in Oxides—Proc Symp, US Dept of Commerce, p 79-84

Cooper AR Jr (1974) Vector space treatment of multicomponent diffusion. In: Hofmann AW, Giletti BJ, Yoder HS Jr, Yund RA (eds) Geochemical transport and kinetics, Carnegie Inst Washington Publ 634:15-30, Washington, DC

Cooper AR Jr (1982) W.H. Zachariasen—The melody lingers on. J Non-Cryst Solids 49:1-17

Cooper AR Jr, Heasley JH (1966) Extension of Darken's equation to binary diffusion in ceramics. J Am Ceram Soc 49:280-284

Crank J (1975) The Mathematics of Diffusion. Clarendon-Oxford, London

Cullinan HT Jr (1965) Analysis of the flux equations of multicomponent diffusion. Ind Eng Chem Fundamentals 4:133-139

Cussler EL (1976) Multicomponent Diffusion. Elsevier, Amsterdam

Cussler EL (1984) Diffusion: Mass Transfer in Fluid Systems. Cambridge University Press, Cambridge

Darken LS (1948) Diffusion, mobility and their interrelation through free energy in binary metallic systems. Am Inst Mining Metall Engineers Trans 175:184-201

Davis JC (1973) Statistics and Data Analysis in Geology. John Wiley & Sons, New York

Dayananda MA (1985) Zero-flux planes, flux reversals and diffusion paths in ternary and quaternary diffusion. In: Dayananda MA, Murch GE (eds) Diffusion in solids: Recent developments. AIME, Pennsylvania, p 195-230

de Groot SR, Mazur P (1984) Non-equlibrium thermodynamics. Dover Publications, New York

Dingwell DB (1990) Effects of structural relaxation on cationic tracer diffusion in silicate melts. Chem Geol 82:209-216

Dingwell DB, Knoche R, Webb SL, Pichavant M (1992) The effect of B_2O_3 on the viscosity of haplogranitic melts. Am Mineral 77:457-461

Dong W, Baros F, André JC (1990) Non-Markovian effect on diffusion-controlled reactions. Ber Bunsenges Phys Chem 94:269-274

Donth E (1982) The size of cooperatively rearranging regions at the glass transition. J Non-Cryst Solids 52:325-330

Dosdale T, Brook RJ (1983) Comparison of diffusion data and of activation energies. J Am Ceram Soc 66:392-395

Duffy JA, Kamitsos EI, Chryssikos GD, Patsis AP (1993) Trends in local optical basicity in sodium borate glasses and relation to ionic mobility. Phys Chem Glasses 34:153-157

Dunn T (1982) Oxygen diffusion in three silicate melts along the join diopside-anorthite. Geochim Cosmochim Acta 46:2293-2299

Dunn T (1986) Diffusion in silicate melts: An introduction and literature review. In: Scarfe CM (ed) Silicate melts, Short Course Handbook 12:57-92, Mineral Assoc Canada

Dunn T, Scarfe CM (1986) Variation of the chemical diffusivity of oxygen and viscosity of an andesite melt with pressure at constant temperature. Chem Geol 54:203-215

Elphick SC, Ganguly J, Loomis TP (1985) Experimental determination of cation diffusivities in aluminosilicate garnets I. Experimental methods and interdiffusion data. Contrib Mineral Petrol 90:36-44

Farnan I, Stebbins JF (1990) Observation of slow atomic motion close to the glass transition using 2-D ^{29}Si NMR. J Non-Cryst Solids 124:207-215

Farnan I, Stebbins JF (1994) The nature of the glass transition in a silica-rich oxide melt. Science 265:1206-1209

Fick AE (1855) Über Diffusion. Poggendorff's Annalen der Physik 94:59-86

Finnila AB, Hess PC, Rutherford MJ (1994) Assimilation by lunar mare basalts: Melting of crustal material and dissolution of anorthite. J Geophys Res 99E:14677-14690

Fisher GW (1973) Nonequilibrium thermodynamics as a model for diffusion-controlled metamorphic processes. Am J Sci 273:897-924

Fiske PS, Stebbins JF (1993) The structural role of Mg in silicate liquids: a high-temperature ^{25}Mg, ^{23}Na, and ^{29}Si NMR study. Am Mineral 79:848-861

Freer R (1980) Self-diffusion and impurity diffusion in oxides. J Mater Sci 15:803-824

Freer (1981) Diffusion in silicate minerals and glasses: A data digest and guide to the literature. Contrib Mineral Petrol 76:440-454

Frischat GH (1975) Ionic diffusion in oxide glasses. Trans Tech Publications, Aedermannsdorf, Switzerland

Franke W, Heitjans P (1992) ^{7}Li-NMR study of diffusion-induced spin-lattice relaxation in glassy and crystalline LiAlSi$_2$O$_6$. Ber Bunsenges Phys Chem 96:1674-1677

Fujita H, Gosting LJ (1956) An exact solution of the equations for free diffusion in three-component systems with interacting flow, and its use in evaluation of the diffusion coefficients. J Am Chem Soc 78:1099-1106

Funke K (1991) Ion transport and relaxation studied by high-frequency conductivity and quasi-elastic neutron scattering. Phil Mag A 64:1025-1034

Gan H, Hess PC (1992) Phosphate speciation in potassium aluminosilicate glasses. Am Mineral 77:495-506

Ganguly J, Bhattacharya RN, Chakraborty S (1988) Convolution effect in the determination of compositional profiles and diffusion coefficients by microprobe step scans. Am Mineral 73:901-909

Ghez R (1988) A Primer of Diffusion Problems. John Wiley & Sons, New York

Ghiorso MS (1987) Chemical mass transfer in magmatic processes III. Crystal growth, chemical diffusion and thermal diffusion in multicomponent silicate melts. Contrib Mineral Petrol 96:291-313

Glasstone S, Laidler KJ, Eyring H (1941) The Theory of Rate Processes. McGraw Hill, New York-London.

Greaves GN (1985) EXAFS and the structure of glass. J Non-Cryst Solids 71:203-217

Greaves GN, Gurman SJ, Catlow CRA, Chadwick AV, Houde-Walter S, Henderson CMB, Dobson, BR (1991) A structural basis for ionic diffusion in oxide glasses. Phil Mag A 64:1059-1072

Greenwood JP, Hess PC (1995) Constraints on flash heating from melting kinetics. Lunar Planetary Science Conf XXVI, NASA, Houston, Texas, p 471-472

Gupta PK, Cooper AR (1971) The [D] matrix for multicomponent diffusion. Phys 54:39-59

Hanson GN (1978) The application of trace elements to the petrogenesis of igneous rocks of granitic composition. Earth Planet Sci Lett 38:26-43

Hart SR (1981) Diffusion compensation in natural silicates. Geochim Cosmochim Acta 45:279-291

Harrison TM, Watson EB (1983) Kinetics of zircon dissolution and zirconium diffusion in granitic melts of variable water content. Contrib Mineral Petrol 84:66-72

Harrison TM, Watson EB (1984) The behavior of apatite during crustal anatexis: Equilibrium and kinetic considerations. Geochim Cosmochim Acta 48:1567-1577

Heinemann I, Frischat GH (1993) The sodium transport mechanism in Na$_2$O·2SiO$_2$ glass determined by the Chemla experiment. Phys Chem Glasses 34:255-260

Heitjans P, Faber W, Schirmer A (1991) ß-NMR studies of spin-lattice relaxation in glassy and layer-crytalline compounds. J Non-Cryst Solids 131-133:1053-1062

Henderson J, Yang L, Derge G (1961) Self-diffusion of aluminum in CaO-SiO$_2$-Al$_2$O$_3$ melts. Trans Metallurgical Soc AIME 221:56-60

Henderson P, Nolan J, Cunninham GC, Lowry RK, (1985) Structural controls and mechanisms of diffusion in natural silicate melts. Contrib Mineral Petrol 89:262-272

Hofmann AW (1980) Diffusion in natural silicate melts: A critical review. In: Physics of Magmatic Processes. Hargraves RB (ed) Princeton Univ Press, Princeton, NJ, p 385-417

Hofmann AW, Hart SR (1978) An assessment of local and regional isotopic equilibrium in the mantle. Earth Planet Sci Lett 38:44-62

Hofmann AW, Margaritz M (1977) Diffusion of Ca, Sr, Ba, and Co in a basalt melt: Implications for the geochemistry of the mantle. J Geophys Res 82:5432-5440

Ingram MD (1987) Ionic conductivity in glass. Phys Chem Glasses 28:215-234

Ingram MD, Moynihan CT, Lesikar AV (1980) Ionic conductivity and the weak electrolyte theory of glass. J Non-Cryst Solids 38/39:371-376

Ingram MD, Maass P, Bunde A (1991) Ionic conductivity and memory effects in glassy electrolytes. Ber Bunsenges Phys Chem 95:1002-1006

Isard JO (1992) Distributions of activation energies in ionic glasses. Phil Mag A 66:213-228

Jambon A (1983) Diffusion dans les silicates fondus: Un bilan des connaissances actuelles. Bull Minéral 106:229-246

Jambon A, Semet MP (1978) Lithium diffusion in silicate glasses of albite, orthoclase, and obsidian composition: An ion microprobe determination. Earth Planet Sci Lett 37:445-450

Jambon A, Zhang Y, Stolper EM (1992) Experimental dehydration of natural obsidian and estimation of D$_{H2O}$ at low water contents. Geochim Cosmochim Acta 56:2931-2935

Joesten R (1977) Evolution of mineral assemblage zoning in diffusion metasomatism. Geochim Cosmochim Acta 41:649-670

Jost W (1960) Diffusion in Solid State Physics. Academic Press, New York

Kahnt H, Kaps Ch, Offermann J (1988) A new method of simultaneous measurement of tracer diffusion coefficient and mobility of alkali ions in glasses. Solid State Ionics 00:215-220

Kalen JD, Boyce RS, Cawley JD (1991) Oxygen tracer diffusion in vitreous silica. J Am Ceram Soc 74:203-209

Keppler H, Bagdassarov NS (1993) High-temperature FTIR spectra of H_2O in rhyolite melt to 1300°C. Am Mineral 78:1324-1327

Kieffer J, Borchardt G, Scherrer S, Weber S (1986) Tracer diffusion of cobalt and silicon in cobaltous oxide-silica liquid mixtures. Mater Sci Forum 7:243-256

Kieffer J, Borchardt G (1987) A kinetic model of silicate melts (silicon tracer diffusion). Chem Geol 62:93-101

Kirkaldy JS (1959) Diffusion in multicomponent metallic systems: IV. A general theorem for construction of multicomponent solutions from solutions of the binary diffusion equation. Can J Phys 37:30-34

Kirkaldy JS, Young DJ (1987) Diffusion in the Condensed State. The Institute of Metals, London

Kirwan AD Jr, Kump LR (1987) Models of geochemical systems from mixture theory: Diffusion. Geochim Cosmochim Acta 51:1219-1226

Knoche RK, Dingwell DB, Webb SL (1992) Non-linear temperature dependence of liquid volumes in the system albite-anorthite-diopside. Contrib Mineral Petrol 111:61-73

Knoche RK, Dingwell DB, Webb SL (1995) Melt densities for leucogranites and granitic pegmatites: Partial molar volumes for SiO_2, Al_2O_3, Na_2O, K_2O, Li_2O, Rb_2O, Cs_2O, MgO, CaO, SrO, BaO, B_2O_3, P_2O_5, F_2O_{-1}, TiO_2, Nb_2O_5, Ta_2O_5 and WO_3. Geochim Cosmochim Acta, in press

Kress VC, Ghiorso, MS (1993) Multicomponent diffusion in $MgO-Al_2O_3-SiO_2$ and $CaO-MgO-Al_2O_3-SiO_2$ melts. Geochim Cosmochim Acta 57:4453-4466

Kress VC, Ghiorso MS (1995) Multicomponent diffusion in basaltic melts. Geochim Cosmochim Acta 59:313-324

Kreuzer HJ (1981) Nonequilibrium thermodynamics and its statistical foundations. Monographs on the physics and chemistry of Materials. Clarendon Press, Oxford

Kubicki JD, Lasaga AC (1988) Molecular dynamics simulations of SiO_2 melt and glass: Ionic and covalent models. Am Mineral 73:945-955

Kubicki JD, Lasaga AC (1991) Molecular dynamics and diffusion in silicate melts. In: Ganguly J (ed) Diffusion, atomic ordering, and mass transport—Selected topics in geochemistry. Advances in Physical Geochemistry 8:1-5, Springer-Verlag, New York

Kubicki JD, Lasaga AC (1993) Molecular dynamics simulation of interdiffusion in $MgSiO_3-Mg_2SiO_4$ melts. Phys Chem Minerals 20:255-262

Kushiro I (1976) Changes in viscosity and structure of melt of $NaAlSi_2O_6$ composition at high pressures. J Geophys Res 81:6347-6350

Kushiro I (1983) Effects of pressure on the diffusivity of network-forming cations in melts of jadeitic compositions. Geochim Cosmochim Acta 47:1415-1422

Laborde P, Kaps Ch, Kahnt H, Feltz A (1989) Simultaneous estimation of drift mobility and tracer diffusion coefficient in the glass $Na_2O \cdot SiO_2$. Mat Res Bull 24:921-930

Lange RL, Carmichael ISE (1990) Thermodynamic properties of silicate liquids with emphasis on density, thermal expansion and compressiblity. In: Nicholls J, Russell JK (eds) Modern Methods of Igneous Petrology: Understanding Magmatic Processes. Rev Mineral 24:25-64

LaTourette T, Fahey AJ, Wasserburg GJ (1995) The effects of ionic radius and charge on diffusion of Mg, Ca, Ba, Ti, Zr, Nd, and Yb in basaltic melt. Lunar Planetary Sci Conf XXVI, NASA, Houston, Texas, p 829-830

Lasaga AC (1979a) Multicomponent exchange and diffusion in silicates. Geochim Cosmochim Acta 43:455-469

Lasaga AC (1979b) The treatment of multi-component diffusion and ion pairs in diagenetic fluxes. Am J Sci 279:324-346

Lasaga AC (1982) Toward a master equation in crystal growth. Am J Sci 282:1264-1288

Lasaga AC (1995) Kinetic theory and application to Geochemistry. Princeton Univ Press, Princeton, NJ

Lasaga AC, Kirkpatrick RJ (1981) Kinetics of Geochemical Processes. Reviews in Mineralogy, vol 8. Mineral Soc Am, Washington, DC

Lesher CE (1994) Kinetics of Sr and Nd exchange in silicate liquids: Theory, experiments, and applications to uphill diffusion, isotopic equilibration, and irreversible mixing of magmas. J Geophys Res B 99:9585-9605

Lesher CE, Walker D (1991) Thermal diffusion in petrology. In: Ganguly J (ed) Diffusion, Atomic Ordering, and Mass Transport—Selected Topics in Geochemistry. Advances in Physical Geochemistry 8: 396-451, Springer-Verlag, New York

Liang Y, Richter FM, Watson EB (1994) Isothermal multicomponent convection in silicate melts. Nature 369:390-392

Liu S-B, Pines A, Brandriss, M, Stebbins JF (1987) Relaxation mechanisms and effects of motion in albite (NaAlSi$_3$O$_8$) liquid and glass: A high temperature NMR study. Phys Chem Minerals 15:155-162

Loomis TP, Ganguly J, Elphick SC (1985) Experimental determination of cation diffusivities in aluminosilicate garnets II. Multicomponent simulation and tracer diffusion coefficients. Contrib Mineral Petrol 90:45-51

Manning JR (1968) Diffusion Kinetics for Atoms in Crystals. Van Nostrand, Princeton Univ Press, Princeton, NJ

Martin SW, Angell CA (1986) Dc and ac conductivity in wide composition range Li$_2$O-P$_2$O$_5$ glasses. J Non-Cryst Solids 83:185-207

Mazurin OV, Streltsina MV, Shvaiko-Shvaikovskaya TP (1983) Handbook of Glass Data. Part A - Silica glass and binary silicate glasses. Elsevier, Amsterdam

Mazurin OV, Streltsina MV, Shvaiko-Shvaikovskaya TP (1987) Handbook of Glass Data. Part C - Ternary silicate glasses. Elsevier, Amsterdam

McMillan PF, Poe BT, Gillet Ph, Reynard B (1994) A study of SiO$_2$ glass and supercooled liquid to 1950 K via high-temperature Raman spectroscopy. Geochim Cosmochim Acta 58:3653-3664.

Meier M, Frischat GH (1990) Comparison between micro- and macro-g interdiffusion experiments in glass melts. In: Proc 7th Eur Symp on Materials and Fluid Sciences in Microgravity. Oxford, UK, ESA SP-295, p 253-256.

Meier M, Frischat GH (1993) A high temperature centrifuge method to study ion exchange in glass melts avoiding convective processes. Phys Chem Glasses 34:71-76

Meier M, Braetsch V, Frischat GH (1990) Self diffusion in Na$_2$O-Rb$_2$O-SiO$_2$ glass melts as obtained by microgravity experiments. J Am Ceram Soc 73:2122-2123

Miller DG, Vitagliano V, Roberto S (1985) Some comments on multicomponent diffusion: Negative main term diffusion coefficients, second law constrains, solvent choices, and reference frame transformations. J Phys Chem 90:1509-1519

Misener DJ (1974) Cationic diffusion in olivine to 1400°C and 35 kbar. In: Hofmann AW, Giletti BJ, Yoder HS Jr., Yund RA (eds) Geochemical transport and kinetics, Carnegie Inst Washington Publ 634:117-130, Washington, DC

Nanba M, Oishi Y (1983) Determination of the Si diffusing unit in the system Na$_2$O-CaO-SiO$_2$ using a multiatomic ion model. J Am Ceram Soc 66:714-716

Neuville DR, Richet P (1991) Viscosity and mixing in molten (Ca, Mg) pyroxenes and garnets. Geochim Cosmochim Acta 55:1011-1019

Oishi Y, Terai R, Ueda H (1975) Oxygen diffusion in liquid silicates and relation to their viscosity. In: Cooper AR, Heuer AH (eds) Mass Transport Phenomena in Ceramics. Plenum Press, New York, p 297-310

Oishi Y, Nanba M, Pask JA (1982) Analysis of liquid-state interdiffusion in the system CaO-Al$_2$O$_3$-SiO$_2$ using multiatomic ion models. J Am Ceram Soc 65:247-253

Okongwu DA, Lu W-K, Hamielec AE, Kirkaldy JS (1973) Diffusion interactions in glasses arising from discontinuities in anion concentration. J Chem Phys 58:777-787

Onsager L (1931a) Reciprocal relations in irreversible processes. I. Phys Rev 37:405-426

Onsager L (1931b) Reciprocal relations in irreversible processes. II. Phys Rev 38:2265-2279

Onsager L (1945) Theories and problems of liquid diffusion. Annals New York Acad Sci 46:241-265

Onsager L, Fuoss RM (1932) Irreversible processes in electrolytes. Diffusion, conductance, and viscous flow in arbitrary mixtures of strong electrolytes. J Phys Chem 36:2689-2778

Petuskey W, Schmalzried H (1980) Ionic transport in PbO-SiO$_2$-melts (II) Transference number measurements. Ber Bunsenges Phys Chem 84:218-222

Philibert J (1991) Atom movements - diffusion and mass transport in solids. Monographies de physique. Les editions de physique, Les Ulis, France

Poe BT, McMillan PF, Cote B, Massiot D, Coutures JP (1994) Structure and dynamics in Calcium Aluminate liquids: High temperature ^{27}Al NMR and Raman spectroscopy. J Am Ceram Soc 77:1832-1838

Poe BT, Rubie DC, McMillan PF, Diefenbacher J (1994) Oxygen and silicon self diffusion in Na$_2$Si$_4$O$_9$ liquid to 15 GPa. EOS Trans Am Geophys Union 75:713

Prigogine I, Defay R (1954) Chemical thermodynamics. Longman, Green, London and Harlow

Prinzhofer A, Allègre CJ (1985) Residual peridotites and the mechanisms of partial melting. Earth Planet Sci Lett 74:251-265

Qin Z (1992) Disequilibrium partial melting model and its implications for trace element fractionations during mantle melting. Earth Planet Sci Lett 112:75-90

Rálková J (1965) Diffusion of radioisotopes in glass and melted basalt. Glass Techn 6: 40-45

Ravaine D, Souquet JL (1977) A thermodynamic approach to ionic conductivity in oxide glasses. Part 1. Correlation of the ionic conductivity with the chemical potential of alkali oxide in oxide glasses. Phys Chem Glasses 18:27-31

Richet P (1984) Viscosity and configurational entropy of silicate melts. Geochim Cosmochim Acta 48:471-483

Richter FM (1993) A method for determining activity-composition relations using chemical diffusion in silicate melts. Geochim Cosmochim Acta 57:2019-2032

Richter FM, Liang Y, Watson EB, Davis AM (1994) Experimental tests of empirical models for multicomponent diffusion in molten CaO-Al$_2$O$_3$-SiO$_2$ at 1500°C and 10 kb. EOS Trans Am Geophys Union 75:702

Rosing MT, Lesher CE, Bird DK (1989) Chemical modification of east Greenland tertiary magmas by two-liquid interdiffusion. Geology 17:626-629

Rubie DC, Brearley AJ (1990) A model for rates of disequilibrium melting during metamorphism. In: Ashworth JR, Brown M (eds) High Temperature Metamorphism and Anatexis. Unwin-Hyman, London, p 57-86

Rubie DC, Ross II CR, Carroll MR, Elphick SC (1994) Oxygen self-diffusion in Na$_2$Si$_4$O$_9$ liquid up to 10 GPa and estimation of high-pressure melt viscosities. Am Mineral 78:574-582

Rustad JR, Yuen DA, Spera FJ (1990) Molecular dynamics simulation of liquid SiO$_2$ under high pressure. Phys Rev A 42:2081-2089

Ryerson FJ (1987) Diffusion measurements: Experimental methods. In: Sammis CG, Henyey T (eds) Methods Exp Phys 24:89-130

Sato H (1975) Diffusion coronas around quartz xenocrysts in andesite and basalt from tertiary volcanic region in northeastern Shikoku, Japan. Contrib Mineral Petrol 50:49-64

Sauer F, Freise V (1962) Diffusion in binären Gemischen mit Volumenänderung. Zeitschr Elektrochemie 66:353-363

Sawyer EW (1991) Disequilibrium melting and the rate of melt-residuum separation during migmatization of mafic rocks from the Grenville front, Quebec. J Petrol 32:701-738

Scamehorn CA and Angell CA (1991) Viscosity-temperature relations and structure in fully polymerized aluminosilicate melts from ion dynamics simulations. Geochim Cosmochim Acta 55:721-730

Scarfe CM, Brearley M (1987) Mantle xenoliths: Melting and dissolution studies under volatile-free conditions. In: Nixon PH (ed) Mantle Xenoliths. Wiley and Sons, New York, p 599-608

Schmalzried, H (1995) Chemical Kinetics of Solids. VCH Verlag, Weinheim

Schut RJ, Cooper AR (1982) A method for determination of [D] in ternary systems from a single experiment. Acta Metall 30:1957-1959

Semkow KW, Rizzo RA, Haskin LA, Lindstrom DJ (1982) An electrochemical study of Ni^{2+}, Co^{2+}, and Zn^{2+} ions in melts of composition CaMgSi$_2$O$_6$. Geochim Cosmochim Acta 46:1879-1889

Semkow KW, Haskin LA (1985) Concentrations and behavior of oxygen and oxide ion in melts of composition CaO·MgO·xSiO$_2$. Geochim Cosmochim Acta 49:1897-1908.

Shaw HR (1972) Viscosities of magmatic silicate liquids: An empirical method of prediction. Am J Sci 272:870-893

Shewmon PG (1963) Diffusion in solids. McGraw-Hill Book Company, New York

Shimizu N, Kushiro I (1984) Diffusivity of oxygen in jadeite and diopside melts at high pressures. Geochim Cosmochim Acta 48:1295-1303

Shimizu N, Kushiro I (1991) The mobility of Mg, Ca, and Si in diopside-jadeite liquids at high pressures. In: Perchuk LL, Kushiro I (eds) Physical Chemistry of Magmas, Advances in Physical Geochemistry 9:192-212, Springer, New York

Sidebottom DL, Green PF, Brow RK (1995) Anomalous-diffusion model of ionic transport in oxide glasses. Phys Rev B 51:2770-2776

Spallek M, Hertz HG, Funsch M, Herrmann H, Weingärtner, H (1990) Ternary diffusion in the aqueous solutions of MgCl$_2$ + KCl, CdCl$_2$ + KCl, ZnCl$_2$ + KCl and Onsager's reciprocity relations. Ber Bunsenges Phys Chem 94:365-376

Spera FJ, Trial AF (1993) Verification of the Onsager reciprocal relations in a molten silicate solution. Science 259:204-206

Stanton TR, Tyburczy JA, Holloway JR, Petuskey WT (1990) Electro-migration of water in silicate glass: Resolution of the valence charge of the diffusion species. Trans Am Geophys Union EOS 71:652

Stebbins JF (1988) NMR spectroscopy and dynamic processes in mineralogy and geochemistry. In: Hawthorne FC (ed) Spectroscopic Methods in Mineralogy and Geology. Rev Mineral 18:405-430

Stein DJ, Spera FJ (1995) Molecular dynamics simulations of liquids and glasses in the system $NaAlSiO_4$-SiO_2: methodology and melt structures. Am Mineral 80:417-431

Sugawara H, Nagata K, Goto KS (1977) Interdiffusivities matrix of CaO-Al_2O_3-SiO_2 melt at 1723 K to 1823 K. Metall Trans B 8B:605-612

Takada Y, Schmalzried H (1983) Ionic transport in PbO-SiO_2 melts (IV). Zeitschrift Phys Chem 138:61-75

Takahashi K, Miura Y (1980) Electrochemical studies on diffusion and redox behavior of various metal ions in some molten glasses. J Non-Cryst Solids 38/39:527-532

Tarento RJ (1989) Influence of evaporation and exchange reactions at the surface on the evolution of an arbitrary tracer distribution by diffusion. Revue Phys Appl 24:11-16

Thompson MS, Morral JE (1986) The effect of composition on interdiffusion in ternary alloys. Acta Metall 34:339-346

Toor HL (1964) Solution of the linearized equations of multicomponent mass transfer: II. Matrix methods. J Am Inst Chem Eng 10:460-465

Toramaru A (1995) Numerical study of nucleation and growth of bubbles in viscous magmas. J Geophys Res 100:1913-1931

Trial AF, Spera FJ (1988) Natural convection boundary layer flows in isothermal ternary systems: Role of diffusive coupling. Int J Heat Mass Transfer 31:941-955

Trial AF, Spera FJ (1994) Measuring the multicomponent diffusion matrix: Experimental design and data analysis for silicate melts. Geochim Cosmochim Acta 58:3769-3783

Tsekov R, Ruckenstein E (1994) Brownian dynamics in amorphous solids. J Chem Phys 101:7844-7849

van der Laan S, Zhang Y, Kennedy AK, Wyllie PJ (1994) Comparison of element and isotope diffusion of K and Ca in multicomponent silicate melts. Earth Planet Sci Lett 123:155-166

Vitagliano PL, Ambrosone L, Vitagliano V (1992) Gravitational instabilities in multicomponent free-diffusion boundaries. J Phys Chem 96:1431-1437

Wagner C (1969) The evaluation of data obtained with diffusion couples of binary single-phase and multiphase systems. Acta Metall 17:99-107

Wasserburg GJ (1988) Diffusion of water in silicate melts. J Geol 96:363-367

Watson EB (1976) Two-liquid partition coefficients: Experimental data and geochemical implications. Contrib Mineral Petrol 56:119-134

Watson EB (1979): Calcium diffusion in a simple silicate melt to 30 kbar. Geochim Cosmochim Acta 43:313-322

Watson EB (1982) Basalt contamination by continental crust: Some experiments and models. Contrib Mineral Petrol 80:73-87

Watson EB (1994) Diffusion in volatile-bearing magmas. Rev Mineral 30:371-411

Watson EB, Baker DR (1991) Chemical diffusion in magmas: An overview of experimental results and geochemical applications. In: Perchuk LL, Kushiro I (eds) Physical chemistry of magmas, Advances in Physical Geochemistry 9:120-151, Springer Verlag, New York

Wendlandt RF (1991) Oxygen diffusion in basalt and andesite melts: Experimental results and discussion of chemical versus tracer diffusion. Contrib Mineral Petrol 108:463-471

Weinberg W (1987) Kationen-Interdiffusion im Glastransformationsbereich mineralischer Gläser unter hohen Drücken im System Jadeit-Albit-Anorthit. PhD dissertation, Georg-August-Universität, Göttingen, Germany

Winchell P (1969) The compensation law for diffusion in silicates. High Temp Science 1:200-215

Winchell P, Norman JH (1969) A study of the diffusion of radioactive nuclides in molten silicates at high temperature. In: Third Int'l Symp High-Temp Techn Proc, Asilomar, California. Butterworth, London, p 479-492

Woodcock LV, Angell CA, Cheeseman P (1976) Molecular dynamics studies of the vitreous state: Simple ionic systems and silica. J Chem Phys 65:1565-1577

Xue X, Stebbins JF, Kanzaki M, McMillan PF, Poe B (1991) Pressure-induced silicon coordination and tetrahedral structural changes in alkali oxide-silica melts up to 12 GPa: NMR, Raman, and infrared spectroscopy. Am Mineral 76:8-26

Yinnon H, Cooper AR Jr (1980) Oxygen diffusion in multicomponent glass forming silicates. Phys Chem Glasses 21:204-211

Young TF, Kieffer J, Borchardt G (1994a) Tracer diffusion of oxygen in CoO-SiO_2 melts. J Phys: Condens Matter 6:9825-9834

Young TF, Kieffer J, Borchardt G (1994b) Application of a kinetic model to tracer diffusion of silicon and oxygen in silicate melts. J Phys Condens Matter 6:9835-9852

Zallen R (1983) The physics of amorphous solids. John Wiley & Sons, New York

Zhang Y (1993) A modified effective binary diffusion model. J Geophys Res B 98:11901-11920

Zhang Y (1994) Reaction kinetics, geospeedometry, and relaxation theory. Earth Planet Sci Lett 122:373-391

Zhang Y, Walker D, Lesher CE (1989) Diffusive crystal dissolution. Contrib Mineral Petrol 102:492-513

Zhang Y, Stolper EM, Wasserburg GJ (1991a) Diffusion of water in rhyolitic glasses. Geochim Cosmochim Acta 55:441-456

Zhang Y, Stolper EM, Wasserburg GJ (1991b) Diffusion of a multi-species component and its role in oxygen and water transport in silicates. Earth Planet Sci Lett 103:228-240

Chapter 11

PRESSURE EFFECTS ON SILICATE MELT STRUCTURE AND PROPERTIES

George H. Wolf and Paul F. McMillan

Department of Chemistry and Biochemistry
Arizona State University
Tempe, Arizona 85287 U.S.A.

INTRODUCTION

The high-pressure properties of silicate melts are likely to have played the single most important role in shaping the physical and chemical evolution of the earth and other terrestrial planets. Magma ascent and emplacement, as well as crystal nucleation, growth and segregation, are all principally controlled by the pressure-dependent densities and transport properties of silicate melts. In the pressure range of the Earth's mantle, silicate melts display high compressibilities relative to their crystalline counterparts, as a result of the great diversity in structural compression mechanisms that occur in these molten systems. Magmatic buoyancy forces, which are determined by the density contrast between the melt and surrounding rock matrix, can strongly decrease as a function of depth and can even invert in the deep interior of a planet (Stolper et al., 1981). Thus, gravitationally stable magma oceans can exist at depth within a planet, profoundly influencing its thermal and chemical evolution (Nisbet and Walker, 1982; Ohtani, 1984; Rigden et al., 1984; Ohtani et al., 1986; Takahashi, 1986; Ohtani and Sawamoto, 1987; Ryan, 1987; Agee and Walker, 1988, 1993; Miller et al., 1991a,b). Pressure can also have an equally large, and sometimes surprising (see below), effect on the viscosities and other transport properties of magmatic liquids. In turn, these effects can significantly influence the segregation dynamics and crystal fractionation pathways in magmas.

There is currently much interest in understanding the microscopic atomistic processes that control the bulk physical properties of magmatic liquids. Experimental and theoretical investigations of aluminosilicate and related liquids and glasses over the last few decades, together with a growing body of data from fundamental studies on the viscoelastic behavior and structural properties of network forming liquids, have led to the emergence of a more complete understanding of the interrelations between the thermodynamic and transport properties of magmatic liquids and their microscopic structure and dynamical processes. This chapter will focus on the pressure variable, both as a critical thermodynamic variable that affects the physical properties of silicate melts and hence magmatic processes within the earth, and as a powerful experimental and theoretical variable that has yielded new fundamental insights into the microscopic origin of silicate melt properties.

It is well known that the effects of volatiles on melt properties is dramatic. Moreover, pressure has a profound influence on the physical properties of hydrous and other volatile-bearing magmas, as underscored by the strong increase in solubility of CO_2 and H_2O that occurs in silicate melts with increasing pressure (see Burnham, 1979; Spera and Bergman, 1980; Stolper and Holloway, 1988; Lange and Carmichael, 1990).

However, the discussion in this chapter is limited to *anhydrous* melts and glasses. A more complete discussion on the properties of volatile-bearing silicate melts was the topic of a recent Mineralogical Society of America short course (Carroll and Holloway, 1994).

GENERAL HIGHLIGHTS

One of the most intriguing aspects of aluminosilicate liquids is the rich diversity in phenomenology which they display. Lower-silica, highly depolymerized aluminosilicate melts behave like normal ionic liquids in their response to pressure. These liquids undergo a gradual, continuous densification with increasing pressure, and display a gradual decrease in their fluidity and atomic diffusivities. However, higher-silica liquids, with compositions near those of crystalline tektosilicates, display more unusual behavior in their physical properties. Most significantly, the viscosity of many of these liquids *decreases* with increasing pressure (Fig. 1) (Kushiro, 1976, 1977, 1978a,b, 1980, 1986; Kushiro et al., 1976; Fujii and Kushiro, 1977; Scarfe et al., 1979, 1987). This is contrary to the behavior expected from free-volume theory, and that observed for most "normal" liquids, where applied pressure leads to an increase in viscosity (see chapters by Richet and Dingwell in this volume). A similar anomaly in the pressure effect on viscosity occurs in water at low temperatures. The microscopic structural and energetic factors which are responsible for the anomalous transport properties in water may also be connected to its well known density maximum at low temperatures. This unusual behavior has been rationalized in terms of the free energy competition between the high energetic stability and low configurational entropy of an open ice-like tetrahedral framework structure in water. Similar explanations have been presented for silica, which itself displays a density maximum in its supercooled liquid state (Bruckner, 1970).

High-silica liquids also display anomalous pressure-dependent behavior in their atomic diffusivities. Experiments reveal that, while the diffusivities of network-modifying ions generally decrease with increasing pressure for aluminosilicate liquids, the diffusivities of the network ions (i.e., Si, Ge, Ga, Al, O) actually *increase* with pressure in many highly polymerized liquids near tektosilicate compositions (see Fig. 2a) (Watson, 1979; Fujii, 1981; Kushiro, 1983; Shimizu and Kushiro, 1984; Dunn and Scarfe, 1986; Rubie et al., 1993; Poe et al., 1994, 1995). These observations were first predicted, and are generally supported, by theoretical molecular dynamics simulations (see Fig. 2b) (Woodcock et al., 1976; Angell et al., 1982, 1987; Kubicki and Lasaga, 1988, 1990, 1991).

The general inverse relation between oxygen diffusion and viscosity suggests that oxygen exchange is the rate limiting step in the viscous flow mechanism of aluminosilicate melts (Scarfe et al., 1987). Results from molecular dynamics simulations on aluminosilicate, and related, liquids have been used to propose that five-coordinated silicon ([5]Si) and aluminum ([5]Al) species are important reactive intermediate states in transport mechanisms and act to facilitate ion exchange (Brawer, 1981; Angell et al., 1982, 1983; Kubicki and Lasaga, 1988, 1991). In the simulations, there is a strong correlation between the network ion diffusivities and the abundance of five-coordinatedd Si and Al species in the melt (Angell et al., 1982, 1983; Kubicki and Lasaga, 1988). The existence of significant abundances of [5]Si (Stebbins and McMillan, 1989; Xue et al., 1991) and [5]Al (Yarger et al., 1995) species in partially depolymerized silicate and aluminosilicate melts has now been confirmed experimentally from MAS NMR studies of glasses quenched from high-pressure.

Although moderately high abundances of high-coordinated Si and Al species can occur in depolymerized composition alkali silicate and aluminosilicate glasses quenched

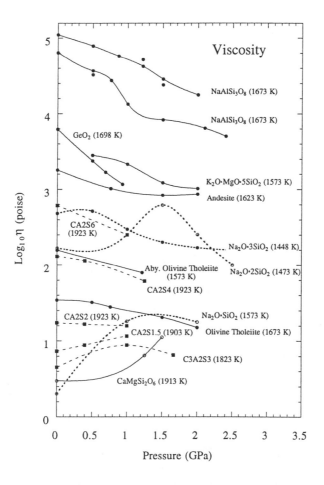

Figure. 1. Viscosity of silicate and aluminosilicate melts as a function of pressure. Temperatures of the measurements are given in parentheses. Data for the calcium aluminosilicate system along the charge-balanced CaAl₂O₄-SiO₂ join: CA2S6, CA2S4, Anorthite (CA2S2), and CA2S1.5 taken from Kushiro (1981) and for the depolymerized Ca₃Al₂Si₃O₁₂ (Gross) composition melt from Mysen et al. (1983). Viscosity data for sodium disilicate (Na₂O·2SiO₂), sodium metasilicate (Na₂O·SiO₂), and diopside (CaMgSi₂O₆) composition melts taken from Scarfe et al. (1979). Data for GeO₂, K₂O·MgO·5SiO₂, Na₂O·3SiO₂, NaAlSi₃O₈, and NaAlSi₂O₆ taken from Kushiro (1976, 1977; 1978a,b). Viscosity data for natural systems from Kushiro et al. (1976) and Fujii and Kushiro (1977).

from high-pressure melts, the experimental evidence for the retention of high-coordinated species in fully polymerized tektosilicate composition glasses quenched from high-pressure melts is much less definitive. However, there is compelling in situ spectroscopic evidence that pressure-induced coordination changes do occur in the fully polymerized silicate and aluminosilicate glasses and melts. High-pressure infrared absorption studies on anorthite composition glass (Williams and Jeanloz, 1988) and spectroscopic studies of silica glass (Hemley et al., 1986) and germania glass (Itie et al., 1989; Durben and Wolf, 1991), a structural analog to silica, indicate that network forming cations of fully polymerized glasses form high-coordinated species at elevated pressures. However, for these fully polymerized systems the high-coordinated species revert to tetrahedral states on decompression. Perhaps the most definitive documentation of this behavior is the in situ high-pressure X-ray absorption edge study of germania glass by Itie et al (Itie et al., 1989, 1990). The data indicate that germanium undergoes a four- to six-fold coordination change between 4 and 12 GPa, and, on decompression the high-coordinated species revert entirely to tetrahedral species over a very narrow pressure range below 4 GPa.

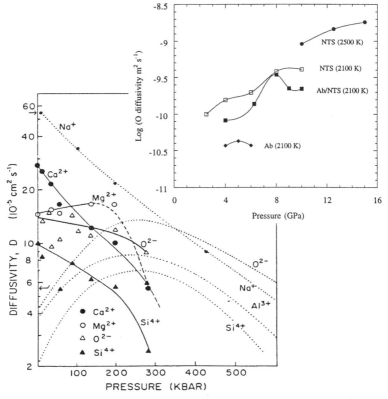

Figure 2. (a) Experimental values for the pressure dependence of oxygen diffusivity in sodium tetrasilicate (NTS) and albite (Ab) system melts as a function of pressure. Data for NTS at 2100 K taken from Rubie et al. (1993). Data for NTS at 2500 K, for Ab at 2100 K and for an equimolar Ab/NTS mixture at 2100 K were taken from Poe et al. (1995). (b) Molecular dynamics simulation results for the pressure dependence of ion diffusivities in diopside (CaMgSiO$_4$, solid lines) and jadeite (NaAlSi$_2$O$_6$, dotted lines) at 6000 K taken from Angell et al. (1987). [Used by permission of the editor of *Chemical Geology*, from Angell et al. (1987), Fig. 4, p. 89.]

In silica glass, analyses of the pressure-dependent Raman (Hemley et al., 1986; Sugiura and Yamadaya, 1992), infrared (Williams and Jeanloz, 1988; Williams et al., 1993) and X-ray diffraction (Meade et al., 1992) spectra suggest that the pressure-induced increase in the silicon coordination takes place above about 15 GPa at room temperature. However, in partially depolymerized alkali silicate glasses, an increase in the silicon coordination takes place at much lower pressures (Stebbins and McMillan, 1989; Xue et al., 1989; Wolf et al., 1990; Xue et al., 1991; Durben, 1993). In these systems, high-coordinated silicon species can be retained in samples quenched from high-pressure melts.

On the basis of the above studies, it has been suggested that there exist two distinct mechanisms for coordination changes in silicate liquids, involving either the bridging or non-bridging oxygen atoms (Wolf et al., 1990; Xue et al., 1991). In the slightly depolymerized, high-silica systems, these mechanisms have very different activation energies and operate over separate pressure ranges. At low pressures, tetrahedral silicon (or aluminum) can be readily attacked by the non-bridging oxygen atoms to form high-coordinated [5]Si and [6]Si species. It is only at much higher pressures that the network bridging oxygens become involved in the formation of high-coordinated species. The differing energetics and reversibility of these two mechanisms can be rationalized by an inspection of the cation arrangement around the oxygen anion. The higher pressure mechanism, involving bridging oxygens, requires the energetically costly formation of oxygen anions bonded to three silicon atoms. However, the lower pressure mechanism involves non-bridging oxygens and requires formation of oxygen anions bonded to only two silicon atoms. The relative energetics of these two mechanisms appears to be strongly affected by the degree of depolymerization of the network structure (Xue et al., 1991; Durben, 1993). With increasing alkali or akaline earth oxide component, the pressure range for the coordination reaction involving non-bridging oxygen atoms is shifted to higher pressures, apparently due to the increasing steric restrictions imposed by the basic metal cations.

The in situ spectroscopic studies unequivocally demonstrate that significant local structural relaxations can occur in the glassy state on decompression from high pressure. It remains unclear as to whether these structural relaxations are equally significant in glasses that are formed from high-pressure melts. The room-pressure structural properties of glasses that have been pressure-cycled at 300 K can be quite different from that of samples which, in addition to pressure cycling, were also annealed or melted at high pressures (Mackenzie, 1963a,b; McMillan et al., 1984; Sugiura and Yamadaya, 1992; Yarger et al., 1995). For this reason, there is a growing emphasis on the development and application of in situ methods to investigate the structural properties of aluminosilicate liquids as a function of pressure and temperature. Although experimental investigations on the high-pressure structure of aluminosilicate melts have to date been limited either to ambient pressure studies on glasses quenched from high-pressure melts or to in situ high-pressure studies of glasses at room temperature, a few pioneering vibrational spectroscopic studies on the combined pressure and temperature effects have been made on alakli germanate (Farber and Williams, 1992) and alkali silicate systems (see chapter by McMillan and Wolf in this volume).

Although in situ spectrocopic investigations of the combined high-pressure and high-temperature effects on silicate melt structure are just beginning, measurements on the physical properties of silicate melts at high pressure have been made for some time and continue to be extended to higher pressures. A number of viscosity (Kushiro, 1976, 1977, 1978a,b, 1980, 1986; Kushiro et al., 1976; Fujii and Kushiro, 1977; Scarfe et al., 1979, 1987) and diffusivity (Watson, 1979; Fujii, 1981; Kushiro, 1983; Shimizu and Kushiro, 1984; Dunn and Scarfe, 1986) measurements have now been made on natural and synthetic composition aluminosilicate melts to pressures up to 2.5 GPa. Most recently, diffusion measurements on aluminosilicates melts have been extended to 15 GPa (Rubie et al., 1993; Poe et al., 1994 , 1995). The electrical conductivity of basaltic and andesitic melts has been examined to 2.5 GPa (Tyburczy and Waff, 1983; Tyburczy and Waff, 1985). Shock wave experiments have provided important fundamental information on the pressure-density equations of state of model aluminosilicate and magmatic liquids to pressures in excess of 30 GPa (Rigden et al., 1984, 1988, 1989; Schmitt and Ahrens, 1989; Miller et al., 1991a,b). Crystal flotation experiments have provided further constraints on the densities of silicate melts at high pressures and have been used to directly test petrogenesis models of crystal segregation in mantle magmas (Agee and Walker, 1988, 1993; Suzuki et al., 1995).

PHYSICAL PROPERTIES

Density is perhaps the most fundamental physical property of a silicate melt or glass. In addition to the critical role that melt densities play in igneous petrogenesis and planetary evolution, the compositional and pressure dependences of silicate melt and glass densities can yield insights into their microscopic structure and deformational mechanisms (Bottinga and Weill, 1970). Equations for estimating densities of multicomponent magmas from acoustic velocity data (Manghnani et al., 1986; Rivers and Carmichael, 1987; Kress et al., 1988) at ambient pressure have recently been summarized by Lange and Carmichael (1990). Since these equations are based on ambient compressibilities, estimates of the high-pressure densities of silicate melts are valid only if there is no significant change in the compression mechanisms at high pressures.

Some of the first investigations of the pressure-dependent structural properties of glasses were directed toward understanding Birch and Dow's (1936) observation that silica glass can undergo a permanent densification under some conditions of pressure cycling. Subsequent studies established that this densification occurs for a number of highly polymerized tetrahedral network glasses and that the magnitude of this effect depends on the precise history of sample stress conditions and temperature (Bridgman and Simon, 1953; Boyd and England, 1963; Mackenzie, 1963a,b; Cohen and Roy, 1965; Uhlmann, 1973; Primak, 1975; Kushiro, 1976; Arndt, 1983; McMillan et al., 1984; Grimsditch, 1986; Hemley et al., 1986; Walrafen and Hokmabadi, 1986; Meade and Jeanloz, 1987; Grimsditch et al., 1988).

An understanding of the mechanisms responsible for the permanent densification and the volume relaxation of densified silicate glasses, and their possible relationship to mechanisms of viscous flow in the liquid near the glass transition temperature, has been of longstanding interest. Early studies identified two or more distinct structural mechanisms for the densification of silica glass (Mackenzie, 1963a; Kimmel and Uhlmann, 1969). Under non-hydrostatic compression, Mackenzie proposed that silica can suffer permanent densification without bond breaking via a polymer entanglement mechanism with a low activation energy (<30 kJ/mol). The second mechanism, involving bond breaking and rearrangement, occurs with much higher activation energies (>300 kJ/mol) and can even occur under completely hydrostatic conditions. Studies by Hsich et al. (1971) suggested that volume relaxation in densified silica glass is characterized by a very broad range of relaxation times, all with approximately the same activation enthalpy (~300 kJ/mol). Although, this value approaches that of the activation energy for viscous flow in silica liquid (400 to 600 kJ/mol), Hsich concluded that the molecular processes associated with viscous flow and volume relaxation were different since the width of the relaxation time distribution for volume relaxation in the glass is enormously broader than that of viscous flow.

Some of the earliest attempts at measurement of the pressure-density equations-of-state of anhydrous aluminosilicate melts were based on the falling-sphere method of Fujii and Kushiro (1977; Scarfe et al., 1979). These early experiments, carried out in a piston-cylinder apparatus, were limited to pressures below 2.5 GPa and were subject to large uncertainties. Using a finite strain expression for the equation of state, and employing the available room-pressure ultrasonic compressibility data on silicate liquids, Stolper et al. (1981) estimated the densities of basaltic liquids to high pressures. Despite the uncertainties in these extrapolations (i.e., without knowledge of the pressure derivatives of compressibilities) they came to the conclusion that because of the high compressibility of liquid silicates relative to that of mantle minerals, ultrabasic liquids may actually be denser than the principal residual crystals in mantle source regions at high pressure. Consequently, magmatic liquids may become neutrally buoyant deep within the terrestrial planets, hence limiting the depths at which magmas will segregate from their source and the degrees of partial melting that can be achieved in these source regions before melt segregation occurs (Stolper et al., 1981).

Using a different approach based on a thermodynamic analysis of melting curve data, Ohtani et al. (1983, 1984) were also able to estimate the equations-of-state of basic to ultrabasic magmatic liquids throughout the pressure range of the upper mantle. From these estimates, they concluded that picritic silicate liquids generated by partial melting of the upper mantle become denser than olivine and pyroxenes at pressures higher than 7 GPa. The density estimates further support the idea that in the early Archaean, when extensive partial melting of Earth's upper mantle was likely, melt segregation could lead to the development of chemically stratified upper mantle composed of an upper residual layer rich in olivine underlain by a garnet-rich layer.

Shock wave experiments have confirmed that the compressibilities of many basaltic composition melts remain high, relative to their crystalline counterparts, throughout the pressure range of the upper mantle (Rigden et al., 1984, 1988, 1989; Boslough et al., 1986; Schmitt and Ahrens, 1989; Miller et al., 1991a,b). Moreover, the shock experiments, and the high-pressure olivine flotation experiments of Agee and Walker (1988; 1993) (see Fig. 3a), generally support earlier speculations on melt segregation in the upper mantle and the possibility for gravitationally stable deep magma source regions. These ideas have been further refined by Miller et al. (1991a,b) from shock experiments on a molten komatiite. The komatiite melt data indicate that olivine and clinopyroxene become neutrally buoyant in a komatiite melt near 8 GPa while garnet-majorite becomes buoyant in ultrabasic melts in the 20 to 24 GPa interval (Fig. 3b). From an extrapolation of their shock data to higher pressures, Miller et al. (1991a) reach the same conclusion made earlier by Ohtani (1983), that liquidus perovskite crystals may become buoyant in lower mantle ultrabasic liquids below 70 GPa. Consequently, a downward migration of a partial melt, with incompatible elements, could occur in the lower mantle at depths below ~1000 km.

The high compressibilities of aluminosilicate melts relative to those of their low pressure crystalline phases, indirectly support earlier speculations (Waff, 1975) and computer simulation results (Matsui and Kawamura, 1980; Angell et al., 1982, 1983, 1987; Matsui et al., 1982; Matsui and Kawamura, 1984) that the framework cations in aluminosilicate melts undergo pressure-induced coordination changes with increasing pressure. Waff (1975) first suggested, by analogy with the high-pressure behavior of crystalline aluminosilicates, that the tetrahedral network cations in aluminosilicate melts will also undergo transformations to six-coordinated species at high pressures. He further speculated that these coordination reactions would take place over a narrow pressure range and thus produce dramatic changes in the densities and viscosities of aluminosilicate melts at mantle conditions. However, shock wave estimates of the equations of state of aluminosilicate liquids reveal more gradual changes in density with increasing pressure, a result more in line with the computer simulations.

There is some indirect evidence that the compressibilities of several fully polymerized (tektosilicate composition) aluminosilicate liquids exhibit anomalous behavior at very low pressures, which is not likely to be resolvable in shock experiments. By employing a thermodynamic analysis of high-pressure melting data, Bottinga et al.

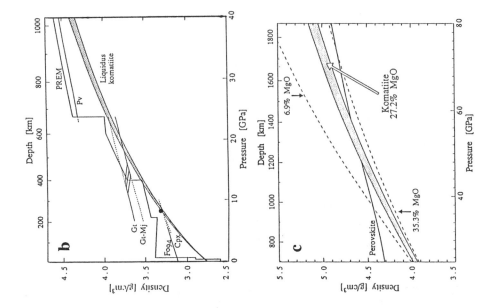

Figure 3 (opposite page). (a) Photomicrograph documenting the flotation of olivine spherules in a fayalite-enriched komatiite liquid (65% komatiite/35% fayalite) at 1.4 GPa and 1525°C. Denser Fe-rich equilibrium olivine crystals have segregated to the bottom of the capsule (taken from Agee and Walker, 1988). These and subsequent flotation results (Agee and Walker, 1993) indicate that equilibrium olivine crystals become buoyant at about 8 GPa in komatiite liquids along their liquidi. (b) The pressure-density relationships of komatiite liquid (27.2% MgO) and its liquidus phases (Fo₉₄, olivine; Cpx, clinopyroxene; Gt, garnet; Gt-Mj, garnet-majorite; Pv, perovskite) along the high-pressure liquidus. Estimate for the bulk mantle (PREM model) is also shown. The data suggests that olivine would be neutrally buoyant near 8.2 GPa (252 km). Figure taken from Miller et al. (1991a). (c) Extrapolation of the pressure-density relationships to lower mantle conditions. Shaded curve is the liquidus for komatiite liquid with 27.2% MgO and dotted lines represent komatiite liquids with 6.9% and 35.3% MgO. These results suggest that komatiite melts ranging from basic to ultrabasic compositions would become denser than perovskite at depths greater than 1000 to 1700 km, respectively. Figure taken from Miller et al. (1991a).

(1985) found that the compressibilities of jadeite and pyrope composition melts show a divergent increase in their compressibilities on decreasing pressure below 5 GPa (Fig. 4). (Compressibility estimates could not be derived at pressures below the termini of their respective congruent melting lines at 2.9 and 3.7 GPa.) Since the magnitude and correlation length of density fluctuations must also diverge as the compressibility diverges, this suggests that there may exist an underlying instability in the network structures of both of these liquids approaching ambient pressure on decompression.

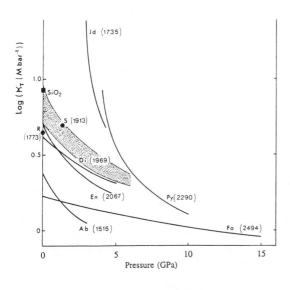

Figure 4. Logarithmic isothermal compressibilities of liquid jadeite (Jd), pyrope (Py), albite (Ab), forsterite (Fo), enstatite (En), and diopside (Di) derived by Bottinga et al. (1985) from a thermodynamic analysis of high-pressure melting data. Also plotted are data for liquid diopside, (S) and (R), from Scarfe et al. (1979) and Rivers and Carmichael (1987), respectively. Temperatures, in Kelvins, are given in parenthesis. The shaded area represents the range of compressibility values of liquid basalt proposed by Stolper et al. (1981). Data and Figure taken from Bottinga et al. (1985). [Used by permission of the editor of *Earth and Planetary Science Letters* from Bottinga (1985), Fig. 2, p. 354.]

Although the physical origin of this instability is not yet known, Bottinga et al. (1985) speculated that an aluminum coordination change could cause the anomalous compressibility and thermodynamic behavior. At first glance this argument appears compelling, since the anomalous behavior of the liquids occurs at about the same pressure as the aluminum coordination change in the stable crystalline phases of these same compositions. However, there is not yet any definitive structural data which supports this assertion. Furthermore, the pressure range of the aluminum coordination change is completely different to that which has been obtained from computer simulations (Angell et al., 1987). Nearly all ion dynamic simulations find that the coordination changes of the framework cations in aluminosilicate liquids take place over a large pressure range. In fact, the pressure ranges are sufficiently large that the volume changes accompanying the coordination change are effectively smeared out. Angell et al. (1987) proposed an alternative origin for the anomalous low pressure compressibility behavior in jadeite melts. They suggested that the diverging compressibility resulted from an impending mechanical instability or "spinodal" which occurs in the tetrahedral network at slightly negative (i.e., tensile) pressures.

In the remaining sections, we focus more specifically on spectroscopic and structural investigations of the pressure-induced deformational mechanisms in simplified aluminosilicate and analog liquid and glass systems. The discussion is separated into classes of materials starting from the most polymerized systems (i.e., silica and germania) and progressing through to the most depolymerized systems (orthosilicates).

STRUCTURAL PROPERTIES

Fully polymerized systems

Silica. Silica is an important compositional component of essentially all magmatic liquids, and represents the reference structural archetype in the description of aluminosilicate glass and melt structures. At room pressure silica liquid and glass exist in a fully polymerized three-dimensional framework of corner-linked silica tetrahedra. The short range structure of the glass is characterized by highly regular SiO_4 tetrahedra with an average Si-O bond length of about 1.61 Å. At ambient pressure, the spread in the distribution of Si-O-Si inter-tetrahedral angles is large (from 120° to 180°) and the average value is near 144° (Mozzi and Warren, 1969; Dupree and Pettifer, 1984). In spite of the prominence of SiO_2 as a component in both technological glasses and magmatic liquids, fundamental details in the description of the intermediate and long range structure of silica glass remain controversial. Models ranging from those based on a continuous random network (Zachariasen, 1932; Warren 1933) to those based on microcrystallite building units (Lebedev, 1921; Valenkov and Porai-Koshits, 1936) have been proposed (see 1988 review by Galeener and chapter by Brown et al. in this volume).

It is well recognized that the extreme flexibility in the inter-tetrahedral angle of Si-O-Si can accommodate an enormous diversity of structures in response to chemical, temperature and pressure changes (Tossell and Gibbs, 1978; Navrotsky et al., 1985; Hemley et al., 1994). For example, the molar volumes of stable and metastable silica phases that are based on a tetrahedral framework structure range from 20.64 and 22.69 cm^3/mol in coesite and quartz to greater than 47 cm^3/mol in some zeolites. Over this enormous range in density the formation enthalpy varies by only 13.6 kJ/mol (Navrotsky, 1994). Room-temperature investigations of the high-pressure metastable behavior of quartz and other silica polymorphs (Hemley et al., 1988; Hazen et al., 1989; Halvorson and Wolf, 1990; Kingma et al., 1993a,b) suggest that the compressional deformation limit for the tetrahedral silica network is very near 17 cm^3/mol, roughly corresponding to the volume where the oxygen anions are in a cubic body-centered packing (Sowa, 1988; Hazen et al., 1989; Chelikowsky et al., 1990) and where the Si-O-Si intertetrahedral angle is close to 120° and the Si atoms are in non-bonded contact. Approaching this density, the SiO_4 tetrahedra become severely distorted (Hazen et al., 1989). Compression beyond this limit leads to bond rearrangement with a disruption of the tetrahedral framework and an increase in the silicon coordination (Verhelst-Voorhees et al., 1994). A mechanical instability of the tetrahedral network has been proposed to underlie the pressure-induced crystal-to-amorphous transition observed in many tetrahedral framework silicate and related phases (Hemley et al., 1988; Tse and Klug, 1991; Binggeli and Chelikowsky, 1992; Wolf et al., 1992; Chaplot and Sikka, 1993; Binggeli et al.,

1994). Chelikowsky et al. (1990) have suggested, on the basis of theoretical investigations, that this instability is driven by strong interpolyhedral oxygen-oxygen repulsions which occur in the superpressed tetrahedral framework at high pressure.

Studies on pressure compacted silica glass have found that residual irreversible densifications of up to 20% can be realized (Bridgman and Simon, 1953; Mackenzie, 1963a,b; Walrafen and Hokmabadi, 1986; Devine et al., 1987; Susman et al., 1990, 1991; Xue et al., 1991). The highest degree of compaction was obtained in glass samples pressure cycled up to 20 GPa and heated above 1600 K. In none of these samples was there evidence for the retention of high-coordinated silicon species. Diffraction and spectroscopic studies on compacted silica samples indicate that the major structural change is the shift in distribution of Si-O-Si intertetrahedral angles to smaller values with a decrease in the overall width of this distribution (Couty and Sabatier, 1978; Grimsditch, 1984; McMillan et al., 1984; Hemley et al., 1986; Devine and Arndt, 1987; Devine et al., 1987; Susman et al., 1990, 1991; Xue et al., 1991; Davoli et al., 1992). Furthermore, high-temperature annealing at high pressures tends to increase this effect. A slight increase in the average Si-O bond length can also be inferred from X-ray diffraction studies of compacted samples (Devine and Arndt, 1987).

In situ Raman (Hemley et al., 1986; Suguira and Yamadaya, 1992), infrared (Williams and Jeanloz, 1988; Williams et al., 1993), ultrasonic (Kondo et al., 1981; Sasakura et al., 1989), Brillouin (Zha et al., 1994; Rau, 1995) and X-ray diffraction (Meade et al., 1992) studies of silica glass have provided important information on the basic structural response of the SiO_2 network to changes in pressure. These studies have served to identify several primary deformation modes in silica glass that dominate over different but overlapping pressure regimes. Up to about 8 GPa, most of the compression of the tetrahedral network is taken up by a concerted rotational motion of the highly regular SiO_4 tetrahedra resulting in a gradual and reversible decrease in the Si-O-Si intertetrahedral angle. Above 8 GPa, increasing distortions of the SiO_4 tetrahedra become an important deformation mechanism. At higher pressures, possibly starting at pressures as low as 14 to 15 GPa, there is a gradual increase in the average silicon coordination number. Silicon is likely distributed over a range of four-, five-, and six-coordinated sites above 15 GPa with the most probable species progressing from four to five and to six with increasing pressure. Most of the four-coordinated silicon species appear to be consumed by about 30 GPa and have reacted to form either five- or six-coordinated species. However, some small concentration of four-coordinated species likely persists in the glass to 40 GPa. The completion of the coordination transformation to octahedral silicon likely occurs at much higher pressures, perhaps even above 50 GPa.

In situ infrared absorption studies (Williams and Jeanloz, 1988; Williams et al., 1993) provide compelling evidence for a pressure-induced coordination change in silica glass above 20 GPa (see Fig. 5b). A similar coordination change is also inferred to occur in metastable crystalline quartz and coesite under compression to pressures well above their thermodynamic stability fields (Williams et al., 1993; Verhelst-Voorhees et al., 1994). The primary spectroscopic signature for the loss of SiO_4 species is the reduction in intensity of the infrared tetrahedral stretching peak near 1100 cm^{-1} relative to absorption in the 600 to 900 cm^{-1} region. This progression is observed for silica glass at high pressures and is most evident in the spectra above 17 GPa (Williams and Jeanloz, 1988; Williams et al., 1993). At 39 GPa only a weak relatively diffuse absorption band is observed in the spectral region near 1100 cm^{-1} for silica glass. A similar loss of a distinct tetrahedral stretching band is also observed for metastable quartz at 39 GPa and for coesite above 47 GPa (see Fig. 6). The infrared absorption spectrum of silica glass at 39

GPa is qualitatively similar to that of stishovite at the same pressure (Williams et al., 1993).

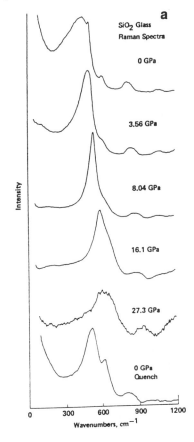

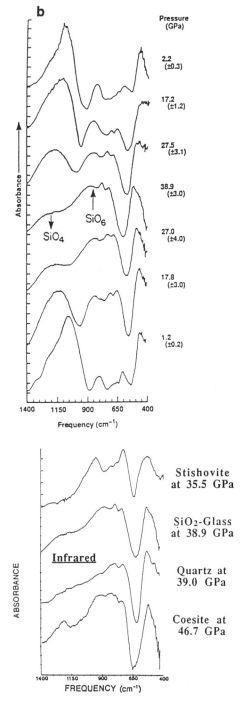

Figure 5 (above). Pressure dependence of the vibrational spectrum of silica glass. (a) Raman spectrum on increasing pressure and on recovered, pressure-cycled sample (Hemley et al., 1986). (b) Infrared absorbance spectrum on increasing and decreasing pressure (Williams and Jeanloz, 1988). Figure taken from spectra redrawn by Hemley et al. (1994).

Figure 6 (right). Comparison of the high-pressure infra-red absorbance spectra of silica glass and several crystalline silica polymorphs (Williams et al., 1993).

Further details of the compressional behavior in silica glass can be obtained from the pressure-dependent changes in the Raman spectrum. At ambient pressure, the Raman spectrum of silica glass in the low frequency region is characterized by a strongly polarized, diffuse band centered near 430 cm[-1] with two relatively sharp bands on the high frequency shoulder at 492 and 606 cm[-1] (see Fig. 5a) The diffuse band is typically associated with a symmetric bending motion of the Si-O-Si linkages largely involving the motion of oxygen atoms in the plane bisecting the Si...Si line. The frequency of this mode is very sensitive to the value of the intertetrahedral angle and increases with decreasing SiOSi angles. The broad, diffuse nature of the scattering in this region is consistent with the broad distribution of intertetrahedral angles and high intermediate range disorder in silica glass. The sharp polarized bands at 492 and 606 cm[-1] have been denoted as "defect" bands and are thought to represent symmetric oxygen breathing motions from, respectively, four- and three-membered siloxane rings which are vibrationally decoupled from the rest of the network (for a more complete description of the Raman bands in silica and other aluminosilicate glasses see McMillan and Wolf's chapter on vibrational spectroscopy in this volume).

Hemley et al. (1994) have measured the in situ Raman spectrum of silica glass to 30 GPa (Fig. 5a). Detailed measurements of the Raman spectra of silica glass have recently been obtained in the 0-20 GPa region by Sugiura and Yamadaya (1992). At pressures up to about 8 GPa, the primary effect of compression on the Raman spectrum is the gradual shift of the main symmetric stretching band to higher frequencies with a strong decrease in width and increase in absolute intensity. These spectral changes are consistent with a significant decrease in the Si-O-Si intertetrahedral angle and a gradual increase in the intermediate range order (Hemley et al., 1994). At pressures near 1 to 15 GPa there begins a marked decrease in intensity and apparent broadening of the main symmetric stretching band with increasing pressure (Sugiura and Yamadaya, 1992). Spectral intensity in the region near the 606 cm[-1] defect band does not appear to decrease. At 27 GPa, the absolute intensity of the Raman scattering is very low and the spectrum is characterized by a weak, very diffuse band centered near 600 cm[-1].

Previous interpretations of the Raman scattering data have concluded that the onset of the silicon coordination change in SiO_2 glass occurs above 20 GPa (Williams et al., 1993; Hemley et al., 1994). We suggest, however, that the strong reduction in the intensity of the main symmetric stretching band in the Raman spectrum that begins near 14 to 15 GPa may be an indication of the onset of this transition. Although the precise pressure of the onset of the silicon coordination change is difficult to unambiguously constrain from the vibrational spectroscopic data, this interpretation is generally consistent with the in situ X-ray diffraction and Brillouin data on silica glass, discussed below, and is also consistent with interpretations of similar pressure-induced spectral changes observed in GeO_2 glass which we discuss in a latter section.

Meade et al. (1992) have measured, in situ, the X-ray diffraction spectrum of silica glass as a function of pressure to 42 GPa (Fig. 7). Their analysis of the diffraction data is generally consistent with conclusions based on the vibrational spectroscopic data. The derived pair correlation functions indicate that at 8 GPa the average Si-O bond length is consistent with complete tetrahedral coordination of silicon. At 28 GPa the average Si-O bond length is about halfway between those expected for tetrahedral and octahedral silicon at the same pressure, and implies that the average silicon coordination is approximately five. No data points between 8 and 28 GPa were reported; thus, the onset pressure of the silicon coordination change could not be more tightly constrained. At 42 GPa, the average Si-O bond length is about 0.03 Å smaller than that of stishovite at this same pressure, consistent with an average silicon coordination in the glass of about 5.5.

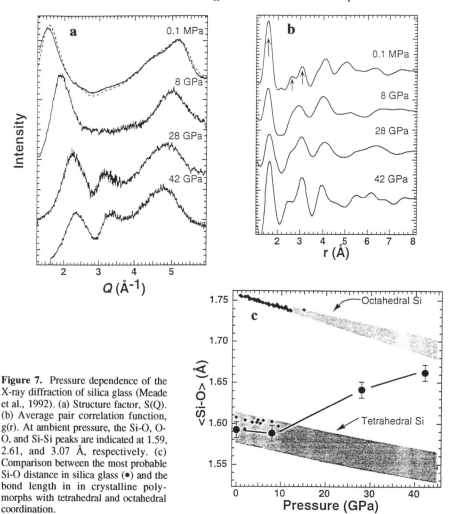

Figure 7. Pressure dependence of the X-ray diffraction of silica glass (Meade et al., 1992). (a) Structure factor, S(Q). (b) Average pair correlation function, g(r). At ambient pressure, the Si-O, O-O, and Si-Si peaks are indicated at 1.59, 2.61, and 3.07 Å, respectively. (c) Comparison between the most probable Si-O distance in silica glass (•) and the bond length in in crystalline poly-morphs with tetrahedral and octahedral coordination.

A critical evaluation of deformational models inferred from the spectroscopic data can be obtained from an analysis of data on the high-pressure compressibility (Kondo et al., 1981; Schroeder et al., 1982; Grimsditch, 1984, 1986; Sasakura et al., 1989; Suito et al., 1992; Polian and Grimsditch, 1993; Zha et al., 1994; Rau, 1995) and equation of state (Bridgman, 1948; Meade and Jeanloz, 1987) of silica glass. Kondo et al. (1981) measured the ultrasonic acoustic velocities and attenuation in silica glass to 3 GPa. A minimum in the acoustic wave velocities was observed near 2.5 GPa, in good agreement with earlier static compression measurements (Birch and Dow, 1936; Bridgman, 1948). Kondo et al. also found that the ultrasonic acoustic waves were highly attenuated near the pressure of the minimum in compressibility. Ultrasonic measurements were later extended to 6 GPa by Sasakura et al. (1989; Suito et al., 1992) and are in good agreement with earlier measurements (Fig. 8).

Vukcevich (1972) showed that it was possible to rationalize most of the observed anomalous behavior of silica glass in terms of a simple two-state model. The states

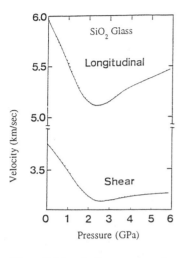

Figure 8 (above). Pressure dependence of ultrasonic acoustic wave velocities in silica glass (Suito et al., 1992).

Figure 9 (right). High-pressure Brillouin scattering data of silica glass.
(a) Pressure dependence of the hypersonic longitudinal, V_p, and shear, V_s, velocities of silica glass. Data from Zha et al. (1994) (filled symbols) and Schroeder et al. (1982) (open symbols).
(b) Comparison of the pressure dependence of the hypersonic bulk sound velocity for silica glass (filled circles; from Zha et al., 1994) with that estimated for quartz, coesite, and stishovite. Bulk sound velocities of the crystalline polymorphs estimated from

$$V_B = \sqrt{K/\rho}$$

using a 3rd-order Birch-Murnaghan equation of state. Equations of state for quartz and coesite from Hemley et al. (1988), for stishovite from Ross et al. (1990).
(c) A comparison of the static (Bridgman, 1948; Meade and Jeanloz, 1987) and shock (Marsh, 1980) compression data for silica glass with that derived from Brillouin data of Zha et al. (1994).

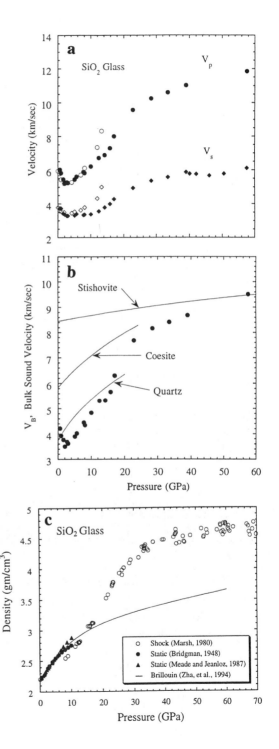

represent two local minima in the silica potential energy surface with distinct intertetrahedral angles and volumes, analogous to that of the α and β states in quartz, tridymite, and cristobalite. Anomalous temperature and pressure-dependent behavior in silica glass arises from a change in the distribution of these states. In this model, the primary contribution to the increase in compressibility of silica glass with pressure arises from the negative volume change associated with the transition between these two states. At pressures above 3 GPa, this reaction is nearly saturated and the compressibility behaves as that of a normal solid.

The elastic properties of silica glass have been investigated to much higher pressures using Brillouin scattering. Grimsditch (1984, 1986) and Polian and Grimsditch (1993) measured the Brillouin scattering spectrum of silica glass up to 25 GPa in a diamond anvil cell. Because of the backscattering geometry used in their studies, index of refraction corrections are needed in order to extract the acoustic wave velocities from the Brillouin data. These corrections were applied in the latest work of Polian and Grimsditch (1993). Schroeder et al. (1982) and Zha et al. (1994) also employed Brillouin scattering, using an equal angle scattering geometry to eliminate the index of refraction correction, to directly obtain the acoustic velocities of silica glass to 13 and 57 GPa, respectively.

At pressures up to about 6 GPa, the ultrasonic and Brillouin scattering data are generally in accord, although, the shear velocities obtained by Schroeder et al. (1982) appear high above 3 GPa, in comparison with the ultrasonic and other Brillouin scattering measurements. At pressures above 6 GPa, the velocities obtained by Schroeder et al. increase much more rapidly with pressure than those obtained from the other Brillouin experiments. The velocity data from Zha et al. (1994) and Polian and Grimsditch (1993) are in good agreement with each other over the entire range of comparison.

One of the most significant aspects of the high-pressure behavior of the acoustic velocities in silica glass is the marked increase in the pressure dependence of both the longitudinal and transverse acoustic velocities which occurs near 15 GPa (see Fig. 9a). This strong pressure dependence saturates near 30 GPa, above which the increase in the velocities is more gradual on up to 57 GPa. The rapid increase in the acoustic velocities above 15 GPa is consistent with the formation of dense, highly rigid domains composed of high-coordinated silicon species. We further suggest that the saturation of the strong pressure dependence of the acoustic velocities, which occurs near 30 GPa, is an indication of a near-complete conversion of tetrahedral silicon sites to either five- or six-coordinated silicon. Some tetrahedral species, however, remain resistant to conversion even at these pressures, perhaps those bound in the smaller siloxane ring structures. The more gradual increase in velocities above 30 GPa, up to 57 GPa, indicates that the completion of the coordination transformation, to all six-coordinated silicon, occurs more gradually.

These arguments can be further validated by a comparison of the bulk sound speed, V_B, ($V_B^2 = V_P^2 - \frac{4}{3}V_S^2$) of silica glass, derived from the Brillouin scattering measurements, with those estimated for stishovite and metastable quartz and coesite (Fig. 9b). The metastable compression limits for quartz and coesite occur at about 21 and 26 GPa, respectively (Hemley et al., 1988; Kingma et al., 1993a), so any extrapolation above these limits is not meaningful. At 22 GPa, the bulk sound velocity of silica glass is similar to that of metastable coesite at this same pressure. With increasing pressure, the sound velocity gradually tends toward that of stishovite. It is interesting to note that there is not much difference between the bulk sound velocities of coesite and stishovite at high pressures. This coincidence is a result of the tradeoff between rigidity and density in determining the sound velocity ($V_B = \sqrt{K/\rho}$, where K is the adiabatic bulk modulus and ρ is the density).

Further insights into the deformational behavior of silica glass has been obtained from both static and dynamic compression measurements of the equation of state. Static compression measurements on the room temperature equation of state of silica glass have been made to about 10 GPa by Bridgman (1948) and by Meade and Jeanloz (1987) (Fig. 9c). Above 6 GPa there is a growing discrepancy between the compression results of these two studies, the origin of which is difficult to evaluate. In Meade and Jeanloz's experiments, the compression measurements were made on samples in a diamond anvil cell employing a novel optical method for measuring linear strains. In these experiments the sample was immersed in a liquid medium which remained completely hydrostatic to 10 GPa. Bridgman carried out his compression experiments in a modified piston-cylinder type apparatus using a liquid pentane/isopentane pressure medium which is known to remain hydrostatic to 7 GPa. In his experiments, the pressure and volume were derived from piston force and displacement measurements.

In the absence of strong viscoelastic (or anelastic) effects, the acoustic wave velocities can also be used to estimate the equation of state (see chapter by Dingwell in this volume). Estimates of the equation of state obtained in this manner are based on the premise that the structural response of the glass which occurs over the timescale of the static compression measurements ($\sim 10^4$ sec) will also occur over hypersonic timescales (10^{-11} sec). Although some viscoelastic behavior is observed in the acoustic velocities at MHz (Kondo et al., 1981) and GHz (Rau, 1995) frequences for silica glass near 2-3 GPa, the relaxing part of the acoustic moduli is not large in this pressure region in comparison to the overall pressure dependencies of the static moduli. In Figure 9c, we compare the static compression data of Bridgman (1948) and Meade and Jeanloz (1987) to estimates based on the ultrasonic and Brillouin scattering measurements. All of the compressional curves are in reasonable agreement up to about 6 GPa. The equation of state estimate derived from Schroeder et al.'s (1982) Brillouin scattering measurements remains in very good agreement with the static compression data of Meade and Jeanloz to 10 GPa. The Brillouin scattering results of Polian and Grimsditch (1983) and Zha et al (1994) show an increasing deviation with the static compression data of Meade and Jeanloz above about 7 GPa. However, these latter Brillouin scattering studies are more in line with Bridgman's original compression data.

Shock compression investigations have been made on silica glass both in the elastic regime (Barker and Hollenback, 1970) and to pressures in excess of 100 GPa (Wackerle, 1962; Anan'in et al., 1974; Marsh, 1980; Sugiura et al., 1981; Lyzenga and Ahrens, 1983; Chhabildas and Grady, 1984; Schmitt and Ahrens, 1989; McQueen, 1992). However, in comparing the properties of shocked silica glass with those obtained from static measurements, it is important to emphasize that the state of silica under dynamic and static compressions is quite different. The behavior of silica glass under shock compression is complex and the interpretation of the shock wave data has been controversial. Early shock recovery experiments (Anan'in et al., 1974) found that silica glass undergoes heterogeneous deformation between 10 and 30 GPa, very similar to that of quartz. These observations are consistent with the high radiative temperature measurements found for shocked silica glass between 10 and 30 GPa (Schmitt and Ahrens, 1989) and indicate that localized hot spots occur within the sample in this shock regime that are nearly 2000 K hotter than the bulk. Shock heating estimates, based on a homogeneous continuum, predict a temperature rise of less than 10 K up to 26 GPa, increasing to 495 K at 30 GPa (Wackerle, 1962). At higher shock pressures, between 30 and 70 GPa, the deformation is largely homogeneous. Lyzenga and Ahrens (1983) have estimated, on the basis of their own thermal emission studies, that the temperature of shocked silica glass at 50 GPa is nearly 4500 K. They suggest that, under these conditions, the silica exists as superheated crystalline stishovite and interpret the sudden decrease in thermal emission at 70 GPa to indicate the metastable melting of this phase.

Figure 9c includes a plot of the shock Hugoniot data for silica glass obtained by Marsh (1980). For comparison, the static compressional data for silica glass, together with the high-pressure equations of state of quartz, coesite, and stishovite, are also included in this figure. The dynamic and static compression data are in general agreement in the elastic regime below 10 GPa. Above 15 GPa the shock Hugoniot strongly deviates from that of the equation of state derived from the hypersonic acoustic velocities. The density of the glass sample under shock becomes greater than that of quartz and coesite above 15 GPa and approaches that of stishovite above 30 GPa. The decoupling of the Hugoniot and hypersonic equation of state above 15 GPa, indicates that there is a significant structural rearrangement that occurs under shock at these conditions that is essentially frozen at hypersonic timescales. This is likely due to both the higher temperature and the longer characteristic timescale of the shock experiment. The characteristic timescale in the dynamic compression experiments can be estimated by the shock rise time and is several orders of magnitude longer than that for the Brillouin experiments (Rigden et al., 1988).

Germania. Germania (GeO$_2$) is a useful chemical and structural analog for silica and exhibits a number of the same structural motifs and anomalous physical properties as SiO$_2$ (Ringwood, 1978). Most useful is the fact that many of the inferred structural deformation modes in silica glass also occur in germania glass but at much lower pressures. In addition, the pressure-induced coordination changes occurring in germania glass have been well characterized using EXAFS (Itie et al., 1989) and Raman spectroscopy (Durben and Wolf, 1991; Smith and Wolf, to be published). At ambient pressure germania glass exhibits a tetrahedral framework structure similar to that of silica. One important difference, however, is that the germania network possesses a much greater degree of intermediate range order compared to silica. In germania, the average intertetrahedral angle is only about 133° at room pressure and the width of the angle distribution is much smaller (~10°) than that found in silica (Leadbetter and Wright, 1972; Sayers et al., 1972; Desa and Wright, 1988). The structure of germania glass at ambient pressure has been likened to that of the silica glass structure at about 8 GPa (Hemley et al., 1986) based on a comparison of the Raman spectra.

High pressure X-ray absorption (XANES and EXAFS) (Itie et al., 1989) studies on germania glass indicate that germanium undergoes a four-fold to six-fold coordination change between about 5 and 10 GPa (see Fig. 10a). The relatively low pressure of this coordination change in GeO$_2$ has made this system particularly amenable to experimental inquiries aimed at understanding the nature of the structural transition and the manner in which this structural change is manifested in the physical and spectroscopic properties.

Durben and Wolf (1991) and Smith and Wolf (1995, to be published) have made extensive investigations of the changes in the Raman spectrum of germania through the pressure-induced coordination change in germania glass (Fig. 11). Perhaps the most noticeable change exhibited in these spectra is the dramatic reduction in the absolute intensity of the main symmetric bending mode near 500 cm^{-1} beginning just above 4 GPa. This pressure only slightly precedes that inferred from the X-ray absorption data for the onset of germanium coordination change (Itie et al., 1989) (although the onset pressures may be consistent within resolution of the X-ray experiment).

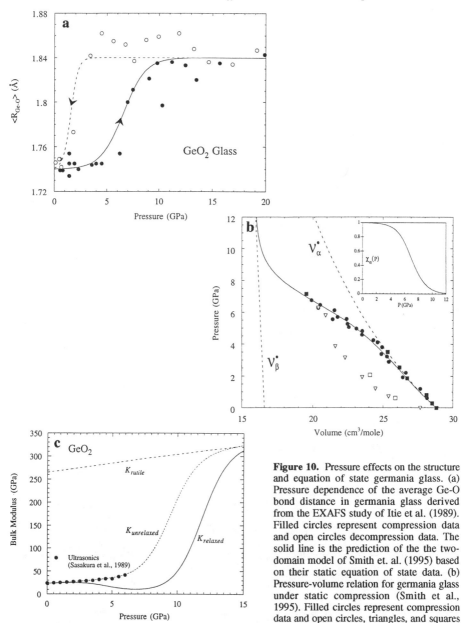

Figure 10. Pressure effects on the structure and equation of state germania glass. (a) Pressure dependence of the average Ge-O bond distance in germania glass derived from the EXAFS study of Itie et al. (1989). Filled circles represent compression data and open circles decompression data. The solid line is the prediction of the the two-domain model of Smith et. al. (1995) based on their static equation of state data. (b) Pressure-volume relation for germania glass under static compression (Smith et al., 1995). Filled circles represent compression data and open circles, triangles, and squares represent decompression data from peak pressures of 3.5, 6.5, and 7.1 GPa, respectively. Dashed lines are the model equations of state for the pure $^{[4]}Ge$ and $^{[6]}Ge$ domains, v_α and v_β. Solid line is a fit of the compression data to the two-domain equation of state model. The inset is a plot of the pressure dependence of the mole fraction of $^{[4]}Ge$, χ_α. (c) Pressure dependence of the bulk modulus of germania glass. Solid circles are ultrasonic data from Suito et al. (1992). Solid line is the bulk modulus derived from the static compression data of Smith et al. (1995). Dotted line is the high-frequency (unrelaxed) bulk modulus predicted from the two-domain model of Smith et al. (1995). Dashed line is the pressure-dependent bulk moduli of the crystalline rutile phase of germania extrapolated from the equation of state parameters of Hazen and Finger (1981).

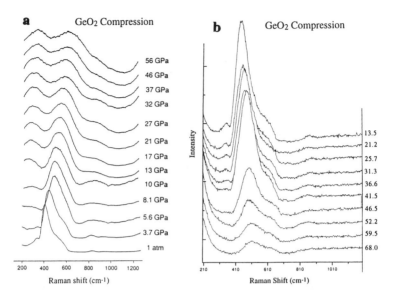

Figure 11. Pressure dependence of the Raman spectrum of germania glass. (a) Relative Raman intensities between ambient and 56 GPa (Durben and Wolf, 1991). (b) Absolute Raman intensities between ambient and 7 GPa (Smith and Wolf, to be published).

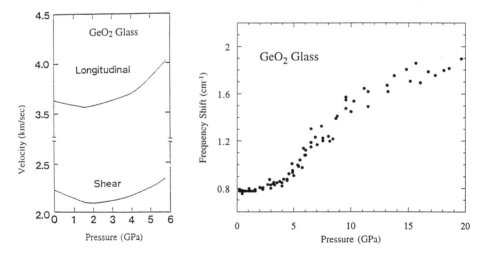

Figure 12. Pressure dependence of the the ultrasonic acoustic wave velocities in germania glass (Suito et al., 1992).

Figure 13. Pressure dependence of the Brillouin frequency shift (backscattering geometry) of the longitudinal acoustic mode in germania glass (Grimsditch et al., 1988).

The pressure-dependent acoustic wave velocities in germania glass have been obtained from both ultrasonic (Fig. 12) (Suito et al., 1992) and Brillouin scattering (Schroeder et al., 1990; Wolf et al., 1992) measurements up to about 6 GPa. Brillouin frequency shifts in germania glass (without index of refraction corrections) have been obtained up to 20 GPa (Fig. 13) (Grimsditch et al., 1988). All of the data reveal the onset of a rapid increase in both the acoustic velocities and Brillouin frequency shifts near 4 GPa (this is most pronounced for the longitudinal waves), similar to that which occurs in silica glass near 15 GPa. Acoustic wave attenuation studies (Wolf et al., 1992) reveal that at room temperature the inferred germanium coordination change above 4 GPa occurs over much longer timescales than that sampled by waves propagating at hypersonic frequencies (i.e., $\tau_c \gg 10^{-11}$ sec). This observation validates the inference that acoustic wave velocities of silicates derived from Brillouin measurements at room temperature sample a frozen coordination configuration. Hence, equations of state directly derived from the acoustic velocities will decouple (be less compressible than) from the equations of state derived from a static compression measurement in the pressure region where these coordination transformations occur.

Very recently, Smith et al. (1995) have measured the equation of state of germania glass under static compression up to 7 GPa (Fig. 10b). The data indicate a marked increase in the static compressibility of the glass that also begins near 4 GPa. Up to 4 GPa the compressional behavior of germania glass is completely reversible in a hydrostatic medium. At higher pressures, a residual densification is always observed, even when the sample is maintained under completely hydrostatic conditions.

Smith et al. (1995) were able to quantitatively fit their equation of state data to a simple two-domain model in which the glass is assumed to be made up of a mixture of low and high density domains. The structural interpretation of this model is that the low density domains are composed of tetrahedral germanium while the high density domains are rutile-like with germanium in six-fold coordination. Within this *ansatz*, the pressure dependence of the mole fraction of tetrahedral germanium could be extracted from the static equation of state data and was used to predict the pressure dependences of the average Ge-O bond length and compressibility of germanium glass. The behavior of the average Ge-O bond length with pressure predicted with this model was in quantitative agreement with the X-ray results of Itie et al. (1989, 1990) (see Fig. 10a).

In the two-domain model, the static compressibility is given by the volume-averaged compressibilities of the low and high density domains, together with a term representing the volume change associated with the coordination change (Vukcevich, 1972). Employing the earlier observation that high-frequency acoustic waves do not access this compression mechanism, Smith et al. (1995) were able to quantitatively reproduce the pressure-dependent behavior of the bulk sound velocity of germania glass obtained from the ultrasonic measurements of Suito et al. (1992) (Fig. 10c).

The consistency of the two-domain model for the description of the high-pressure properties of germania glass does not eliminate the possibility for the existence of five-coordinated Ge species at high pressures. More complex models (with additional parameters) could certainly be found which will also provide a consistent interpretation of the data. For silica, molecular dynamics simulations on the glass and melt predict that a distribution of four-, five-, and six-coordinated Si species occur at high pressures (Angell et al., 1982, 1983, 1987; Rustad et al., 1991). Moreover, experimental investigations of the more "depolymerized" aluminosilicate systems, which we discuss below, have unequivocally established that five-coordinated silicon and aluminum species do occur at high pressures in the glass and liquid states. However, the existence of [5]Ge species in GeO$_2$ has not been established from spectroscopic studies on either in situ or quench samples. Furthermore, G. Calas (personal communication to PM) has indicated that the XANES data of Itie et al. (1989, 1990) would be most consistent with only [4]Ge and [6]Ge species.

The extensive in situ structural data on germania glass provides an unusual opportunity to develop semi-quantitative "calibrations" of vibrational spectral bands. Smith and Wolf (to be published) have recently made a detailed analysis of the pressure-dependent absolute Raman band intensities of germania glass. The Raman data show (Fig. 11b) that between 4 and 7 GPa there is a near 85% loss in the total intensity of the main symmetric tetrahedral stretching mode. This intensity reduction can be qualitatively interpreted by a model in which approximately half of the [4]Ge species are converted to high-coordinated ([5]Ge or [6]Ge) species, effectivley eliminating nearly all Q^4-Q^4 linkages (which are associated with the main symmetric stretching band). This would result in an average Ge coordination of about 4.5 to 5.0, depending on whether five- or six-coordinated germanium is formed and assuming a random spatial distribution. By comparison, the results of Itie et al. (1989, 1990) suggest that at 7 GPa the average coordination is also in this same range (on the basis of bond length inferences). This general consistency between the Raman absolute intensity data and the XANES data suggest that absolute Raman intensity data may be a useful marker for inferring coordination changes in network glasses and melts.

"Charge-balanced" aluminosilicates. Spectroscopic investigations of the ambient structure of aluminosilicate glasses with compositions along the "charge-balanced" SiO_2–M^+AlO_2 or SiO_2–$M^{2+}Al_2O_4$ joins (M^+ = alkali metal; M^{2+} = alkaline earth) generally indicate that Al enters the framework structure in tetrahedral coordination substituting for Si as a "network former." At ambient pressure the structures of glasses along these joins are well described in terms of fully polymerized networks of tetrahedral silicon and aluminum (Taylor and Brown, 1979a,b; Taylor et al., 1980; Engelhardt and Michel, 1987; Oestrike et al., 1987), in similarity to the crystalline structures they form at ambient pressure. A possible exception to this generalization has been suggested from NMR data on ambient glasses very near the charge-balanced SiO_2-$MgAl_2O_4$ join which are consistent with the presence of small amounts of [5]Al and [6]Al species (McMillan and Kirkpatrick, 1992).

Waff (1975) was the first to suggest that aluminum and other tetrahedral network cations in aluminosilicate melts would undergo transformations to six-coordinated species at high pressures. The basis for his speculation was from analogy with the well-documented transformations of this type in the crystalline state of these same systems. He further suggested that these coordination transformations would take place over relatively narrow pressure ranges producing dramatic changes in melt properties. Kushiro (1976) and Velde and Kushiro (1978) later rationalized their observations of the anomalous pressure dependence of viscosity in a number of highly polymerized aluminosilicate liquids in terms of Waff's original ideas. In part, this rationalization was based on their interpretation of infrared absorption data and wavelength shifts in the aluminum K_α and K_β X-ray absorption lines of jadeite ($NaAlSi_2O_6$) composition glasses quenched from high-pressure melts (Velde and Kushiro, 1978). From this data they concluded that in melts of this composition aluminum converts to six-coordination at pressures below 3 GPa.

The spectral interpretations of Velde and Kushiro were later criticized by Sharma et al. (1979) who could find no evidence for an aluminum coordination change in Raman spectroscopic studies of similar glasses. This negative result was borne out in a number of subsequent Raman (McMillan and Graham, 1980; Mysen et al., 1982, 1985; Seifert et al., 1982) and X-ray (Hochella and Brown, 1985) studies on jadeite ($NaAlSi_2O_6$), albite ($NaAlSi_3O_8$), and other fully polymerized aluminosilicate glasses that were quenched from melts as high as 4 GPa.

Later, Ohtani et al. (1985) reported observation of spectral lines indicative of six-coordinated aluminum in the ^{27}Al MAS NMR spectra of albite glasses quenched from 6 and 8 GPa melts. This claim was again questioned, however, by Stebbins and Sykes (1990), who were not able reproduce the results of Ohtani et al. in their own ^{27}Al MAS NMR studies of high-pressure albite composition melts. In their study, spectrally well-defined [5]Al or [6]Al bands were not present in NMR spectra of melt glasses quenched from pressures as high as 10 GPa (Fig. 14a). These authors suggested that the well resolved peak at -16 ppm present in the spectra of Ohtani et al. may be due to contamination or the presence of poorly crystalline jadeite nuclei. The spectra of Stebbins and Sykes did, however, exhibit a small feature on the high-field side of the main [4]Al resonance, which could indicate the presence of small quantities (< 5%) of [5]Al and [6]Al sites in the high-pressure glass. Moreover, these features were found to be independent of the magnetic field strength and so are not likely to originate from quadrupole broadened, highly distorted tetrahedral sites. Furthermore, Stebbins and Sykes concluded that ~50%

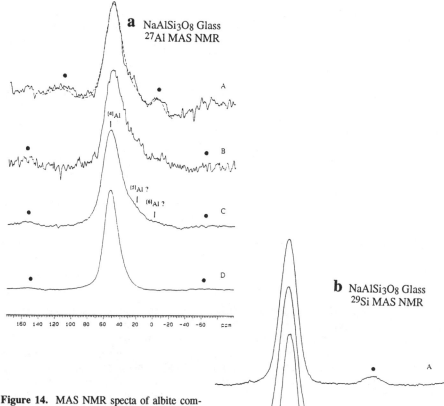

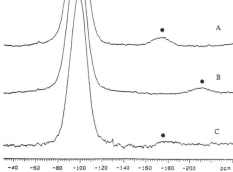

Figure 14. MAS NMR specta of albite composition glass quenched from high-pressure melts (Stebbins and Sykes, 1990). (a) [27]Al MAS spectrum of same glasses. A: 8 GPa sample (130.3 MHz). B: 10 GPa sample (104.2 MHz). C: 8 GPa sample (104.2 MHz). D: 1 bar sample (104.2 MHz). Spinning sidebands are indicated by dots. (b) [29]Si MAS spectrum (79.46 MHz). A: 10 GPa sample. B: 8 GPa sample. C: 1 bar sample.

of the total ^{27}Al signal intensity could not be accounted for in their spectral bands. They suggested that the loss of signal in the NMR spectrum of these high-pressure glasses is an indication of the presence of highly distorted Al polyhedra with large quadrupolar coupling constants. Very similar ^{27}Al NMR spectroscopic results were also obtained in a more recent study on orthoclase ($KAlSi_3O_8$) composition glasses quenched from high-pressure melts (Sykes et al., 1993).

For silicon, the resolving power of ^{29}Si MAS NMR for different coordination species is much greater than for ^{27}Al which has a quadrupolar lineshape. In the Stebbins and Sykes (1990) study, no evidence was found for the presence of five- or six-coordinated Si species in the high-pressure albite glasses above the detection limit of 0.5% (see Fig. 14b).

Unambiguous evidence for the existence of high-coordinated Al and Si species in aluminosilicate composition glasses along the charge-balanced joins has been difficult to establish from studies of quenched samples at ambient pressure. This can be contrasted to the results from some recent studies on fully-polymerized aluminate melts. Daniel et al. (1995) have inferred, from an in situ Raman study, that the aluminum in $CaAl_2O_4$ glass increases in coordination above 11 GPa. Very recently, this was confirmed by Yarger et al. (1995) who found NMR evidence for the retention of high-coordinated Al species in $CaAl_2O_4$ glasses quenched from high pressures. In their study, well resolved features occur in the ^{27}Al MAS NMR spectra of glasses quenched from high pressure that are consistent with relatively high abundances (tens of %) of [5]Al and [6]Al species.

Contrary to ambient studies on fully polymerized aluminosilicate composition glasses quenched from high-pressure melts, in situ spectroscopic studies on silica and germania glasses (discussed in the previous sections) provide compelling evidence that high-coordinated species are formed in fully polymerized silicate and related systems at high pressures, even though these species are not retained on decompression (Hemley et al., 1986; Williams and Jeanloz, 1988; Itie et al., 1990; Durben and Wolf, 1991). Similar results in an aluminosilicate system were found by Williams and Jeanloz (1988). In their study, the infrared absorbance spectrum of anorthite ($CaAl_2Si_2O_8$) composition glass was measured as a function of pressure to 31 GPa (see Fig. 15). In general the spectral changes observed for anorthite glass are similar to that observed for silica glass, but occur at lower pressures. The infrared data suggest that the tetrahedral coordination change in anorthite glass commences between 3 and 11 GPa. By 22 GPa the high frequency spectral bands associated with tetrahedral species are essentially gone.

The spectral interpretations of Williams and Jeanloz (1988) are remarkably consistent with molecular dynamic simulations on jadeite ($NaAlSi_2O_6$) melts by Angell et al (Angell et al., 1982, 1983, 1987). In these simulations it was found that the tetrahedral network cations (Si^{4+}, Al^{3+}) undergo coordination changes over a wide range in pressure (at least at simulation temperatures, 4000 to 6000 K). Furthermore, the coordination changes for Al^{3+} take place at considerably lower pressures than for Si^{4+}. In jadeite, the average coordination of Al^{3+} is 5 at about 10 GPa. However, for Si^{4+} cations, this average coordination state is not realized until pressures of about 20 GPa.

One of the more interesting results of the experimental and theoretical investigations on the fully polymerized silicate and germanate systems, has been the diversity in character that the coordination transformations can display in the amorphous state. It appears that these coordination changes can, in some systems, resemble first-order transformation (i.e., GeO_2) with relatively sharp changes in properties and significant hysteresis, while in silica and the few other fully polymerized aluminosilicates that have been studied, the transformations appear to occur with minimal hysteresis over a much larger pressure range. These generalities must be modified, however, in describing the high-pressure behavior of the more "depolymerized" aluminosilicates. As we discuss below, additional coordination change mechanisms are expressed that can display extreme hysteresis, resulting in the retention of high-coordinated species in the decompressed samples.

Depolymerized silicate systems

Alkali and alkaline earth oxides are known to act as depolymerizing agents on the silicate tetrahedral network structure. This action results from the chemical disruption of the strong Si-O-Si linkages and formation of much weaker and less directional $Si-O^-$-M^+ and $Si-O^-$-M^{2+} interactions (Brawer and White, 1975; Hess, 1977; Murdoch et al., 1985; Dupree et al., 1986; Schneider et al., 1987). The degree of disruption of the tetrahedral framework is strongly evident by its effect on the viscosity and thermodynamic properties of melts in these systems (Bockris and Lowe, 1954; Charles, 1967; Bottinga and Weill, 1972; Mysen et al., 1980; Urbain et al., 1982; Hochella and Brown, 1984; Richet, 1984; Stebbins et al., 1984; Richet and Bottinga, 1985).

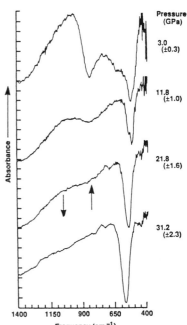

$CaAl_2Si_2O_8$ Glass

Figure 15. Pressure dependence of the infrared absorbance of anorthite glass (Williams and Jeanloz, 1988). [Used with permission of the American Association of the Advancement of Science, from Williams and Jeanloz (*Science*, 1988, Fig. 1, p. 903.]

In describing the structure of "depolymerized" aluminosilicate networks, it is convenient to distinguish between two different chemical environments of the oxygen atoms. Oxygen atoms that are contained within the strong T-O-T (T = Si or Al) linkages are typically referred to as *bridging* oxygens or *BOs* while those directly bonded to only one silicon atom are called *nonbridging* oxygens or *NBOs*. It is also useful to distinguish silicate tetrahedral units in terms of the number of NBOs they contain. In this description, the increasing depolymerization of the silicate framework is expressed by the successive appearance of tetrahedral silicate units with 1, 2, 3, and 4 NBOs. These tetrahedral units are often labeled as Q^n species, where n ($0 \leq n \leq 4$) indicates the number of BOs in that unit (Engelhardt et al., 1975). In amorphous SiO_2, all of the oxygen atoms are bridging and hence each silicon tetrahedral unit is a Q^4 species. In alkali and alkaline earth aluminosilicates, the glass and melt structures exhibit a distribution of Q^n species, where the abundances of lower n value species proportionally increases with increasing alkali or alkaline earth oxide concentrations. The presence and compositional dependence of these different structural species has been extensively documented using Raman scattering (Brawer and White, 1975; Sharma et al., 1978; Mysen et al., 1980, 1982; Virgo et al., 1980; Furukawa et al., 1981; Matson et al., 1983; McMillan, 1984b; Mysen, 1988, 1990) and NMR spectroscopy (Murdoch et al., 1985; Dupree et al., 1986; Schneider et al.,

1987; Kirkpatrick, 1988; Stebbins, 1988; Stebbins et al., 1992) (see also the chapter in this volume by McMillan and Wolf).

Binary alkali silicates. Spectroscopic measurements on glasses quenched from high-pressure melts have been used to investigate the effect of pressure on the equilibrium tetrahedral species distributions in several high silica binary alkali silicates (Dickinson and Scarfe, 1985; Stebbins and McMillan, 1989; Xue et al., 1989, 1991; Dickinson et al., 1990). These studies have generally concluded that increasing pressure results in a greater distribution of Q species in these melts. The observed changes in the tetrahedral species distribution have been discussed in terms of pressure shifts in the equilibrium for disproportionation reactions of the type

$$2 Q^3 \leftrightarrow Q^2 + Q^4 \tag{1}$$

$$2 Q^2 \leftrightarrow Q^3 + Q^1 \tag{2}$$

Xue et al. (1989, 1991) used ^{29}Si MAS NMR, Raman and infrared spectroscopies in an attempt to quantify the pressure dependence of the species distributions in potassium tetrasilicate ($K_2Si_4O_9$) and sodium disilicate ($Na_2Si_2O_5$) and tetrasilicate ($Na_2Si_4O_9$) composition glasses that were quenched from melts at pressures up to 12 GPa. Below about 4 to 5 GPa the quenched glasses displayed only minor changes in their vibrational and NMR spectra. However, at higher pressures, significant changes were observed in the NMR, Raman and infrared reflectance spectra of all of the quenched glasses (Fig. 14). By modeling the ^{29}Si MAS NMR spectra, Xue et al were able to estimate the pressure-induced changes in the tetrahedral speciation abundances in these glasses. In general, the abundances of the Q^3 species decreased and the Q^2 and Q^4 species increased with increasing pressure. For example, in sodium disilicate melts, essentially no resolvable change in the species distribution was found in glasses quenched from 5 GPa compared to that of a normal ambient sample. However, for the same composition glass quenched from 8 GPa, the relative abundance of Q^3 species decreased from about an ambient value of 84% to a value of 72%, while the Q^4 and Q^2 species abundances increased from 8% to 16% and from ~0 to 3%, respectively.

It is interesting that the changes inferred in the tetrahedral species distribution of alkali silicate glasses quenched from high-pressure melts are qualitatively similar to those observed to occur with increasing temperature at ambient pressure (Seifert et al., 1981; Stebbins, 1987; Liu et al., 1988; Stebbins, 1988; Mysen, 1990). It is generally assumed that the structure of a glass represents the structure of the liquid at the glass transition temperature, T_g (Gibbs and DiMarzio, 1958) (see also chapters in this volume by Richet and Bottinga, and Dingwell). Thus, it might be expected that if T_g increases with pressure in these systems, then the inferred pressure-dependent structural changes might instead result from a temperature effect. It is known that the viscosity of high-silica alkali silicates decreases with increasing pressure. Thus, the glass transition temperature in these high silica systems will also likely decrease with increasing pressure. This conclusion of the relation between the pressure dependences of viscosity and T_g is supported by low pressure (< 0.7 GPa) T_g measurements on silicate melts (Rosenhauer et al., 1979). On the basis of these arguments, Xue et al. (1991) concluded that their observed changes in the tetrahedral speciation distribution would likely underestimate the true pressure effects.

Dickinson et al. (1990) have suggested that the increase in the distribution of tetrahedral species in alkali silicate melts with pressure could significantly contribute to the observed decrease in viscosity of these systems with increasing pressure. Their argument is based on the Adam-Gibbs (1965) configurational entropy theory of relaxation processes and the expectation that an increase in the speciation distribution would result in an increase in the configurational entropy and a decrease in size of polymeric flow units (see chapter by Richet and Bottinga). Dickinson et al. (1990) further speculated that the increase in the Q^3 disproportionation with pressure could be the primary compression mechanism of alkali silicate melts at low pressures.

Much more significant than the observation of the effect of pressure on the tetrahedral speciation distribution was Xue et al.'s (1989) report of spectral evidence for the existence of [6]Si species (~ 1.5% abundance) in a sodium disilicate glass quenched from 8 GPa and 1500 °C. The ^{29}Si MAS NMR data presented by these authors was the first clear evidence for the existence of [6]Si species in a silicate or aluminosilicate melt; although, [6]Si species were previously known to occur in phosphosilicate glasses prepared at ambient pressure (Dupree et al., 1987, 1988).

Stebbins and McMillan (1989) reported evidence for the existence of both five- and six-coordinated silicon species (0.4% and 0.2% abundances, respectively) in a potassium disilicate glass quenched from 1.9 GPa and 1200°C. This was the first report for the existence of a [5]Si species in a silicate, a species postulated years before by Angell et al. (1982) to play a key role in the viscous flow mechanism of silicate liquids. The presence of very small amounts of [5]Si species even in ambient pressure alkali silicate melts was later documented by Stebbins (1991) for potassium tetrasilicate. It was found that the abundance of [5]Si increased with increasing fictive temperature, ranging from 0.06% in slow quenched melts to 0.10% in fast quenched samples. Although the observation of five-coordinated silicon in a silicate is novel, this species does occur, typically as a distorted trigonal bipyramidal structure, in organic molecules (Marsmann, 1981; Coleman, 1983; Tandura et al., 1986) and has also been matrix stabilized in a SiF$_5^-$ complex (Ault, 1979).

Although the abundances of [5]Si and [6]Si species in silicate glasses reported in these early studies were extremely low, the study by Xue et al. (1991) demonstrated that much higher fractions of five- and six-coordinated silicon could be retained in alkali silicate glasses quenched from higher pressures (see Fig. 16). Maximum abundances of about 8.5% [5]Si and 6.3% [6]Si were recorded in a sodium tetrasilicate sample quenched from a melt at 12 GPa. Xue et al. also found a marked compositional dependence to the fractional abundances of the [4]Si, [5]Si, and [6]Si species at all pressures (see also Stebbins and McMillan, 1993 and chapter in this volume by Stebbins).

One of the most significant trends supported by Xue et al.'s data is that the proportion of high-coordinated species ([5]Si + [6]Si) is greatest near the tetrasilicate composition and decreases strongly both toward SiO_2 and toward higher alkali concentrations (Fig. 17). As these authors have discussed, this observation is consistent with observations made in the alkali germanate system at ambient pressure. Spectroscopic analyses suggest that alkali germanate glasses formed at ambient pressure contain both four- and six-coordinated germanium species (Ueno et al., 1983). The abundance of high-coordinated germanium is strongly sensitive to the alkali concentration and is maximized near the tetrasilicate composition (Ueno et al., 1983). Studies on the alkali germanate system further suggest that the high-coordinated germanium species are formed through a consumption of the non-bridging oxygens rather than by an attack from bridging oxygens. A similar explanation has been made to rationalize the compositional dependence of boron coordination species in alkali borate glasses at atmospheric pressure (Bray and O'Keefe, 1963; Jellison et al., 1978) and is also consistent with trends observed for the composition dependence of silicon speciation in alkali silicophosphate glasses at 1 bar (Dupree et al., 1987).

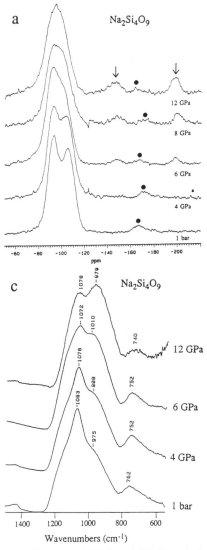

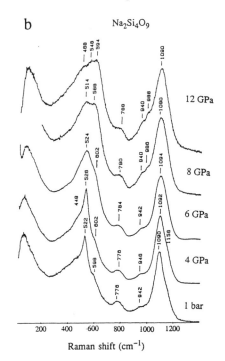

Figure 16. Spectroscopic data of sodium tetra-silicate glasses quenched from high-pressure melts (Xue et al., 1991). (a) ^{29}Si MAS NMR spectra of $Na_2Si_4O_9$ glasses quenched from liquids at 1 atm (1200°C), 4 GPa (1800°C), 6 GPa (190°C), 8 GPa (2000°C), and 12 GPa (2100°C). Solid dots show spinning side bands, arrow show positions of peaks due to $^{[5]}$Si and $^{[6]}$Si. (b) Raman spectra of same $Na_2Si_4O_9$ glasses. (c) Micro-infrared specular relectance spectra of same $Na_2Si_4O_9$ glasses.

Xue et al. (1991) and Wolf et al. (1990) have suggested that the formation of high-coordinated silicon species in alkali silicate melts at low pressures also occurs through a consumption of non-bridging oxygens via mechanisms of the type

$$Q^3 + Q^4 \rightarrow {}^{[5]}Si + Q^{4*} \tag{3}$$

or

$$2Q^3 + Q^4 \rightarrow {}^{[6]}Si + 2Q^{4*} . \tag{4}$$

where Q^{4*} is a SiO_4 species that has three $^{[4]}$Si neighbors and one $^{[5]}$Si or $^{[6]}$Si neighbor (see Fig. 18). The greater reactivity of the non bridging oxygens to formation of high-coordinated Si species over that of the bridging oxygens is also consistent with the known crystalline transformation of $K_2Si_4O_9$ near 2.5 GPa to a wadeite-structured phase (Kinomura et al., 1975; Kanzaki et al., 1989). In this phase, 1/4 of the silicon atoms occur in octahedral sites and the remaining 3/4 occur in tetrahedral sites in a fully polymerized network with no non-bridging oxygens.

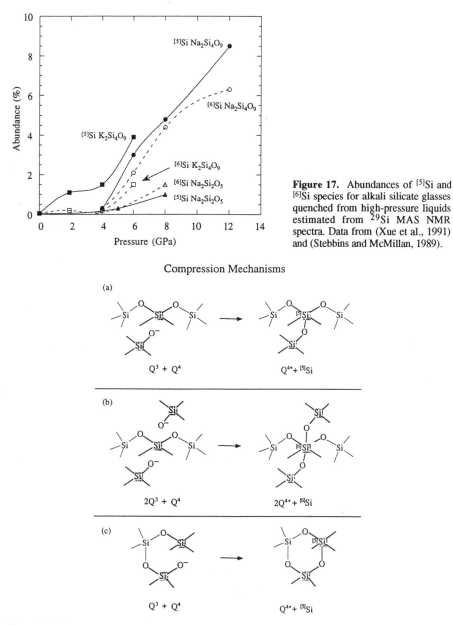

Figure 17. Abundances of [5]Si and [6]Si species for alkali silicate glasses quenched from high-pressure liquids estimated from ^{29}Si MAS NMR spectra. Data from (Xue et al., 1991) and (Stebbins and McMillan, 1989).

Compression Mechanisms

Figure 18. Compression mechanisms proposed by Wolf et al. (1990) and Xue et al. (1991) for the formation of high-coordinate Si species through the reaction non-bridging oxygen atoms. (a) Formation of [5]Si species. (b) Formation of [6]Si species. (c) Formation of small ring structures with high-coordinated Si.

Raman scattering, like NMR spectroscopy, has also proven to be useful for investigating the degree of polymerization in silicate networks. One important advantage of Raman scattering is that in situ studies can now be routinely made on extremely small samples contained at high pressures (and high temperatures) in the diamond anvil optical cell. Unlike NMR however, the Raman intensity of any particular mode is not in itself quantitative. (The Raman intensity depends on the modulation amplitude of the sample polarization that accompanies the mode atomic displacements. Since these displacement patterns can, in general, change with pressure, temperature, or composition, the scattering cross section can also change.) Nevertheless, with sufficient cross referencing with other probes, Raman scattering provides a useful qualitative, and sometimes semiquantitative, description of pressure or temperature induced changes in the short and medium range structure of glasses and melts.

With increasing alkali or alkaline earth content, the increasing depolymerization of the silica tetrahedral network is characterized in the Raman spectrum by the successive appearance of strong, highly polarized modes at frequencies near 1100 cm^{-1}, 950 cm^{-1}, 900 cm^{-1}, and 850 cm^{-1} which have been assigned to highly localized symmetric Si-O stretching vibrations of Q^3, Q^2, Q^1 and Q^0 species, respectively (Brawer and White, 1975; Sharma et al., 1978; Mysen et al., 1980, 1982; Virgo et al., 1980; Furukawa et al., 1981; Matson et al., 1983; McMillan, 1984b; Mysen, 1988, 1990). Antisymmetric stretching vibrations of the fully polymerized Q^4 units also appear in the frequency region between 1000 and 1250 cm^{-1} but are considerably weaker than the symmetric stretching vibrations of the depolymerized units. At the tetrasilicate and disilicate compositions, there are on average 0.5 and 1.0 NBOs per SiO_4 tetrahedron, respectively. For alkali tetrasilicate glasses the Raman spectrum in the high frequency region consists of a strong asymmetric band near 1100 cm^{-1}, corresponding to vibrations of Q^3 units, and a weaker band near 950 cm^{-1}, more evident for the higher field strength cations, corresponding to Q^2 vibrations (see Fig. 16b). For the disilicate glasses, both the Q^2 and Q^3 vibrational bands are considerably stronger, indicative of the greater depolymerization of the network.

The interpretation of the Raman spectra of alkali silicate glasses in the frequency range below 700 cm^{-1} is less clear. At ambient pressure, the Raman spectra of high silica alkali silicate glasses in this frequency range is characterized by two relatively sharp, polarized bands near 600 cm^{-1} and 520 cm^{-1} with weaker, diffuse scattering at lower frequencies. The diffuse scattering below 500 cm^{-1} is generally assigned to bending deformations associated with fully polymerized regions within the glass structure. Matson et al. (1983) have suggested that the 600 cm^{-1} band in alkali silicates has a similar origin to the 606 cm^{-1} defect band of vitreous silica and have assigned it to a symmetric oxygen breathing motion from three-membered siloxane rings vibrationally decoupled from the glass network. The weak dependence of the position of this band both on the nature of the alkali cation and on temperature is consistent with this interpretation. Furthermore, the retention of this band to higher alkali contents indicates that depolymerized tetrahedral species can be incorporated in tri-siloxane ring structures. The 520 cm^{-1} band has been assigned to delocalized symmetric bending deformations of structural units that contain both Q^3 and Q^4 species (Matson et al., 1983). This assignment is partly based on the observation that the intensities of the 520 cm^{-1} band and 1100 cm^{-1} are highly correlated with changes in alkali content (Verweij and Konijnendijk, 1976; Verweij, 1979; Mysen et al., 1980; Furukawa et al., 1981).

Wolf et al. (1990) and Durben (1993) used Raman scattering to investigate the in situ high-pressure behavior of binary alkali tetrasilicate and disilicate glasses under room temperature compression. The in situ Raman spectra of the tetrasilicate and disilicate glasses clearly reveal a reduction in intensities of both the 1100 cm^{-1} and 520 cm^{-1} bands with increasing pressure (see Figs. 19 and 20). For all of the tetrasilicate glasses studied, the 520 cm^{-1} band shows a marked reduction in intensity beginning at about 3 GPa. In

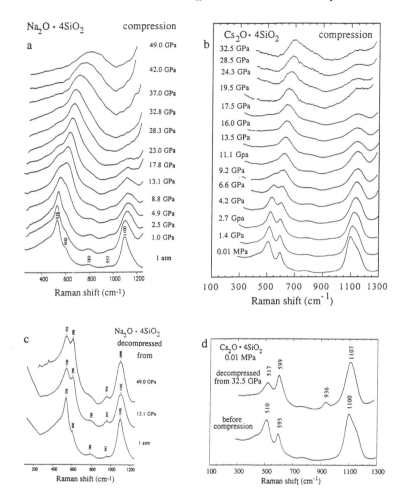

Figure 19. Pressure dependence of the Raman spectra of sodium and cesium tetrasilicate glasses (Wolf et al., 1990; Durben, 1993). (a) In: situ Raman spectrum of $Na_2Si_4O_9$ glass. (b) In situ Raman spectrum of $Cs_2Si_4O_9$ glass. (c) Comparison of the ambient Raman spectra of $Na_2Si_4O_9$ glasses pressure-cycled to 49 and 13 GPa with the spectrum of an unpressurized sample. (d) Comparison of the ambient Raman spectrum of $Cs_2Si_4O_9$ glass pressure-cycled to 32.5 GPa with the spectrum of an unpressurized sample.

sodium and potassium tetrasilicate glasses, the disappearance of the 520 cm^{-1} band near 16 and 23 GPa, respectively, closely coincides with the disappearance of the 1100 cm^{-1} band. However, for Cs tetrasilicate glass, the 520 cm^{-1} band disappears at about 11 GPa while the 1100 cm^{-1} band, although weak, can still be resolved to at least 32 GPa.

For all of the disilicate glasses studied, the 520 cm^{-1} band disappears at a much lower pressure (~ 5 GPa) than that found for the tetrasilicate glasses. Moreover, although the Q^2 and Q^3 bands near 950 cm^{-1} and 1100 cm^{-1} show a gradual reduction in their absolute intensities (not apparent in the scaled spectra shown in Figures 19 and 20) with increasing pressure, scattering in this frequency envelope persists to the highest pressures of the experiments. In this spectral region, the ratio of the relative band intensity, Q^2:Q^3,

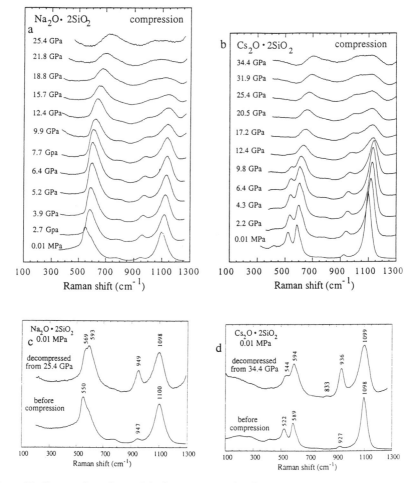

Figure 20. Pressure dependence of the Raman spectra of sodium and cesium disilicate glasses (Durben, 1993). (a) In situ Raman spectrum of Na₂Si₂O₇ glass. (b) In situ Raman spectrum of Cs₂Si₂O₇ glass. (c) Comparison of the ambient Raman spectrum of Na₂Si₂O₇ glasses pressure-cycled to 25.4 GPa with the spectrum of an unpressurized sample. (d) Comparison of the ambient Raman spectrum of Cs₂Si₂O₇ glass pressure-cycled to 34.4 GPa with the spectrum of an unpressurized sample.

appears to increase with pressure. In addition, there is also a buildup of relative scattering intensity between the 950 cm^{-1} and 1100 cm^{-1} bands resulting in a loss in resolution of these bands above 10 GPa.

Common to both the tetrasilicate and disilicate glasses, there is a gradual broadening and shift of the 600 cm^{-1} band to higher frequencies which begins near ambient pressure and continues to the highest pressures of these experiments. For the tetrasilicate glasses, the 600 cm^{-1} band is shifted to about 680 cm^{-1} near 28 GPa, independent of the nature of the alkali cation. Data at higher pressures from Wolf et al. (1990) on sodium tetrasilicate glass show that above 28 GPa the pressure dependence of the width and position of this band markedly increases. At 49 GPa, only a single band centered near 820 cm^{-1}, with a band width at half height of nearly 300 cm^{-1}, can be resolved in the Raman spectrum of sodium tetrasilicate glass. For the disilicate glasses, the pressure shift of the 600 cm^{-1} band is greater and exhibits a more significant alkali effect. Near 20 GPa, the 600 cm^{-1} band is shifted to about 700 cm^{-1} for sodium disilicate glass but appears at about 680 cm^{-1} and 660 cm^{-1} for the potassium and cesium disilicate glasses, respectively, at the same pressure.

The pressure-induced changes in the in situ Raman spectra of the tetrasilicate and disilicate glasses can be interpreted in terms of two distinct silicon coordination change reactions which operate over different, but possibly overlapping, pressure intervals (Wolf et al., 1990; Durben, 1993). At low pressure, the loss in the 520 cm^{-1} band and the reduction in intensity of the 1100 cm^{-1} band in both the tetrasilicate and disilicate glasses are consistent with the mechanisms (described above in Eqns. 3 and 4) where high-coordinated Si species are formed through the reaction of non-bridging oxygen atoms associated with Q^3 species. At much higher pressures, as discussed more fully below, the marked increase in the frequency shift and width of the 600 cm^{-1} band are consistent with the formation of high-coordinated Si species through the reaction of bridging oxygen atoms with neighboring silicon atoms. The differing energetics of these two mechanisms can be rationalized by an inspection of the cation arrangement around the oxygen anion The higher pressure mechanism, involving bridging oxygens, requires the energetically costly step of forming [3]O species where oxygen is bonded to three silicon atoms. However, the lower pressure mechanism involves non-bridging oxygens and the formation of [2]O species where oxygen is bonded to only two silicons.

The detailed interpretation of the pressure-induced changes in the Raman spectra of the tetrasilicate glasses is more straightforward than that of the disilicate glasses. For the tetrasilicate glasses, the reaction of the non-bridging oxygen atoms associated with Q^3 species to form high-coordinated Si species shows a significant alkali effect. In $K_2Si_4O_9$ glass the consumption of non-bridging oxygen atoms to form high-coordinated Si species largely occurs in the pressure range between 3 and 16 GPa. For $Na_2Si_4O_9$ glass the pressure range extends to about 23 GPa and for $Cs_2Si_4O_9$ glass to at least 30 GPa. This alkali effect, inferred from the in situ Raman measurements, is consistent with that found in the NMR studies of alkali tetrasilicate glasses quenched both from high-pressure and ambient pressure melts (Xue et al., 1991; Stebbins and McMillan, 1993). In the NMR studies, the abundance of [5]Si retained in the quenched glasses is greatest for potassium silicate (Stebbins and McMillan, 1993).

For the sodium and potassium tetrasilicate glasses, the 520 cm^{-1} and 1100 cm^{-1} bands display a nearly simultaneous decrease in intensity with pressure, and disappear at about the same pressure for each glass. The concomitant intensity reduction of these bands generally supports their assignment to vibrations involving Q^3 units in the glass structure. However, for cesium tetrasilicate glass and, more significantly for all of the disilicate glasses studied, the 1100 cm^{-1} band persists to higher pressures than that at which the 520 cm^{-1} band disappears. These observations suggest that the 1100 cm^{-1} band results from a localized vibration of all the Q^3 species whereas the 520 cm^{-1} band results from only a subset of medium ranged structural units that contain Q^3 species. One possible interpretation for this observation is that the 520 cm^{-1} band arises from symmetric bending vibrations of Q^3-O-Q^4 linkages. Linkages involving two Q^3 species may, instead, contribute to the band envelope at higher frequencies, and would be more dominant in the disilicate and lower silica glasses. In this interpretation, regions that contain isolated non-bridging oxygens (i.e., structural regions related to the 520 cm^{-1} band) would generally be expected to be involved in the formation of high-coordinated Si species more easily (at lower pressures) than regions where non-bridging oxygen and alkali metal atoms are clustered, especially for the larger alkali metal ions. The increased steric restrictions imposed by the cations in the clustered regions would substantially reduce the efficiency of the Si coordination transformation.

The increase in relative intensity of the 600 cm^{-1} band in the tetrasilicate and disilicate glasses with pressure is reminiscent of the behavior observed for vitreous silica and germania. In silica glass, the relative intensity of the 606 cm^{-1} "defect" mode increases most significantly at pressures above 15 GPa (Hemley et al., 1986). Similarly, the relative intensity of the related 520 cm^{-1} "defect" band in germania glass increases above 4 GPa (Durben and Wolf, 1991). We have speculated above that these spectral changes are related to the formation of three-ring structures through high-coordinated species in these fully polymerized networks. The in situ Raman data on the alkali silicate glasses is consistent with the interpretation that related structural changes may be taking place in these glasses as well, but beginning at lower pressures. Under the interpretation that vibrations associated with three-membered rings strongly contribute to the scattering intensity of the 600 cm^{-1} band in alkali silicate glasses, then the increase in intensity of this band at low pressures suggests a relatively easy pathway for the formation of three-membered rings in silicates containing non-bridging oxygens. Wolf et al. (1990) proposed a mechanism for the formation of small ring structures in alkali silicate glasses (Fig. 18). In this reaction a non-bridging oxygen associated with a tetrahedral Q^3 species binds to a Q^4 species in the same chain to form $^{[5]}Si$ and $^{[2]}O$ species and a three-membered ring. In vitreous silica, since all of the oxygen atoms are *bridging* the formation of a small ring requires either the formation of an $^{[3]}O$ species and $^{[5]}Si$ species, or Si-O bond breaking and rearrangement of the tetrahedral network. The available pathways for three-ring formation in fully polymerized systems would be expected to have a higher activation energy than that for depolymerized systems since, in the latter, neither high-coordinated $^{[3]}O$ species are formed nor are any Si-O bonds broken.

Wolf et al. (1990) suggested that scattering associated with bending vibrations of $Q^{4*}-{^{[5]}Si}$ and $Q^{4*}-{^{[6]}Si}$ linkages may also contribute to the observed increase in the relative intensity and broadening of the 600 cm^{-1} band in alkali silicate glasses with increasing pressure. Raman spectra of the high-pressure $MgSiO_3$ garnet and $K_2Si_4O_9$ wadeite phases provide some support for this interpretation. Both of these phases, which contain corner-shared linkages between SiO_4 and SiO_6 polyhedra, have modes in their Raman spectra near 600 cm^{-1}.

As discussed above, the Raman data indicate that there are essentially no non-bridging oxygens in sodium and potassium tetrasilicate glass above 20 GPa. If we assume that only the non-bridging oxygens are incorporated in higher coordinated silicon species, this would imply an average Si coordination of 4.5, similar to that of the high-pressure wadeite phase of $K_2Si_4O_9$. Any further transformation of the remaining tetrahedral silicon atoms in the tetrasilicate glass to high-coordinated states requires the involvement of bridging oxygens and the formation of $^{[3]}O$ species. Wolf et al. (1990) suggested that the marked increase in the frequency shift and broadening of the 600 cm^{-1} band in sodium tetrasilicate glass above about 28 GPa is an indication of the formation of high-coordinated Si species through reactions involving non-bridging oxygen atoms. It is difficult to establish the onset pressure of this higher pressure reaction mechanism since the formation of high-coordinated species via non-bridging oxygens also results in a shift and broadening of the 600 cm^{-1} band.

An interesting comparison can be made between the high-pressure Raman spectra of the alkali silicate glasses with that of silica glass. At about 27 GPa, the Raman spectra of silica glass shows a broad weak feature centered at about 620 cm^{-1} with a spectral width of about 200 cm^{-1} at half height (see Fig. 5a). At about this same pressure, the in situ diffraction measurements of silica glass by Meade et al. (1992) suggest an average Si coordination of about 5. Hemley et al. (1986) report that at higher pressures the main Raman band of silica glass further broadens and becomes indistinct, perhaps reflecting the complete reaction of all remaining tetrahedral Si to high-coordinated species. In the tetrasilicate glasses the main Raman band is centered near 680 cm^{-1} GPa at 28 GPa. Above 28 GPa, this band markedly broadens and weakens, and displays a stronger shift to high frequencies without disappearing.

In all of the disilicate glasses, residual intensity persists in the 800 to 1200 cm^{-1} region to the highest pressures obtained in these experiments. Moreover, the spectral intensity in this region is considerably greater for the cesium disilicate glass than for sodium disilicate. These observations suggest that tetrahedral species can persist to much higher pressures in more depolymerized glasses than in the highly polymerized systems. Thus, whereas NBOs can easily react to form high-coordinated Si species in highly polymerized silicates, this reaction pathway becomes progressively less favored with increasing depolymerization. The shift in the energetics of this reaction likely arises from the increasing steric restrictions imposed by the increasing concentration of alkali metals.

Additional important inferences can be made from the in situ Raman data regarding the pressure dependence of the Q^n speciation distribution in alkali silicate glasses. For the disilicate glasses, the Raman data indicate that the relative intensity ratio of the Q^2:Q^3 bands increases strongly with increasing pressure in these systems. This observation, by itself, is consistent with earlier conclusions, based on ambient pressure studies of glasses quenched from high-pressure melts, that there is an increase in the Q^n speciation distribution in alkali silicate glasses with increasing pressure (Dickinson and Scarfe, 1985; Xue et al., 1989, 1991; Dickinson et al., 1990; Mysen, 1990). In the earlier studies the changes in the Q^n speciation distribution were discussed in terms of pressure shifts in tetrahedral speciation reactions. However, the in situ Raman measurements suggest that the pressure-induced formation of high-coordinated Si species may play a more significant role in altering the Q^n speciation distribution in alkali silicate liquids. In particular, the Raman data indicate that the non-bridging oxygens associated with the Q^3 species display a much greater reactivity to formation of high-coordinated Si species in comparison to those associated with the Q^2 species. These inferences are further supported by the in situ Raman data on the tetrasilicate glasses. In these systems there appears to be no significant increase in the relative intensity of the Q^2 band with increasing pressure, but only a gradual reduction in the intensity of the Q^3 band.

The greater reactivity of the Q^3 species compared to the that of Q^2 species (and presumably Q^1 and Q^0 species) is also consistent with the conjecture that increased cation clustering decreases the efficiency of the reaction of non-bridging oxygens to form high-coordinated species. The increase in abundance of high-coordinated Si species observed by Xue et al. (1991) in sodium tetrasilicate composition glasses quenched from high-pressure melts compared to that found in the sodium disilicate systems further supports this idea. It would thus be expected, that Si coordination changes would be much more difficult in the highly depolymerized metasilicate systems despite the increase in the concentration of non-bridging oxygens.

The Q^n speciation reactions inferred from studies on quenched glasses were based on the assumption that structural relaxations do not occur in the glass on decompression to ambient pressure. The in situ Raman measurements on the alkali tetrasilicate and disilicate glasses clearly demonstrate that significant local structural changes do occur along the decompression route (Wolf et al., 1990; Durben, 1993). Examples of this behavior can be seen in comparing the high-pressure Raman spectra of the tetrasilicate and disilicate glasses with that of the initial starting materials and the pressure-cycled samples (Figs. 17 and 18). On decompression of sodium tetrasilicate glass, the 1100 cm^{-1} band is essentially absent until about 5 GPa. It is only below 1.3 GPa that this band, and the 520 cm^{-1} band, recover a significant fraction of their original intensities.

There are important differences in the Raman spectra of the normal uncompressed glasses in comparison with glasses that have been pressure-cycled (Figs. 19c,d and 20c,d). In particular, the pressure-cycled glasses display an increase in the relative intensities of the Q^2 band near 950 cm^{-1} and the "defect" band near 600 cm^{-1}. The enhanced intensity of the 600 cm^{-1} band indicates that a greater fraction of three-membered siloxane ring structures is retained in the pressure cycled samples. As discussed above, a similar enhancement of the intensity of the defect band of silica and germania glass is also observed under pressure cycling.

Wolf et al. (1990) have proposed that much of the increase in the Q^2 species abundance observed in decompressed alkali silicate samples could result from non-equilibrium reactions that take place along the decompression route. They propose that these highly depolymerized species are formed on the reversion of the high-coordinated silica species to tetrahedral coordination via reaction pathways that differ from those which occur under compression. An example of these differing reaction pathways is illustrated in Figures 21a,b. Under compression, the reaction of Q^4 and Q^3 species can

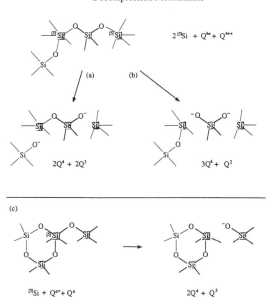

Figure 21. Decompression reversion mechanisms of high-coordinated Si species proposed by Wolf et al. (1990). (a) Formation of Q^4 and Q^3 species. (b) Formation of Q^2 and Q^4 species. (c) Formation of a three-membered siloxane ring.

result in the formation of Si tetrahedral species, labeled Q^{4**}, which are linked to two $^{[5]}Si$ species. Under pressure release, the $^{[5]}Si$ species can revert via two alternate pathways:

$$2\ ^{[5]}Si + Q^{4*} + Q^{4**} \rightarrow 2\ Q^4 + 2\ Q^3 \tag{5}$$

or
$$2\ ^{[5]}Si + Q^{4*} + Q^{4**} \rightarrow 3\ Q^4 + Q^2. \tag{6}$$

In one reaction the Q^{4**} species reverts to a Q^3 species whereas in the other reaction it reverts to a Q^2 species. In either mechanism, atomic diffusion is not required and the resultant distribution of tetrahedral species depends only on which particular Si–O bonds break under decompression. Alternate reaction pathways can also occur for the reversion of $^{[6]}Si$ species.

Finally, it is important to point out that differences may occur in the decompression pathways of glasses that were thermally equilibrated at high pressures above T_g compared to that of glasses which are pressure-cycled at room temperature. Significant structural differences are evident in comparing the Raman spectra of room temperature pressure-cycled glasses with that of glasses at 1 bar that were quenched from high-pressure melts. Compare, for example, the 1 bar Raman spectrum of a sodium tetrasilicate glass sample obtained from a melt at 12 GPa (Xue et al., 1991) with those of the room temperature glass sample cycled to a peak pressure of 13 GPa (Figs. 14 and 17). This underscores the general belief that pressure-induced structural changes that occur in glasses below T_g are likely to occur at much lower pressures in samples that are simultaneously annealed at high temperatures. Furthermore, it may also be that high-pressure samples that are also annealed at high temperatures may be less susceptible to local structural relaxations on decompression to ambient pressure. These questions also indicate the importance of obtaining in situ spectroscopic measurements on silicate liquids under simultaneous high-pressure/high-temperature conditions.

Alkali aluminosilicate systems. Mysen et al. (1990) have used Raman spectroscopy to investigate pressure-induced structural changes in depolymerized alkali aluminosilicate melts. These studies were limited to a peak pressure of 3 GPa and were made on glass samples at 1 bar that were quenched from melts at high pressure. Small changes in the tetrahedral Q^n speciation distribution in these systems were inferred from the Raman spectra. These authors attempted to make a quantitative analysis of the speciation distribution by deconvoluting the Raman band envelope in the 800 cm^{-1} to 1200 cm^{-1} frequency region and calibrating the Raman band cross sections with available NMR data. Through this analysis, Mysen et al. obtained estimates for the equilibrium constant of the Q^3 disproportionation reaction (given in Eqn. 1) for a variety of depolymerized alkali aluminosilicate systems as a function of pressure. They concluded that the Q^3 disproportionation reaction shifts in the direction of increasing disproportionation both with increasing pressure and with increasing Al/(Al+Si) ratio. Although the general conclusion that disproportionation increases with pressure is consistent with Xue et al.'s (1991) interpretations, Mysen et al. find that for high silica aluminosilicates (Al/(Al+Si) = 0.05) the most significant pressure effects occur at low pressure (<1 GPa) whereas Xue et al. (1991) found no significant change in disproportionation below 5 GPa for alumina-free alkali silicate melts.

Yarger et al. (1995) have recently investigated the pressure-dependent structural properties of a slightly depolymerized alkali aluminosilicate melt intermediate along the albite-sodium tetrasilicate join (Al/(Al+Si) = 0.125 with an average of 0.25 NBOs per network cation). [The degree of depolymerization and silica content of this system are much closer to basaltic or andesitic magmas than either albite or sodium tetrasilicate.] In their study, glass samples quenched from melts up to 12 GPa were characterized by vibrational spectroscopy and MAS NMR on ^{29}Si, ^{27}Al, and ^{23}Na nuclei. One very interesting result from this study was that, in contrast to the behavior observed for the depolymerized alkali silicates with the same silica content, no spectral evidence (< 1.0% detection limit) was found for the presence of high-coordinated Si species in these quenched samples to at least 12 GPa (Fig. 22a). Instead, a broad distribution of Al coordination species, including $^{[4]}$Al, $^{[5]}$Al, and $^{[6]}$Al, were inferred from the ^{27}Al MAS spectra of all the high-pressure samples (Fig. 22b). In the normal glass quenched from an ambient pressure melt, the ^{27}Al MAS spectrum shows a single resonance that indicates that essentially all of the aluminum atoms are in four-coordinated sites. With increasing pressure, the average Al coordination gradually increases with the appearance of peaks for five- and six-coordinated aluminum. The concentration of $^{[5]}$Al reaches a maximum value of about 28% at 8 GPa, and then decreases at higher quench pressures. The concentration of $^{[6]}$Al continues to increase with increasing pressure, reaching a value of 48% at 12 GPa. The average Al coordination at 12 GPa is about 5.1. No substantial changes are found in the ^{23}Na MAS NMR spectra with increasing pressure (see Fig. 22c).

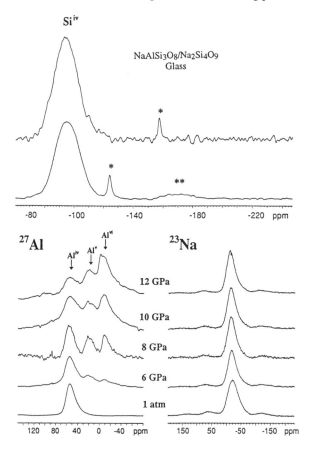

Figure 22. MAS NMR spectra of Ab$_{50}$/NTS$_{50}$ glass quenched from high-pressure melts at temperatures between 1900 and 2200°C (Yarger et al., 1995). (a) ^{29}Si MAS NMR spectrum of Ab/NTS glass quenched from 10 GPa (top spectrum) and 0.1 MPa (bottom spectrum). Samples were enriched in ^{29}Si (92%). Spinning sidebands from the Si$_3$N$_4$ rotor (*) and the $^{[4]}$Si resonance (**) are also indicated. (b) ^{27}Al MAS NMR spectrum of Ab/NTS glasses quenched from high pressures. Resonances for $^{[4]}$Al, $^{[5]}$Al, and $^{[6]}$Al are indicated. (c) ^{23}Na MAS NMR spectrum of Ab/NTS glasses quenched from high pressures.

Raman spectra of the pressure-quenched Ab_{50}/NTS_{50} glass samples (Fig. 23a) show a strong increase in the intensity of the 600 cm^{-1} band and a marked weakening of the 1100 cm^{-1} band (Yarger et al., to be published). These effects become enhanced with increasing peak pressure. As in the alkali tetrasilicate glasses, there is also enhanced scattering intensity in the 950 cm^{-1} region and a buildup of intensity in the region between the 950 cm^{-1} Q^2 and 1100 cm^{-1} Q^3 bands with increasing pressure.

Yarger et al. (to be published) also measured the in situ Raman spectrum of the same albite-sodium tetrasilicate glass as a function of pressure (Fig. 23b). As in the high-silica alkali silicate glasses, the Raman spectrum of this glass shows a concerted loss of the 520 cm^{-1} and 1100 cm^{-1} bands with increasing pressure. Over this same pressure interval, the 600 cm^{-1} band broadens and displays a marked increase in both its absolute and relative intensities. At 12.5 GPa, the 520 cm^{-1} and 1100 cm^{-1} bands have nearly disappeared.

The results of Yarger et al. (1995) support the early suggestion by Waff (1975) that network aluminum atoms are much more susceptible to the formation of high-coordinated species under pressure than silicon. As discussed above, earlier studies on fully polymerized aluminosilicate systems (no NBOs) have found no definitive evidence for the retention of high-coordinated Al or Si species in decompressed samples. It is interesting, however, that in such a weakly depolymerized alkali aluminosilicate melt (0.25 NBOs per network cation), that such a large fraction of high-coordinated Al species can be retained in the decompressed sample. In the melt composition studied by Yarger et al., if all of the NBOs were bound up in the formation of high-coordinated Al species, then a maximum aluminum coordination of six could be obtained. Thus, the energetically costly reaction mechanism involving bridging oxygens is not required to account for the pressure-dependent coordination trends observed in this low alumina (Al/(Al+Si) = 0.125) system. If the coordination change does take place via reaction pathways involving NBOs, these results further suggest that the NBOs in this sodium aluminosilicate system reside on the silicon atoms with at least one aluminum atom present in the surrounding cosphere.

The in situ Raman scattering results of Yarger et al. suggest that at 12.5 GPa, nearly all of the NBOs have been consumed in the formation of high-coordinated Al or Si species. Furthermore, there is no evidence in the in situ Raman spectra that bridging oxygens become involved in the formation of high-coordinated species up to 12.5 GPa at room temperature. The in situ Raman spectra of the pressure-cycled sample shows that a large fraction of the high-coordinated Al species revert to tetrahedral coordination on decompression. Further ^{27}Al MAS NMR measurements by Yarger et al. (to be published) show that far fewer high-coordinated Al species are actually retained in decompressed samples that are pressure cycled without heating than in samples that were heated above T_g at the same pressure. For example, in a glass sample pressure-cycled at room temperature to 12 GPa, a quantitative analysis of the MAS NMR spectra shows that only 5% [5]Al and 6% [6]Al species are recovered compared to 17% and 48%, respectively, for a glass sample quenched obtained from a melt at the same peak pressure. In addition, the increased relative intensity of the 600 cm^{-1} band in the room temperature pressure-cycled sample (Fig. 23c) is consistent with the previous suggestion that three-membered tetrahedral siloxane-type rings are formed on the reversion of high-coordinated species under decompression (see Fig. 21c) (Wolf et al., 1990). Also similar to the alkali silicate glasses is the inference from the in situ Raman data that a significant fraction of Q^2 species are formed through a reversion of high-coordinated species along the decompression path.

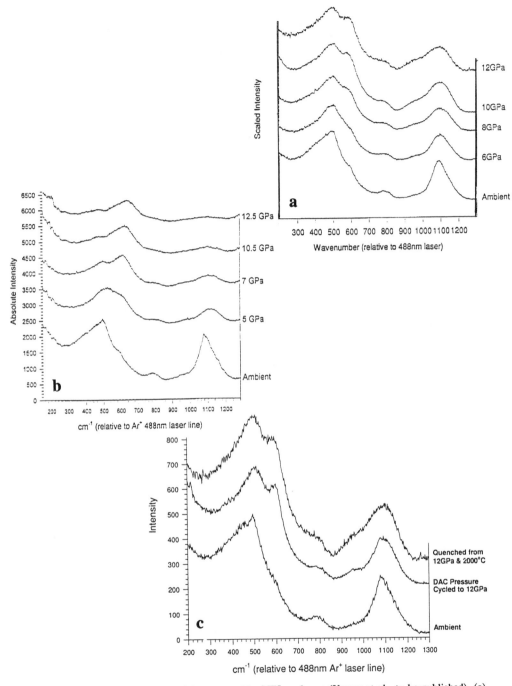

Figure 23. Raman spectra of high-pressure Ab$_{50}$/NTS$_{50}$ glasses (Yarger et al., to be published). (a) Ambient Raman spectra of glasses quenched from high-pressure melts. (b) *In situ* Raman spectrum of glass obtained as a function of pressure. (c) Comparison of the Raman spectra of glasses quenched from 12 GPa and 2000°C, pressure-cycled at room temperature to 12 GPa, and quenched from a melt at 0.1 Mpa.

A final important observation can be made in comparing the 1 bar Raman spectra of the room temperature pressure-cycled sample with that of the high-pressure melt quench, both obtained from a peak pressure of about 12 GPa (Fig. 23c). The Raman spectrum of the high-pressure thermally annealed sample displays an enhanced scattering intensity in the region between the Q^2 and Q^3 high frequency stretching bands (i.e., 950 to 1100 cm^{-1}) compared to that of the room temperature pressure-cycled sample. Alkali silicate glass samples quenched from melts at high pressures also show an enhanced Raman scattering intensity in the spectral region between 950 cm^{-1} and 1100 cm^{-1} which is not observed to the same extent in glass samples that are pressure-cycled below T_g (Dickinson et al., 1990; Xue et al., 1991) (see Fig. 16b). Dickinson et al. (1990) have suggested that scattering in this same region may arise from vibrational modes related to high-coordinated Si (or Al) species. Further work on cross-calibrating the Raman spectrum with NMR data is necessary to more fully interpret these spectral changes.

Alkaline earth metasilicates. In alkaline earth silicates, the structural distinction between "network former" and "network modifier" cations is not as precise as in the alkali silicates (Navrotsky et al., 1982, 1985; McMillan, 1984a). Analysis of X-ray diffraction data on MgSiO$_3$ glass indicates that the Mg ion is surrounded by four oxygen atoms at a distance of 2.08 Å with two additional oxygens at 2.50 Å (Yin et al., 1983). Molecular dynamic simulations generally find that the Mg^{2+} ion coordination is ill defined in alkaline earth silicate melts at low pressures, and occurs in both tetrahedral and octahedral coordinations (Matsui and Kawamura, 1980, 1984; Matsui et al., 1982; Angell et al., 1987; Kubicki and Lasaga, 1991). In contrast Ca^{2+} and the larger, more basic alkaline earth cations primarily exist in octahedral sites. EXAFS and neutron scattering experiments on CaSiO$_3$ glass indicate that Ca has a well-defined octahedral coordination (Eckersley et al., 1988a,b).

Williams and Jeanloz (1988) investigated the pressure-induced structural changes in diopside (CaMgSi$_2$O$_6$) composition glass in situ to pressures above 30 GPa using infrared absorption spectroscopy. The high-pressure spectral changes observed for diopside glass closely parallel the behavior they found for silica glass (Fig. 24). In the diopside glass spectra above 20 GPa, there is a noticeable reduction in intensity of the high frequency tetrahedral stretching band near 1000 cm^{-1} relative to absorption in the 600-900 cm^{-1} region. By 37 GPa, only a broad, featureless absorption band is observed over the entire spectral range from 700 to 1200 cm^{-1}, very similar to that observed for silica glass at the same pressure. On decompression, the pressure-induced spectral changes of diopside glass are largely reversible, again reminiscent of the behavior of silica glass. Williams and Jeanloz concluded that the changes in the infrared absorption spectra of diopside glass are a result of a pressure-induced Si coordination change. This coordination transformation takes place over a large pressure range and, even at 37 GPa, there is probably a significant distribution of Si over the different coordination states.

The spectral interpretations of Williams and Jeanloz (1988) are generally consistent with molecular dynamic simulations on diopside (CaMgSi$_2$O$_6$) melts by Angell et al. (1987). In these simulations it was found that tetrahedral Si^{4+} and Mg^{2+} cations increase coordination over a wide range in pressure, and that at 20 GPa the average coordination state of both the Si^{4+} and Mg^{2+} cations is about five.

Kubicki et al. (1992) used both infrared and Raman vibrational spectroscopies to study in situ the pressure-induced structural changes in MgSiO$_3$ (enstatite), CaMgSi$_2$O$_6$, (diopside), and CaSiO$_3$ (wollastonite) composition glasses to pressures up to 45 GPa. The in situ infrared absorption and Raman spectra of these glasses are shown in Figure 25. All of the glasses display a gradual decrease with pressure in their IR absorbance between 900 cm^{-1} and 1250 cm^{-1} with a simultaneous increase in absorbance in the 600 cm^{-1} to 900 cm^{-1} region. These spectral changes appear to occur at lower pressures in enstatite glass than for either the diopside or wollastonite glasses. At 24 GPa, the infrared absorption spectra of MgSiO$_3$ glass shows a relatively flat absorption between 700 and 1200 cm^{-1}, and is similar to that of silica glass at about 38 GPa (see Fig. 5b). At 35 GPa, the absorption of CaSiO$_3$ glass becomes very broad in this same spectral region, however a distinct band in the 1100 cm^{-1} region can still be resolved at this pressure.

At ambient pressure the Raman spectra of the MgSiO$_3$-CaSiO$_3$ glasses show two relatively strong bands centered near 630 cm^{-1} and 1000 cm^{-1}. The broad band envelope at high frequency can be deconvoluted into four bands at about 850 cm^{-1}, 900 cm^{-1}, 950 cm^{-1}, and 1100 cm^{-1} which, as in the alkali silicates correspond to symmetric Si-O stretching vibrations in Q^0, Q^1, Q^2 and Q^3 species, respectively. The band near 650 cm^{-1} probably corresponds to symmetric bending deformations of Si-O-Si linkages in the highly depolymerized network.

With increasing pressure there is a marked positive frequency shift and broadening of the 630 cm^{-1} Raman band. This band weakens and disappears in all three samples near 20 GPa. Simultaneous to these changes, there is a gradual reduction in intensity and loss of resolution in the high frequency band envelope between 800 and 1150 cm^{-1}. However, even at the highest pressures of the Raman measurements, 35 to 45 GPa, some residual scattering intensity persists in this spectral region.

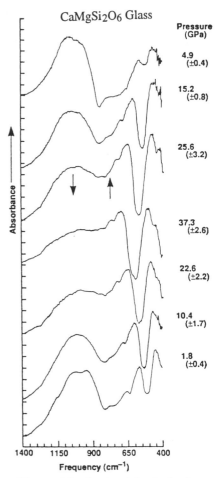

Figure 24. In situ infrared absorbance spectrum of diopside composition glass on increasing and decreasing pressure, top to bottom (Williams and Jeanloz, 1988). [Used by permission of the editor of *Science*, from Williams and Jeanloz (1988), Fig. 1, p. 903.]

The infrared absorbance data for the CaSiO$_3$-MgSiO$_3$ glasses are consistent with the interpretation that there is a gradual increase in the Si and Mg coordinations with increasing pressure. The interpretation of the Raman data is less clear. The reduction in intensity of the Raman bands in the 800 cm^{-1} to 1150 cm^{-1} region with increasing pressure, more marked for MgSiO$_3$ glass, suggests that NBOs are consumed in forming high-coordinated species. However, the decrease in intensity in this spectral region is not as dramatic as in the alkali tetrasilicate and disilicate glasses, and occurs over a broader pressure range. This observation is in line with the trend inferred from data on alkali

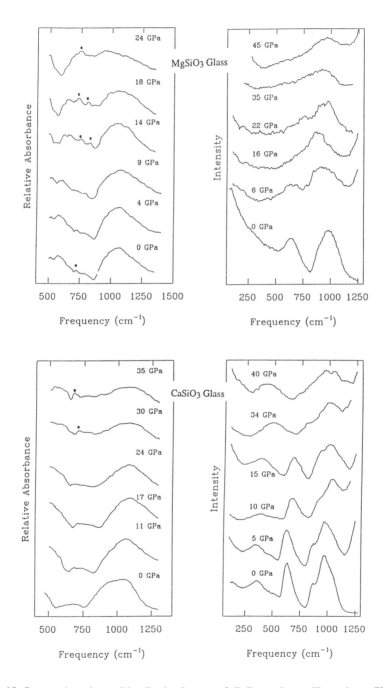

Figure 25. Pressure dependence of the vibrational spectra of alkaline earth metasilicate glasses (Kubicki et al., 1992). (a) Infrared absorbance spectrum of $MgSiO_3$ glass. (b) Infrared absorbance spectrum of $CaSiO_3$ glass. (c) Raman spectrum of $MgSiO_3$ glass. (d) Raman spectrum of $CaSiO_3$ glass.

silicate glasses that the activation energy for the formation of high-coordinated species through NBOs increases with increasing depolymerization of the tetrahedral network. The onset of the coordination transformations in the metasilicate glasses is difficult to establish from the data. However, there is a distinct reduction in intensity in both the 630 cm^{-1} and 1000 cm^{-1} Raman band envelopes of MgSiO$_3$ glass even in the 6 GPa spectrum and suggests that coordination transformations could begin in MgSiO$_3$ glass at pressures as low as 6 GPa. A similar degree of change in the Raman spectrum of CaSiO$_3$ glass does not occur until about 15 ($\pm$ 3) GPa.

Kubicki et al. (1992) concluded that the residual Raman intensity observed in the 800 cm^{-1} to 1100 cm^{-1} region for the metasilicate glasses indicates that tetrahedral species with NBOs are retained in the glass structures to at least 45 GPa. However, this conclusion must be considered tentative since, as discussed above, high-pressure Raman studies on depolymerized alkali silicate and aluminosilicate glasses suggest that scattering in this spectral region may also arise from vibrational modes of structural units incorporating high-coordinated network species (Dickinson et al., 1990; Xue et al., 1991; Yarger et al., to be published). In addition, if ordered regions containing edge-shared SiO$_6$ octahedra are present in the glass structure at this pressure (formed via the reaction on bridging oxygens), then Raman active vibrational modes could occur in this frequency region. At 45 GPa, the A$_{1g}$ Raman mode of stishovite occurs at about 900 cm^{-1}, and the much weaker E$_g$ and B$_{2g}$ modes occur near 680 cm^{-1} and 1120 cm^{-1}, respectively (Kingma et al., 1995).

The experimental high-pressure studies on MgSiO$_3$ glass can be compared to several molecular dynamic simulations on this same system. Kubicki and Lasaga (1991; see also Kubicki et al., 1992) carried out MD simulations on MgSiO$_3$ glass at 300 K using potentials derived from molecular orbital theory and electron-gas calculations. In these simulations, the transformation of Si^{4+} to octahedral coordination occurred over an extremely broad range in pressure, and at much higher (estimated) pressures than that suggested by the in situ spectroscopic measurements. In the simulation experiments, only about 8% of the Si^{4+} ions are in high-coordinated ($^{[5]}$Si + $^{[6]}$Si) sites at 30 GPa; whereas between 60 GPa and 90 GPa the abundance of high-coordinated Si species changes more significantly, from 27% to 65%. The Mg^{2+} coordination change occurs at much lower pressures. By 30 GPa, essentially all of the Mg^{2+} ions in the simulation runs occur in high-coordinated sites. A precise comparison with the simulation results of Kubicki et al. is complicated by the fact that the simulation pressures were not directly obtained but, instead, were estimated from the simulation densities and an extrapolation of the equation of state of MgSiO$_3$ glass (based on the ambient experimental compressibility). A perhaps more meaningful comparison is that the most pronounced change in the Si coordination in the simulations occurs over the same density range as that which occurs in the solid state on the transformation from tetrahedral to octahedral Si near 25 GPa (Ito and Matsui, 1977, 1978; Yagi et al., 1978). Qualitatively similar results are also found in the MD simulation experiments by Matsui and Kawamura (1980, 1984) and Matsui et al. (1982) on MgSiO$_3$ glass and by Wasserman et al. (1993) on MgSiO$_3$ melts. In these studies, empirical pair potential models were used with parameters derived from fitting to structural data at ambient pressure.

Alkaline earth orthosilicates. A characterization of the ultra-high pressure melt behavior in highly depolymerized silicate systems is crucial for an understanding of mantle melting relations in ultrabasic rock assemblages and the properties of deep-origin ultramafic magmas. However, very little data currently exists on the high-pressure structural changes in orthosilicate or other highly depolymerized composition glasses and melts. At ambient pressure, the structure of orthosilicate glasses are thought to be very similar to that of their related crystalline phases. In crystalline orthosilicates, the silica tetrahedra are completely depolymerized and the structure is composed entirely of isolated Q^0 tetrahedral species.

Considerable insights into the compressional mechanisms and high-pressure behavior of orthosilicate liquids and glasses has resulted from studies on the high-pressure metastable behavior of crystalline orthosilicates. Fayalite (Fe_2SiO_4) and forsterite (Mg_2SiO_4) are two of the most extensively studied orthosilicate minerals. It is found that near 35 to 40 GPa, fayalite becomes X-ray amorphous under static, room-temperature compression (Richard and Richet, 1990; Williams et al., 1990). Near this same pressure there is significant collapse in the volume along the shock Hugoniot (Mashimo et al., 1980). In situ infrared absorption measurements of fayalite by Williams et al. (1990) indicate that, beginning near 20 to 25 GPa, there is a dramatic and reversible weakening of the high frequency tetrahedral stretching band and an emergence of a strong band between 600 and 800 cm^{-1} (Fig. 26). These authors have interpreted the spectral changes to indicate that silicon undergoes a continuous transition from fourfold toward sixfold coordination in metastable fayalite beginning near 20 GPa and extending to pressures above 46 GPa. Furthermore, they have speculated that the crystal-to-amorphous transition in this system is precipitated by a pressure-induced instability of the tetrahedral silicon coordination.

Shock and static compression experiments suggest that similar behavior occurs in crystalline forsterite, but at much higher pressures. Shock measurements on crystalline

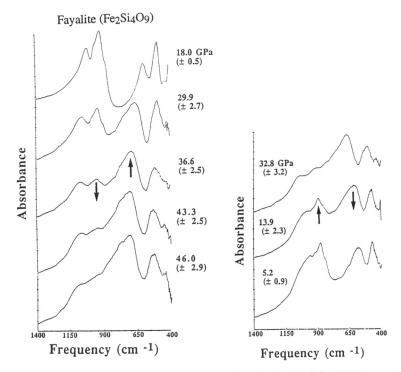

Figure 26. In situ infrared absorbance spectrum of crystalline fayalite (Fe_2SiO_4) (Williams et al., 1990). (a) Increasing pressure. (b) Decreasing pressure.

forsterite indicate a significant volume collapse of the structure occurs along the Hugoniot near 50 GPa (Jackson and Ahrens, 1979; Furnish and Brown, 1986; Brown et al., 1987) and that samples recovered above this pressure show diaplectic glass formation (Jeanloz et al., 1977; Jeanloz, 1980). Experiments (Guyot and Reynard, 1992) suggest that crystalline olivine, $(Mg,Fe)_2SiO_4$, transforms to an amorphous phase under static compression near 70 GPa.

There have been few high-pressure structural studies on orthosilicate liquids and glasses. Williams (1990) made one of the first attempts at investigating the ultra-high pressure structure of an orthosilicate composition melt. In his study, 1-bar infrared absorption spectra were measured of quenched Mg_2SiO_4 glasses which were synthesized by laser fusion at pressures near 50 GPa. The infrared spectrum of these glasses was consistent with a structure that was completely composed of tetrahedral silicon. Williams concluded that if high-coordinated Si species are formed at high pressures in orthosilicate liquids then they are completely unstable under decompression to ambient pressure.

Durben et al. (1993) used Raman spectroscopy to investigate the high-pressure structural properties of Mg_2SiO_4 glass at room temperature as a function of pressure. The in situ Raman spectra they obtained for forsterite glass to 50 GPa are shown in Figure 27. At ambient pressure, the Raman spectrum of the glass is characterized by a single broad asymmetric band centered near 870 cm^{-1} which is assigned to vibrations principally involving symmetric and antisymmetric stretching motions of isolated Q^0 tetrahedral species (Williams et al., 1989; Cooney and Sharma, 1990). The small band at somewhat lower frequencies (near 720 cm^{-1}) has been assigned to stretching vibrations of Si_2O_7 dimers in the structure (Williams et al., 1989; Cooney and Sharma, 1990).

Under initial compression, the frequency of the main tetrahedral stretching band of Mg_2SiO_4 glass increases at a similar rate to the pressure shift of the Q^0 tetrahedral stretching band of crystalline forsterite (Durben et al., 1993). The 720 cm^{-1} "dimer" band has a slightly greater positive frequency shift which is consistent with a tightening of the Si-O-Si angles within these units. The intensity of this band decreases with increasing pressure and cannot be followed to above about 20 GPa. However, it is difficult to distinguish whether this apparent loss in intensity is real or is due to a shift of the dimer band into the broadened manifold of the main Raman band. No other major changes in the spectrum are evident up to 20 GPa, and all frequency shifts are fully reversible over this pressure range. At pressures above 20 GPa there is a gradual but marked increase in the breadth of the main Raman band in forsterite glass, primarily arising from an increase in intensity on the low frequency side of this band between 700 cm^{-1} and 900 cm^{-1}. These changes continue gradually to at least 51 GPa where the main Raman band is centered near 950 cm^{-1}. On decompression, the increased scattering intensity in the 700 to 900 cm^{-1} region is lost and the Si_2O_7 dimer band near 720 cm^{-1} returns.

At ambient pressure, the Raman spectrum of the pressure-cycled glass sample shows some small but significant differences from that of the normal glass sample. For the pressure-cycled glass sample there is a well-defined shoulder near 950 cm^{-1}, which does not appear in the normal glass spectrum, and a small increase in the intensity of the Si_2O_7 dimer band near 720 cm^{-1}.

The pressure-induced changes in the Raman spectrum of Mg_2SiO_4 glass can be related to similar changes that occur in crystalline forsterite at high pressure (Fig. 27) (Durben et al., 1993). In forsterite, two new bands emerge in the Raman spectrum near

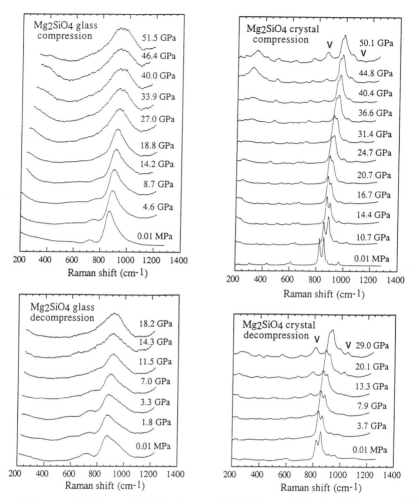

Figure 27. Pressure dependence of the Raman spectra of Mg_2SiO_4 crystal and glass (Durben et al. 1993). (a) Glass Raman spectrum on compression. (b) Crystal Raman spectrum on compression. (c) Glass Raman spectrum on decompression. (d) Crystal Raman spectrum on decompression.

30 GPa: a relatively strong band at 825 cm^{-1} and a weaker band at 1060 cm^{-1}. Both of these bands increase in intensity and shift to higher frequencies with increasing pressure. On decompression the new bands are retained in the Raman spectrum to about 8 GPa, below which they can no longer be resolved. From an extrapolation of the pressure data, estimated values of 750 cm^{-1} and 960 cm^{-1} are obtained for the positions of these bands at ambient pressure. The Raman spectrum of the pressure-cycled crystal sample is very similar to that of uncompressed crystalline forsterite except for a residual broadening of all of the Raman bands and an enhanced scattering intensity on the high frequency side of the main Raman bands.

Durben et al. (1993) have interpreted the spectral changes of both Mg_2SiO_4 glass and crystal to indicate that pressure-induced Si-O-Si linkages begin to form in both of these materials above about 20 and 30 GPa, respectively. The extrapolated position of the new Raman band in crystalline forsterite at 750 cm^{-1} occurs in a frequency range typical of symmetric bending vibrations of Si-O-Si dimers in β-Mg$_2$SiO$_4$ and other pyrosilicate structures (McMillan and Akaogi, 1987). The assignment of the extrapolated 960 cm^{-1} band is more ambiguous but may be related to vibrations of terminal Q^1 species on silica dimers, or possibly Q^2 species on more polymerized silica units. In β-Mg$_2$SiO$_4$, a similar band, assigned to the terminal Q^1 vibrations, appears near 918 cm^{-1} (McMillan and Akaogi, 1987). The observation of excess spectral intensity in the Raman spectrum of pressure-cycled forsterite glass in the 750 cm^{-1} and 950 cm^{-1} region suggests that similar silica dimer, or more polymerized, silica species may be retained in the glass structure of the decompressed sample.

Whether these polymerized species involve tetrahedral or higher coordinated silicon can not be distinguished from the spectral data. Formation of Si$_2$O$_7^{-6}$ dimers would retain four-fold Si coordination. However, this could only occur in the olivine stoichiometry if a Si-O bond were broken leaving one oxygen atom (per dimer) nonbonded to silicon. Formation of Si$_2$O$_8^{-8}$ dimers would not require Si-O bond breakage and would result in the formation of either one or two five-coordinated silicon species (depending on whether the polyhedra are corner- or edge-shared). The assignment of the 750 cm^{-1} band to vibrations involving higher coordinated silicon polyhedral species is consistent with Williams et al.'s (1990) interpretation of a new band which appears in the same spectral region of the infrared absorption spectrum of crystalline fayalite (Fe$_2$SiO$_4$) above 20 GPa.

Durben et al. (1993) have proposed that the formation of dimer defects in metastable forsterite at lower pressures is a prelude to the crystal-amorphous transition. The high-pressure shock and static experiments on forsterite suggest that a more complete polymerization of the silica network, through an extensive formation of high-coordinated silicon species at higher pressures, may be required to produce a significant volume collapse and irreversible disordering of the structure. Because of the inferred similarity in the pressure-dependent structural properties of crystalline and amorphous Mg$_2$SiO$_4$, analogous high-pressure polymerization reactions would be expected to occur in the glass in a similar pressure regime.

ACKNOWLEDGMENTS

The authors are indebted to Jeffrey Yarger, Brent Poe, Kenneth Smith, David Rubie, Dan Durben, Mary VerHelst-Voorhees, Isabelle Daniel, and Jason Diefenbacher for allowing us to include in this review some of their most recent data which is not yet in press. This most recent and highly seminal work has greatly increased our general understanding of pressure-effects on aluminosilicate melt structure and properties and has added immensely to the scope of this review. We are also grateful to Jonathan Stebbins, Kathleen Kingma, Andrew Chizmeshya, Don Dingwell and two anonymous reviewers for constructive comments on the manuscript and to A. Chizmeshya for his help in preparing several of the figures. We owe a special thanks to Paul Ribbe, Series Editor, without whom there would be no Short Course series. Our research on silicate melt studies has been supported by grants from the National Science Foundation, Experimental Geochemistry program.

REFERENCES

Adam G, Gibbs JH (1965) On the temperature dependence of cooperative relaxation properties in glass-forming liquids. J Chem Phys 43:139-146

Agee CB, Walker D (1988) Static compression and olivine flotation in ultrabasic silicate liquid. J Geophys Res 93:3437-3449

Agee CB, Walker D (1993) Olivine flotation in mantle melts. Earth Planet Sci Lett 114:315-324

Anan'in AV, Breusov ON, Dremin AN, Pershin SV, Rogacheva AI, Tatsii VF (1974) Action of shock waves on silicon dioxide, II, quartz glass. Combust Explos Shock Waves 10:504-508

Angell CA, Cheeseman PA, Kadiyala R (1987) Diffusivity and thermodynamic properties of diopside and jadeite melts by computer simulation studies. Chem Geol 62:85-95

Angell CA, Cheeseman PA, Tamaddon S (1983) Water-like transport property anomalies in liquid silicates investigated at high T and P by computer simulation techniques. Bull Minéral 1 -2:87-9 9

Angell CA, Cheeseman PA, Tammadon S (1982) Pressure enhancement of ion mobilities in liquid silicates from computer simulation studies to 800 kbar. Science 218:885-887

Arndt J (1983) Densification of glasses of the system TiO_2-SiO_2 by very high static pressures. Phys Chem Glasses 24:104-110

Ault BS (1979) Infrared matrix isolation studies of $M+SiF_5^-$ ion pair and its chlorine-fluorine analogs. Inorg Chem 18:3339-3343

Barker LM, Hollenbach RE (1970) Shock-wave studies of PMMA, fused silica, and sapphire. J Appl Phys 41:4208-4226

Binggeli N, Chelikowsky JR (1992) Elastic instability in α-quartz under pressure. Phys Rev Lett 69:2220-2223

Binggeli N, Keskar NR, Chelikowsky JR (1994) Pressure-induced amorphization, elastic instability, and soft modes in α-quartz. Phys Rev B 49:3075-3081

Birch F, Dow RB (1936) Compressibility of rocks and glasses at high temperatures and pressures. Bulletin of the Geological Society of America 47:1235-1255

Bockris JOM, Lowe DC (1954) Viscosity and the structure of molten silicates. Proceedings of the Royal Society London A 226:423-435

Boslough MB, Rigden SM, Ahrens TJ (1986) Hugoniot equation of state of anorthite glass and lunar anorthosite. Geophys J R Astron Soc Geophys Res Letts 84:455-473

Bottinga Y (1985) On the isothermal compressibility of silicate liquids at high pressure. Earth Planet Sci Lett 74:350-360

Bottinga Y, Weill D (1970) Densities of liquid silicate systems calculated from partial molar volumes of oxide components. Am J Sci 269:169-182

Bottinga Y, Weill DF (1972) The viscosity of magmatic silicate liquids: a model for calculation. Am J Sci 272:438-475

Boyd FR, England JL (1963) Effect of pressure on the melting of diopside, $CaMgSi_2O_6$, and albite, $NaAlSi_3O_8$, in the range up to 50 kilobars. J Geophys Res 68:311-323

Brawer SA (1981) Defects and fluorine diffusion in sodium fluoroberyllate glass: a molecular dynamics study. J Chem Phys 75:3522-3541

Brawer SA, White WB (1975) Raman spectroscopic investigation of the structure of silicate glasses. I. The binary alkali silicates. J Chem Phys 63:2421-2432

Bray PJ, O'Keefe JG (1963) Nuclear magnetic resonance investigations of the structure of alkali borate glasses. Phys Chem Glasses 4:37-46

Bridgman PW (1939) The high pressure behavior of miscellaneous minerals. Am J Sci 237:7-18

Bridgman PW (1948) The compression of 39 substances to 100,000 kg/cm^2. Proc Am Acad Arts Sci 76:55-70

Bridgman PW, Simon J (1953) The effects of very high pressures on glass. J Appl Phys 24:405-413

Brown JM, Furnish MD, McQueen RG (1987) Thermodynamics for $(Mg,Fe)_2SiO_4$ from the Hugoniot. In: Manghnani MH, Syono Y (eds) High Pressure Research in Mineral Physics, p. 373-384, AGU, Washington D. C.

Bruckner R (1970) Properties and structure of vitreous silica. I. J Non-Cryst Solids 5:123-175

Burnham CW (1979) The importance of volatile constituents. In: Yoder HS (ed) The Evolution of Igneous Rocks. Princeton Univ Press, Princeton, p 439-482.

Chaplot SL, Sikka SK (1993) Molecular dynamics simulation of pressure-induced crystalline-to-amorphous transitions in some corner-linked polyhedral compounds. Phys Rev B47: 5710-5714

Charles RJ (1967) Activities in the Li_2O-, Na_2O- and K_2O-SiO_2 solutions. J Am Ceram Soc 50:631-640

Carroll MR, Holloway Jr (eds) (1994) Volatiles in Magmas. Rev Mineral Vol 30, 517 p, Mineralogical Soc America, Washington, DC

Chelikowsky JR, King HE, Jr., Troullier N, Martins JL, Glinnemann J (1990) Structural properties of α-quartz near the amorphous transition. Phys Rev Lett 65:3309-3312

Chhabildas LC, Grady DE (1984) Shock loading behavior of fused quartz. In: Asay JR, Graham RA, Straub GK (eds) Shock Waves in Condensed Matter—1983. p 155-178, Elsevier, New York

Cohen HM, Roy R (1965) Densification of glass at very high pressure. Phys Chem Glasses 6:149-161

Coleman B (1983) Applications of silicon-29 NMR spectroscopy. In: Lazlo P (ed) NMR of Newly Accessible Nuclei, 2:192-228, Academic Press, New York

Couty R, Sabatier G (1978) Contribution a l'étude de l'enthalpie du verre de silice densifie. J Chimie Phys 75:843-848

Daniel I, Gillet Ph, Poe BT, McMillan PF (1995) Raman spectroscopic study of structural changes in calcium aluminate ($CaAl_2O_4$) glass at high pressure and high temperature. Chem Geol, submitted

Davoli I, Paris E, Stizza S, Benfatto M, Fanfoni M, Gargano A, Bianconi A, Seifert F (1992) Structure of densified vitreous silica: silicon and oxygen XANES spectra and multiple scattering calculations. Phys Chem Min 19:171-175

Desa JAE, Wright AC, Sinclair RN (1988) A neutron diffraction investigation of the structure of vitreous germania. J Non-Cryst Solids 99:276-288

Devine RAB, Arndt J (1987) Si-O bond length modification in pressure-densified amorphous SiO_2. Phys Rev B 35:

Devine RAB, Dupree R, Farnan I, Capponi JJ (1987) Pressure-induced bond-angle variation in amorphous SiO_2. Phys Rev B 35:2560-2562

Dickinson JE, Scarfe CM (1985) Pressure induced structural changes in $K_2Si_4O_9$ silicate melt. EOS Trans Am Geophys Union 66:395

Dickinson JE, Scarfe CM, McMillan P (1990) Physical properties and structure of $K_2Si_4O_9$ melt quenched from pressures up to 2.4 GPa. J Geophys Res 95:15675-15681

Dunn T, Scarfe CM (1986) Variation of the chemical diffusivity of oxygen and viscosity of an andesite melt with pressure at constant temperature. Chem Geol 54:203-215

Dupree E, Pettifer RF (1984) Determination of the Si-O-Si- bond angle distribution in vitreous silica by magic angle spinning NMR. Nature 308:523-525

Dupree R, Holland D, Mortuza MG (1987) Six-coordinated silicon in glasses. Nature 328:416

Dupree R, Holland D, Mortuza MG, Collins JA, Lockyer MWG (1988) An MAS study of network-cation coordination in phosphosilicate glasses. J Non-Cryst Solids 106:403-407

Dupree R, Holland D, Williams DS (1986) The structure of binary alkali silicate glasses. J Non-Cryst Solids 81:185-200

Durben DJ (1993) Raman Spectroscopic Studies of the High Pressure Behavior of Network Forming Tetrahedral Oxide Glasses. Ph.D Dissertation, Arizona State University

Durben DJ, McMillan PF, Wolf GH (1993) Raman study of the high-pressure behavior of forsterite (Mg_2SiO_4) crystal and glass. Am Mineral 78:1143-1148

Durben DJ, Wolf GH (1991) Raman spectroscopic study of the pressure-induced coordination change in GeO_2 glass. Phys Rev B 43:2355-2363

Eckersley MC, Gaskell PH, Barnes AC, Chiex P (1988a) The environment of Ca ions in silicate glasses. J Non-Cryst Solids 106:132-136

Eckersley MC, Gaskell PH, Barnes AC, Chiex P (1988b) Structural ordering in a calcium silicate glass. Nature 335:525-527

Engelhardt G, Michel D (1987) High-Resolution Solid-State NMR of Silicates and Zeolites. John Wiley and Sons, New York

Engelhardt G, Zeigan D, Jancke H, Hoebbel D, Weiker W (1975) Zu abhangigkeit der struktur der silicatanionen in wassrigen natriumsilicatlosungen vom Na:Si verhaltnis. Z Anorg Allg Chem 418:17-28

Farber DL, Williams Q (1992) Pressure-induced coordination changes in alkali-germanate melts: an in-situ spectroscopic investigation. Science 256:1427-1430

Fujii T (1981) Ca-Sr chemical diffusion in melt of albite at high temperature and pressure. EOS Trans Am Geophys Union 62:429

Fujii T, Kushiro I (1977) Density, viscosity and compressibility of basaltic liquids at high pressures. Carnegie Inst Washington Yearb 76:419-424

Furnish MD, Brown JM (1986) Shock loading of single-crystal olivine in the 100-200 GPa range. J Geophys Res 91:4723-4729

Furukawa T, Fox KE, White WB (1981) Raman spectroscopic investigation of the structure of silicate glasses. III. Raman intensities and structural units in sodium silicate glasses. J Chem Phys 75:3226-3237

Galeener FL (1988) Current models for amorphous SiO_2. In: Devine RAB (ed) The Physics and Technology of Amorphous SiO_2, p 1-13, Plenum Press, New York

Gibbs JH, DiMarzio EA (1958) Nature of the glass transition and the glassy state. J Chem Phys 28:373-383

Grimsditch M (1984) Polymorphism in amorphous SiO_2. Phys Rev Lett 52:2379-2381

Grimsditch M (1986) Annealing and relaxation in the high-pressure phase of amorphous SiO_2. Phys Rev B 34:4372-4373

Grimsditch M, Bhadra R, Meng Y (1988) Brillouin scattering from amorphous materials at high pressures. Phys Rev B 38:7836-7838

Guyot F, Reynard B (1992) Pressure-induced structural modifications and amorphization in olivine compounds. Chem Geol 96:411-420

Halvorson K, Wolf GH (1990) Pressure-induced amorphization of cristobalite: structural and dynamical relationships of crystal-amorphous transitions and polymorphic glass transitions in silica polymorphs. EOS Trans Am Geophys Union 71:1671

Hazen RM, Finger LW (1981) Bulk moduli and high-pressure crystal structures of rutile-type compounds. J Phys Chem Solids 42:143-151

Hazen RM, Finger LW, Hemley RJ, Mao HK (1989) High-pressure crystal chemistry and amorphization of α-quartz. Solid State Comm 72:507-511

Hemley RJ, Jephcoat AP, Mao HK, Ming LC, Manghnani MH (1988) Pressure-induced amorphization of crystalline silica. Nature 334:52-54

Hemley RJ, Bell PM, Mysen BO (1986) Raman spectroscopy of SiO_2 glass at high pressure. Phys Rev Lett 57:747-750

Hemley RJ, Prewitt CT, Kingma KJ (1994) High-pressure behavior of silica. In: Heaney PJ, Prewitt CT, Gibbs GV (eds) Silica: Physical Behavior, Geochemistry, and Materials Applications. Rev Mineral 29:41-81

Hess PC (1977) Structure of silicate melts. Can Mineral 15:162-178

Hochella MF, Brown GE (1984) Structure and viscosity of rhyolitic composition melts. Geochim Cosmochim Acta 48:2631-2640

Hochella MF, Brown GE (1985) The structures of albite and jadeite composition glasses quenched from high pressure. Geochim Cosmochim Acta 49:1137-1142

Houser B, Alberding N, Ingalls R, Crozier ED (1988) High-pressure study of α-quartz GeO_2 using extended X-ray absorption fine structure. Phys Rev B 37:6513-6516

Hsich S-Y, Montrose CJ, Macedo PB (1971) The annealing dynamics of fused silica. J Non-Cryst Solids 6:37-48

Itie JP, Polian A, Calas G, Petiau J, Fontaine A, Tolentino H (1989) Pressure-induced coordination changes in crystalline and vitreous GeO_2. Phys Rev Lett 63:398-401

Itie JP, Polian A, Calas G, Petiau J, Fontaine A, Tolentino H (1990) Coordination changes in crystalline and vitreous GeO_2. High Press Res 5:717-719

Ito E, Matsui Y (1977) Silicate ilmenites and the post-spinel transformations. In: Manghnani MH, Akimoto S (eds) High-Pressure Research - Applications to Geophysics, p 193-209, Academic Press, New York

Ito E, Matsui Y (1978) Synthesis and crystal-chemical characterization of $MgSiO_3$ perovskite. Earth Planet Sci Lett 38:443-450

Jackson I, Ahrens TJ (1979) Shock-wave compression of single crystal forsterite. J Geophys Res 84:3039-3048

Jeanloz R (1980) Shock effect in olivine and implications for the Hugoniot data. J Geophys Res 85:3163-3176

Jeanloz R, Ahrens TJ, Lally JS, Nord Jr. GL, Christie JM, Heuer AM (1977) Shock produced olivine glass: first observation. Science 197:457-459

Jellison GE, Feller SA, Bray PJ (1978) A re-examination of the fraction of 4-coordinated boron atoms in lithium borate glass system. Phys Chem Glasses 19:52-53

Jorgensen JD (1978) Compression mechanisms in α-quartz structures—SiO_2 and GeO_2. J Appl Phys 49:5473-5478

Kanzaki M, Xue X, Stebbins F (1989) High pressure phase relations in $Na_2Si_2O_5$, $Na_2Si_4O_9$ and $K_2Si_4O_9$ up to 12 GPa. EOS Trans Am Geophys Union 70:1418

Kimmel RM, Uhlmann DR (1969) On the energy spectrum of densified silica glass. Phys Chem Glasses 10:12-17

Kingma KJ, Cohen RE, Hemley RJ, Mao HK (1995) Transformation of stishovite to a denser phase at lower mantle pressures. Nature 374:243-245

Kingma KJ, Hemley RJ, Mao H-K, Veblen DR (1993a) New high-pressure transformation in α-quartz. Phys Rev Lett 70:3927-3930

Kingma KJ, Meade C, Hemley RJ, Mao HK, Veblen DR (1993b) Microstructural observations of α-quartz amorphization. Science 259:666-669

Kinomura N, Kume S, Koizumi M (1975) Synthesis of $K_2SiSi_3O_9$ with silicon in 4- and 6-coordination. Mineral Mag 40:401-404

Kirkpatrick RJ (1988) MAS NMR spectroscopy of minerals and glasses. In: Spectroscopic Methods in Mineralogy and Geology. Hawthorne FC (ed) Rev Mineral 18: 341-404

Kondo KI, Ito S, Sawaoka A (1981) Nonlinear pressure dependence of the elastic moduli of fused quartz up to 3 GPa. J Appl Phys 52:2826-2831

Kress VC, Williams Q, Carmichael ISE (1988) Ultrasonic investigation of melts in the system Na_2O-Al_2O_3-SiO_2. Geochim Cosmochim Acta 52:283-293

Kubicki JD, Hemley RJ, Hofmeister AM (1992) Raman and infrared study of pressure-induced structural changes in $MgSiO_3$, $CaMgSi_2O_6$, and $CaSiO_3$ glasses. Am Mineral 77:258-269

Kubicki JD, Lasaga AC (1988) Molecular dynamics simulations of SiO_2 melt and glass: ionic and covalent models. Am Mineral 73:941-955

Kubicki JD, Lasaga AC (1990) Molecular dynamics and diffusion in silicate melts. In: Ganguly J (ed) Diffusion, atomic ordering, and mass transport: selected problems in geochemistry, p 1-50, Springer-Verlag, New York

Kubicki JD, Lasaga AC (1991) Molecular dynamics simulation of pressure and temperature effects on $MgSiO_3$ and Mg_2SiO_4 melts and glasses. Phys Chem Min 17:661-673

Kushiro I (1976) Changes in viscosity and structure of melts of $NaAlSi_2O_6$ composition at high pressure. J Geophys Res 81:6347-6350

Kushiro I (1977) Phase transformation in silicate melts under upper-mantle conditions. In: Manghnani MH, Akimoto S (eds) High-Pressure Research: Applications in Geophysics, p 25-37, Academic Press,

Kushiro I (1978a) Density and viscosity of hydrous calc-alkalic andesite magma at high pressures. Carnegie Inst Washington Year Book 77:675-678

Kushiro I (1978b) Viscosity and structural change of albite ($NaAlSi_3O_8$) melt at high pressure. Earth Planet Sci Lett 41:87-90

Kushiro I (1980) Viscosity, density, and structures of silicate melts at high pressures, and their petrological applications. In: Hargraves RB (ed) Physics of Magmatic Processes, p 92-120, Princeton University Press, Princeton

Kushiro I (1983) Effect of pressure on the diffusivity of network-forming cations in melts of jadeitic compositions. Geochimica Cosmochimica Acta 47:1415-1422

Kushiro I (1986) Viscosity of partial melts in the upper mantle. J Geophys Res 91:9343-9350

Kushiro I, Yoder HS, Mysen BO (1976) Viscosities of basalt and andesite melts at high pressures. J Geophys Res 81:66351-6356

Lange RL, Carmichael ISE (1990) Thermodynamic properties of silicate liquids with emphasis on density, thermal expansion and compresibility. In: Nicholls J, Russell JK (eds) Modern Methods of Igneous Petrology: Understanding Magmatic Processes. Rev Mod Mineral 24:25-64.

Leadbetter AJ, Wright AC (1972) Diffraction studies of glass structure II. The structure of vitreous germania. J Non-Cryst Solids 7:37-52

Lebedev AA (1921) Proc State Opt Inst Leningr 2:10

Liu SB, Stebbins JF, Schneider E, Pines A (1988) Diffusive motion in alkali silicate melts: an NMR study at high temperature. Geochim Cosmochim Acta 52:527-538

Lyzenga GA, Ahrens TJ (1983) Shock temperatures of SiO_2 and their geophysical implications. J Geophys Res 88:2431-2444

Mackenzie JD (1963a) High-pressure effects on oxide glasses: I, Densification in rigid state. J Am Ceram Soc 46:461-470

Mackenzie JD (1963b) High-pressure effects on oxide glasses: II, Subsequent heat treatment. J Am Ceram Soc 46:470-476

Marsh SP (1980) LASL Shock Hugoniot Data. University of California Press, Berkeley

Manghnani MH, Sato H, Rai CS (1986) Ultrasonic velocity and attenuation measurements on basalt melts to 1500 °C: role of composition and structure in the viscoelastic properties. J Geophys Res 91:9333-9342

Marsmann H (1981) Silicon-29 NMR. In: Diehl P, Fluck E, Kosfeld R (eds) NMR Basic Principles and Progress, 17, p 65-235, Springer-Verlag, Berlin

Mashimo T, Kondo KI, Sawaoka A, Syono Y, Takei H, Ahrens TJ (1980) Electrical conductivity measurement of fayalite under shock compression up to 56 GPa. J Geophys Res 85:1876-1881

Matson DW, Sharma SK, Philpotts JA (1983) The structure of high-silica alkali-silicate glasses: a Raman spectroscopic investigation . J Non-Cryst Solids 58:323-352

Matsui Y, Kawamura K (1980) Instantaneous structure of an $MgSiO_3$ melt simulated by molecular dynamics. Nature 285:648-649

Matsui Y, Kawamura K (1984) Computer simulation of structures of silicate melts and glasses. In: Sunagawa I (ed) Materials Science of the Earth's Interior, p 3-23, Terra Scientific, Tokyo

Matsui Y, Kawamura K, Syono Y (1982) Molecular dynamics calculations applied to silicate systems: Molten and vitreous $MgSiO_3$ and Mg_2SiO_4 under low and high pressures. In: Akimoto S, Manghnani MH (eds) High Pressure Research in Geophysics, p 511-524, Center for Academic Publications, Tokyo

McMillan P (1984a) A Raman spectroscopic study of glasses in the system CaO-MgO-SiO_2. Am Mineral 69:645-659

McMillan P (1984b) Structural studies of silicate glasses and melts--applications and limitations of Raman spectroscopy. Am Mineral 69:622-644

McMillan P, Piriou B, Couty R (1984) A Raman study of pressure-densified vitreous silica. J Chem Phys 81:4234-4236

McMillan PF, Akaogi M (1987) The Raman spectra of β-(modified spinel) and γ-(spinel) Mg_2SiO_4. Am Mineral 72:361-364

McMillan PF, Kirkpatrick RJ (1992) Al coordination in magnesium aluminosilicate glasses. Am Mineral 77:898-900

McMillan PR, Graham CM (1980) The Raman spectra of quenched albite and orthoclase glasses from 1 atm to 40 kb. In: Ford CE (ed) Progress in Experimental Petrology, p 112-115, Eaton Press,

McQueen RG (1992) The velocity of sound behind strong shocks in SiO_2. In: Schmidt SC (ed) Shock Compression of Condensed Matter - 1991, p 75-78, Elsevier, New York

Meade C, Hemley RJ, Mao HK (1992) High-pressure X-ray diffraction of SiO_2 glass. Phys Rev Lett 69:1387-1390

Meade C, Jeanloz R (1987) Frequency-dependent equation of state of fused silica to 10 GPa. Phys Rev B 35:236-244

Miller GH, Stolper EM, Ahrens TJ (1991a) The equation of state of a molten komatiite, 1, shock wave compression to 36 GPa. J Geophys Res 96:11831-11848

Miller GH, Stolper EM, Ahrens TJ (1991b) The equation of state of a molten komatiite, 2, application to komatiite petrogenesis and the Hadean mantle. J Geophys Res 96:11849-11864

Mozzi RL, Warren BE (1969) The structure of vitreous silica. J Appl Cryst 2:164-172

Murdoch JB, Stebbins JF, Carmichael ISE (1985) High-resolution ^{29}Si NMR study of silicate and aluminosilicate glasses: the effect of network-modifying cations. Am Mineral 70:332-343

Mysen B (1990) Effect of pressure, temperature and bulk composition on the structure and species distrubution in depolymerized alkali aluminosilicate melts and quenched melts . J Geophys Res 95:15733-15744

Mysen BO (1988) Structure and Properties of Silicate Melts. Elsevier, Amsterdam

Mysen BO, Virgo D, Scarfe CM (1980) Relations between the anionic structure and viscosity of silicate melts - a Raman spectroscopic study. Am Mineral 65:690-710

Mysen BO, Virgo D, Seifert FA (1982) The structure of silicate melts: implications for chemical and physical properties of natural magma. Rev Geophys Space Phys 20:353-383

Mysen BO, Virgo D, Seifert FA (1985) Relationships between properties and structure of aluminosilicate melts. Am Mineral 70:88-105

Navrotsky A (1994) Thermochemistry of crystalline and amorphous silica. In: Heaney PJ, Prewitt CT, Gibbs GV (eds) Silica: Physical Behavior, Geochemistry, and Materials Applications. Rev Mineral 29:309-329

Navrotsky A, Geisinger KL, McMillan P, Gibbs GV (1985) The tetrahedral framework in glasses and melts - Inferences from molecular orbital calculations and implications for structure, thermodynamics, and physical properties. Phys Chem Min 11:284-298

Navrotsky A, Peraudeau G, McMillan P, Coutures JP (1982) A thermochemical study of glasses and crystals along the joins silica-calcium aluminate and silica-sodium aluminate. Geochim Cosmochim Acta 46:2093-2047

Nisbet EG, Walker D (1982) Komatiites and the structure of the Archaean mantle. Earth Planet Sci Lett 60:105-113

Oestrike R, Yang W-H, Kirkpatrick RJ, Hervig RL, Navrotsky A, Montez B (1987) High-resolution ^{23}Na, ^{27}Al, and ^{29}Si NMR spectroscopy of framework aluminosilicate glasses. Geochim Cosmochim Acta 51:2199-2210

Ohtani E (1983) Melting temperature distribution and fractionation in the lower mantle. Phys Earth Planet Inter 33:12-25

Ohtani E (1984) Generation of komatiite magma and gravitational differentiation in the deep mantle. Earth Planet Sci Lett 67:261-272

Ohtani E, Kato T, Sawamoto H (1986) Melting of a model chondritic mantle to 20 GPa. Nature 322:352-353

Ohtani E, Sawamoto H (1987) Melting experiment on a model chondritic mantle composition at 25 GPa. Geophys Res Letts 14:733-736

Ohtani E, Taulle F, Angell CA (1985) Al^{3+} coordination changes in liquid aluminosilicates under pressure. Nature 314:78-81

Poe BT, Rubie DC, McMillan PF, Diefenbacher J (1994) Oxygen and silicon self diffusion in $Na_2Si_4O_9$ liquid to 15 GPa. EOS Trans Am Geophys Union 75:713

Poe BT, Rubie DC, Yarger J, Diefenbacher J, McMillan PF (1995) Silicon and oxygen self diffusion in $Na_2Si_4O_9$ liquids to 15 GPa and 2500. Nature (submitted)

Polian A, Grimsditch M (1993) Sound velocities and refractive index of densified a-SiO$_2$ to 25 GPa. Phys Rev B 47:13979-13982

Primak W (1975) The Compacted States of Vitreous Silica. Gordon and Breach, New York

Rau S, Baebler S, Kasper G, Weiss G, Hunklinger S (1995) Brillouin scattering of vitreous silica. Ann Physik 4:91-98

Richard G, Richet P (1990) Room-temperature amorphization of fayalite and high-pressure properties of Fe$_2$SiO$_4$ liquid. Geophys Res Letts 17:2093-2096

Richet P (1984) Viscosity and configurational entropy of silicate melts. Geochim Cosmochim Acta 48:471-483

Richet P, Bottinga Y (1985) Heat capacity of aluminum-free liquid silicates. Geochim Cosmochim Acta 49:471-486

Rigden SM, Ahrens TJ, Stolper EM (1984) Densities of liquid silicates at high pressures. Science 226:1071-1074

Rigden SM, Ahrens TJ, Stolper EM (1988) Shock compression of molten silicate: results for a model basaltic composition. J Geophys Res 93:367-382

Rigden SM, Ahrens TJ, Stolper EM (1989) High-pressure equations of state of molten anorthite and diopside. J Geophys Res 94:9508-9522

Ringwood AE (1978) Composition and Petrology of the Earth's Mantle. McGraw and Hill, New York

Rivers ML, Carmichael ISE (1987) Ultrasonic studies of silicate melts. J Geophys Res 92:9247-9270

Rosenhauer M, Scarfe CM, Virgo D (1979) Pressure dependence of the glass transition temperature in glasses of diopside, albite, and sodium trisilicate composition. Carnegie Inst Washington Yearb 78:556-559

Ross NL, Shu J-F, Hazen RM, Gasparik T (1990) High-pressure crystal chemistry of stishovite. Am Mineral 75:739-747

Rubie DC, Ross CR, Carroll MR, Elphick SC (1993) Oxygen self-diffusion in Na$_2$Si$_4$O$_9$ liquid up to 10 GPa and estimation of high-pressure melt viscosities. Am Mineral 78:574-582

Rustad JR, Yuen DA, Spera FJ (1991) Molecular dynamics of amorphous silica at very high pressures (135 GPa): Thermodynamics and extraction of structures through analysis of Voronoi polyhedra. Phys Rev B 44:2108-2121

Ryan MP (1987) Neutral buoyancy and the mechanical evolution of magmatic systems. In: Mysen, BO (ed) Magmatic Processes: Physiochemical Principles. Geochem Soc Special Pub 1:259-287

Sasakura T, Suito K, Fujisawa H (1989) Measurement of ultrasonic wave velocities in fused quartz under hydrostatic pressures up to 6.0 GPa. In: Novikov NV, Chistyakov YM (eds) Proceedings of the XIth AIRAPT Int. Conf. on High Pressure Science and Technology, 2, p 60-72, Naukova Dumka, Kiev

Sayers DE, Lytle FW, Stern EA (1972) Structure determination of amorphous Ge, GeO$_2$, and GeSe by Fourier analysis of extended X-ray absorption fine structure (EXAFS). J Non-Cryst Solids 8-10:401-407

Scarfe CM, Mysen BO, Virgo D (1979) Changes in viscosity and density of sodium disilicate, sodium metasilicate, and diopside composition with pressure. Carnegie Inst Washington Yearb 78:547-551

Scarfe CM, Mysen BO, Virgo D (1987) Pressure dependence of the viscosity of silicate melts. In: Mysen BO (ed) Magmatic Processes: Physicochemical Principles, p 59-68, The Geochemical Society, University Park, PA

Schmitt DR, Ahrens TJ (1989) Shock temperatures in silica glass: implications for modes of shock-induced deformation, phase transformation, and melting with pressure. J Geophys Res 94:5851-5872

Schneider E, Stebbins FJ, Pines A (1987) Speciation and local structure in alkali and alkaline earth silicate glasses: constraints from ^{29}Si NMR spectroscopy. J Non-Cryst Solids 89:371-383

Schroeder J, Bilodeau TG, Zhao X-S (1990) Brillouin and Raman scattering from glasses under high pressure. High Press Res 4:531-533

Schroeder J, Dunn KJ, Bundy F (1982) Brillouin scattering from amorphous silica under hydrostatic pressures up to 133 kbar. In: Blackman CK, Johannison T, Tegner L (eds) High Pressure in Research and Industry, Proceedings of the 8th AIRAPT Conference, p 259-267, Arkitektkopia, Uppsala

Seifert FA, Mysen BO, Virgo D (1981) Structural similarities of glasses and melts relevant to petrological processes. Geochim Cosmochim Acta 45:1879-1884

Seifert FA, Mysen BO, Virgo D (1982) Three-dimensional network structure of quenched melts (glass) in the systems SiO$_2$-NaAlO$_2$, SiO$_2$-CaAl$_2$O$_4$ and SiO$_2$-MgAl$_2$O$_4$. Am Mineral 67:696-717

Sharma SK, Virgo D, Mysen BO (1978) Structure of glasses and melts of Na$_2$O-xSiO$_2$ (x=1,2,3) composition from Raman spectroscopy. Carnegie Inst Wash Yearb 77:649-652

Sharma SK, Virgo D, Mysen BO (1979) Raman study of the coordination of aluminum in jadeite melts as a function of pressure. Am Mineral 64:779-787

Shimizu N, Kushiro I (1984) Diffusivity of oxygen in jadeite and diopside melts at high pressures. Geochim Cosmochim Acta 48:1295-1303

Smith KH, Shero E, Chizmeshya A, Wolf GH (1995) The equation of state of polyamorphic germania glass: two-domain description of the viscoelastic response. J Chem Phys 102:6851-6857

Sowa H (1988) The oxygen packings of low-quartz and ReO_3 under high pressure. Zeits Kristal 184:257-268

Spera FJ, Bergman SC (1980) Carbon dioxide in igneous petrogenesis: I. Aspects of the dissolution of CO_2 in silicate liquids. Contrib Mineral Petrol 74:55-66

Stebbins J, Sykes D (1990) The structure of $NaAlSi_3O_8$ liquid at high pressure: new constraints from NMR spectroscopy. Am Mineral 75:943-946

Stebbins JF (1987) Identification of multiple structural species in silicate glasses by ^{29}Si NMR. Nature 330:465-467

Stebbins JF (1988) Effects of temperature and composition on silicate glass structure and dynamics: Si-29 NMR results. J Non-Cryst Solids 106:359-369

Stebbins JF (1991) NMR evidence for five-coordinated silicon in a silicate glass at atmospheric pressure. Nature 351:638-639

Stebbins JF, Carmichael ISE, Moret LK (1984) Heat capacities and entropies of silicate liquids and glasses. Contrib Mineral Petrol 86:131-148

Stebbins JF, Farnan I, Xue X (1992) The structure and dynamics of alkali silicate liquids: a view from NMR spectroscopy. Chem Geol 96:371-385

Stebbins JF, McMillan PF (1989) Five- and six-coordinated Si in $K_2Si_4O_9$ glass quenched from 1.9 GPa and 1200°C. Am Mineral 74:965-968

Stebbins JF, McMillan PF (1993) Compositional and temperature effects on five-coordinated silicon in ambient pressure silicate glasses. J Non-Cryst Solids 160:116-125

Stolper E, Holloway JR (1988) Experimental determination of the solubility of carbon dioxide in molten basalt at low pressure. Earth Planet Sci Lett 87:397-408

Stolper E, Walker D, Hager BH, Hays JF (1981) Melt segregation from partially molten source regions: the importance of melt density and source region size. J Geophys Res 86:6261-6271

Sugiura H, Kondo K, Sawaoka A (1981) Dynamic response of fused quartz in the permanent densification region. J Appl Phys 52:3375-3382

Sugiura H, Yamadaya T (1992) Raman scattering in silica glass in the permanent densification region. J Non-Cryst Solids 144:151-158

Suito K, Miyoshi M, Sasakura T, Fujisawa H (1992) Elastic properties of obsidian, vitreous SiO_2, and vitreous GeO_2 under high pressure up to 6 GPa. In: Syono Y, Manghnani MH (eds) High-Pressure Research: Application to Earth and Planetary Sciences, p 219-225, Am. Geophysical Union, Washington D.C.

Susman S, Volin KJ, Liebermann RC, Gwanmesia GD, Wang Y (1990) Structural changes in irreversibly densified fused silica: implications for the chemical resistance of high level nuclear waste glasses. Phys Chem Glasses 31:144-150

Susman S, Volin KJ, Price DL, Grimsditch M, Rino JP, Kalia RK, Vashishta P, Gwanmesia G, Wang Y, Liebermann RC (1991) Intermediate-range order in permanently densified vitreous SiO_2: a neutron-diffraction and molecular-dynamics study. Phys Rev B 43:1194-1197

Suzuki A, Ohtani E, Kato T (1995) Flotation of diamond in mantle melt at high pressure. Science 269:216-218

Sykes D, Poe B, McMillan PF, Luth RW, Sato RK (1993) A spectroscopic investigation of the structure of anhydrous $KAlSi_3O_8$ and $NaAlSi_3O_8$ glasses quenched from high pressure. Geochim Cosmochim Acta 57:3574-3584

Takahashi E (1986) Melting of a dry peridotite KLB-1 up to 14 GPa: implications on the origin of peridotite upper mantle. J Geophys Res 91:9367-9382

Tandura SN, Alekseev NV, Voronkov MG (1986) Molecular and electronic structure of penta- and hexa-coordinate silicon compounds. Topics in Current Chemistry 131:99-186

Taylor M, Brown GE (1979a) Structure of mineral glasses. I. The feldspar glasses, $NaAlSi_3O_8$, $KAlSi_3O_8$, $CaAl_2Si_2O_8$. Geochim Cosmochim Acta 43:61-77

Taylor M, Brown GE (1979b) Structure of mineral glasses. II. The SiO_2-$NaAlSiO_4$ join. Geochim Cosmochim Acta 43:1467-1473

Taylor M, Brown GE, Fenn PM (1980) Structure of mineral glasses. III. $NaAlSi_3O_8$ supercooled liquid at 805°C and the effects of thermal history. Geochim Cosmochim Acta 44:109-119

Tossell JA, Gibbs GV (1978) The use of molecular-orbital calculations on model systems for the prediction of bridging-bond-angle variations in siloxanes, silicates, silicon nitrides and silicon sulphides. Acta Crystallogr A34:463-472

Tse JS, Klug DD (1991) Mechanical instability of α-quartz: a molecular-dynamics study. Phys Rev Lett 67:3559-3562

Tyburczy JA, Waff HS (1983) Electrical conductivity of molten basalt and andesite to 25 kilobars pressure: Geophysical significance and implications for charge transport and melt structure. J Geophys Res 88:2413-2430

Tyburczy JA, Waff HS (1985) High pressure electrical conductivity in molten natural silicates. In: Schock RN (ed) Point Defects in Minerals, 31, p 78-87, American Geophysical Union, Washington, D. C.

Ueno M, Misawa M, Suzuki K (1983) On the change in coordination of Ge atoms in Na_2O-GeO_2 glasses. Physica 120B:347-351

Uhlmann DR (1973) Densification of alkali silicate glasses at high pressure. J Non-Cryst Solids 13:89-99

Urbain G, Bottinga Y, Richet P (1982) Viscosity of liquid silica, silicates and alumino-silicates. Geochim Cosmochim Acta 46:1061-1072

Valenkov N, Porai-Koshits EA (1936) X-ray investigation of the glassy state. Z Krist Kristall Kristallphys Kristallchem 95:195-229

Velde B, Kushiro I (1978) Structure of sodium alumino-silicate melts quenched at high pressure: infrared and aluminum K-radiation data. Earth Planet Sci Lett 40:137-140

Verhelst-Voorhees M, Yarger J, Diefenbacher J, Poe BT, Wolf GH, McMillan PF (1994) ^{29}Si NMR, HREM, and Raman study of pressure-vitrified quartz: new evidence for high-coordinate silicon. EOS Trans Am Geophys Union 75:635

Verweij H (1979) Raman study of the structure of alkaligermanosilicate glasses II. lithium, sodium and potassium digermanosilicate glasses. J Non-Cryst Solids 33:55-69

Verweij H, Konijnendijk WL (1976) Structural units in K_2O-PbO-SiO_2 glasses by Raman spectroscopy. J Am Ceram Soc 59:517-521

Virgo D, Mysen BO, Kushiro I (1980) Anionic constitution of 1-atmosphere silicate melts: implications for the structure of igneous melts. Science 208:1371-1373

Vukcevich MR (1972) A new interpretation of the anamolous properties of vitreous silica. J Non-Cryst Solids 11:25-63

Wackerle J (1962) Shock-wave compression of quartz. J Appl Phys 33:922-937

Waff HS (1975) Pressure-induced coordination changes in magmatic liquids. Geophys Res Letts 2:193-196

Walrafen GE, Hokmabadi MS (1986) Raman structural correlations from stress-modified and bombarded vitreous silica. In: Walrafen GE, Revesz AG (eds) Structure and Bonding in Noncrystaline Solids, p 185-202, Plenum Press, New York

Warren, BE (1933) X-ray diffraction of vitreous silica. Z Krist Kristall Kristallphys Kristallchem 86:349-352

Wasserman EA, Yuen DA, Rustad JR (1993) Molecular dynamics study of the transport properties of perovskite melts under high temperature and pressure conditions. Earth Planet Sci Lett 114:373-384

Watson EB (1979) Calcium diffusion in a simple silicate melt to 30 kbar. Geochim Cosmochim Acta 43:313-322

Williams Q (1990) Molten $(Mg_{0.88}Fe_{0.12})_2SiO_4$ at lower mantle conditions: melting products and structure of quenched glasses. Geophys Res Letts 17:635-638

Williams Q, Hemley RJ, Kruger MB, Jeanloz R (1993) High pressure infrared spectra of a-quartz, coesite, stishovite, and silica glass. J Geophys Res 98:22157-22170

Williams Q, Jeanloz R (1988) Spectroscopic evidence for pressure-induced coordination changes in silicate glasses and melts. Science 239:902-905

Williams Q, Knittle E, Reichlin R, Martin S, Jeanloz R (1990) Structural and electronic properties of Fe_2SiO_4-fayalite at ultrahigh pressures: amorphization and gap closure. J Geophys Res 95:21549-21564

Wolf GH, Durben DJ, McMillan PF (1990) High pressure Raman spectroscopic study of sodium tetrasilicate ($Na_2Si_4O_9$) glass. J Chem Phys 93:2280-2288

Wolf GH, Wang S, Herbst CA, Durben DJ, Oliver WF, Kang ZC, Halvorson K (1992) Pressure induced collapse of the tetrahedral framework in crystalline and amorphous GeO_2. In: Syono Y, Manghnani MH (eds) High-Pressure Research: Application to Earth and Planetary Sciences, p 503-517, Am. Geophysical Union, Washington D.C.

Woodcock LV, Angell CA, Cheeseman P (1976) Molecular dynamics studies of the vitreous state: simple ionic systems and silica. J Chem Phys 65:1565-1577

Xue X, Stebbins J, Kanzaki M, Poe B, McMillan P (1991) Pressure-induced silicon coordination and tetrahedral structural changes in alkali oxide-silica melts up to 12 GPa: NMR, Raman and infrared spectroscopy. Am Mineral 76:8-26

Xue X, Stebbins JF, Kanzaki M, Tronnes RG (1989) Silicon coordination and speciation changes in a silicate liquid at high pressures. Science 245:962-964

Yagi T, Mao HK, Bell P (1978) Structure and crystal chemistry of perovskite-type $MgSiO_3$. Phys Chem Min 3:97-110

Yarger JL, Poe BT, Diefenbacher J, Smith KH, Wolf GH, McMillan PF (1995) Al^{3+} coordination changes in high pressure aluminosilicate liquids. submitted to Science

Yin CD, Okuno M, Morikawa H, Marumo F (1983) Structure analysis of MgSiO$_3$ glass. J Non-Cryst Solids 55:131-141

Zachariasen WH (1932) The atomic arrangement in glass. J Am Chem Soc 54:3841-3851

Zha CS, Hemley RJ, Mao HK, Duffy TS, Meade C (1994) Acoustic velocities and refractive index of SiO$_2$ glass to 57.5 GPa by Brillouin scattering. Phys Rev B 50:13105-13112

Chapter 12

COMPUTER SIMULATIONS OF SILICATE MELTS

Peter H. Poole

Department of Applied Mathematics
The University of Western Ontario
London, Ontario, N6A 5B7 Canada

Paul F. McMillan and George H. Wolf

Department of Chemistry and Biochemistry
Arizona State University
Tempe, Arizona 85287 U.S.A.

INTRODUCTION

Silicate melts form refractory systems, which are often difficult to study under the high-pressure, high-temperature conditions developed within the Earth's interior. In addition, the highly associated, non-periodic and time-dependent character of the liquid state represents a challenge to the theoretical understanding of the melt properties, and the interpretation of the results of spectroscopic or diffraction experiments in terms of the atomic-level structure. For these reasons, workers in liquid state chemistry and physics have long had recourse to computer simulation techniques to model the structure and dynamics of melts, glasses and dense fluids (March and Tosi, 1976; Hansen and McDonald, 1976, 1986; Brawer, 1985). These simulations were initially rendered possible by the appearance of the digital computer (Metropolis et al., 1953), and advances in the field remain inherently tied to developments in computational algorithms and hardware.

As a tool of scientific research, the modern computer is unique. Although its limits are continually being tested and extended in a range of highly specialized scientific applications, the utility and performance of computers are growing mainly due to mass-market economic forces, which have little to do with the desires or designs of scientists. The result is that researchers using computational techniques can rely on the fact that their basic research tool will grow in power with time, irrespective of their efforts. The rate of this growth is enormous. Currently, the doubling time for the performance of workstations is approximately 18 months. Such growth is expected to continue over the foreseeable future, probably during the entire careers of the graduate students now reading this review.

The implications of such explosive growth are at once exhilarating and sobering for the practitioners of computer modelling. There is obviously a tremendous opportunity to model more and more complex systems, with increasing accuracy. On the other hand, it requires that we pay careful and continuous attention to our research methods and strategies, which can quickly become obsolete or inappropriate due to rapidly advancing computational technology. It even raises questions for the dissemination of results of computer simulation research. It is already difficult to communicate three-dimensional, time-dependent visualizations of the simulated liquid systems, which are often the principal result of a series of simulation studies, via normal publication channels.

Although computer modeling has now become a valuable and even essential component of almost all scientific research fields, our understanding of the liquid state of matter has been particularly influenced by computer simulation techniques. The Monte Carlo (MC) and molecular dynamics (MD) methods for evaluation of the properties of statistical mechanical systems, such as fluids, liquids and solids, appeared shortly after the development of fast digital computers (Metropolis et al., 1953; Wood and Parker, 1957; Alder and Wainwright, 1957: these and other classic papers are collected together in an annotated bibliography by Ciccotti et al., 1987). These methods were soon applied to study the structures and thermodynamic and dynamic properties of molten silicates (Woodcock et al., 1976; Borgianni and Granati, 1979).

In simulation studies it is possible to examine the local and medium-range structure around any chosen atom within the liquid, and to identify the individual atoms or groups of atoms which play a deciding role in diffusion or structural relaxation processes (Woodcock et al., 1977; Kubicki and Lasaga, 1988, 1991a,b). In addition, the vibrational properties associated with particular groups of atoms in the liquid can be calculated, which will be invaluable for interpreting experimental spectra under high temperature-high pressure conditions (see chapters by McMillan and Wolf, and Wolf and McMillan, this volume).

Computer simulation studies applied to geochemically important liquids have been reviewed by Matsui et al. (1982), Matsui and Kawamura (1984), Erikson and Hostetler (1987), Kubicki and Lasaga (1988, 1991a), Rustad et al. (1991a, 1992), and Stein and Spera (1995). A great advantage of simulation studies over experiments is that the structure and properties of the melt can be examined in situ under the high-pressure and high-temperature conditions of the Earth's deep interior (Matsui et al., 1982; Kubicki and Lasaga, 1991b; Wasserman et al., 1993a). There have been extensive experimental studies of densification processes and coordination changes linked to magma rheology at depth within the Earth (Kushiro et al., 1976; Stolper et al., 1981; Scarfe et al., 1987; Miller et al., 1991a,b: see chapters by Stebbins, and by Wolf and McMillan, this volume). These phenomena have been studied in detail using computer simulations. The simulations can readily be carried out under extreme conditions which are impossible, or extremely difficult, to attain within the laboratory. Such conditions include states of highly non-hydrostatic compression (J Badro and JL Barrat, pers comm: recent work on SiO_2), metastable "stretched" states described by *negative* values of P (Angell et al., 1987; Green et al., 1990; Poole et al., 1992; Hemmati et al., 1995), states of arbitrarily large thermal excitation (Scamehorn and Angell, 1987), or extremely rapid dynamic compression or thermal quenching. This last property permits a leisurely examination of the glassy states of materials that can not be easily vitrified in the laboratory, such as H_2O (Poole et al., 1992, 1993, 1994) and Mg_2SiO_4 (Matsui et al., 1982; Williams et al., 1989; Kubicki and Lasaga, 1991b).

The drawback in these studies is that a simulation only provides a model of the real system, limited by the appropriateness of the modeling parameters, including the type of method chosen, the system size and shape, and the interparticle potential function used to simulate interactions. The problem facing the simulationist is to design a system which will adequately model the liquid phenomena of interest and allow interpretation or extrapolation of experimental results. The usual choice of a simulation modeller is to devise a potential function which allows the system to be identified with the melt of, say, SiO_2.

However, it is important to bear in mind that the potential parameters can be varied continuously. The "Si" or "O" "atoms" can be made arbitrarily large or small, light or heavy, with any type of interaction with their neighbors. Atomic configurations can be arbitrarily fixed in one desired state as pressure and temperature is changed. This use of simulation represents a great advantage over laboratory studies, in that "thought" experiments can be carried out, unhampered by real experimental constraints or the behavior of known liquids (Ciccotti et al., 1979). Any number of "what if" experiments can be carried out to probe the phenomenology of liquid state behavior in general. In fact, although most attention is usually paid to trying to get the potential function "right", i.e. to give results which are in agreement with experiment, the whole family of other solutions gives the true power to simulation studies. In simulation experiments, it is possible to explore what a silicate liquid is *not*, and why. Alternatively, such experiments can yield insights into previously unobserved types of behavior which might be expected from silicate melts, under certain conditions.

For example, in studies of simple amorphous networks (H_2O, SiO_2, GeO_2), it has been suggested that increased coordination of the network-forming ions with pressure leads to distinct forms of the glass which may form separate phases (Mishima, 1985, 1994; Tse and Klein, 1987; Jeanloz, 1990; Poole et al., 1992, 1993, 1994; Wolf et al., 1992; Stanley et al., 1994; Smith et al., 1995). Studies of such vitreous "polyamorphism" (Wolf et al., 1992; Angell, 1995; Poole et al., 1995a,b) relate to the potential occurrence of density- and entropy-driven transitions between multiple glassy, or even liquid, phases of the same composition. Such transitions have been suggested to occur in many simple metallic or semiconducting systems (Thompson et al., 1984; Donovan et al., 1985; Ponyatovsky and Barkalov, 1992). Observation of such a first-order phase transition, between two liquid phases with different structure and density but the same composition, has recently been observed within a *multicomponent* aluminate liquid (Aasland and McMillan, 1994). This type of transition could possibly occur within aluminosilicate liquids, under the high pressure conditions of the deep Earth. Computer simulations are currently being used to explore the phase behavior of liquid silicates under high P-T conditions (Belonoshko, 1994a,b; Shao and Angell, 1995; P Poole, M Hemmati, CA Angell, in prep.).

In this chapter, our goal is to present an overview of computer simulation as applied to the study of molten silicates. These techniques can now be readily applied to problems of interest to geochemistry and igneous petrology, even by workers with little prior background, although there are obvious pitfalls to avoid. The range of useful information which can be gained from such studies is discussed in the final section, which could be read first by students and researchers with little interest in the detailed methodology, at present. In the following section, we present the basic techniques used in computer simulation of liquids. This section echoes several recent reviews which have appeared on the application of computer simulation methods to silicate melts studies (Erikson and Hostetler, 1987; Kubicki and Lasaga, 1988, 1991a; Rustad et al., 1991a, 1992; Stein and Spera, 1995). Students requiring further details of the methodology of computer simulation of liquids should consult one of several excellent texts and reviews (Binder, 1979; Kalos and Whitlock, 1986; Allen and Tildesley, 1987, 1993; Ciccotti et al., 1987; Haile, 1992).

MODELLING AND MEASUREMENT

Despite continuing research efforts over the past several decades, the liquid state remains much less well understood than gases or solids. The characteristic features of a liquid - solid-like density, lack of long-range order, and often chaotic motions on the atomic scale - combine to make liquid systems extremely difficult to treat analytically. The intrinsic many-body nature of the liquid state problem means that the solution is fixed by well-defined but intractable integral equations (Hansen and McDonald, 1986). The problem is not particularly of *what* to do, but rather, of *how* to do it. To address these issues, we first outline the nature of the problem, and then describe how computers are used in its solution.

The basic problem

Let us choose to model a classical liquid in the following way. We consider N particles (atoms, ions or molecules) confined within a container of fixed volume V, in thermodynamic equilibrium at temperature T. The potential energy is given by a function $\Phi(\{r\})$, which depends upon the set of particle positions $\{r\}$ (and any external field applied to the simulation system), and is usually expressed as an interaction function between the particles. The particles are free to change their positions.

In classical mechanics, the properties of a system is fully specified if the positions and momenta of all of the particles are known. The set of instantaneous positions is

$$\{r\} = \{r_1, r_2, r_3, \dots r_i, \dots r_N\} \tag{1a}$$

and the momenta, defined as $p_i = m_i dr_i/dt$ (m_i is the particle mass), are

$$\{p\} = \{p_1, p_2, p_3, \dots p_i, \dots p_N\} \tag{1b}$$

A particular set of $\{r\}$ and $\{p\}$ serves to define a *configuration* (*k*) of the system. The total energy (E) for a given configuration is the sum of the kinetic (L) and potential (Φ) energies. The usual form for the kinetic energy is

$$L = \sum_{i=1}^{N} \sum_{\alpha=1}^{3} \frac{p_{i\alpha}^2}{2m_i} \tag{2}$$

where $p_{i\alpha}$ is the magnitude of one of the three (usually Cartesian) vector components of the particle momentum (the kinetic energy of a particle i travelling in a component direction α is $L_{i\alpha} = p_{i\alpha}^2/2m_i = 1/2\, m_i v_{i\alpha}^2$, where $v_{i\alpha}$ is the particle velocity in that direction). Use of this expression implies that the motion of the "atomic" particles is being treated classically, not by quantum mechanics. The potential energy function is specified for the system being studied. A common approximation is to express the potential energy as a function of the separation between pairs of particles (the *pair potential* approximation), so that

$$\Phi = \frac{1}{2} \sum_{i=1}^{N} \sum_{\substack{j=1 \\ j \neq i}}^{N} \phi_{ij}(r_i, r_j) \equiv \sum_{i} \sum_{j<i} \phi_{ij}(r_i, r_j) \ , \tag{3}$$

The choice of potential energy function is discussed below.

With this information, we can calculate other properties of the system, such as the pressure P inside the container, or the internal energy U of the system, using results derived from classical and statistical mechanics (Hill, 1956; McQuarrie, 1976; March and Tosi, 1976; Hansen and McDonald, 1976, 1986; Landau and Lifshitz, 1980; Kittel and Kroemer, 1980; Goldstein, 1980; Friedman, 1985; Callen, 1985; Chandler, 1987). In general, the bulk thermodynamic properties can be calculated as a weighted average over all possible configurations of the system. Each configuration *k* is defined by one set of $\{r, p\}$. The thermodynamic state of the system is defined by the values of a small set of thermodynamic variables or parameters: the number of particles N, the volume V, the temperature T, pressure P, etc.. The set of particle position and momentum values corresponding to each configuration can be thought of as "coordinates", defining a point Γ in a multi-dimensional *phase space*. This space has dimension (3N positions + 3N momenta) = 6N. The instantaneous value of some property *A* of the system is dependent upon the particular set of particle positions and momenta: i.e. *A* is a function of *k* or Γ. *A* could be the internal energy of the system, the probability that any two atoms are separated by some distance, or the intensity of scattering of light or neutrons at a specified wavevector by the system.

In general, the particle positions and momenta in the system will change with time (i.e. the system will *evolve*, following some *path in phase space*): $\Gamma(t)$. If the evolution time is sufficiently long, the time average of the property of interest, $\langle A \rangle_{time}$, can be compared with the experimentally observed value of that property, A_{obs}. (This and the following statements are actually concerned with a fundamental assumption of statistical thermodynamics, called the *ergodic hypothesis*. This is discussed further below.) The property average over an observation time t is defined by:

$$\langle A \rangle_{time} = \langle A(\Gamma(t)) \rangle_{time} = \lim_{t \to \infty} \frac{1}{t} \int_0^t A(\Gamma(t')) dt' \tag{4}$$

More generally, the property average is taken over the configurations sampled during the simulation. The property depends on the particular configuration: $A(\Gamma) \equiv A(k)$. The statistical average over particle configurations is "weighted", in the sense that all configurations do not contribute equally to the average. If the "weight", or relative probability, for the system to be in a specific configuration k is denoted $P(k)$, then any quantity $A(k)$ which is defined for a specific k, will have a (configuration) average value $<A>_k$ given by

$$\langle A \rangle_k = \frac{\sum_k A(k) P(k)}{\sum_k P(k)} \tag{5}$$

Because such a large number of configurations are sampled, A can be taken to depend continuously on k, and the sum written as an integral:

$$\langle A \rangle_k = \frac{\int_k A(k) P(k) dk}{\int_k P(k) dk} \tag{6}$$

The role of computer simulation methods is to numerically evaluate integrals such as (4) and (6), which can not be treated analytically. Two principal techniques have evolved, which address respectively the solution of Equations (6) and (4). These are known as the *Monte Carlo* (MC) and *molecular dynamics* (MD) methods.

The Monte Carlo method

This method was that devised to solve integrals of the type in Equation (6), in the first application of computer simulation techniques to this kind of problem (Metropolis et al., 1953). The application closely followed the development of the first generation of fast digital computers (specifically, the MANIAC, at Los Alamos National Laboratory). The name "Monte Carlo" derives from the use of random numbers in the method, such as might be generated by a (fair) roulette wheel. The basic idea is that for high-dimensional integrals, it is extremely demanding computationally to evaluate the integrand systematically over the full domain of the integral. Instead, the domain is sampled randomly, and the integral is estimated as the average value of the integrand over a random sample of configurations, multiplied by the "volume" of the domain. Although the number of possible configurations k of the system is very large, many of these are not very likely. For example, under normal circumstances, one would not expect to find all of the air molecules in a room to occupy the same point in space. It is then necessary to evaluate those configurations which contribute substantially (i.e. for which $P(k)$ is significant) to the integral. This is done by using the result from statistical mechanics:

$$P(k) = e^{-\frac{E_k}{k_B T}} \tag{7}$$

where E_k is the energy of the kth configuration, k_B is Boltzmann's constant, and T is the temperature. This biased sampling procedure, known as "*importance sampling*," was introduced in the earliest Monte Carlo calculation (Metropolis et al., 1953), and is implemented in the following way. We start from some liquid configuration k which is chosen to be "reasonable", and a new configuration $k+1$ is generated by imposing upon each particle position a random displacement in a random direction, usually not too far from its current position. The difference between the relative weights of the two configurations, ΔP, is obtained from the difference in the configurational energy, $\Delta E = E(k+1) - E(k)$:

$$\Delta P = e^{-\frac{\Delta E}{k_B T}}$$ (8)

The new configuration then contributes to the estimate of the integral in (6) with a probability proportional to ΔP. In practice this is achieved by comparing the value of ΔP to a random number between 0 and 1. The contribution from the new configuration is accepted if the random number is less than ΔP, and rejected otherwise. If $k+1$ is rejected, the original configuration k is taken to be the new configuration, $k+1$. Each step in the process generates a specific contribution to the estimation of the integral, and is usually termed a MC "time step," although physical time does not appear in the simulation. The sequence of configurations generated is termed a *Markov chain*, and they trace a path followed by the system in phase space $\{\Gamma_1, \Gamma_2, \Gamma_3, ..., \Gamma_M\}$ (Allen and Tildesley, 1987). Relying on the validity of the importance sampling procedure, the true configuration average value of the property of interest $<A>_k$ is taken to be the average value of $A(k)$ over the set of configurations sampled:

$$\langle A \rangle_k = \frac{1}{M} \sum_{k=1}^{M} A(k)$$ (9)

This is the Monte Carlo estimate for the integral specified in Equation (6).

It is important to note that the MC method only samples that portion of configuration space that depends on the 3N particle coordinates $\{\mathbf{r}\}$; the particle momenta $\{\mathbf{p}\}$ do not appear in the above treatment. As a result, the MC method only gives estimates of system properties which depend only on particle positions; i.e. "static" properties. For such properties, that portion of the integration in Equation (6) which deals with the momenta can be factored out and evaluated analytically (Allen and Tildesley, 1987). For this reason, it is sufficient for the MC method to sample only the position-space portion of the domain of the integral in Eq. (6), in order to make an estimate of the value of the full expression.

Despite the simplicity of the MC technique, only a few Monte Carlo simulation studies of silicate liquids and glasses have been carried out (Borgianni and Granati, 1979; Hostetler, 1982 (see Erikson and Hostetler, 1987); Stixrude and Bukowinski, 1988, 1989, 1990, 1991), although it has been used extensively in studies of simple liquids (March and Tosi, 1976; Hansen and McDonald, 1976, 1986; Binder 1979, 1984). Apart from questions of taste and practice, the main reason for this neglect has been the great interest in time-dependent properties in silicate studies. These are most easily investigated using the molecular dynamics technique, described below. However, there are certain advantages to the MC method, especially for the calculation of thermodynamic properties and for investigations of the system very close to the glass transition (see chapters by Dingwell, and by Richet and Bottinga, in this volume), and Monte Carlo simulations should be pursued in future work on silicates.

The molecular dynamics method

In molecular dynamics (MD) simulations, the computer is used to obtain approximate solutions to the *time average* of the system properties ($\langle A \rangle_{time}$: Eqn. 4), and in doing so, information is obtained on the time dependence of the property ($A(t)$) as the system evolves along its phase space trajectory through time, $\Gamma(t)$. In a practical simulation, the study is carried out over a large number (n) of finite *time steps*, of duration Δt, determined by various properties of the system being investigated, and the computing power available. The number in the sequence of time steps can be denoted by an integer τ, and the total simulation run time is $t = n\Delta t$. Expression (4) reduces to a sum:

$$\langle A \rangle_{time} = \frac{1}{n} \sum_{\tau=1}^{n} A(\Gamma(\tau)) \tag{10}$$

In order to allow the particles to change their positions and momenta and generate new configurations, the force (f_i) on each particle is evaluated using the classical relation:

$$\mathbf{f}_i = -\mathbf{grad}_i \, \Phi(r) = -\nabla_i \Phi \tag{11}$$

In this expression, the operator ∇_i ("del") is the differential vector operator for particle i, ($\partial/\partial x_i$, $\partial/\partial y_i$, $\partial/\partial z_i$), where x_i, y_i, and z_i are the Cartesian components of the particle position vector, $\mathbf{r}_i$. Φ is the many-body potential energy function, which in general, depends upon the simultaneous positions of the all the atoms (simplified models for the potential function, as used in practical simulations, are discussed below). The trajectory of each atom in space can be evaluated through application of the classical equations of motion, which form N coupled differential equations (these are coupled through the potential energy function, which depends on the instantaneous *relative* positions of the particles):

$$\mathbf{f}_i = m_i \frac{d^2 \mathbf{r}_i}{dt^2} \tag{12}$$

This can not be solved analytically, but a numerical estimate of the particle positions as a function of time, $\mathbf{r}_i(t)$, can be obtained using a finite-difference method. The typical strategy is to estimate a future position $\mathbf{r}_i(t + \Delta t)$ for a time Δt ahead of the current position. One simple algorithm due to Verlet (1967) is obtained by adding truncated Taylor expansions for $\mathbf{r}_i(t + \Delta t)$ and by adding $\mathbf{r}_i(t - \Delta t)$, which yields

$$\mathbf{r}_i(t + \Delta t) = -\mathbf{r}_i(t - \Delta t) + 2\mathbf{r}_i(t) + \mathbf{a}_i(t)(\Delta t)^2 \tag{13}$$

This estimates the new particle position at time ($t + \Delta t$) from the current and immediately preceding positions, and the current acceleration $\mathbf{a}_i(t)$, obtained from the force law ($\mathbf{a}_i = f_i/m_i$) (Eqn. 12). Another algorithm in common use in silicate simulations is that of Schofield (1973), with

$$\mathbf{r}_i(t + \Delta t) = \mathbf{r}_i(t) + \mathbf{v}_i(t)\Delta t + \frac{1}{6}\left(4\mathbf{a}_i(t) - \mathbf{a}_i(t - \Delta t)\right)(\Delta t)^2 \tag{14}$$

and $$\mathbf{v}_i(t + \Delta t) = \mathbf{v}_i(t) + \frac{1}{6}\left(2\mathbf{a}_i(t + \Delta t) + 5\mathbf{a}_i(t) - \mathbf{a}_i(t - \Delta t)\right)(\Delta t) \tag{15}$$

$v_i(t)$ is the particle velocity at time t. Using algorithms such as these, each particle is moved step-wise through time along its path, with Δt being the size of the time-step.

In such finite-difference methods, a key common feature is that Δt must be chosen to be sufficiently small that the solution is numerically stable. In practice, the fact that the total energy E must be conserved over the course of the simulation can be use to test this stability. For typical potential energy functions chosen to model atomic and molecular interactions, this means choosing a time step on the order of 1 to 10 fs (1 femtosecond = 10^{-15} s), which typically results in a conserved energy to within 1 part in 10,000, if the numerical integrations have been done properly. The maximum size of the time step, determined by the potential function, results in real computing constraints for the simulation study. The simulation must be carried out over a sufficiently large number of time steps that the configuration achieves an equilibrium value, for the time average of the property of interest to be comparable with an observed thermodynamic quantity. A typical vibrational mode has a period of $\sim 10^{-13}$ s, so that an MD simulation must be run for several thousand time steps to obtain information on the vibrational spectrum, corresponding to "real" system run times on the order of 10^{-10} s, or ~ 100 ps. For most silicate liquids at temperatures exceeding ~ 3000 to 4000 K, the relaxation time for the structure to achieve an equilibrium configuration is on the order of a few vibrational periods (Angell, 1985, 1991; Stein and Spera, 1995), so that runs on the order of a few hundred ps are usually sufficient. Depending on the computing resources available, the number of particles studied, and the complexity of the potential energy function, this can take minutes, hours (usually), days or even weeks and months of real computing time. However, relaxation times increase exponentially with decreasing temperature (see chapter by Dingwell and Webb, this volume), so that it is usually not possible to carry out simulations of silicate systems at realistic magmatic temperatures; hence the run must be carried out at extremely high temperature in order to achieve reasonable equilibration in reasonable time (Scamehorn and Angell, 1987). The problem becomes particularly severe for highly polymerized liquids such as SiO_2. In this case, the activation energy for structural relaxation is very large (~ 400 kJ/mol) and follows an Arrhenian temperature dependence. Structural equilibration within $\sim 10^{-12}$ to 10^{-13} s is not achieved until temperatures in excess of ~ 6000 K, so that exploration of liquid properties at lower temperatures requires much longer runs. In contrast, the very "fragile" liquids (i.e. with very non-Arrhenian viscosity-temperature relations: Angell, 1985, 1991) along the CaO-Al_2O_3 join have very low viscosities and short structural relaxation times to quite low temperatures, so that simulations of these liquids could be readily carried out at temperatures in the range of a recent experimental NMR investigation (~ 2000 K) (Poe et al., 1992b, 1993a,b; Coté et al., 1993).

By moving each particle step-wise along its classical trajectory, a sequence of distinct configurations of the liquid system is generated which resembles the sequence of configurations generated in a MC simulation. Due to the energy conservation principle, a collection of atoms all moving according to the classical equations of motion must have constant total energy E as a function of time. As a result, the chain of configurations generated in MD will lie on a hypersurface of constant E in the 6N-dimensional configuration space of the system, and hence $P(k)$ will be the same for all configurations (Eqn. 7). In this sense, MD generates a sequence of configurations that all have the same value of $P(k)$. Under these circumstances, the property averages described by equations (4) and (6) become equivalent, if the configuration space has been adequately sampled:

$$\langle A \rangle = \langle A \rangle_k = \langle A \rangle_{time} \tag{16}$$

This equality provides a formal statement of an assumption in statistical mechanics known as the "*ergodic hypothesis*" (McQuarrie, 1976; Landau and Lifshitz, 1980; Chandler, 1987). The ergodic hypothesis states that the trajectory through phase space generated by the classical equations of motion visits (in the limit of long time) all the configurations of the statistical ensemble, with the result that the time average over the trajectory is equivalent to the ensemble average of statistical mechanics. The history of the development of this hypothesis began with the work of Maxwell, Boltzmann and Gibbs on the kinetic gas theory and the foundations of modern thermodynamics (Brush, 1964; Wood, 1968; Brawer, 1980).

In the MD approach, ergodicity will generally be achieved when the simulation is carried out for a sufficient number of time steps that a relaxed (equilibrium) configuration is achieved. If the simulation is stopped before this point, because the structural relaxation time is too long for equilibrium to be obtained, then the system studied will no longer exhibit ergodic behavior: i.e. it will "fall out of equilibrium". This *ergodicity-breaking* is intimately related to the occurrence of the (kinetic) glass transition in the simulated system (Angell, 1981; Brawer, 1985; Scamehorn and Angell, 1987; Barrat et al., 1990; Barrat and Klein, 1991; Stein and Spera, 1995). It is of interest that different subsets of configurations in the liquid will often "freeze out" at different temperatures (Angell, 1991). For example, at temperatures well below the structural glass transition for a silicate framework, it is known that alkali ions are still highly mobile; i.e. their behavior is largely ergodic, although the silicate part of the structure has become a glass (Dingwell and Webb, 1990). MD simulation techniques have been used extensively to study the conduction mechanisms of alkali metals and other highly mobile cations in glasses (Angell, 1990).

"The Devil is in the details..."

Initial conditions. For both the MC and MD methods, the initial conditions of the system must be chosen at the beginning of the calculation: in both approaches, there needs to be a "starting point" for the sequence of configurations generated by the given method. This means identifying initial positions in space for all the atoms. In MD, the velocities of each atom must also be specified. This issue is not usually a serious one, if only because the system must always be brought into thermodynamic equilibrium (or "aged"), no matter what choice of initial conditions is made. Hence the simplest choice is probably best (see however, Erikson and Hostetler, 1987, for SiO_2), which usually means starting out with the atoms placed on a regular lattice in space (such as a convenient crystal structure), and for MD, choosing velocities randomly from a Maxwell-Boltzmann distribution. In many studies, it is common practice to initiate new runs using the last configuration of a previous run, which typically reduces the required equilibration time.

System size and periodic boundary conditions. A more serious issue is the finite size, in terms of the number of particles (N), of the simulated system. Historically, due to limitations of processor speed and computer memory capacity, simulations were limited to a few tens or hundreds of particles. Current computer performance permits the simulation of much larger system sizes, commonly in the range of 10^3 to 10^5 particles. Modern "cell multipole" methods can increase particle numbers into the millions (Greengard, 1988; Ding et al., 1992). The current "world record" for the largest MD simulation is apparently a Lennard-Jones simulation consisting of N=600,000,000 atoms (Tamayo, 1995: in prep for Ann Rev Comp Phys, vol 3), though this number will almost certainly have been exceeded by the time the present work appears in print. Nakano et al. (1993) have used a parallel computing architecture to carry out a simulation of SiO_2 with 41,472 particles. Obviously, such a large system size requires considerable expense of computing power. For most applications, however, it is sufficient (and necessary) to define the system size required to adequately model the properties of interest, in a reasonable time. Early studies on simple liquids obtained reasonable results for systems containing less than 100 atoms (Wood and Parker, 1957). However, the simulationist must generally beware the onset of system size effects: i.e. properties of the simulated liquid which become critically dependent on the number of particles. Whenever possible, this should be tested. The dependence of calculated properties on system size has begun to be investigated in MD simulations on silicates (Rustad et al., 1990; Stein and Spera, 1995; Diefenbacher et al., 1995).

The finite size of simulated liquid systems leads to one of the most significant approximations used in simulation: the periodic boundary condition (PBC) (Born and Von Karman, 1912; Metropolis et al., 1953). In this approach, the atoms move within a specified finite region of space (the simulation "box"), usually taken to be cubic with edge length L. (Simulation boxes having any symmetry compatible with filling space by being repeated in a regular array can be used: e.g. Adams (1979). This is particularly useful when liquid state results need to be compared with a non-cubic crystalline state under the same conditions). When the motion of an atom in any of the three Cartesian directions takes it outside this region, the atom is treated as if it has just entered the box from the opposite side. [The same procedure is used in many video games.] Distances between particles are handled in a similar way, using what is termed the "nearest image convention", illustrated in Figure 1 (Metropolis et al., 1953; Brush et al., 1966).

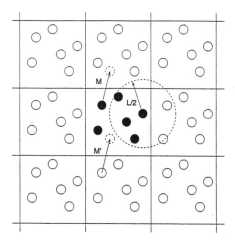

Figure 1. Illustration (in two dimensions) of periodic boundary conditions and the nearest-image convention. The simulation consists of six particles (filled circles) moving in the central simulation cell. The effect of periodic boundary conditions on the particle displacement M is shown: the moving particle "disappears" as it passes out of the central cell, "reappearing" on the other side of the box via path M'. In the nearest-image convention, a given particle only interacts with neighbors within a distance L/2 (indicated by the large dashed circle), regardless of whether those neighbors are in the central cell, or the surrounding "image" cells.

The simulated system thus has no surface, in the sense that no boundary with a container, or free surface to vacuum, exists within the model. Formally, the system is a periodic array of identical cells, extending to infinity in all directions (so that, in fact, the simulated "liquid" is really a "crystal" with an arbitrarily large "unit cell" of length L, and an internal structure which fluctuates in time). The advantage of PBCs is clear: the approach allows one to simulate an approximation of an infinitely large system using a relatively small number of atoms. Further, each atoms "sees" a local environment which consists only of other atoms. The simulated liquid is thus pure, homogeneous and surface-free. These properties can pose some problems in the study of first order phase changes, such as crystallization, melting or unmixing, because there are no convenient nucleation sites.

It must be borne in mind that the use of PBCs sets a limit on the maximum distance in space that can be considered physical. Due to the nearest image convention, no pair of atoms is ever more than L/2 apart. Hence, any phenomenon occurring on a longer length scale (or shorter than the corresponding wave-vector in reciprocal space) will not be observable. In general, phenomena which involve a physical length scale on the order of the simulation box size are inevitably strongly affected. For example, a fixed number of atoms in a simulation box of fixed V sets the bulk density to which the simulation corresponds. However, there is an additional, hidden, constraint on the problem. The use of PBCs means that locally there cannot be any density fluctuations with a wavelength larger than L/2. In the study of thermodynamic states where such fluctuations are particularly important, as near a liquid-gas critical point, the effect of PBCs must be taken into account. The kinetics of crystal nucleation in simulations of simple liquids become strongly anomalous (even unphysical) when the simulation box is smaller than the size of the critical nucleus (typically, several thousands of atoms). Conversely, simulations of small crystalline systems display an enhanced and unphysical stability against melting and solid state amorphization, since the small size of the simulation box suppresses the longer wavelength instabilities that would normally destroy the crystalline order (Hemmati et al., 1995), and the lack of surfaces eliminates the most efficient sites for melt or glass nucleation. As advanced computing capabilities are becoming more available, the effects of system size should be investigated as a matter of course in simulation studies. This may even give valuable additional information on the system, such as the wave-vector dependence of calculated properties. This can be invaluable, for example, in the interpretation of experimental neutron diffraction data on liquids and glasses (Buchenau et al., 1984, 1986).

Constraints and ensembles. Fixing the density of a simulation by fixing N and V is an example of one of a family of constraints on the physical properties of the model system that inevitably appear in computer simulations. These constraints arise as a direct result of the methods (i.e. MC or MD) used to generate a set of configurations $\{\Gamma\}$ to evaluate the integrals in Equations (4) or (6). Such constraints establish the particular thermodynamic *ensemble* in which the averaging will take place.

In the simplest MC simulation, N and V are fixed. Further, since the temperature T occurs in the expression for $P(k)$, it too must be fixed at a constant value. The condition of fixed N, V and T establishes that such a simulation models a system in the *canonical ensemble*.

Of course, there are other thermodynamic ensembles possible, each having a different function $P(k)$ weighting the configurations of the ensemble. As a result, sequences of sampled configurations can be generated in any ensemble by using the appropriate $P(k)$ function. For example, instead of constraining V to a fixed value, its conjugate thermodynamic variable P (pressure), can be constrained by replacing the $P(k)$ function for the canonical ensemble (i.e. fixed (N,V,T)) with that for the "*isothermal-isobaric*" ensemble (i.e. fixed (N,P,T)). In this case,

$$P(k) = e^{-\frac{E+PV}{k_B T}} \tag{17}$$

Here, the pressure P is now fixed at a constant value, while V for the simulation box is randomly varied in successive configurations, analogous to the way in which atomic positions are varied randomly. The sequence of configurations is thus one in which both atomic positions and V vary from step to step.

Of particular significance are the set of MC techniques used to sample the *grand canonical ensemble* , with constant (μ,V,T) (μ is the chemical potential). For this to occur, the number of particles N must be allowed to vary under the constraint that the chemical potential μ of the system is fixed (Norman and Filinov, 1969). This is achieved by adding a step to the procedure for generating a new configuration in which either (1) an attempt is made to insert a new particle into the system at a randomly chosen location, or (2) an attempt is made to annihilate a randomly chosen particle. The functional form of $P(k)$ used to evaluate the relative probablity of acceptance of the new configuration thus generated is given in Table 1. Simulations of the grand canonical ensemble are especially useful because the chemical potential (equivalent to the Gibbs free energy per particle) is fixed at the outset. Grand canonical simulations are thus the only commonly simulated ensemble in which the Gibbs free energy is known immediately. In all other ensembles, neither the free energy nor the entropy are immediately available for evaluation. There are however, indirect methods to estimate these quantities, as described below.

Table 1. Summary of thermodynamic ensembles explored in computer simulations. The * denotes the fact that MD simulation of the indicated ensemble is only possible using modified (non-Newtonian) equations of motion.

Ensemble	Fixed parameters	Weighting function for configuations	Simulation methods
microcanonical	N,V,E	$\delta(E-E_0)$	MD
canonical	N,V,T	$\exp(-E/k_BT)$	MC, MD*
isothermal-isobaric	N,P,T	$\exp[-(E+PV)/k_BT]$	MC, MD*
grand canonical	μ,V,T	$\exp[-(E-\mu N)/k_BT]$	MC

In the case of MD simulations, the simple approach described above for integrating the equations of motion yields a sampling of the microcanonical ensemble (i.e. constant (N,V,E)), since the total energy is conserved. However, the MD approach can be modified to generate a trajectory through phase space that explores other ensembles. For example, the Newtonian equations of motion can be readily modified so that only the kinetic energy (and hence the temperature) is conserved, rather than the total energy (Hoover et al., 1982; Evans 1983). This is achieved by adding a dissipative, velocity-dependent term to the expression for the force on each particle. A prefactor of this term is calculated at each time-step such that the resulting modified forces are just those that keep the kinetic energy fixed (Evans and Morris, 1990). It can be shown that these new non-Newtonian equations of motion generate a phase space trajectory that explores the canonical ensemble (constant (N,V,T)). In another (cruder) approach, the so-called "velocity re-scaling" method, which maintains the temperature of an MD simulation near a desired value, the velocities at each time step are are multiplied by a common prefactor so as to restore the kinetic energy to a specific value. This method is in common use as a means to bring a MD run close to a desired T, but is not well behaved and is less physical than the above alternatives.

Similar approaches can be used to constrain many other macroscopic variables in MD. For example, an equation of motion for the box volume V can be added to the system of differential equations for the particles, such that P is held constant. An important generalization of this approach was achieved by Parrinello and Rahman (1981, 1982), who demonstrated the utility of adding equations of motion for each of the six parameters that specify the geometry of the simulation box (i.e. three lengths and three direction-cosines, equivalent to the strain tensor). The result is that conditions can be simulated in which the symmetry of the simulation box itself is allowed to vary in response to non-hydrostatic (i.e. anisotropic) stress.

The ability to choose the particular thermodynamic ensemble for a given simulation (either in MD or MC) is a powerful feature. Which ensemble is "best" is determined mostly by the particular property or phenomenon of interest, but also to some degree by personal taste and expertise. Constant pressure ensembles are often useful to match most closely a set of laboratory measurements, which are almost always carried out at constant P rather than at constant V. In many MD simulations of silicates, an (N,V,E) ensemble is used, and the system size and box dimensions chosen to match the (approximate) experimental density. Constant μ ensembles are particularly well-suited for gas phase simulations, but their use in the study of dense liquids requires special techniques. It also must be remembered that dynamic properties can, strictly speaking, only be evaluated in MD, and even then, only during those portions of a simulation run in which an equilibrated microcanonical ensemble is being explored. In all other ensembles, the corresponding simulation methods require adjustments of atomic positions and/or velocities which do not arise directly from interparticle forces. Hence the motion of particles in any MD run not carried out in the microcanonical ensemble will not reflect the dynamic behavior of a real system.

Interaction potentials

General features. Up to now, we have described the modelling strategy and techniques used in liquid state simulation without saying much at all about the model liquid itself. That part of the simulation which makes it a model *of something* is the the function Φ describing the potential energy of interaction between particles in the system. The issue of interaction potentials is both subtle and important, especially in the context of silica and silicate melt simulations (Woodcock, 1975; Angell et al., 1981).

The potential energy function of a system can be expanded generally in a series of the form

$$\Phi = \sum_i \phi_i(\mathbf{r}_i) + \sum_i \sum_{j<i} \phi_{ij}(\mathbf{r}_i,\mathbf{r}_j) + \sum_i \sum_{j<i} \sum_{k<j<i} \phi_{ijk}(\mathbf{r}_i,\mathbf{r}_j,\mathbf{r}_k) + ... \tag{18}$$

The first term in this series is the particle *self energy*, and appears in first principles calculations of the potential energy function (Chizmeshya et al., 1994), or if an external field is applied to the simulation box. The second term is the pair potential, giving the contribution to the potential energy function from the interaction of pairs of particles. The third and higher terms represent the *many-body* contributions to the potential energy. It should be emphasized that in this formal expansion, each successive higher-order term contains no contribution from lower-order terms, otherwise over-counting occurs. Because of the additional computational effort required to carry out simulations with many-body potentials, most simulations, especially of silicate liquids, have been carried out within the two-body central pair potential approximation. In the absence of external fields,

$$\Phi = \sum_i \sum_{j<i} \phi_{ij}^{eff}(r_{ij}) \tag{19}$$

in which r_{ij} is the magnitude $|\mathbf{r}_j-\mathbf{r}_i|$. The potential here is termed an "effective" two-body potential (ϕ_{ij}^{eff}), because the effects of many-body interactions have generally been included in the formulation of the potential function and its adjustable parameters, for example, by comparing with experimental data.

Several types of such pair potentials are in common use, for modelling particle interactions in different types of liquid systems. The earliest simulations of model liquid systems used *hard-sphere* potentials, in which the repulsive contact between spheres was set at a fixed distance (Fig. 2). Note that because of the non-analytic nature of such a pair potential function, it is usually only employed in MC simulations. Conventional MD simulations require that the derivative of the pair potential (with respect to interparticle distance) exist at all points in space. A variant of the hard-sphere potential, that can be used in MD, is the soft-sphere potential, in which the repulsive part no longer rises infinitely steeply with interparticle contact. Many profitable studies have been carried out to investigate the behavior of simple liquids and general phenomenology of the liquid state using this class of potentials (Hansen and McDonald, 1976, 1986; March and Tosi, 1976; Ciccotti et al., 1987; Barrat and Klein, 1991).

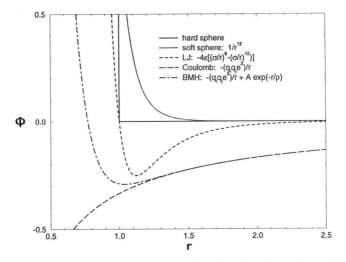

Figure 2. Plot of several pair potential functions used in simulations. In the Lennard-Jones potential (LJ), $\varepsilon = 1/4$ and $\sigma = 1$. In the Coulomb and BMH potentials, $(q_i q_j e^2)$ is set to unity, while $A = e^8$ and $\rho = 0.1$. Note the long-range nature of the Coulomb and BMH interactions.

The next most "sophisticated" type of potential is the Lennard-Jones (or "6-12") pair potential (Maitland et al., 1981):

$$\phi_{ij} = 4\varepsilon\left[\left(\frac{\sigma}{r_{ij}}\right)^{12} - \left(\frac{\sigma}{r_{ij}}\right)^{6}\right]$$ (20)

This potential function is used to approximate inter-particle interactions for simple, weakly interacting atoms or molecules which are approximately spherical in shape, such as the rare gases, methane, or molecular organic liquids such as ortho-terphenyl. Here ε is a parameter characterizing the strength of the interaction, while σ specifies a characteristic length for the interaction. These parameters are fitted to experimental data, such as the compressibility or melting temperature (Allen and Tildesley, 1987). In this potential, the $1/r^{12}$ term represents the short-ranged repulsion due to the overlap of closed-shell electron orbitals. The $1/r^{6}$ "attractive" term has the theoretical form for an induced dipole-dipole (or van der Waals) interaction. The exponent in the repulsive term is chosen purely for computational convenience, i.e. $1/r^{12} = (1/r^{6})^{2}$. The Lennard-Jones potential is an archetypal simulation model, in that it represents a compromise between physical accuracy, and computational utility and contains empirically adjustable parameters.

Coloumbic interactions are commonly encountered in a wide variety of simulated liquids, especially in the context of silicate and related systems (Brawer, 1985). Several potential types have been used to simulate this class of liquids. All contain a long-ranged Coloumbic interaction, and a short-ranged repulsive term. Also, additional terms are often added to include the effects of induced-dipole, and higher electrostatic multipole, interactions. Obviously, when electrostatic potentials are used, the model liquid necessarily consists of two or more ionic species, in a stoichiometric proportion consistent with overall charge neutrality. Again, it is emphasized that these are semi-empirical potentials, in that the parameters are chosen by the simulationist, usually to reproduce experimental results. Note that in many cases, the charges of the ions themselves are not taken to be their formal values; instead, non-integral "partial charges" can be optimized to improve the performance of the model (Mitra, 1982; Mitra et al., 1981; Lasaga and Gibbs, 1987).

Although simulations with pair potentials have achieved quite spectacular results in the case of silicates (Woodcock et al., 1976; Rustad et al., 1990; Stein and Spera, 1995), it is well known that spherically symmetric ionic potentials can not adequately simulate many structural features of the liquid. For example, the calculated SiOSi angle is always too large, and the frequencies of the SiOSi and OSiO bending vibrations are not reproduced well (Garofalini, 1982; Erikson and Hostetler, 1987; Kubicki and Lasaga, 1988; Stixrude and Bukowinski, 1988; Rustad et al., 1991a). One suggested solution to this problem has been to supplement the ionic potential with "covalent" terms for OSiO or SiOSi angle bending (Lasaga and Gibbs, 1987; Kubicki and Lasaga, 1988; Stixrude and Bukowinski, 1988). However, in these "mixed" potentials, it is not easy to avoid over-counting of terms. For example, the OSiO (θ) angle is governed by $O^{2-}...O^{2-}$ repulsion in a purely ionic calculation. If an additional term is added for the covalent OSiO angle bending, in terms of an angle bending force constant f_θ, the angle could become over-constrained (i.e. too stiff), if the ionic repulsion is not decreased accordingly. Such "over-counting" of contributions to the potential function can be avoided if careful attention is paid to the choice of the "mixed" potential used (Rappe and Goddard, 1991; Rappe et al., 1992). In addition, the particle charges used should be made as realistic as possible (Breneman and Wiberg, 1990; Chizmeshya et al., 1994).

Even for simple oxides such as MgO and CaO, it is well known that two-body central pair potentials do not give a realistic representation of the forces in the system, from the Cauchy violations observed for the crystals (Cohen et al., 1987; Chizmeshya et al., 1994; Anderson, 1995). There have been several attempts to incorporate non-central and many-body terms into potential functions for silicates, with excellent results for the simulated properties (Sanders et al., 1984; Feuston and Garofalini, 1988; Vessal et al., 1989; Vashishta et al, 1990; Zirl and Garofalini, 1990). In general, the three body terms are usually introduced in the form of an angular dependence to the potential function, giving interaction potentials which include terms that depend on the simultaneous positions of triplets of atoms. Great care must be taken not to include an additional two-body part into the three-body function, because of the trigonometric relation between bond lengths (described by the two-body interaction), and the included angle.

Despite the improvement in agreement between simulation results and certain experimental data when potential functions containing more terms are used, there is a caveat to use of more and more sophisticated potentials. As more potential parameters are included, their physical justification becomes less clear. There is a real danger that potential functions with many variable parameters become simply digital representations of the data with which the simulation results are compared, imparting them with limited value for extrapolation to regions where no data are present. With the simplest ionic pair potentials, there are few parameters, and the sources of error and limits to realistic interpretation and extrapolation of the results are obvious.

It has been suggested that it is possible to avoid the parameter fitting required in the development of semi-empirical potentials, through the use of *ab initio* methods to determine the parameters describing interparticle interactions (Lasaga and Gibbs, 1987; Tsuneyuki et al., 1988; van Beest et al., 1990; Tijskens et al., 1995). In this approach, *ab initio* calculations are carried out for molecular clusters designed to model structural features within the glass or melt; in particular, the SiO_4^{4-} (or $Si(OH)_4$) tetrahedral unit and the Si-O-Si linkage. These clusters are deformed and the total energy plotted as a function of the deformation coordinate. For example, if the SiO_4^{4-} tetrahedral unit is chosen as a cluster, one obvious deformation is the symmetric stretching of the four Si-O bonds about their equilibrium value, and a second is the symmetric deformation of the O-Si-O angles (Lasaga and Gibbs 1987, Tsuneyuki et al., 1988; van Beest et al., 1990). These *ab initio* deformation energies are then fitted to an appropriate potential function form, for later use in a molecular dynamics calculation. The *ab initio* potentials thus derived have the advantage that they are constructed entirely without direct reference to experimental data. However, there are problems associated with the construction of such potentials, which have to do with the choice of the deformational coordinates used to obtain the *ab initio* energy surface. For example, redundancy in the deformation modes of the SiO_4 tetrahedron has not yet been taken into account, when obtaining the *ab initio* energies to construct the potential (McMillan and Hess, 1990; Hess et al., 1994). (A tetrahedron has 4 bond stretching + 6 angle bending parameters, for a total of 10 structural variables, but only $3 \times 5 - 6 = 9$ degrees of freedom: it is impossible to deform a single tetrahedral angle, leaving all the others fixed). Other potential problems have to do with the adequacy of the basis set used in the *ab initio* calculation, neglect of electron correlation, and the choice of the cluster used to model the condensed system (Sauer, 1989; Catlow and Price, 1990; Hess et al., 1994).

The obvious future of simulation studies lies in incorporating more and more of an *ab initio* approach in the simulation, to eliminate the need for empirically derived interaction potentials. Two families of methods for carrying out "*ab initio* MD" have been proposed and are now being tested for work on real liquids, including oxides. In the Car-Parrinello "simulated annealing" approach, the electronic degrees of freedom, which are represented in a plane wave expansion, are given an explicit time dependence. Not only the nuclear positions, but also the electronic wave functions, relax as a function of time (Car and Parrinello, 1985). Grumbach (1993) and Grumbach and Martin (1995) have used the Car-Parrinello approach to investigate the properties of liquid carbon to extremely high pressures (and have investigated the occurrence of density-driven phase transitions within this elemental liquid). This type of approach has now been implemented in *ab initio* studies of crystalline framework SiO_2 compounds (Allan and Teter, 1987, 1990; Demkov et al., 1995), and it is only a matter of time before true *ab initio* MD simulations are carried out for silicates using a variant of the Car-Parrinello method. This approach has recently been applied to crystalline $MgSiO_3$ perovskite, in an investigation of the high-pressure properties of this important mineral (Wentzcovitch et al., 1993, 1995).

A similar approach using local orbitals, coupled with an efficient but accurate approximate density functional method, has been developed by Sankey and co-workers (Sankey and Allen, 1986; Sankey and Niklewski, 1989; Drabold et al., 1991). This has been used to investigate a wide range of molecular crystals with large unit cells, including fullerenes, Si clathrates, and SiO_2 polymorphs (Sankey et al., 1990, 1993; Adams et al., 1994; Demkov et al., 1994, 1995). The beauty of these methods is that information is obtained on the electronic structure, as well as the nuclear positions.

An alternative approach, based on a real-space description of the system electronic charge density in terms of component ion densities, derives from the approximate electron-gas model developed by Gordon and Kim (1972). This approach has been tested in MD simulations of molecular systems (LeSar and Gordon, 1982, 1983; LeSar, 1984). Variants of this model, which incorporate deformable ion densities to model many-body effects, have been used extensively in *ab initio* calculations of silicate minerals (Cohen et al., 1987; Wolf and Bukowinski, 1988; Chizmeshya et al., 1994). These should be readily adapted to large-scale simulations of oxide liquids (Cohen and Gong, 1994; Inbar and Cohen, 1995; Chizmeshya et al., in prep).

Long-range interactions. From the standpoint of a practical computer simulation, one of the most important features of an interaction potential is the manner in which the potential energy goes to zero as the interparticle separation goes to infinity. For short-range functions like the soft-sphere or Lennard-Jones potentials, the magnitude of the potential energy rapidly approaches zero after a few particle diameters. Hence the potential energy and force between a pair of particles separated by the maximum physical distance accessible to the simulation (L/2) is already a small fraction of the total, and only minor adjustments are needed to account for the absence of direct interactions over distances larger than L/2 (Allen and Tildesley, 1987). Truly long range interactions, such as those described by a Coulombic potential, present a more serious difficulty. In this case, the contribution to the total energy arising from interactions between particles separated by more than L/2 cannot be safely ignored.

Formally, the response to this problem is to carry out an evaluation of the Madelung potential energy of the entire infinite array of repeated simulation boxes using an *Ewald summation* technique (Ewald, 1921; Bertaut, 1952; Adams and Dubey, 1987). The Coulombic interaction potential between pairs of ions is of the general form

$$\phi_{ij} = \frac{z_i z_j e^2}{r_{ij}} \tag{21}$$

where z_i and z_j are ionic charges, e is the unit of electronic charge, and r_{ij} is the interparticle separation. This is a very long-ranged interaction, which falls off only as $1/r$ from any given ion. The interaction is repulsive (positive) between particles with the same sign charge, and attractive (negative) between particles of different sign. The total Coulombic contribution to the potential energy is the sum of all of these pair contributions:

$$\Phi^{Coul} = \sum_i \sum_{j<i} \frac{z_i z_j e^2}{r_{ij}} \tag{22}$$

Because the positive and negative contributions to this sum generally form alternating terms (i.e. positive ions are generally surrounded by negative ions, and vice versa), this is a very slowly convergent sum (Boeyens and Gafner, 1969). Ewald (1921) first showed that rapid convergence could be obtained for ionic crystals by using a Fourier transform method. In the Ewald method, two sums are obtained, one which converges in direct space, and one in which the summation is carried out in reciprocal space, and the two are added to obtain the electrostatic force acting on a given ion. Although a detailed description of this elegant method is beyond the scope of this review, a careful study of the Ewald sum and its properties is a "must" for anyone intending to carry out computer simulations on silicate systems. Details of the Ewald method and its implementation can be found in review papers (Tosi, 1964; Woodcock, 1975; Anastasiou and Fincham, 1982) or texts such as Allen and Tildesley (1986) and Haile (1992). In many early calculations, where computing power was limited, approximate forms of the Ewald summation were used in which the reciprocal space sum was neglected (Woodcock, 1975; Soules, 1979). Erikson and Hostetler (1987) have discussed the effect of such approximations on calculations for silicates, and have also considered the effects of finite simulation box dimensions. In modern simulations, there is no need for such approximations, and a full Ewald summation is generally carried out. In particular, recent advances in algorithm design, including the so-called "cell multipole methods", can greatly increase the efficiency of the evaluation of long-range forces (Adams and Dubey, 1987; Greengard, 1988).

Potentials for silica and silicates. All of the potentials contain a long range Coulombic part, along with a short-ranged repulsive term. In the pioneering work by Woodcock et al. (1976) on SiO_2, a Born-Mayer-Huggins type of pair potential (Fumi and

Tosi 1964; Tosi and Fumi, 1964; Busing, 1970) was chosen:

$$\phi_{ij}(r_{ij}) = \frac{z_i z_j e^2}{r_{ij}} + \left(1 + \frac{z_i}{n_i} + \frac{z_j}{n_j}\right) be^{\frac{\sigma_i + \sigma_j - r_{ij}}{\rho}} \tag{23}$$

in which z_i is the charge on ion i, n_i is the number of valence shell electrons for the ion i, σ_i is a distance parameter which defines the "size" (or ionic radius) of ion i, and b and ρ are empirical constants. The parameter ρ determines the steepness of the short range repulsive potential, and is known as a "softness" parameter. The σ parameters for both Si^{4+} and O^{2-} were adjusted so that the first peak in each of the three possible radial distribution functions (RDFs) due to Si-O, O-O and Si-Si pairs for the simulated liquid fit (approximately) the experimentally determined RDFs determined by Mozzi and Warren (1969). Although Woodcock et al. (1976) expressed concern with retaining full ionic charges for the Si^{4+} and O^{2-} ions in a partially covalent material, they found that the calculated dissociation energy for their simulated amorphous SiO_2 agreed to within 5% with that for crystalline quartz, obtained from experimental data via a Born-Haber cycle. This indicated that the form and parameters of the potential energy function represented at least a good *effective* potential for carrying out simulations on silicates. This potential, with minor variations, has been used in many later studies from this group on silicate melts (Angell et al., 1982, 1983, 1987; Scamehorn and Angell, 1991; Poe et al., 1992a,b; 1993a,b; Coté et al., 1993; Diefenbacher et al., 1994, 1995). In this later work, additional potential parameters for interactions involving ions such as Na^+, Ca^{2+}, Mg^{2+} and Al^{3+} were developed in a similar way to those for Si^{4+} and O^{2-}. In general, potentials of this kind, constructed of a sum of repulsive and Coulombic terms, are refered to as "*rigid-ion*" potentials.

Following the development of methods which permitted the study of larger systems, Mitra et al. (1981) and Mitra (1982) repeated the calculations on SiO_2 glass with a system containing 375 particles. These authors chose to use a form for the interatomic forces suggested by Pauling (Fumi and Tosi, 1964):

$$F(r_{ij}) = \frac{z_i z_j e^2}{r_{ij}^2} \left(1 + sign(z_i z_j)\left(\frac{s_i + s_j}{r_{ij}}\right)^n\right) \tag{24}$$

Here, s_i is a measure of the ionic radius of ion i with charge z_i, n determines the steepness of rise, or the "hardness" of the repulsive potential, and the "sign" function takes the value +1 when $z_i z_j$ is positive, and -1 for $z_i z_j$ negative. These authors originally took +4 and -2 as charges for Si and O ions, and the values suggested by Pauling for the ionic radii, but found poor agreement with the Mozzi and Warren structure for SiO_2 glass. Allowing these parameters to vary gave a set which fit the experimental RDF best, with n = 10, q_{Si} = +2.272, q_O = -1.136, s_{Si} = 0.02374 nm, and s_O = 0.12 nm (Mitra et al., 1981).

The behavior of all such early empirical potentials, and a comparison of their performance in calculations for SiO_2, were reviewed by Erikson and Hostetler (1987). They further concluded that Mitra's (1981) potential, using partial charges, gave the best agreement with experiment for the compressibility of crystalline polymorphs of SiO_2. These authors also studied and discussed the effect of approximations used in the Ewald sum evaluation of the long-range electrostatic interactions in earlier simulation studies. For example, in the original work of Woodcock et al. (1976), the Ewald evaluation was "abbreviated" to decrease the computational effort required in this initial exploratory effort. Similarly, Soules (1979) chose a modified form for the Born-Mayer-Huggins potential for his simulations on sodium silicate glass, so as to simultaneously account for some of the contribution of the Ewald sum, while avoiding its most computationally demanding terms.

Erikson and Hostetler (1987) demonstrated the difficulty that arises when trying to compare results obtained with different potentials and different approximation methods. They showed that approximations to the full Ewald sum, though giving internal energies close to those found with a full Ewald treatment, magnify the disorder of the local structure of the melt. They also pointed to the importance of care in the choice of starting configuration, especially in simulations conducted at the limit of computational run-time resources.

More recent developments of empirical potentials for SiO_2 have focused on the inclusion of three-body interaction terms (Sanders et al., 1984; Feuston and Garofalini, 1988; Vessal et al., 1989, 1991; Vashishta et al, 1990). These potentials do generally result in much better agreement between the simulation results and experimental data (Vashishta et al., 1990; Nakano et al., 1993, 1994; Jin et al., 1993, 1994), but they involve considerably more computational effort, and the potentials are not easily transferable to multicomponent aluminosilicate systems. Zirl and Garofalini (1990) have experimented with a three-body potential for sodium aluminosilicates.

Several groups have pioneered the development of *ab initio*-derived potentials for MD and MC calculations on silicates (Lasaga and Gibbs 1987, Stixrude and Bukowinski, 1988; Tsuneyuki et al., 1988; van Beest et al., 1990; Tijskens et al., 1995). A Morse-type potential, containing added terms for covalent OSiO angle bending and SiOSi (expressed as Si...Si repulsion) terms was developed by Stixrude and Bukowinski (1988) for their Monte Carlo simulations of SiO_2 glass at high P (Stixrude and Bukowinski, 1990; 1991). This model used covalent force field parameters obtained from the *ab initio* cluster calculations of O'Keeffe and McMillan (1986) and Hess et al. (1986, 1987), for H_6Si_2O and $Si(OH)_4$. Lasaga and Gibbs (1987) developed a potential based on Hartree-Fock calculations on the neutral siloxane molecules $Si(OH)_4$ and $H_6Si_2O_7$, to obtain potential parameters for the SiO_4 unit and the SiOSi linkage. The static and dynamic properties of the Lasaga-Gibbs (LG) potential were explored by Kubicki and Lasaga (1988), and compared with results using a simple ionic potential. Tsuneyuki et al. (1988) developed their potential based on deformation of the SiO_4^{4-} ion only.

The Tsuneyuki et al. (1988) potential (TTAM) has the Born-Mayer-Huggins form with an added short-range van der Waals'-like term (i.e. in $1/r^6$):

$$\phi_{ij}(r_{ij}) = \frac{z_i z_j e^2}{r_{ij}} + A_{ij} e^{\frac{-r_{ij}}{\rho_{ij}}} - \frac{C_{ij}}{r_{ij}^6} \tag{25}$$

The parameters defining the TTAM potential for SiO_2 are listed in Tsuneyuki et al. (1988) and Rustad et al. (1991a). This potential has been used to calculate properties of crystalline SiO_2 polymorphs (Tsuneyuki et al. 1989, 1990), to investigate melting of stishovite at high pressures (Belonoshko, 1994a), and has been tested for silicate melt studies by Della Valle and Andersen (1992) and Rustad et al. (1991a). Rustad et al. (1991a) compared the performance of the TTAM potential with those of Mitra (1981) and Lasaga and Gibbs (1987), and concluded that the TTAM potential gave the best results for densification of SiO_2. Rustad et al. (1990) did note a problem with the TTAM potential in calculations of transport properties. The inclusion of the term in $1/r^6$ allows particles to coalesce onto the same point in space due to the infinitely strong attraction at small interparticle separations (i.e. nuclear fusion occurs!). Care should be taken when using this type of potential in any high temperature simulation.

The potentials derived by van Beest et al. (1990) for Si-O, Al-O and P-O interactions were based on *ab initio* calculations for the $Si(OH)_4$, $Al(OH)_4^-$ and $P(OH)_4^+$ molecules, and are similar in form to TTAM. However, the parameters differ, and better agreement with experiment was found for structural parameters of crystalline silicates using the van Beest et al. potential. The performance of this potential function has been tested thoroughly by Tse and Klug (1991). A further potential for simulations of SiO_2 polymorphs based on *ab initio*-derived parameters has recently been developed by Tijskens et al. (1995).

To date, most effort has been expended to develop adequate potentials to model SiO_2 polymorphs. For interactions involving alkali or alkaline earth ions, simple ionic pair potentials have been used to date. While the results seem to be quite adequate (i.e. the coordinated numbers and metal-oxygen bond lengths, and metal ion diffusion coefficients appear to be in reasonable agreement with experiment (Borgianni and Granati, 1979; Soules, 1980, 1982; Soules and Busby, 1981; Angell et al., 1982, 1983; Matsui et al., 1984, 1987, 1988, 1991, 1992; Kubicki and Lasaga, 1993; Wasserman et al., 1993), and the potential parameters appear to be quite transferable (Lienenweber and Navrotsky, 1988), we know that such rigid ion potentials are inadequate (Cohen et al., 1987; Chizmeshya et al., 1994). As we develop more techniques for the study of the "modifier" cation environment in silicate melts, and more information becomes available on the local environment of such ions in glasses and melts (Greaves et al., 1981, 1992; Xue et al., 1991; Fiske and Stebbins, 1994; Xu and Stebbins, 1995; see chapters by Stebbins, by Brown et al., and by McMillan and Wolf, in this volume), more attention will be paid to refining the potentials used to model the interactions involving these ions.

Making "measurements"

Having covered what the simulation model consists of, we now turn to the matter of what results can be obtained from the simulation, and how this is achieved. The central role played by the time variable in liquid simulations, whether it be the "physical" time (i.e. simulation run clock) that appears in MD runs, or the number of "time" steps in a MC run, cannot be overemphasized at this stage. A typical liquid simulation consists of an initial equilibration phase, followed by a "production" phase, during which the properties of interest are actually evaluated. The equilibration period starts from whatever initial (inevitably out-of-equilibrium) state has been specified and proceeds until a plausible thermodynamic equilibrium state has been achieved. In the particular case for which the production phase is to be a microcanonical MD run, it is common that the equilibration include an initial period during which the liquid is brought to a desired T and/or P through the use of a different ensemble, such as the canonical or the isothermal-isobaric ensemble.

Establishing the achievement of an equilibrium state with any confidence can be problematic. As a minimum requirement, all properties of interest should have relaxed from their initial non-equilibrium values, and be observed to be fluctuating around an average value which is not changing systematically over time. However, this "steady state" should be treated with some skepticism, unless it is certain that the thermodynamic state simulated is particularly innocuous. Simulated states near phase transitions, or at very low T, can get trapped in metastable states, which give every appearance of thermodynamic equilibrium, despite the fact that the true equilibrium state may be radically different. Metastable states may have very long characteristic lifetimes before decaying to the true equilibrium state. Also, the kinetics of the non-equilibrium transformation from the metastable to the equilibrium state may be very slow compared to the simulation timescale. In the absence of a precise knowledge of the phase diagram of the system, it is very difficult to be sure that a given state is not metastable: a simulation simply may not probe the timescale required to reveal the metastability. The general behavior of a system in equilibrium is illustrated in Figure 3.

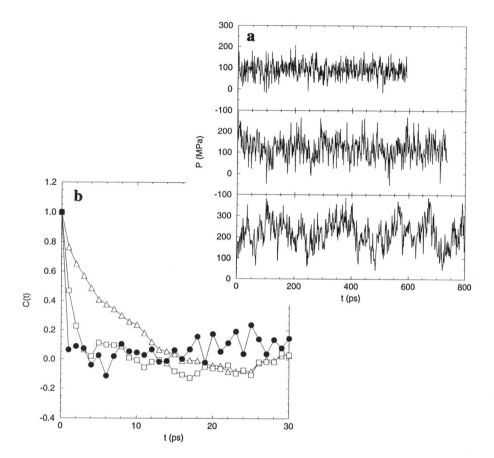

Figure 3. (a) Time series of calculated pressure (P) in a typical MD simulation of a liquid (NVT ensemble), at high T (top panel), moderate T (middle), and low T (bottom). (b) Autocorrelation functions for the three time series of P given in (a). At high T (filled circles) the time correlations die out quickly, within a few ps; at moderate T (squares) correlations persist for slightly longer, while at low T (triangles) the values of P are correlated for 10 to 20 ps.

The question of how long to run for equilibration also bears directly on how long to pursue the production phase of the run. The production phase must be sufficiently long so that enough *independent* configurations of the ensemble are sampled to provide well-averaged property values. Hence the time required to establish and assure equilibrium will be comparable to the time required *in equilibrium* to sample phase space sufficiently for such an average to be obtained, for the property of interest. It is important to note that different quantities will generally have different relaxation times. For example, structure factor values for small wavevector will take longer to average than for large wavevector, since the size of the region that must relax in equilibrium is larger for small wavevector (i.e. a longer wavelength fluctuation). To be sure, one should directly measure the *autocorrelation function* ($C(t)$) of the quantity of interest $A(t)$ with respect to time; defined by:

$$C(t) = \frac{\langle A(t)A(0)\rangle - \langle A\rangle^2}{\langle A^2\rangle - \langle A\rangle^2} \tag{26}$$

A(0) is the value of the property at t = 0. The time it takes for this function to decay to zero establishes a reasonable time between truly independent contributions to the corresponding thermodynamic or structural average (Fig. 3b). This is a useful normalized form of the more conventional autocorrelation functions used to date in many geochemical simulation studies, which take the simpler form $C(t) = \langle A(t)A(0)\rangle$. In the form of Equation (26), the second term in the numerator guarantees that the function will eventually decay to zero, and the denominator normalizes the function so that it it unity at t = 0. This generalized form is particularly useful in comparing correlation functions for different quantities (Hansen and McDonald, 1986).

A production run to evaluate the average should proceed over many multiples of this basic relaxation time. Newcomers to simulation studies should also pay careful attention to the proper calculation of uncertainties for evaluated averages (see, for example, Allen and Tildesley (1987) and Haile (1992)).

Static properties. The values of some thermodynamic properties will be fixed by the choice of ensemble within which the simulation is carried out. Some of those that are not fixed can be evaluated from the simulation results. For example, in a canonical MC or MD simulation, N, V and T are fixed parameters. The thermodynamic variables conjugate to N and T, respectively μ and S, are not directly measurable. The internal energy (U) is the sum of average values of the potential energy (Φ) and the kinetic energy (L), obtained by summing over the momenta of the particles (Eqn. 2):

$$U = \langle \Phi\rangle + \langle L\rangle \tag{27}$$

In a Monte Carlo simulation, there is no explicit kinetic energy term. In constant V simulations it is common to evaluate the pressure, P, from the interparticle distances and forces through the average value of the *virial* ($\langle W\rangle$). This is defined as

$$\langle W\rangle = \frac{1}{3}\left\langle \sum_{i=1}^{N} \mathbf{r}_i \cdot \mathbf{f}_i \right\rangle \tag{28}$$

where the force ($\mathbf{f}_i$) is given by Equation (11). The second line in the expression for the virial expresses the sums in terms of inter-particle separations, $\mathbf{r}_{ij}$. Using the equipartition and virial theorems (Goldstein, 1980), P is then evaluated from

$$P = \frac{Nk_BT + \langle W\rangle}{V} \tag{29}$$

Note that this calculation should be done in a manner independent of the simulation coordinate system. (See also the appendix in Haile (1992) for some subtleties on the definition and evaluation of P in simulations.)

As indicated earlier, the chemical potential μ and entropy S are not directly available in any single simulation run, except in the case of a grand canonical MC simulation. However, given some care, these quantities can be referenced to some known standard state by performing a *set* of simulation runs along a path connecting the reference state to the state of interest. Such "thermodynamic integration" methods rely on the fact that an evaluation of the equation of state over a range of thermodynamic states can be numerically integrated to give an accurate estimate of the free energy, and consequently yield μ and S (Allen and Tildesley, 1987).

One of the unique features of simulation data, as compared to experimental data, is the fact that the positions and velocities (in MD) of all particles are known in complete detail. As a result, information on the atomic or molecular structure of the liquid is immediately available and readily analyzed, and can be compared with experimental data (Wright, 1993). For example, it is common to evaluate the pair correlation function g(r), which gives the probability density of finding a particle within an interval δr at a distance r from a specified particle. For small values of the distance r, this function is found to display the oscillations characteristic of the local order found in dense liquids (Fig. 4). The first maximum corresponds to the shell of first neighbors found around each particle. The function g(r) is defined as

$$g(r) = \frac{V}{N^2} \left\langle \sum_{i=1}^{N} \sum_{\substack{j=1 \\ j \neq i}}^{N} \delta(r - r_{ij}) \right\rangle \tag{30}$$

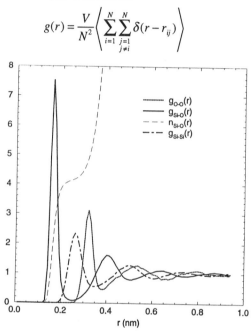

Figure 4. The g(r) and n(r) functions for SiO_2 liquid at a density (ρ) of 2.2 g/cm^3 and T = 6000 K, as modeled by the BMH potential.

From a practical viewpoint, this average is not evaluated as a continuous function of r, but as a histogram. The space around each particle is partitioned into spherical shells of a specified thickness, and a histogram is built up by counting the number of neighboring particles that appear in each shell over time. The average number of particles in each shell, plotted against the distance of the shell from the central particle, gives an estimate of g(r), once the normalization by the prefactor V/N^2 is carried out. Also shown in Figure 4 is the "*running coordination number*", n(r), which is an integral over g(r):

$$n(r) = \frac{4\pi N}{V} \int_0^r dr'(r')^2 g(r') \tag{31}$$

The form of n(r) is such that it gives the average number of particles found inside a sphere of radius r centered on each particle. The first plateau in n(r) as a function of r indicates the first (nearest neighbor) coordination number characteristic of the central particle.

This wealth of microscopic structural information actually leads to a problem in the presentation of simulation results. Often data on atomic structure are generated and discussed which are difficult or impossible to compare with experimental results, simply because the experimental study of the corresponding real system does not permit measurement of the appropriate information. This issue can be a serious one (more so for non-simulationists than simulationists) when it results in predictions that cannot be confirmed or refuted experimentally. As a general rule, it is probably best to focus attention on simulation data that can be directly compared to existing or future experiments, presenting other results only as aids to intuition.

Dynamics. A simple and commonly measured dynamic quantity is the diffusion coefficient or constant D (see chapter by Chakraborty, in this volume). D is commonly evaluated by plotting the *mean squared displacement* (MSD) of the X ions of a given type considered in the simulation (X ≤ N),

$$\left\langle r^2(t) \right\rangle = \frac{1}{X} \left\langle \sum_{i=1}^{X} \left| \mathbf{r}_i(t) - \mathbf{r}_i(t_0) \right|^2 \right\rangle \tag{32}$$

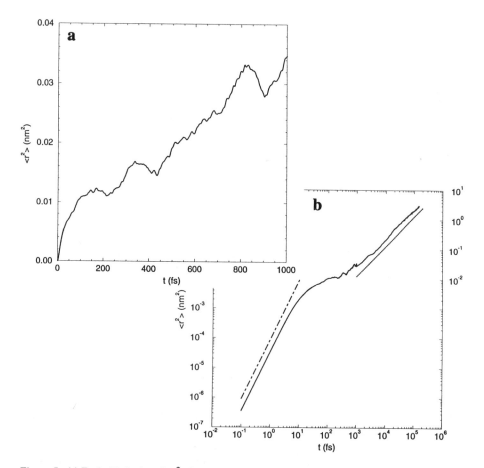

Figure 5. (a) Typical behavior of <r²> for oxygen ions as a function of time, from an MD simulation of SiO₂ (BMH potential) at ρ = 2.2 g/cm³ and T = 6000 K. (b) Log-log plot of $\langle r^2 \rangle$ versus t for the same system described in (a). In (b), The ballistic regime is visible as the initial part of the curve having slope 2; the diffusive regime appears at long times, where the slope becomes unity. (The dot-dashed and dotted lines have slopes 2 and 1 respectively.) Note the size of the cross-over regime. Though it might have been tempting to extract a value for D by fitting the curve in (a) between (say) t = 200 and 1000 fs, to a straight line, the data in (b) show that the true diffusive regime does not even begin until after 1000 fs.

of the atoms from their positions at some specified time t_0, as a function of time (t). A schematic plot of the typical behavior observed is shown in Figure 5. Three regimes can be identified in the behavior of $\langle r^2 \rangle$. For very short times, before a significant number of collisions takes place, there is an initial "ballistic" regime in which the particle motion is predominantly characterized by inertial flight through space; in such motion the distance travelled is proportional to the time, and hence $\langle r^2 \rangle$ varies parabolically (i.e. like t^2) with respect to time. At the opposite extreme of very long time, a particle's trajectory has the form of Brownian motion, or a "random walk", for which $\langle r^2 \rangle$ varies linearly with t. Between these two extremes is a cross-over regime in which the two effects are mixed are identified in the behavior of $\langle r^2 \rangle$. For very short times, before a significant number of collisions takes place, there is an initial "ballistic" regime in which the particle motion is predominantly characterized by inertial flight through space; in such motion the distance (Hansen and McDonald, 1976; March and Tosi. 1976). The value of D is obtained from the slope of the curve *in the limit of large t*, as indicated by the Einstein expression for D:

$$D = \lim_{t \to \infty} \frac{\langle r^2(t) \rangle}{6t} \tag{33}$$

When evaluating D from such a plot, care must be taken to be sure that the long-time regime has been reached in the simulation run. The cross-over to truly asymptotic behavior can be very slow, especially at low T (Fig. 5b). Also, $\langle r^2 \rangle$ data should be averaged over multiple "time origins" (different t_0 values), as well as over the diffusing particles (Kubicki and Lasaga, 1991a). Note that spurious results can occur if the net momentum of the simulated system is not set and maintained to be zero (this can be particularly problematic for small system sizes). Even an extremely small center-of-mass motion in the system will eventually dominate the long time behavior by superimposing a parabolic component on the asymptotic linear behavior. The diffusion coefficient D provides a crude, but convenient, measure of the point at which the behavior of the system becomes non-ergodic. Typical nearest-neighbor inter-particle separations ("bond lengths") are on the order of 1 to 2 Å. Sufficient particle diffusion must occur in order for the system to adequately sample configuration space, for ergodicity to be achieved. If we take a minimum value of $\langle r^2 \rangle = 1$ Å^2 for this to occur, for a simulation run time of $\sim 10^{-11}$ s, $D > 0.2 \times 10^{-5}$ cm^2/s (Brawer, 1985; Scamehorn and Angell, 1987). If the diffusion coefficient for a given ion or set of ions is smaller than this, it is unlikely that the system will have achieved structural equilibrium, for the configurations involving those ions. As already indicated, ergodicity-breaking sets in at much higher temperatures, or for shorter run times, for the framework ions (Si^{4+}, Al^{3+}, O^{2-}) in silicates, than for modifier ions such as Li^+ or Na^+, which behave as ionic conductors. Note that measured diffusion coefficients larger than this value are no guarantee of achievement of equilibrium.

D can also be calculated from the time correlation function for the velocity of individual particles, or "*velocity autocorrelation function*" $\varphi(t)$, defined as

$$\varphi(t) = \langle \mathbf{v}(t) \cdot \mathbf{v}(0) \rangle \tag{34}$$

It can be shown that D is simply the integral of this function (Boon and Yip, 1980):

$$D = \frac{1}{3} \int_0^\infty \varphi(t) dt \tag{35}$$

In simulations, $\varphi(t)$ can evaluated directly and the integral taken after the run has ended. Note that the upper limit of the integral in Equation (35) must necessary be replaced by the time over which $\varphi(t)$ was evaluated; once more, the run must therefore be long enough to approximate the asymptotic behavior. Note that, in general, any macroscopic transport property (such as D, thermal conductivity, shear and bulk viscosity, etc.) can be written in the form of an integral over the time correlation function for an appropriate quantity, as in Equation (35) (Hansen and McDonald, 1976, 1986; Allen and Tildesley, 1987), and that for each such expression (called "*Green-Kubo formulae*") there corresponds a relation analogous to the Einstein relation of Equation (33), valid in the limit of long time (Boon and Yip, 1980). The Green-Kubo formalism has been applied to calculate the shear viscosity in molten silicate systems (Ogawa et al., 1990; Wasserman et al., 1993a).

The time correlation function $\varphi(t)$ is also a useful starting point for the consideration of how to compare simulation results to experimental spectra obtained by, for example, neutron scattering or vibrational (infra-red or Raman) spectroscopy (McQuarrie, 1976; Allen and Tildesley, 1987). $\varphi(t)$ contains all the information about particle vibrations, due to the fact that the particle velocities (along with positional coordinates) are necessarily periodic during vibrational motions. By Fourier transforming $\varphi(t)$ from a function of t to one of frequency ω, a *power spectrum* of vibrational modes $\varphi(\omega)$ is found which is proportional to the vibrational density of states $D(\omega)$, in the limit of low T (March and Tosi, 1976; Garofalini, 1982; Brawer, 1983, 1985; Rustad et al., 1991a; Wasserman et al., 1993a,b; Alavi et al., 1995; Jin et al., 1993). This transformation should be carried out at low temperature, where the vibrational behavior is nearly harmonic (Brawer, 1983; Allen and Tildesley, 1987). As illustrated in the next section, $\varphi(\omega)$ as calculated from simulation can therefore be compared to frequency-domain experimental spectra, such as inelastic neutron scattering, or infrared and Raman spectra, since all such probes measure some component of $D(\omega)$. Correspondences between peaks in the simulation and experimental spectra can be studied, as well as changes in the strengths and positions of these peaks as a function of external parameters such as P and T. In these studies, however, the simulated box size (and shape) and number of particles determine the correlation wavelength, or wave vector, of the vibrational modes studied. This should be carefully examined, when comparing with experimental data. Some vibrational modes of silicate glasses do show a dependence on wave vector, which can be studied by carrying out simulations with varying box size (Buchenau et al., 1984, 1986).

With modest additional effort, specific types of spectral functions can be calculated from simulation data (Hansen and McDonald, 1976, 1986; Allen and Tildesley, 1987), allowing more precise comparison with experimental results. For example, the dynamic structure factor $S(\mathbf{k},\omega)$, measured in inelastic neutron scattering experiments, can be evaluated from simulation data by noting its connection with the van Hove correlation function, $G(\mathbf{r},t)$, defined by

$$G(r,t) = \frac{1}{N}\left\langle \sum_{i=1}^{N}\sum_{j=1}^{N}\delta(\mathbf{r} + \mathbf{r}_i(0) - \mathbf{r}_j(t))\right\rangle \qquad (36)$$

(Jin et al., 1993). $G(\mathbf{r},t)$ simply expresses the relative probability that if particle i is at position $\mathbf{r}_i$ at t = 0, then particle j will be at position $\mathbf{r}_j$ at time t. Given that appropriate care is taken in the storage of simulation configurations, $G(\mathbf{r},t)$ can readily be evaluated in MD, using a generalization of the histogram approach employed to evaluate g(r) above. $S(\mathbf{k},\omega)$ is then found by Fourier transforming, both from position-space to momentum-space (the wave vector in reciprocal space is $\mathbf{k}$), and from the time to the frequency (ω) domain:

$$S(\mathbf{k},\omega) = \frac{1}{2\pi}\int e^{i\omega t}dt\int e^{-i\mathbf{k}\cdot\mathbf{r}}G(\mathbf{r},t) \qquad (37)$$

In practice, the $\mathbf{r}$-to-$\mathbf{k}$ Fourier transform can be accounted for during the simulation run, which is a useful approach when it is known beforehand which values of **k** are of interest. Also, there are complications involved in Fourier transforming any discrete data set known only over a finite domain, to which heed must be paid (Press et al., 1986).

In a similar way, an IR spectrum $I(\omega)$ (I is the infrared absorption intensity) can be found from simulation data by first evaluating the time correlation function for the total electric dipole moment of the system, and then Fourier transforming the result (Brawer, 1983, 1985; Boulard et al., 1992; Wasserman et al., 1993a,b). In practice, $I(\omega)$ is usually calculated from the time correlation function for the electric flux, since this quantity carries the same information, and facilitates a coordinate-system-independent result (Woodcock, 1975; Brawer, 1983, 1985; Boulard et al., 1992). Related time-correlation functions can be devised to calculate Raman scattering intensities, but these are more difficult to devise and interpret (Brawer, 1983, 1985; Boulard et al., 1992; Jin et al., 1993). The main problem is in modelling that fluctuation of the system which will best mimic the interaction of light with amorphous systems in a Raman scattering experiment. In particular, the effect of simulation box size appears to significantly affect the results (Boulard et al., 1992). In these simulations of the vibrational spectra of fluoride and oxide glasses, simple ionic potentials seem to do well for the high frequency stretching vibrations (Be-F, Si-O) (Woodcock et al., 1976; Brawer, 1983, 1985; Boulard et al., 1992). This is not, however, surprising, because the potential parameters are chosen to model the bond length. Poorer agreement is found for lower frequency vibrations involving OSiO or SiOSi bending vibrations, unless a more sophisticated potential function is used (Rustad et al., 1991a; Jin et al., 1993).

SIMULATIONS OF SILICATE LIQUIDS

Mostly SiO_2

Most simulations of silicate materials to date have focused on the structure and properties of SiO_2 glass and melt, because it forms the simplest prototype silicate system, and also because of its technological significance.

Structure. In their pioneering study, Woodcock et al. (1976) calibrated their potential function against the experimentally determined RDF for SiO_2 glass of Mozzi and Warren (1969). In this sense, the agreement found between the simulated and the experimental structure was to some degree "built-in" to the resulting model. This is a common characteristic of simulations carried out using empirical potentials. However, the quality of the agreement that was achieved in this first attempt should not be overlooked. It is by no means obvious *a priori* that a classical, two-body potential should do as well as was found. The first peak due to Si-O interactions was modelled very well, though the O-O pair function was slightly too small, and the Si-Si separation was too large (Fig. 4). It is now well known that simple ionic potentials result in Si-O-Si angles which are too wide (Erikson and Hostetler, 1987; Stixrude and Bukowinski, 1988; Kubicki and Lasaga, 1988; Vessal et al., 1989; Vashishta et al., 1990; Rustad et al., 1991a). Woodcock et al. (1976) did in fact speculate that the lack of covalent terms allowed for greater distortion of the SiO_4 tetrahedra. In spite of these deficiencies, their simulation correctly found that SiO_2 liquid was composed of corner-shared SiO_4 tetrahedral groups. They also found the internal energy to be consistent with that of α-quartz (Woodcock et al., 1976). All of the simple two-body ionic potentials share similar problems in accurately reproducing structural details of SiO_2 melt and glass, especially the SiOSi bond angle. That proposed by Mitra et al. (1981) appears to give the best fit with the experimental RDF (Erikson and Hostetler, 1987; Rustad et al., 1991).

When evaluating differences in structure between simulated and real SiO_2, it should be recognized that the "glass" is formed via the MD runs under quite different conditions than any laboratory-synthesized sample. Specifically, due to the relatively much shorter simulation observation time, the MD glass appears on the computer time scale at a much higher fictive temperature (on the order of 4000 K) than the real glass, which passes through its glass transition near 1500 K (Angell and Torell, 1983; Scamehorn and Angell, 1987). Hence, detailed comparison between simulation and experiment may not be strictly valid, because the structural details may differ considerably, simply due to the temperature dependence of the structure (Scamehorn and Angell, 1991). For this reason, it is difficult to decide that a given potential is "better", based on comparing the simulated structure with experimental observations on a glass formed with normal laboratory cooling rates.

In reproducing the glass and crystal structure, it seems clear that the potentials developed by Tsuneyuki et al. (1988) and van Beest et al. (1990) - which contain a short range dispersive (C_{ij}/r^6) term in addition to the ionic part, and are fit to results of *ab initio* cluster calculations - do a much better job in reproducing the structural parameters of SiO_2 polymorphs, in particular, the O-Si-O and Si-O-Si angles (Tsuneyuki et al., 1990; Rustad et al., 1991a; Tse and Klug, 1991; Della Valle and Andersen, 1991, 1992; Keskar and Chelikowsky, , 1992, 1995). However, care must be taken in use of these potentials to investigate the properties of SiO_2 liquid at high temperature (>5000 K), in particular the O^{2-} diffusivity, because of the "particle coalesence" problem associated with the presence of the (C_{ij}/r^6) term (Rustad et al., 1990).

The potential developed for Monte Carlo calculations by Stixrude and Bukowinski (1988) contains covalent angle bending terms derived from *ab initio* calculations, and also does an excellent job of reproducing silicate structures (Stixrude and Bukowinski, 1988, 1989). However, this potential constrains the silicate unit to tetrahedral geometry (because a harmonic tetrahedral OSiO bond-bending is used), so that densification and structural relaxation processes which involve formation of more highly coordinated species can not be examined. This does allow the interesting "thought exepriment" to be carried out: how well can a constrained tetrahedral model account for the compression behavior of SiO_2 glass and liquid (Stixrude and Bukowinski, 1990, 1991)?

The many-body potentials developed by Sanders et al. (1984), Feuston and Garofalini (1988), Vessal et al. (1989) and Vashishta et al. (1990) also do an excellent job in reproducing the structures and the vibrational properties of crystalline and amorphous polymorphs of SiO_2, and simulations with such potentials will provide a powerful tool for future studies on liquid silicates (Nakano et al., 1993, 1994; Jin et al., 1993, 1994).

Vibrational spectrum. The simplest type of information on the vibrational properties of silica is obvious in the results of Woodcock et al. (1976) (Fig. 6). In their plot of radial distance versus time for oxygen atom displacements about an Si atom, it is seen that the period of the oxygen motion is $\sim 10^{-13}$ s, corresponding to a vibrational frequency ~ 1000 cm^{-1} (Soules, 1982). This is the approximate value of the observed Si-O stretching frequency in SiO_2 glass. Subsequently, several workers have used the results of MD simulations to directly calculate the vibrational spectrum of SiO_2 glass or melt, by using an appropriate autocorrelation function and performing a time Fourier transform (Garofalini, 1982; Rustad et al., 1990, 1991a; Alavi et al., 1992; Jin et al., 1993). These calculations are particularly interesting and useful, in that the velocity autocorrelation functions associated with the Si^{4+} or O^{2-} ions are evaluated separately, so that the degree of participation of each ion in the atomic motions can be investigated. This leads to an assignment of the particular vibrational mode, which is especially useful if the potential function allows a reliable calculation of the vibrational spectrum (Jin et al., 1993). In general, the simple rigid ion potentials permit calculation of the Si-O bond stretching vibrations, but SiOSi and OSiO bending modes are not well reproduced (Garofalini, 1982). The more sophisticated potential TTAM and van Beest potentials appear to do a better job (Rustad et al., 1990; Della Valle and Andersen, 1991). The three-body potentials appear to give a vibrational mode distribution in excellent agreement with experiment (Jin et al., 1993). This type of study is in its infancy, but will prove invaluable for the interpretation of infrared and Raman spectra for silicate liquids and glasses, obtained in situ at high temperature and/or high pressure (see chapters by Wolf and McMillan, and McMillan and Wolf). In particular, the simulations could be used to obtain the vibrational signature associated with, for example, the stretching vibrations of SiO_5 or AlO_6 groups within the structure, a point of intense interest in current melt structure studies.

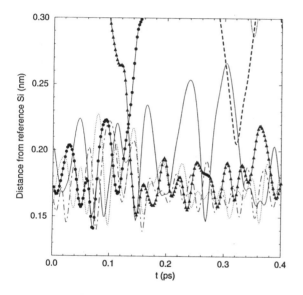

Figure 6. Plot of distance of neighboring O^{2-} ions from a selected Si ion as a function of time, from an MD simulation of SiO_2 liquid at $\rho = 2.2$ g/cm^3 and T = 8000 K, as modeled by the BMH potential. Initially there are four O^{2-} ions nearby, executing anharmonic vibrations around the central Si ion. At t = 0.13 ps, one of these O^{2-} ions (●) leaves the first coordination shell, and is simultaneously replaced by a new O^{2-} ion (▲), illustrating a "rattle-and-hop" mechanism for diffusion. Note that at a later time t = 0.31 ps, there is a unsuccessful attempt to exchange O^{2-} ions.

Expanded silica. The growing power of computer simulations is well illustrated by two studies of porous silica, a form of SiO_2 formed by supercritical drying of silica gels. The result is a highly ramified and porous amorphous structure, containing cavities of up to several nanometers in size. Simulations by Kieffer and Angell (1988) showed how this material could be produced in simulations. In their work, a silica glass at normal density was subjected to isotropic tensile stress by successively increasing the simulation box size in small increments. The silica glass was observed initially to support the imposed stress, but eventually began to "cavitate" through the formation of pores over a range of sizes. This trend continued until the simulated system fractured completely, but before this eventual mechanical failure, a well-defined fractal structure was observed. This initial result has been exhaustively confirmed in a set of impressive simulations carried out by Nakano et al. (1993). They simulated a silica system containing 41,472 ions—this may yet be the largest silica simulation ever conducted—and subjected a typical silica glass configuration to careful successive expansions of the box volume. Again a well-defined fractal pore-size distribution was observed. Owing to the large size of their simulation, this work was able to model the full range of length scales important to the characterization of this important material.

Phase relations. Improvements in computer performance have not only led to an increase in the number of particles and/or time steps that can be realized in silicate simulations. An additional significant development is the ability to carry out a great number of separate simulation runs of the same model liquid, in order to evaluate in detail the functional dependence of measurable properties on the simulation conditions. Such a research program is particularly suited to a workstation cluster, or "farm", in that each state point can be simultaneously run on a separate processor. The proliferation of workstation clusters at many institutions represents a valuable opportunity to realize such results.

The recent work of Belonoshko and Dubrovinsky (1995) is a good example of this research strategy. In this work the melting curve of stishovite was determined over a wide range of T and P. Their method consisted of initiating a series of simulations, in which the initial configuartion was a simulation box half-filled with a stishovite crystal arrangement, and half-filled with a liquid arrangement of ions. A number of trial values of T and P were chosen, and the system allowed to evolve until one or the other phase completely filled the simulation box. The boundary in the plane of T and P separating the runs that became totally liquid and those that become totally crystal represents an estimate of the melting line in the phase diagram. The results are consistent with experimental measurements at low P, but extend to higher P than the experimental data.

Also, a series of simulations can be used to evaluate the $P(V,T)$ equation of state of a melt, simply by executing a number of canonical ensemble runs at various T and V, and monitoring P. Experimental measurements of $P(V,T)$ over a broad range can be particularly difficult in high T silicate melts, and recent simulation studies have demonstrated that unexpected and important features may exist in the equation of state of several silicate and silicate-like liquids. As an example, consider the $P(V,T)$ data for the BMH model of liquid SiO_2, presented in Figure 7 (Stanely et al., 1994; Poole et. al., 1995b). Though the isotherms of P versus V are relatively innocuous at the highest T, as T decreases a "double-

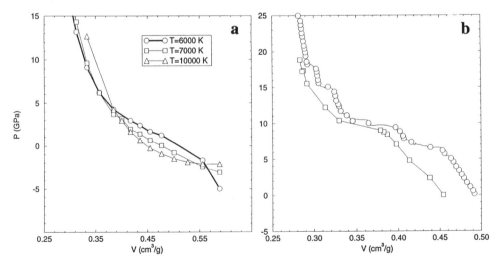

Figure 7. (a) Equation of state of simulated liquid SiO_2 (from Poole et al., 1995b): P as a function of V for various T (6000 K, 7000 K, and 10,000 K) from simulations using the rigid-ion BMH potential. (b) Comparison of simulation and experimental data for shock compression of SiO_2 glass. The MD data (circles) were generated by quenching an ambient P liquid state configuration to 100 K, and then subjecting it to successively higher P over a 1 ns run (Poole, et al., 1995b). The experimental data (squares) are from shock compression of fused silica (adapted from Sugiura, Kondo and Sawaoka, 1982).

inflection" appears in the isotherms. There are wide-spread implications of such behavior if it occurs in the real system. It demonstrates, for example, that a third-order singularity (i.e. the inflections of the isotherms) in the free energy of the liquid has developed as T decreases. An implication of this is that if the isothermal compressibility κ_T of liquid SiO_2 is measured as a function of P at sufficiently low T, a maximum will occur in κ_T as the region of the inflection in the equation of state is traversed. Furthermore, the trend in the simulation data shows that the inflections grow in strength as T decreases. Though the temperature of the simulations is not low enough to directly confirm this, there is the possibility that the inflections eventually become "van der Waals loops", indicating the existence of a second-order critical point, followed at lower T by a first-order phase transition between two *distinct* phases of liquid SiO_2.

Though exotic, such behavior would be consistent with that of a number of other liquids dominated by tetrahedral coordination at the atomic level, including liquid Si and water (Thompson et al., 1984; Donovan et al., 1985; Mishima, 1985, 1994; Ponyatovsky and Barkalov, 1992; Stanley et al., 1994; Poole et al., 1992; 1994, 1995a,b). Furthermore, this behavior is consistent with, and explanatory of, the behavior of amorphous solid SiO_2 when subjected to compression, since a similar double-inflection has been observed in the mechanical response. The behavior of the glass under compression can then be understood, not only in terms of microscopic atomic rearrangements and instabilities, but also as a "frozen in" reflection of the equilibrium behavior of the liquid. Such an interpretation has also been proposed to explain the quite similar behavior of liquid and amorphous solid water. Also, a number of tetrahedral amorphous solids (like GeO_2) exhibit similar behavior when compressed (see chapter by Wolf and McMillan). There is therefore a need for experimental P(V,T) data on the liquid states of these materials to determine if the simulation prediction is plausible.

Diffusion coefficients and mechanism. In their original MD study, Woodcock et al. (1976) obtained diffusion coefficients (D) for Si^{4+} and O^{2-} from their simulation runs, which agreed with the expected values estimated from measured viscosities (extrapolated to the temperature of the simulation). Within the uncertainty of the simulation, these diffusion coefficients were found to be the same. This agrees with results of more recent simulation studies of SiO_2 liquid at low pressure (Kubicki and Lasaga, 1988; Rustad et al., 1990; Frenkel and Vasserman, 1991; Della Valle and Andersen, 1992). The viscosity (η) can be obtained from the Stokes-Einstein expression:

$$\eta = \frac{k_B T}{6\pi D a} \tag{38}$$

in which a is taken to be the radius, or the jump distance (taken to be 2.5 Å in studies on silicate species) of the diffusing ion (Scamehorn and Angell, 1991). For this expression to yield viscosities which are comparable with experimental values, it is important to determine that the diffusing ion is involved in the structural relaxation mechanism of the aluminosilicate framework. This is the case for Si^{4+}, Al^{3+} and O^{2-}, but not for Na^+, in sodium aluminosilicates (Angell et al., 1982a,b, 1987). Scamehorn and Angell (1991) noted that, although the Eyring equation has been recommended over the Stokes-Einstein expression for silicate and related systems, based on analysis of experimental data (Yinnon and Cooper, 1980; Dingwell and Webb, 1990), the simulations are carried out at such high T that their fluidities are much closer to those of the molten salt systems for which the Stokes-Einstein equation is appropriate (Scamehorn and Angell, 1991). The variation in diffusion constants with temperature (Kubicki and Lasaga, 1988; Vessal et al., 1989; Rustad et al., 1990; Scamehorn and Angell, 1991; Frenkel and Wasserman, 1991; Stein and Spera, 1995) is consistent with highly Arrhenius behavior for liquid SiO_2 liquid (Scamehorn and Angell, 1991), in agreement with experiment (Angell, 1985, 1991).

In order to study the atomic mechanism of the diffusion process, Woodcock et al. (1976) followed the movement of the oxygen ions around one silicon (Fig. 6). For most of the observation time, the oxygens executed vibrational motions about their silicon, but occasionally, one oxygen ion was observed to leave the coordination sphere and enter that of a neighboring silicon. At the same time, the original oxygen is replaced by one entering from another silicate tetrahedron. This "rattle and hop" mechanism provides the basis for O^{2-} diffusion in simulated SiO_2 liquid. This study demonstrates the power of simulation studies in elucidating the microscopic atomic mechanisms for diffusion and structural relaxation in liquids (Brawer, 1985).

In a series of simulation studies on BeF_2 liquid, a structural analog to SiO_2, Brawer (1981, 1985) observed the presence of structural defects consisting of a five-fold coordinated Be atom ([5]Be) and a three-coordinated fluorine ([3]F). Brawer suggested that these defects played a central role in F^- diffusion in the melt, acting as a short-lived intermediate state, and proposed a detailed model for structural relaxation in the liquid based on these species (Brawer, 1985). He also suggested that such species might play an analogous role in the diffusion process in SiO_2 melt, especially as the melt density is increased, a view shared by Angell based on the results observed by Woodcock et al. (1976) "in private conversation" (Brawer, 1985, p. 129). Such five-coordinated silicon species have now been identified in silicate glasses from NMR spectra (Stebbins and McMillan, 1989, 1993; Xue et al., 1991; Stebbins, 1992: see chapter by Stebbins). This mechanism appears to be confirmed in the recent MD simulations by Kubicki and Lasaga (1988) and Vessal et al. (1989). In their study, Kubicki and Lasaga (1988) clearly showed the role of first a three-coordinated Si atom, as the initial Si-O bond breaking step is completed, followed by formation of an intermediate five-coordinated Si species (Angell et al., 1982a,b; McMillan et al., 1994). However, Rustad et al. (1990) have suggested that the importance of this mechanism may need to be re-assessed. In their study, they investigated the pressure dependence of O^{2-} diffusivity in SiO_2, using the TTAM potential, with systems containing between 252 to 1371 particles. These authors found that the calculated diffusivities depended strongly on the system size. Although they did find a correlation between the ion diffusivity, melt density and proportion of [5]Si species, they suggested that this might be over-printed by system size effects. However, Rustad et al. (1990) did point out the problem of particle coalescence associated with the use of the TTAM potential, which could affect the diffusivity results.

Diffusion maximum in highly polymerized silicates at high pressure. One of the most interesting findings from MD simulations on SiO_2 liquid has been the unusual pressure dependence of the ion diffusivities. Woodcock et al. (1976) calculated the diffusivity as a function of density, up to compactions corresponding to application of pressure of nearly 200 GPa. The initial effect of increasing P was to *increase* the O^{2-} and Si^{4+} diffusivities, which would considerably lower the viscosity, with a pronounced maximum in O^{2-} diffusivity near 20 to 30 GPa at 6000 K. Woodcock et al. (1976) noted the analogy with the behavior of water, and suggested that this anomalous diffusivity behavior might be a characteristic of any open network liquid. The behavior for SiO_2 has been confirmed in all subsequent simulations (Kubicki and Lasaga, 1988; Vessal et al., 1989; Rustad et al., 1990; Frenkel and Wasserman, 1991), suggesting that it is not simply an artifact of the simulation studies and will be found in the real liquid (Fig. 8). Although no measurements for SiO_2 are available, work on the structural analogue GeO_2 (Sharma et al., 1979a), and highly polymerized aluminosilicate melts (Kushiro, 1976, 1978, 1981; Kushiro et al., 1976; Scarfe et al., 1987; Rubie et al., 1993; Poe et al., 1995), indicates that an anomalous diffusivity increase or viscosity decrease with increasing pressure is indeed observed. This constitutes an interesting and useful prediction of a silicate melt property, which is serving to stimulate further experimental work.

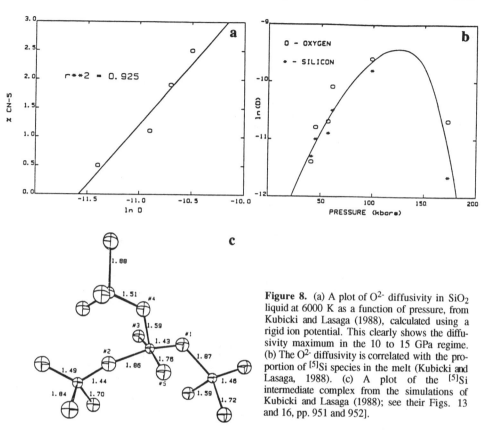

Figure 8. (a) A plot of O^{2-} diffusivity in SiO_2 liquid at 6000 K as a function of pressure, from Kubicki and Lasaga (1988), calculated using a rigid ion potential. This clearly shows the diffusivity maximum in the 10 to 15 GPa regime. (b) The O^{2-} diffusivity is correlated with the proportion of $[5]Si$ species in the melt (Kubicki and Lasaga, 1988). (c) A plot of the $[5]Si$ intermediate complex from the simulations of Kubicki and Lasaga (1988); see their Figs. 13 and 16, pp. 951 and 952].

Woodcock et al. (197M) proposed that the effect of pressure was to break down the four-coordinated silicate network structure, reducing the energetic barrier to ionic diffusion. Later studies on sodium silicate and aluminosilicate compositions showed a similar diffusivity maximum with increasing pressure (Angell et al., 1982a,b, 1983, 1987). Angell et al. (1982a) made the connection with the diffusivity mechanism proposed by Brawer (1981), and suggested that the diffusivity maximum in $NaAlSi_2O_6$ liquid occurred at a pressure for which the proportion of $[5]Si$ species was maximized. Angell et al. (1987) noted that, in the earlier work (Angell et al. 1982a, 1983), they had not documented the coordination behavior of the Al^{3+} ions in these aluminosilicate compositions. They found that the Al^{3+} coordination in the liquid increased much more rapidly than the Si^{4+} coordination, consistent with observations of Kushiro (1983). Consistent with this simulation result, Ohtani et al. (1985) suggested that six-coordinated Al^{3+} species ($[6]Al$) were present in $NaAlSi_3O_8$ melt above 6 GPa, based on ^{27}Al NMR spectroscopy of glasses quenched from high pressure. However, Stebbins and Sykes (1990) have repeated this experiment, and did not find the same obvious peak for $[6]Al$ species, in glasses quenched from up to 10 GPa. Instead, they did find weak, broad features, which likely indicate the presence of small amounts (< 5%) of five- and six-coordinated Al species, retained in the quenched glass (see chapters by Stebbins, and by Wolf and McMillan). This topic is pursued further below.

Silicon coordination at high pressure. The effect of pressure on the silicon coordination behavior, and its relationship to network ion diffusion and structural relaxation in highly polymerized silicate compositions, has been addressed in several experimental and simulation studies. In their simulation of SiO_2 using a rigid ion potential, Kubicki and Lasaga (1988) found a clear correlation between the proportion of five-coordinated Si species and the O^{2-} diffusivity, in simulations at 6000 K. They also indicate that, for pressures above 10 GPa, the simulated system has changed to a melt dominated by silicon in six-fold coordination. (The oxygen coordination is then also III). This pressure corresponds to the region in which the diffusivity maximum is observed to occur (Woodcock et al., 1977; Kubicki and Lasaga, 1988). Although Rustad et al. (1990) do find a general correlation between the increased diffusivity and the [5]Si proportion, they find that this relation is apparently over-printed by effects of system size. These authors further indicate (Rustad et al., 1991a) that the TTAM potential is the only one to exhibit a change in silicon coordination to six-fold at high pressure, although this clearly disagrees with the findings of Kubicki and Lasaga (1988).

A series of high pressure simulations for SiO_2 liquid (6000 to 10,000 K) using a rigid ion (Born-Mayer-Huggins) potential have recently been carried by Poole et al. (1995b), which clearly show an increase in the proportion of [6]Si species above a density of ~3.0 g/cm^3 (Fig. 9a) which corresponds to a pressure near 10 GPa (Fig. 6a). Shao and Angell (1995) have carried out a simulation of SiO_2 glass (i.e. at 300 K), and these show a gradual increase in the average Si coordination (Fig. 9). This reaches a value of approximately 5 near 8 GPa, 5.5 near 15 GPa, and tends toward 6 above ~20 GPa. This is in qualitative agreement with structural changes inferred from the X-ray diffraction data of Meade et al. (1992), and with in situ IR and Raman spectra on silica glass (Hemley et al., 1986; Williams and Jeanloz, 1988; Williams et al., 1993). These spectra show the disappearance of the bands due to the tetrahedral network, along with growth of a broad feature in the 600 to 900 cm^{-1} region of the IR spectra, which has been taken as indicative of formation of highly coordinated species (see chapter by Wolf and McMillan). However, these changes are reversible upon decompression, and the identification of highly coordinated silicate species in SiO_2 glass and their onset pressure still remain to be demonstrated. No highly coordinated species have yet been directly observed in densified SiO_2 glass (Xue et al., 1991).

Stixrude and Bukowinski (1988; 1989, 1990, 1991) have carried out an interesting Monte Carlo study of the densification of SiO_2 glass, in which they constrained the silicate framework to remain *tetrahedrally* coordinated. They showed that compression mechanisms involving changes in the deformation, relative arrangement and connectivity of the silicate tetrahedra could account for the compressibility of SiO_2 to pressures in excess of 50 GPa, and would even result in crystal-liquid density inversions. They suggest that these processes could provide efficient densification mechanisms for silicate melts over much of the pressure range before occurrence of highly coordinated Si species becomes important.

Other silicate melt studies

Overview. Shortly after the pioneering study by Woodcock et al. (1976) on SiO_2, Soules (1979) carried out the first MD study of the structures of a series of sodium silicate glasses (as well as a few other compositions), and Borgianni and Granati (1979) carried out a Monte Carlo simulation of alkaline earth silicate and aluminosilicate melt structure. These were followed by a series of MD simulations of alkali silicate and borosilicate glasses and melts by Soules and co-workers (Soules, 1980, 1982; Soules and Busbey, 1981; Soules and Varshneya, 1981). During this period, Brawer carried out MC and MD simulations for BeF_2 and fluoroberyllate glasses and liquids (Brawer, 1980, 1981, 1982, 1983, 1985; Brawer and Weber, 1981). BeF_2 has a tetrahedral network structure analogous to SiO_2, with bridging [4]Be-F-[4]Be bonds Addition of modifier fluorides (NaF,

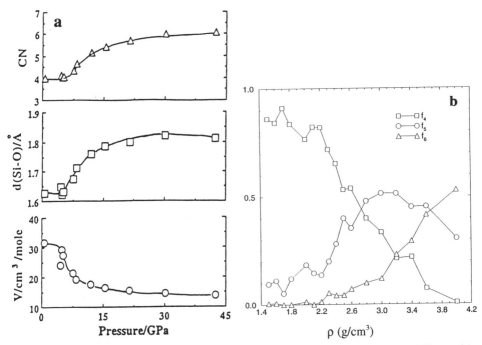

Figure 9. (a) The pressure dependence of the average Si coordination calculated using a BMH potential (from Shao and Angell, 1995, to be published. Used by permission of the authors). (b) Plots of f_n, the fraction of Si ions coordinated by n O ions, as a function of density, evaluated from a MD simulation of a BMH model of liquid SiO_2 at T = 8000 K (Poole, et al., 1995b). Note the appearance of a significant population of 6-coordinated Si^{4+} ions at the highest density. In each case, f_n was calculated as the value of the running coordination number n(r) for r at the first minimum in the Si-O g(r) function.

CaF_2) results in creation of non-bridging fluorine atoms (Be-F⁻), like the corresponding binary silicates (Brawer, 1985). Angell and co-workers began a series of MD simulations on silicates (Angell et al., 1982a,b, 1983, 1987, 1988; Scamehorn and Angell, 1991; Poe et al., 1992a). These simulations have most recently been used to help interpret the ^{27}Al NMR spectra obtained in situ at very high temperatures, on aluminosilicate and aluminate liquids (Poe et al., 1992b, 1993a,b; Coté et al., 1993), and new measurements on network ion diffusivities in silicate melts at high pressure (Rubie et al., 1993; Poe et al., 1994, 1995; Diefenbacher et al., 1994, 1995). The use of three-body potentials in aluminosilicates has been pioneered by Zirl and Garofalini (1990). Matsui and Kawamura (1980) and Matsui et al. (1981, 1982, 1984) used the MD method to investigate the structure of molten $MgSiO_3$ and Mg_2SiO_4 melts, and then followed up with MD studies of crystalline high pressure magnesium silicates (Matsui, 1988; Matsui and Busing, 1984; Matsui et al., 1987), including a study of the pre-melting behavior of $MgSiO_3$ perovskite (Matsui and Price, 1991). Other simulation studies on crystalline minerals have been summarized by Catlow and Cormack (1987) and Catlow and Price (1990). The behavior of $MgO-SiO_2$ liquids has also been investigated by Kubicki and Lasaga (1991a,b; 1993) and Wasserman et al. (1993a,b), including a study of the melting behavior of crystalline Mg_2SiO_4 and $MgSiO_3$ perovskite (Kubicki and Lasaga, 1992). Most recently, an extensive series of MD simulations of melts along the SiO_2-$NaAlSiO_4$ join has appeared (Stein and Spera, 1995). These complement an earlier study by Dempsey and Kawamura (1984). In other recent work, Huang and Cormack (1990), Vessal et al. (1992) and Greaves (1992) have used a three-body potential to investigate local structure and cation clustering around the non-bridging oxygens in binary silicate glasses. As noted by Huang and Cormack (1990), the alkali ions link the non-bridging oxygens (NBO) into a network distinct from the silicate framework. This work has important consequences for understanding the medium range order of silicate glasses, and has been used to understand the ion mobility, including the mixed alkali effect (Greaves et al., 1991; Greaves, 1992; Vessal et al., 1992)

Structure and properties of binary silicates. In the early studies of binary silicate liquids and glasses, the comforting result was found that the silicate units remained tetrahedral (at least, to a first approximation: see Brawer, 1985), forming corner-shared polymer units. The alkali atoms were not randomly distributed, but were found to be clustered around the non-bridging oxygens. The Na^+ coordination in sodium silicate melts has an average value between five and six, with a range from approximately three to eight (Soules, 1979). This agrees generally with the later work of Newell et al. (1989), Huang and Cormack (1990), Greaves (1992) and Vessal et al. (1992), using three-body potentials. In general, the distribution of oxygen ions around the alkali or alkaline earth cations is much broader than the coordination around silicon, as expected for the weaker metal-oxygen bonding (Soules, 1982). In the later work, significant ordering in the alkali ions around the non-bridging oxygens was recognized, with implications for the medium-range order in the glass (Huang and Cormack, 1990; Greaves, 1992; Vessal et al., 1992). These observations agree well with diffraction and spectroscopic evidence for the melt and glass structures (Greaves et al., 1981, 1991; Soules, 1982; Navrotsky et al., 1985; Xue and Stebbins, 1993).

Advanced simulation techniques were used by Ogawa et al. (1990) to carry out a systematic comparison of three different methods for determining the shear viscosity in liquid sodium disilicate. They evaluated the viscosity via (a) integration of the autucorrelation function of the components of the microscopic stress tensor (i.e. a Green-Kubo relation), (b) measurement of the diffusion constant followed by use of the Stokes-Einstein equation, and (c) exploitation of the methods of "non-equilibrium molecular dynamics" (NEMD) (Evans and Morris, 1990), in which shear stress is imposed directly on the simulated liquid through a moving boundary condition. They found that the first two approaches gave order of magnitude agreement with experimental measurements. However the high shear rates that had to be imposed on the system to realize the NEMD approach in the available computation time proved to be too large to allow extrapolation to the experimental regime. A similar study has been carried out for magnesium silicate liquids at high pressure by Wasserman et al. (1993a).

The mixing properties of silicate melts are of considerable importance in determining the properties of magmas and silicate glasses. Kieffer and Angell (1989) have used MD simulations to investigate the enthalpy of mixing in K_2O-SiO_2 and MnO-SiO_2 liquids. In a related study, Angell et al. (1988) studied the mixing properties of neutral gas particles, in a simulated silicate liquid, with implications for the dissolution of these gases in magmas at high P and temperature. Mixing in MgO-SiO_2 liquids has been studied by Wasserman et al. (1993b).

Alkali ion mobility. The alkali and alkaline earth ions are much more mobile than the network-forming ions within both the melt and the glass structure. The MD method has been used to carry out detailed studies of ionic diffusion within glass and melt structures, with diffusion coefficients in excellent agreement with experiment (Soules, 1979, 1982; Soules and Busbey, 1981; Angell et al., 1982b; Angell, 1990). The mixed alkali effect has recently been studied by Vessal et al. (1992) (see also Greaves et al., 1991). With increased pressure, the diffusivity of non-framework metal ions decreases normally with increasing pressure, even for highly polymerized aluminosilicates (Angell et al., 1982a, 1987)

Concerning the mobility of non-framework ions, an extremely interesting observation by Angell and Torell (1983) was that the high temperature Arrhenius plot of relaxation associated with ionic motion in $(Ca,K)NO_3$ liquids extrapolated naturally to the lattice vibrational frequency in the far infrared spectrum. In this study, the authors examined the mechanical and dielectric response of the simulated and real liquids as a function of relaxation timescale, and showed the identity of anharmonic vibrational absorption with the short timescale (high frequency) limit of ionic diffusion. This property has considerable application in the study of ionic conductors (Angell, 1990), and has been applied to the analysis of alkali ion motion in silicate liquids and glasses (Angell et al., 1982b). This phenomenon is also of particular current interest, in determining the fundamental short timescale anharmonic vibrational event which eventually triggers the transition to the supercooled liquid on heating a glass (Angell, 1981, 1995; Frick and Richter, 1995). This phenomenon has recently been studied in detail via MD methods for SiO_2 liquids at low and high density by Shao and Angell (1995).

Diffusion of network-forming ions. The Brawer (1981, 1985) mechanism for network ion diffusion in highly polymerized fluoroberyllates and silicates was described above. Soules (1980, 1982) did find that some of the non-bridging oxygen ions in simulations of binary silicate liquids "moved into the face of other silicate tetrahedra making some of the silicon atoms five-coordinated" (Soules, 1982). This mechanism agrees well with that proposed on the basis of NMR and Raman spectroscopic studies, for the densification mechanism responsible for formation of [5]Si species (Wolf et al., 1990; Xue et al., 1991: see chapters by Stebbins, and by Wolf and McMillan). Soules (1982) did indicate that fewer [5]Si groups are present at larger volume, consistent with this as a densification process. However, Soules also indicates that the [5]Si species disappear with increasing temperature, which is not consistent with the experimental observation (Stebbins, 1992).

Based on in situ high temperature ^{29}Si NMR measurements, Stebbins and co-workers have built up a detailed picture of the O^{2-} diffusion mechanism in high silica alkali silicate melts, which also appears to serve as the fundamental structural relaxation process (Farnan and Stebbins, 1990a,b, 1994; Stebbins et al., 1992: see chapter by Stebbins, this volume). In this mechanism, the intermediate step in the O^{2-} transfer reaction is the formation of a [5]Si species, as postulated by Brawer (1985). The proposed mechanism involves attack of a non-bridging oxygen on an adjacent tetrahedral Si atom, forming a new [4]Si-O-[5]Si linkage involving a five-coordinated intermediate (Fig. 10). The O^{2-} transfer

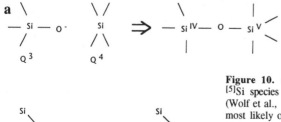

Figure 10. (a) The proposed reaction for formation of [5]Si species in the presence of non-bridging oxygens (Wolf et al., 1990; Xue et al., 1991). This mechanism most likely operates at lower pressure in alkali silicates (but see chapter by Wolf and McMillan). (b) At higher pressure, the highly coordinated [6]Si) species are formed by attack of a briding oxygen on neighboring Si. Creating of a three-coordinated oxygen requires the simultaneous formation of [6]Si.

is achieved by breaking a different [5]Si-O bond, from that formed in the original reaction. Because the [5]Si species are also formed in the same way during densification of alkali silicates (Wolf et al., 1990; Xue et al., 1991: chapters by Stebbins, and Wolf and McMillan), this rationalizes the increased O^{2-} diffusivity with increasing pressure (Rubie et al., 1993; Poe et al., 1994, 1995). Poe et al. (1994, 1995) have recently found that the Si^{4+} diffusivity also increases with pressure for $Na_2Si_4O_9$ liquid. The reduced viscosity observed for these high silica alkali silicates (Scarfe et al., 1987; Dickinson et al., 1990) at high pressure is related to the increased oxygen diffusivity, and also the greater configurational entropy due the presence of highly coordinated species (Dickinson et al., 1990: see chapter by Richet and Bottinga).

The diffusion mechanism of network forming ions (O^{2-}, Si^{4+}) and structural relaxation in alkali trisilicate and tetrasilicate melts at high pressure has been studied in detail via molecular dynamics simulation (Angell et al., 1982b, 1983, 1987; Diefenbacher et al., 1993, 1995). In the study by Diefenbacher et al. (1993, 1995) for $Na_2Si_4O_9$, the Born-Mayer-Huggins potential used in the studies by Angell and co-workers was used. A Birch-Murnaghan fit to the simulation P-V relation gave $K = 7.5$ GPa for the zero-pressure bulk modulus and $K' = 9.0$ for its pressure derivative at 6000 K, not unreasonable for a high temperature silicate liquid (Bottinga, 1985). Using an expression for the compressibility of silicate liquids in terms of oxide components, derived from ultrasonic measurements (Rivers and Carmichael, 1987; Lange and Carmichael, 1990) gives $K = 13.7$ GPa at 1673 K.

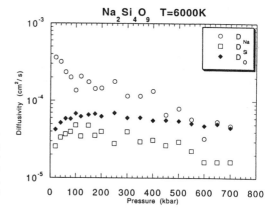

Figure 11. The diffusivity maximum observed for Si^{4+} and O^{2-} ions in $Na_2Si_4O_9$ melt with increasing pressure by Diefenbacher et al. (1993, 1995). The Na^+ diffusivity decreases normally with pressure. A BMH potential was used, and system size was N = 345 particles.

The Na^+ diffusivity decreases rapidly as P increases, as expected. However, Si^{4+} and O^{2-} diffusivities increase with increasing pressure, as found in the earlier work (Angell et al., 1982a,b, 1983, 1987), reaching a maximum value above approximately 10 GPa (Fig. 11). [This maximum has not yet been observed in the experimental studies: Rubie et al., 1993; Poe et al., 1994, 1995: see chapter by Wolf and McMillan.] These simulations were carried out using 345 particles, comparable with the previous studies from Angell et al. Motivated by the observations of Rustad et al. (1990) and Stein and Spera (1995), the effect of system size on diffusivity at a few pressures was evaluated, for systems containing 345 to 1500 particles (Fig. 12). At low pressure (large box volume), there was no obvious system size dependence of the O^{2-} diffusivity. For smaller box volumes, the O^{2-} diffusivity showed a gradual increase with increasing particle number, opposite to the effect observed by Rustad et al. (1990) and Stein and Spera (1995), for fully polymerized liquids. The diffusivities all agreed to within a factor of approximately 3 (half a decade).

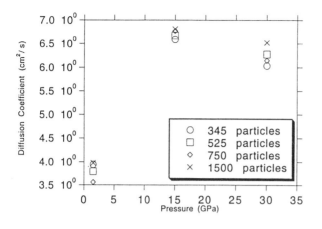

Figure 12. The effect of system size on the diffusivity maximum in Figure 11. The O^{2-} diffusivity in systems containing between 345 and 1500 particles is shown.

In order to gain a structural understanding of the densification process, the Si coordination numbers obtained from this study are plotted in Figure 13a. The four-coordinated sites decrease, and the five-coordinated species increase, up to a pressure near 15 GPa. The proportion of six-coordinated Si remains low and practically constant throughout this pressure range. This is consistent with the mechanism (a) shown in Figure 10, in which five-coordinated Si is formed by attack of a non-bridging oxygen on a neighboring four-coordinated species. Above this pressure, the proportion of six-coordinated Si begins to increase, corresponding to mechanism (b), in which bridging oxygens are used in the densification mechanism. Formation of three-coordinated oxygen (Fig. 13b) results in the simultaneous formation of [6]Si species. The saturation of the initial mechanism (a) and the onset of the (energetically less favorable: Wolf et al., 1990; McMillan et al., 1994) mechanism (b) appears to coincide with the broad maximum in the network ion diffusivities (Fig. 11).

Al and Si coordination in high temperature melts. The question of Al coordination in aluminosilicate liquids at high temperature has a long history. There have been extensive series of experimental studies on glasses in the systems $M_2O-Al_2O_3-SiO_2$ and $MO-Al_2O_3-SiO_2$ (M is an alkali or alkaline earth cation) (see chapters by Stebbins, and by McMillan and Wolf). For compositions with lower alumina content than the "charge balanced join" (SiO_2-MAlO_2 or $SiO_2-MAl_2O_4$), the general conclusion is that all Al is in tetrahedral coordination to oxygen. Highly coordinated Al species (five- and six-coordinated) have been reported in glass samples along the $SiO_2-Al_2O_3$ binary, or other high alumina compositions (e.g. Sato et al., 1991). This general picture is consistent with the results of the pioneering MC study by Borgianni and Granati (1979), of melts in the $CaO-Al_2O_3-SiO_2$ system.

In their MD study of diffusivity in $NaAlSi_2O_6$ liquid at high pressure, Angell et al. (1987) remarked that the Al coordination increased more rapidly than that of Si with increasing pressure, but did not comment on the coordination number in the room pressure liquid. Matsui et al. (1982, 1984) studied structure and coordination in $CaAl_2Si_2O_8$ melt. They observed that the AlO_4 tetrahedra are more irregular than those of SiO_4, and that 5-coordinated Al^{3+} ions were observed when the melt was compressed. Dempsey and Kawamura (1984) have carried out a structural study of three liquids along the $SiO_2-NaAlSiO_4$ join, using Born-Mayer-Huggins pair potential parameters optimized from runs on aluminosilicate crystal structures. The calculated TOT (T = Si, Al) angle distributions

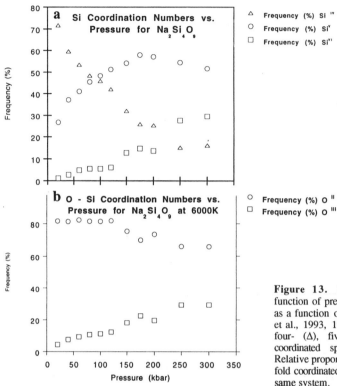

Figure 13. (a) Si coordination as a function of pressure in $Na_2Si_4O_9$ liquid as a function of pressure (Diefenbacher et al., 1993, 1995). The proportion of four- (Δ), five- (O) and six- (□) coordinated species is shown. (b) Relative proportions of two- and threefold coordinated oxygen species for the same system.

were in good general agreement with the range of values observed in crystals, and the running Si and Al coordination numbers were given. The results suggested that the Si and Al coordinations were both equal to four. The Na^+ coordination showed a range of values, with an average coordination near six (slightly greater than for the alkali silicates studied by Soules, 1979, 1982).

Scamehorn and Angell (1991) have carried out a series of MD simulations of SiO_2, $NaAlSi_3O_8$, $NaAlSiO_4$, $CaAl_2Si_2O_8$, and $MgAl_2Si_2O_8$ liquids at 1 atm pressure as a function of temperature. These authors obtained the O^{2-} diffusivities, and used them to calculate the viscosity via the Stokes-Einstein relationship. The viscosity-temperature relations showed an increasing "fragility" (Angell, 1985, 1991), or degree of departure of from Arrhenian behavior, as the silica content is reduced along these charge-balanced aluminosilicate joins. This agrees generally with experiment. These authors found that all of the simulated liquids contained substantial proportions (20 to 40%) of [5]Si and [5]Al, along with a few per cent six-coordinated species, and that the percentage of five-coordinated species present in the liquid increased rapidly with decreasing alumina content, and in the order Na < Ca < Mg. This prediction was generally borne out by the experimental observation of [5]Al and [6]Al species present in glasses along the SiO_2-$MgAl_2O_4$ join, prepared by rapid quenching from the melt (McMillan and Kirkpatrick, 1992). Similar studies of other alkali or alkaline earth glasses have shown no evidence for highly coordinated species (see Stebbins, this volume).

The simulations of Scamehorn and Angell (1991) indicated that substantial quantities of [5]Al and [5]Si, along with some six-coordinated species, might be present in "normal" aluminosilicate liquid compositions at magmatic temperatures. The work of Stebbins (1992), in which the formation enthalpy of [5]Si species was measured as approximately 30 kJ/mol for $K_2Si_4O_9$ liquid, places some constraint on the likely proportion of five-coordinated silicon at low pressure: at 2000 K, the concentration of [5]Si is unlikely to exceed 1% (although this will become very much larger, as the pressure is increased). Experiments on glasses along the SiO_2-Al_2O_3 join prepared with different quench rates indicated that the formation of [5]Al species from [4]Al and [6]Al is endothermic (Poe et al., 1992a). A series of MD simulations for these liquids, using the same potential parameters as Scamehorn and Angell (1991), gave Al coordination distributions which agreed generally with the NMR results on the glasses, although the proportion of highly coordinated species was higher in high-temperature simulated liquids (Poe et al., 1992a).

The results of the MD simulations for the SiO_2-Al_2O_3 system were used to interpret high temperature ^{27}Al NMR data, obtained directly on the liquid at temperatures close to those of the simulation (2500 to 3000 K) (Poe et al., 1992b) (see chapter by Stebbins). This situation is unusual in MD simulation studies, where the simulation run temperature is usually far in excess of available experimental data. The NMR data showed a single Lorentzian line for the [27]Al resonance, which was interpreted as due to chemical exchange averaging over four-, five- and six-coordinated aluminate species. The timescale for this process was estimated from the NMR linewidth, and was found to agree quite well with the timescale for structural relaxation, obtained from viscosity data (Poe et al., 1992b). This observation was used to suggest that the O^{2-} diffusion and structural relaxation mechanism in the aluminosilicate liquids was analogous to that for binary silicates, with the [5]Al species providing an intermediate for the O^{2-} exchange reaction. A similar combined high temperature NMR and MD study was carried out for liquids along the CaO-Al_2O_3 join, and $MgAl_2O_4$ (Poe et al., 1993a,b; Coté et al., 1993). In this case, the average [27]Al chemical shift in the liquid agreed well with the estimated value, obtained by taking the average Al coordination number from the MD simulation, and characteristic shifts for four-, five- and six-species from crystals. This approach did not work so well for the SiO_2-Al_2O_3 liquids, which was thought to be due to competing effects of Al coordination number and Si nearest neighbors on the chemical shifts (Poe et al., 1992b). In these studies, the MD simulations have been invaluable in helping decipher the structural sites present in the high temperature liquid, in cases for which only a single time-averaged signal is observed (Poe et al., 1992a,b, 1993a,b).

However, details of the potential function used in the simulation can substantially change the calculated coordination number distribution. Stein and Spera (1995) have recently recalculated Si and Al coordination numbers in SiO_2-$NaAlSiO_4$ liquids, comparing the potentials of Scamehorn and Angell (1991) ("SA91") and Dempsey and Kawamura (1984) ("DK84"). They have found that the "SA91" potential consistently yields higher average coordination numbers (larger proportions of five- and six-coordinated species) (Fig. 14). Zirl and Garofalini (1990) have recently carried out simulations of sodium aluminosilicate glasses using a three-body potential, adapted from their previous work on SiO_2 (Garofalini, 1982), and found only tetrahedrally coordinated Al over the entire composition range studied. However, the focus in this study was on Na^+ mobility in the low temperature glassy state, rather than the Al coordination in a high temperature liquid, so highly coordinated Al states may not have been sampled. Stein and Spera (1995) also found differences in the O^{2-} diffusivities for SiO_2-$NaAlSiO_4$ liquids, calculated using the "SA91" and "DK84" parameter sets (Fig. 14). The $\langle r^2 \rangle$ values at 4000 K were approximately one order of magnitude larger for the "SA91" potential than for the "DK84" potential function. However, the $\langle r^2 \rangle$ values calculated using the "DK84" potential were very small, on the order of 1 Å^2 after 40 ps, suggesting that diffusion is only just taking place. It is obvious that these points must be examined by further simulations using more sophisticated potentials for the aluminosilicate systems (Zirl and Garofalini, 1990), taking care to evaluate system-size-dependent effects, coupled with additional experiments to determine the structure of the liquids.

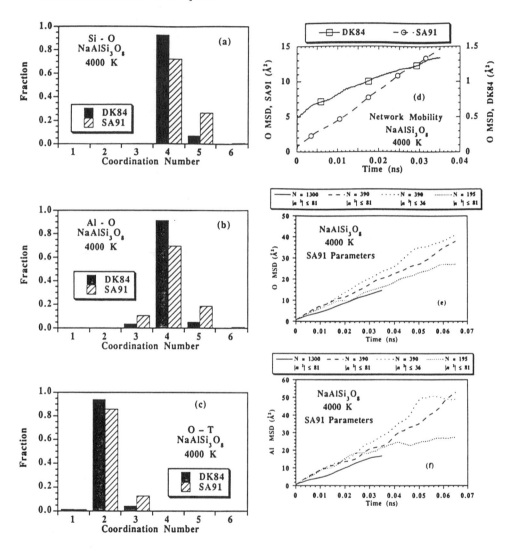

Figure 14. Comparison of Si (a), Al (b) and O (c) coordination number distribution from simulations of NaAlSi₃O₈ liquid using the potentials of Scamehorn and Angell (1991) (SA91) and Dempsey and Kawamura (1984) (DK84), taken from Stein and Spera (1995). (d) Comparison of the O^{2-} diffusivities ($\langle r^2 \rangle$ = MSD) at 4000 K obtained for the same liquid, with the two potentials (Stein and Spera, 1995). (e) and (f) Effects of system size on the O^{2-} and Al^{3+} diffusivities, using the SA91 potential (Stein and Spera, 1995, Fig. 2, p. 422).

The effect of pressure on Si and Al coordination. As noted in the previous section, the observed diffusivity maximum with increasing pressure for network forming ions (Woodcock et al., 1976; Angell et al., 1982a,b, 1983, 1987) was interpreted in terms of a coordination increase in Si and Al in the melt. This agreed with the suggestion made by Waff (1975), and was used to rationalize the viscosity relations observed by Kushiro (1976, 1978, 1981) and Kushiro et al. (1976), on highly polymerized melts. However, experimental studies on glasses quenched from high pressure found no evidence for highly coordinated Al species (Sharma et al., 1979b; McMillan and Graham, 1981; Mysen et al., 1983), and the viscosity changes were re-interpreted as due to changes in the TOT (T = Si, Al) angle. Ohtani et al. (1985) then obtained ^{27}Al NMR spectra of albite glasses quenched from high pressure and observed a peak due to six-coordinated Al, but this result was shown to be in error (Stebbins and Sykes (1990): see chapters by Stebbins, and by Wolf and McMillan). At this stage, there is still no definitive experimental demonstration of the presence of large quantities of highly coordinated Al or Si species in fully polymerized high pressure melts, as indicated by the MD simulations. Highly polymerized alkali silicates containing non-bridging oxygens do form highly coordinated silicate groups at high pressure, and some of the $^{[5]}$Si and $^{[6]}$Si species are retained on decompression (Wolf et al., 1990; Xue et al., 1991). However, in situ infrared spectroscopic studies of fully polymerized silicates do indicate the breakdown of the tetrahedral structure, and the appearance of broad bands which may be associated with highly coordinated species (Williams and Jeanloz, 1988), but these changes are largely non-quenchable. Weak broad bands in the ^{27}Al NMR study of Stebbins and Sykes (1990) may indicate the presence of trace quantities of $^{[5]}$Al and $^{[5]}$Al species in the decompressed sample. A very recent study on fully polymerized $CaAl_2O_4$ glass indicates the formation of highly coordinated species above 11 GPa (Daniel et al., 1995), and five- and six-coordinated Al species are retained on decompression, clearly indicated by NMR (Yarger et al., in prep; see chapter by Wolf and McMillan). A series of high pressure glasses containing large amounts of $^{[5]}$Al and $^{[6]}$Al groups has been prepared along the join between $Na_2Si_4O_9$ and $NaAlSi_3O_8$, containing non-bridging oxygens (Yarger et al., 1994, 1995; Poe et al., 1995). These compositions are being studied via MD simulation (Diefenbacher et al., 1993; Poe et al., 1995; Diefenbacher et al., in prep).

Mantle melts. The structure and properties of basic melts in the system MgO-FeO-CaO-SiO_2 at high P and T are most relevant to the likely behavior of partial melts within the mantle. There have been several studies of these melts via MD simulation. Matsui and Kawamura (1980, 1984), Matsui et al. (1982), and Dempsey et al. (1984) carried out studies of the structure of melts in the MgO-SiO_2 and MgO-CaO-SiO_2 systems. Angell et al. (1987) investigated the diffusivity in diopside melt at high pressure. More recent studies have been carried out by Kubicki and Lasaga (1991b, 1992, 1993) and Wasserman et al. (1993a,b). In general, at low pressure, the melts consisted of monomer, dimer, and chain silicate polymer units, as expected from spectroscopic studies. The average Mg^{2+} coordination (4 to 7) is smaller than that for Ca^{2+}, and both are much more distorted than the tetrahedral Si sites. At high pressure, the Si coordination increases to an average value near 5.3 at 78 GPa in $MgSiO_3$ melt (Wasserman et al., 1993a). The average Mg^{2+} coordination for the same composition increases from 4.2 at 5 GPa to over seven at high pressure. In diopside melt, Angell et al. (1987) found that the Si^{4+}, Ca^{2+} and O^{2-} diffusivities decreased with increasing pressure, as expected for a highly depolymerized system, but that the Mg^{2+} diffusivity appeared to show an initial increase and maximum near 15 GPa. They suggested that this might be due to the formation of five-fold coordinated Mg^{2+} species, consistent with the Brawer model for diffusion and discussed previously. Kubicki and Lasaga (1993) have carried out a detailed MD study of the interdiffusion mechanisms in Mg_2SiO_4 and $MgSiO_3$ melts. Si^{4+} diffusion proceeds via an initial Si-O (bridging oxygen) bond breaking step, followed by an inversion of the resulting three-coordinated silicon, and formation of a new Si-O bond to an available non-bridging oxygen on the far side of the complex (Fig. 15). The much higher Mg^{2+} diffusivities are associated with a "rattle and hop" mechanism within the more loosely bound metal sites in

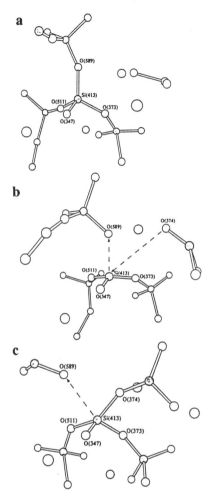

the melt. These authors found no correlation between particular Mg^{2+} coordination numbers and diffusivity.

An $MgSiO_3$ phase with the perovskite structure likely dominates the mineralogy of the lower mantle (Hemley and Cohen, 1992). Matsui and Price (1991) have simulate the behavior of this phase at high temperatures, in the "pre-melting" regime, via MD methods. Belonoshko (1994b) has recently used MD simulation to investigate melting of $MgSiO_3$ perovskite. Wasserman et al. (1993a) have investigated the transport properties of $MgSiO_3$ melt at mantle pressures via MD simulation, and Wasserman et al. (1993b) have calculated mixing properties for melts in the MgO-SiO_2 system. Wasserman et al. (1993a,b) have also calculated the infrared spectra for MgO-SiO_2 melts at low and high pressures. These spectra show the effect of the Si coordination changes, in general agreement with the experimental results of Williams and Jeanloz (1988) and Kubicki et al. (1992). This type of study will prove invaluable for the interpretation of spectroscopic data, as experimental in situ studies of melts under mantle conditions are pursued (see chapters by McMillan and Wolf, and by Wolf and McMillan, in this volume).

Figure 15. Illustration of the network ion diffusion mechanism in $MgSiO_3$-Mg_2SiO_4 liquids, taken from Kubicki and Lasaga (1993). [From Kubicki and Lasaga (1993), Fig. 7, p. 261].

CONCLUSIONS

In this chapter, we believe that we have demonstrated the immense power of computer simulation studies for the study of silicate melts. As available computing power continues to increase, the calculations will become more and more sophisticated and more and better-constrained information on larger systems will become available. It is not inconceivable that useful melt properties, such as thermal and chemical diffusion profiles, thermodynamic state, and viscosity, can be calculated on a realistic model of a magma chamber, at some point in the near future.

The interpretability of simulation results in terms of real systems will always be limited by the appropriateness of the potential, as long as empirical potentials are used. The development of fast and accurate simulation methods based on first principles or *ab initio* approaches will eventually address this problem, but these will take some time to become available. Quantum MD methods can currently simulate about as many particles, and about as long time scales, as classical MD could model about two decades ago. However, empirical simulations will most likely remain a valuable tool for the foreseeable future.

Much emphasis is often placed on developing empirical potentials to reproduce experimental data for a given system as well as possible. Highly calibrated potential functions can be used for interpretation and interpolation of data, within the range for which the potential was calibrated. However, extrapolations can be subject to the same uncertainties as a simpler potential. In addition, the limits of validity of a simple potential are usually quite clear: this is often not so obvious for a sophisticated potential function. Considerable effort remains to be made to calibrate and test even simple potential functions for liquid simulations of multicomponent aluminosilicate systems. In future simulations on silicate systems, the utility of carrying out calculations with potentials that are obviously "incorrect" should not be overlooked, to test the limits of the liquid behavior and establish phenomenological relations.

In modern simulation studies, the increased computing power makes it possible to study quite large systems of particles. The dependence of calculated properties on system size should be investigated, and used to study the true wave vector dependence of liquid state fluctuations. This will give much valuable information on the behavior of silicate liquids at high pressures and temperatures.

Simulation studies by their nature find themselves at the forefront of discussions concerning the dissemination of scientific results in this age of electronic communication (Winograd and Zare, 1995). Many of the simulation results do not lend themselves to the two-dimensional static representation of the printed page. The use of graphics in silicate simulations has been described by Kubicki and Lasaga (1991a). The geophysical community already takes advantage of large scale electronic networks for the transfer and dissemination of data. This could easily be extended to simulation studies of silicate melt structure and dynamics. Some thought should be given to how this can be implemented, and how credit should be assigned, and scientific quality control achieved.

Computer simulation "experiments" have yielded much valuable information about the structure and properties of silicate melts, in particular for the detailed investigation of local structure and diffusion mechanisms, under P-T conditions which have not always been achieved in the laboratory. The range of potential future applications to the field is enormous.

ACKNOWLEDGMENTS

PHP thanks NSERC (Canada) for financial support. PFM and GHW thank the National Science Foundation for continued support of their work on silicate liquids. The authors are indebted to C. Austen Angell for many discussions on simulation, liquid structure, and glassy behavior. Mahin Hemmati is thanked for her contributions to the work on the equation of state of liquid SiO_2. Jason Diefenbacher carried out the work on sodium silicate liquids, presented here. Thanks are also due to Jonathan Stebbins for his editorial assistance, as well as to an anonymous reviewer for their helpful comments at short notice.

REFERENCES

Aasland S, McMillan PF (1994) Density-driven liquid-liquid phase separation in the system Al_2O_3-Y_2O_3. Nature 369:633-636

Adams DJ (1979) Computer simulations of ionic systems: the distorting effects of the boundary conditions. Chem Phys Lett 62:329-332

Adams DJ, Dubey GS (1987) Taming the Ewald sum in the computer simulation of charged systems. J Comp Phys 72:156-176

Adams GB, O'Keeffe M, Demkov AA, Sankey OF, Huang YM (1994) Wide-band-gap Si in open fourfold-coordinated structures. Phys Rev B 49:8048-8054

Alavi A, Alvarez LJ, Elliott SR, McDonald IR (1992) Charge-transfer molecular dynamics. Phil Mag B 65:489-500

Alder BJ, Wainwright TE (1957) Phase transition for a hard sphere system. J Chem Phys 27:1208-1209

Alder BJ, Wainwright TE (1959) Studies in molecular dynamics. I. General method. J Chem Phys 31:459-466

Allen DC, Teter MP (1987) Nonlocal pseudopotentials in molecular-dynamical density-functional theory: application to SiO_2. Phys Rev Lett 59:1136-1139

Allen DC, Teter MP (1990) Local density approximation total energy calculations for silica and titania structure and defects. J Am Ceram Soc 73:3247-3250

Allen MP, Tildesley DJ (1987) Computer Simulation of Liquids. Clarendon Press, Oxford.

Allen MP, Tildesley DJ, eds (1993) Computer Simulations in Chemical Physics. Kluwer Press, Dordrecht, Netherlands

Anastasiou N, Fincham D (1982) Programs for the dynamic simulation of liquids and solids. II. Rigid ions using the Ewald sum. Comp Phys Commun 25:159-176

Angell CA (1981) The glass transition: Comparison of computer simulation and laboratory studies. Ann New York Acad Sci 371:136-150

Angell CA (1985) Strong and fragile liquids. In: K Ngai, GB Wright (eds) Relaxation in Complex Systems, p 3-11 National Technical Information Service, US Dept of Commerce, Springfield, IL

Angell CA (1990) Dynamic processes in ionic glasses. Chem Rev 90:523-542

Angell CA (1991) Relaxation in liquids, polymers and plastic crystals-strong/fragile patterns and problems. J Non-Cryst Solids 131/133:13-31

Angell CA (1995) Formation of glasses from liquids and biopolymers. Science 267:1924-1935

Angell CA, Torell LM (1983) Short time structural relaxation processes in liquids, comparison of experimental and computer simulation glass transition on picosecond timescales. J Chem Phys 78:937-945

Angell CA, Clarke JHR, Woodcock LV (1981) Interaction potentials and glass formation, a survey of computer experiments. Adv Chem Phys 48:397-453

Angell CA, Cheeseman PA, Tamaddon S (1982a) Computer simulation studies of migration mechanisms in ionic glasses and liquids. J Phys Coll 43:C9/381-C9/385

Angell CA, Cheeseman PA, Tammadon S (1982b) Pressure enhancement of ion mobilities in liquid silicates from computer simulation studies to 800 kbar. Science 218:885-887

Angell CA, Cheeseman PA, Tamaddon S (1983) Water-like transport property anomalies in liquid silicates investigated at high T and P by computer simulation techniques. Bull Minéral 106:87-99

Angell CA, Cheeseman PA, Kadiyala R (1987) Diffusivity and thermodynamic properties of diopside and jadeite melts by computer simulation studies. Chem Geol 62:85-95

Angell CA, Scamehorn CA, Phifer CC, Kadyala RR, Cheeseman PA (1988) Ion dynamics studies of liquid and glassy silicates and gas-in-liquid solutions. Phys Chem Minerals 15:221-227

Barrat JL, Roux JN, Hansen JP (1990) Diffusion, viscosity and structural slowing down in soft sphere alloys near the kinetic glass transition. Chem Phys 149:197-208

Barrat JL, Klein ML (1991) Molecular dynamics simulations of supercooled liquids near the glass transition. Ann Rev Phys Chem 42:23-53

Belonoshko AB (1994a) Molecular dynamics of silica at high pressures: equation of state, structure, and phase transitions. Geochim Cosmochim Acta 58:1557-1566

Belonoshko AB (1994b) Molecular dynamics of $MgSiO_3$ perovskite at high pressure: equation of state, structure, and melting transition. Geochim Cosmochim Acta 58:4039-4047

Belonoshko AB, Dubrovinsky LS (1995) Molecular dynamics of stishovite melting. Geochimica et Cosmochimica Acta 59:1883-1889

Bertaut F (1952) L'énergie électrostatique de réseaux ioniques. J Phys Radium 13:499-505

Binder K, ed (1979) Monte Carlo Methods in Statistical Physics. Springer-Verlag, Berlin

Binder K, ed (1984) Applications of the Monte Carlo Method in Statistical Physics. Springer-Verlag, Berlin

Boltzmann L (1896, 1898) Vorlesungen über Gastheorie (Parts I and II), JA Barth, Leipzig, Germany. English translation (SG Brush translator) (1964)—Lectures on Gas Theory, University of California Press, Berkeley. Reprinted by Dover Publications, New York, 1995

Borgianni C, Granati P (1979) Monte Carlo calculations of ionic structure in silicate and alumino-silicate melts. Metall Trans B 10:21-25

Born M, Von Karman T (1912) Uber Schwingungen in Raumgittern. Physik Z 13:297-309

Boon JP, Yip S (1980) Molecular Hydrodynamics. Dover, New York

Bottinga Y (1985) On the isothermal compressibility of silicate liquids at high pressure. Earth Planet Sci Lett 74:350-360

Boulard B, Keiffer J, Phifer CC, Angell CA (1992) Vibrational spectra in fluoride crystals and glasses at normal and high pressures by computer simulation. J Non-Cryst Solids 140:350-358

Boeyens JCA, Gafner G (1969) Direct summation of Madelung energies. Acta Cryst A 25:411-414

Brawer SA (1980) The glass transition in BeF$_2$: a Monte Carlo study. J Chem Phys 72:4264

Brawer SA (1981) Defects and fluorine diffusion in sodium fluoroberyllate glass: a molecular dynamics study. J Chem Phys 25:3516-3521

Brawer SA (1983) Ab initio calculation of the vibrational spectra of BeF$_2$ glass simulated by molecular dynamics. J Chem Phys 79:4539-4544

Brawer S, Relaxation in Viscous Liquids and Glasses (1985) American Ceramic Society, Columbus, Ohio.

Breneman CM, Wiberg KB (1990) Determining the atom-centered monopoles from molecular electrostatic potentials. The need for high sampling density in formamide conformational analysis. J Comp Chem 11:361-373

Brush SG (1964) Translator's introduction to English edition of L Boltzmann (1896, 1898), Lectures on Gas Theory, p 1-17. University of California Press, Berkeley. Reprinted by Dover Publications, New York, 1995.

Brush SG, Sahlin HL, Teller E (1966) A Monte Carlo study of a one-component plasma. J Chem Phys 40:2102-2118

Buchenau U, Nucker N, Dianoux AJ (1984) Neutron scattering study of the low-frequency vibrations in vitreous silica. Phys Rev Lett 53:2316-2319

Buchenau U, Prager M, Nucker N, Dianoux AJ, Ahmad N, Phillips WA (1986) Low-frequency modes in vitreous silica. Phys Rev B 34:5665-5673

Busing WR (1970) An interpretation of the structures of alkaline earth chlorides in terms of interionic forces. Trans Am Cryst Assoc 6:57-72

Callen HB (1985) Thermodynamics and an Introduction to Thermostatistics. Wiley, New York

Catlow CRA, Cormack AN (1987) Computer modelling of silicates. Int Rev Phys Chem 6:227-250

Catlow CRA, Price GD (1990) Computer modelling of solid-state inorganic materials. Nature 347:243-248

Chandler D (1987) An Introduction to Modern Statistical Mechanics. Oxford University Press, New York

Chizmeshya A, Zimmermann FM, LaViolette RA, Wolf GH (1994) Variational charge relaxation in ionic crystals: an efficient treatment of statics and dynamics. Phys Rev B 50:15559-15574

Ciccotti G, Jacucci G, McDonald IR (1979) "Thought experiments" by molecular dynamics. J Stat Phys 21:1-22

Ciccotti G, Frenkel D, McDonald IR (eds) (1987) Simulation of Liquids and Solids. Molecular Dynamics and Monte Carlo Methods in Statistical Mechanics. North-Holland, Amsterdam

Cohen RE, Gong Z (1994) Melting and melt structure of MgO at high pressures. Phys Rev B 50:12301-12311

Cohen RE, Boyer LL, Mehl MJ (1987) Theoretical studies of charge relaxation effects on the static and lattice dynamics of oxides. Phys Chem Minerals 14:294-302

Coté B, Massiot D, Poe B, McMillan P, Taulelle F, Coutures JP (1993) Liquids and glasses: structural differences in the CaO-Al$_2$O$_3$ system as evidenced by ^{27}Al NMR spectroscopy. J Phys IV Coll C2:223-226

Della Valle RG, Andersen HC (1991) Test of a pairwise additive ionic potential model for silica. J Chem Phys 94:5056-5060

Della Valle RG, Andersen HC (1992) Molecular dynamics simulation of silica liquid and glass. J Chem Phys 97:2682-2689

Demkov AA, Sankey OF, Schmidt KE, Adams GB, O'Keeffe M (1994) Theoretical investigation of alkali-metal doping in Si clathrates. Phys Rev B 50:17001-17008

Demkov AA, Ortega J, Sankey OF, Grumbach MP (1995) Electronic structure approach for complex silicas. Phys Rev B 52:1618-1630

Dempsey MJ, Kawamura K (1984) Molecular dynamics simulation of the structure of aluminosilicate melts. In: CMB Henderson (ed) Progress in Experimental Petrology, Sixth Prog Rep of Research supported by NERC, p 49-56, Natural Environment Research Council, Manchester, UK

Dempsey MJ, Kawamura K, Henderson CMB (1984) Molecular dynamics modelling of akermanite- and diopside-composition melts. In: CMB Henderson (ed) Progress in Experimental Petrology, Sixth Prog Rep of Research supported by NERC, p 57-59, Natural Environment Res Council, Manchester, UK

Dickinson JE, Scarfe CM, McMillan P (1990) Physical properties and structure of K$_2$Si$_4$O$_9$ melt quenched from pressures up to 2.4 GPa. J Geophys Res B95:15675-15681

Diefenbacher J, McMillan PF, Poe BT, Rubie D (1993) Diffusivities of Si and O in Na$_2$Si$_4$O$_9$ liquid at high pressure: experiments and ion dynamics simulation. EOS Trans Am Geophys Union 74:630

Diefenbacher J, Poe BT, McMillan PF (1995) Ion dynamics simulations of Si^{4+} and O^{2-}: diffusivity and structure of Na$_2$Si$_4$O$_9$ liquid at high pressure. Am Mineral, submitted

Ding HQ, Karasawa N, Goddard WA (1992) Atomic level simulations on a million particles: the cell multipole method for Coulombic and London non-bonded interactions. J Chem Phys 97:4309-4315

Dingwell DB, Webb SL (1990) Relaxation in silicate melts. Eur J Mineral 2:427-449

Donovan EP, Spaepen F, Turnbull D, Poate JM, Jacobson DC (1985) Calorimetric studies of crystallization and relaxation of amorphous Si and Ge prepared by ion implantation. J Appl Phys 57:1795-1804

Drabold DA, Wang R, Klemm S, Sankey OF, Dow JD (1991) Efficient *ab initio* molecular-dynamics simulations of carbon. Phys Rev B 43:5135-5137

Durben DJ, Wolf GH (1991) Raman spectroscopic study of the pressure-induced coordination change in GeO_2 glass. Phys Rev B 43:2355-2363

Erikson RL, Hostetler CJ (1987) Application of empirical ionic models to SiO_2 liquid, Potential model approximations and integration of SiO_2 polymorph data. Geochim Cosmochim Acta 51:1209-1218

Evans DJ (1983) Computer experiment for nonlinear thermodynamics of Couette flow. J Chem Phys 78:3297-3302

Evans DJ, Morriss GP (1990) Statistical Mechanics of Nonequilibrium Liquids. Academic Press, San Diego

Ewald PP (1921) Die Berechnung optischer und elektrostatischer Gitterpotentiale. Ann Phys 64:253-287

Farnan I, Stebbins JF (1990a) A high temperature ^{29}Si NMR investigation of solid and molten silicates, J Am Chem Soc 112:32-39

Farnan I, Stebbins JF (1990b) Observation of slow atomic motions close to the glass transition using 2-D ^{29}Si NMR. J Non-Cryst Solids 124:207-215

Farnan I, Stebbins JF (1994) The nature of the glass transition in a silica-rich oxide melt. Science 265:1206-1209

Fincham D, Makrodt WC, Mitchell PJ (1994) MgO at high temperatures and pressures: shell-model lattice dynamics and molecular dynamics. J Phys C: Cond Matter 6:393-404

Feuston BP, Garofalini SH (1988) Empirical three-body potential for vitreous silica. J Chem Phys 89:5818-5824

Fiske P, Stebbins JF (1994) The structural role of Mg in silicate liquids: a high-temperature ^{25}Mg, ^{23}Na and ^{29}Si NMR study. Am Mineral 79:848-861

Frenkel MY, Vasserman YA (1991) Modeling molten silica by molecular-dynamics. Geokhim 8:1194-1203

Frick B, Richter D (1995) The microscopic basis of the glass transition in polymers from neutron scattering studies. Science 267:1939-1945

Friedman HL (1985) A Course in Statistical Mechanics. Prentice-Hall, Englewood Cliffs, NJ

Fumi FG, Tosi MP (1964) Ionic sizes and Born repulsive parameters in the NaCl-type alkali halides-I. The Huggins-Mayer and Pauling forms. J Phys Chem Solids 25:31-43

Garofalini SH (1982) Molecular dynamics simulation of the frequency spectrum of amorphous silica. J Chem Phys 76:3189-3192

Garofalini SH (1983) Pressure variation in molecular dynamics simulated vitreous silica. J Non-Cryst Solids 55:451-454

Gillet P, Badro J, McMillan PF (1995) High pressure behavior in α-$AlPO_4$, Amorphization and the memory-glass effect. Phys Rev B 51:11262-11269

Goldstein H (1980) Classical Mechanics. Addison-Wesley, Reading, Massachusetts

Gordon RG, Kim YS (1972) Theory for the forces between closed-shell atoms and molecules. J Chem Phys 56:3122-3133

Greaves GN (1992) Glass structure and ionic transport. In: LD Pye, WC Lacourse, HJ Stevens (eds) Physics of Non-Crystalline Solids, p 453-459. Taylor and Francis, London

Greaves GN, Fontaine A, Lagarde P, Raoux D, Gurman SJ (1981) Local structure of silicate glasses. Nature 293:611-616

Greaves GN, Gurman SJ, Catlow CRA, Chadwick AV, Houde-Walter S, Henderson CMB, Dobson BR (1991) A structural basis for ionic diffusion in oxide glasses. Phil Mag A 64:1059-1072

Green, JL, Durben DJ, Wolf GH, Angell CA (1990) Water and solutions at negative pressure: Raman spectroscopic study to -80 megapascals. Science 249:649-652

Greengard, L (1988) The Rapid Evaluation of Potential Fields in Many Particle Systems. MIT Press, Cambridge, MA

Grumbach MP (1993) First principles molecular dynamics simulation of carbon at high pressures and temperatures. PhD dissertation, University of Illinois at Urbana-Champaign

Grumbach MP, Martin RM (1995) Properties of liquid carbon at high pressure: implications for the phase diagram. Nature, submitted

Haile JM (1992) Molecular Dynamics Simulation: Elementary Methods. Wiley, New York

Hansen JP, McDonald IR (1976) Theory of Simple Liquids. Academic Press, London.

Hansen JP, McDonald IR (1976) Theory of Simple Liquids (2nd Edn). Academic Press, London.

Hemley RJ, Mao HK, Bell PM, Mysen BO (1986) Raman spectroscopy of SiO_2 glass at high pressure. Phys Rev Lett 57:747-750

Hemley RJ, Cohen RE (1992) Silicate perovskite. Ann Rev Earth Planet Sci 20:553-600

Hemmati M, Chizmeshya A, Wolf GH, Poole P, Shao J, Angell CA (1995) The crystalline-amorphous transition in silicate perovskites. Phys Rev B 51:14841-14848

Hess AC, McMillan PF, O'Keeffe M (1986) Force fields for SiF_4 and H_4SiO_4: Ab initio molecular orbital calculations. J Phys Chem 90:5661-5665

Hess AC, McMillan PF, O'Keeffe M (1987) Ab initio force field for H_4SiO_4 in S_4 conformation. J Phys Chem 90:5661-5662

Hess AC, McCarthy MI, McMillan PF (1994) Ab-initio methods in geochemistry and mineralogy. In: TH Dunning Jr (ed) Advances in Electronic Structure Theory 2:143-201 Jai Press, Greenwich, Connecticut

Hill TL (1956) Statistical Mechanics. McGraw-Hill, New York

Hoover WG, Ladd AJC, Moran B (1982) High strain rate plastic flow studied via nonequilibrium molecular dynamics. Phys Rev Lett 48:1818-1820

Hostetler CJ (1982) Equilibrium properties of some silicate materials: a theoretical study. PhD dissertation, University of Arizona, Tucson, AZ

Huang C, Cormack AN (1990) The structure of sodium silicate glass. J Chem Phys 93:8180-8186

Inbar I, Cohen RE (1995) High pressure effects on thermal properties of MgO. Geophys Res Lett 22:1533-1536

Jeanloz R (1990) Thermodynamics and evolution of the Earth's interior: high pressure melting of silicate perovskite as an example. 1990 Gibbs Symp, Am Math Soc, p 211-226, Washington, DC

Jin W, Vashista P, Kalia RK, Rino JP (1993) Dynamic structure factor and vibrational properties of SiO_2 glass. Phys Rev B 48:9359-9368

Jin W, Kalia RK, Vashishta P (1993) Structural transformation, intermediate-range order and dynamical behavior of SiO_2 glass at high pressures. Phys Rev Lett 71:3146-3149

Jin W, Kalia RK, Vashishta P, Rino JP (1994) Structural transformation in densified silica glass, A molecular-dynamics study. Phys Rev B 50: 118-131

Kalos MH, Whitlock PA (1986) Monte Carlo Methods. Volume I: Basics. John Wiley & Sons, New York.

Keskar NR, Chelikowsky JR (1992) Structural properties of nine silica polymorphs. Phys Rev B 46:1-13

Keskar NR, Chelikowsky JR (1995) Calculated thermodynamic properties of silica polymorphs. Phys Chem Minerals 22:233-240

Kieffer J, Angell CA (1988) Generation of fractal structures by negative pressure rupturing of SiO_2 glass. J Non-Cryst Solids 106:336-342

Kieffer, J, Angell CA (1989) Structural incompatibilities and liquid-liquid phase separation in molten binary silicates, a computer simulation. J Chem Phys 90:4982-4991

Kittel C, Kroemer H (1980) Thermal Physics. WH Freeman and Company, San Francisco

Kubicki JD, Lasaga AC (1988) Molecular dynamics simulations of SiO_2 melt and glass, ionic and covalent models. Am Mineral 73:941-955

Kubicki JD, Lasaga AC (1991a) Molecular dynamics and diffusion in silicate melts. In: I Kushiro, L Perchuk (eds) Advances in Physical Geochemistry 8:1-50, Springer-Verlag, Berlin

Kubicki JD, Lasaga AC (1991b) Molecular dynamics simulation of pressure and temperature effects on $MgSiO_3$ and Mg_2SiO_4 melts and glasses. Phys Chem Minerals 17:661-673

Kubicki JD, Lasaga AC (1992) Ab initio molecular dynamics simulations of melting in forsterite and $MgSiO_3$ perovskite. Am J Sci 292:153-183

Kubicki JD, Lasaga AC (1993) Molecular dynamics simulations of interdiffusion in $MgSiO_3$-Mg_2SiO_4 melts. Phys Chem Minerals 20:255-262

Kubicki JD, Hemley RJ, Hofmeister AM (1992) Raman and infrared study of pressure-induced structural changes in $MgSiO_3$, $CaMgSi_2O_6$, and $CaSiO_3$ glasses. Am Mineral 77:258-269

Kushiro I (1976) Changes in viscosity and structure of melts of $NaAlSi_2O_6$ composition at high pressure. J Geophys Res 81:6347-6350

Kushiro I (1978) Viscosity and structural change of albite ($NaAlSi_3O_8$) melt at high pressure. Earth Planet Sci Lett 41:87-90

Kushiro I (1981) Viscosity, density and structure of silicate melts at high pressures, and their petrological applications. In: RB Hargraves (ed) Physics of Magmatic Processes, p 93-120

Kushiro I (1983) Effect of pressure on the diffusivity of network-forming cations in melts of jadeite compositions. Geochim Cosmochim Acta 47:1415-1422

Kushiro I, Yoder HS, Mysen BO (1976) Viscosities of basalt and andesite melts at high pressure. J Geophys Res 81:6351-6357

Landau LD, Lifshitz EM (1980) Statistical Physics (Course of Theoretical Physics, Vol 5), 3rd Edn revised by EM Lifshitz, LP Pitaevskii, Pergamon Press, Oxford

Lange RL, Carmichael ISE (1990) Thermodynamic properties of silicate liquids with emphasis on density, thermal expansion and compressibility. In: J Nicholls, JK Russell (eds) Modern Methods of Igneous Petrology: Understanding Magmatic Processes. Rev Mineral 24:25-64

Lasaga AC, Gibbs GV (1987) Applications of quantum mechanical potential surfaces to mineral physics calculations. Phys Chem Minerals 14:107-117

Leinenweber K, Navrotsky A (1988) A transferable interatomic potential for crystalline phases in the system MgO-SiO$_2$. Phys Chem Minerals 15:588-596

LeSar R (1984) Improved electron-gas model calculations of solid N$_2$ to 10 GPa. J Chem Phys 81:5104-5108

LeSar R, Gordon TG (1982) Density functional theory for the solid alkali cyanides. J Chem Phys 77:3682-3692

LeSar R, Gordon TG (1983) Density functional theory for solid nitrogen and carbon dioxide at high pressure. J Chem Phys 78:4991-4996

Maitland GC, Rigby M, Smith EB, Wakeham WA (1981) Intermolecular Forces: their Origin and Determination. Clarendon Press, Oxford

March NH, Tosi MP (1976) Atomic Dynamics in Liquids. General Publishing Co., Toronto, Canada. Reprinted (1991) by Dover Publications, New York.

Matsui M (1988) Molecular dynamics study of MgSiO$_3$ perovskite. Phys Chem Minerals 16:234-238

Matsui M, Kawamura K (1980) Instantaneous structure of an MgSiO$_3$ melt simulated by molecular dynamics. Nature 285:648-649

Matsui M, Kawamura K (1984) Computer simulation of structures of silicate melts and glasses. In: I Sunagawa (ed) Materials Science of the Earth's Interior, p 3-23, D. Reidel, Boston

Matsui M, Busing WR (1984) Computational modelling of the structure and elastic constants of the olivine and spinel forms of Mg$_2$SiO$_4$. Phys Chem Minerals 11:55-59

Matsui M, Price GD (1991) Simulation of the pre-melting behavior of MgSiO$_3$ perovskite at high pressures and temperatures. Nature 351:735-737

Matsui M, Kawamura K, Syono Y (1982) Molecular dynamics calculations applied to silicate systems: molten and vitreous MgSiO$_3$ and Mg$_2$SiO$_4$. In: S Akimoto, MH Manghnani (eds) High Pressure Research in Geophysics, p 511-524, D. Reidel, Boston

Matsui M, Akaogi M, Matsumoto T (1987) Computational model of the structural and elastic properties of the ilmenite and perovskite phases of MgSiO$_3$. Phys Chem Minerals 14:101-106

Matsui Y, Tsuneyuki S (1992) Molecular dynamics study of rutile-CaCl$_2$-type phase transition of SiO$_2$. In: Y Syono, MH Manghnani (eds) High-Pressure Research, Application to Earth and Planetary Sciences, p 433-439 Terra Scientific Publishing Co/Am Geophysical Union, Tokyo/Washington, DC

McMillan PF, Hess AC (1990) Ab initio force field calculations for quartz. Phys Chem Minerals 17:97-107

McMillan PF, Kirkpatrick RJ (1992) Aluminium coordination in magnesium aluminosilicate glasses. Am Mineral 77:898-900

McMillan PF, Poe BT, Gillet P, Reynard B (1994) A study of SiO$_2$ glass and supercooled liquid to 1950 K via high-temperature Raman spectroscopy. Geochim Cosmochim Acta 58, 3653-3664

McQuarrie DA (1976) Statistical Mechanics. Harper and Row, New York.

Meade C, Hemley RJ, Mao HK (1992) High-pressure x-ray diffraction of SiO$_2$ glass. Phys Rev Lett 69:1387-1390

Metropolis N, Rosenbluth AW, Rosenbluth MN, Teller AH, Teller, E (1953) Equation of state calculations by fast computing machines. J Chem Phys 21:1087-1092

Miller GH, Stolper EM, Ahrens TJ (1991a) The equation of state of a molten komatiite, 1, shock wave compression to 36 GPa. J Geophys Res 96:11831-11848

Miller GH, Stolper EM, Ahrens TJ (1991b) The equation of state of a molten komatiite, 2, application to komatiite petrogenesis and the Hadean mantle. J Geophys Res 96:11849-11856

Mishima O (1994) Reversible first-order transition between two H$_2$O amorphs at 0.2 GPa and 135 K. J Chem Phys 100:5910-5912

Mishima O, Calvert LD, Whalley E (1985) Apparently first-order phase transition in amorphous ice. Nature 314:76-79

Mitra SK, Amini M, Fincham D, Hockney RW (1981) Molecular dynamics simulation of silicon dioxide glass. Phil Mag B 43:365-372

Mitra SK (1982) Molecular dynamics simulation of silicon dioxide glass. Phil Mag B 45:529-548

Mitra SK, Parker JM (1984) Molecular dynamics simulation of a soda-silica glass containing fluorine. Phys Chem Glasses 25:95-99

Mozzi RL, Warren BE (1969) The structure of vitreous silica. J Appl Cryst 2:164-172

Mysen BO, Virgo D, Danckwerth P, Seifert FA, Kushiro I (1983) Influence of pressure on the structure of melts on the joins NaAlO$_2$-SiO$_2$, CaAl$_2$O$_4$-SiO$_2$, MgAl$_2$O$_4$-SiO$_2$. N Jahrb Mineral Abh 147:281-303

Nakano A, Bi L, Kalia RK, Vashishta P (1993) Structural correlations of porous silica, molecular dynamics simulation on a parallel computer. Phys Rev Lett 71:85-88

Nakano A, Kalia RK, Vashishta P (1994) First sharp diffraction peak and intermediate-range order in amorphous silica: finite-size effects in molecular dynamics simulations. J Non-Cryst Solids 171:157-163

Navrotsky A, Geisinger K, McMillan P, Gibbs GV (1985) The tetrahedral framework in glasses and melts - inferences from molecular orbital calculations and implications for structure, thermodynamics and physical properties. Phys Chem Minerals 11:284-298

Newell RG, Feuston BP and Garofalini SH (1989) The structure of sodium trisilicate glass via molecular dynamics employing three-body potentials. J Mater Res 4:434-439

Norman GE, Filinov VS (1969) Investigation of phase transitions by a Monte Carlo method. High Temp (USSR) 7:216-222

O'Keeffe M, McMillan PF (1986) The Si-O-Si force field: ab initio MO calculations. J Phys Chem 90:541-542

Ohtani E, Taulelle F, Angell CA (1985) Al^{3+} coordination changes in liquid aluminosilicates under pressure. Nature 314:78-81

Parrinello M, Rahman A (1981) Polymorphic transitions in single crystals: a new molecular dynamics method. J Appl Phys 52:7182-7190

Parrinello M, Rahman A (1982) Strain fluctuations and elastic constants. J Chem Phys 76:2662-2666

Poe BT (1993) An Investigation of CaO-MgO-Al_2O_3-SiO_2 Liquids by Nuclear Magnetic Resonance Spectroscopy, Vibrational Spectroscopy, and Ion Dynamics Simulation. PhD dissertation, Arizona State University, Tucson, AZ

Poe BT, McMillan PF, Sato RK, Angell CA (1992a) Al and Si coordination in SiO_2-Al_2O_3 glasses and liquids: a study by NMR and IR spectroscopy and MD simulations. Chem Geol 96, 333-349

Poe BT, McMillan PF, Cote B, Massiot D, Coutures JP (1992b) SiO_2-Al_2O_3 liquids: in situ study by high-temperature ^{27}Al NMR spectroscopy and molecular dynamics simulations. J Phys Chem 96, 8220-8224

Poe BT, McMillan PF, Cote B, Massiot D, Coutures JP (1993a) The Al environment in $MgAl_2O_4$ and $CaAl_2O_4$ liquids: an in situ study via high temperature NMR. Science 259:786-788

Poe BT, McMillan PF, Cote B, Massiot D, Coutures JP (1993b) Structure and dynamics in CaO-Al_2O_3 liquids: Investigation by high-temperature ^{27}Al NMR spectroscopy. J Am Ceramic Soc 77:1832-1838

Poe BT, Rubie DC, McMillan PF, Diefenbacher J (1994) Oxygen and silicon self diffusion in $Na_2Si_4O_9$ liquid to 15 GPa. EOS Trans Am Geophys Union 75:713

Poe BT, Rubie DC, Yarger J, Diefenbacher J, McMillan PF (1995) Silicon and oxygen self diffusion in aluminosilicate liquids to 15 GPa and 2500. Nature (submitted)

Ponyatovsky EG, Barkalov OI (1992) Pressure-induced amorphous phases. Mat Sci Rep 8:147-191

Poole PH, Sciortino F, Essmann U, Stanley HE (1992) Phase behavior of metastable water. Nature 360:324-328

Poole PH, Essmann U, Sciortino F, Stanley HE (1993) Phase diagram for amorphous solid water. Phys Rev E 48:4605-4610

Poole PH, Sciortino F, Grande T, Stanley HE, Angell CA (1994) Effect of hydrogen bonds on the thermodynamic behavior of liquid water. Phys Rev Lett 73:1632-1635

Poole PH, Grande T, Sciortino F, Stanley HE, Angell CA (1995a) Amorphous polymorphism. J Comp Mat Sci (submitted)

Poole PH, Hemmati M, Angell CA (1995b) Equation of state of liquid SiO_2. Phys Rev, submitted

Press WH, Flannery BP, Teukolsky SA, Vetterling WT (1986) Numerical Recipes. Cambridge University Press, Cambridge, UK

Rappe AK, Goddard WA (1991) Charge equilibration for molecular dynamics simulations. J Phys Chem 95:3358-3363

Rappe AK, Casewit CJ, Colwell KS, Goddard WA, Skiff WM (1992) UFF, a full periodic table force field for molecular mechanics and molecular dynamics simulations. J Am Chem Soc 114:10024-10035

Rigden SM, Ahrens TJ, Stolper EM (1984) Densities of liquid silicates at high pressures. Science 226:1071-1074

Rigden SM, Ahrens TJ, Stolper EM (1988) Shock compression of molten silicate: results for a model basaltic composition. J Geophys Res 93:367-382

Rigden SM, Ahrens TJ, Stolper EM (1989) High-pressure equations of state of molten anorthite and diopside. J Geophys Res 94:9508-9522

Rivers ML, Carmichael ISE (1987) Unltrasonic studies of silicate melts. J Geophys Res 92:9247-9270

Rubie DC, Ross II CR, Carroll MR, Elphick SC (1993) Oxygen self-diffusion in $Na_2Si_4O_9$ liquid up to 10 GPa and estimation of high-pressure melt viscosities. Am Mineral 78:574-582

Rustad JR, Yuen DA, Spera FJ (1990) Molecular dynamics of liquid SiO_2 under high pressure. Phys Rev A 42:2081-2089

Rustad JR, Yuen DA, Spera FJ (1991a) The sensitivity of physical and spectral properties of silica glass to variations in interatomic potentials under high pressure. Phys Earth Planet Int 65:210-230

Rustad JR, Yuen DA, Spera FJ (1991b) Molecular dynamics of amorphous silica at very high pressures (135 GPa): Thermodynamics and extraction of structures through analysis of Voronoi polyhedra. Phys Rev B 44:2108-2121

Rustad JR, Yuen DA, Spera FJ (1991c) The statistical geometry of amorphous silica at lower mantle pressures: implications for melting slopes of silicates and anharmonicity. J Geophys Res 96:19665-19673

Rustad JR, Yuen DA, Spera FJ (1992) Coordination variability and the structural components of silica glass under high pressures. Chem Geol 96:421-437

Sanders MJ, Leslie M, Catlow CRA (1984) Interatomic potentials for SiO_2. J Chem Soc Chem Commun:1271-1273

Sankey OF, Allen RE (1986) Atomic forces from electronic energies via the Hellmann-Feynman theorem, with application to semiconductor (110) surface relaxation. Phys Rev B 33:7164-7171

Sankey OF, Niklewski DJ (1989) *Ab initio* multicenter tight-binding model for molecular-dynamics simulations and other applications in covalent systems. Phys Rev B 40:3979-3995

Sankey OF, Niklewski DJ, Drabold DA, Dow JD (1990) Molecular-dynamics determination of electronic and vibrational spectra, and equilibrium structures of small Si clusters. Phys Rev B 41:12750-12759

Sankey OF, Demkov AA, Petuskey WT, McMillan PF (1993) Energetics and electronic structure of the hypothetical cubic zincblende form of GeC. Modelling Simulation Mater Sci Eng 1:1-14

Sato RK, McMillan PF, Dennison P, Dupree R (1991) High resolution ^{27}Al and ^{29}Si MAS NMR investigation of SiO_2-Al_2O_3 glasses. J Phys Chem 95:4483-4489

Sauer J (1989) Molecular models in ab initio studies of solids and surfaces: from ionic crystals and semiconductors to catalysts. Chem Rev 89:199-205

Scamehorn CA, Angell CA (1991) Viscosity-temperature relations and structure in fully polymeized aluminosilicate melts from ion dynamics simulations. Geochim Cosmochim Acta 55:721-730

Scarfe CM, Mysen BO, Virgo D (1987) Pressure dependence of the viscosity of silicate melts. Geochem Soc Spec Pub No 1:59-67

Shao J, Angell CA (1995) Vibrational anharmonicity and the glass transition in strong and fragile vitreous polymorphs. Proc Int'l Conf Glass, Beijing, China, submitted

Schofield P (1973) Computer simulation studies of the liquid state. Comp Phys Commun 5:17-23

Sharma SK, Virgo D, Kushiro I (1979a) Relationship between density, viscosity and structure of GeO_2 melts at low and high pressures. J Non-Cryst Solids 33:235-248

Sharma SK, Virgo D, Mysen BO (1979b) Raman study of the coordination of aluminum in jadeite melts as a function of pressure. Am Mineral 64:779-787

Smith KH, Shero E, Chizmeshya A, Wolf GH (1995) The equation of state of germania glass: evidence for vitreous polymorphism and spinodal behavior. J Chem Phys, in press

Somayazulu MS, Sharma SM, Garg N, Chaplot SL, Sikka SK (1993) The behavior of alpha-quartz and pressure-induced SiO_2 glass under pressure: a molecular dynamics study. J Phys: Condens Matter 5:6345-6356

Soules TF (1979) A molecular dynamic calculation of the structure of sodium silicate glasses. J Chem Phys 71:4570-4578

Soules TF (1980) A molecular dynamic calculation of the structure of B_2O_3 glass. J Chem Phys 73(8):4032-4036

Soules TF (1982) Molecular dynamic calculations of glass structure and diffusion in glass. J Non-Cryst Solids 49:29-52

Soules TF, Varshneya AK (1981) Molecular dynamic calculations of a sodium borosilicate glass structure. J Am Ceram Soc 64:145-150

Soules TF, Busbey RF (1981) Sodium diffusion in alkali silicate glass by molecular dynamics. J Chem Phys 75:969-975

Stanley HE, Angell CA, Essmann U, Hemmati M, Poole PH, Sciortino F (1994) Is there a second critical point in liquid water? Physica A 205:122-139

Stebbins, JF (1991) NMR evidence for five-coordinated silicon in a silicate glass at atmospheric pressure. Nature 351:638-639

Stebbins JF, McMillan P (1989) Five- and six-coordinated Si in $K_2Si_4O_9$ glass quenched from 1.9 GPa and 1200 °C. Am Mineral 74:965-968

Stebbins JF, McMillan PF (1993) Compositional and temperature effects on five-coordinated silicon in ambient pressure silicate glasses. J Non-Cryst Solids 160:116-125

Stebbins JF, Sykes D (1990) The structure of $NaAlSi_3O_8$ liquid at high pressure: New constraints from NMR spectroscopy. Am Mineral 75:943-946

Stebbins JF, Farnan I, Xue X (1992) The structure and dynamics of alkali silicate liquids: a view from NMR spectroscopy. Chem Geol 96:371-385

Stein DJ, Spera FJ (1995) Molecular dynamics simulations of liquids and glasses in the system $NaAlSiO_4$-SiO_2: methodology and melt structures. Am Mineral 80L417-431

Stixrude L, Bukowinski MST (1988) Simple covalent potential models of tetrahedral SiO_2: Applications to a-quartz and coesite at pressure. Phys Chem Minerals 16L199-206

Stixrude L, Bukowinski MST (1989) Compression of tetrahedrally bonded SiO_2 liquid and silicate liquid-crystal density inversion. Geophys Res Lett 16:1403-1406

Stixrude L, Bukowinski MST (1990) A novel topological compression mechanism in a covalent liquid. Science 250:541-543

Stixrude L, Bukowinski MST (1991) Atomic structure of SiO_2 glass and its response to pressure. Phys Rev B 44:2523-2534

Stolper EM, Walker D, Hager BH, Hays JF (1981) Melt segregation from partially molten source regions: the importance of melt density and source region size. J Geophys Res 86:6261-6271

Sugiura H, Kondo K, Sawaoka A (1982) in High Pressure Research in Geophysics. Akimoto S, Manghnani MH (eds) p 551-561, Reidel, Dordrecht, Netherlands

Sykes D, Poe BT, McMillan PF, Luth, R Sato, RK (1993) A spectroscopic investigation of anhydrous $KAlSi_3O_8$ and $NaAlSi_3O_8$ glasses quenched from high pressure. Geochim Cosmochim Acta 57:1753-1759

Thompson MO, Galvin GJ, Mayer JW, Peercy PS, Poate JM, Jacobson DC, Cullis AG, Chew NG (1984) Melting temperature and explosive crystallization of amorphous silicon during pulsed laser irradiation. Phys Rev Lett 52:2360-2363

Tijskens E, Viaene WA, Geerlings P (1995) The ionic model: extension to spatial charge distributions, derivation of an interaction potential for silica polymorphs. Phys Chem Minerals 22:186-199

Tosi MP (1964) Cohesion of solids in the Born model. Solid State Phys 16:1-120

Tosi MP, Fumi FG (1964) Ionic sizes and Born repulsive parameters in the NaCl-type alkali halides-II. The generalized Huggins-Mayer form. J Phys Chem Solids 25:45-52

Tse JS, Klein ML (1987) Pressure-induced phase transformations in ice. Phys Rev Lett 58:1672-1675

Tse JS, Klug DD (1991) Mechanical instability of α-quartz: a molecular-dynamics study. Phys Rev Lett 67:3559-3562

Tse JS, Klug DD (1991) The structure and dynamics of silica polymorphs using a two-body effective potential model. J Chem Phys 95:9176-9185

Tse JS, Klug DD (1992) Structural memory in pressure-amorphized $AlPO_4$. Science 255:1559-1561

Tsuneyuki S, Tsukada M, Aoki H, Matsui Y (1988) First-principles interatomic potential of silica applied to molecular dynamics. Phys Rev Lett 61:869-872

Tsuneyuki S, Matsui Y, Aoki H, Tsukada M (1989) New pressure-induced structural transformations in silica obtained by computer simulation. Nature 339:209-211

Tsuneyuki S, Aoki H, Tsukada M, Matsui Y (1990) Molecular-dynamics study of the α to β structural phase transition of quartz. Phys Rev Lett 64:776-779

van Beest BWH, Kramer GJ, van Santen RA (1990) Force fields of silicas and aluminophosphates based on ab initio Calculations. Phys Rev Lett 64:1955-1958

Vashishta, P, Kalia RK, Rino JP (1990) Interaction potential for SiO_2: a molecular dynamics study of structural correlations. Phys Rev B 41:12197-12209

Verlet L (1967) Computer "experiments" on classical fluids. I. Thermodynamical properties of Lennard-Jones molecules. Phys Rev 159:98-103

Vessal B, Amini M, Fincham D, Catlow CRA (1989) Water-like melting behavior of SiO_2 investigated by the molecular dynamics simulation technique. Phil Mag B 60:753-775

Vessal B, Amini M, Catlow CRA, Leslie M (1991) Simulation studies of silicate glasses. Trans Am Cryst Assoc 27:15-34

Vessal B, Greaves GN, Marten PT, Chadwick AV, Mole R, Houde-Walter S (1992) Cation micro-segregation and ionic mobility in mixed alkali glasses. Nature 356:504-506

Waff HS (1975) Pressure-induced coordination changes in magmatic liquids. Geophys Res Lett 2:193-196

Wasserman EA, Yuen DA, Rustad JR (1993a) Molecular dynamics study of the transport properties of perovskite melts under high temperature and pressure conditions. Earth Planet Sci Lett 114:373-384

Wasserman EA, Yuen DA, Rustad JR (1993b) Compositional effects on the transport and thermodynamic properties of MgO-SiO_2 mixtures using molecular dynamics. Phys Earth Planet Int 77:189-203

Wentzcovitch RM, Martins JL, Price GD (1993) Ab initio molecular dynamics with variable cell shape: application to $MgSiO_3$. Phys Rev Lett 70:3947-3950

Wentzcovitch RM, Ross NL, Price GD (1995) Ab initio study of $MgSiO_3$ and $CaSiO_3$ perovskites at lower-mantle pressures. Phys Earth Planet Int 90:101-111

Williams Q, Jeanloz R (1988) Spectroscopic evidence for pressure-induced coordination changes in silicate glasses and melts. Science 239:902-905

Williams Q, McMillan P, Cooney T (1989) Vibrational spectra of orthosilicate glasses. The Mg-Mn join. Phys Chem Minerals 16:352-359

Williams, Q, Hemley RJ, Kruger MB, Jeanloz R (1993) High pressure vibrational spectra of quartz, coesite, stishovite and amorphous silica. J Geophys Res B 98:22157-22170

Winograd S, Zare RN (1995) "Wired" science or whither the printed page? Science 269:615

Wolf GH, Bukowinski MST (1988) Variational stabilization of the ionic charge densities in the electron gas theory of crystals: applications to MgO and CaO. Phys Chem Minerals 15:209-220

Wolf GH, Wang S, Herbst CA, Durben DJ, Oliver WF, Kang ZC, Halvorson K (1992) Pressure induced collapse of the tetrahedral framework in crystalline and amorphous GeO_2. In: Y Syono, MH Manghnani (eds) High-Pressure Research: Application to Earth and Planetary Sciences, p 503-517, Terra Scientific Publishing Co/Am Geophysical Union, Tokyo/Washington, DC

Wolf GH, Durben DJ, McMillan PF (1990) High pressure Raman spectroscopic study of sodium tetrasilicate ($Na_2Si_4O_9$) glass. J Chem Phys 93:2280-2288

Wood WW (1968) Monte Carlo studies of simple liquid models. In: HNV Temperley, JS Rowlinson, GS Rushbrooke (eds) Physics of Simple Liquids, p 115-230, North-Holland, Amsterdam

Woodcock LV (1975) Molecular dynamics calculations on molten salts. In: Braunstein, Mamantor, Smith (eds) Molten Salt Chemistry 3:1-74. Plenum Press, New York

Woodcock LV, Angell CA, Cheeseman P (1976) Molecular dynamics studies of the vitreous state: simple ionic systems and silica. J Chem Phys 65:1565-1577

Wright AC (1993) The comparison of molecular dynamics simulations with diffraction experiments. J Non-Cryst Solids 159:264-268

Xu Z, Stebbins JF (1995) ^6Li and ^7Li NMR chemical shifts, relaxation, and coordination in silicates. Solid State Mag Res, in press

Xue X, Stebbins JF (1993) ^{23}Na NMR chemical shifts and local Na coordination environments in silicate crystals, melts and glasses. Phys Chem Minerals 20:297-307

Xue X, Stebbins J, Kanzaki M, Tronnes RG (1989) Silicon coordination and speciation changes in a silicate liquid at high pressures. Science 245:962-964

Xue X, Stebbins J, Kanzaki M, Poe B, McMillan PF (1991) Pressure-induced silicon coordination and tetrahedral structural changes in alkali oxide-silica melts up to 12 GPa, NMR, Raman and infrared spectroscopy. Am Mineral 76:8-26

Yarger J, Diefenbacher J, Poe BT, Wolf GH, McMillan PF (1994a) High coordinate Si and Al in sodium aluminosilicate glasses quenched from high pressure. EOS Trans Am Geophys Union 75:713

Yarger J, Diefenbacher J, Poe BT, Wolf GH, McMillan PF (1995) High coordinate Si and Al in sodium aluminosilicate glasses quenched from high pressure. Science (submitted)

Yinnon, H, Cooper A Jr (1980) Oxygen diffusion in multicomponent glass forming systems. Phys Chem Glasses 21:204-211

Zirl DM, Garofalini SH (1990) Structure of sodium aluminosilicate glasses. J Am Ceram Soc 73:2848-2856